THE OCEAN BASINS AND MARGINS

Volume 7A

The Pacific Ocean

THE OCEAN BASINS AND MARGINS

THE OCEAN BASINS AND MARGINS

Edited by

Alan E. M. Nairn

Earth Science and Resources Institute
University of South Carolina
Columbia, South Carolina

Francis G. Stehli

College of Geosciences
University of Oklahoma
Norman, Oklahoma

and

Seiya Uyeda

Earthquake Research Institute
University of Tokyo
Tokyo, Japan

Volume 7A

The Pacific Ocean

PLENUM PRESS · NEW YORK AND LONDON

Library of Congress Cataloging in Publication Data

Nairn, A. E. M.
   The ocean basins and margins.

   Vol. 5 edited by A. E. M. Nairn, M. Churkin, Jr., and F. G. Stehli.
   Includes bibliographies.
   Contents: c. 1. The South Atlantic. — v. 2. The North Atlantic. — [etc.] — v. 7A. The
Pacific Ocean.
   1. Submarine geology. 2. Continental margins. I. Stehli, Francis Greenough, joint
author. II. Title.
QE39.N27   551.4′608   72-83046

ISBN 978-1-4612-9440-5   ISBN 978-1-4613-2351-8 (eBook)
DOI 10.1007/978-1-4613-2351-8

Additional material to this book can be downloaded from http://extras.springer.com.

© 1985 Plenum Press, New York
Softcover reprint of the hardcover 1st edition 1985
A Division of Plenum Publishing Corporation
233 Spring Street, New York, N.Y. 10013

CONTRIBUTORS TO THIS VOLUME

Luis Aguirre
Department of Geology
University of Liverpool
Liverpool, England

John W. Baldock
British Geological Survey
Nottingham, England

Biq Chingchang
Geology Department
Chinese Cultural University
Yangmingshan, Taiwan

Sam Boggs, Jr.
Department of Geology
University of Oregon
Eugene, Oregon

J. N. Carney
British Geological Survey
Nottingham, England

J. C. Chen
Institute of Oceanography
National Taiwan University
Taipei, Taiwan

David A. Clague
United States Geological Survey
Menlo Park, California

Edwin John Cobbing
British Geological Survey
Nottingham, England

Frank I. Coulson
British Geological Survey
Nottingham, England

Thomas A. Davies
Institute for Geophysics
The University of Texas at Austin
Austin, Texas

Gabriel Dengo
Centro de Estudios Geológicos de
    America Central
Guatemala City, Guatemala

Robert A. Duncan
College of Oceanography
Oregon State University
Corvallis, Oregon

v

Helios S. Gnibidenko

Institute of Marine Geology and
    Geophysics
Sakhalin, USSR

E. Honza

Geological Survey of Japan
Ibaraki, Japan

Kazuo Kobayashi

Ocean Research Institute
University of Tokyo
Tokyo, Japan

A. Macfarlane

British Geological Survey
Nottingham, England

D. I. J. Mallick

British Geological Survey
Nottingham, England

Floyd W. McCoy

Lamont-Doherty Geological
    Observatory
Palisades, New York

John Milsom

Department of Geological
    Sciences
University of London
London, England

Constance Sancetta

Lamont-Doherty Geological
    Observatory
Palisades, New York

C. T. Shyu

Institute of Oceanography
National Taiwan University
Taipei, Taiwan

K. Tamaki

Geological Survey of Japan
Ibaraki, Japan

CONTENTS

Chapter 7. The Southern Andes

Luis Aguirre

Chapter 10. **The Bonin Arc**

E. Honza and K. Tamaki

Chapter 11. **Taiwan: Geology, Geophysics, and Marine Sediments**

Biq Chingchang, C. T. Shyu, J. C. Chen and Sam Boggs, Jr.

Chapter 14. **The Vanuata Island Arc: An Outline of the Stratigraphy,
         Structure and Petrology**

J. N. Carney, A. Macfarlane, D. I. J. Mallick

Chapter 1

NORTH PACIFIC SEDIMENTS

Floyd W. McCoy and Constance Sancetta
Lamont–Doherty Geological Observatory
Palisades, New York 10964

I. INTRODUCTION

A blanket of sediment mantles the floor of the world ocean. It forms the largest geological deposit on the earth's surface in terms of volume, areal extent, conterminous characteristics, or syndepositional processes. It is the result of contributions from the hydrosphere, lithosphere, atmosphere, and solarsphere, in about that order of input—an amazing assortment of debris reflecting the activity and residue from the interplay of biological, geological, chemical, and oceanographic processes interacting within the boundaries of these four environmental spheres.

Geophysical processes have constantly reshaped the boundaries within which these sedimentation processes have been active. The size, geographic configuration and physiography of the ocean basins have changed and continue to change. These physical changes have influenced climatic conditions, oceanographic circulation, and biological evolution, three variables of considerable influence on marine sedimentation processes. Clearly, understanding the genesis of the sediment blanket in the modern ocean is necessary for interpreting the geological record; the environmental and industrial concerns of civilization also require this knowledge.

For an area as vast as the North Pacific Ocean our discussion of oceanic sedimentation must be confined to general concepts. At the map scale used here only the abyssal environment can be considered. The hemipelagic and shelf sediments along continental and insular margins as well as within the marginal seas surrounding the North Pacific Basin, with their complex lithologies, must be ignored.

A few types of sedimentary particles dominate in the abyssal milieu. These particles are usually fine-grained, in the silt and clay sizes. This simplifies classification schemes involving such textural parameters and laboratory methods for their analysis. By assigning only two general categories describing the origin of these particles, biogenic or nonbiogenic, classification schemes are further simplified because minimal genetic criteria are considered. Terms such as turbidite, contourite, homogenite, glacial sediment, etc., are excluded. Simplification of classification schemes and laboratory analysis methods is necessary to provide a framework for efficiently comparing and merging previous sediment-mapping studies into one coherent format. Such a framework must be created from a data base that is both extensive for adequate regional coverage and internally consistent in terms of sampling procedures. For these reasons the core collection and laboratory techniques at the Lamont–Doherty Geological Observatory have been used as the foundation for summarizing the sediments of the North Pacific Ocean.

II. FRAMEWORK OF SEDIMENTATION

A. Physiography

Four large basins dominate the North Pacific seafloor physiography: the Northeast Pacific Basin off the North American continent, a smaller western basin forming the Phillipine Sea, and, between these, the Northwest–Central Pacific and Marianas basins (Fig. 1).

The Northeast Basin extends west from North America to the long lineament of seafloor mountains and ridges formed by the Emperor Seamounts–Hess Rise, the Hawaiian Ridge, the Mid-Pacific Mountains, and the Line Islands Ridge (Fig. 2). This is an almost continuous feature bisecting the North Pacific Ocean, except for the offset between the Hawaiian Rise and the Mid-Pacific Mountains and Line Islands Ridge. The Aleutian Trench forms a northern boundary; another trench system, the Middle America Trench, forms a southeastern boundary along Central America. The North American continental margin between these two trenches is the only area of the North Pacific where fans of hemipelagic sediment extend out onto the abyssal seafloor without the intervention of a trench system. One major river system lies along this margin, the Columbia River.

Within this northeastern basin are the large east–west fracture zones of which only four, the Mendocino, Murray, Clarion, and Clipperton, have

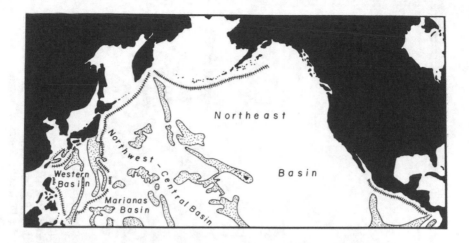

Fig. 1. Major physiographic provinces of the North Pacific Ocean. Trenches are traced with a hatched line. Ridges, rises, and seamount chains are broadly outlined by stippled patterns. Regional basins are identified.

sufficient topographic expression to be noticeable in Fig. 2. Much of the seafloor between these large tectonic features—fracture zones, trenches, seamounts, and ridges—has low relief and includes the three major abyssal plains of the North Pacific Ocean: the Alaska, Aleutian, and Tufts abyssal plains with their deposits of pelagic clays. At the southeastern corner of the North Pacific Basin is the East Pacific Rise and Cocos Ridge, with the adjoining Guatemala and Panama basins (Fig. 2).

The Northwest–Central Basin is separated from the Northeastern Basin by the Emperor Seamounts–Hawaiian Rise–Mid-Pacific Mountains–Line Islands Ridge feature. On the northern and western boundaries of the basin are the extensive circum-Pacific trench systems that extend from the Aleutian Trench through the Kamchatka, Kuril, Japan, and Bonin trenches. This nearly continuous system of trenches, in conjunction with the various marginal seas, prevents the dispersal of large volumes of terrigenous detritus onto the Central Pacific seafloor by mass-movement erosional mechanisms. Thus, the Kuril Trench as well as the marginal sea west of the trench prevent significant terrigenous influx from the largest fluvial system emptying into the Northwest Basin, the Amur River of central Asia. The Central Basin lies between the Line Islands and Mid-Pacific Mountains, and the Marshall Islands (Fig. 2).

West of the Central Basin is the Marianas Basin. The Marianas–Caroline–Yap trenches form a western boundary, the Marshall Islands an eastern boundary. The lowest point mapped on the earth's surface, 10,915 m

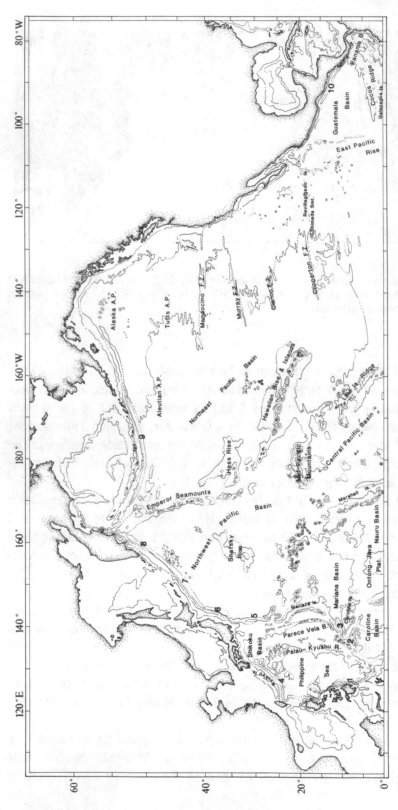

Fig. 2. Bathymetry and major physiographic features of the North Pacific Ocean. Isobaths are 1 km, 3 km, and 5 km; bathymetry is from Chase (1975). Abbreviations are: A. P., abyssal plain; Smt. seamount; Is., island(s); Plat., plateau; R., ridge or rise; B., basin; F. Z., fracture zone. Numbers identify major trench and island arc features as follows: 1, Phillipine Trench; 2, Caroline Trench; 3, Mariana Trench; 4, Ryukyu Trench; 5, Bonin Trench and Islands; 6, Japan Trench; 7, Kuril Trench and Islands; 8, Kamchatka Trench; 9, Aleutian Trench and Islands; 10, Middle America Trench.

(35,810 ft) below sea level, occurs in the southernmost portion of the Marianas Trench. Numerous seamounts and island chains lie within this basin
and are the sites of coral and algal reefs that provide calcareous detritus to
the surrounding seafloor. Small basins are located between the island chains,
such as the Caroline and Nauru basins (Fig. 2).

The small western basin is the Phillipine Sea. It is surrounded by
trenches and island arcs—the Bonin–Marianas–Caroline–Yap trenches
along the eastern boundary and the Ryukyu–Phillipine trenches along the
western boundary. The latter trenches and adjacent marginal seas isolate
the Phillipine Sea from any large-volume detrital input from the two major
drainage systems off eastern Asia, the Hwang Ho (Yellow) and the Yangtze
rivers. The Palau–Kyushu Ridge separates the Phillipine Sea into two
smaller basins. Much of the seafloor is marked by rugged relief, reflecting
the active volcanism and tectonism of this area, that provides a foundation
for reefs and a provenance for calcareous–volcanic pelagic sediments.

B. Oceanography

Surface circulation in the North Pacific Ocean is the result of nine major
currents (Fig. 3a). Velocities of current flow vary from only a few centi-
meters per second (cm/sec) up to 90 cm/sec in parts of the Kuroshio. Surface
currents are the primary distribution mechanism for particles originating
from air-fall or pelagic sources, and are the most influential factor in oceanic
productivity in the euphotic zone. The northern edge of the North Pacific
Current roughly defines the limits of drift by icebergs into the North Pacific
Ocean and thus the southerly extent of ice-rafted sedimentary debris.

Four major water masses can be defined in the upper 300 by regional
temperature and salinity characteristics (Fig. 3b). The cyclonic subarctic
gyre (Dodimead et al., 1963) is characterized by a surface salinity minimum
and permanent halocline caused by an excess of precipitation over evapo-
ration, with surface salinities of 33.00–32.00% and surface temperatures less
than 12°C. The anticyclonic subtropical gyre is characterized by a surface
salinity maximum and permanent thermocline, with surface salinities of
34.00–35.00% and temperatures of 20°–25°C. Between the two gyres is a
transitional zone of steep latitudinal temperature and salinity gradients
(34.00–33.00%, 12°–20°C). The transitional zone extends southward along
the coast of North America, following the California Current. Within the
equatorial region, the interaction of currents and countercurrents results in
an east–west subdivision of this generally warm, salty system, with lower
temperatures (20°–25°C) in the eastern equatorial Pacific. This is due to
upwelling of the Peru Current, which is fed by the equatorial countercurrent
and the Chile Current. The western equatorial Pacific is characterized by

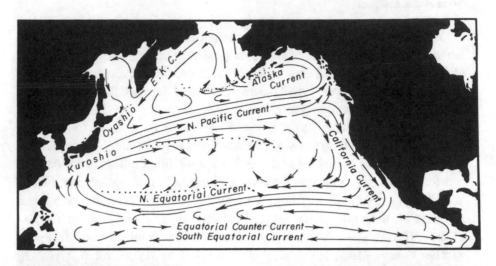

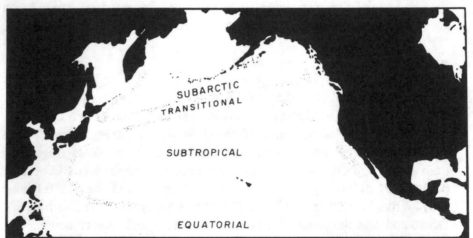

high salinity (35.00%) and the highest temperature of the ocean (25°–28°C). The interaction of these water masses and their boundary currents controls the distribution of nutrients in surface waters, which in turn determines rates of planktonic productivity and the distribution of microfossil populations.

Bottom-water movement (Fig. 3c) is considerably more sluggish than surface-water motion, with velocities on the order of 1–10 cm/sec. An influx of bottom water occurs into the North Pacific basin at about 170°–175°W in the equatorial region (Gordon and Gerard, 1970; Edmond et al., 1971) where it flows northward in two streams, mixes, then separates south of the Line Islands Ridge to flow east and west. Eastward flow at low latitudes continues into the Northeast Pacific basin north of the Clipperton Fracture Zone. A westward branch continues south of the Marianas through the Palau–Kyushu Ridge via the Parece Vela and Shikoku basins, into the Phillipine Sea (Mantyla and Reid, 1983). Eventually this water moves up to the middle latitudes of the Northeast Pacific and Central Pacific basins.

Deep circulation extends about 2000 m above the seafloor, and therefore influences the dispersal of particles settling through the water column. Bottom-sediment erosion by low-velocity flow north of the Clipperton Fracture Zone has been well documented (van Andel et al., 1975; Johnson, 1972; Craig, 1979). Where velocities are high due to constriction of flow between or over topographic features on the seafloor, such as through the Samoan passage in the South Pacific (Reid and Lonsdale, 1974; Hollister et al., 1974) or over the Cocos Ridge (Dowding, 1977), erosion and transport of pelagic deposits occurs. Acoustic mapping of sediments in the northwestern Pacific, however, suggest more erosion and redeposition activity by bottom currents than has been previously recognized, especially around and between seamounts (Damuth et al., 1983). A near-bottom water layer with increased concentrations of suspended sediments, known as a nepheloid layer, that is anywhere from 100 to 1000 m thick, occupies most of the deeper basins below depths of about 3500 m. This layer is a significant factor in the distribution of fine-grained clay-sized particles (Biscaye and Eittreim, 1977) particularly near the continental margins of North America and Japan; the latter within a western boundary undercurrent, where suspended-sediment

←————————————————————————————————————

Fig. 3. (a) Major surface currents in the North Pacific Ocean; E.K.C. identifies the East Kamchatka Current; the northern dotted line delineates the Arctic Convergence (subarctic boundary), and the southern dotted line delineates the Subtropical Convergence. (b) Upper water masses and the approximate boundaries between them (stippled pattern). (c) Bottom-water circulation pattern in the North Pacific Ocean with the stippled area outlining the approximate extent of bottom-water distribution as defined from the 1°C potential temperature isotherm (modified from Sverdrup et al., 1942; Dodimead et al., 1963; Gordon and Gerard, 1970; Lonsdale et al., 1972; Johnson, 1972; Moore et al., 1973; Nemoto and Korenke, 1981; Mantyla and Reid, 1983; Kadko, 1983).

concentrations are considerably higher than they are in mid-basin areas (Ewing and Connary, 1970).

C. Meteorology

Summer meteorological conditions (Fig. 4a) are dominated by a large high-pressure system centered over the northeast Pacific. Clockwise circulation around this high extends out over most of the northern area, resulting in onshore winds over Asia, the latter also being influenced by thermal advection over the Asian interior. South of about 20° latitude, easterly trade winds dominate surface-air circulation across the entire ocean. In the equatorial area the convergence of these two wind regimes produces a region of light, variable winds known as the doldrums. Upper atmospheric circulation is dominated by the jet stream flowing off Asia at 10–12 km altitudes with velocities up to 18 m/sec that increase to 22 m/sec over the open ocean; this is a significant mechanism of particulate transport offshore.

Winter conditions are considerably different (Fig. 4b). A large low-pressure cell off the Kamchatka Peninsula extends across much of the northern Pacific and produces surface winds up to about 600 m in altitude blowing off central and northern Asia. At higher altitudes, about 3000 m or so, zonal westerlies also flow off central and southern Asia. A smaller high-pressure cell exists off Mexico; easterly circulation south of this cell and along the equatorial area are trades that are strongest during the winter. The doldrums persist throughout the year. The jet stream axis shifts south of its summer track and follows a variable path across the central and eastern Pacific depending on ocean surface temperatures. Velocities of this upper atmospheric flow increase during the winter to as much as 54 m/sec off Asia. Both surface winds and jet stream flow during the winter carry detritus off the continents into the North Pacific. They are also the primary mechanism for dispersal of volcanic particles following ash-producing eruptions.

III. PRIOR MAPPING AND STUDIES

In 1854 the first scientific sample of seafloor sediment was taken at a water depth of 3870 m by a young midshipman in the U.S. Navy, J. M. Brooke, from the USS *Vincennes* by tying a quill to the lead at the end of a sounding line. The quill came back covered with foraminifera tests of the genus *Globigerina,* which Brooke looked at under $50\times$ magnification and described in the ship's log as "three roughly spherical balls . . . closely bunched together . . . with filaments radiating from each rosy ball like the thistledown of dusty blue."

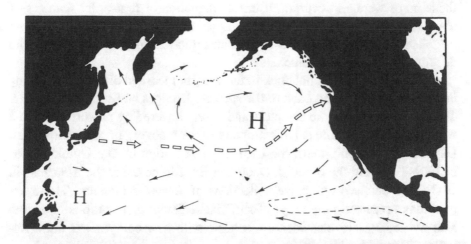

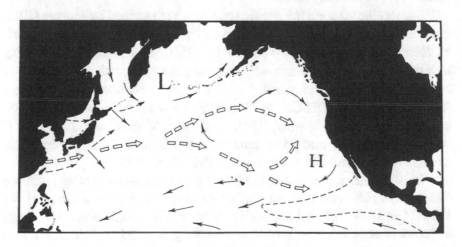

Fig. 4. Generalized atmospheric circulation over the North Pacific Ocean during (a) summer and (b) winter. Small arrows indicate surface wind directions (0–1000 m altitude). Dashed arrows indicate the direction of the westerlies off Asia during the winter (approx. 3000 m altitude). Large open arrows trace the jet stream axis (10–12 km altitude), whose track varies during the winter. Dashed line outlines the doldrums. High pressure cells are marked by an H; low pressure cells are marked by an L; relative sizes of letters suggest cell strength or magnitude (modified from Barry and Chorley, 1976).

Subsequent improvements of Brooke's technique used sounding weights coated with tallow, hollowed-out weights containing beeswax, and dredges. Using samples collected by such equipment while making spot soundings, the first maps depicting sediment distributions in the Pacific Ocean were published in the reports following England's *Challenger* Expedition during 1872–1876 by Murray and Renard (1891). Later work has confirmed that these maps were an accurate, if crude, depiction of first-order features of abyssal sediment distributions and seafloor relief. The classification scheme established by Murray and Renard remains today as the basic format for the description of marine sediments.

Wet clays often did not stick to the wax, but washed off as the sounding lines were slowly raised back to the surface. Dredges had similar problems. The development of grab samplers and corers lowered by powerful and fast winches with steel cable led to better and easier recovery of pelagic material. Using this new equipment, the *Carnegie* Expedition by the United States in 1909–1928, and the Swedish Deep-Sea Expedition using the *Albatross* in 1947–1948 obtained extensive collections of pelagic sediments. These expeditions presented new maps of surficial sediment distributions using the Murray and Renard classification scheme with new bathymetric data (Revelle, 1944; Arrhenius, 1952).

Substantial relief on the Pacific seafloor was described by Murray (1941, 1945, 1946), summarizing soundings recorded by spot measurements, then by Hess (1946, 1948), Menard and Dietz (1951), and Hamilton (1956) using new data collected by continuously recording electronic echosounders—the pelagic sediment blanket mantled a very irregular seafloor. Modern maps (e.g., Chase, 1975; GEBCO, 1975) portray an even more complex bathymetry; for example, Murray (1941, 1945) and Menard and Dietz (1951) mapped between 30 and 35 seamounts in the Gulf of Alaska, whereas at least 300 now have been mapped (e.g., Bek-Bulat, 1982).

In 1948, Shepard published the first book on submarine geology. Kuenen followed in 1950, and Menard concentrated on the marine geology of the Pacific Ocean in 1964. All three of these books discussed pelagic sediments in terms of sedimentation processes involving physical and biological processes within the framework of a seafloor containing significant relief. Thus the foundation was laid for the marine geological research of the past few decades. Systematic sampling of Pacific sediments by the Lamont–Doherty Geological Observatory (L-DGO) began in 1953 from the *Vema* and later in 1962 from the *Robert D. Conrad*. The result was an unparalleled collection of seafloor samples—1200 in the North Pacific Ocean—collected by one type of sampler, the piston corer.

The L-DGO collection was the basis for a colored map depicting the sedimentary provinces of the North Pacific by Horn *et al.* (1970). Sediment types were portrayed using a combination of textural criteria, based on the median diameter of the dominant particle size in terms of gravel/sand/silt/ clay, and genetic nomenclature for volcanic ash, turbidites, and pelagic muds. Later versions employed different schemes involving pelagic muds, pelagic oozes, terrigenous sediment, turbidites, and "sediment related to topographic highs."

In 1972 sediment distribution maps were published by Frazer *et al.* that used core data from Scripps Institution of Oceanography (SIO) and L-DGO (coarse-fraction data) in addition to information from the published literature. A complex graphic scheme for portraying lithologies involved color and black-and-white patterns using a mixed textural and genetic nomenclature for sediment classification. Also in 1972, Lisitzin summarized pacific sediment data from the extensive sample collection in the Soviet Union.

In 1978 a sediment distribution map covering the entire world ocean was prepared by Rawson and Ryan which also depicted the distribution and geochemical characteristics of polymetallic nodules on the seafloor. Sediment data were primarily from L-DGO, with additional information from other major oceanographic institutions. Sediment characteristics were mapped using a simple color scheme combining textural and genetic terms.

New maps of surficial sediments for the Pacific Ocean and its marginal seas were published by McCoy (1983a,b, 1986) as part of the Circum-Pacific Council for Energy and Mineral Resources mapping effort. The primary contribution of these maps has been their use of new analyses of L-DGO core sediments, allowing a systematic framework to be established for regional mapping, one to which older information can be compared and evaluated. This is important, as noted previously, due to: (1) the diverse types of sampling equipment used at sea, with their varying efficiencies in collecting the soft and unconsolidated sediment covering most of the seafloor (McCoy, 1980); (2) the variety in shipboard and laboratory techniques applied to sediment analyses; and (3) the varying schemes used to classify sediments. Thus, a framework was created based on one type of sampler [a piston corer with a trigger-weight (TW) corer] a consistent laboratory technique (smear slides and measurements of total carbonate content), and a uniform sediment classification scheme. These maps and the data derived from their preparation form the basis for this summary.

An extensive literature exists on seafloor sediments and sedimentary processes in the North Pacific Ocean. Every attempt has been made to summarize the important and appropriate literature, but a complete literature review is not attempted here. The more current literature has been favored over the older literature and should serve as a starting point for the interested reader.

IV. SEDIMENT CHARACTERIZATION

Analysis methods and mapping techniques for marine sediments rely on certain assumptions: First, the 2–5 cc volume of sediment collected in the top centimeter of each trigger-weight and piston core is presumed representative of a large area of nearby seafloor. The core tops represent a minuscule fraction of the true surface sediments, despite the appearance of extensive coverage shown in Figure 5. A subsample of one-tenth to one-thousandth of this volume for smear slide observations is presumed equally representative. Second, in mapping sediment types, contacts are drawn as distinct boundaries although in reality they describe vague and indistinct zones of variable width on the seafloor where mixing between lithologies occurs. Extrapolation of mapped contacts between data points relies on well-known physical or chemical controls on sedimentation, such as calcareous sediments mantling physiographic features above regional calcite compensation depth (CCD) levels. Third, surficial sediments sampled and mapped here are not necessarily of Holocene age, nor need they be the result of Holocene sedimentary or biological processes.

A. Sediment Data Base

A primary data base was formulated from approximately 1200 cores in the L-DGO archive from the North Pacific Ocean (Fig. 5). Most samples were taken from the tops of trigger-weight (TW) cores, with a few coming from the tops of piston cores or from other types of samplers such as box corers. Station data (cruise, core number, equipment used, water depth, etc.) are available from L-DGO. For a discussion of bottom sampling equipment, techniques, and constraints on interpreting the recovered cores, see McCoy (1980).

Additional sediment information came from published smear slide analyses of sediments obtained in hydraulic piston cores (HPC) taken from the *Glomar Challenger* during five legs of the Deep Sea Drilling Project (DSDP) in the eastern equatorial Pacific: Leg 64 (Curray *et al.*, 1982), Leg 66 (Watkins *et al.*, 1981), Leg 68 (Prell *et al.*, 1982), Leg 69 (Cann *et al.*, 1983), and Leg 70 (Honnorez *et al.*, 1983). HPC data from other North Pacific DSDP holes remains unpublished as of this writing. Rotary drilling techniques used by DSDP do not collect an undisturbed sediment sample from the sediment–water interface.

A secondary data set incorporated mapped lithology contacts from Horn *et al.*, (1970), Frazer *et al.*, (1972), and Rawson and Ryan (1978) as well as information from the published and unpublished literature up to 1982 (listed in McCoy, 1983a, and McCoy *et al.*, 1985). The latter two include historical data on sediments collected by the *Challenger* (Murray and Renard, 1891), *Carnegie* (Revelle, 1944), and *Albatross* (Arrhenius, 1952).

Both the primary and secondary data sets were used for plotting the distributions of surface sediment types and total calcium carbonate ($CaCO_3$) content, with the primary data set forming a framework for evaluating the

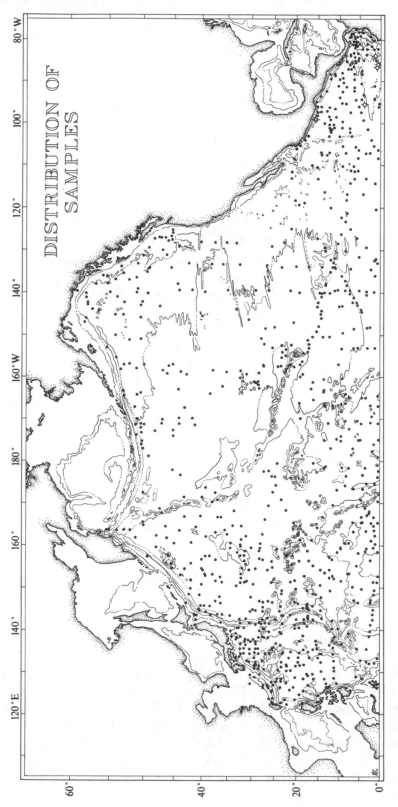

Fig. 5. Locations of piston and trigger-weight cores (solid circles) from the Lamont–Doherty Geological Observatory archive, and hydraulic piston cores (open circles with crosses) from the Deep-Sea Drilling Project, used in mapping lithologic types and sediment components.

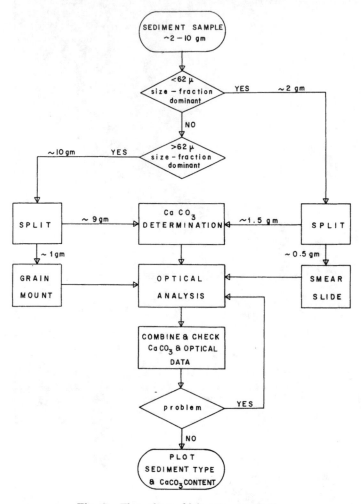

Fig. 6.   Flow chart of laboratory methods.

secondary data. Distributions of sedimentary components were based on the primary data set with selected additions from the secondary data set.

B. Sediment Analysis Methods

Smear slide and coarse-fraction analyses combined with measurements of total calcium carbonate ($CaCO_3$) content were used to identify sediment types and sedimentary components (Fig. 6). Coarse-fraction analysis involves the separation of sand-sized particles (>62 μm) by sieving, then examination of the sands for their mineralogical and microfossil content using a binocular microscope. This technique does not emphasize fine-grained (<62 μm) particles. Sediment criteria from this method were used on the

Horn *et al.* (1970) map, and, in part, on the Rawson and Ryan (1978) map where older (pre-1980) L-DGO core description data were incorporated.

Smear slide techniques provide a quick and reproducible method for descriptions of fine-grained sediments. It is an old technique used during oceanographic expeditions of the last century that has received additional recognition after its use by DSDP and L-DGO. A smear slide is a thin film of about $\frac{1}{2}$ g or less of sediment gently smeared onto a glass slide with distilled water, which is dried, then mounted using a conventional mounting medium such as Canada balsam. A cover slip is then applied. Analyses of sedimentary particle compositions are done using conventional optical techniques with a petrographic microscope.

Relative abundances of mineral and biogenic particles were estimated to the nearest 5–10% in smear slides, the precision limits of the technique. Visual approximations of particle concentrations used the following categories: dominant (>60%), abundant (30–60%), common (15–30%), present (5–15%), and rare (<5%). Quantitative measures were needed as checks on these visual estimates, the most convenient and important measure being the total content of calcium carbonate ($CaCO_3$) in the sediment. For the data summarized here, $CaCO_3$ analyses of approximately every tenth sample provided this check. Additional $CaCO_3$ data from the literature also has been used. Smear slide data from other sources were used only if $CaCO_3$ measurements were available.

C. Sediment Nomenclature and Classification

The classification scheme used here is based on the Murray and Renard (1891) system because the latter is descriptive, applicable to fine-grained material, and easily adapted to smear slide work by defining a one-third/two-thirds subdivision between nonbiogenic/biogenic particles (Fig. 7a,b). This scheme has been modified into a two-component format comparing particle composition to particle size (McCoy, 1983a,b) with size definitions following the Wentworth (1922) grade scale.

Nonbiogenic components predominantly are allochthonous from land—terrigenous—and are classified by the median diameter of the dominant grain size according to the Wentworth subdivisions of gravel, sand, silt, and clay. Pelagic clays are included with the clay category because these abyssal deposits are allochthonous clay mineral residues remaining after dissolution of calcareous matter, with subordinate amounts of authigenic components. Differentiation between allogenic and authigenic clay minerals by smear slide techniques is impossible.

Biogenic components are subdivided into the two dominant types of such debris in deep-sea deposits—siliceous and calcareous particles. For

(a) PARTICLE COMPOSITION

NON-BIOGENIC   **BIOGENIC**

% BIOGENIC (CaCO$_3$ or Biosiliceous) COMPONENTS
% NON-BIOGENIC (Terrigenous, Volcanic, Authigenic, etc.) COMPONENTS

PARTICLE SIZE	MAJOR SEDIMENT TYPES	MINOR SEDIMENT TYPES	DOMINANT GRAIN-SIZE	0% (100%)		15% (85%)		30% (70%)		60% (40%)	[30%]		100% (10%)	SPECIAL SEDIMENT TYPES	
			2mm	GRAVEL	SANDY GRAVEL / SILTY GRAVEL / CLAYEY GRAVEL	VOLCANIC GRAVEL –									VOLCANIC / CALCAREOUS / DIATOM PARTICLES 30-100%
			62 μm	SAND	GRAVEL SAND / SILTY SAND / CLAYEY SAND / SILTY-CLAYEY SAND	SAND –		CALCAREOUS GRAVEL – SAND – SILT							
			4 μm	SILT	GRAVEL SILT / SANDY SILT / CLAYEY SILT / SANDY-CLAYEY SILT	SILT		DIATOM SILT							
				CLAY		CALCAREOUS	FORAMINIFERA CLAY / NANNO-FORAM. CLAY / FORAM-NANNO. CLAY	MARL	FORAMINIFERA MARL / NANNO-FORAM. MARL / FORAM-NANNO. MARL	CALCAREOUS	FORAMINIFERA OOZE / NANNO-FORAM. OOZE / FORAM-NANNO. OOZE			CALCAREOUS > BIOSILICEOUS	(RELATIVE AMOUNTS) 0 — 30% FORAMINIFERA — 60% — 100% CALC. / 0 — 30% NANNOFOSSILS — 60% / TERRIGENOUS PARTICLES 10-30%
					GRAVEL CLAY	CLAY	NANNOFOSSIL CLAY		NANNOFOSSIL MARL	OOZE	NANNOFOSSIL OOZE				
					SANDY CLAY	SANDY CALCAREOUS CLAY / SILTY CALCAREOUS CLAY		SANDY MARL / SILTY MARL		SANDY CALCAREOUS OOZE / SILTY CALCAREOUS OOZE					
					SILTY CLAY	CALCAREOUS SILTY CLAY / BIOSILICEOUS CALCAREOUS CLAY		BIOSILICEOUS MUD-MARL		BIOSILICEOUS-CALCAREOUS OOZE					
				PELAGIC	SAND-SILT-CLAY	CALCAREOUS - BIOSILICEOUS CLAY / BIOSILICEOUS SILTY CLAY	BIOSILICEOUS	RADIOLARIA CLAY / DIATOM-RAD. CLAY / RAD.-DIATOM CLAY	MARL-BIOSILICEOUS MUD	BIOSILICEOUS	RADIOLARIA MUD / DIATOM-RAD. MUD / RAD.-DIATOM MUD	CALCAREOUS - BIOSILICEOUS OOZE	BIOSILICEOUS	RADIOLARIA OOZE / DIATOM-RAD. OOZE / RAD.-DIATOM OOZE	BIOSILICEOUS > CALCAREOUS	(RELATIVE AMOUNTS) 0 — 30% RADIOLARIA — 60% — 100% / 0 — 30% DIATOMS — 60% / TERRIGENOUS PARTICLES 10-30%
				CLAY	McCOY 1983	CLAY	DIATOM CLAY	MUD	DIATOM MUD	OOZE	DIATOM OOZE				
						SANDY BIOSILICEOUS CLAY / SILTY BIOSILICEOUS CLAY		SANDY BIOSILICEOUS MUD / SILTY BIOSILICEOUS MUD		SANDY BIOSILICEOUS OOZE / SILTY BIOSILICEOUS OOZE					

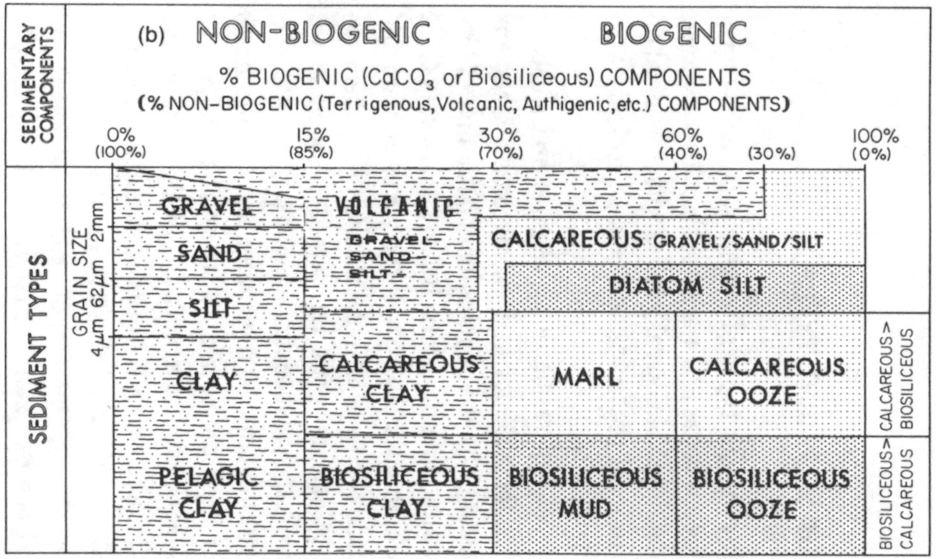

Fig. 7. Sediment classification scheme: (a) detailed 73-category format used for smear slide data and CaCO$_3$ data. (b) Simplified format of 12 sediment types used for initial plotting, and from which the three sediment types mapped in Fig. 8 are derived.

biogenic debris of siliceous composition, the term "biosiliceous" is used synonymously with "biogenic siliceous," the prefix "bio" being applied to prevent ambiguity with nonorganic siliceous detritus such as clay and volcanic ash.

Smear slide data were input initially into a 73-category classification format (Fig. 7a). These data then were summarized in a 12-category scheme (Fig. 7b) and subsequently mapped as only three sediment types (Fig. 8). Control in mapping sediment contacts was provided by the more complex 73-category format.

V. DISTRIBUTION OF MAJOR SEDIMENT TYPES

The distribution of the three major types of marine sediments—terrigenous and pelagic clays, calcareous oozes and marls, and biosiliceous oozes and muds—is a function of planktonic biological productivity, chemical effects, terrestrial input, redepositional processes and benthic biological activity in about that order of significance.

Planktonic biological productivity results in calcareous and siliceous microfossils that accumulate on the seafloor as biogenic oozes. Chemical effects are expressed predominantly by the postmortem dissolution of biogenic microfossils; those of calcareous composition below a mid-water CCD and those of siliceous composition both throughout the water column and within the sediment. Authigenic chemical activity results in a varied assortment of minerals within pelagic deposits that are present in minor amounts, such as zeolites, phosphates, glauconite, dolomite, clays, etc. Extensive accumulations of polymetallic (or ferromanganese, or manganese) nodules and crusts also result from authigenic reactions.The terrigenous contribution predominantly comes from the denudation of the surrounding continents, with a minor contribution from mid-ocean islands and seamount-ridge features underwater, and leaves a clay blanket over the entire basin. Redepositional processes occur through erosion of seafloor sediments by bottom currents, slumping and benthic faunal activity. The latter usually results in vertical mixing of lithologies rather than significant lateral displacement of material, and is a function of net sedimentation rates (Berger and Killingley, 1982). Net sedimentation rates are, in turn, a function of biological productivity, terrigenous influx, chemical dissolution, and reworking of sediment.

The interrelationship of these processes is obvious. The rain rate of particles through the water column usually exceeds their accumulation rate on the seafloor in terms of both short-term and long-term geologic time

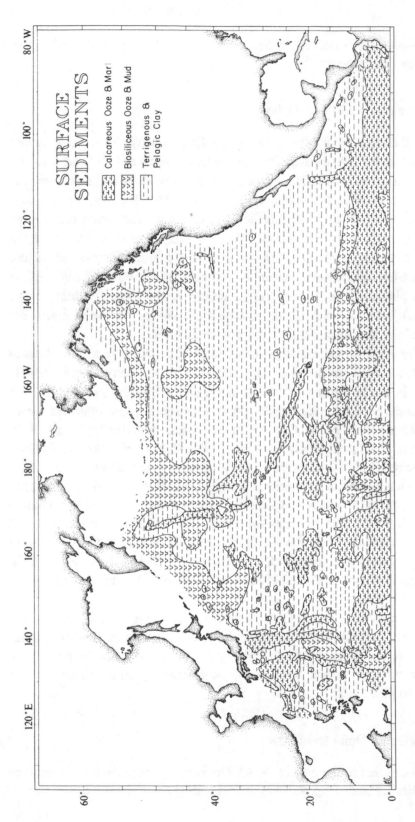

Fig. 8. Distribution of surficial sediments on the seafloor of the North Pacific Ocean in terms of the three major types of pelagic sediments: terrigenous and pelagic clays, calcareous oozes and marls, and biosiliceous oozes and muds. See Fig. 7 for definitions.

scales—much is dissolved, eaten, or otherwise destroyed or modified in transit or after deposition.

A. Terrigenous and Pelagic Clays

Of the three major types of pelagic sediments, clays form the dominant sediment on the North Pacific seafloor (Fig. 8). Much of it is "red" clay on the abyssal plains, reflecting the vast area of the seafloor that lies below the CCD where calcareous components are dissolved leaving a residue of clays. The bulk of these clays are detrital, having been transported offshore as suspended sediments. Significant volumes of clays and silts also are blown off the continents by winds. Minor amounts of iron manganese oxides and zeolites often are present, usually in association with volcanic material, as well as phosphatic (e.g., fish teeth), chitinous, and meteoric debris (e.g., tectites.). These trace minerals become more obvious in the sediment because of the lack of dilution by calcareous material.

Silty clays are prevalent proximal to continental areas where turbidity currents flow into trenches or out onto the abyssal plains. Sands usually are left in the hemipelagic deposits along the continental margins and are prevented from dispersing farther offshore by trenches that intercept downslope movement of detritus, except in the northeastern Pacific (Neeb, 1943; Hamilton, 1967; Anikouchine and Ling, 1967; Horn et al., 1969, 1970, 1971, 1974; Scholl, 1974). Sediments of glacial–marine origin show an increase of coarse clastic content primarily in hemipelagic sediments along the continental margins of Alaska and Kamchatka where late Quaternary glaciation occurred. Today, few icebergs survive and drift through the Bering Strait and Sea into the North Pacific (Hibler, 1980); this, plus the trapping of detritus by trenches, prevents a significant input of glacial–marine material to pelagic sediments. The distribution of hemipelagic sediments cannot be adequately portrayed here.

Sedimentation rates vary considerably, from a meter or more per thousand years in turbidites and slumps along continental and insular margins or in trenches, to less than a millimeter per thousand years for pelagic red clays in the open ocean. Average physical properties of a pelagic red clay are listed in Table I.

B. Calcareous Oozes and Marls

Calcareous sediments are the next most abundant pelagic sediment type in the North Pacific (Fig. 8). They reflect biologic productivity by the calcite-

TABLE I

Average Values for the Physical Properties of North Pacific Pelagic Sediments[a]

Sediment type	Grain size	Biogenic content	Wet bulk density (gm/cc)	Bulk grain (mineral) density (gm/cc)	Water content	Porosity
Pelagic red clay	1 μm: (clay)	<15% $CaCO_3$	1.35–1.40	2.70	125%	75–80%
Nannofossil marl/ooze	2–6 μm: (clay/fine silt)	30–100% $CaCO_3$	1.35–1.70	2.61–2.71	120%	65–80%
Radiolarian mud/ooze	2–5 μm: (clay/fine silt)	30–100% SiO_2	1.15	1.90–2.30	340–390%	80–90%
Diatomaceous mud/ooze	2–3 μm: (clay/fine silt)	30–100% SiO_2	1.20–1.40	2.45	150%	70–90%

[a] From Baas-Becking and Moore, 1959; Hamilton, 1974; Horn et al., 1974; Hurd and Theyer, 1977; Hamilton et al., 1982; Hamilton and Bachman, 1982.

and aragonite-secreting organisms that produce nannofossils (coccolithophores), microfossils (primarily foraminifera and pteropods), and, less commonly, larger fossil remains of shells, corals, etc. While whole fossils are obvious, broken fragments can be very difficult to identify, and thus many fossil components may be underestimated in smear slides. Coccolithophores and foraminifera as sedimentary components in pelagic deposits are discussed in Section VI. Calcareous productivity is dominantly planktonic, occurring in the euphotic zone of the water column with deposition of the carbonate remains on the seafloor above CCD levels (Sliter *et al.*, 1975; Broecker and Peng, 1982). Calcareous sediments thus mantle ridges and seamounts, the usual analogy being to snow capping mountains on land, with the snowline being the underwater CCD level; the corresponding simile to the latter has resulted in the term "carbonate line."

These factors are reflected in the distribution of $CaCO_3$ values in sediment (Fig. 9). High values reflect the equatorial belt of productivity by carbonate-secreting pelagic organisms. High values also outline islands and near-surface seamounts where coral and algal reefs supply carbonate-rich material to oceanic sediments, such as in the Panama Basin where bottom currents winnow and transport fine-grained carbonates into deeper water from surrounding ridges (Moore *et al.*, 1973). Almost any seafloor area above CCD levels, removed from significant influx of terrigenous detritus, has calcareous sediment reflecting zooplankton/phytoplankton productivity in the open ocean. Details of these patterns cannot be shown here; thus the hundreds of seamounts, guyots, and islands that are speckled across the Pacific Basin and capped with carbonate-rich sediments are impossible to portray. The extent of carbonates on major ridges and rises has been overemphasized for cartographic clarity.

CCD levels in mid-ocean typically are at 4–5 km water depth and generally follow regional isobaths, but rise to 3 km or less along the continental margins (Moore *et al.*, 1973; Berger *et al.*, 1976; Thunell *et al.*, 1981). Below these depths calcareous particles are dissolved, their rate of supply to the seafloor being balanced by dissolution rates within both the deep-water masses and the pore waters within the sediment.

Average physical properties for a nannofossil marl and ooze are summarized in Table I. Other types of calcareous sediments such as foraminiferal silts and sands, which do not form significant deposits on the deep-ocean floor, have considerably different physical property values. Grain size or porosity, for example, can have much higher values, particularly where a micritic or clay matrix is not present to fill interstitial spaces or foraminiferal tests, or where much biogenic burrowing and mixing occurs (Rhoads

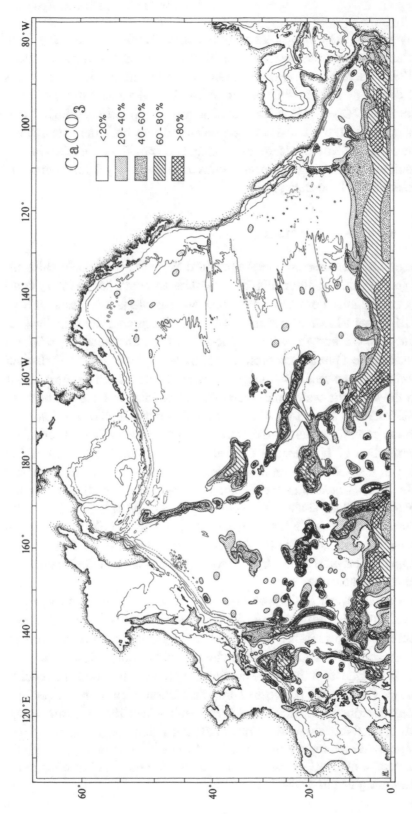

Fig. 9. Concentration of CaCO₃ in seafloor sediments of the North Pacific Ocean expressed as weight percent of dry sediment. Data points are most of those shown in Fig. 5, with added information from the literature. The >80% isopleth is mapped only where discernable at this scale.

and Boyer, 1982). In two areas of the North Pacific, the Ontong Java Plateau and the southern area of the Northeast Pacific Basin, Hamilton *et al.* (1982) describe a decrease in mean grain size within calcareous oozes lying at depths of 1600 to 4900 m due to breakage (fragmentation) of foraminiferal tests. Mayer (1979) has noted a correlation between higher calcareous concentrations and increased values of wet bulk densities. Sedimentation rates for calcareous pelagic deposition, exclusive of redepositional events such as biocalcareous turbidites, vary from millimeters to decimeters per thousand years.

C. Biosiliceous Oozes and Muds

Biogenic siliceous sediments are the third major type of pelagic sediment on the North Pacific seafloor (Fig. 8). Their characteristic sedimentary components are the microfossils from diatoms and radiolaria (discussed in Sections VI.B.1 and VI.B.2). Less abundant, but widespread, are spicules broken from sponges and remains of silicoflagellates. Distributions of these microfossils reflect planktonic productivity in the euphotic zone with distinct affinities to water masses. Diatomaceous oozes are deposited on the seafloor beneath the subarctic water mass. Diatom and radiolarian oozes lie beneath the transitional and equatorial water masses, and radiolarian oozes dominate beneath the central portion of the equatorial water mass (Figs. 3b and 8).

Distribution of biosiliceous deposits in the eastern equatorial region often may be masked by calcareous particles and thus are obvious only below the CCD. Mixed calcareous–biosiliceous sediments are not common on the North Pacific seafloor since they occur only in regions of high productivity in the narrow depth range between the lysocline and CCD. In the northeastern Pacific, the extent of biosiliceous sediments is not as extensive as that mapped by Nayudu and Enbysk (1964) due to masking of the biosiliceous material by volcanic shards.

Because the water in the world ocean is undersaturated with respect to opal, significant dissolution of biosiliceous debris occurs. Although dissolution rates are considerably lower than those for carbonates in the deep waters (Berger, 1976; Hurd and Birdwhistell, 1983), only a few percent of the opaline residue from the total standing crop in the euphotic zone reaches the seafloor (Heath, 1974). Most of the dissolution occurs on the seafloor and accumulations of biosiliceous oozes result where the rain rate is significantly higher than dissolution rates (Broecker and Peng, 1982).

Sedimentation rates are on the order of 1 cm/10^3 yr in high-productivity areas. Average physical properties for radiolarian and diatomaceous muds and oozes are given in Table I.

VI. DISTRIBUTION OF DOMINANT SEDIMENTARY COMPONENTS

A. Biogenic Calcareous Particles

Two types of biogenic calcareous particles dominate in deep-sea sediments, tests of foraminifera and platelets from disintegrated coccospheres of calcareous nannoplankton. The distribution of pelagic calcareous sediment types formed by concentrations of these particles in excess of 30% (Fig. 7), foraminiferal and nannofossil oozes and marls, are mapped in Fig. 10. The distribution of foraminifera and calcareous nannofossils in terms of relative abundance estimates from smear slides is mapped in Figs. 11 and 12.

1. *Foraminifera*

Foraminifera are heterotrophic protozoa with either calcareous or agglutinated shells that occupy both planktonic (primarily euphotic) and benthonic habitats. Shell sizes are usually between about 50 and 500 μm, within the coarse silt to coarse sand size on the Wentworth grade scale (see Fig. 7). Larger shells of benthic foraminifera up to a centimeter (pebble size) are often found in shallow-water deposits near reefs, but are rare in pelagic deposits. Calcareous planktonic foraminifera are overwhelmingly dominant in deep-sea sediments; benthic species are present only in trace amounts, although they are more common near continental margins where the input of organic carbon is higher.

Foraminifera are most common in calcareous deposits mantling the shoal portions of major rises, plateaus, and island chains (Fig. 11). This distribution reflects a planktonic productivity in the warmer surface waters of the tropical and western subtropical water masses (Fig. 3b) as well as good preservation above the CCD. With increasing depth below the lysocline, the less-resistant species become fragmented and disappear, so that only the most robust forms occur near the CCD. Planktonic foraminiferal tests sink at rates of 0.3 to 4 cm/sec., the higher rate applying only to rare thick-walled tests measured under nonturbulent laboratory conditions (Berger, 1971). Accordingly, a test would settle to the deep seafloor (5 km) within a day to a month depending on the test size. Thus, distributions of tests in seafloor sediments reflect their biological habitat and provenance.

The taxonomic composition of planktonic foraminiferal assemblages shows distinct variations corresponding to surface water mass properties and to the depth of the seafloor. Coulbourn *et al.* (1980), Thompson (1981), Sverdlove (1983), and Bé *et al.* (1985) have reported on the distribution and abundance of species in various regions of the North Pacific. In the

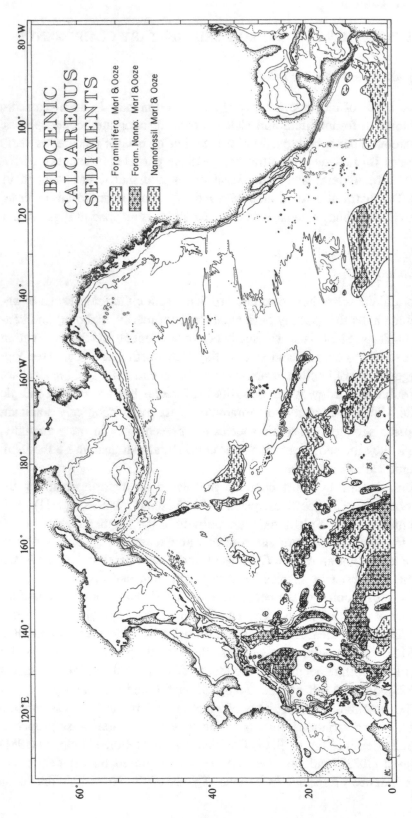

Fig. 10. Distribution of the two major types of calcareous biogenic pelagic sediments on the seafloor of the North Pacific Ocean—foraminifera and nannofossil oozes/marls. These sediments are defined in Fig. 7b; overlapping areas of their distributions define foraminifera–nannofossil oozes and marls.

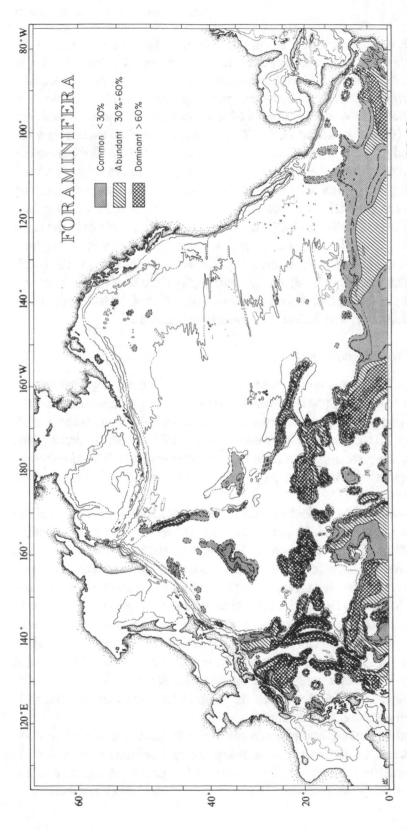

Fig. 11. Distribution of foraminifera in North Pacific pelagic sediments. Relative abundances are based on smear slide analyses and CaCO₃ measurements on sediment from core tops. No distinction has been made between planktonic and benthonic types. Blank areas are where foraminifera are not present or where data are inadequate; dashed line within the pattern representing <30% concentrations ("common") approximately delineates the 15% isopleth of foraminifera abundances.

eastern Pacific, and especially in the Panama Basin (Bé *et al.*, 1985), diversity is fairly high in the water column, but strong dissolution results in a depauperate sediment assemblage composed primarily of *Neogloboquadrina dutertrei*. The most shallow equatorial sites, particularly in the western Pacific, show greater occurrences of *Globigerinoides, Pulleniatina,* and *Globorotalia* (*G. menardii.*and *G. tumida*). Seamounts and plateaus below the subtropical gyre and in the Phillipine Sea are dominated by *Globigerinoides* in the shallowest samples, and by *Globigerinita glutinata* in deeper samples (Thompson, 1981; Coulbourn *et al.,* 1980). Calcareous sediments underlying the transition zone (Hess and Shatsky rises) have rare *Globorotalia inflata* and *Neogloboquadrina pachyderma* (Thompson, 1981), while only the latter species occurs on the Emperor Seamount chain below the subarctic gyre.

There has been no systematic study of benthic foraminifera in the North Pacific. The vast abyssal regions are essentially devoid of all foraminifera. In the equatorial Pacific, workers such as Vincent *et al.* (1981) report *Planulina, Oridorsalis,* and *Nuttalides,* among other genera.

2. *Calcareous Nannofossils*

Coccolithophores are a photoautotrophic, unicellular alga that secrete shieldlike calcareous platelets which completely or partially surround the cell. The complete organism forms a coccosphere holding an average of 20 platelets, with as many as 150 platelets (Honjo, 1976). With the death of the organism, the platelets separate and settle to the seafloor. The morphology of platelets, most of which are elliptical or circular in shape, is characteristic of a species.

An individual nannofossil platelet, or coccolith, has a diameter of about 1 to 20 μm, typically 5–10 μm in the clay to fine silt size range. Settling rates for these tiny particles are on the order of 100 m/yr, so that 50 years would be needed for a coccolith to reach a 5 km depth. However, zooplankton graze on coccolithophores, so that the predominant mechanism of transport for platelets to the seafloor is within fecal pellets (Bishop *et al.,* 1977). These pellets settle at rates of 40 to 400 m/day, with a mean rate of about 200 m/day for 100-μ sized (very fine sand) pellets (Fowler and Small, 1972; Honjo, 1976). Accordingly, anywhere from half a month to a few months, or slightly less than a month at the mean rates, is required for a coccolith encased in a fecal pellet to reach the deep (5 km) seafloor, considerably faster than an individual coccolith.

Sheathing by the fecal pellets protects the nannofossils from dissolution and this factor, in combination with their planktonic habitat and rapid settling rate through the water column, produces a widespread pattern to their distribution in deep-sea sediments (Fig. 12). Considering that each fecal pellet

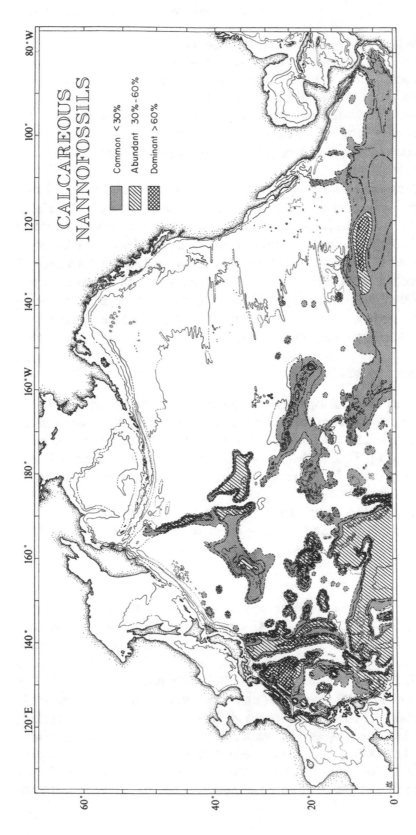

Fig. 12. Distribution of calcareous nannofossils in North Pacific pelagic sediments. Relative abundances are based on smear slide analyses and $CaCO_3$ measurements on sediment from core tops. Blank areas are where calcareous nannofossils are not present or where data are inadequate. Dashed line within the pattern representing <30% concentrations ("common") approximately delineates the 15% isopleth of calcareous nannofossil abundances.

could contain 100,000 coccoliths, the significance of the calcareous nannofossils as a sedimentary component is obvious.

The small size of coccoliths results in a frequent underestimation of
their abundance in smear slide analyses, which has been taken into account
here. Their distribution pattern (Fig. 12) is similar to that of the total carbonate content in pelagic sediments (Fig. 9), which reflects the relatively
greater resistance to dissolution of nannofossils and their abundance in the
plankton.

Geitzenauer *et al.* (1976) have reported the regional distribution of coccoliths in Pacific sediments. *Gephyrocapsa* is overwhelmingly dominant
throughout the tropical and western subtropical Pacific underlying the North
Pacific Current and the Kuroshio Current. *Umbellosphaera* and *Rhabdos-
phaera* are more common below the subtropical gyre, while *Cyclococcoli-
thina* occurs in the northern part of the subtropical area and in the transition
zone. Within the subarctic gyre *Coccolithus pelagicus* constitutes a monospecific assemblage, apparently due to its greater resistance to dissolution
rather than to any ecologic limitation.

B. Biogenic Siliceous Particles

Diatoms and radiolaria are the source of most biosiliceous particles.
Where concentrations of these particles are greater than 30%, the resulting
pelagic sedimentary accumulations form radiolarian or diatomaceous muds
and oozes (Fig. 7). Distribution of these two types of biogenic siliceous
sediments is shown in Figure 13. Relative abundances of diatoms and radiolaria are shown in Figures 14 and 15, respectively. Distributions of other
silica-secreting organisms, such as silicoflagellates and sponge spicules, are
not mapped because they are very minor components of pelagic assemblages.

1. *Diatoms*

Diatoms are algae which occupy the euphotic zone and tolerate a wide
range of environments. High abundances of diatoms, however, are restricted
to sediments underlying areas of high productivity in upper water masses
such as coastal areas, regions of upwelling, or nutrient-rich waters along
oceanographic convergence zones. All three situations are obvious in Figures 13 and 14: (1) the very high abundance of diatoms off coastal northeastern Asia; (2) the diatom oozes and muds marking equatorial upwelling
areas in the eastern equatorial Pacific, as well as the absence of diatoms in
much of the central Pacific basin sediments where there is no upwelling;
and (3) the diatom oozes forming a band across much of the northern Pacific
seafloor south of the Aleutians. Diatoms also can form a dominant sedi-

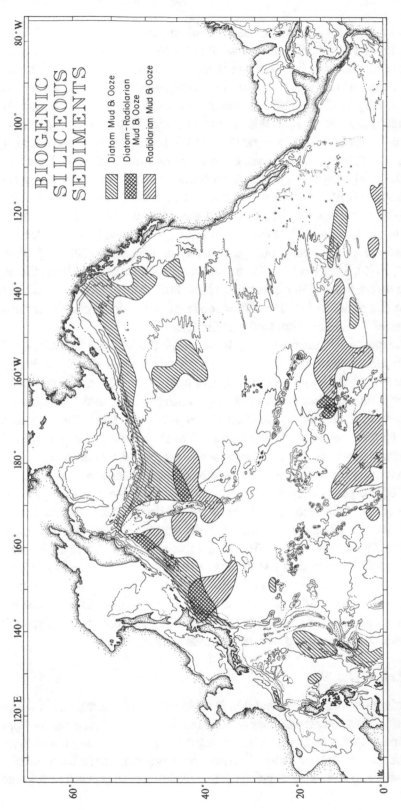

Fig. 13. Distribution of the two major types of biosiliceous pelagic sediments in the North Pacific Ocean—diatom and radiolarian oozes/muds. These sediments are defined in Figure 7b; overlapping areas of their distributions define diatom—radiolarian oozes and muds.

mentary constituent in seafloor deposits as residues where carbonates have been removed by dissolution below CCD levels, or as local concentrations produced by mass-wasting processes such as slumps into small, deep basins.

Diatoms secrete a perforate frustule composed of two valves of opaline silica, usually in the 10–100 μm size range, or in the fine silt to very fine sand size range. Because of this minute size and the perforations, frustules settle to the seafloor slowly at rates of only 0.1–0.6 cm/sec (Smayda, 1971). At these rates, a diatom can take up to a few months to settle to 5 km; thus they are widely dispersed by ocean currents. While an organic coating may protect some frustules from dissolution during settling, many are dissolved within the upper few hundred meters of the water column and it is only the more robust types—with fewer perforations, thicker-walled frustules, and lower degree of opal hydration—that survive in pelagic sediments (Hurd and Thayer, 1975). Diatoms can also be incorporated into fecal pellets, thereby settling through the water column at rates three times that of an individual frustule (Smayda, 1971). This latter mode of transport probably accounts for most of the diatom flux to the sediments.

Kanaya and Koizumi (1966), Jousé et al. (1971), and Sancetta (1979) have discussed the taxonomic distribution of diatoms in the North Pacific. The eastern equatorial diatom oozes and muds represent a diverse assemblage of *Thalassiosira, Coscinodiscus, Thalassionema,* and *Thalassiothrix* species, with numerous others (Burckle, unpublished data). The concentration of diatoms in the western equatorial Pacific represents a monospecific ooze of the resistant genus *Ethmodiscus.* Along the course of the highly productive Oyashio Current, the sediment assemblage is dominated by spores of the genus *Chaetoceros,* common in the plankton of coastal upwelling areas. The oozes and muds of the subarctic Pacific are dominated by the pennate genus *Denticulopsis,* with lesser numbers of large *Coscinodiscus.* The overall distribution of diatoms shown in Figure 14 is similar to mapped distributions of absolute numbers of diatom frustules presented by Jousé et al. (1971) and Lisitzin (1971, 1972). The latter are quantitative data calculated on a carbonate-free basis, whereas our qualitative smear slide data were estimated on a relative basis including carbonates; discrepancies thus are most pronounced in high-carbonate areas.

2. Radiolaria

Radiolaria form the second dominant biosiliceous microfossil in pelagic sediments. They are planktonic protozoa that live predominantly in the upper water masses with certain species having the capability to live over a wider depth range into deeper water. Skeletal size variations range from about 50 μm (coarse silt) to 400 μm (medium sand). Skeletons are generally perforated

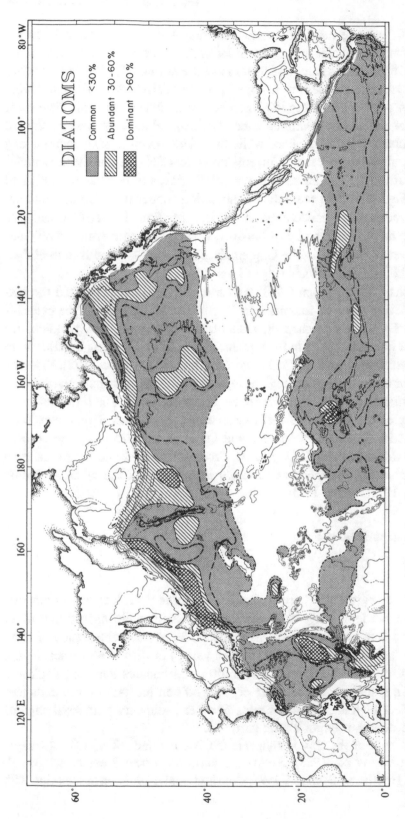

Fig. 14. Distribution of diatoms in North Pacific pelagic sediments. Relative abundances are based on smear slide analyses of sediment from core tops. Blank areas are where diatoms are not present or where data are inadequate; dashed line within the pattern representing <30% concentrations ("common") approximately delineates the 15% isopleth of diatom abundances.

and show varying rates of postmortem dissolution. Gravitational settling rates through seawater are on the order of 15–200 m/day, with a maximum rate of 416 m/day. As much as a year or as little as a couple of weeks therefore would be required for settling to abyssal depths (Takahashi and Honjo, 1983).

Two distinct zones of radiolarian accumulations are clear in Figure 15, a tropical and a middle-latitude zone. The tropical zone extends northward to the subtropical convergence while the middle-latitude zone occurs along the Kuroshio–North Pacific Currents down to 40°N (Nigrini, 1970; Lisitzin, 1972; Burnett, 1975; and Kruglikova, 1981). Areas of radiolarian muds and oozes (Fig. 13) reflect sea floor accumulations under these tropical and subarctic radiolarian productivity zones. Perturbations in distributional patterns are due to masking by carbonates, which is most pronounced off Central America, such as in the Guatemala Basin, on the Line Islands Ridge, and in the Marshall Islands (Fig. 13 and 15).

Sachs (1973), Nigrini (1968), and Moore (1978) have reported the taxonomic occurrence of radiolaria in North Pacific sediments. Spumellarids such as *Euchitonia* (Y-shaped), *Ommatartus* (elongate), and pylonids, are the most common forms in the equatorial and tropical zones, together with some nassellarids such as *Theoconus* (Nigrini, 1968; Moore, 1978). Pylonids persist into the subtropical region, together with circular spumellarids such as *Ommatodiscus* and *Stylochlamydium* (Moore, 1978). The transition zone is characterized by a mixture of nassellarids (*Theoconus, Siphocampe*) and spumellarids (elliptical *Phorticium* and *Lithelius,* spherical large-pored *Haliomma*) (Moore, 1978; Sachs, 1973). *Stylochlamydium* and the nasselarian *Siphocampe, Theocalyptra* and *Tristylospyris* dominate the subarctic Pacific (Moore, 1978; Sachs, 1973).

C. Terrigenous Detritus

1. *Clays*

Clay-sized particles are the major constituent in deep-sea sediments. They are predominantly clay minerals but also include micrite and other clay-sized mineral grains such as quartz. The relative abundance of clay-sized (2–4 μm) debris (Fig. 16) reflects a variety of natural processes: higher abundances might be due to dissolution of carbonates leaving pelagic red clays as a residue or to the influx of clays off continents; lower abundances may reflect high-accumulation rates of other sedimentary material such as biogenic debris or larger clastic particles.

The bulk of the clay is allogenic (El Wakeel and Riley, 1961; Biscaye, 1965; Griffen *et al.,* 1968). Authigenic clays form near areas of active volcanism (Heath and Dymond, 1977; Cole and Shaw, 1983), in association with

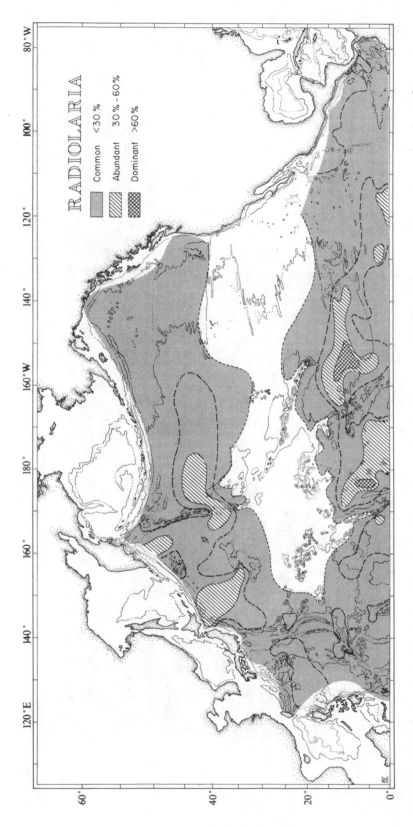

Fig. 15. Distribution of radiolaria in North Pacific pelagic sediments. Relative abundances are based on smear slide analyses of sediment from core tops. Blank areas are where radiolaria are not present or where data are inadequate. Dashed line within the pattern representing <30% concentrations ("common") approximately delineates the 15% isopleth of radiolaria abundance.

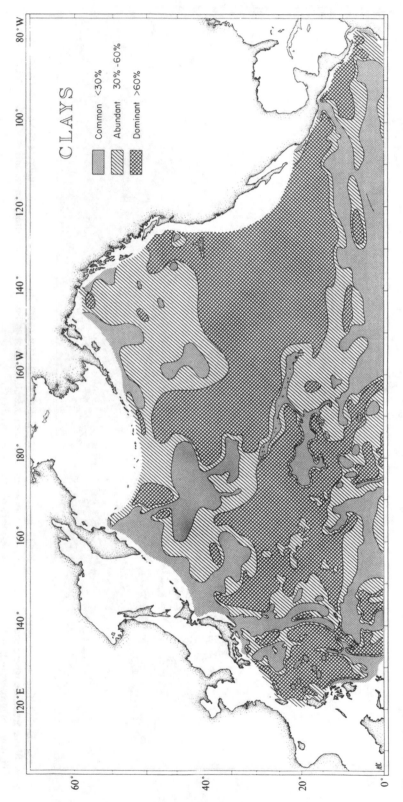

Fig. 16. Distribution of clay-sized detritus in North Pacific pelagic sediments. Relative abundances are based on smear slide analyses of sediment from core tops.

altered volcanic debris (Aoki *et al.*, 1974), from the low-temperature combination of detrital metalliferous hydroxides and biogenic silica (Heath and
Dymond, 1977; Hein *et al.*, 1979b), or where subareal weathering of volcanic
terrains in tropical climates contributes amorphous aluminosilicates for authigenic clay mineral formation (Moberly *et al.*, 1968; Moberly, 1963).

Aggregation of fine-grained detritus, through flocculation or incorporation into fecal pellets, is the dominant method by which clays settle to the
seafloor, rather than by individual particle deposition (McCave, 1975). Such
aggregates have much higher settling velocities than individual clay or silt-
sized particles. Data derived from mid-ocean sediment traps suggests that
most of the vertical mass flux of fine-grained particles (<53 μm) to pelagic
deposits through the upper 400 m in equatorial areas is composed of large
(>53 μm) aggregated particles, primarily fecal pellets (Bishop *et al.*, 1977).
The flux of these settling aggregates has been shown to be commensurate
with sedimentation rates in abyssal areas (McCave, 1975; Bishop *et al.*,
1977).

As individual, nonaggregated particles, clays can be distributed as suspended sediment over wide areas by ocean currents because of their small
size and settling velocities that require years or decades for deposition at
abyssal depths. Clays also can be transported in traction currents as mud
turbidites into trenches or onto hemipelagic aprons (Piper, 1972, 1978; Rupke
and Stanley, 1974), such as off the Columbia River (Kulm and Scheidegger,
1979).

Erosion, resuspension, and transport of pelagic clays by strong bottom
currents has been documented in various areas of the Pacific: within the
Samoan Passage just south of the Central Pacific Basin (Hollister *et al.*,
1974), around seamounts and guyots in the central Pacific region (Lonsdale
et al., 1972), on Hess Rise (Nemoto and Kroenke, 1981), over the abyssal
seafloor between the Clarion and Clipperton fracture zones (Johnson and
Johnson, 1970; Johnson, 1972, 1976; Marchig, 1978; Craig, 1979), and in
portions of the Panama Basin (Kowsmann, 1973; Dowding, 1977; Heezen
and Rawson, 1977). Seafloor erosion due to bottom currents has been inferred from acoustic records throughout much of the northwestern Pacific,
especially around and between seamounts, ridges, and rises (Damuth *et al.*,
1983). Resuspension and transport of sediment on a regional scale is implied
by the nepheloid layer (Ewing and Connary, 1970; McCave and Swift, 1976;
Biscaye and Eittreim, 1977), which primarily carries allogenic clay-sized
particles (Carder *et al.*, 1971).

Significant volumes of clay are transported out to sea by aeolian processes from both continents. Airborne dusts are predominantly illite over
the North Pacific (Griffin *et al.*, 1968; Windom, 1975) and other minerals,

usually quartz. Such dust is almost completely (90%) composed of clay-sized particles (Ferguson *et al.*, 1970; Rahn *et al.*, 1981). The residence time for aerosols in the atmosphere, or the duration of their transport and settling (gravitational or by incorporation into rain drops) through the atmosphere before arriving at the sea surface, is a function of grain size and altitude. Particles that are fine silts and coarse clays can remain in the troposphere, which is the level of the jet streams, for about a month, while clays higher in the stratosphere might remain suspended for up to two years (Prospero *et al.*, 1983).

Most of the dust is present in the troposphere, where residence times are ample for extensive dispersal out to sea. Trajectories of Asian dust at this altitude extend as much as 12,000 to 15,000 km from source areas (Rahn *et al.*, 1981). Input of clay-sized aerosols higher into the stratosphere probably occurs only by thunderheads and volcanic plumes, thus transport at this level is not usually significant. High concentrations of quartz in North Pacific sediments define an east–west zone at middle-latitudes (Rex and Goldberg, 1958; Windom, 1975; Leinen and Heath, 1981) that directly underlies the jet stream flow, particularly the summer flow pattern (Fig. 4a). Peak dust loads in Hawaii occur between April and May (Parrington *et al.*, 1983). An influx of airborne particulates occurs into the Panama Basin from South America (Kowsmann, 1973).

Rex and Goldberg (1958) estimated that 50% of the particulate nonbiogenic debris in pelagic sediments was aeolian material. Windom (1969, 1975), in noting the similarity between the mineralogy of pelagic sediments and airborne dusts, calculated that about half of the less than 2 μm-sized (clay size) particulates in these sediments was of aeolian derivation, and that all of the particles of greater than 2 μm size represented aeolian derivation. These calculations indicated that aeolian material represented 75% of the total particulate contribution to central North Pacific abyssal sediments. The flux of airborne debris at the sea surface has been measured at between 15–35 $mg/cm^2/10^3$ years in the northern Line Islands (Eniwetok Atoll) (Duce *et al.*, 1980; Turekian and Cochran, 1981), to 180 $mg/cm^2/10^3$ years in deep-sea sediments north of the Hawaiian Islands (Janecek and Rea, 1983). A single storm off Asia carried 240 tons of dust per hour, or 6000 tons per day (Rahn *et al.*, 1981).

2. Detrital Clastic Minerals

The input of clastics from land is particularly noticeable in the distribution of detrital minerals (Fig. 17), which primarily indicate resedimentation processes by mass-wasting along continental and insular margins. Abundant concentrations of the common detrital minerals, quartz, feldspar, mica and heavy minerals, occur in the silty clays within trenches such as the Alaska, Japan, Bonin, and Middle America trenches (Fig. 17). Where large land masses are not proximal to trenches to form a detrital source of clastics, such as in the Ryukyu Trench, or where mid-slope basins trap turbidity currents before they reach the trench floor as occurs in the Japan–Kuril–Kamchatka trenches (Uyeda, 1974; Scholl, 1974), the sediment fill within the trench is clay-rich (Fig. 16). Off North America, sediments containing higher proportions of clastics extend offshore onto the Alaska and Tufts abyssal plains (Figs. 16 and 17), the result of active turbidite deposition off the continent unconstrained by physiographic barriers like trenches (Hamilton, 1967; Horn et al., 1969, 1971, 1972, 1974).

Detrital mineral distributions in sediments between the Kamchatka Trench and the Shatsky Rise–Emperor Seamounts (Fig. 17) approximately define the area of ice-rafted material described by Connolly and Ewing (1970). Ice-rafted debris forms only a small proportion of the clastic material in the eastern Aleutian Trench (Piper et al., 1973). Feldspar is abundant in the detritus off Central America and within the Marianas Trench, probably reflecting volcanism (Peterson and Goldberg, 1962). Detrital minerals in the western Pacific represent erosional debris from islands and aeolian contributions from Asia.

Extraterrestrial particles (microtectites, cosmic spherules) have been found in surficial sediments throughout the North Pacific (Glass et al., 1979; Blanchard et al., 1980; Han-chang et al., 1982; Parkin et al., 1983). These glassy, stony, iron spherules are usually black (rarely white), rounded, spherical or dumbbell-shaped, and occur in such small numbers that their presence on smear slides is unusual. They were first identified in *Challenger* expedition samples (Murray and Renard, 1883). Annual worldwide flux of this material has been estimated to be on the order of 30 to 5000 metric tons/year (Bruun et al., 1955; Pettersson and Fredriksson, 1958).

D. Volcanic Detritus

The distribution of volcanic detritus (Fig. 18) reflects the distribution of major tephra-producing volcanoes within and around the Pacific Basin that have been active during the Quaternary. This detritus is usually recognizable as glass shards in the finer sand sizes and coarser silt sizes. Rare grains of volcanic rock fragments may be present, and much of the feldspar included with the detrital minerals (Fig. 17) is probably of volcanic origin (Peterson and Goldberg, 1962; Rehm, 1983).

An input of volcanic debris occurs as a finite influx from an eruption via a tephra-laden cloud or by the mechanical abrasion of floating pumice fragments. Erosion of volcanic terrains above and below sea level also con-

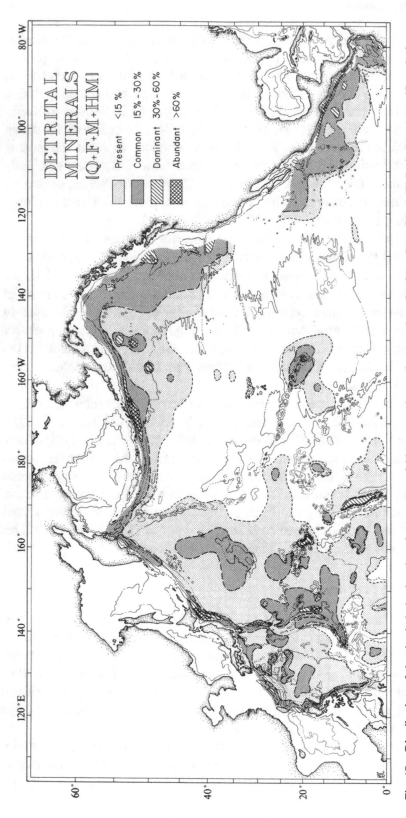

Fig. 17. Distribution of the detrital clastic minerals, quartz + feldspar + mica + heavy minerals (Q + F + M + HM), in North Pacific pelagic sediments. Relative abundances are based on smear slide analyses of sediment from core tops. Blank areas are where these detrital components are present in trace quantities, absent, or where data are inadequate.

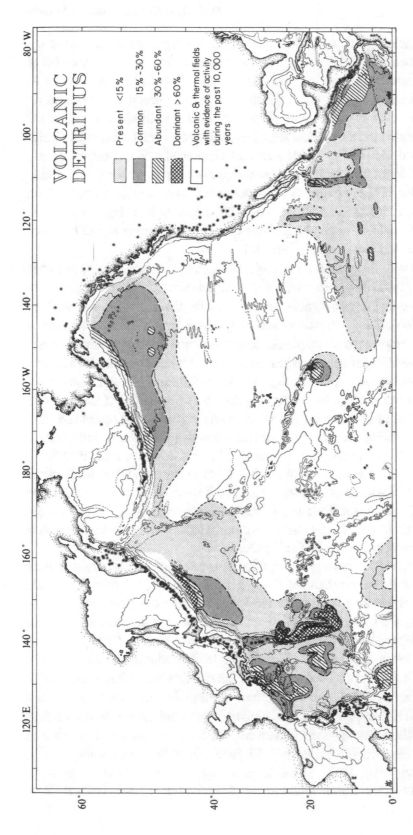

Fig. 18. Distribution of volcanic debris, predominantly glass shards, in North Pacific pelagic sediments. Relative abundances are based on smear slide analyses of sediment from core tops. Blank areas are where volcanic debris is present in trace quantities, absent, or where data are inadequate. Volcanoes, fumerolic fields, and submarine hydrothermal vents with known or mapped activity during the late Quaternary are indicated by asterisks. Data are from Klitgord and Mudie, 1974; Lonsdale, 1977; Francheteau et al., 1979; Spiess et al., 1980; Simkin et al., 1981; Circum-Pacific Plate Tectonic Map, 1981a,b; Malahoff et al., 1982; Normark et al., 1982; Rehm and Halbach, 1982; Hekinian et al., 1983.

tributes volcanic detritus to hemipelagic and pelagic sediments, although subaerial weathering in subtropical and tropical climatic zones may be of less importance in supplying volcanic particles to pelagic deposits due to chemical weathering processes, which preferentially produce amorphous aluminosilicates or clay minerals (Moberly, 1963). Volcanic gravels, sands and silts mantle the submarine slopes of most volcanic islands, such as Hawaii (Moore and Fiske, 1969; Fan and Grunwald, 1971; Fornari et al., 1979a,b), and are the product of both erosional and eruptive mechanisms. Redeposition of these hemipelagic deposits by mass-wasting processes can make a significant and frequent contribution locally to pelagic sediments, the best example perhaps being abundant volcanic detritus in the back-arc basin of the Marianas Islands (Fig. 18).

Because of their small size, shape characteristics, and density contrast with water, the settling velocities of glass shards through water are low, 0.2–2 cm/sec, so that days or weeks are required for them to reach the seafloor (Fisher, 1965). They are thus prone to dispersal by ocean currents, which often is subsequent to atmospheric dispersal within volcanic plumes. While settling velocities for ash particles through air are considerably higher than in water (Walker et al., 1971), the turbulence within eruption clouds tends to maintain ash in suspension for long distances downwind of the eruption center. Postdepositional mixing processes on the seafloor by oceanographic, tectonic, or biological activity disperse glass shards (or any sedimentary particle) within sediment (Berger and Heath, 1968; McCoy, 1981; Ruddiman and Glover, 1982) at rates on the order of 100 to 200 $cm^2/10^3$ years (Peng et al., 1979; DeMaster and Cochran, 1982). A thick ash blanket causes mortality of benthic life (Cita and Podenzani, 1980) and prevents stirring by organisms. Volcanic ash layers are not distinctive nor widespread as an unmixed surficial sediment in the North Pacific Ocean.

Floating rafts of silicic pumice fragments release a constant rain of glass shards through the mechanical disintegration of pumice within these rafts as a result of abrasion from wave motion. Such rafts can be extensive. One raft drifting west from a volcanic eruption of Isla San Benedicto, south of Baja California, was 345 nautical miles long with pumice fragments more than a meter in size (Richards, 1958). A 1973 submarine eruption in the northern Tonga Islands produced a pumice raft that was a few meters thick and was tracked across the Pacific into the Indian Ocean by satellites and commercial airline pilots (Simkin, personal communication). Rounded drift pumice has been found washed up on numerous islands and atolls throughout the Pacific (Sachet, 1955; Emery, 1956; Richards, 1958; Coombs and Landis, 1966; Bryan, 1968, 1971). Breakdown of pumice produces the typical Y-shaped bubble-wall shards found in pelagic sediments (Fisher, 1963; Heiken,

1972). Basaltic pumice does not float for long and thus does not form extensive rafts (Macdonald, 1972).

In the northern Pacific, the extent of volcanic particles mixed in surficial sediments, as determined from core top samples, is similar to the extent of air-fall Neogene and Quaternary volcanic ash layers found downsection in piston cores (Nayudu, 1964; Horn *et al.*, 1969; Hays and Ninkovich, 1970). This similarity indicates air-fall deposition from eruption clouds as the primary transport mechanism for the disseminated volcanic particles in modern pelagic sediments in this area, presuming minimal ice-rafting of ash during glacial stages. Both summer and winter atmospheric wind patterns would contribute to transporting Aleutian and Alaskan tephra seaward (Figs. 4a,b). The contour tracing 15–30% concentrations on Figure 18 in the northeastern Pacific, for example, outlines the distribution of ash from the Katmai eruption in 1912, as identified by Nayudu (1964). Off Kamchatka, only winter wind directions act to disperse tephra seaward (Fig. 4b), thus the extent of volcanics in seafloor deposits east of the peninsula is less than that southeast of the Aleutians.

East of Japan, higher concentrations of volcanics in a distinct southerly and easterly pattern, emphasized by the abundant amounts off northern Japan (Fig. 18), suggest surface-current transport of ash and pumice from Kamchatka by the southerly-flowing East Kamchatka Current and the Oyashio Current (Fig. 3a). Lower concentrations in the Japan Trench are due to masking of volcanics by terrigenous detritus.

In the western equatorial Pacific, the presence of volcanics in seafloor sediments reflects an input by atmospheric winds and ocean currents from the numerous volcanoes surrounding and within the Phillipine Sea, particularly those in the southern Japan–Ryukyu islands and Bonin–Marianas islands. Near the latter islands, higher abundances of volcanic detritus within the back-arc basin probably also represent contributions by mass-wasting from the island arc (Hussong and Fryer, 1983). High concentrations off the northern Ryukyu Islands occur near a long chain of submarine volcanoes (Geological Survey of Japan, 1977). On the Ontong–Java Plateau, volcanics probably are derived from the volcanoes of the Bismarck Archipelago–Solomon Islands–New Britain to the south. An eruption about 536 A.D. on Rabaul in New Britain, for example, may have been one of the largest known in recorded history (Stothers, 1984). The central area within this pattern apparently indicating a lack of volcanic detritus is an effect of dilution by calcareous sediments capping the plateau.

Volcanic detritus in sediments southeast of Hawaii are derived from eruptions there. Ash-producing phreatomagmatic eruptions are not frequent in the Hawaiian Islands (Macdonald, 1967), thus volcanic detritus does not extend far downwind of the islands. The southerly extent of this detritus is based on Hein *et al.* (1979a) and Rehmand Halbach (1982).

In the eastern equatorial Pacific, volcanic detritus distributions surround island volcanoes such as those of the Islas Revillagigedo and Galapagos. Another area to the west is centered on Shimada volcanic seamount (Gardner *et al.*, 1984), with a southerly dispersal pattern in response to the dominant oceanic and atmospheric circulation patterns. Distribution patterns also outline portions of the East Pacific Rise, an area of exposed volcanic features well documented by submersible exploration and acoustic side-scan surveys (e.g., Fornari *et al.*, 1983).

Volcanic detritus extending west off Central and South America in the Guatemala and Panama basins is derived from tephra-producing eruptions on the adjacent continent, the most dramatic example being the 1982 eruption of El Chichon in southern Mexico. As in the northern Pacific, these patterns in surficial sediments are similar to those indicated by the distribution of air-fall Quaternary and Neogene ash layers downsection in cores (Worzel, 1959; Bowles *et al.*, 1973; Ninkovich and Shackleton, 1975; Hahn and Rose, 1979; Drexler *et al.*, 1980) indicating the importance of aeolian dispersal processes in these basins. Extension of this distribution pattern west of the East Pacific Rise suggests deposition of volcanics from drifting pumice rafts driven westerly by both trade winds and surface currents from Central and South America.

E. Authigenic Particles, Nodules, and Crusts

Two types of authigenic components are discussed here: zeolites and polymetallic deposits occurring as nodules or crusts (Figs. 19 and 20). Zeolites are hydrous aluminosilicates that form silt-sized mineral grains in deep-sea sediments. Because of their small size, in addition to the fact that they are present only in trace amounts and are optically difficult to recognize, zeolites often are underestimated in smear slide analyses. Phillipsite, the dominant type of zeolite in North Pacific sediments, is an alteration product of glassy volcanic material (Murray and Renard, 1891; Bonatti, 1963, 1965; Kastner and Stonecipher, 1978; Bernat and Church, 1978). This relationship is clear in the concentration of zeolites around volcanic terrains and in areas of ash deposits on the East Pacific Rise, in the eastern equatorial seafloor off Central America, in portions of the Tufts–Alaska–Aleutian abyssal plains, and in the Phillipine Sea (Fig. 19). Zeolites also are associated with sediments containing detrital components, such as on the Tufts–Alaska–Aleutian abyssal plains (Fig. 17) (Bonatti, 1963), and with red clays in the Northeast Pacific and Central Pacific basins where sedimentation rates are low l($<$1mm–5mm/10^3 years) (Fig. 21 and 22).

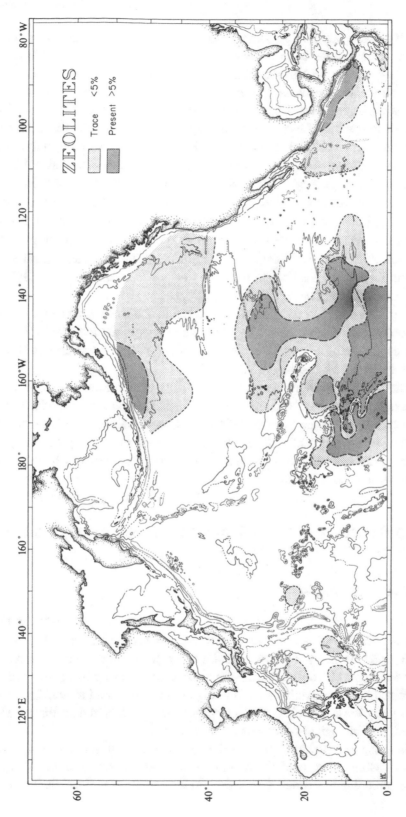

Fig. 19. Distribution of zeolites in North Pacific pelagic sediments. Relative abundances are based on smear slide analyses of sediment from core tops. Blank areas are where zeolites are present in trace amounts, absent, or where data are inadequate.

Polymetallic nodules, also known as ferromanganese or manganese nodules, cover vast areas of the North Pacific seafloor (Fig. 20). The density of nodule coverage on the sediment surface is usually less than 10%, with higher abundances between 25% and 50% occurring in small areas of the northeastern basin abyssal plains and on ridges and seamounts (McCoy *et al.*, 1985). In general, these are regions of pelagic red clays and biosiliceous oozes where sedimentation rates are low (Fig. 21 and 22). Nodules may also be concentrated by reworking of sediments and/or nodules; Lonsdale (1980), for example, has photographed high concentrations of 10 cm nodules arranged into bedforms by strong bottom currents on the Carnegie Ridge in the Panama Basin. Nodules grow at rates of only 1 mm/10^6 yr through adsorption and diffusion of metals from the surrounding sediment with some assistance from benthic biological mixing processes, and from diagenetic reactions involving biosiliceous particles and volcanic ash (Cronan, 1977; Heath, 1979, 1982; Broecker and Peng, 1982).

Ferromanganese crusts are usually associated with sediment-free areas where bottom-current scour occurs, steep slopes where sediments or nodules cannot accumulate, or hydrothermal fields along volcanic lineaments (Halbach *et al.*, 1982; Halbach, 1982).

Other authigenic minerals are found in North Pacific deep-sea sediments but are not discussed or mapped here. These include phosphorites (see Baturin, 1982), barites (see Cronan, 1974), and iron sulphides. The latter were present in smear slides of sediment from the central portions of the pelagic red clay areas and from areas in some proximity to volcanic hydrothermal fields along the East Pacific Rise.

VII. SUMMARY: THE SEDIMENTARY PROCESSES AND PROVINCES OF THE NORTH PACIFIC OCEAN

A. Sedimentary Processes

Input of nonbiogenic, allogenic detritus into and within the North Pacific Ocean occurs by six sedimentary processes of regional extent (Fig. 21): slumping and turbidity current deposition, suspended sediment transport, aeolian transport off continents, air-fall dispersal of volcanics following eruptions, dispersal of volcanics from floating pumice rafts, and ice-rafting processes. Only these principal physical processes are depicted in Figure 21; chemical and biological processes are not shown.

Slumping and turbidity current deposition are the dominant processes of mass-wasting along continental and insular margins, and provide a large volume of sediment primarily into the circum-Pacific trenches. Suspended

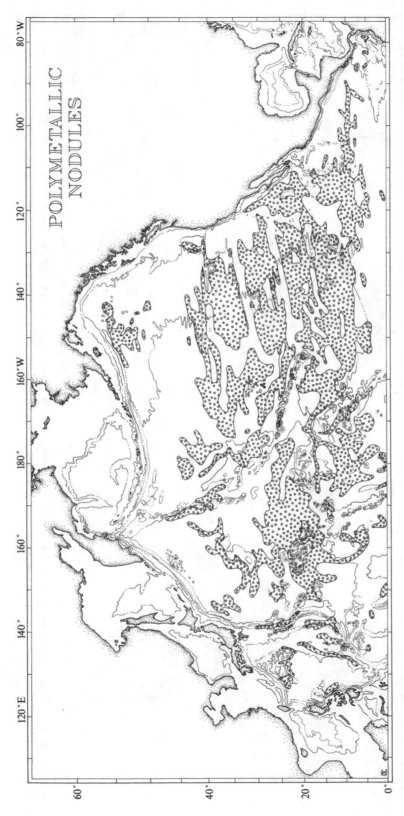

Fig. 20. Distribution of polymetallic (manganese) nodules on the North Pacific seafloor (modified from Rawson and Ryan, 1978; Craig *et al.*, 1982; Exon, 1983). Boundaries of nodule fields are approximate, not nearly as abrupt as suggested here by boundary outlines, which are used only for visual clarity. Nodule density within fields is usually only a few percent coverage of the seafloor.

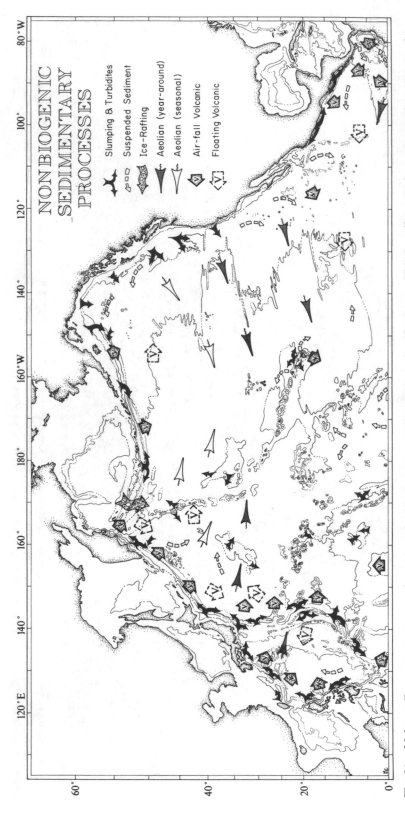

Fig. 21. Major sedimentary processes of nonbiogenic pelagic sediments in the North Pacific Ocean. This is a diagramatic representation of regional processes. Relative intensities of erosion or transportation is not suggested. Suspended sediment transport by bottom currents is indicated by shorter arrows, transport by surface currents by longer arrows.

sediment movement disperses terrigenous sediment by ocean currents into abyssal areas beyond the trenches, particularly where fluvial input of clays is significant such as along much of western North America. Bottom currents also disperse fine-grained material throughout the abyssal basin as suspended sediment often forming a nepheloid layer, and may be more significant in the erosion and redeposition of pelagic deposits than indicated in Fig. 21.

Aeolian processes transport sediment into the North Pacific from the surrounding continents as well as from volcanic eruptions. This influx varies with seasonal tropospheric wind patterns at higher latitudes, but is less variable at lower latitudes in the trade winds over the eastern North Pacific. Stratospheric jet stream flow patterns shift with seasons but remain a west-to-east flow transporting fine-grained particulate debris from Asia. Injection and transport of volcanic ash at tropospheric levels from circum-Pacific volcanic eruptions is more prevalent than at stratospheric levels (Simkin *et al.*, 1981). Extensive pumice rafts produced by some of these eruptions drift across the North Pacific Ocean following surface currents and deposit volcanic particles as pumice fragments sink or are abraded.

Ice-rafting is not an important sedimentary process in the North Pacific today, but may have been during glacial stages. Ice-rafted detritus is noticeable only in surface sediments off Kamchatka.

B. Sedimentation Rates

Pelagic sedimentation rates are highest (>30 mm/10^3 yr) near continental areas and lowest (< 1 mm/10^3 yr) on the abyssal plains (Fig. 22). High rates are due to terrigenous influx by slumping and turbidity currents into trenches and onto abyssal areas where trenches are absent in the northeastern Pacific; to aeolian influx into the northwestern Pacific; and to biogenic contributions in high productivity areas in the latter region as well as in the eastern equatorial Pacific. In general, high rates coincide with increased sediment thicknesses and may depict paleosedimentary patterns, the best example being the southeasterly bulge of higher sedimentation rates south of the Aleutian Trench (Opdyke and Foster, 1970).

Sedimentation rates shown in Fig. 22 are net rates, in that sediment removal by chemical dissolution or by erosion is considered. They are calculated from a combination of biostratigraphic, magnetostratigraphic, and tephrostratigraphic levels identified in cores, thus representing an integration over varying portions of the Quaternary. For example: rates from Morley *et al.* (1982) are based on a variety of biostratigraphic levels ranging from 50,000 to 400,000 yr BP; Opdyke and Foster (1970), Kobayashi *et al.* (1971), and Malahoff and Hammond (1971) present rates based on the Brunhes pa-

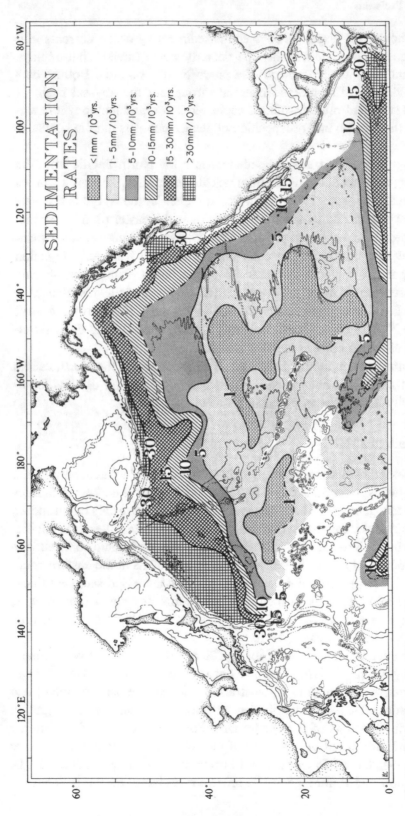

Fig. 22. Regional pelagic sedimentation rates in the North Pacific Ocean (dashed where inferred). Blank areas are where data are inadequate or where local complexities make depiction of regional patterns difficult. Data are from: Uchupi and Emery, 1963; Opdyke and Foster, 1970; Heath *et al.*, 1970; Kobayashi *et al.*, 1971; Lisitzin, 1972; Piper *et al.*, 1973; Saito *et al.*, 1975; van Andel *et al.*, 1975; Corliss and Hollister, 1979; Embley and Johnson, 1980; Morley *et al.*, 1982; Berger and Killingley, 1982; Scholl *et al.*, 1982; Prell *et al.*, 1982; Sancetta, 1983; Honnorez *et al.*, 1983; Carson and Arcaro, 1983; McCoy *et al.*, 1985.

leomagnetic epoch, or the past 700,000 years; rates from van Andel *et al.* (1975) represent calculations over the past million years. Lisitzin (1972) based sedimentation rates on the total sediment thickness from seismic reflection profiles thus integrating over a wide range of crustal ages.

C. Sedimentary Provinces

Six sedimentary provinces can be identified on the North Pacific seafloor (Fig. 23). These are large regions where pelagic sediments have common, mappable characteristics. Provinces overlap. Cartographic clarity, however, dictates a boundary line between them and does not allow the inclusion of smaller sedimentary provinces that occur on seamounts, rises, etc. Hemipelagic sedimentary provinces are not shown.

1. Pelagic Red Clay Province

The pelagic red clay province is the largest sedimentary province in the North Pacific (Fig. 23). Sediments here are the residues after dissolution of calcareous and biosiliceous particles below CCD levels in areas of low-biologic productivity and lie most distal from continental sources. It is an accumulation of allogenic and authigenic clays, the latter being a minor component, within which authigenic minerals such as zeolites and iron/manganese oxides are obvious due to a lack of dilution by biogenic material. Biogenic siliceous microfossils may be present in areas closer to regions of high productivity. An important contribution comes from aeolian transported detritus (Fig. 21). Sedimentation rates typically are less than 1 mm/10^3 years (Fig. 22). Much of this province is carpeted with manganese nodule fields (Fig. 20).

2. Calcareous and Biosiliceous Provinces

Sediments forming these provinces (Fig. 23) result from biologic productivity in the upper water masses and, in the case of calcareous debris, also from the numerous algal and coral reefs on submerged platforms and around islands in the southwestern Pacific. These provinces are the second largest in the North Pacific Ocean. The principal limitation on their distribution is the susceptibility of calcite and opal to dissolution in the water column or in surface sediments. The resulting provincal boundaries reflect seafloor physiography as well as zones of productivity; they follow the distribution of calcareous and biosiliceous sediments (Fig. 8).

Pelagic sedimentation rates in the calcareous and biosiliceous sedimentary provinces are a function of planktonic and benthonic productivity rates, as much as they are of rates of reef degradation and of the erosion of insular

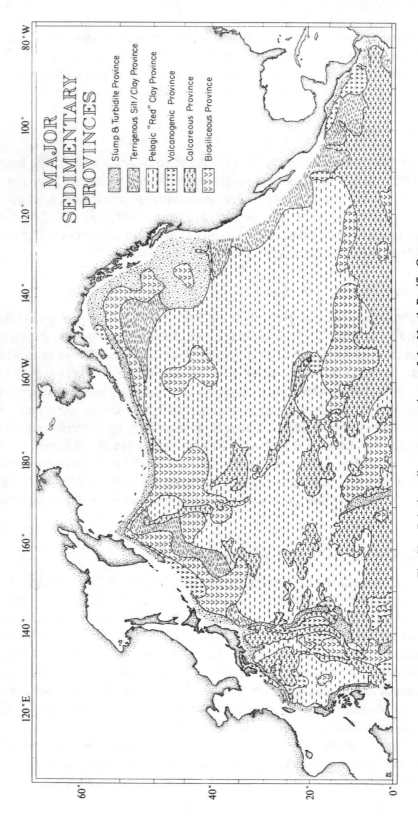

Fig. 23. Pelagic sedimentary provinces of the North Pacific Ocean.

or continental margins. Pelagic sedimention rates in the southeastern equatorial Pacific are 5–30 mm/10^3 yr (Fig. 22) as a result of planktonic carbonate productivity.

3. *Slump and Turbidite Province*

A slump and turbidite province surrounds the North Pacific Basin adjacent to continental areas (Fig. 23). In general, this province outlines the circum-Pacific trench system. In the single area where trenches do not exist, the northeastern Pacific, the slump and turbidite province extends well offshore, particularly in the vicinity of the Columbia River north of the Mendocino Fracture Zone. Along much of the continental margin north of here, the Queen Charlotte Sound and Alaskan inland waterway serve as traps retaining much of the terrigenous sediment washing off the continent, and the province does not extend as far offshore. Offshore dispersal of hemipelagic sediment from this continental margin occurs through the numerous leveed channels leading onto the abyssal plains (e.g., Hamilton, 1967).

Sedimentation rates are high in this province (Fig. 22). Turbidites include mud turbidites or lutite flows as described by Piper (1972, 1978) and Rupke and Stanley (1974), as well as bioclastic and volcanoclastic turbidites in trenches removed from continents. Small areas of the slump and turbidite province surround almost every seafloor feature with significant relief. These may be only a few hundred meters in size, and cannot be mapped in Figure 23. Damuth *et al.* (1983) have noted the widespread distribution of such areas in the northwestern Pacific Basin. Other areas have been described by Safonov (1978) around western Pacific seamounts, Fornari *et al.* (1979a,b) around Hawaii, Nemoto and Kroenke (1981) on the Hess Rise, Hussong and Fryer (1983) in the Marianas, Lonsdale *et al.* (1972) on Horizon Guyot, Heezen and Rawson (1977) on Cocos Ridge, and many others.

4. *Terrigenous Silt/Clay Province*

The terrigenous silt/clay province (Fig. 23) defines seafloor areas where the input of terrestrial detritus is primarily from suspended sediment dispersal by ocean currents or from atmospheric sources (Fig. 21). Turbidites, including mud turbidites, are present but are minor sedimentary features. Sedimentation rates are on the order of 5 to 15 mm/10^3 yr (Fig. 22).

The largest area occurs off North America from south of the Mendocino Fracture Zone to the Guatemala Basin. Clays and fine silts are transported south from North America by the California Current, and offshore by other surface currents across the Middle America Trench from South and Central America.

In the northwestern Pacific, this province partially outlines the relict turbidite province described by Hamilton (1967) and Horn *et al.* (1970). Surficial sediments here indicate pelagic sedimentation characteristic of the terrigenous silt/clay province as described above; turbidites occur downcore and represent a subsurface facies.

Aeolian influx of fine-grained detritus, in addition to suspended sediment input, contributes to formation of the terrigenous silt/clay province in the northwestern Pacific. The outline of the province here partially follows the path of the western boundary undercurrent and its peripheral components suggesting some dispersal of sediment by these currents.

5. *Volcanogenic Province*

Volcanic debris present in pelagic deposits in quantities adequate to define volcanogenic sedimentary provinces outline areas proximal to volcanic fields on land and under water (Fig. 23). Ash dispersal is primarily by tropospheric wind transport of eruption plumes, with additional dispersal by abrasion of drift pumice in rafts; slumping and redeposition of volcanics may also be significant, as in the Marianas back-arc basin (Fig. 21).

ACKNOWLEDGMENTS

Preliminary analyses and compilation of sediment information were supported by the Circum-Pacific Council for Energy Resources to the senior author. The continued encouragement by W. B. F. Ryan to the senior author has been a valuable and appreciated contribution. His review and additional comments from J. Morley and T. Janecek (L-DGO), and H. Zimmerman (Union College) are gratefully acknowledged. Computer data manipulation, map plotting and preparation, and drafting were provided by S. Coughlin, D. Kiel, S. Lewis, and S. Porta (L-DGO). Clerical help was provided by M. J. Foster and R. Lotti. Assistance from F. Mills and W. Addicott of the U. S. Geological Survey is appreciated. This compilation is the result of support by the National Science Foundation and the Office of Naval Research to the L-DGO curatorial facility. This is L-DGO contribution number 3616.

REFERENCES

Anikouchine, W. A., and Ling, H.-Y., 1967, Evidence for turbidite accumulation in trenches in the Indo-Pacific region, *Mar. Geol.,*v. 5, p. 141–154.
Aoki, S., Kohyama, N., and Sudo, T., 1974, An iron-rich montmorillonite in a sediment core from the northeast Pacific, *Deep Sea Res.*, v. 21, p. 865–875.

Arrhenius, G., 1952, Sediment cores from the east Pacific, *Reports of the Swedish Deep-Sea Expedition*, v. 5, p. 1–227.

Baas-Becking, L. G. M., and Moore, D., 1959, Density distribution in sediments, *J. Sed. Petrol.*, v. 29, p. 47–55.

Barry, R. G., and Chorley, R. J., 1976, *Atmosphere, Weather, and Climate* London: Methuen.

Baturin, G. N., 1982, Phosphorites on the seafloor: origin, composition and distribution, in: *Developments in Sedimentology, No. 33*, New York: Elsevier.

Bé, A. W. H., Sverdlove, M. S., Bishop, J. K. B., and Gardner, W. D., 1985, Standing stock, vertical distribution, and flux of planktonic foraminifera in the Panama Basin. *Mar. Micropaleontol.* v. 9, p. 307–333.

Bek-Bulat, B. Z., 1982, Seamounts of the Gulf of Alaska and the adjacent Pacific Ocean, *Oceanology*, v. 22, p. 731–734.

Berger, W. H., 1971, Sedimentation of planktonic foraminifera, *Mar. Geol.*, v. 11, p. 325–358.

Berger, W. H., 1976, Biogenous deep-sea sediments: production, preservation, and interpretation, in: *Treatise on Chemical Oceanography* Riley, J. P., and Chester, R., eds., v. 5, New York: Academic p. 265–388.

Berger, W. H., and Heath, G. R., 1968, Vertical mixing in pelagic sediment, *J. Mar. Res.* v. 26, p. 134–143.

Berger, W. H., and Killingley, J. S., 1982, Box cores from the equatorial Pacific: [14]C sedimentation rates and benthic mixing, *Mar. Geol.*, v. 45, p. 93–125.

Berger, W. H., Adelseck, C. G., and Mayer, L. A., 1976, Distribution of carbonate in surface sediments of the Pacific Ocean, *J. Geophys. Res.*, v. 81, p. 2617–2627.

Bernat, M., and Church, T. M., 1978, Deep-sea phillipsites: trace geochemistry and modes of formation, in: *Natural Zeolites: Occurrence, Properties, Use* Sand, L. B., and Mumpton, F. A., eds., New York: Pergamon, p. 259–267.

Biscaye, P. E., 1965, Mineralogy and sedimentation of Recent deep sea clay in the Atlantic Ocean and adjacent seas and oceans, *Geol. Soc. Am. Bull.*, v. 76, p. 803–832.

Biscaye, P. E., and Eittreim, S. L., 1977, Suspended particulate loads and transports in the nepheloid layer of the abyssal Atlantic Ocean, *Mar. Geol.*, v. 23, p. 155–172.

Bishop, J. K. B., Edmond, J. M., Ketten, D. R., Bacon, M. P., and Silker, W. B., 1977, The chemistry, biology, and vertical flux of particulate matter from the upper 400 meters of the equatorial Atlantic Ocean, *Deep Sea Res.*, v. 24, p. 511–548.

Blanchard, M. B., Brownlee, D. E., Bunch, T. E., Hodge, P. W., and Kyte, F. T., 1980, Meteoroid ablation spheres from deep-sea sediments, *Earth Planet. Sci. Lett.*, v. 46, p. 178–190.

Bonatti, E., 1963, Zeolites in Pacific pelagic sediments, *N.Y. Acad. Sci. Trans. Ser. II*, v. 25, p. 938–948.

Bonatti, E., 1965, Palagonite, hyaloclastites, and alteration of volcanic glass in the ocean, *Bull. Volcanol.*, v. 28, p. 1–15.

Bowles, F. A., Jack, R. N., and Carmichael, I. S. E., 1973, Investigation of deep-sea volcanic ash layers from equatorial Pacific cores, *Geol. Soc. Am. Bull.*, v. 84, p. 2371–2388.

Broecker, W. S., and Peng, T.-H., 1982, *Tracers in the Sea,* Palisades: Eldigio.

Bruun, A. F., Langer, E., and Pauly, H., 1955, Magnetic particles found by raking the deep sea bottom, *Deep Sea Res.*, v. 2, p. 230–246.

Bryan, W. B., 1968, Low-potash dacite drift pumice from the Coral Sea, *Geol. Mag.*, v. 105, p. 431–439.

Bryan, W. B., 1971, Coral Sea drift pumice stranded on Eua Island, Tonga, in 1969, *Geol. Soc. Am. Bull.*, v. 82, p. 2799–2812.

Burnett, W. C., 1975, Trace element geochemistry of biogenic sediments from the western equatorial Pacific, *Pac. Sci.*, v. 29, p. 291–225.

Cann, J. R., Langseth, M. G., Honnorez, J., Von Herzen, R. P., White, S. M., *et al.*, 1983, *Initial Reports of the Deep Sea Drilling Project,* v. 69, Washington, D. C.: U. S. Government Printing Office.

Carder, K. L., Beardsley, G. F., and Pak, H., 1971, Particle size distributions in the eastern equatorial Pacific, *J. Geophys. Res.*, v. 76, p. 5070–5077.

Carson, B., and Arcaro, N. P., 1983, Control of clay mineral stratigraphy by selective transport in late Pleistocene–Holocene sediments of northern Cascadia Basin–Juan de Fuca Abyssal Plain: implications for studies of clay mineral provenance, *J. Sed. Petrol.*, v. 53, p. 0395–0406.

Chase, T. E., 1975, Topography of the Oceans, 1:26,640,000, *Scripps Inst. Oceanog., Mercator Proj.*

Circum-Pacific Map Project, 1981a, Plate Tectonic Map, Northeast Quadrant, 1:10,000,000, *Cirum-Pacific Council for Energy and Mineral Resources.*

Circum-Pacific Map Project, 1981b, Plate Tectonic Map, Northwest Quadrant, 1:10,000,000, *Circum-Pacific Council for Energy and Mineral Resources.*

Cita, M. B., and Podenzani, M., 1980, Destructive effects of oxygen starvation and ash falls on benthic life: a pilot study, *Quat. Res.*, v. 13, p. 230–241.

Cole, T. G., and Shaw, H. F., 1983, The nature and origin of authigenic smectites in some Recent marine sediments, *Clay Min.*, v. 18, p. 239–252.

Conolly, J. R., and Ewing, M., 1970, Ice-rafted detritus in northwest Pacific deep-sea sediments, in: *Geological Investigations of the North Pacific*, Hays, J. D., ed., *Geol. Soc. Am. Mem.* 126, p. 219–232.

Coombs, D. S., and Landis, C. A., 1966, Pumice from the South Sandwich eruption of March 1962 reaches New Zealand, *Nature*, v. 209, p. 289–290.

Corliss, B. H., and Hollister, C. D., 1979, Cenozoic sedimentation in the North Pacific, *Nature*, v. 282, p. 707–709.

Coulbourn, W. T., Parker, F. L., and Berger, W. H., 1980, Faunal and solution patterns of planktonic foraminifera in surface sediments of the North Pacific, *Mar. Micropaleo.*, v. 5, p. 329–399.

Craig, J. D., 1979, Geological investigation of the equatorial North Pacific seafloor: a discussion of sediment redistribution, in: *Marine Geology and Oceanography of the Pacific Manganese Nodule Province*, Bischoff, J. L., and Piper, D. Z., eds., New York: Plenum, p. 529–558.

Craig, J. D., Andrews, J. E., and Meylan, M. A., 1982, Ferromanganese deposits in the Hawaiian archipelago, *Mar. Geol.*, v. 45, p. 127–157.

Cronan, D. S., 1974, Authigenic minerals in deep-sea sediments, in: *The Sea, Marine Chemistry*, Goldberg, E. D., ed., New York: Wiley-Interscience.

Cronan, D. S., 1977, Deep-sea nodules: Distribution and geochemistry, in: *Marine Manganese Deposits*, Glasby, G. P., ed., Amsterdam: Elsevier, p. 11–44.

Curray, J. R., Moore, D. G., *et al.*, 1982, *Initial Reports of the Deep Sea Drilling Project*, v. 64, Washington, D. C.: U. S. Government Printing Office.

Damuth, J. E., Jacobi, R. D., and Hayes, D. E., 1983, Sedimentation processes in the Northwest Pacific Basin revealed by echo-character mapping studies, *Geol. Soc. Am. Bull.*, v. 94, p. 381–395.

DeMaster, D. J., and Cochran, J. K., 1982, Particle mixing rates in deep-sea sediments determined from excess ^{210}Pb and ^{32}Si profiles, *Earth Planet. Sci. Lett.*, v. 61, p. 257–271.

Dodimead, A. V., Favorite, F., and Hirano, T., 1963, Review of oceanography of the subarctic Pacific region, *Int. North Pac. Fish. Comm. Bull.*, v. 13, p. 1–176.

Dowding, L. G., 1977, Sediment dispersal within the Cocos Gap, Panama Basin, *J. Sed. Petrol.*, v. 47, p. 1132–1156.

Drexler, J. W., Rose, W. I., Jr., Sparks, R. S. J., and Ledbetter, M. T., 1980, The Los Chocoyos ash, Guatemala: a major stratigraphic marker in Middle America and in three ocean basins, *Quat. Res.*, v. 13, p. 327–345.

Duce, R. A., Unni, C. K., Ray, B. J., Prospero, J. M., and Merril, J. T., 1980, Long-range atmospheric transport of soil dust from Asia to the tropical North Pacific: temporal variability, *Science*, v. 209, p. 1522–1524.

Edmond, J. M., Chung, Y., and Sclater, J. G., 1971, Pacific bottom water: penetration east around Hawaii, *J. Geophys. Res.,* v. 76, p. 8089–8097.

El Wakeel, S. K., and Riley, J. P., 1961, Chemical and mineralogical studies of deep-sea sediments, *Geochim. Cosmochim. Acta.,* v. 25, p. 110–146.

Embley, R. W., and Johnson, D. A., 1980, Acoustic stratigraphy and biostratigraphy of Neogene carbonate horizons in the north equatorial Pacific, *J. Geophys. Res.,* v. 85, p. 5423–5437.

Emery, K. O., 1956, Marine geology of Johnston Island and its surrounding shallows, central Pacific Ocean, *Geol. Soc. Am. Bull.,* v. 67, p. 1505,–1519.

Ewing, M., and Connary, S. D., 1970, Nepheloid layer in the North Pacific, in: *Geological Investigations of the North Pacific* Hays, J .D., ed., *Geol. Soc. Am. Mem.,* v. 126, p. 41–82.

Exon, N. F., 1983, Manganese nodule deposits in the Central Pacific Ocean and their variation with latitude, *Mar. Min.,* v. 4, p. 70–107.

Fan, P.-F., and Grunwald, R. R., 1971, Sediment distribution in the Hawaiian Archipelago, *Pac. Sci.,* v. 25, p. 484–488.

Ferguson, W. S., Griffin, J. J., and Goldberg, E. D., 1970, Atmospheric dusts from the North Pacific—a short note on a long-range eolian transport, *J. Geophys. Res.,* v. 75, p. 1137–1139.

Fisher, R. V., 1963, Bubble-wall texture and its significance, *J. Sed. Petrol.,* v. 33, p. 224–235.

Fisher, R. V., 1965, Settling velocity of glass shards, *Deep Sea Res.,* v. 12, p. 345–353.

Fornari, D. J., Malahoff, A., and Heezen, B. C., 1979(a), Submarine slope micromorphology and volcanic substructure of the island of Hawaii inferred from visual observations made from U. S. Navy deep-submergence vehicle (DSV) "Sea Cliff", *Mar. Geol.,* v. 32, p. 1–20.

Fornari, D. J., Moore, J. G., and Calk, L., 1979(b), A large submarine sand-rubble flow on Kiluaea Volcano, Hawaii, *J. Volcanol. Geotherm. Res.,* v. 5, p. 239–256.

Fornari, D. J., Ryan, W. B. F., and Fox, P. J., 1983, Sea MARC 1 side-scan sonar imaging near the East Pacific Rise, *EOS Trans. Am. Geophys. Union,* v. 64, p. 482.

Fowler, S. W., and Small, L. F., 1972, Sinking rates of Euphausiid fecal pellets, *Limnol. Oceanogra.,* v. 17, p. 293–296.

Francheteau, H. D., Needham, H. D., Choukroune, P., Juteau, T., Séguret, M., Ballard, R. D., Fox, P. J., Normark, W., Carranza, A., Cordoba, D., Guerrero, J., Rangin, C., Bougault, H., Cambon, P., and Hekinian, R., 1979, Massive deep-sea sulphide ore deposits discovered on the East Pacific Rise, *Nature,* v. 277, p. 523–528.

Frazer, J. Z., Hawkins, D. L., and Arrhenius, G., 1972, Surface sediments and topography of the North Pacific, 1:3,630,000, *Scripps Inst. Of Oceanog., Geol. Data Ctr.,* Charts 1–10.

Gardner, J. V., Dean, W. E., and Blakely, R. J., 1984, Shimada Seamount: an example of Recent mid-plate volcanism, *Geol. Soc. Am. Bull.,* v. 95, p. 855–862.

GEBCO, General Bathymetric Chart of the Oceans, 1975ff., 1:10,000,000, *International Hydrographic Organization, Intergovernmental Oceanographic Commission,* Sheets 5-02, 5-03, 5-06, 5-07.

Geitzenauer, K. R., Roche, M. B., and McIntyre, A., 1976, Modern Pacific coccolith assemblages: derivation and application to late Pleistocene paleotemperature analysis, in: *Investigation of late Quaternary Paleoceanography and Paleoclimatology,* Cline, R. M., and Hays, J. D., eds., *Geol. Soc. Am. Mem.,* v. 145, p. 423–448.

Geological Survey of Japan, 1977, Geological Map around Ryukyu Arc, 1:10,000,000. *Mar. Geol. Map. Ser.* 7.

Glass, B. P., Swincki, M. B., and Zwart, P. A., 1979, Australasian, Ivory Coast, and North American tektite-strewn fields: size, mass, and correlation with geomagnetic reversals and other earth events, *Proc. Tenth Lunar Planet Sci. Conf.,* p. 2535–2545.

Gordon, A. L., and Gerard, R. D., 1970, North Pacific bottom potential temperature, in: *Geological Investigations of the North Pacific,* Hays, J. D., ed., *Geol. Soc. Am. Mem.* 126, p. 23–40.

Griffen, J. J., Windom, H., and Goldberg, E. D., 1968, The distribution of clay minerals in the world oceans, *Deep Sea Res.,* v. 15, p. 433–459.

Hahn, G. A., and Rose, W. I., 1979, Geochemical correlation of genetically related rhyolitic ash-flow and air-fall ashes, central and western Guatemala and the equatorial Pacific, *Geol. Soc. Am. Spec. Pap.,* 180, p. 101–111.

Halbach, P., 1982, Co-rich ferromanganese seamount deposits of the central Pacific Basin, in: *Marine Mineral Deposits—New Research Results and Economic Prospects,* Halbach, P., and Winter, P., eds., *Marine Rohstoffe und Meerestechnik,* v. 6, p. 60–85. Essen: Verlag Glückauf.

Halbach, P., Manheim, F. T., and Otten, P., 1982, Co-rich ferromanganese deposits in the marginal seamount regions of the central Pacific Basin—results of the Midpac '81, *Erzemetall.,* v. 35, p. 447–453.

Hamilton, E. L., 1956, Sunken islands of the Mid-Pacific Mountains, *Geol. Soc. Am. Mem.,* 64.

Hamilton, E. L., 1967, Marine geology of abyssal plains in the Gulf of Alaska, *J. Geophys. Res.,* v. 72, p. 4189–4213.

Hamilton, E. L., 1974, Prediction of deep-sea sediment properties: state-of-the-art, in: *Deep Sea Sediments,* Inderbitzen, A. L., ed., New York: Plenum,. p. 1–43.

Hamilton, E. L., and Bachman, R. T., 1982, Sound velocity and related properties of marine sediments, *J. Acoust. Soc. Am.,* v. 72, p. 1891–1904.

Hamilton, E. L., Bachman, R. T., Berger, W. H., Johnson, T. C., and Mayer, L. A., 1982, Acoustic and related properties of calcareous deep-sea sediments, *J. Sed. Petrol.,* v. 52, p. 733–753.

Han-chang, P., Kui-huan, Z., and Sui-tian, C., 1982, The evidence of "choi" in sediment core samples of Pacific Ocean, *J. Geophys. Res.,* v. 87, p. 5563–5565.

Hays, J. D., and Ninkovich, D., 1970, North Pacific deep-sea ash chronology and age of present Aleutian underthrusting, in: *Geological Investigations of the North Pacific,* Hays, J. D., ed., *Geol. Soc. Am. Mem.,* 126, p. 263–290.

Heath, G. R., 1974, Dissolved silica and deep-sea sediments, in: *Studies in Paleo-Oceanography,* Hay, W. W., ed., *Soc. Econ. Paleontol. Mineral. Spec. Publ* 20, p. 77–93.

Heath, G. R., 1979, Burial rates, growth rates, and size distributions of deep-sea manganese nodules, *Science,* v. 205, p. 903–904.

Heath, G. R., 1982, Deep-sea ferromanganese nodules, in: *The Environment of the Deep Sea,* Ernst, W. G., and Morin, J. G., eds., Englewood Cliffs: Prentice-Hall, p. 105–153.

Heath, G. R., and Dymond, J., 1977, Genesis and transformation of metalliferous sediments from the East Pacific Rise, Bauer Deep, and Central Basin, northwest Nazca plate, *Geol. Soc. Am. Bull.,* v. 88, p. 723–733.

Heath, G. R., Moore, T. C., Jr., Somayajula, B. L. K., and Cronan, D. S., 1970, Sediment budget in a deep-sea core from the central equatorial Pacific, *J. Mar. Res.,* v. 28, p. 225–234.

Heezen, B. C., and Rawson, M., 1977, visual observations of contemporary current erosion and tectonic deformation on the Cocos Ridge crest, *Mar. Geol.,* v. 23, p. 173–196.

Heiken, G., 1972, Morphology and petrography of volcanic ashes, *Geol. Soc. Am. Bull.,* v. 83, p. 1961–1987.

Hein, J. R., Ross, C. R., Alexander, E., and Yeh, H.-W., 1979, Mineralogy and diagenesis of surface sediments from DOME areas A, B, and C, in: *Marine Geology and Oceanography of the Pacific Manganese Nodule Province,* Bischoff, J. L., and Piper, D. Z., eds., New York: Plenum. p. 365–396.

Hein, J. R., Yeh, H.-W., and Alexander, E., 1979, Origin of iron-rich montmorillonite from the manganese nodule belt of the north equatorial Pacific, *Clays Clay Miner.,* v. 27, p. 185–194.

Hekinian, R., Fevrier, M., Avedik, F., Cambon, P., Charlou, J. L., Needham, H. D., Raillard, J., Boulegue, J., Merlivat, L., Moinet, A., Manganini, S., and Lange, J., 1983, East Pacific Rise near 13°N: geology of new hydrothermal fields, *Science,* v. 219, p. 1321–1324.

Hess, H. H., 1946, Drowned ancient islands of the Pacific Basin, *Am. J. Sci.*, v. 244, p. 722–791.

Hess, H. H., 1948, Major structural features of the western North Pacific, *Geol. Soc. Am. Bull.*, v. 59, p. 417–446.

Hibler, W. D., 1980, Sea ice growth, drift, and decay, in: *Dynamics of Snow and Ice Masses*, Colbeck, S., ed., New York: Academic, p. 141–209.

Hollister, C. D., Johnson, D. A., Lonsdale, P. F., 1974, Current-controlled abyssal sedimentation; Samoa Passage, equatorial west Pacific, *J. Geol.*, v. 82, p. 275–300.

Honjo, S., 1976, Coccoliths: production, transportation and sedimentation, *Mar. Micropaleo.*, v. 1, p. 65–79.

Honnorez, J., Von Herzen, R. P., *et al.*, 1983, *Initial Reports of the Deep Sea Drilling Project*, v. 70, Washington, D. C.: U. S. Government Printing Office.

Horn, D. R., Delach, M. N., and Horn, B. M., 1969, Distribution of volcanic ash layers and turbidites in the North Pacific, *Geol. Soc. Am. Bull.*, v. 80, p. 1715–1724.

Horn, D. R., Delach, M. N., and Horn, B. M., 1974, Physical properties of sedimentary provinces, North Pacific and North Atlantic Oceans, in: *Deep-Sea Sediments*, Inderbitzen, A. L., ed., New York: Plenum, p. 417–442.

Horn, D. R., Ewing, J., and Ewing, M., 1972, Graded-bed sequences emplaced by turbidity currents north of 20°N in the Pacific, Atlantic and Mediterranean, *Sedimentology*, v. 18, p. 247–275.

Horn, D. R., Horn, B. M., and Delach, M. N., 1970, Sedimentary provinces of the North Pacific, in: *Geological Investigations of the North Pacific*, Hays, J. D., ed., *Geol. Soc. Am. Mem.*, 126, p. 1–22.

Horn, D. R., Ewing, M., Delach, M. N., and Horn, B. M., 1971, Turbidites of the Northeast Pacific, *Sedimentology*, v. 16, p. 55–69.

Hurd, D. C., and Birdwhistell, S., 1983, On producing a more general model for biogenic silica dissolution, *Am. J. Sci.*, v. 283, p. 1–28.

Hurd, D. S., and Theyer, F., 1975, Changes in the physical and chemical properties of biogenic silica from the central equatorial Pacific, *Adv. Chem. Ser.*, v. 147, p. 211–230.

Hurd, D. S., and Theyer, F., 1977, Changes in the physical and chemical properties of biogenic silica from the central equatorial Pacific: Part II, refraction index density and water content of acid-cleaned samples, *Am. J. Sci.*, v. 277, p. 1168–1202.

Hussong, D. M., and Fryer, P., 1983, Back-arc seamounts and the SeaMARC II seafloor mapping system, *EOS. Am. Geophys. Union*, v. 64, p. 627–632.

Janecek, T. R., and Rea, D. K., 1983, Eolian deposition in the northeast Pacific Ocean: Cenozoic history of atmospheric circulation, *Geol. Soc. Am. Bull.*, v. 94, p. 730–738.

Johnson, D. A., 1972, Oceanfloor erosion in the equatorial Pacific, *Geol. Soc. Am. Bull.*, v. 83, p. 3121–3144.

Johnson, D. A., and Johnson, T. C., 1970, Sediment redistribution by bottom currents in the central Pacific, *Deep Sea Res.*, v. 17, p. 157–170.

Johnson, T. C., 1976, Biogenic opal preservation in pelagic sediments of a small area in the eastern tropical Pacific, *Geol. Soc. Am. Bull.*, v. 87, p. 1273–1282.

Jousé, A. P., Kozlova, O. G., and Muhina, V. V., 1971, Distribution of diatoms in the surface layer of sediment from the Pacific Ocean, in: *The Micropalaeontology of Oceans*, Funnell, B. M., and Riedel, W. R., eds., Cambridge, England: Cambridge University, p. 263–270.

Kadko, D., 1983, A multitracer approach to the study of erosion in the northeast equatorial Pacific, *Earth Planet. Sci. Lett.*, v. 63, p. 13–33.

Kanaya, T., and Koizumi, I., 1966, Interpretation of diatom thanatocoenoses from the North Pacific applied to a study of core V20-130 (Studies of a deep-sea core V20-130, Part IV), *Sci. Rep. Tohoku Univ.*, Ser. 2, v. 37, p. 89–130.

Kastner, M., and Stonecipher, S. A., 1978, Zeolites in pelagic sediments of the Atlantic, Pacific, and Indian Oceans, in: *Natural Zeolites: Occurrence, Properties, Use*, Sand, L. B., and Mumpton, F. A., eds., New York: Pergammon. p. 199–220.

Klitgord, K. D., and Mudie, J. D., 1974, The Galapagos spreading centre: a near-bottom geophysical survey, *Geophys. J. R. Astron. Soc.*, v. 38, p. 563–586.

Kobayashi, K., Kitazawa, K., Kanaya, T., and Saka, T., 1971, Magnetic and micropaleontological study of deep-sea sediments from the west-central euqatorial Pacific, *Deep Sea
Res.*, v. 18, p. 1045–1062.

Kowsmann, R. O., 1973, Coarse components in surface sediments of the Panama Basin, eastern
equatorial Pacific, *J. Geol.*, v. 81, p. 473–494.

Kruglikova, S. B., 1981, Radiolarians in the surface layer of bottom sediments in the eastern
part of the Pacific Ocean's tropical zone, *Oceanology*, v. 21, p. 359–364.

Kuenen, Ph. H., 1950, *Marine Geology*, New York: Wiley.

Kulm, L. D., and Scheidegger, K. F., 1979, Quaternary sedimentation on the tectonically active
Oregon continental slope in: *Geology of Continental Slopes*, Doyle, L. J., ed., *Soc. Econ.
Paleontol. Mineral. Spec. Publ.*, v. 27, p. 247–263.

Leinen, M., and Heath, G. R., 1981, Sedimentary indicators of atmospheric activity in the
northern hemisphere during the Cenozoic, *Palaeogeogra. Palaeoclimatol. Palaeoecol.*, v.
36, p. 1–21.

Lisitzin, A. P., 1972, *Sedimentation in the World Ocean, Soc. Econ. Paleontol. and Mineral.
Spec. Publ. No. 17*, p. 218.

Lisitzin, A. P., 1971, Distribution of siliceous microfossils in suspension and in bottom sediments, in: *The Micropalaeontology of Oceans*, Funnel, B. M., and Riedel, W. R., eds.,
Cambridge, England: Cambridge University. p. 173–196.

Lonsdale, P., 1977, Deep-tow observations at the mounds abyssal hydrothermal field, Galapagos
Rift, *Earth Planet. Sci. Lett.*, v. 36, p. 92–110.

Lonsdale, P., 1980, Manganese-nodule bedforms and thermohaline density flows in a deep-sea
valley on Carnegie Ridge, Panama Basin, *J. Sed. Petrol.*, v. 50, p. 1033–1048.

Lonsdale, P., Normark, W. R., and Newman, W. A., 1972, Sedimentation and erosion on
Horizon guyot, *Geol. Soc. Am. Bull.*, v. 83, p. 289–316.

Macdonald, G. A., 1967, Forms and structures of extrusive basaltic rocks, in: *Basalts: The
Poldervaart Treatise on Rocks of Basaltic Composition*, Hess, H. H., and Poldervaart, A.,
eds., New York: Wiley-Interscience. p. 1–61.

Macdonald, G. A., 1972, *Volcanoes*, Englewood Cliffs: Prentice-Hall.

Malahoff, A., and Hammond, S. R., 1971, Sediment accumulation rates along the Murray
Fracture Zone between 131°W and 167°W longitude, *Geol. Soc. Am. Bull.*, v. 82, p. 1429–
1432.

Malahoff, A., McMurty, G. M., Wiltshire, J. C., and Yeh, H.-W., 1982, Geology and chemistry
of hydrothermal deposits from active submarine volcano Loihi, Hawaii, *Nature*, v. 298,
p. 234–239.

Mantyla, A. W., and Reid, J. L., 1983, Abyssal characteristics of the World Ocean waters,
Deep Sea Res., v. 30, p. 805–833.

Marchig, V., 1978, Brown clays from the central Pacific—metalliferous or not? *Geol. Jahrb.
Reihe D*, v. 30, p. 3–25.

Mayer, L. A., 1979, Deep sea carbonates: acoustic, physical and stratigraphic properties, *J.
Sed. Petrol.*, v. 49, p. 819–836.

McCave, I. N., 1975, Vertical flux of particles in the ocean, *Deep Sea Res.*, v. 22, p. 491–502.

McCave, I. N., and Swift, S. A., 1976, A physical model for the rate of deposition of fine-
grained sediments in the deep sea, *Geol. Soc. Am. Bull.* v. 87, p. 541–546.

McCoy, F. W., 1980, Photographic analysis of coring, *Mar. Geol.*, v. 38, p. 263–282.

McCoy, F. W., 1981, Areal distribution, redeposition and mixing of tephra within deep-sea
sediments of the Eastern Mediterranean Sea, in: *Tephra Studies*, Self, S., and Sparks, R.
S. J., eds., New York: D. Reidel. p. 245–254.

McCoy, F. W., 1983a, Seafloor sediments: Geologic map of the Circum-Pacific region, Northeast Quadrant, 1:10,000,000 *Circum-Pacific Council for Energy and Mineral Resources,
Circum-Pacific Map Project*.

McCoy, F. W., 1983b, Surficial sediments of the northeastern Pacific Ocean, in: *Proceedings of the 34th Circum-Pacific Energy and Mineral Resources Conference,* Halbouty, M., ed., Honolulu: American Association of Petroleum Geologists.

McCoy, F. W., 1986, Seafloor sediments: Geologic map of the Circum-Pacific region, Northwest Quadrant, 1:10,000,000 Circum-Pacific Council for Energy and Mineral Resources, Circum-Pacific Map Project.

McCoy, F. W., Swint, T., and Piper, D., 1985, Manganese nodules, seafloor sediment, and sedimentation rates of the Circum-Pacific region, 1:17,000,000 *Circum-Pacific Council for Energy and Mineral Resources.*

Menard, H. W., 1964, *Marine Geology of the Pacific,* New York: McGraw-Hill.

Menard, H. W., and Dietz, R. S., 1951, Submarine geology of the Gulf of Alaska, *Geol. Soc. of Am. Bull.,* v. 62, p. 1263–1265.

Moberly, R., 1963, Amorphous marine muds from tropically weathered basalts, *Am. J. Sci.,* v. 261, p. 767–772.

Moberly, R., Kimura, H. S., and McCoy, F. W., 1968, Authigenic marine phyllosilicates near Hawaii, *Geol. Soc. Am. Bull.,* v. 79, p. 1449–1460.

Moore, J. G., and Fiske, R. S., 1969, Volcanic substructure inferred from dredge samples and ocean-bottom photographs, Hawaii, *Geol. Soc. Am. Bull.,* v. 80, p. 1191–1202.

Moore, T. C., 1978, The distribution of radiolarian assemblages in the modern and ice-age Pacific, *Mar. Micropaleo.,* v. 3, p. 229–266.

Moore, T. C., Heath, G. R., and Kowsmann, R. O., 1973, Biogenic sediments of the Panama Basin, *J. Geol.,* v. 81, p. 458–472.

Morley, J. J., Hays, J. D., and Robertson, J. H., 1982, Stratigraphic framework for the late Pleistocene in the northwest Pacific Ocean, *Deep Sea Res.,* v. 29, p. 1485–1499.

Murray, H. W., 1941, Submarine mountains in the Gulf of Alaska, *Geol. Soc. Am. Bull.,* v. 52, p. 333–362.

Murray, H. W., 1945, Profiles of the Aleutian Trench, *Geol. Soc. Am. Bull.,* v. 56, p. 757–782.

Murray, H. W., 1946, Submarine relief of the Aleutian Trench, *Trans. Am. Geophys. Union.,* v. 27, p. 871–875.

Murray, J., and Renard, A. F., 1883, On the measurement characters of volcanic ashes and cosmic dust and their distribution in deep-sea deposits, *Proc. R. Soc., Edinburgh,* v. 9, p. 247–261.

Murray, J., and Renard, A. F., 1891, Report on deep-sea deposits based on the specimens collected during the voyage of HMS *Challenger* in the years 1872 to 1876, in: *Report on the Scientific Results of the Voyage of HMS* Challenger *during the Years 1873–1876,* Thomson, C. W., and Murray, J., eds., London, His Majesty's Stationery Office.

Nayudu, Y. R., 1964, Volcanic ash deposits in the Gulf of Alaska and problems of correlation of deep-sea ash deposits, *Mar. Geol.,* v. 1, p. 194–212.

Nayudu, Y. R., and Enbysk, B. J., 1964, Biolithology of northeast Pacific surface sediments, *Mar. Geol.,* v. 2, p. 310–342.

Neeb, G. A., 1943, *The Composition and Distribution of the Samples,* Snelluis *Expedition: Scientific Results,* Snelluis *Expedition East Indian Geologic Archipelago, 1929–1930, Bottom Samples.* Leiden: Brill.

Nemoto, K., and Kroenke, L. W., 1981, Marine geology of the Hess Rise 1: bathymetry, surface sediment distribution and environment of deposition, *J. Geophys. Res.,* v. 86, p. 10734–10752.

Nigrini, C., 1970, Radiolarian assemblages in the North Pacific and their application to a study of Quaternary sediments in core V20-130, in: *Geological Investigations of the North Pacific,* Hays, J. D., ed., *Geol. Soc. Am. Mem.,* 126, p. 139–184.

Nigrini, C. A., 1968, Radiolaria from eastern tropical Pacific sediments, *Micropaleontology,* v. 14, p. 51–63.

Ninkovitch, D., and Shackleton, N. J., 1975, Distribution, stratigraphic position and age of ash layer "L" in the Panama Basin, *Earth Planet. Sci. Lett.,* v. 27, p. 20–34.

Normark, W. R., Morton, J. L., and Delaney, J. R., 1982, Geologic setting of massive sulfide deposits and hydrothermal vents along the southern Juan de Fuca Ridge, *U. S. Geol. Surv. Open-File Rep.* 82-200A, p. 1–21.

Opdyke, N. D., and Foster, J. H., 1970, Paleomagnetism of cores from the North Pacific, in: *Geological Investigations of the North Pacific,* Hays, J. D., ed., *Geol. Soc. Am. Mem.,* 126, p. 83–121.

Parkin, D. W., Sullivan, R. A. L., and Bull, R. K., 1983, Cosmic spherules and asteroidal collisions, *Geophys. J. R. Astron. Soc.,* v. 75, p. 473–491.

Parrington, J. R., Zoller, W. H., and Aras, N. K., 1983, Asian dust: seasonal transport to the Hawaiian Islands, *Science,* v. 220, p. 195–197.

Peng, T.-H., Broecker, W. S., and Berger, W. H., 1979, Rates of benthic mixing in deep-sea sediment as determined by radioactive tracers, *Quat. Res.,* v. 11, p. 141–149.

Peterson, M. N. A., and Goldberg, E. D., 1962, Feldspar distributions in South Pacific pelagic sediments, *J. Geophys. Res.,* v. 67, p. 3472–3492.

Pettersson, H., and Fredriksson, K., 1958, Magnetic spherules in deep sea deposits, *Pac. Sci.,* v. 12, p. 71–81.

Piper, D. J. W., 1972, Turbidite origin of some laminated mudstones, *Geol. Mag.,* v. 109, p. 115–126.

Piper, D. J. W., 1978, Turbidite muds and silts on deep sea fans and abyssal plains, in: *Sedimentation in Submarine Caynons, Fans, and Trenches,* Stanley, D. J., and Kelling, G., eds., Stroudsburg, Pennsylvania: Dowden, Hutchinson, and Ross, p. 163–176.

Piper, D. J. W., Von Huene, R., and Duncan, J. R., 1973, Late Quaternary sedimentation in the active eastern Aleutian Trench, *Geol. Mag.,* v. 110, p. 19–22.

Prell, W. L., Gardner, J. V., *et al.,* 1982, *Initial Reports of the Deep Sea Drilling Project,* v. 68, Washington, D. C.: U. S. Government Printing Office.

Prospero, J. M., Charlson, R. J., Mohnen, V., Jaenicke, R., Delany, A. C., Moyers, J., Zoller, W., and Rahn, K., 1983, The atmospheric aerosol system: an overview, *Rev. Geophys. Space Phys.,* v. 21, p. 1607–1629.

Rahn, K. A., Borys, R. A., and Shaw, G. E., 1981, Asian desert dust over Alaska: anatomy of an Arctic haze episode, in: *Desert Dust: Origin, Characteristics, and Effect on Man,* Pewe, T. L., ed., *Geol. Soc. Am. Spec. Pap.,* 186, p. 37–70.

Rawson, M. D., and Ryan, W. B. F., 1978, Ocean Floor Sediment and Polymetallic Nodules, 1:23,230,000, *U. S. Dept. State.*

Rehm, E., 1983, Coarse-grained volcanic detritus in deep-sea sediments of the northeastern equatorial Pacific, *Mar. Geol.,* v. 51, p. 347–361.

Rehm, E., and Halbach, P., 1982, Hawaiian-derived volcanic ash layers in equatorial northeastern Pacific sediments, *Mar. Geol.,* v. 50, p. 25–40.

Reid, J. L., and Lonsdale, P. F., 1974, On the flow of water through the Samoan Passage, *J. Phys. Oceanogr.,* v. 4, p. 58–73.

Revelle, R. R., 1944, Marine bottom samples collected in the Pacific Ocean by the *Carnegie* on its seventh cruise, *Carnegie Inst. Washington Publ.* vol. 556, p. 168–174.

Rex, R. W., and Goldberg, E. D., 1958, Quartz contents of pelagic sediments of the Pacific Ocean, *Tellus,* v. 10, p. 153–159.

Rhoads, D. C., and Boyer, L. F., 1982, The effects of marine benthos on physical properties of sediments, a successional perspective, in: *Animal-Sediment Relations: The Biogenic Alterations of Sediments,* McCall, P. L., and Tevesz, M. J. S., eds., New York: Plenum, p. 3–52.

Richards, A. F., 1958, Transpacific distribution of floating pumice from Isla San Benedicto, Mexico, *Deep Sea Res.,* v. 5, p. 29–35.

Ruddiman, W. F., and Glover, L. K., 1982, Mixing of volcanic ash zones in subpolar North Atlantic sediments, in: *The Ocean Floor,* Scrutton, R. A., and Talwani, M., eds., New York: Wiley, p. 37–60.

Rupke, N. A., and Stanley, D. J., 1974, Distinctive properties of turbidite and hemipelagic mudlayers in the Algero–Balearic Basin, western Mediterranean Sea, *Smithson. Contrib. Earth. Sci.* No. 13, 40 p.

Sachet, M.-H., 1955, Pumice and other extraneous volcanic materials on coral reefs, *Atoll Res. Bull.*, v. 37, p. 1–27.

Sachs, H. M., 1973, North Pacific radiolarian assemblages and their relationship to oceanographic parameters, *Quat. Res.*, v. 3, p. 73–88.

Safonov, V. G., 1978, Some types of clastic rocks of the Marcus–Necker seamounts, *Oceanology*, v. 18, p. 181–185.

Saito, T., Burckle, L. H., and Hays, J. D., 1975, Late Miocene to Pleistocene biostratigraphy of equatorial Pacific sediments, in: *Late Neogene Epoch boundaries*, Saito, T., and Burckle, L., eds., *Micropaleontology Spec. Publ.* 1, p. 226–244, New York: Micropaleontology.

Sancetta, C. A., 1979, Oceanography of the North Pacific during the last 18,000 years: evidence from fossil diatoms, *Mar. Micropaleo.*, v. 4, p. 103–123.

Sancetta, C. A., 1983, Biostratigraphic and paleoceanographic events in the eastern equatorial Pacific: results of Deep Sea Drilling Project Leg 69, in: *Initial Reports of the Deep Sea Drilling Project*, v. 69, Cann, J. R., Langseth, M. G., *et al.*, eds., Washington, D. C.: *U. S. Government Printing Office*, p. 311–320.

Scholl, D. W., 1974, Sedimentary sequences in the North Pacific trenches, in: *The Geology of Continental Margins* Burk, C. A., and Drake, C. L., eds., New York: Springer-Verlag, p. 493–504.

Scholl, D. W., Vallier, T. L., and Stevenson, A. J., 1982, Sedimentation and deformation in the Amlia fracture zone sector of the Aleutian Trench, *Mar. Geol.*, v. 48, p. 105–134.

Shepard, F. P., 1948, *Submarine Geology* (1st ed.), New York, Harper and Row.

Simkin, T., Siebert, L., McClelland, L., Bridge, D., Newhall, C., and Latter, J. H., 1981, *Volcanoes of the World*, Stroudsburg, Pennsylvania: Hutchinson Ross.

Sliter, W. V., Bé, A. W. H., and Berger, W. H., eds., 1975, Dissolution of Deep-Sea Carbonates, *Cushman Found. Foraminiferal Res., Spec. Publ.* v. 13, 159p.

Smayda, T. J., 1971, Normal and accelerated sinking of phytoplankton in the sea, *Mar. Geol.*, v. 11, p. 105–122.

Spiess, F. N., 1980, East Pacific Rise: hot springs and geophysical experiments, *Science*, v. 207, p. 1421–1433.

Stothers, R. B., 1984, Mystery cloud of AD 536, *Nature*, v. 307, p. 344–345.

Stonecipher, S. A., 1976, Origin, distribution and diagenesis of phillipsite and clinoptilolite in deep-sea sediments, *Chem. Geol.*, v. 17, p. 307–318.

Sverdlove, M. S., 1983, *Planktonic Foraminiferal Ecology of the Eastern Equatorial Pacific Ocean Including a Paleoceanographic Reconstruction of the Panama Basin for the Last 320,000 Years*, Ph.D. Dissertation, University of Cincinnati, Ohio.

Sverdrup, H. U., Johnson, M. W., and Fleming, R. H., 1942, *The Oceans*, Englewood Cliffs: Prentice-Hall.

Takahashi, K., and Honjo, S., 1983, Radiolarian skeletons: size, weight, sinking speed, and residence time in tropical pelagic oceans, *Deep Sea Res.*, v. 30, p. 543–568.

Thompson, P. R., 1981, Planktonic foraminifera in the western North Pacific during the past 150,000 years: comparison of modern and fossil assemblages, *Palaeogeog., Paleoclimat., Palaeoecol.*, v. 35, p. 241–279.

Thunnell, R. C., Keir, R. S., and Honjo, S., 1981, Calcite dissolution: an *in situ* study in the Panama Basin, *Science*, v. 212, p. 659–661.

Turekian, K. K., and Cochran, J. K., 1981, ^{210}Pb in surface air at Eniwetok and the Asian dust flux to the Pacific, *Nature*, v. 292, p. 522–524.

Uchupi, E., and Emery, K. O., 1963, The continental slope between San Francisco, California, and Cedros Island, Mexico, *Deep Sea Res.*, v. 10, p. 397–447.

Uyeda, S., 1974, Northwest Pacific trench margins, in: *The Geology of Continental Margins*, Burk, C. A., and Drake, C. L., eds., New York: Springer-Verlag, p. 473–504.

van Andel, T. H., Heath, G. R., and Moore, T. C., 1975, Cenozoic history and paleoceanography of the central equatorial Pacific Ocean, *Geol. Soc. Am. Mem.*, v. 143, p. 134.

Vincent, E., Killingley, J. S., and Berger, W. H., 1981, Stable isotope composition of benthic foraminifera from the equatorial Pacific, *Nature*, v. 289, p. 639–643.

Walker, G. P. L., Wilson, L., and Bowell, E. L. G., 1971, Explosive volcanic eruptions I. The rate of fall of pyroclasts, *Geophys. J. R. Astr. Soc.*, v. 22, p. 377–383.

Watkins, J., Moore, C., *et al.*, 1981, *Initial Reports of the Deep Sea Drilling Project*, v. 66, Washington, D. C.: U. S. Government Printing Office.

Wentworth, C. K., 1922, A scale of grade and class terms for clastic sediments, *J. Geol.*, v. 30, p. 377–392.

Windom, H. L., 1969, Atmospheric dust records in permanent snowfields: implications to marine sedimentation, *Geol. Soc. Am. Bull.*, v. 80, p. 761–782.

Windon, H. L., 1975, Eolian contributions to marine sediments, *J. Sed. Pet.*, v. 45, p. 520–529.

Worzel, J. L., 1959, Extensive deep-sea sub-bottom reflections identified as white ash, *Proc. Nat. Acad. Sci.*, v. 45, p. 349–355.

Chapter 2

MESOZOIC AND CENOZOIC SEDIMENTATION IN THE PACIFIC OCEAN BASIN

Thomas A. Davies

Institute for Geophysics
The University of Texas at Austin
Austin, Texas 78712

I. INTRODUCTION

The Pacific Ocean is the largest and oldest of the ocean basins. It represents approximately half of the world's oceanic regions and more than one-third of the total surface area of the earth. It is also the most truly "oceanic" of the oceans in that the deep trenches that surround it effectively isolate the basin from external terrigenous sediment input, except in the northeast where the trench has been filled to overflowing with sediments eroded from Alaska and the Canadian Rocky Mountains. As a result, the nature and distribution of the sediments in the Pacific Basin are largely functions of processes occurring within the basin and are relatively independent of the geology and geologic history of the surrounding regions, which are the subjects of separate chapters in this volume. McCoy (Chapter 1) has discussed the distribution of surficial sediments in the Pacific. This chapter addresses briefly the sedimentary record.

Much of our present knowledge of the pre-Neogene sediments of the Pacific comes from the results of deep-sea drilling by *Glomar Challenger*. Since 1969, when the ship first entered the Pacific, well over 200 holes have been drilled. These are almost all located north of 20° south, or in the Tasman Sea and the area south of New Zealand. The thin sediment cover and isolation from major port facilities have so far precluded drilling in the central and eastern South Pacific. The results of deep-sea drilling are presented in the series *Initial Reports of the Deep Sea Drilling Project*. Among the volumes published to date, volumes 5–9, 16–21, 29–35 and 54–68 are relevant to the Pacific Ocean. An up-to-date, brief but valuable summary of our present knowledge of the history of sedimentation in the Pacific has been provided by Kennett (1982). Other, more extensive, reviews include van Andel *et al.* (1975), Fischer *et al.* (1970), and Worsley and Davies (1979). There are also a number of recently published detailed studies of particular regions, e.g., Kulm *et al.* (1981), von Huene (1981), etc.

Given the vast amount of information that has been gathered in the past decade, and which continues to grow apace, I have not attempted either a detailed region-by-region account of the stratigraphy or a comprehensive synthesis. Nor have I addressed the sedimentary record of the ocean margins and marginal seas that can more properly be considered in the context of the geology of the circum-Pacific regions. Rather, what follows is a short overview of the salient features of sedimentation within the Pacific Basin in the context of geologic time.

II. FRAMEWORK OF SEDIMENTATION

A. Geographic and Tectonic Setting

Figure 1 shows the principal topographic features of the Pacific Basin. Surrounding the basin almost entirely are the deep-ocean trenches and associated island arcs or volcanic mountain ranges. The eastern part of the Pacific is shallower and dominated by the East Pacific Rise and, farther north, the scarps of the major fracture zones. The western Pacific basin floor is deeper, but marked by many guyots and elevated rises. Linear island chains, of which the Hawaiian chain is probably the best known, are a notable feature of the Pacific, although these are also found in other oceans. Similar to the linear island chains are linear chains of guyots, e.g., the Emperor Seamounts. These linear chains are generally interpreted as the result of the passage of the lithosphere over a hot spot in the mantle (Wilson, 1963; Clague, 1981).

Figure 2 shows the principal tectonic features of the Pacific Basin. Unlike the other oceans of the world, the Pacific is almost entirely surrounded by convergent margins. As a result, unlike the margins of the Atlantic and Indian oceans, which record the continuing history of those oceans, the margins of the Pacific record a complex history not easily related to events within the ocean basin or to the adjacent land masses. Most of the Pacific

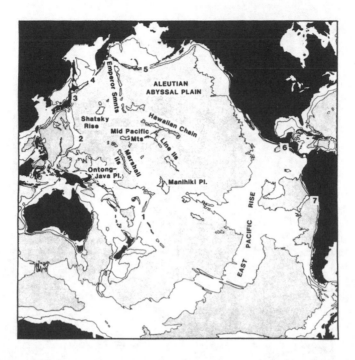

Fig. 1. Major bathymetric features of the Pacific Ocean Basin as defined by the 1000 m and 4000 m contours. Shaded areas are regions deeper than 4000 m. Principal deep-ocean trenches are numbered as follows: 1, Tonga; 2, Mariana; 3, Japan; 4, Kuril; 5, Aleutian; 6, Middle America; 7, Peru–Chile. (Modified from Davies and Gorsline, 1976. Reproduced by permission of Academic Press.)

floor is made up of the lithospheric plate, now moving in a generally northwesterly direction away from the major spreading center, the East Pacific Rise. Typical rates of spreading are on the order of 10 cm/yr or more (Handschumacher, 1976). East and south of the East Pacific Rise lie the Cocos, Nazca, and Antarctic plates. The Juan de Fuca plate in the eastern North Pacific, off Oregon and Washington, is a tiny remnant of a once much larger plate that has now been overridden by North America (Atwater, 1970). The geophysics and plate tectonics of the Pacific are discussed in detail in Volume 7B of this title by Hussong as well as in the excellent monograph edited by Sutton *et al.* (1976).

The inferred ages of the Pacific Ocean floor, generally confirmed by drilling results, are shown in Figure 3. Almost half the crust beneath the Pacific is pre-Cenozoic. The oldest parts of the basin lie in the west, and available data suggest that the oldest sediments are no more than about 160 m.y. old. Much of the sedimentary history of the Pacific has been destroyed in the subduction zones, and little record of the early Mesozoic remains. The remaining ocean crust of Jurassic age, for example, represents only a

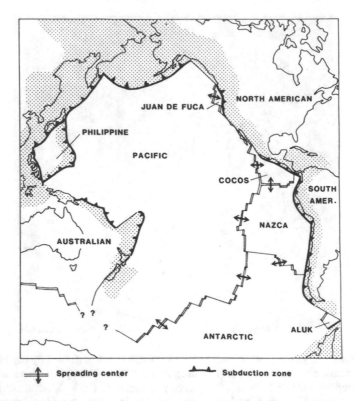

Fig. 2. Simplified map of the principal tectonic features of the Pacific basin. Shaded area represents Mesozoic–Cenozoic megasutures. (Based on data from many sources.)

small part of the Jurassic ocean floor. In Fig. 4 we have used paleogeographic reconstructions from Whitman (1981) to compare the remnant crustal area of a given age or older with an estimate of the area of the basin at that time. Less than 40% of the pre-Cenozoic and probably less than 10% of the pre-Cretaceous basin floor now remain.

The tectonic setting of the Pacific makes paleogeographic reconstruction exceedingly difficult. There are few constraints on any model that might be proposed. There is general agreement that the other oceans have grown over the past 200 m.y. at the expense of the Pacific, which has become smaller. Most researchers also agree that the Pacific has retained much of its original shape and geographic location. Using these boundary conditions, the approach to global paleogeographic reconstruction has usually been to make reconstructions of the Atlantic and Indian oceans and consider the Pacific to be what remains, e.g., Smith and Briden (1977), Barron *et al.* (1981). One of the most comprehensive attempts to reconstruct the Cenozoic

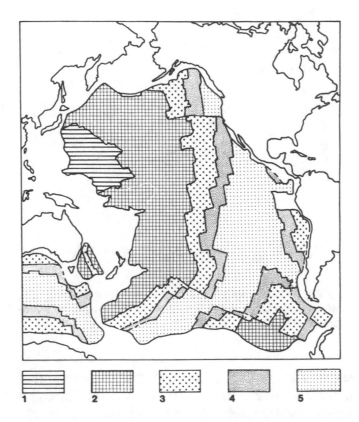

Fig. 3.   Age of the Pacific Ocean crust based on magnetic anomalies. 1, Jurassic; 2, Cretaceous; 3, Paleocene and Eocene; 4, Oligocene; 5, Miocene–Recent. (Based on Pitmann *et al.*, 1974.)

history of the Pacific, using marine magnetic anomalies and other data from within the basin and the immediately adjacent regions, is that of Whitman (1981).

The general scheme for the history of the Pacific (from Barron *et al.*, 1981) is depicted in Fig. 5. Paleogeographic reconstructions for early Jurassic time show a single supercontinent surrounded by a single superocean, a markedly different configuration from the present. In the middle Jurassic the supercontinent began to separate into northern and southern halves, with the opening of the Tethys Sea in the east and the Gulf of Mexico–Caribbean– North Atlantic region in the west. The Tethys continued to extend westward so that by middle to late Cretaceous time a continuous east–west low latitude seaway was established, linking the Pacific, North Atlantic, and Indian oceans. The connection between the Pacific and Indian oceans became severely restricted by middle Cenozoic time with the blocking of the Indo-

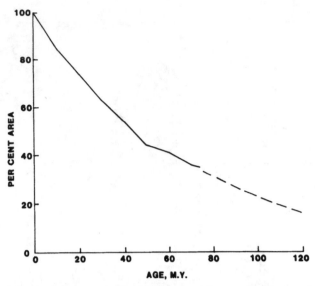

Fig. 4. Area of Pacific Ocean crust of a given age or older now remaining as a percentage of the total area of the Pacific Ocean at that time. Paleo-oceanic areas are estimated from paleo-geographic reconstructions (Whitman, 1981).

nesian Seaway which resulted from the northward movement of Australia. However, the low-latitude connection between the Atlantic and Pacific oceans persisted until the middle Miocene and was finally severed when the Isthmus of Panama separated the equatorial Pacific from the Gulf of Mexico in the Pliocene. Sub-Antarctic connection between the Pacific and Indian oceans became established in the middle Cenozoic as the Australian–Antarctic Seaway south of Australia was opened, but a connection with the South Atlantic, permitting complete circum-Antarctic circulation was not made until the Drake Passage opened in the late Oligocene.

B. Ocean Circulation, Sediment Distribution, and Tectonics

The basic surface circulation pattern in the Pacific, which influences most oceanic sedimentation, is relatively simple. Large central gyres north and south of the equator are separated by an equatorial current system. The equatorial current system is the locus of intense upwelling, especially in the eastern part of the region. In the far north is a less well-developed subpolar gyre. The far south is dominated by the eastward flowing circum-Antarctic system.

Bottom circulation in the Pacific is probably relatively sluggish, except where the topography of the ocean floor influences local circulation. The principal source of cold bottom water in the South Pacific is in the Ross

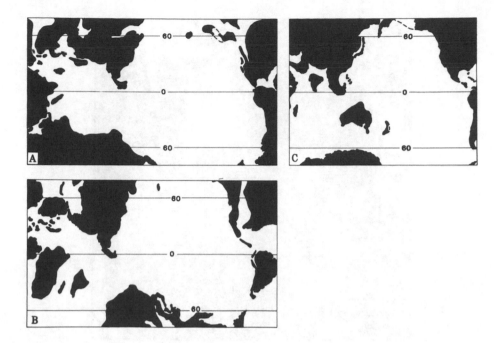

Fig. 5. The changing geography of the Pacific Ocean. A, 160 m.y. BP; B, 80 m.y. BP; C, 20 m.y. BP. All maps are Mercator projection at the same scale. (Based on Barron *et al.*, 1981.)

Sea, south of New Zealand. From here it flows northward through the Tasman Sea and through the Samoan Passage (Johnson and Johnson, 1970), into the western North Pacific.

Figure 6 shows the generalized distribution of the major surface sediment types in the Pacific. Beneath the central water masses of the northern and southern Pacific gyres are extensive areas of slowly accumulating deep-sea clay. The shallow areas of the East Pacific Rise and the guyots and plateaus of the southwestern Pacific are the sites of accumulating carbonate sediment. The upwelling associated with the equatorial current system gives rise to a belt of high-biogenous sediment accumulation, dominated by siliceous sediment. Around the periphery of the basin sedimentation is dominantly hemipelagic and detrital, and, except in the North Pacific, is limited in areal extent by the trenches, as noted earlier. As might be expected, glacial and siliceous sediments dominate the far southern part of the basin. Volcanic ash and fine-grained windblown sediments are ubiquitous, but are most significant near volcanic islands and in the western Pacific, downwind of the major volcanic chains and deserts of eastern and southeastern Asia. The crest of the mid-ocean ridges is the site of much hydrothermal activity,

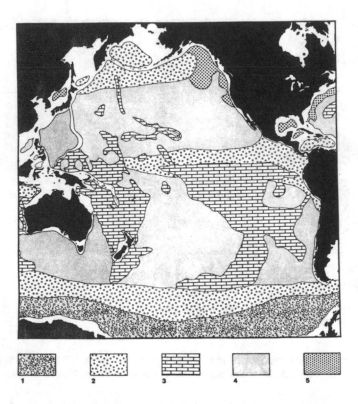

Fig. 6. Distribution of principal surficial sediment types. 1, Glacial sediments; 2, siliceous ooze; 3, calcareous ooze; 4, deep-sea clay; 5, terrigenous sediment; Blank, ocean margin sediments. (From Davies and Gorsline, 1976. Reproduced by permission of Academic Press.)

leading to the accumulation of metal-rich sediments (Edmond *et al.*, 1979). For further details of present day sediments, see Chapter 1 of this volume.

Sediment accumulation rates over the past 3 m.y., both total sediment and carbonate, are shown in Figure 7. The high accumulation under the equatorial high-productivity belt and on the rises and platforms of the West Pacific is immediately apparent. In the Pacific, as in the other oceans, sediment accumulation has been far from continuous and the sedimentary record, especially in the western and southwestern Pacific, is interrupted by hiatuses. Moore *et al.* (1978) and other writers attribute these hiatuses to changes in bottom-water characteristics (circulation patterns, current intensity, temperature).

Because the Pacific Ocean has retained much of its original shape and geographic location, it has also retained its basic, relatively simple circulation pattern. In marked contrast, the circulation patterns of the Atlantic and Indian oceans have changed significantly as those oceans have grown in size and changed in configuration. We might therefore expect the basic

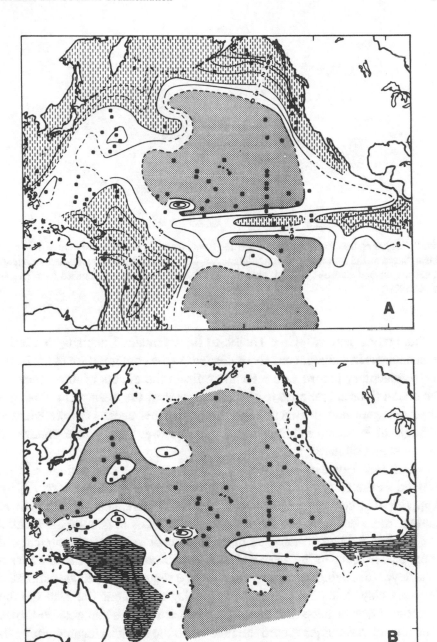

Fig. 7. Pliocene to Recent (0–3 m.y.) sediment accumulation rates expressed in $g/cm^2/10^3$ yr. Stippled areas indicate areas receiving less than 0.25 $g/cm^2/10^3$ yr. Vertical dashes or brickwork patterns indicate areas receiving $\geq$ 1.0 $g/cm^2/10^3$ yr. A, total sediment; B, carbonate. Solid squares are DSDP sites. (From Worsley and Davies, 1979. Reproduced by permission of the Society of Economic Paleontologists and Mineralogists.)

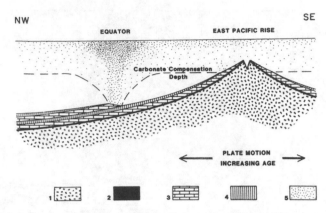

Fig. 8.   A generalized schematic cross section from northwest to southeast across the Pacific Basin, based on the concept of plate stratigraphy. 1, Oceanic basement; 2, metal-rich sediments; 3, calcareous and siliceous ooze; 4, abyssal clay; 5, organic detritus. (Modified from Heezen *et al.*, 1973.)

sedimentation pattern of the Pacific to have remained unchanged. Rather than changes in sediment supply or changes in the configuration of the basin, the sedimentary record in the Pacific reflects the effects of movements of the ocean floor in response to seafloor spreading, the opening or closing of important gateways connecting the Pacific to other oceans (Drake Passage, Isthmus of Panama, Australian–Antarctic Seaway) and global changes in sea level and climate.

The link between sedimentary sequences and plate tectonics is expressed in the concept of *plate stratigraphy* (Berger and Winterer, 1974). Applying this concept to the Pacific, both the increasing water depth resulting from cooling of the lithosphere and changes in geographic location due to plate movement must be taken into account. Thus, the typical sedimentary sequence in the western North Pacific consists of a metal-enriched basal layer (overlying basement) overlain by a sequence of pelagic carbonate, deep-sea clay, calcareous and siliceous pelagic sediments, and, at the top, a second layer of deep-sea clay. The metal-enriched sediments and lower sequence of carbonates record the time when the site was located above the carbonate-compensation depth, near the crest of the East Pacific Rise. The upper sequence of carbonate and siliceous sediments records the passage of the site beneath the equatorial high-productivity belt (Fig. 8). Lancelot and Larson (1975) applied this general concept to the sedimentary record formed at Deep Sea Drilling Project (DSDP) sites 303–313 in the northwest Pacific and were able to use the results interpreted from the sedimentary sequence to considerably refine the tectonic history deduced from the magnetic data alone.

III. MESOZOIC SEDIMENTATION

A. Jurassic (pre–135 m.y. BP)

The Jurassic rocks of the western Pacific represent the only remaining record of the early Mesozoic superocean. Unfortunately, these rocks have been sampled at only a few locations. Those samples that have been obtained consist of short fragments of core from widely separated locations. The lithologies present include cherts, porcellanites, limestones, and marls, and are indicative of open-ocean conditions not unlike those of the present-day (Heezen *et al.*, 1973; Larson *et al.*, 1975). Many of the samples are without identifiable fossils. Thus a more complete interpretation is impossible. Indeed one of the priorities for future scientific ocean drilling is the thorough investigation of the pre-Cretaceous record of the oceans, including the western Pacific (JOIDES, 1981).

B. Cretaceous (135–65 m.y. BP)

Cretaceous rocks have been widely sampled in the western Pacific both in the deep ocean and on rises and guyots. Notable localities where ocean drilling has permitted extensive sampling of the stratigraphic record include the Nauru Basin (Larson *et al.*, 1980), the Manihiki Plateau and Line–Tuamoto Islands (Schlanger *et al.*, 1976), the Mid-Pacific Mountains and Hess Rise (Thiede *et al.*, 1981), and the Shatsky Rise (Larson *et al.*, 1975). The Lower Cretaceous is characterized by clays, siltstones, marls, and marly limestone. Carbonaceous and siliceous deposits are also found. In the Upper Cretaceous pelagic limestones and reef limestones become dominant.

Organic carbon contents of Lower Cretaceous sediments of the Pacific are above those typical for a late Cenozoic or present-day open-ocean setting, but the laminated "black shales" characteristic of the Upper Cretaceous (Aptian) sediments of the Atlantic and Indian oceans are found only at paleodepths of 2000–3000 m on the rises and plateaus (Weissert, 1981). In some localities the organic-rich sediments are associated with volcaniclastic deposits. Much of the organic matter in these sediments is plant material derived from nearby land areas (Thiede *et al.*, 1981). The restricted depth-range suggests that, at least in the Pacific, the black shales reflect an expansion of the oxygen-minimum zone resulting from high-organic input coupled with sluggish ocean circulation. [Such an explanation will not hold for the black shales of the Atlantic and Tethys, however, and the global significance of early Cretaceous organic-rich sediments remains the subject of debate (Weissert, 1981).]

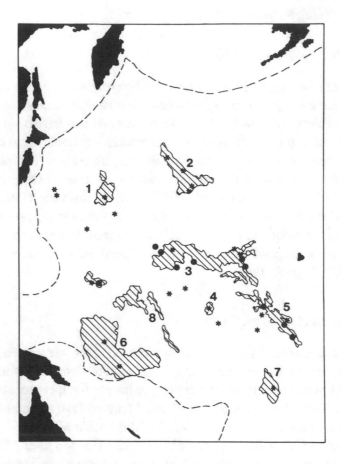

Fig. 9. Cretaceous volcanism. Diagonal shading indicates major edifices as follows: 1, Shatsky Rise; 2, Hess Rise; 3, Mid-Pacific Mountains; 4, Magellan Rise; 5, Line–Tuamoto chain; 6, Ontong–Java plateau; 7, Manihiki Plateau; 8, Marshall Islands. Broken line represents the western and northern limits of Mesozoic crust. Stars indicate locations at which volcanogenic sediments have been sampled; solid circles, locations where middle to late Cretaceous reefs have been sampled. (Based largely on Schlanger, 1981.)

The Cretaceous is generally regarded as a period of warm, equable conditions. It was also, especially in the middle Cretaceous, a period of extraordinarily voluminous and widespread mid-plate volcanism (Fig. 9). The results of dredging and deep-sea drilling show that most of the sea-mounts, plateaus, and rises of the western Pacific are underlain by volcanic edifices of Cretaceous age. The foundations of these seem to have been formed during a period of intense volcanic activity in the Early Cretaceous (pre-Barremian), but volcanic activity over a very wide area continued with pulses in Aptian and Campanian times (Rea and Vallier, in press). Indeed many of the guyots and rises were built above sea level to form islands, as evidenced in the abundance of plant debris found in neighboring basins. Even after subsiding they persisted as shoal areas throughout the Cretaceous and much of the Cenozoic. The sedimentary record in adjacent deep-sea basins shows abundant evidence of displaced shallow-water debris, generally introduced in discrete episodes coinciding with low stands of sea level (Thiede *et al.*, 1981; Schlanger, 1981).

Rises and guyots now situated north of the equator (e.g., Shatsky Rise, Hess Rise) were formed south of the equator, and their passage through the equatorial region is marked by thicknesses of rapidly accumulated middle and late Cretaceous pelagic limestones. Some of these limestones contain chert formed from the diagenesis of opaline silica tests indicative of the equatorial high-productivity belt (Thiede *et al.*, 1981).

Guyots in the Mid-Pacific Mountains, Line Islands, and Marshall Islands show evidence of middle to late Cretaceous reefs, indicating they were close to the equator and above or close to sea level at that time (Heezen *et al.*, 1973; Matthews *et al.*, 1974; Schlanger, 1981). Some of these reefs are of "Urgonian" facies and represent a segment of the world-encircling Tethyan "Urgonian belt" (Ferry and Schaaf, 1981). These reefs ceased to exist near the end of the Cretaceous. Matthews *et al.* (1974) attribute this to drowning resulting from a rapid eustatic rise of sea level.

IV. CENOZOIC SEDIMENTATION

A. Paleocene and Eocene (65–38 m.y. BP)

Available data from this time interval are still somewhat limited. However, sufficient data exist for the late Eocene that oceanwide patterns of sedimentation are discernable (Fig. 10).

In general, sediment accumulation rates were low, in large part because of extensive dissolution of carbonate and/or substantial accumulation of carbonate outside the Pacific Basin, e.g., the shallow seas of the Tethyan region and North America (van Andel *et al.*, 1975). Carbonate sediments appear to have accumulated only in water depths less than 3000 m (van Andel *et al.*, 1975). The elevation of the carbonate-compensation depth may be interpreted as evidence for a strongly stratified ocean with little or no circulation, permitting a buildup of oxidative carbon dioxide. The equatorial high-productivity belt appears as an ill-defined zone with relatively low accumulation rates. Carbonate sediments are, however, geographically widespread, a result of warm, equable conditions (Kennett, 1982). In fact, carbonate sedimentation continued into the late Eocene in some shallow areas adjacent to Antarctica.

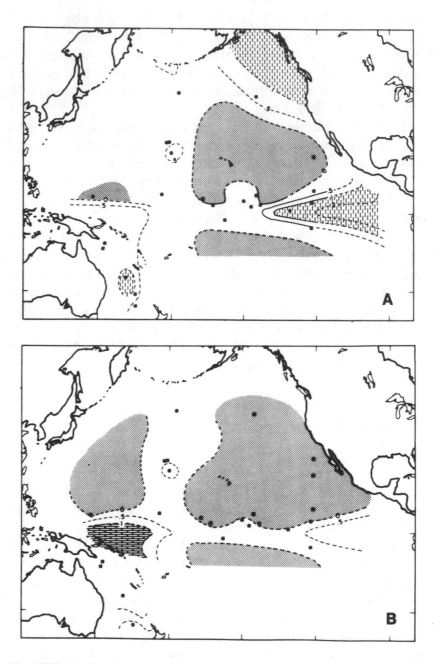

Fig. 10.   Middle Eocene (45–48 m.y. BP) sedimentation rates. DSDP site locations have been reconstructed using a simple model based on Morgan (1972). A, total sediment; B, carbonate. See Fig. 7 for key. (From Worsley and Davies, 1979. Reproduced by permission of the Society of Economic Paleontologists and Mineralogists.)

Terrigenous sedimentation became important in the Gulf of Alaska in the early Eocene, a reflection of tectonic activity on the northwestern margin of North America. High terrigenous sedimentation continued in this region until the middle Oligocene, building the Aleutian abyssal plain (Scholl and Creager, 1973).

B. Oligocene (38–23 m.y. B.P.)

The early Oligocene shows a sharpening of dissolution gradients and a drop in the depth of the carbonate-compensation depth at low latitudes. Carbonate sedimentation became focused in the equatorial region and the southwestern Pacific and ceased at high latitudes. By the late Oligocene carbonate sedimentation on the Emperor Seamounts and in the region to the west of them had ceased. Accumulation rates in the early Oligocene are low (Fig. 11), but hiatuses in the record are few, at least in the eastern Pacific. By contrast, in the late Oligocene, we see high accumulation rates (Fig. 12) and a sharp focusing of sedimentation in the peripheral and equatorial regions, perhaps resulting from the beginning of bottom circulation associated with opening of the Australian–Antarctic Seaway and the Drake Passage to deep water, circum-Antarctic circulation, and the beginning of glaciation in Antarctica (Kennett, 1977).

Increased volcanic and tectonic activity on the western margins of the Pacific in the late Oligocene is recorded in the increase in the number and thickness of ash layers in cores from those regions, e.g., DSDP Site 292 in the Philippine Basin (Donnelly, 1975; Kennett et al., 1977), and in the general increase in noncarbonate sedimentation rates near the marginal basins. By the late Oligocene, the Aleutian region was no longer the site of significant terrigenous sediment accumulation with the cessation of deposition on the Zodiac Fan (Stevenson et al., in press).

C. Miocene to Recent (23–0 m.y. B.P.)

In the Miocene (Fig. 13) we begin to see substantial effects of global cooling. The isolation of Antarctica by the circum-Antarctic current system led to extensive glaciation by Miocene time and the development of the present bottom circulation system of the oceans (Kennett, 1977). At the same time, low-latitude connections between the oceans were severed with the closing of the Tethys (early Miocene) and the uplift of the Isthmus of Panama (late Pliocene). The development of strong bottom circulation is reflected in the abundance of hiatuses in the middle Miocene record. Carbonate sedimentation was severely restricted and siliceous sediments (diatom and radiolarian oozes) become significant components of pelagic sedimentation,

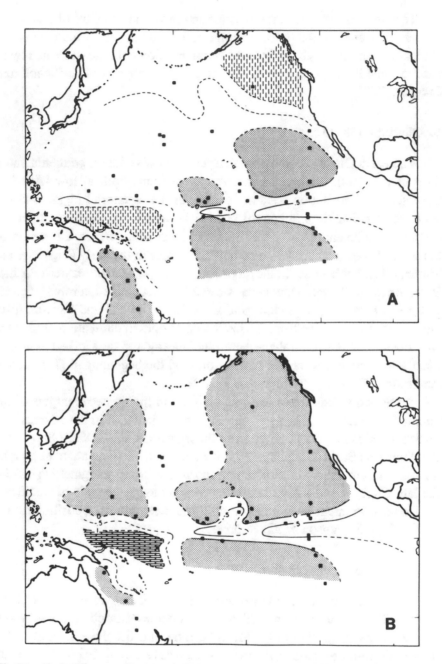

Fig. 11.   Early Oligocene (33–36 m.y. BP) sedimentation rates. DSDP site locations have been reconstructed according to a simple model based on Morgan (1972). A, total sediment; B, carbonate; See Fig. 7 for key. (From Worsley and Davies, 1979. Reproduced by permission of the Society of Economic Paleontologists and Mineralogists.)

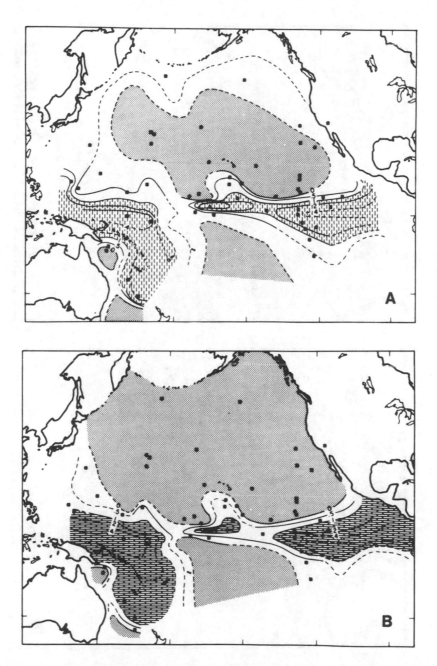

Fig. 12.   Late Oligocene (24–27 m.y. BP) sedimentation rates. DSDP site locations have been reconstructed according to a simple model based on Morgan (1972). A, total sediment; B, carbonate. See Fig. 7 for key. (From Worsley and Davies, 1979. Reproduced by permission of the Society of Economic Paleontologists and Mineralogists.)

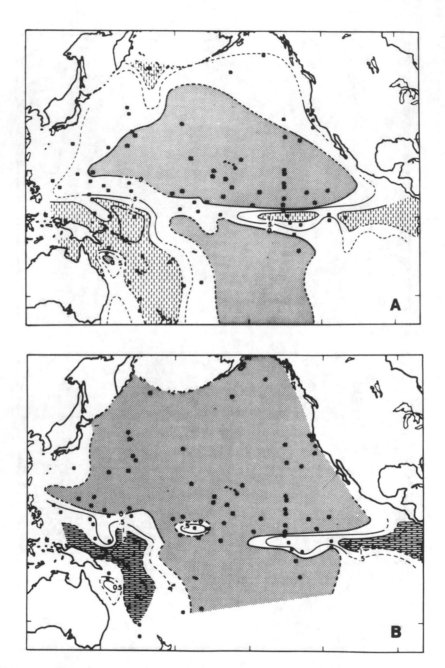

Fig. 13.   Late middle Miocene (9–12 m.y. BP) sedimentation rates. DSDP site locations have been reconstructed according to a simple model based on Morgan (1972). A, total sediment; B, carbonate. See Fig. 7 for key. (From Worsley and Davies, 1979. Reproduced by permission of the Society of Economic Paleontologists and Mineralogists.)

particularly at high latitudes, a consequence of increased upwelling at these latitudes resulting from the deterioration of climate (Brewster, 1980; Ingle and Garrison, 1977). Clay sediments on the northern Emperor Seamounts, for example, give way to rapidly deposited diatom oozes (Fullam *et al.*, 1973). In the late Miocene and Pliocene, terrigenous sediments again became a significant component in the northeast Pacific, reflecting renewed tectonic activity along those margins (von Huene and Kulm, 1973).

An important aspect of present-day sedimentation in the eastern Pacific is the accumulation of metal-rich sediments associated with hydrothermal activity at the crests of the mid-ocean ridges. Iron-enriched sediments form a basal layer in the sedimentary section in many parts of the ocean basins (Bonatti, 1981). Their occurrence has long been attributed to some sort of hydrothermal or other enrichment process operating at the mid-ocean ridge crests (Boström and Peterson, 1969; Boström, 1970). Recently an area of intense hydrothermal activity in the Galapagos rift zone has been investigated in detail by submersibles (Corliss *et al.*, 1979). The region supports a unique biological community and is the site of active deposition of iron- and manganese-enriched sediments. The latter form mounds several meters high. Deep-sea drilling in the area has yielded a complete section through the hydrothermal deposits and provides insight as to how the composition of the precipitates has changed through time (Natland *et al.*, 1979). In the Galapagos region, the sediments are rich in iron and manganese oxides. Near the mouth of the Gulf of California (Francheteau *et al.*, 1979) and, more recently, on the Juan de Fuca Ridge (Koski *et al.*, 1982), deposits of crystalline aggregates of zinc, iron, copper, and lead sulfides have now been found associated with hydrothermal activity. These findings, besides being indicative of the fact that hydrothermal activity is one of the important processes associated with spreading centers, are of considerable significance in the interpretation of ophiolites and massive sulfide ore deposits.

D. Some Mass-Balance Considerations

The abundance of Cenozoic sediment accumulation-rate data, coupled with our knowledge of the changing geography of the ocean basins, permits some estimates of the total mass of sediments deposited in the oceans as a function of time. Order-of-magnitude calculations for carbonate sediments have been presented by Davies and Worsley (1981). The method used requires a number of simplifying assumptions, but the results for the long-term flux of carbonate to the ocean floor (10.3–12.5×10^{14} g/yr) are surprisingly close to estimates made from considerations of riverine input (12.2×10^{14} g/yr; Garrels and MacKenzie, 1971). An important conclusion by Davies and Worsley is that, as a first approximation, the amount of carbonate

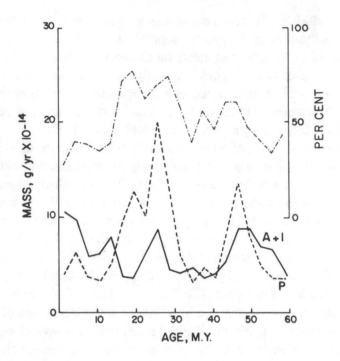

Fig. 14. Estimates of total carbonate sedimentation in the oceans through the Cenozoic. A + I, Atlantic and Indian oceans combined; P, Pacific Ocean. The uppermost line (—·—) shows the percentage of total carbonate sediment accumulating in the Pacific. (Modified from Davies and Worsley, 1981. Reproduced by permission of the Society of Economic Paleontologists and Mineralogists.)

deposited in the deep oceans is proportional to the exposed land-area of the continents.

Figure 14 shows the results for the Pacific compared to the Atlantic and Indian oceans combined. Three distinct periods can be recognized; 0–15 m.y. BP (middle Miocene to Recent), 15–30 m.y. BP (middle Oligocene to middle Miocene), and prior to 30 m.y. BP (pre-middle Oligocene). Prior to 30 m.y. BP, the exposed land area was less than the present (high sea level) and circulation between the oceans was dominantly latitudinal. All of the oceans behaved in concert and the Pacific, representing more than half the world's oceans, accumulated 40–60% of the total carbonate. At 30 m.y. BP sea level dropped markedly (Vail *et al.*, 1977). This exposed a larger area of the continents, increasing the total carbonate transported into the deep oceans, and at the same time the Pacific became the repository for 60–80% of the total. This was probably a consequence of the existence of the many seamounts and oceanic rises of the central and western Pacific which provide suitable sites for carbonate accumulation. A consideration of the hypsometry of the ocean basins (Menard and Smith, 1966) shows that at times of lowered sea level the proportion of shallow seafloor in the Pacific would increase significantly more than in the Atlantic. The period 12–15 m.y. BP corresponds to the closure of the Tethys and major glaciation in the Antarctic (Kennett, 1977). It also marks the incursion of cold water into the Atlantic from the Norwegian Sea (Thiede, 1979). These events lead to the establishment of the present circulation regime in the oceans which, as Berger (1970) has shown, tends to favor accumulation of carbonate in the Atlantic at the expense of the Pacific.

V. SUMMARY AND CONCLUSIONS

The Pacific Ocean holds the longest record of oceanic sedimentation, extending from the Jurassic to the present. Terrigenous sediments are of relatively minor significance, the dominant sediment types being pelagic oozes and clays and volcanogenic sediments. The distribution of sediments in the Pacific can be explained in terms of the interaction of ocean circulation (sediment supply and distribution) and plate tectonics (sediment preservation).

The Jurassic record is poorly known but appears to consist predominantly of clays and porcellanites. The Lower Cretaceous is marked by shales and marly limestones. Organic carbon contents of the sediments are high, but the early Cretaceous black shales found in the other oceans are represented at only a few localities. Evidence of extensive mid-plate volcanism in the Cretaceous is common. The late Cretaceous was a period of widespread carbonate sedimentation and reef building, presumably a reflection of warm, equable climates and extensive shallow-water environments.

The Cenozoic record has been widely sampled and shows the gradual evolution of the present pattern of sediment distribution. The deterioration in climate, the isolation of Antarctica, and the destruction of low-latitude connections with the Atlantic and Indian oceans lead to sharp climatic gradients and the development of the present meridianal circulation. These changes are reflected in the sedimentary record. Mass-balance considerations show the effects of changing sea level and progressive isolation of the world's oceans.

ACKNOWLEDGMENTS

In preparing this review, I have drawn heavily on the ideas and observations of many colleagues, particularly those who over the years have participated in the voyages of the *Glomar Challenger*. I have attempted to give the appropriate recognition in every case and apologize for any omissions or misinterpretations which may have crept in. J. A. Austin, Jr., and T. L. Vallier kindly reviewed the manuscript and provided helpful criticism. S. L. Thompson assisted in checking references and suggested a number of editorial improvements. This is contribution No. 548 from the Institute for Geophysics of the University of Texas at Austin.

REFERENCES

Atwater, T., 1970, Implications of plate tectonics for the Cenozoic tectonic evolution of western North America, *Geol. Soc. Am. Bull.*, v. 81, p. 3513–3566.

Barron, E. J., Harrison, C. G. A., Sloan, J. L., and Hay, W. W., 1981, Paleogeography, 180 million years ago to the present, *Eclogae Geol. Helv.*, v. 74, p. 443–470.

Berger, W. H., 1970, Biogenous deep-sea sediments, fractionation by deep-sea circulation, *Geol. Soc. Am. Bull.*, v. 81, p. 1385–1402.

Berger, W. H., and Winterer, E. L., 1974, Plate stratigraphy and the fluctuating carbonate line, in: *Pelagic Sediments on Land and under the Sea*, Hsü, K. J., and Jenkyns, H. C., eds, *Int. Assoc. Sediment. Spec. Publ. No. 1*, Oxford: Blackwell p. 11–98.

Bonatti, E., 1981, Metal deposits in the oceanic lithosphere, in: *The Oceanic Lithosphere, The Sea: Ideas and Observations on Progress in the Study of the Seas*, v. 7, Emiliani, C., ed., New York: Wiley, p. 639–686.

Boström, K., 1970, Geochemical evidence for ocean floor spreading in the South Atlantic Ocean, *Nature*, v. 227, p. 1041.

Boström, K., and Peterson, M. N. A., 1969, Origin of aluminum-poor ferromanganoan sediments in areas of high heat flow on the East Pacific Rise, *Mar. Geol.*, v. 7, p. 427–447.

Brewster, N. A., 1980, Cenozoic biogenic silica sedimentation in the Antarctic Ocean based on two Deep Sea Drilling Project sites, *Geol. Soc. Am. Bull.*, v. 91, p. 337–347.

Clague, D. A., 1981, Linear island and seamount chains, aseismic ridges and intraplate volcanism, in: *Deep Sea Drilling Project: A Decade of Progress*, Douglas, R. G., and Winterer, E. L., eds., *Soc. Econ. Paleontol. Mineral. Spec. Publ.* 32, p. 7–22.

Corliss, J. B., Dymond, J., Gordon, L. I., Edmond, J. M., von Herzen, R. P., Ballard, R. D., Green, K., Williams, D., Bainbridge, A., Crane, K., and van Andel, Tj. H., 1979, Submarine thermal springs on the Galapagos rift, *Science*, v. 203, p. 1073–1083.

Davies, T. A., and Gorsline, D. S., 1976, Oceanic sediments and sedimentary processes, in: *Chemical Oceanography*, Riley, J. P., and Chester, R., eds., v. 5, London: Academic, p. 1–80.

Davies, T. A., and Worsley, T. R., 1981, Paleoenvironmental implications of oceanic carbonate sedimentation rates, in: *The Deep Sea Drilling Project: A Decade of Progress*, Douglas, R. G., and Winterer, E. L., eds., *Soc. Econ. Paleontol. Mineral. Spec. Publ.* 32, p. 169–179.

Donnelly, T. N., 1975, Neogene explosive volcanic activity of the western Pacific: Sites 292 and 296, DSDP Leg 31, in: *Initial Reports of the Deep Sea Drilling Project*, Karig, D. E., Ingle, J. C., *et al.*, eds. v. 31, Washington, D.C.: United States Government Printing Office, p. 577–597.

Edmond, J. M., Measures, C., Mangum, B., Grant, B., Sclater, J. G., Collier, R., Hudson, A., Gordon, L. I., and Corliss, J. B., 1979, On the formation of metal-rich deposits at ridge crests, *Earth Planet. Sci. Lett.*, v. 46, p. 19–30.

Ferry, S., and Schaaf, A., 1981, The early Cretaceous environment at Deep Sea Drilling Project site 463 (Mid Pacific Mountains), with reference to the Vocontian Trough (French Subalpine Ranges), in: *Initial Reports of the Deep Sea Drilling Project*, Thiede, J., Vallier, T. L., *et al.*, eds., v. 62, Washington, D.C.: United States Government Printing Office, p. 669–682.

Fischer, A. G., Heezen, B. C., Boyce, R. E., Bukry, D., Douglas, R. G., Garrison, R. E., Kling, S. A., Krasheninnikov, V., Lisitzin, A. P., and Pimm, A. C., 1970, Geological history of the western North Pacific, *Science*, v. 168, p. 1210–1214.

Francheteau, J., Needham, H. D., Choukroune, P., Juteau, T., Seguret, M., Ballard, R. D., Fox, P. J., Normark, W. R., Carranza, A., Cordoba, D., Guerrero, J., Rangin, C., Bougault, H., Cambon, P., and Hekinian, R., 1979, Massive deep-sea sulfide ore deposits discovered on the East Pacific Rise, *Nature*, v. 277, p. 523–528.

Fullam, T. J., Supko, P. R., Boyce, R. E., and Stewart, R. W., 1973, Some aspects of late Cenozoic sedimentation in the Bering Sea and North Pacific Ocean, in: *Initial Reports of the Deep Sea Drilling Project*, Creager, J. S., Scholl, D. W., *et al.*, eds., v. 19, Washington, D. C.: United States Government Printing Office, p. 887–896.

Garrels, R. M., and MacKenzie, F. T., 1971, *Evolution of Sedimentary Rocks*, New York: Norton.

Handschumacher, D. W., 1976, Post Eocene plate tectonics of the eastern Pacific, in: The Geophysics of the Pacific Ocean Basin and its Margins, Sutton, G. H., Manghnani, M. H., and Moberly, R., eds., Washington, D.C.: American Geophysical Union, p. 177–202.

Heezen, B. D., MacGregor, I., *et al.*, 1973, *Initial Reports of the Deep Sea Drilling Project*, v. 20, Washington, D.C.: United States Government Printing Office.

Ingle, J. C., and Garrison, R. E., 1977, Origin, distribution, and diagenesis of Neogene diatomites around the North Pacific rim, *Proceedings of the 1st International Congress on Pacific Neogene Stratigraphy*, Tokyo: Kaiyo Shuppan Co., p. 348–350.

Johnson, D. A., and Johnson, T. C., 1970, Sediment redistribution by bottom currents in the Central Pacific, *Deep-Sea Res.*, v. 17, p. 157–170.

Joides, 1981, *Report of the Conference on Scientific Ocean Drilling*, Washington, D.C.: JOI, Inc., 110 p.

Kennett, J. P., 1977, Cenozoic evolution of Antarctic glaciation, the circum-Antarctic ocean and their impact on global paleoceanography, *J. Geophys. Res.*, v. 82, p. 3843–3860.

Kennett, J. P., 1982, *Marine Geology*, Englewood Cliffs: Prentice-Hall.

Kennett, J. P., McBirney, A. R., and Thunnell, R. C., 1977, Episodes of Cenozoic volcanism in the circum-Pacific region, *J. Volcanol. Geotherm. Res.*, v. 2, p. 145–163.

Koski, R. A., Normark, W. R., Morton, J. L., and Delaney, J. R., 1982, Metal sulfide deposits on the Juan de Fuca Ridge, *Oceanus*, v. 25, p. 42–48.

Kulm, L. D., Dymond, J., Dasch, E. J., and Hussong, D. M., eds., 1981, *Nazca Plate: Crustal Formation and Andean Convergence*, Geol. Soc. Am. Mem. v. 154.

Lancelot, Y., and Larson, R. L., 1975, Sedimentary and tectonic evolution of the northwestern Pacific, in: *Initial Reports of the Deep Sea Drilling Project*, Larson, R. L., Moberly, R., *et al.*, eds. v. 32, Washington, D.C.: United States Government Printing Office, p. 925–939.

Larson, R., Moberly, R., *et al.*, 1975, *Initial Reports of the Deep Sea Drilling Project*, v. 32, Washington, D.C.: United States Government Printing Office.

Larson, R., Schlanger, S. O., *et al.*, 1980, *Initial Reports of the Deep Sea Drilling Project*, v. 61, Washington, D.C.: United States Government Printing Office.

Matthews, J. L., Heezen, B. C., Catalano, R., Coogan, A., Tharp, M., Natland, J., and Rawson, M., 1974, Cretaceous drowning of reefs on mid-Pacific and Japanese guyots, *Science*, v. 184, p. 462–464.

Menard, H. W., and Smith, S. M., 1966, Hypsometry of ocean basin provinces, *J. Geophys. Res.*, v. 71, p. 4305–4325.

Moore, T. C., van Andel, Tj. H., Sancetta, C., and Pisias, N., 1978, Cenozoic hiatuses in pelagic sediments, *Micropaleontology*, v. 24, p. 113–138.

Morgan, J., 1972, Deep mantle convection plumes and plate motions, *Am. Assoc. Pet. Geol. Bull.*, v. 56, p. 203–213.

Natland, J. H., Rosendahl, B., Hekinian, R., Dmitriev, Y., Fodor, R. V., Goll, R. M., Hoffert, M., Humphris, S., Mattey, D. P., Peterson, N., Roggenthen, W., Schrader, E. L., Srivastava, R. K., and Warren, N., 1979, Galapagos hydrothermal mounds: Stratigraphy and chemistry revealed by deep sea drilling, *Science*, v. 204, p. 613–616.

Pittman, W. C., Larson, R. L., and Herron, E. M., 1974, The age of the ocean basins, *Geol. Soc. Amer. Map Chart Series*, MC-6.

Rea, D. K., and Vallier, T. L., 1983, Two Cretaceous volcanic episodes in the western Pacific Ocean, *Geol. Soc. Am. Bull.* v. 19, p. 1430–1437.

Schlanger, S. O., 1981, Shallow water limestones in oceanic basins as tectonic and paleoceanographic indicators, in: *The Deep Sea Drilling Project: A Decade of Progress*, Douglas, R. G., and Winterer, E. L., eds., *Soc. Econ. Paleontol. Mineral. Spec. Publ.* 32, p. 209–266.

Schlanger, S. O., Jackson, E. D., *et al.*, 1976, *Initial Reports of the Deep Sea Drilling Project*, v. 33, Washington, D.C.: United States Government Printing Office.

Scholl, D. W., and Creager, J. S., 1973, Geological synthesis of Leg 19 (DSDP) results: far North Pacific, and Aleutian Ridge, and Bering Sea, in: *Initial Reports of the Deep Sea Drilling Project*, Creager, J. S., Scholl, D. W., *et al.*, eds., v. 19, Washington, D.C.: United States Government Printing Office, p. 897–913.

Smith, A. G., and Briden, J. C., 1977, *Mesozoic and Cenozoic Paleocontinental Maps*, Cambridge, England: Cambridge University.

Stevenson, A. J., Scholl, D. W., and Vallier, T. L., 1983, Tectonic and geologic implications of the Zodiac Fan, Aleutian abyssal plain, northeast Pacific, *Geol. Soc. Am. Bull.* v. 94, p. 259–273.

Sutton, G. H., Manghnani, M. H., Moberly, R., eds., 1976, The geophysics of the Pacific Ocean basin and its margin, *Geophys. Monogr. Am. Geophys. Union.* v. 19.

Thiede, J., 1979, History of the North Atlantic Ocean: evolution of an asymmetric zonal paleoenvironment in a latitudinal basin, in: *Deep Drilling Results in the Atlantic Ocean: Continental Margins and Paleoenvironment*, Talwani, M., Hay, W. W., and Ryan, W. B. F., eds., Washington, D.C.: American Geophysical Union, p. 275–296.

Thiede, J., Dean, W. E., Rea, D. K., Vallier, T. L., and Adelseck, C. G., 1981, The geologic history of the Mid-Pacific Mountains in the central North Pacific Ocean—a synthesis of deep sea drilling studies, in: *Initial Reports of the Deep Sea Drilling Project*, Thiede, J., Vallier, T. L., *et al.*, eds. v. 62, Washington, D.C.: United States Government Printing Office, p. 1073–1120.

Thiede, J., Vallier, T., *et al.*, 1981, Initial Reports of the Deep Sea Drilling Project, v. 62, Washington, D.C.: United States Government Printing Office.

Vail, P. R., Mitchum, R. M., Todd, R. G., Widmier, J. M., Thompson, S., Sangree, J. B., Bubb, J. N., and Hatlelid, W. G., 1977, Seismic stratigraphy and global changes in sea level, in: *Seismic Stratigraphy—Applications to Hydrocarbon Exploration*, Payton, C. E., ed., *Am. Assoc. Pet. Geol. Mem.* v. 26, p. 49–212.

van Andel, Tj. H., Heath, G. R., and Moore, T. C., 1975, Cenozoic Tectonics Sedimentation and Paleoceanography of the Central Equatorial Pacific, *Geol. Soc. Am. Mem.* 143.

von Huene, R., 1981, Review of early results from drilling of the IPOD-1 active margin transects across the Japan, Mariana, and Middle America convergent margins, in: *The Deep Sea Drilling Project: A Decade of Progress*, Douglas, R. G. and Winterer, E. L., eds., *Soc. Econ. Paleontol. Mineral. Spec. Publ.* 32, p. 57–66.

von Huene, R., and Kulm, L. D., 1973, Tectonic survey of Leg 18, in: *Initial Reports of the Deep Sea Drilling Project*, Kulm, L. D., von Huene, R., *et al.*, eds., v. 18, Washington, D.C.: United States Government Printing Office, p. 961–975.

Weissert, H., 1981, The environment of deposition of black shales in the early Cretaceous: an ongoing controversy, in: *The Deep Sea Drilling Project: A Decade of Progress*, Douglas R. G., and Winterer, E. L., eds., *Soc. Econ. Paleontol. Mineral. Spec. Publ.* 32, p. 547–560.

Whitman, J. M., 1981, Tectonic and bathymetric evolution of the Pacific Ocean Basin since 74 Ma, M.S. Thesis, University of Miami, Miami, Florida.

Wilson, J. T., 1963, A possible origin of the Hawaiian Islands, *Can. J. Phys.*, v. 41, p. 863–870.

Worsley, T. R., and Davies, T. A., 1979, Cenozoic sedimentation in the Pacific Ocean: steps toward a quantitative evaluation, *J. Sediment. Petrol.*, v. 49, p. 1131–1146.

Chapter 3

PACIFIC PLATE MOTION RECORDED BY LINEAR VOLCANIC CHAINS

Robert A. Duncan

College of Oceanography
Oregon State University
Corvallis, Oregon 97331

and

David A. Clague

United States Geological Survey
345 Middlefield Rd.
Menlo Park, California 94025

I. INTRODUCTION

Seamounts and islands are quite common features of the deep-ocean basins, particularly the Pacific Basin. A conspicuous but relatively small number of these volcanoes are arranged in long linear chains (Figure 1). The vast majority, however, occur as isolated edifices. Recent work on a small number of these isolated volcanoes shows that many were formed at or very close to mid-ocean spreading axes (Clague and Dalrymple, 1975; Batiza, 1980) but that others are significantly younger than the underlying oceanic crust (G. B. Dalrymple, D. A. Clague, H. G. Greene, unpublished data). These isolated volcanoes require much additional study before their origin is understood. The volcanoes arranged in linear chains have been the focus of a great deal of work, and much has been learned about their origin and evolution. Most of the linear volcanic chains in the Pacific Basin are oriented roughly west-northwesterly and apparently formed sequentially above stationary hot spots during the past 42 m. y. as the Pacific plate rotated clockwise about a pole located near 69°N, 68°W (Clague and Jarrard, 1973a). Another group of linear chains exhibit roughly north-trending orientations and apparently were formed by the same mechanism between at least 80 and 42 m. y. ago as the Pacific plate rotated about a pole of rotation located near 17°N, 107°W (Clague and Jarrard, 1973b). Earlier age data available from linear island and seamount chains were summarized by Jackson (1976), Jarrard and Clague (1977), and McDougall and Duncan (1980); we will briefly review the information presented there and update it with more recent data.

We have focused our attention on the age relations along and among the linear island chains that occur in the Pacific Ocean. A detailed summary and bibliography of the geology, petrology, and geophysics of the Hawaiian–Emperor volcanic chain is presented in Jackson et al. (1980). No similar summaries or detailed bibliographies exist for the other linear volcanic chains. Detailed information of the geology, petrology, and geophysics of these volcanoes can be found through the references cited in the various articles that report radiometric age data.

Jarrard and Clague (1977) discussed the reliability of the available age data using several specific examples. We emphasize here that nearly all submarine volcanic rocks are altered to some degree, which makes the interpretation of the analytically determined radiometric ages less than straightforward. We do not consider an age to be reliable unless we have obtained several concordant ages, have dated fresh mineral separates, or have obtained $^{40}Ar/^{39}Ar$ incremental heating data showing a well-defined plateau and isochron. Clague et al. (1975), Dalrymple and Clague (1976), and Dalrymple, et al. (1980) have shown that total-fusion $^{40}Ar/^{39}Ar$ ages of somewhat altered submarine rocks commonly give ages consistent with K-Ar ages on feldspars and with ages based on $^{40}Ar/^{39}Ar$ incremental heating experiments. We urge caution, however, since this is not always the case, particularly for tholeiitic basalt samples.

Another factor that affects the interpretation of available age data is the finite period of time during which volcanoes form. The sequence and duration of volcanic events are relatively well known in the Hawaiian Islands, where the time from the pretholeiitic-stage alkalic lavas (Moore et al., 1982) to the last posterosional eruptions of strongly alkalic lavas can be as long as 4.5 m. y. (Clague et al., 1983). Whereas for Hawaiian volcanoes the tholeiitic shield stage was probably less than 1 m.y. in duration (McDougall, 1964; McDougall and Duncan, 1980), and the formation of the volcano up to and including the postcaldera alkalic stage probably took less than 2 m.y., evidence is mounting that the duration of volcanic activity is considerably longer for some of the volcanoes in the Austral–Cook Islands (Dalrymple

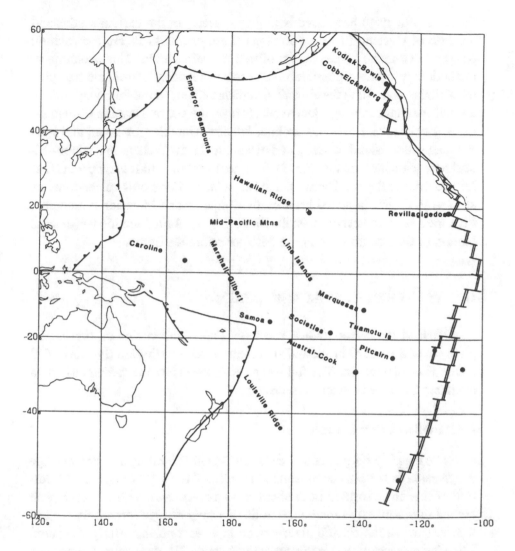

Fig. 1.   Location map showing the Pacific Basin and the linear volcanic chains discussed in the text. Dots mark the location of hot spots.

et al., 1975; Duncan and McDougall, 1976) and in the Pratt–Welker Seamount chain in the Gulf of Alaska (Turner *et al.*, 1980).

Plate rotation models of the absolute motion of the Pacific plate relative to the underlying hot spots are based on two types of data: (1) the geometry of the volcanic chains, and (2) the rate of volcanic migration along the chains. Early models used the geometry of the chains to determine the location of the rotation pole and, separately, the rate of volcanic propagation along the chains to determine the rate of rotation about that pole (e.g., Clague and Jarrard, 1973a,b). More recent models (e.g., McDougall and Duncan, 1980;

Turner *et al.*, 1980) have used both the geometry of the various chains and the rate of volcanic propagation along those chains to locate the rotation pole and determine a least-squares-fitted rate of rotation. This evolution of methodology became possible due to the availability of reliable age data from chains other than the Hawaiian volcanic chain. In the following section we will summarize the age data available from the coeval Hawaiian, Austral–Cook, Society, Marquesas, Caroline, Pitcairn–Gambier, Samoan, and Islas Revillagigedos island chains, and from the Pratt–Welker and Cobb–Eickelberg seamount chains (Fig. 1). The next section will review chains that formed prior to the bend between the Hawaiian and Emperor volcanic chains at 42 m.y. Chains included here are the Emperor and Musician seamounts, the Line Islands, and the Louisville Ridge (Fig. 1). A final section will discuss motion of the Pacific plate in the hot spot reference frame.

II. LATE TERTIARY VOLCANIC CHAINS

Most of the better-studied seamount and island chains on the Pacific plate are late Tertiary in age and are oriented west-northwesterly subparallel to the Hawaiian chain. The following sections review the radiometric data available from these volcanic lineaments.

A. Hawaiian Volcanic Chain

The classic example of a linear island chain exhibiting a systematic age progression is the Hawaiian–Emperor chain. The Emperor Seamount portion of this chain formed from about 70 to 42 m.y. ago, as the Pacific plate moved in a northerly direction, and the Hawaiian Ridge formed from 42 to 0 m.y. ago, as the Pacific plate moved in a west-northwesterly direction. The Emperor chain will be discussed in Section III, describing Cretaceous and early Tertiary chains. The Hawaiian chain includes the high volcanic Hawaiian Islands, the low pinnacles and atolls of the Leeward Islands, and numerous seamounts and guyots. Age data are available from 31 volcanoes in the chain. The only major gap in ages is for the far western end of the chain. Work in progress (Duncan and Clague, 1984) on dredged samples from Colahan and Abbott seamounts has determined the age of volcanism in this area to be between 36 and 40 m.y.

Table I lists "best" ages for each edifice. All data have been converted to current decay constants (Steiger and Jager, 1977). Distances have been recalculated from Loihi Seamount instead of from Kilauea because Loihi is the youngest Hawaiian volcano (Moore *et al.*, 1982). For many of the dated volcanoes, particularly those older than Northampton Bank, tholeiitic lavas

suitable for K-Ar dating have not been recovered. The ages of these volcanoes are based on ages of alkalic lavas. These alkalic lavas were thought
to all postdate the shield-building tholeiitic stage. The recent discovery that
young Loihi Seamount is erupting both tholeiitic and alkalic lavas (Moore
et al., 1982) suggests that at least some of the alkalic lavas dredged along
the chain, particularly from small volcanoes, may represent the early rather
than the late alkalic eruptive phase.

New age data include a near-zero age for Loihi Seamount, based on
palagonite thicknesses (Moore *et al.*, 1982). New radiocarbon ages on char-
coal beneath surficial flows on Hualalai volcano on Hawaii are all less than
12,000 years old (R.B. Moore, personal communication, 1983), and tholeiitic
lavas dredged from the northwest rift of Hualalai still have fresh glassy rinds
indicating a probable age of less than 100,000 years (Clague, 1982). New age
data from Haleakala, Kahoolawe, West Maui, Lanai, and East and West
Molokai (Naughton *et al.*, 1980; Bonhommet *et al.*, 1977) confirm the age
relations for these volcanoes published by McDougall (1964). McDougall
(1979) has added new data for Kauai, while Dalrymple (cited in Dalrymple
et al., 1980a) and Grooms (1980) have added new data for Niihau Island and
Kaula Island, respectively. Grooms (1980) has also added limited data for
an unnamed seamount between Nihoa and Necker islands, for Brooks Bank,
and for Gardner Pinnacles. Only the data for Gardner Pinnacles are based
on K-Ar ages of more than one sample; the others must be considered less
reliable. New data provide a reliable age of 19.9 ± 0.3 m.y. for Laysan
Island, and a less reliable age on one sample of 26.6 ± 2.7 m.y. for Northampton Bank (Dalrymple *et al.*, 1981). Dalrymple *et al.* (1977) redeter-
mined the age of Midway Atoll using less altered alkalic cobbles from the
Midway drill core. The new age of 27.7 ± 0.6 m.y. demonstrates that the
previously reported age of 17.9 ± 0.6 m.y. (Dalrymple *et al.*, 1974) represents a minimum age due to the highly altered condition of the tholeiitic
flows. This example serves to illustrate the difficulties inherent in dating
altered submarine volcanic rocks.

Estimates of the rate of volcanic migration along the Hawaiian chain
vary considerably. McDougall and Duncan (1980) used a linear regression
of all the data available at that time and calculated a rate of 9.66 cm/yr ±
0.27 cm/yr. Using the more complete data set presented in Table I, we have
recalculated the best-fit volcanic migration rate as 9.03 ± 0.4 cm/yr for all
dated volcanoes older than Mauna Kea. We excluded the youngest volcanoes from the calculations because they are still active, and extrapolation
based on Loihi Seamount indicates they will continue to be active for up to
another 0.9 m.y. If these volcanoes were sampled several million years from
now, it would be these still-to-be erupted lavas that would be sampled and

TABLE I
Summary of Age Data from the Hawaiian–Emperor Volcanic Chain

Name	Number[a]	Distance from Loihi along Hawaiian–Emperor Trend (km)	Best K-Ar age ($\times 10^6$ yrs)	Source[c]	Remarks
Hawaiian Ridge					
Loihi Seamount	0	0	Active	1	Age estimated from palagonite
Kilauea	1	32	Active	1	Historic tholeiitic eruptions
Mauna Loa	2	56	Active	1	Historic tholeiitic eruptions
Mauna Kea	4	66	0.375 ± 0.05	2	Samples from tholeiitic shield
Hualalai	3	103	0.01	3	^{14}C ages on oldest alkalic flows
Kohala[b]	5	107	0.40 ± 0.02	4	Samples from tholeiitic shield
Haleakala[b]	6	184	0.75 ± 0.04	5, 8	Samples from tholeiitic shield
Kahoolawe	7	213	1.03 ± 0.18	5	Alkalic samples from upper member
W. Maui[b]	8	230	1.63 ± 0.03	5, 8	Samples from tholeiitic shield
Lanai[b]	9	258	1.28 ± 0.04	6	Samples from tholeiitic shield
E. Molokai[b]	10	272	1.48 ± 0.07	5, 8	Samples from tholeiitic shield
W. Molokai[b]	11	302	1.84 ± 0.07	5, 8	Samples from tholeiitic shield
Koolau[b]	12	369	2.3 ± 0.1	7, 8	Samples from tholeiitic shield
Waianae[b]	13	404	2.9 ± 0.1	7	Samples from tholeiitic shield
Kauai[b]	14	549	5.1 ± 0.2	9	Samples from tholeiitic shield
Niihau[b]	15	595	5.5 ± 0.2	10	Samples from tholeiitic shield
Kaula	15	615	4.1 ± 0.1	11	Phonolite blocks in tuff
Nihoa	17	810	7.2 ± 0.3	12	Samples from tholeiitic shield
(unnamed)	21	950	10.1 ± 0.6	11	Dredged hawaiite
Necker[b]	23	1099	10.3 ± 0.4	12	Samples from tholeiitic shield
LaPerouse Pinnacle[b]	26	1239	12.0 ± 0.4	12	Samples from tholeiitic shield

Brooks Bank	28	1330	17.6 ± 0.5	11	Dredged hawaiite
Gardner Pinnacle[b]	30	1420	16.0 ± 1.2	11	Dredged tholeiite and alkalic basalt
Laysan Island[b]	36	1848	19.9 ± 0.3	13	Dredged hawaiite and mugearite
Northampton Bank	37	1871	26.6 ± 2.7	13	Dredged tholeiitic basalt
Pearl and Hermes Reef	50	2311	20.6 ± 0.5	14	Dredged phonolite, hawaiite and alkalic basalt
Midway Islands[b]	52	2462	27.7 ± 0.6	15	Drilled cobbles of hawaiite and mugearite
(unnamed)	57	2630	28.0 ± 0.4	14	Dredged alkalic basalt
(unnamed)	56, 58	2702	28–31	16	Fossils, turbidites, DSDP 311
(unnamed)	63	2855	27.4 ± 0.5	14	Dredged alkalic basalt
Emperor Seamounts					
Kammu	65	3440	41 ± 3	17	Large foraminifera
Daikakuji	67	3523	42.4 ± 2.3	18	Dredged alkalic basalt
Yuryaku	69	3550	43.4 ± 1.6	14	Dredged alkalic basalt
Kimmei	72	3698	39.9 ± 1.2	18	Dredged alkalic basalt
Koko	74	3788	48.1 ± 0.8	19	Dredged alkalic basalt, trachyte
Ojin	81	4132	55.2 ± 0.7	20	Hawaiite and tholeiite, DSDP 430
Jingu	83	4205	55.4 ± 0.9	21	Dredged hawaiite and mugearite
Nintoku	86	4482	56.2 ± 0.6	20	Alkalic basalt, DSDP 432
Suiko	90	4824	59.6 ± 0.6	22	Single dredged mugearite
Suiko	91	4890	64.7 ± 1.1	20	Alkalic basalt and tholeiite DSDP site 433
Meiji	108	5860	74 ± 3 (61.9 ± 5.0)	23, 24	Nanoplankton, (minimum age on altered tholeiite) DSDP 192

[a] Volcano number from Bargar and Jackson (1974).

[b] Age data used to calculate "best" volcanic migration rate (see text).

[c] *Sources:* 1. Moore et al. (1982); 2. Porter et al. (1977); 3. R. Moore (unpublished ^{14}C date); 4. McDougall and Swanson (1972); 5. Naughton et al. (1980); 6. Bonhommet et al. (1977); 7. Doell and Dalrymple (1973); 8. McDougall (1964); 9. McDougall (1979); 10. G. B. Dalrymple (unpub. data); 11. Grooms (1980); 12. Dalrymple et al. (1974); 13. Dalrymple et al. (1981); 14. Clague et al. (1975); 15. Dalrymple et al. (1977); 16. Larson, et al. (1975); 17. Sachs, quoted in Clague and Jarrard (1973); 18. Dalrymple and Clague (1976); 19. Clague and Dalrymple (1973); 20. Dalrymple et al. (1980a); 21. Dalrymple and Garcia (1980); 22. Saito and Ozima (1975, 1977); 23. Dalrymple et al. (1980b); 24. Worsley (1973); 25. McDougall and Aziz-ur-Rahman (1972).

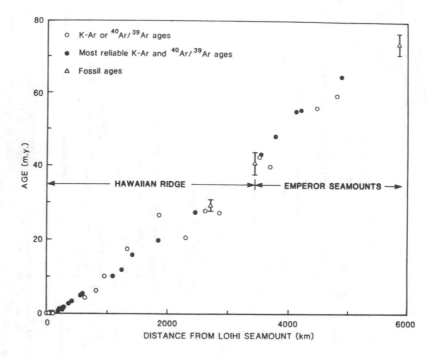

Fig. 2.   Age–distance plot of volcanoes in the Hawaiian–Emperor chain. Age data and distance are from Table I.

dated. The best-fit line has a correlation coefficient of 0.979 and a y intercept of 0.93 m.y. This correlation line predicts an age of 37.2 m.y. for the Hawaiian–Emperor bend, somewhat younger than the K-Ar and fossil age determinations for volcanoes in the southernmost Emperor Seamounts.

If we recalculate the volcanic migration rate, using only the most reliable age data along the entire Hawaiian chain, we obtain a rate of 8.6 ± 0.2 cm/yr, a correlation coefficient of 0.998, a y intercept of −1.44 m.y., and a predicted age for the Hawaiian–Emperor bend of 38.6 m.y. (Fig. 2)

This supposes that the volcanic migration rate has been constant since the change in Pacific plate motion at the Hawaiian–Emperor bend. It is possible that Pacific plate velocity was slower during the early stages of the Hawaiian Ridge, and a faster rate might be more appropriate for the younger part of the ridge. The best-fit volcanic migration rate for dated volcanoes from LaPerouse Pinnacles to Kohala volcano on Hawaii is 9.49 ± 0.40 cm/yr. This portion of the Hawaiian Ridge spans the last 12 m.y. of hot spot activity. Hence, this rate, rather than that determined for the entire Hawaiian chain, is appropriate for comparison with other Pacific volcanic lineaments that are much less extensive or whose pre-Miocene seamounts have not been as well documented.

B. Gulf of Alaska Volcanic Chains

Two major linear chains of seamounts and guyots occur in the Gulf of Alaska: The Pratt–Welker (Turner *et al.*, 1973, 1980) or Kodiak–Bowie (Silver *et al.*, 1974) chain to the north, and the Cobb–Eickelberg chain to the south (Figure 1). The Pratt–Welker chain extends at least 1000 km across the Gulf of Alaska from Kodiak Seamount in the northwest to Bowie Seamount in the southeast. It is possible that the chain extends an additional 600 km to the southeast and that the present-day hot spot is located at or near Dellwood Knolls (Silver *et al.*, 1974). Dredged lavas include transitional and alkalic basalt and trachyte (Turner *et al.*, 1980).

Turner *et al.* (1980) presented new K-Ar age determinations for four seamounts in the chain and reviewed earlier data from Kodiak and Giacomini seamounts (Turner *et al.*, 1973). Table II lists their data and one additional $^{40}Ar/^{39}Ar$ incremental heating age determination for Welker Seamount (Dalrymple, written communication, 1982). Turner *et al.* (1980) proposed that the data for the chain can best be explained if Denson, Davidson, and Hodgkins seamounts, which are composed of transitional basalt, formed at a spreading axis, while Kodiak, Giacomini, Dickens, and Bowie seamounts represent a hot spot trace. Hodgkins Seamount apparently had a phase of renewed volcanism as it passed over the hot spot as well. This complex interpretation is further complicated by the new 14.9 ± 0.3 m.y. age determination for Welker Seamount, in as much as the model presented by Turner *et al.* (1980) predicts an age of only 10.1 m.y. for this edifice. The calculated volcanic propagation rate for the 600 km length of chain between Kodiak and Welker seamounts is 6.7 cm/yr, which would place the present-day hot spot about 1000 km southeast of Welker Seamount and close to Dellwood Knolls rather than near Bowie Seamount. If this model is correct, Dickens, Hodgkins, and Bowie seamounts have experienced rejuvenated volcanism within the past 4 m.y. If the hot spot is located at Dellwood Knolls, the average rate of volcanic migration for the entire chain is 6.1 cm/yr.

The Cobb–Eickelberg volcanic chain extends from an axial seamount on the Juan de Fuca Ridge (Delaney *et al.*, 1981), through Cobb, Eickelberg, and perhaps Horton seamounts (Turner *et al.*, 1980). The only age data are from Cobb, Horton, Miller, and Murray seamounts (Table II). No simple age relations are evident along the chain except to say that Cobb Seamount to the southeast is much younger than the other seamounts. Calculated volcanic migration rates range from around 5 to 6.4 cm/yr, depending on whether the age of Horton, Miller, or Murray seamount is used. This chain, like the Pratt–Welker chain, appears to have a complex history that will only be resolved with additional sampling and $^{40}Ar/^{39}Ar$ age dating.

TABLE II
Radiometric Ages from Pacific Island and Seamount Chains

Island or seamount	Position		Age range[a]	Source
	Latitude (°N)	Longitude (°W)		
Pratt–Welker				
Bowie	53.3	135.6	0.075–<0.7	Herzer (1971)
Hodgkins	53.5	136.0	2.5–14.4	Turner *et al.* (1980)
Davidson	53.7	136.5	>17.4	Turner *et al.* (1980)
Dickens	54.6	136.9	3.8–4.1	Turner *et al.* (1980)
Denson	54.0	137.4	16.8–19.7	Turner *et al.* (1980)
Welker	55.2	140.3	14.9[b]	Dalrymple, personal communication (1982)
Giacomini	56.5	146.6	20.6–21.4	Turner *et al.* (1973)
Kodiak	56.9	149.2	23.4–24.8	Turner *et al.* (1973)
Cobb–Eickelberg				
Cobb	46.8	130.8	1.6	Dymond *et al.* (1968)
Horton	50.3	142.6	19.4–23.2	Turner *et al.* (1980)
Miller	53.5	144.3	25.2	Dalrymple, personal communication (1982)
Murray	53.9	148.5	25.7	Dalrymple, personal communication (1982)
Islas Revillagigedos				
San Benedicto	19.3	110.8	0.0	Bryan (1966)
Socorro	18.7	111.0	0.0–0.3	Dymond, personal communication (1983)
Clarion	18.3	114.7	1.1–2.4	Dymond, personal communication (1983)

Marquesas Islands

Fatu Hiva	-10.5	138.6	1.3–1.4	Duncan and McDougall (1974)
Tahuata	-10.0	139.1	1.8–2.1	Duncan and McDougall (1974)
Hiva Oa	-9.8	139.0	1.6–2.5	Duncan and McDougall (1974)
Ua Huka	-8.9	139.5	2.7–2.8	Duncan and McDougall (1974)
Nuku Hiva	-8.9	140.1	3.0–4.3	Duncan and McDougall (1974)
Eiao	-8.0	140.7	5.2–8.8	Brousse and Bellon (1974)
			5.0	McDougall and Duncan (1980)

Pitcairn–Gambier Islands

Pitcairn	-24.1	130.1	0.5–0.9	Duncan et al. (1974)
Gambier	-23.2	135.0	5.2–7.2	Brousse et al. (1972)
Mururoa	-22.0	139.0	8.0	Chevallier (1973)

Society Islands

Mehetia	-17.9	148.3	0.0	Talandier and Kuster (1976)
Tahiti–iti	-17.8	149.2	0.4–0.5	Duncan and McDougall (1976)
Tahiti	-17.6	149.5	0.5–1.2	Duncan and McDougall (1976)
Moorea	-17.5	149.8	1.5–1.7	Duncan and McDougall (1976)
Huahine	-16.7	151.0	2.0–2.6	Duncan and McDougall (1976)
Raiatea	-16.8	151.5	2.4–2.6	Duncan and McDougall (1976)
Tahaa	-16.6	151.5	2.6–3.2	Duncan and McDougall (1976)
Bora Bora	-16.5	151.8	3.1–3.4	Duncan and McDougall (1976)
Maupiti	-16.4	152.2	4.0–4.5	Duncan and McDougall (1976)

Austral–Cook Islands

Macdonald	-29.0	140.2	0.0	B. Keating, personal communication, 1983; Johnson and Malahoff (1971)
Marotiri	-27.6	143.8	3.5–4.0	B. Keating, personal comm. 1983
Rapa	-27.0	144.3	5.9–5.2	Krummenacher and Noetzlin (1966)
Raivavae	-23.9	147.7	5.6–7.6	Duncan and McDougall (1976)
Tubuai	-23.3	149.5	8.5–10.6	Duncan and McDougall (1976)
Rurutu	-22.4	151.3	0.6–12.3	Dalrymple et al. (1975); Duncan and McDougall (1976); Turner and Jarrard (1982)

(continued)

TABLE II. (*continued*)

Rimatara	−22.8	152.7	(4.8–28.6)	Turner and Jarrard (1982)
Mangaia	−21.9	157.9	13.7–19.6	Dalrymple et al. (1975); Turner and Jarrard (1982)
Mauke	−20.1	157.5	4.8–6.1	Turner and Jarrard (1982)
Mitiaro	−19.8	157.8	(12.3)	Turner and Jarrard (1982)
Atiu	−20.0	158.2	7.4–10.3	Turner and Jarrard (1982)
Rarotonga	−21.2	159.8	1.1–2.3	Dalrymple et al. (1975); Turner and Jarrard (1982)
Caroline Islands				
Kusaie	5.4	162.9	1.2–2.6	Keating et al. (1984)
Ponape	6.9	158.3	3.0–8.6	Keating et al. (1984)
Truk	7.3	151.8	4.8–13.9	Keating et al. (1984)
New Hebrides–Samoa Lineament				
Taviuni	−12.3	174.6	5.4	Duncan (1985)
Lalla Rookh	−13.0	175.6	10.0	Duncan (1985)
Combe	−12.7	177.6	13.5	Duncan (1985)
Line Islands				
LI-143	19.5	169.0	88.1[b]	Schlanger, et al. (1984)
LI-142	18.0	169.1	93.4[b]	Schlanger, et al. (1984)
			128[b]	Saito and Ozima (1977)
LI-63	16.5	168.2	86.0[b]	Schlanger, et al. (1984)
LI-61	15.0	167.5	82[b]	Schlanger, et al. (1984)
LI-137	14.5	169.0	56[b]	Saito and Ozima (1977)
LI-59	12.5	167.0	85.0[b]	Schlanger, et al. (1984)
LI-133	12.0	165.8	83[b]	Saito and Ozima (1977)

LI-134	10.3	168.0	48[b]	Saito and Ozima (1977)
LI-128	9.2	160.7	78.7[b]	Schlanger, et al. (1984)
LI-130	8.3	164.3	72[b]	Saito and Ozima (1977)
LI-33	8.2	161.9	39.3[b]	Schlanger, et al. (1984)
LI-123	5.8	160.7	76.4[b]	Schlanger, et al. (1984)
LI-119	2.7	165.0	68[b]	Saito and Ozima (1977)
LI-PC6	2.5	158.5	69.8[b]	Schlanger, et al. (1984)
LI-41	2.1	157.3	35.5[b]	Schlanger, et al. (1984)
LI-43	-0.7	155.3	59.0[b]	Schlanger, et al. (1984)
LI-44	-7.6	151.5	71.9[b]	Schlanger, et al. (1984)
LI-45	-9.1	150.7	70.5[b]	Schlanger, et al. (1984)
LI-52	-15.0	149.0	44.6[b]	Schlanger, et al. (1984)
Louisville Ridge				
LV-VM36-02	-40.8	165.3	45.5[b]	Duncan (unpublished)
LV-VM36-04	-38.3	167.7	44.6[b]	Duncan (unpublished)
LV-VM36-05	-33.9	171.2	53.3[b]	Duncan (unpublished)
Osborn	-26.0	175.0	(30–36)	Ozima et al. (1970)
Musician Seamounts				
Khatchaturian	27.1	162.2	66.9 ± 2.6	Clague and Dalrymple (1975)
Rachmaninoff	28.6	163.5	88.8 ± 5.2	Clague and Dalrymple (1975)
(unnamed)	33.5	166.5	95.6 ± 1.9[b]	M. Pringle, personal communication (1983)

[a] Ages recalculated where necessary using the following decay and abundance constants: $\lambda_e = 0.581 \times 10^{-10}$ yr^{-1}; $\lambda_\beta = 4.962 \times 10^{-10}$ yr^{-1}; $^{40}K/K = 1.167 \times 10^{-4}$ mol/mol.

[b] $^{40}Ar/^{39}Ar$ total-fusion or incremental-heating age. Ages in parentheses are reported as minimum ages.

C. Caroline Islands

The Caroline Island chain consists of the volcanic islands of Truk, Pon-ape, and Kusaie, as well as additional coral atolls and seamounts. The volcanic islands have recently been studied by Mattey (1981, 1982) who distinguished a mildly alkaline Main Lava series, a Transitional Lava series, and a Nephelinite series of post erosional lavas. The sequence of eruptions is similar to that observed in Hawaii with the exception that the shield-building lavas are mildly alkalic rather than tholeiitic basalt, and the post-erosional lavas postdate the shield-building phase by as much as 5 m.y.

The available age data indicate that the Caroline Islands decrease in age to the east with the subaerial shield-building lavas of Truk being 13.9–9.9 m.y., those from Ponape 5.2 m.y., and those of Kusaie ~1–3 m.y. (Keating et al., 1984). The rate of volcanic migration then is 12.1 ± 3.9 cm/yr. Keating et al. (1981) cited evidence that the hot spot that formed the chain is located at 4.8°N, 154.7°E, and is still seismically (and presumably volcanically) active. Mattey (1981, 1982) argued that the Caroline hot spot is declining in activity because Truk, Ponape, and Kusaie have progressively smaller volumes, and the shield-building lavas become progressively more alkaline (i.e., generated by smaller degrees of partial melting).

D. Islas Revillagigedos

This small island group is located near the intersection of the Clarion Fracture Zone with the East Pacific Rise, just south of the Baja California Peninsula (Fig. 1). San Benedicto, Roca Partida, and Socorro islands form a cluster at the northern end of the Mathematician Seamounts, a north-trending lineament thought to be an abandoned spreading axis left as spreading moved to the East Pacific Rise between 11 and 3.5 m.y. (Klitgord and Mammerickx, 1982). Clarion Island lies 400 km to the west. Lavas within this group belong to the alkali olivine basalt association (Bryan, 1966, 1967) and culminate in eruptions of soda-rich rhyolite on Socorro Island, and trach-yte on Clarion Island. Active Barcena Volcano òn San Benedicto Island is a palagonitic ash cone with a central dome of trachyte.

Historic eruptions have occurred at Socorro and San Benedicto islands. No evidence exists for recent volcanism on Clarion Island (Bryan, 1967). Potassium–argon age determinations (Dymond, unpublished data) show that the shield-building alkali basalts at Socorro Island are 0.18–0.30 m.y. old, whereas trachytic rocks are essentially age zero. At Clarion Island the oldest exposed rocks are 2.35–1.43 m.y. old, and later differentiated lavas are as young as 1.05 m.y. old. Volcanism apparently was initiated earlier in the west, at Clarion, and has shifted eastward to Socorro and San Benedicto.

We hesitate to assign a volcano migration rate to this trend on the basis of only two dated volcanoes. Note, however, that the calculated rate of 16.0 cm/yr is much faster than any other Pacific island chain. The position of Clarion Island may be controlled by the Clarion Fracture Zone, since the island is south of a line subparallel with the Hawaiian Ridge, which passes through Socorro or San Benedicto.

E. Island Chains of French Polynesia

After the Hawaiian Ridge, the best-documented age-progressive subparallel volcanic lineaments on the Pacific plate lie in French Polynesia. These include the Marquesas Islands (Duncan and McDougall, 1974), the Society Islands (Duncan and McDougall, 1976, Dymond, 1975), the Pitcairn–Gambier islands (Brousse et al., 1972; Duncan et al., 1974), and the Austral–Cook islands (Dalrymple et al., 1975; Duncan and McDougall, 1976; Turner and Jarrard, 1982; B. Keating, personal communication, 1983). The Tuamotu Islands are coral atolls but undoubtedly have volcanic pedestals. Subaerial lavas comprise alkali olivine basalts, ankaramites, basanites, and their differentiates. Tholeiitic compositions have been reported from the Marquesas only.

Studies of radiometric ages of these island chains have been reported elsewhere. We summarize these data, with recalculated new decay and abundance constants in Table II, and with calculated volcano migration rates in Table III. Historic eruptions have occurred at Mehetia (southeast end of the Society Islands), and volcanism is active at two seamounts near Mehetia (Talandier and Kuster, 1976) and at Macdonald Seamount at the southeast end of the Austral–Cook lineament. The Marquesas Islands and the Society Islands exhibit monotonically decreasing volcano ages towards the southeast, with volcanic migration rates of 10.4 ± 1.8 and 10.9 ± 1.0 cm/yr, respectively (McDougall and Duncan, 1980). A less well-defined but similar age progression exists in the Pitcairn–Gambier islands chain (Duncan et al., 1974).

Volcano ages in the Austral–Cook island chain do not provide such a simple pattern (Dalrymple et al., 1975; Duncan and McDougall, 1976; Turner and Jarrard, 1982; B. Keating, personal communication, 1983). In addition to the presently active Macdonald Seamount, Pleistocene volcanism has occurred at three sites along this lineament: Aitutaki, Rarotonga, and Rurutu. At Aitutaki and Rurutu these young rocks were erupted onto older edifices (Duncan and McDougall, 1976; Turner and Jarrard, 1982). Age determinations from seven volcanoes (Mangaia, Rurutu, Tubuai, Raivavae, Rapa, Marotiri, and Macdonald) do yield an age progression of 10.7 ± 1.6 cm/yr (Duncan and McDougall, 1976), which is in good agreement with other

TABLE III
Rates of Migration of Volcanism for Pacific Island and Seamount Chains

Lineament	Age range	Extent of dated volcanism (km)	Rate ± 1σ (cm/yr)
Phase I			
Pratt–Welker	0–24.8	600	6.7 ± 2.0
Cobb–Eickelberg	0–25.7	1540	5.7 ± 2.0
Hawaiian Ridge	0.–28	2855	9.5 ± 0.4 (8.6 ± 0.2)[a]
Islas Revillagigedos	0.–2.4	400	16.0
Marquesas	1.3–5.0	360	10.4 ± 1.8
Pitcairn–Gambier	0.5–8.1	1100	12.7 ± 5.5
Society	0.–4.5	500	10.9 ± 1.0
Austral–Cook	0.–19.6	2000	10.7 ± 1.6
Caroline	1–13.9	1280	12.1 ± 3.9
New Hebrides-Samoa	0.–27.7	1600	7.2 ± 2.3
Phase II			
Emperor Seamounts	42–64.7	1450	6.3 ± 0.4
Musicians	67–95	830	(3.0)
Line Islands	44–93	4500	8.5 ± 1.3

[a] Rate for entire Hawaiian chain. First figure gives rate for last 12 m.y. of activity.

rates from French Polynesia. Remaining ages in this lineament could be reconciled with the hot spot model by postulating hot spots at Rurutu and Rarotonga, in addition to the Macdonald Seamount activity (Turner and Jarrard, 1982). The alignment of these three possible hot spots together with Samoa, Pitcairn, and proposed hot spots on the Nazca plate to the east (Easter, Sala y Gomez, San Felix), led Turner and Jarrard (1982) to favor a model of "hot line" volcanism (Bonatti and Harrison, 1976) for the islands and seamounts in this long swath. According to this hypothesis, eruptions may occur sporadically along volcanic lineaments in response to either plate boundary stress or longitudinal convective rolls in the mantle (Richter, 1973).

F. New Hebrides–Samoa Lineament

The Samoa Islands have posed a problem for the hot spot model. Pleistocene to Holocene volcanic activity has occurred at both ends of this lineament, which is subparallel to other Pacific island chains that exhibit well-defined age progressions. Natland (1980) proposed that Pleistocene to Holocene undersaturated posterosional lavas overlie alkali basalt shield-building lavas that geomorphologically seem to get younger to the east. This shield-building phase could be related to hot spot activity (now at the eastern end of the Samoa Islands), and the later undersaturated lavas in the west could be a rejuvenated volcanic episode triggered by subduction tectonics near the Tonga Trench. Such a model is easily tested with radiometric ages from these volcanoes, which will be available soon (I. McDougall, personal communication).

Support for a hot spot origin, however, comes from age determinations on rocks dredged from seamounts to the west of the Samoa Islands. The New Hebrides–Samoa Lineament (Hawkins, 1976) is a volcanic swath of seamounts, ridges, and islands that stretches westward from the Samoa Islands to just east of Vanuatu. This volcanic province lies within a region of considerable tectonic complexity, at the northern margin of the Fiji Plateau and the Lau Basin, slightly north of the current transform boundary between the Pacific and Australia–India plates. Some features of this volcanic lineament, then, may be related to plate margin tectonics, either transform-faulting or failed subduction (Halunen, 1979). But radiometric ages ($^{40}Ar/^{39}Ar$) from Taviuni Bank (4.2 ± 0.3 m.y.), Lalla Rookh Bank (9.8 ± 0.3 m.y.), and Combe Bank (14.1 ± 1.1 m.y.) fit well with age-progressive volcanism from zero age in eastern Samoa (Duncan, 1985). The calculated volcano migration rate is 7.7 ± 2.5 cm/yr. Younger volcanoes such as Wallis Island (0.82 ± 0.03 m.y.) and Rotuma Island may be associated with development of the Lau Basin and Fiji Plateau.

III. LATE CRETACEOUS TO EARLY TERTIARY VOLCANIC CHAINS

The older volcanic chains on the Pacific plate are not well documented. Volcanic portions of these lineaments are entirely below sea level, so that samples can be collected only by dredging and drilling operations. Hence the stratigraphic relationship of multiple samples from the same volcano are not often known. In addition all samples have experienced some degree of alteration by seawater. It is not surprising then that only a few late Cretaceous and early Tertiary volcanic chains include enough well-dated seamounts to warrant disucssion. Those that have provided useful information are the Emperor Seamounts, the Line Islands, the Musician Seamounts, and the Louisville Ridge. These lineaments are oriented north-northwest, subparallel to the Emperor Seamounts.

A. Emperor Seamount Chain

The Emperor Seamounts are the older, northward continuation of the Hawaiian Ridge. Age data are now available for 11 volcanoes in the chain including fossil age determinations for Meiji and Kammu Seamounts. Early radiometric studies of the Emperor Seamount chain were based on dredged lavas from the southern end of the chain (Clague and Dalrymple, 1973; Clague *et al.*, 1975; and Dalrymple and Clague, 1976) and from Suiko Seamount (Ozima *et al.*, 1970; Saito and Ozima, 1975, 1977). More recent work is based on drilled samples from DSDP Leg 55 (Dalrymple *et al.*, 1980a) and a few dredged samples from Jingu Seamount (Dalrymple and Garcia, 1980). The data are summarized in Table I. The new data confirm that the Emperor Seamounts increase in age to the north, as predicted by the hot spot model. The only age that is inconsistent with this trend is the 39.9 ± 1.2 m.y. age of altered alkali basalt from Kimmei Seamount (Dalrymple and Clague, 1976). Even this age is only 4 m.y. younger than predicted by a linear age-progression model. The age for the southern portion of Suiko Seamount presented by Saito and Ozima (1975, 1977) is based on a single sample of mugearite selected from among numerous ice-rafted erratics. In addition, Dalrymple *et al.* (1980a) demonstrate that Saito and Ozima (1975, 1977) handled their $^{40}Ar/^{39}Ar$ data in a way that mathematically forced the age-spectrum age to agree with the isochron age. For these reasons, we have not used this age in calculating a rate of volcanic migration along the chain. A best-fit line through the remaining data yields an average rate of volcanic migration of about 6.4 cm/yr and predicts an age of 42 m.y. for the Hawaiian–Emperor bend and an age of 80 m.y. for Meiji Seamount. The lack of radiometric age data for seamounts north of Suiko Seamount will probably not be rectified without additional drilling, inasmuch as these volcanoes are blanketed by ice-rafted erratics.

B. Line Islands

The Line Islands chain is a major bathymetric lineament composed of seamounts, atolls, and elongated submarine ridges. It extends 4200 km from the northwest end of the Tuamotu Archipelago to Horizon Guyot at the eastern end of the Mid-Pacific Mountains. The southern half of this lineament is subparallel to the Emperor Seamounts (Morgan, 1972), whereas the northern half has a more northwesterly bearing (Jarrard and Clague, 1977). Late Cretaceous basement ages (Santonian to Campanian) have been assigned at Deep Sea Drilling Project (DSDP) Sites 165, 315, 316 based on paleontological examination of lowermost sediments. Because these sites are separated by almost 1000 km, Schlanger *et al.* (1976) proposed that the central Line Islands were built by synchronous rather than age-progressive volcanism. Recently, Haggerty *et al.* (1982) reported dredged rocks from the southern Line Islands, 100 km northwest of Caroline Island, containing late Cretaceous fossils associated with volcanic debris that they interpret as evidence for a reef-bearing volcanic edifice of minimum age 70–75 m.y. Those authors extend the synchroneity of late Cretaceous volcanism from DSDP Site 165 to Caroline Island, a distance of 2500 km, to argue against any hot spot origin for the Line Islands or any temporal equivalence with the Emperor Seamounts.

Radiometric ages (K–Ar and ^{40}Ar/^{39}Ar) indicate, however, that no single mechanism may have produced this volcanic province. Volcanic rocks from the Line Islands, like other Cretaceous submarine volcanic rocks, are moderately to severely altered. Attempts to date these rocks by the conventional K–Ar method have produced scattered, irreproducible results (Davis *et al.*, 1980; Duncan, 1983; Schlanger *et al.*, 1984). In all cases the measured ages are minimum estimates. Both total fusion and incremental heating ^{40}Ar/^{39}Ar are more consistent, and we will restrict our discussion to these data (Table II). The majority of these age determinations support an age-progressive volcanic origin. From latitude 18°N (near Horizon Guyot) to 15°S (northwest end of the Tuamotu Archipelago) volcano ages decrease from 93 m.y. to 44 m.y., yielding a migration rate of volcanism of 8.5 ± 1.3 cm/yr. This compares rather favorably with age-progressive volcanism for the Emperor Seamounts. Eocene ages at the north-central Line Islands correspond to the position of the Marquesas–Line Swell (Crough and Jarrard, 1981) which is subparallel to the Hawaiian Ridge and may have been produced by the Marquesas hot spot. The alternative, that the young ages reflect rejuvenated eruptions better explained by "hot line" volcanism, seems less likely in view of the age progression seen in the few dated Eocene seamounts.

The estimated basement age from fossil evidence at DSDP Site 165 agrees with radiometric ages of seamounts in the vicinity. Basement ages at DSDP Sites 315 and 316, and at the late Cretaceous seamounts of the southern Line Islands (Haggerty *et al.*, 1982) are older than the main hot spot trend by 10–30 m.y. They are about the age of the seafloor on which the Line Islands sit (Cretaceous magnetic quiet zone, 83–118 m.y.) and were probably erupted as seamounts or islands close to the Pacific–Farallon spreading ridge and later incorporated into the age-progressive swath. Similar occurrences are documented at Cretaceous seamounts in the Hawaiian Ridge (Clague and Dalrymple, 1975).

Most of the reliable radiometric age data from the Line Islands now seem to fit with age-progressive hot spot volcanism, as predicted by Morgan (1972). The northern Line Islands predate the Emperor Seamounts and give valuable information about the motion of the Pacific plate during the middle and latter part of the Cretaceous period. If we rotate the southernmost point of the Line Islands (44 m.y. old, also the approximate age of the Hawaiian–Emperor Bend) according to the rotation pole for the Hawaiian Ridge and the other parallel lineaments, we find the position of a hot spot which could have generated the Line Islands and younger portions of the Tuamotu Archipelago. This hot spot is at 23°S, 116°W, on the East Pacific Rise and northwest of Easter Island. Henderson and Gordon (1981) have proposed a more complex volcanic history involving several hot spots which leave overprinted traces in the Line Islands. The present age data do not permit us to distinguish between their model and Morgan's (1972) single hot spot model.

C. Louisville Ridge

Three volcanoes along the Louisville Ridge have been dated, although one yields a minimum age in view of seawater alteration. The others yield $^{40}Ar/^{39}Ar$ total-fusion ages of 53, 46 and 45 m.y. (Table II, Duncan, unpublished data from dredged samples collected by A. B. Watts), which places the Louisville Ridge in the Emperor Seamounts's phase of Pacific plate motion, in agreement with Jarrard and Clague's (1977) geometrical analysis. Osborn Seamount was dated at 30 to 36 m.y. by K–Ar methods (Ozima *et al.*, 1970) and we regard these as minimum age estimates.

No radiometric age data for the Marshall–Gilbert chain have been reported, but the azimuth of the chain suggests it is contemporaneous with the Line Islands and Emperor Seamounts (Morgan, 1972).

D. Musician Seamounts

The Musician Seamounts may be subdivided into four distinct provinces: a north-trending chain of volcanoes to the southwest; a northwest-trending chain of volcanoes to the northwest; a group of roughly westerly-oriented ridges to the southeast; and another group of westerly ridges to the northeast (Pringle, personal communication). Of these groups, only the northwest-trending chain of volcanoes appears to be an age-progressive chain. Clague and Dalrymple (1975) report ages of 87 and 65 m.y. for Rachmaninoff and Khachaturian Seamounts, respectively. Pringle (personal communication, 1983) reports an additional age of 95 m.y. for a previously uncharted volcano at the northwest end of the chain. These three seamounts decrease in age to the south-southeast and apparently formed at the same time as the northern Line Islands.

IV. PACIFIC PLATE MOTION IN THE HOT SPOT REFERENCE FRAME

Carey (1958), Wilson (1963), and Morgan (1971, 1972) proposed that linear, intraplate, volcanic features such as the Hawaiian Ridge manifest the movement of the lithosphere over a thermal anomaly (hot spot, plume) that is fixed in the mantle. Previous estimates of inter-hot spot movement have varied (Burke *et al.*, 1973; Molnar and Atwater, 1973; Minster *et al.*, 1974; Molnar and Francheteau, 1975; Minster and Jordan, 1978), but most recent studies (Morgan, 1981; Duncan, 1981) conclude that hot spots move less than 0.5 cm/yr with respect to one another. On this basis, the resulting volcanic lineaments accurately record plate motions. From examination of island and seamount lineaments on the Pacific plate only, Jarrard and Clague (1977) and McDougall and Duncan (1980) found strong support for the Carey–Wilson–Morgan model of fixed hot spots.

If the Pacific hot spots are fixed, or only very slowly moving, the volcanic trails emanating from them should be small circles of rotation about the finite rotation pole which describes the motion of the Pacific plate over the mantle. Also, the rates of migration of volcanism within these lineaments should vary as the sine of the angular distance from that rotation pole. This is analogous to the relation between relative-motion poles and transform-fault azimuths and half-spreading rates (Minster et al., 1974). The same methods of finite-rotation pole location can be applied to defining the Pacific plate over the mantle rotation pole.

Even a casual glance at the orientations of Pacific island and seamount chains would reveal abrupt changes in Pacific plate motion with respect to hot spots. In this review we will restrict our discussion to three phases of Pacific plate motion: 0–42 m.y., 42–100 m.y., and 100–150 m.y. Motion has been west-northwest along the trend of the Hawaiian Ridge in the most recent period. Between 42 m.y. and the middle Cretaceous (about 100 m.y.) motion was more northerly, following the trend of the Emperor Seamounts and Line Islands. From the late Jurassic to the middle Cretaceous the direction of Pacific plate motion was probably defined by west-trending volcanic lineaments such as the Mid-Pacific Mountains and the Magellan Seamounts.

Finer scale changes in plate motion are likely to have occurred. For example, the Marquesas Islands and the Hawaii to Kauai (0–5 m.y.) portion of the Hawaiian Ridge seem to have a slightly more southerly azimuth than the trends of island and seamount chains averaged over the past 42 m.y. (Jarrard and Clague, 1977; Epp, 1978). Other changes, in migration rate particularly, may have occurred in the late Tertiary (Dalrymple *et al.*, 1981). These departures are not large, however, and need to be better documented in more than one lineament before they can be used to add definition to plate motions.

Various finite-rotation poles for Pacific plate motion have been suggested for the period 0–42 m.y., the age of the most recent dramatic change

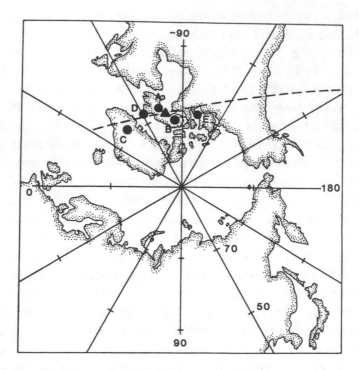

Fig. 3.   Rotation poles for Pacific-plate motion from 0 to 42 m.y. Pole positions shown are those of Morgan (1972), A; Clague and Jarrard (1973b), B; Winterer (1973), C; Minster *et al.* (1974), D; and McDougall and Duncan (1982), E; The triangle denotes our preferred pole position at 68°N, 75°W.

in hot spot lineament azimuths (Morgan, 1972; Clague and Jarrard, 1973b; Winterer, 1973; Minster *et al.*, 1974; McDougall and Duncan, 1980). Figure 3 shows that these poles are quite close to one another, forming an elliptical group whose major axis (greatest variation) lies along a great circle passing from the cluster of poles into the central Pacific. This is so because most of the studied Pacific volcanic lineaments are close to the equator of this group of poles, where the sine function, and hence the volcano migration rate, is relatively insensitive to variations in the angular distance.

The rates found in the Gulf of Alaska seamount chains form an especially strong constraint in determining the distance from the pole to the Pacific hot spot lineaments. Therefore, McDougall and Duncan (1980) proposed a pole position closest to the Pacific region based on a 4.3 cm/yr rate of migration of volcanism for Pratt–Welker Seamount ages (Turner *et al.*, 1973). Recent additional age information (Table II) yields a faster rate of 6.7 cm/yr, which moves the best-fitting pole back away from the Pacific and closer to Morgan's (1972) original suggestion.

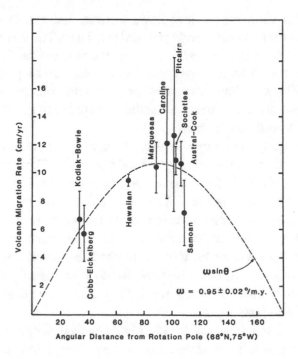

Fig. 4. Volcanic migration rate as a function of angular distance from the rotation pole located at 68°N, 75°W. The dotted curve is a least-squares best-fit with an angular rotation rate of 0.95 ± 0.02°/m.y. Volcano migration rates are from Table III.

In Figure 4 we plot the rate of migration of volcanism in nine hot spot lineaments (Table III) against angular distance from our preferred finite rotation pole at 68°N, 75°W. These data can be fitted to the theoretical relationship $v = \omega \sin \theta$, where v is the linear velocity at an angular distance θ, to estimate the angular velocity of rotation ω. By weighted least-squares linear regression of v on $\sin \theta$ the best-fit estimate of ω is 0.95 ± 0.02°/m.y. The correlation coefficient of the regression is 0.999, indicating a very close fit to the theoretical relationship, which is based on the model of fixed hot spots.

This calculation uses the more rapid rate of migration of volcanism for the Hawaiian Ridge determined for the past 12 m.y., which is appropriate for the other Pacific volcanic chains (Table III). The average Hawaiian rate of migration from 42 m.y. to the present, 8.6 cm/yr, is not consistent with the faster rates shown in Figure 4. It seems that the plate has moved more rapidly since the end of the Oligocene (the precise timing will depend on further geochronological work on seamounts at the older end of the Hawaiian Ridge). But we suspect that the volcano migration rates for most chains are somewhat overestimated due to sampling bias wherein younger volcanoes

have only the youngest surficial flows sampled and dated, while old volcanoes have relatively early lavas sampled and dated. The tendency then is to
obtain ages older than the average age of the volcano on older volcanoes
and ages younger than the average age of the volcano on young and active
volcanoes. This sampling bias causes the calculated volcanic migration rates
to be too rapid. In light of these difficulties, this phase of Pacific plate motion
is remarkably well constrained.

Unfortunately, the same density of data is not available for earlier
phases of Pacific plate motion. Jarrard and Clague (1977) proposed a rotation
pole at 17°N, 107°W to form the Emperor Seamounts. We accept this estimate for the period 65–42 m.y. The rotation rate about this pole was 0.61°/
m.y., calculated from radiometric ages of the Emperor Seamounts and southern Line Islands. The Musician Seamounts and the northern portion of the
Line Islands have a more northwesterly trend, however, and require a different pole of rotation, close to 32°N, 84°W. From age determinations on
northern Line Islands volcanoes, we find the rotation rate to have been about
0.63°/m.y. during this period.

At around 100 m.y. the Pacific plate changed direction dramatically,
from westerly to northwesterly. No radiometric age information for sea-

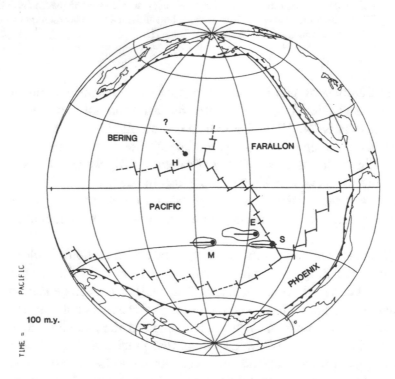

Fig. 5. Configuration of Pacific hot spots and spreading ridges at 100 m.y. Active hot spots
include the Hawaiian (H), Macdonald (M), Easter (E), and Sala y Gomez (S).

Fig. 6. Configuration of Pacific hot spots and spreading ridges at 42 m.y. Active hot spots include the Hawaiian (H), Macdonald (M), Easter (E), and Sala y Gomez (S).

mounts older than 100 m.y. is available, but minimum age estimates come from paleontologic examination of dredged material. Seamount ages in the Mid-Pacific Mountains seem to get younger from the west (Darwin Guyot) to the east (Horizon Guyot) between about 150 and 100 m.y. ago (Heezen *et al.*, 1973; Watts *et al.*, 1980). We pick a high-latitude pole of rotation for this period, reflecting westerly motion at a slow angular velocity of about

TABLE IV
Rotation Poles for the Pacific Plate over a
Hot Spot Reference Frame

Time (m.y.)	Latitude (°N)	Longitude (°W)	Angle (°ccw)
0–42	68.0	75.0	34.0
42–65	17.0	107.0	14.0
65–74	22.0	95.0	7.5
74–100	36.0	76.0	15.0
100–150	85.0	−165.0	24.0

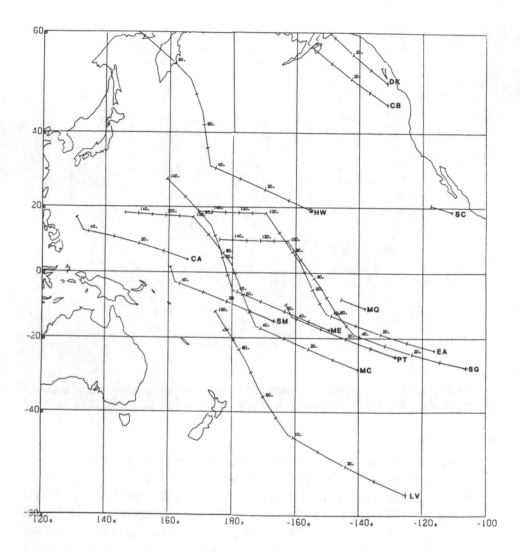

Fig. 7. Pacific plate at present time showing predicted hot spot lineaments and calculated ages in 10-m.y. increments for each volcanic chain assuming fixed hot spots. Hot spots include the Louisville Ridge (LV), Macdonald Seamount in the Austral Islands (MC), Pitcairn Island (PT), Mehetia in the Society Islands (ME), Samoa (SM), the Caroline Islands (CA), the Marquesas Islands (MQ), Easter Island (EA), Sala y Gomez (SG), Hawaii (HW), Socorro Island in the Revillagigedos Islands (SC), Cobb Seamont (CB) and Dellwood Knolls (DK).

0.48°/m.y. (Table IV). Henderson and Gordon (1981) have examined this early phase of Pacific plate motion and suggest that many of the submerged oceanic plateaus (Ontong Java, Manihiki, Hess, Shatsky) were formed at this time by slow plate motion away from hot spots lying beneath spreading ridges.

To summarize, the Pacific plate has undergone two abrupt changes in direction relative to hot spots since Jurassic time: at about 100 m.y. and at 42 m.y. Also, the angular velocity of the plate has steadily increased. Other more subtle changes will be delineated as more age information becomes available. The cause of these dramatic plate motion changes is not clear. Undoubtedly they are related to global plate motion reorganizations, but the causative agent (e.g., new subduction zones, changes in mantle convection patterns, episodes of rapid true polar wander) is unknown. Dalrymple and Clague (1976) argue that the plate reorganization at 42 m.y. that resulted in the change of Pacific plate motion recorded as the Hawaiian–Emperor bend was ultimately caused by the closure of the Tethys seaway and collision of India with Eurasia.

In Figures 5 and 6 we show the position and boundaries of the Pacific plate in the hot spot reference frame at about 100 m.y. and 42 m.y. when changes in plate motion occur. The positions of the Pacific–Farallon, Pacific–Phoenix, and Pacific–Bering spreading ridges are reconstructed using identified magnetic anomalies (interpolated for 100 m.y.) on the Pacific plate (Hilde *et al.*, 1976). Figure 7 illustrates the predicted lineaments and volcano ages that would be left by fixed hot spots with the Pacific plate motions given in Table IV. The correlation with known bathymetric features and measured ages is remarkably good.

V. CONCLUSIONS

Several alternatives to the hot spot model have been proposed to explain age-progressive linear volcanism on the Pacific plate. Propagating fractures (Betz and Hess, 1942; Jackson and Wright, 1970; Green, 1971; McDougall, 1971) might develop in response to tensional stresses resulting from cooling of the plate away from spreading ridges or translatitude changes in the curvature of the earth's surface (Turcotte and Oxburgh, 1973; 1976). As has been pointed out (Jarrard and Clague, 1977; McDougall and Duncan, 1980), however, this hypothesis does not predict the observed congruent set of fractures that accurately reflect lithospheric plate motion over the mantle throughout a large plate like the Pacific. Green (1971) suggested that volcanic lineaments erupted from tensional stress resulting from movement of the plate over stationary bumps in the upper mantle. This model, however, is tantamount to a hot spot model in which the thermal anomaly is replaced by a topological one. Richter's (1973) longitudinal-roll model of upper mantle convection is consistent with observations from Pacific plate island and seamount chains with the advantage that recurrent or simultaneous volcanism along lineaments is allowed. On the other hand, the irregular distribution of parallel volcanic chains does not seem to reflect a regular spacing of upper mantle convection cells.

The Carey–Wilson–Morgan hot spot model for age-progressive linear volcanism seems to explain the geometry and distribution of ages within Pacific island and seamount chains extremely well. This is best documented for the latest phase of Pacific plate motion, 42 m.y. to the present, but older lineaments seem to support the model as well. As more age data become available from late Jurassic, Cretaceous, and early Tertiary seamounts, more precise limits of inter-hot spot motion may be possible. At present, there is no evidence that these thermal anomalies move significantly with respect to one another.

This conclusion is important because the population of hot spots can then be used as a reference frame for precisely reconstructing plate positions through time. If, as Morgan (1972) suggested, hot spots are the tops of narrow, upwelling plumes of deep-mantle material, their stationary nature implies an extremely stable pattern of mantle convection. In addition, volcanic chains of islands and seamounts record the motion of lithospheric plates with respect to the mantle, which constitutes the major portion of the earth (aside from the outer and inner core).

Paleomagnetic and paleoclimatic data record plate motions (latitude only) with respect to the Earth's spin axis. Thus, any difference in plate paleolatitudes between the two reference frames will reveal true polar wander, that is, a motion of the figure of the earth (i.e., mantle) relative to its spin axis. Estimates of the northward late Cretaceous and Cenozoic component of Pacific plate motion from seamount magnetization (Sager, 1983), skewness of seafloor magnetic anomalies (Gordon, 1982; Jarrard and Cande, 1982) and analysis of equatorial sediment facies (van Andel *et al.*, 1975; Hammond *et al.*, 1979; Gordon and Cape, 1981) are significantly less than that determined from the hot spot reference frame during the same period. The difference is on the order of 10° to 15° and is seen also in similar comparisons for other plates (Hargraves and Duncan, 1973; Jurdy, 1981; Morgan, 1981). Whether this inferred true polar wander was rapid or gradual may be determined in the near future with more detailed paleomagnetic polar wander curves.

Much information, then, can be gleaned from Pacific island and seamount chains regarding the direction and velocity of plate motions relative to the Earth's mantle. Many questions remain, and we raise just a few. As noted previously the Pacific plate has changed its direction of motion abruptly at about 100 m.y. and 42 m.y. Is the cause of these sudden changes to be found in plate tectonic processes (i.e., confined to the lithosphere) or in more deep-seated processes? What is the characteristic lifetime of Pacific hot spots and what can be inferred about mantle convection from their distribution? The Pacific oceanic plateaus that formed during the middle Cretaceous have a very different morphology from the island and seamount chains. Did these also originate by hot spot volcanism, or by some different process?

ACKNOWLEDGMENTS

We thank Jack Dymond, Brent Dalrymple, Barbara Keating, and Malcolm Pringle for allowing us to discuss radiometric age determinations in advance of publication. This research was partially supported by NSF grant OCE 82-19189 and the Office of Naval Research (R.A.D.).

REFERENCES

Bargar K. E., and Jackson, E. D., 1974, Calculated volumes of individual shield volcanoes along the Hawaiian–Emperor chain, *J. Res. U. S. Geol. Surv.*, v. 2, p. 545–550.

Batiza, R., 1980, Origin and petrology of young oceanic volcanoes: Are most tholeitic rather than alkalic?, *Geology*, v. 8, p. 477–483.

Betz, F., Jr., and Hess, H. H., 1942, The floor of the north Pacific Ocean, *Geogr. Rev.*, v. 32, p. 99–116.

Bonatti, E., and Harrison, C. G. A., 1976, Hot lines in the earth's mantle, *Nature*, v. 263, p. 402–404.

Bonhommet, N., Beeson, M. H., and Dalrymple, G. B., 1977, A contribution to the geochronology and petrology of the island of Lanai, Hawaii, *Geol. Soc. Am. Bull.*, v. 88, p. 1282–1286.

Brousse, R. and Bellon, H., 1974, Age du volcanisme de lile d'Eiao, au Nord de l'Archipel des Marquises (Ocean Pacifique), *C. R. Acad. Sci. Paris*, v. 278, p. 827–830.

Brousse, R., Phillippet, J.-C., Guille, G., and Bellon, H., 1972, Geochronometrie des iles Gambier (Ocean Pacifique), *C. R. Acad. Sci. Paris*, v. 274, p. 1995.

Bryan, W. B., 1966, History and mechanism of eruption of soda rhyolite and alkali basalt, Socorro Island, Mexico, *Bull. Volcanol.*, v. 29, p. 453–479.

Bryan, W. B., 1967, Geology and petrology of Clarion Island, Mexico, *Geol. Soc. Am. Bull.*, v. 78, p. 1461–1476.

Burke, K., Kidd, W. S. F., and Wilson, J. T., 1973, Relative and latitudinal motion of Atlantic hot spots, *Nature*, v. 245, p. 133–137.

Carey, S. W., 1958, A tectonic approach to continental drift, in: *Continental Drift, a Symposium* (Carey, S. W., ed. Hobart: Univ. of Tasmania, p. 177–355.

Chevallier, J. P., 1973, Geomorphology and geology of coral reefs in French Polynesia, in: Biology and Geology of Coral Reefs, Volume 1. Jones, O. A., and Erdean, R., eds. New York: Academic. p. 113–141.

Clague, D. A., 1982, Petrology of tholeiitic basalt dredged from Hualalai volcano, Hawaii, *Tran. Am. Geophys. Union*, v. 63, p. 1138.

Clague, D. A. and Dalrymple, G. B., 1973, Age of Koko Seamount, Emperor Seamount chain, *Earth Planet. Sci. Lett.*, v. 17, p. 411–415.

Clague, D. A. and Dalrymple, G. B., 1975, Cretaceous K–Ar age of volcanic rocks from the Musician Seamounts and the Hawaiian Ridge, *Geophys. Res. Lett.*, v. 2, p. 305–308.

Clague, D. A., and Jarrard, R. D., 1973a, Tertiary Pacific plate motion deduced from the Hawaiian–Emperor chain, *Geol. Soc. Am. Bull.*, v. 84, p. 1135–1154.

Clague, D. A., and Jarrard, R. D., 1973b, Hot spots and Pacific plate motion (abstract), *Trans. Am. Geophys. Union*, v. 54, p. 238.

Clague, D. A., Dalrymple, G. B., and Moberly, R., 1975, Petrography and K–Ar ages of dredged volcanic rocks from the western Hawaiian–Emperor chain, *Geol. Soc. Am. Bull.*, v. 84, p. 991–998.

Clague, D. A., Dao-Gong, C., Murnane, R., Beeson, M. H., Lanphere, M. A., Dalrymple, G. B., and Holcomb, R. T., 1982, Age and petrology of the Kalaupapa basalt, Molokai, Hawaii, *Pac. Sci.*, v. 36, 411–420.

Crough, S. T., and Jarrard, R. D., 1981, The Marquesas–Line swell, *J. Geophys. Res.*, v. 86, p. 11,763–11,772.

Dalrymple, G. B., and Clague, D. A., 1976, Age of the Hawaiian–Emperor Bend, *Earth Planet. Sci. Lett.*, v.. 31, p. 313–329.

Dalrymple, G. B., and Garcia, M. O., 1980, Age and chemistry of volcanic rocks dredged from Jingu Seamount, Emperor Seamount chain, in: *Initial Reports of the Deep Sea Drilling Project*, Jackson, E. D. and Koizumi, I., eds., v. 55, Washington, D. C.: U. S. Government Printing Office, p. 685–594.

Dalrymple, G. B., Clague, D. A., and Lanphere, M. A., 1977, Revised age for Midway volcano, Hawaii volcano chain, *Earth Planet Sci. Lett.*, v. 37, p. 107–116.

Dalrymple, G. B., Jarrard, R. D., and Clague, D. A., 1975, K–Ar ages of some volcanic rocks from the Cook and Austral Islands, *Geol. Soc. Am. Bull.*, v. 86, p. 1463–1467.

Dalrymple, G. B., Lanphere, M. A., and Clague, D. A., 1980a, Conventional and ^{40}Ar–^{39}Ar ages of volcanic rocks from Ojin (site 430), Nintoku (site 432), and Suiko (site 433) seamounts and the chronology of volcanic propagation along the Hawaiian–Emperor chain, in: *Initial Reports of the Deep Sea Drilling Project*, Jackson, E. D. and Koizumi, I., eds., v. 55, Washington, D. C.: U. S. Government Printing Office, p. 659–676.

Dalrymple, G. B., Lanphere, M. A., and Jackson, E. D., 1974, Contributions to the petrography and geochronology of volcanic rocks from the leeward Hawaiian Islands, *Geol. Soc. Am. Bull.*, v. 85, p. 727–738.

Dalrymple, G. B., Lanphere, M. A., and Natland, J. H., 1980b, K–Ar minimum age for Meiji Guyot, Emperor Seamount chain, in: *Initial Reports of the Deep Sea Drilling Project*, Jackson, E. D. and Koizumi, I., eds., v. 55, Washington, D. C.: U. S. Government Printing Office, p. 677–684.

Davis, G. T., Naughton, J. J., and Philpotts, J. A., 1980, Petrology, geochemistry, and ages of Line Islands basaltic rocks, central Pacific Ocean, *Trans. Am. Geophys. Union*, v. 61, p. 1144.

Delaney, J. R., Johnson, H. D., and J. Karsten, 1981, The Juan de Fuca Ridge hot spot-propagating rift system: New tectonic, geochemical, and magnetic data, *J. Geophys. Res.*, v. 86, p. 11,747–11,750.

Doell, R. R., and Dalrymple, G. B., 1973, Potassium–argon ages and paleomagnetism of the Waianae and Koolau series, Oahu, Hawaii, *Geol. Soc. Am. Bull.*, v. 84, p. 1217–1242.

Duncan, R. A., 1981, Hot spots in the southern oceans—an absolute frame of reference for motion of the Gondwana continents, *Tectonophysics*, v. 74, p. 29–42.

Duncan, R. A., 1983, Overprinted, age-progressive volcanic chains in the Line Islands and Pacific plate Motion since 100 m.y., *Trans. Am. Geophys. Union*, v. 64, p. 345.

Duncan, R. A., 1985, Radiometric ages from volcanic rocks along the New Hebrides–Samoa lineament, in: *Geological Investigations of the Northern Melanesian Borderland, Circum-Pacific Council for Energy and Mineral Resources: Earth Science Series*, Brocher, T. M., ed, *Am. Assoc. Pet. Geol. Mem.* (in press).

Duncan, R. A., and Clague, D. A., 1984, The earliest volcanism on the Hawaiian Ridge, *Trans. Am. Geophys. Union*, v. 65, p. 1076.

Duncan, R. A., and McDougall, I., 1974, Migration of volcanism with time in the Marquesas Islands, French Polynesia, *Earth Planet. Sci. Lett.*, v. 21, p. 414–420.

Duncan, R. A. and McDougall, I., 1976, Linear volcanism in French Polynesia, *J. Volcanol. Geotherm. Res.*, v. 1, p. 197–227.

Duncan, R. A., McDougall, I., Carter, R. M., and Coombs, D. S., 1974, Pitcairn Island—another Pacific hot spot? *Nature*, v. 251, p. 679–682.

Dymond, J., 1975, K–Ar ages of Tahiti and Moorea, Society Islands, and implications for the hot-spot model, *Geology*, v. 3, p. 236–240.

Dymond, J. R., Watkins, N. D., and Nayudu, Y. R., 1968, Age of Cobb Seamount, *J. Geophys. Res.*, v. 73, p. 3977–3939.

Epp, D., 1978, Age and tectonic relationships among volcanic chains on the Pacific plate, Ph.D. Dissertation, Honolulu: Univ. of Hawaii, 235 pp.

Gordon, R. G., 1982, The late Maastrichtian paleomagnetic pole of the Pacific plate, *Geophys. J. R. Astr. Soc.*, v. 70, p. 129–140.

Gordon, R. G., and Cape, C. D., 1981, Cenozoic latitudinal shift of the Hawaiian hot spot and its implications for true polar wander, *Earth Planet. Sci. Lett.*, v. 55, p. 37–47.

Green, D. H., 1971, Composition of basaltic magmas as indicators of origin: application to oceanic volcanism, *Phil. Trans. R. Soc. London, Ser. A.*, v. 267, p. 707–725.

Grooms, D. G., 1980, Contributions to the petrography, geochemistry and geochronology of volcanic rocks along and near the western Hawaiian ridge and Kaula Island, Hawaiian Islands, M. S. Thesis, Honolulu: Univ. of Hawaii, 97 pp.

Haggerty, J. A., Schlanger, S. O., and Premioli-Silva, I., 1982, Late Cretaceous and Eocene volcanism in the southern Line Islands and implications for hot spot theory, *Geology*, v. 10, p. 433–437.

Halunen, A. J., Jr., 1979, Tectonic of the Fiji Plateau, Ph.D. Dissertation, Honolulu: University of Hawaii.

Hammond, S. R., Epp., D., and Theyer, F., 1979, Neogene relative motion between the Pacific plate, the mantle, and the earth's spin axis, *Nature*, v. 278, p. 309–312.

Hargraves, R. B. and Duncan, R. A., 1973, Does the mantle roll? *Nature*, v. 245, p. 361–363.

Hawkins, J. W., 1976, Petrology and geochemistry of basaltic rocks of the Lau Basin, *Earth Planet. Sci. Lett.*, v. 28, p. 283–297.

Heezen, B. C., MacGregor, I. D., *et al.*, 1973, *Initial Reports of the Deep Sea Drilling Project*, v. 20, Washington, D. C.: U. S. Government Printing Office.

Henderson, L. J. and Gordon, R. G., 1981, Oceanic plateaus and the motion of the Pacific plate with respect to the hot spots, *Trans. Am. Geophys. Union*, v. 62, p. 1028.

Herzer, R. H., 1971, Bowie Seamount, a recently active flat-topped seamount in the northeast Pacific Ocean, *Can. J. Earth Sci.*, v. 8, p. 676–687.

Hilde, T. W. C., Uyeda, S., and Kroenke, L., 1976, Tectonic history of the western Pacific, in: *Geodynamics: Progress and Prospects*, Drake, C. L., ed. Washington, D. C.: American Geophysical Union, p. 1–15.

Jackson, E. D., 1976, Linear volcanic chains on the Pacific plate, in: *The Geophysics of the Pacific Ocean Basin and its Margin*, Sutton, G. H., Manghani, M. H., and Moberly, R., eds. *Geophys. Mon.* 19, Washington, D. C.: American Geophysical Union, p. 319–335.

Jackson, E. D., and Wright, T. L., 1970, Xenoliths in the Honolulu volcanic series, Hawaii, *J. Petrol.*, v. 11, p. 405–430.

Jackson, E. D., Koizumi, I., Dalrymple, G. B., Clague, D. A., Kirkpatrick, R. J., and Greene, H. G., 1980, Introduction and summary of results from DSDP Leg 55, the Hawaiian–Emperor hot spot experiment, in: *Initial Reports of the Deep Sea Drilling Project*, Jackson, E. D., Koizumi, I., *et al.* eds., Washington: U. S. Government Printing Office, p. 5–31.

Jarrard, R. D., and Cande, S.C., 1982, A Pacific polar wandering curve based on the skewness of marine magnetic anomalies, *Trans. Am. Geophys. Union*, v. 63, p. 916.

Jarrard, R. D. and Clague, D. A., 1977, Implications of Pacific Island and seamount ages for the origin of volcanic chains, *Rev. Geophys. Space Phys.*, v. 15, p. 57–76.

Johnson, R. H., and Malahoff, A., 1971, Relation of Macdonald volcano to migration of volcanism along the Austral chain, *J. Geophys. Res.*, v. 76, p. 3282–3290.

Jurdy, D. M. 1981, True polar wander, *Tectonophysics*, v. 74, p. 1–16.

Keating, B. H., Mattey, D. P., Naughton, J., and Helsley, C. E., 1983b, Age and origin of Truck Atoll, eastern Caroline Islands; geochemical, radiometric age, and paleomagnetic evidence, *Geol. Soc. Am. Bull.*, v. 95, p. 350–356.

Keating, B. H., Mattey, D., Naughton, J., Epp, D., and Helsley, C. E., 1981, Evidence for a new Pacific hot spot, *Trans. Am. Geophys. Union*, v. 62, p.381.

Klitgord, K. D., and Mammerickx, J., 1982, Northern East Pacific Rise: Magnetic anomaly and bathymetric framework, *J. Geophys. Res.*, v. 87, p. 6725–6750.

Krummenacher, D. and Noetzlin, J., 1966, Ages isotopiques K–Ar de roches prélevées dans les possessions francaises du Pacifique. *Bull. Soc. Geol. France.*, v. 8, p. 173–175.

Larson, R. L., Moberly, R., *et al.*, 1975, *Initial Reports of the Deep Sea Drilling Project*, v. 32, Washington, D. C.: U. S. Government Printing Office.

Mattey, D. P., 1981, Secular geochemical variations in lavas from Truk, Ponope and Kusaie, eastern Caroline Islands, *Trans. Am. Geophys. Union*, v. 62, p. 1068

Mattey, D. P., 1982, Minor and trace element geochemistry of volcanic rocks from Truk, Ponape, and Kusaie, eastern Caroline Islands: evolution of a young hot spot trace across old oceanic crust, *Contr. Min. Petrol.* (in press).

McDougall, I., 1964, Potassium–argon ages from lavas of the Hawaiian Islands, *Geol. Soc. Am. Bull.*, v. 75, p. 107–128.

McDougall, I., 1971, Volcanic island chains and sea-floor spreading, *Nature*, v. 230, p. 141–144.

McDougall, I., 1979, Age of shield-building volcanism of Kauai and linear migration of volcanism in the Hawaiian Island Chain, *Earth Planet. Sci. Lett.*, v. 46, p. 31–42.

McDougall, I., and Aziz-ur-Rahman, 1972, Age of the Gauss–Matuyama boundary and of the Kaena and Mammoth events. *Earth Planet Sci. Lett.*, v. 14, p. 367–380.

McDougall, I., and Duncan, R. A., 1980, Linear island chains—recording plate motions? *Tectonophysics*, v. 63, p. 275–295.

McDougall, I. and Swanson, D. A., 1972, Potassium–argon ages of lavas from the Hawi and Pololu volcanic series, Kohala volcano, Hawaii, *Geol. Soc. Am. Bull.*, v. 83, p. 3731–3738.

Minster, J. B., and Jordan, T. H., 1978, Present-day plate motions, *J. Geophys. Res.*, v. 83, p. 5331–5354.

Minster, J. B., Jordan, T. H., Molnar, P., and Haines, E., 1974, Numerical modeling of instantaneous plate motions, *Geophys. J. R. Astron. Soc.*, v. 36, p. 541–576.

Molnar, P., and Atwater, T., 1973, Relative motions of hot spots in the mantle, *Nature*, v. 246, p. 288–291.

Molnar, P., and Francheteau, J., 1975, The relative motion of hot spots in the Atlantic and Indian oceans during theh Cenozoic, *Geophys. J. R. Astr. Soc.*, v. 43, p. 763–774.

Moore, J. G., Clague, D. A., and Normark, W. R., 1982, Diverse basalt types from Loihi Seamount, Hawaii, *Geology*, v. 10, p. 88–92.

Morgan, W. J., 1971, Convection plumes in the lower mantle, *Nature*, v. 230, p. 42–43.

Morgan, W. J., 1972, Deep mantle convection: plumes and plate motions, *Am. Assoc. Petrol. Geol. Bull.*, v. 56, p. 203–213.

Morgan, W. J., 1981, Hot spot tracks and the opening of the Atlantic and Indian oceans, in: *The Sea, Emiliani, C., ed. v. 7-Wiley-Interscience, New York: p. 443–475.*

Natland, J. H., 1980, The progression of volcanism in the Samoa linear volcanic chain, Am. J. Sci., v. 280-A, p. 709–735.

Naughton, J. J., Macdonald, G. A., and Greenberg, V. A., 1980, Some additional potassium–argon ages of Hawaiian rocks: the Maui volcanic complex of Molokai, Maui, Lanai, and Kahoolawe, *J. Volcanol. Geotherm. Res.*, v. 7, p. 339–355.

Ozima, M., Kaneoka, I., and Aramaki, S., 1970, K–Ar ages of submarine basalts dredged from seamounts in the western Pacific area and discussion of oceanic crust, *Earth Planet. Sci. Lett.*, v. 8, p. 237–249.

Porter, S. C., Stuiver, M., and Yang, I. C., 1977, Chronology of Hawaiian glaciations, *Science*, v. 195, p. 61–63.

Richter, F. M., 1973, Convection and the large-scale circulation of the mantle, *J. Geophys. Res.*, v. 78, p. 8735–8745.

Sager, W. W., 1983, A late Eocene paleomagnetic pole for the Pacific plate, *Earth Planet. Sci. Lett.*, v. 63, p. 408–422.

Saito, K., and Ozima, M., 1975, ^{40}Ar–^{39}Ar isochron age of a mugearite dredged from Suiko Seamount in the Emperor chain, *Rock Magn. and Paleogeophysics*, v. 3, p. 81–84.

Saito, K., and Ozima, M., 1976, ^{40}Ar–^{39}Ar ages of submarine rocks from the line Islands: implications on the origin of the Line Islands, *Am. Geophys. Union, Geophys. Mon.*, v. 19, p. 369–374.

Saito, K. and Ozima, M., 1977, ^{40}Ar–^{39}Ar geochronological studies on submarine rocks from the western Pacific area, *Earth Planet. Sci. Lett.*, v. 33, p. 353–369.

Schlanger, S. O., Jackson, E. D., *et al.*, 1976, *Initial Reports of the Deep Sea Drilling Project*, v. 33, Washington, D. C.: U. S. Government Printing Office.

Schlanger, S. O., Garcia, M. O., Keating, B. H., Naughton, J. J., Duncan, R. A., Haggerty, J., and Philpotts, J., 1984, Geology and geochronology of the Line Islands, *J. Geophys. Res.* (in press).

Silver, E. A., von Huene, R., and Crouch, J. K., 1974, Tectonic significance of the Kodiak–Bowie Seamount chain, northeastern Pacific, *Geology*, v. 2, p. 147–150.

Steiger, R. H. and Jager, E., 1977, Subcommission on geochronology: convention on the use of decay constants in geo- and cosmo-chronology, *Earth Planet. Sci. Lett.*, v. 36, p. 359–362.

Talandier, J. and Kuster, G. T., 1976, Seismicity and submarine volcanism in French Polynesia, *J. Geophys. Res.*, v. 81, p. 939–948.

Turcotte, D. L. and Oxburgh, E. R., 1973, Mid-plate tectonics, *Nature*, v. 244, p. 337–339.

Turcotte, D. L. and Oxburgh, E. R., 1976, Stress accumulation in the lithosphere, *Tectonophysics*, v. 35, p. 183–199.

Turner, D. L., and Jarrard, R. D., 1982, K–Ar dating of the Cook–Austral Island chain: a test of the hot spot hypothesis, *J. Volcanol. Geotherm. Res.*, v. 12, p. 187–220.

Turner, D. L., Forbes, R. B., and Naeser, C. W., 1973, Radiometric ages of Kodiak Seamount and Giacomini Guyot, Gulf of Alaska: implications for circum-Pacific tectonics, *Science*, v. 182, p. 579–581.

Turner, D. L., Jarrard, R. D., and Forbes, R. B., 1980, Geochronology and origin of the Pratt–Welker Seamount chain, Gulf of Alaska: a new pole of rotation for the Pacific plate, *J. Geophys. Res.*, v. 85, p. 6547–6556.

Van Andel, Tj. H., Heath, G. R., and Moore, T. C., 1975, Cenozoic history and paleoceanography of the central equatorial Pacific Ocean, *Geol. Soc. Am. Mem.*, v. 143, p. 1–134.

Watts, A. B., Bodine, J. H. and Ribe, N. M., 1980, Observations of flexure and the geological evolution of the Pacific Ocean basin, *Nature*, v. 283, p. 532–537.

Wilson, J. T., 1963, A possible origin of the Hawaiian Islands, *Can J. Phys.*, v. 41, p. 863–870.

Winterer, E. L., 1973, Sedimentary facies and plate tectonics of equatorial Pacific, *Am. Ass. Pet. Geol. Bull.*, v. 57, p. 265–282.

Winterer, E. L., Ewing, J. I., *et al.*, 1973, *Initial Reports of the Deep Sea Drilling Project*, v. 17, Washington, D. C.: U. S. Government Printing Office.

Worsley, J. R., 1973, Calcareous nannofossils: leg 19 of the Deep Sea Drilling Project, in Initial Reports of the Deep Sea Drilling Project, Creager J. S., and Scholl, D. W., *et al.*, eds., v. 19, Washington, D. C.: U. S. Government Printing Office, p. 741–750.

Chapter 4

MID AMERICA: TECTONIC SETTING FOR THE PACIFIC MARGIN FROM SOUTHERN MEXICO TO NORTHWESTERN COLOMBIA

Gabriel Dengo

Centro de Estudios Geológicos de America Central
Guatemala City, Guatemala

I. INTRODUCTION

This chapter presents a synthesis of the most relevant geological and tectonic features of the land and continental margin that borders the Pacific Ocean from southern Mexico to northwestern Colombia (Fig. 1). The objectives are to emphasize the variations in geological constitution of different segments of the region and to review briefly the major problems in understanding the geological history and the tectonic processes that resulted in its present configuration.

The scope of the paper is limited not only by the large size of the area under discussion, but by its geological variety and complexity. It complements, on the Pacific side, the articles by Banks, Dengo, Helwig, and Shagam, in Volume 3 of this series on the Gulf of Mexico–Caribbean Sea.

The geology and tectonic history of most of the area have been summarized earlier by the author (Dengo, 1968). More detailed studies pertinent to specific parts of the region are given by López Ramos (1979) for Mexico; Weyl (1980) for Central America; Irving (1971) and Duque-Caro (1980) for Colombia; and the regional tectonic syntheses by Butterlin (1977), Bellizzia *et al.* (1981) and Auboin *et al.* (1982) as well as a study based on the concept of geological provinces and tectono-stratigraphic terranes by Case, Hol-

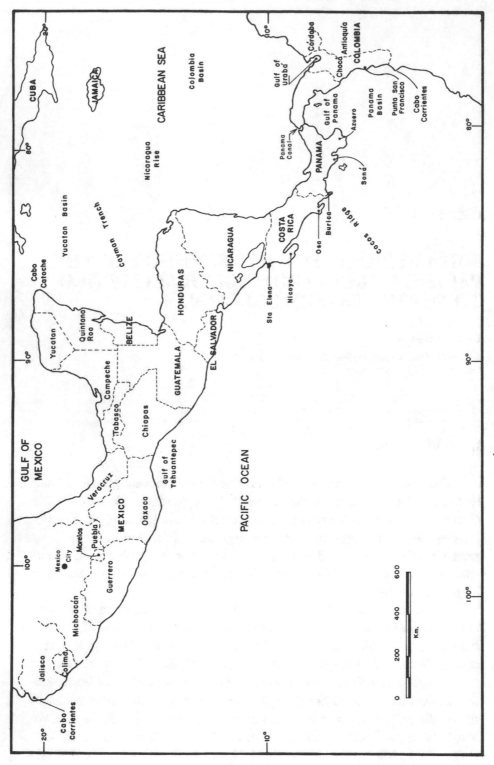

Fig. 1. Index map: political boundaries and localities mentioned in the text.

combe, and Martin (1984). For references to specific topics or small areas the publications cited above present extensive bibliographies.

The seismicity of the region was analyzed in terms of present tectonic activity by Sykes and Ewing (1965) and Molnar and Sykes (1969) and the gravity features by Bowin (1976). These two aspects are mentioned in the discussion only as associated with specific geological and tectonic features.

The selected references accompanying this paper include a list of recent geological and tectonic maps that cover the entire region, or parts of it, and which have been used to support some of the interpretations discussed here. Formational names are given only in those cases where necessary for stratigraphical comparisons between crustal blocks or for the tectonic history discussion.

II. REGIONAL TECTONIC FRAMEWORK

According to present understanding and terminology the region we are discussing covers parts of five large tectonic plates, namely, North America, Caribbean, South America, Cocos, and Nazca. The boundaries between these plates are in some cases well defined by major fault zones, spreading zones, or by subduction zones, while in other parts these are not clear and, therefore, divergent interpretations have been suggested. This latter situation applies in the western portion of the Caribbean and South America plates and also in their respective boundary zones with the eastern Cocos plate and the northern part of the Nazca plate (Panama Basin).

Several of the proposed interpretations, are shown in Fig. 2 with the purpose of presenting the parts for which there is agreement and discrepancy. The interpretation of Molnar and Sykes (1969) is based on earthquake foci distribution and has been widely accepted. The interpretation of Bowin (1976) was inferred from gravity results, while that proposed by Shagam (1975) is one of many different alternatives to show the boundaries of northeastern South America based on a composite synthesis. Finally, the interpretation based on Case and Holcombe (1980, Map) is essentially the same that adapted by Kellog and Bonini (1982).

Based on these models, and on an earlier attempt by the author (Dengo, 1972) supported by the occurrence of ultramafic rocks and characteristics of the crust and tectonic history, a revised version of plate boundaries is presented (Fig. 3) that will be used as the basis for the discussion that follows. Within the major plates, smaller crustal blocks, each with their own geological characteristics, are also identified. The approach followed in this synthesis is to describe: (1) the major geological features of each block, (2) their boundaries, (3) differences and similarities in geological history, and

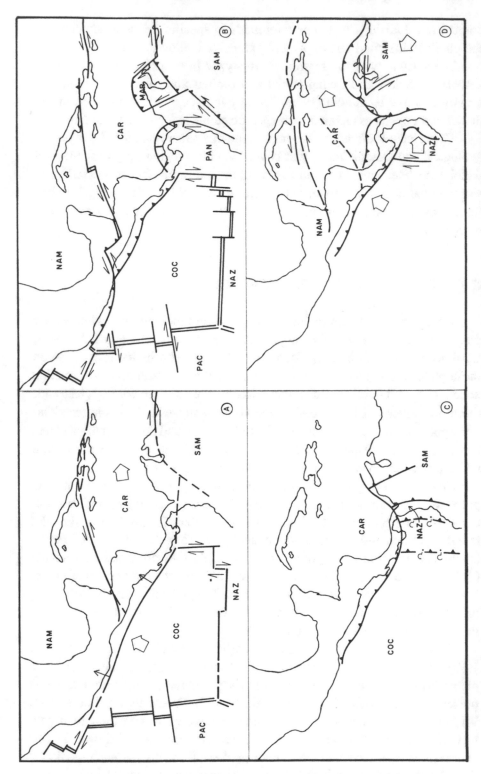

Fig. 2. Comparison of present plate boundary interpretations: A, based on Molnar and Sykes (1969); B, simplified from Bowin (1976); C, after Shagam (1975); D, based on Case and Holcombe (1980, Map) and Kellog and Bonini (1982); NAM, North America Plate; PAC, Pacific Plate; CAR, Caribbean Plate; COC, Cocos Plate; NA2, Nazca Plate; SAM, South America Plate.

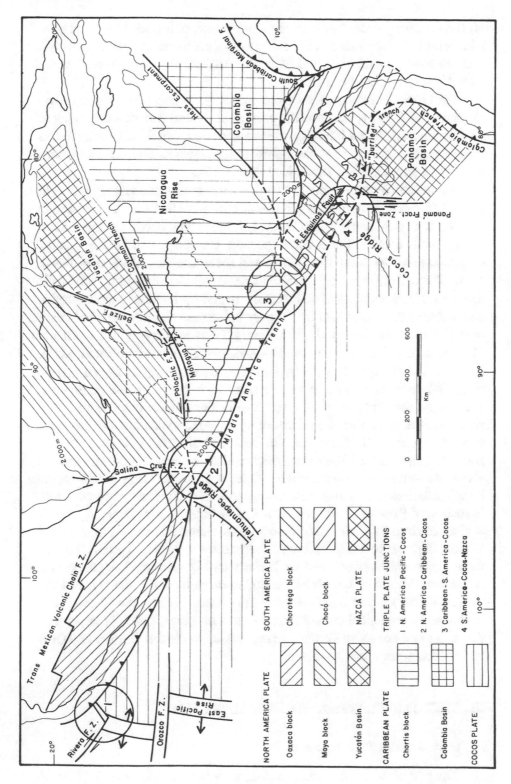

Fig. 3. Plate segments, crustal blocks and major tectonic features.

(4) the boundary zones between the plate segments included in the region, with particular emphasis on the Pacific side of each one. A similar approach was followed by Anderson and Schmidt (1983) for the larger region of the Gulf of Mexico–Caribbean Sea.

The blocks defined are the Oaxaca and Maya as part of the North America plate, the Chortis, which is the portion of Central America belonging to the Caribbean plate, and the Chorotega and Chocó, which, according to this interpretation, form part of the South America plate. The oceanic Cocos and Nazca plates are not subdivided into blocks.

As an introductory summary to these major tectonic elements it is of interest to mention their most prominent geological features. The blocks in the northern part, namely, the Oaxaca, Maya, and Chortis are cratonic in nature, while those in the southern portion, Chorotega and Chocó, display oceanic crustal characteristics. The oldest rocks known in the region, which are of Precambrian age, occur in the Oaxaca block followed by Paleozoic metamorphic, igneous, and sedimentary rocks; by Mesozoic metamorphic and sedimentary rocks; and by Tertiary volcanics. The northern boundary of the block is marked by the Quaternary Trans-Mexican Volcanic Chain. The Maya block has a metamorphic–igneous basement of Precambrian(?), Lower Paleozoic age, overlain by Upper Paleozoic terrigenous and marine rocks different from those of the Oaxaca block, by Mesozoic continental and marine rocks in part similar to those of the Oaxaca block, and by restricted areas of Tertiary and Quaternary volcanic rocks, including at least one active volcano (El Chichón). The basement of the Chortis block is metamorphic, possibly of Paleozoic age, overlain by continental and marine Mesozoic sequences—which differentiates it from the Maya block—some local Tertiary marine sequences, and by extensive Tertiary volcanics, with intrusives of Paleozoic, Mesozoic and Cenozoic ages. The Chorotega block is characterized by a Mesozoic ophiolitic basement that ranges in age from at least Jurassic to early Tertiary, overlain by restricted Tertiary basins, formed as clastic wedges, and by Tertiary and Quaternary volcanic rocks. A common feature of the Chortis and Chorotega blocks is the belt of Quaternary volcanoes that parallels the Pacific coast. Finally, the Chocó block is similar in geological constitution to the Chorotega, but it is larger and contains more extensive Tertiary marine basins, while the Tertiary and Quaternary rocks are more restricted.

III. TECTONIC PLATES AND CRUSTAL BLOCKS

A. North America Plate Segment

The part of the North America plate under discussion extends from the neovolcanic or Trans-Mexican Volcanic Chain that dissects the North American continent approximately along the 19°N parallel, southeastward to the Polochic and Motagua fault zones in central Guatemala, near the 15°N parallel. The boundary between this plate segment and the Cocos plate to the west is the Middle America Trench.

Two major continental crustal blocks are recognized within this segment, the Oaxaca and the Maya, which are separated by an imperfectly defined fault zone in the Isthmus of Tehuantepec that extends from the Pacific to the Gulf of Mexico in a general northerly trend.

1. Oaxaca Block

The part of the Oaxaca block of interest to us is the southern one which corresponds to the one that López Ramos (1979) has defined as the Sierra Madre del Sur and Oaxaca Highlands geological provinces. It has also been defined as a morphotectonic unit in the classification of Guzmán and de Cserna (1963), which was also followed by Dengo (1975) in Volume 3 of this series. The northern portion of the block belongs to the Gulf of Mexico geological realm. It should be pointed out that the geographic name Sierra Madre del Sur is also used in the State of Chiapas (Fig. 1), which under the present interpretation is part of the Maya block.

(a) Stratigraphic Sequence. Most of the Sierra Madre del Sur bordering the Pacific Ocean in the States of Michoacán, Guerrero, and Oaxaca is characterized by Precambrian and Paleozoic (?) metamorphic complexes, the former being the more extensive, mostly in central Oaxaca. (Fig. 4).

According to Ortega-Gutiérrez (1976, 1979, 1980) the basement rocks in this part of Mexico may be divided into three main sequences characterized by their lithology, age, and structural style. The oldest sequence is exposed mainly in the state of Oaxaca (Oaxaca Complex) and is comprised of rocks of the granulite and amphibolite metamorphic facies. They are predominantly layered gneisses, but also include charnockite, anorthosites, and complex pegmatites of Precambrian age, 1050 ± 20 m.y. (Ortega-Gutiérrez and Anderson, 1977), with a N–NW structural trend. The following sequence (Xolapa Complex) is of early Paleozoic age and is exposed in the southern part of the State of Puebla, northern Oaxaca, and northeastern Guerrero. It is formed by a metamorphosed ophiolitic complex that includes amphibolites, possible eclogites, and migmatites, and which probably represents a metamorphosed paleosubduction zone accreted to the Precambrian craton, but now separated from it by a wide mylonitic zone. The Xolapa Complex is poorly known, and it probably also includes Mesozoic rocks (Ortega-Gutiérrez and Guerrero-García, 1982). The third metamorphic sequence (Acatlán Complex) crops out extensively along the Pacific coast and again it is also in tectonic contact along a zone of mylonitization with the Oaxaca Complex. It is characterized by quartzo-feldspathic ortho and paragneisses

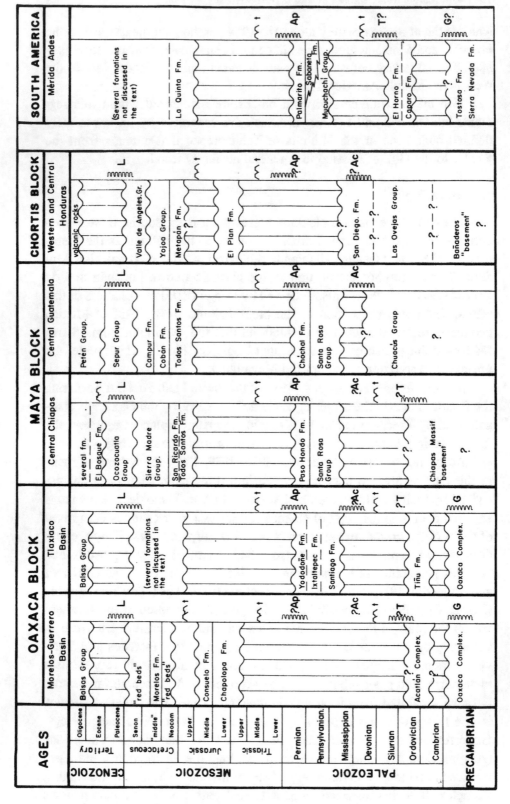

Fig. 4. Simplified stratigraphic and tectonic event table of cratonic blocks in southern Mexico and northern Central America compared with the Mérida Andes, Venezuela. L, Laramide Orogeny; Ap., Appalachian Orogeny; Ac., Acadian Orogeny; T, Taconic Orogeny; G, Grenville Orogeny; t, Taphrogenesis.

and migmatites of the amphibolite metamorphic facies. Rodríguez (1970) has assigned it a Cambro–Ordovician, and Ortega-Gutiérrez (1978, 1979) a pre-Mississippian age.

Overlying the Precambrian basement there are also unmetamorphosed Paleozoic sedimentary rocks in the Tlaxiaco Basin of central Oaxaca (Fig. 4) described by Pantoja Alor and Robinson (1967) and by López Ramos (1969). In general the sequence is fairly thin and consists of Cambro–Ordovician thin-bedded limestone and shale, Mississippian sandstones, shales and conglomerates, and Pennsylvanian–Permian siltstones and sandstones, with a possible Silurian or Devonian unconformity. Because of their restricted area and limited thickness they were considered by Dengo (1975) to belong to an isolated small basin that was integrated into a geosyncline during middle Paleozoic time.

Mesozoic terrigenous and marine sedimentary rocks are generally extensive in Mexico, as summarized by Dengo (1975) and described by Benavides (1956) and López Ramos (1979). In the Oaxaca block, however, they are more restricted in areal extent, or are covered by Tertiary volcanics and younger sediments. They occur mainly in the Morelos–Guerrero and Tlaxiaco basins (Fig. 5) and represent westward embayments of the Mesozoic Mexican Geosyncline that paralleled what is now the Gulf of Mexico. On the Pacific side Mesozoic sedimentary rocks are restricted to small areas.

The Mesozoic sequence in the Pacific side (States of Guerrero and Oaxaca) begins with continental and brackish-water late Triassic–early Jurassic sediments, composed of red-bed conglomerates and tuffaceous sandstones and shales (Chapolopa Formation) that are overlain by plant-bearing middle Jurassic clastic rocks (Consuelo Formation). These in turn are covered by Lower Cretaceous red beds and by Aptian–Albian limestones which become very thick and widespread to the north and east as they pass into the Mexican Geosyncline. Locally, there are andesite flows interbedded in this sequence. In the Mountain of Guerrero it was established recently that some of the red beds originally considered older overlie the Aptian–Albian limestones (Salinas and Ramírez, 1980). This general mesozoic sequence has more similarities with the Chortis block than with the Maya block.

Cenozoic sedimentary basins within the Oaxaca block are restricted to very small areas on the Pacific side (Durham et al., 1981), while larger ones occur on the Gulf of Mexico side. Several intrusive rocks are found in the Oaxaca block, some associated with the Precambrian rocks, and others of known late Cretaceous age. Among the latter, large batholiths dominantly of monzonite and granodiorite composition form a large part of the Sierra Madre del Sur. They are the Jalisco, Michoacán–Guerrero, and Oaxaca batholiths. In the central part of Oaxaca and in other parts of the block there

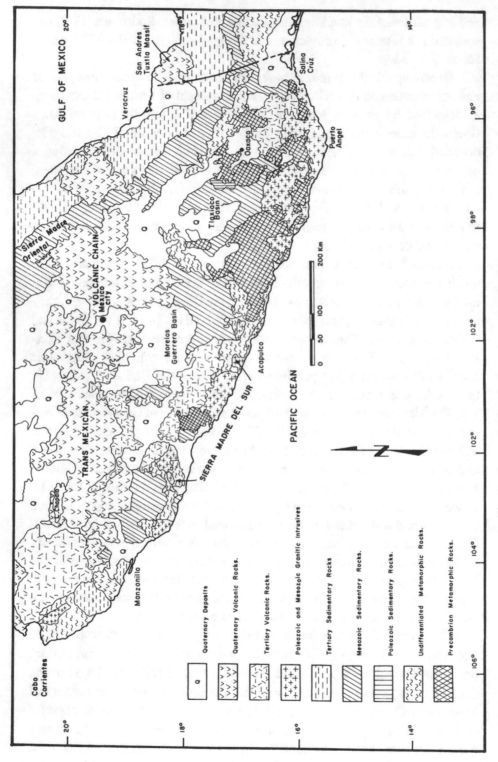

Fig. 5. Generalized geologic map, Oaxaca block.

are extensive volcanic rocks, in part basalts and andesites, but with a great predominance of felsic ignimbrites of Oligocene–Miocene age.

Quaternary volcanics are largely restricted to the nothern part of the block, along the Trans-Mexican Volcanic Chain (Fig. 4 and 9). The volcanic activity along this belt probably started in late Tertiary time, became stronger during Quaternary and has continued until the present. The composition is essentially andesitic. On the basis of paleomagnetic measurements several separate eruptive phases have been identified (Mooser *et al.*, 1974). The belt is continuous across Mexico from the Pacific to the Gulf of Mexico, undoubtedly along a major structural zone that, according to de Cserna (1971), is an old Permo–Triassic feature, and which, according to Mooser *et al.* (1974), from the distribution of the major centers has the pattern indicated in Fig. 3. A separate large volcanic cluster on the Gulf of Mexico side is that of San Andrés Tuxtla (Fig. 9), which could be related either to the Trans-Mexican belt or to a fracture zone across the Isthmus of Tehuantepec.

(b) Tectonic History. From the relationships of the rock sequences briefly described and from those exposed along the margin of the Gulf of Mexico, the following summary of the tectonic history of the Oaxaca block may be presented, and will be used as a basis for comparison with the two other cratonic blocks, the Maya and the Chortis.

The Oaxaca Complex has been considered by several authors, among them Banks (1975), de Cserna (1971), and Kesler and Heath (1970), to be a southward extention of the Grenville Foldbelt of North America that underwent metamorphism and anatexis during the interval 1,000–900 m. y. ago, and that corresponds to the Oaxacan Structural Belt as defined by de Cserna (1971). It is the oldest tectonic episode detected in the region. If the Xolapa and Acatlán complexes are in part Lower Paleozoic (Cambro–Ordovician) protoliths, their metamorphism represents a strong tectonic deformation during or after Silurian that could correspond to the Taconic Orogeny, or, if after the Devonian, to the Acadian Orogeny. This orogeny in central Oaxaca, the Tlaxiaco Basin, must have been of lesser intensity and is indicated by an unconformity between the unmetamorphosed sedimentary rocks of Cambro–Ordovician age and the Mississippian–Pennsylvanian sedimentary sequence.

The absence of late Permian and early Triassic deposits as well as the unconformity between the Paleozoic and the Mesozoic rocks is indicative of a regional tectonic episode, probably not accompanied by regional metamorphism, equivalent to the Appalachian orogeny in North America, also designated Coahuilan Orogeny by Guzmán and de Cserna (1963).

Regional uplift during the Triassic was followed by the long depositional history (Jurassic–Cretaceous–early Tertiary) of the Mexican Geosyncline on the Gulf of Mexico side of the Oaxaca block as well as to the north and east of it, and the embayments in the Morelos–Guerrero and Tlaxiaco basins in its central and southern parts. During most of the Mesozoic the Sierra Madre del Sur must have remained as a landmass.

The late Cretaceous to Eocene (Laramide) Orogeny, also designated Hidalgoan by Guzmán and de Cserna (1963), is geologically well documented in northern Mexico as well as in the northern part of the Oaxaca block and the Maya block, as one of the deformations of broad regional extention. In the Oaxaca block it was accompanied by granitic intrusives along trends essentially parallel to the Pacific coast. Whether there was a Cretaceous subduction zone south of the present coast or the intrusives represent the roots of a volcanic chain parallel to it, is a difficult problem to answer since the evidence is not clear due to the present truncation of the basement rocks at the continental border.

The Tertiary tectonic history is characterized by the formation of small sedimentary basins in the southern part of the block along its present coast, and by the extensive lava and ignimbrite flows in its central portion. The Quaternary tectonics will be dealt with in the discussion of the Middle America Trench.

The Isthmus of Tehuantepec has been recognized as a zone that separates two areas with considerable geological differences as well as similarities. Some authors have inferred that a fault is present in this area with an almost N–S trend (Alvarez, 1958; Viniegra, 1971; Sánchez Barreda, 1981). It is designated the Salina Cruz Fault. If the trend of this fault is taken to be NNW, and is projected into the Gulf of Mexico, the fault corresponds approximately with the position of the San Andrés Tuxtla volcanic massif and, in the gulf itself with a linear boundary that separates a folded sea-bottom area to the west and a salt dome area to the east. This boundary is well defined by gravity features (Moore and Del Castillo, 1974). A zone of Mesozoic metamorphic rocks in parts of the isthmus (López Ramos and Hernández Sánchez-Mejorada, 1976, Map) could represent a suture zone between the Oaxaca and the Maya blocks.

2. *Maya Block*

The Maya block essentially corresponds to the area designated by other authors as the Yucatán block. It seems better, however, to restrict the name Yucatán to its northern part, which is the tectonically stable platform that forms the peninsula of the same name. In terms of the morphotectonic units described for Mexico and Central America by Guzman and de Cserna (1963) and Dengo (1975) it includes in its southern portion the eastern part of the Sierra Madre del Sur in Mexico (Chiapas Massif) and the Central Cordillera of Guatemala, which is the backbone of the Sierras of the Northern Central America unit; in the central part, the Chiapas–Guatemala Foldbelt; and in the north the Yucatán Platform. Based on the interpretation of Bouguer gravity anomalies and seismic data the block has an intermediate to continental thickness in its northern part, ranging from 20 to 25 km in the Yucatán Platform, and 30 to 40 in the southern part (Woollard and Monges Caldera, 1956; Dengo 1968, López Ramos 1975; Case *et al.*, 1983).

(a) *Stratigraphic Sequence*. In general terms, the block contain a basement of the oldest rocks in its southern part, the Chiapas Massif, and the Central Cordillera of Guatemala (Fig. 6). In the central part of the state of Chiapas and northern Guatemala the basement is covered by a thick and extensive Mesozoic sedimentary sequence that starts with continental rocks and is followed by marine rocks, mainly evaporites and limestones in the Chiapas–Guatemala Foldbelt, which were the southeastern extension of the Mexican Geosyncline. Isolated small Tertiary basins occur in the central portion while a large area of thick Tertiary marine rocks is found on the Gulf of Mexico side. The following brief description emphasizes the geological nature of the Pacific side of the block. (Fig. 4).

The Chiapas Massif is still poorly known geologically. Originally it was considered as a large single batholith of possible Paleozoic age because granitic rocks, some partially metamorphosed, form the bulk of the mountain range. Subsequent studies and radiometric age determinations have shown that the massif is formed by rocks of different ages, mostly granitic plutons in nature. According to Pantoja Alor *et al.* (1971) and López Ramos (1979) Precambrian granitic rocks occur in the south central part of the massif near the town of Tonalá and extend northward across it. Most of the massif is still regarded as Paleozoic, while isolated smaller granitic intrusions are dated as Jurassic and Tertiary. More accurate radiometric ages are needed in order to establish a clear sequence of events.

Metasedimentary rocks, schists, and marbles, which occur mainly at the eastern and western parts, are also regarded as Precambrian (López Ramos and Hernández Sánchez-Mejorada, 1979, Map). Other metasedimentary rocks are classified as Lower Paleozoic because they are overlain by Mississippian unmetamorphosed sedimentary rocks.

At the extreme western part, near the Isthmus of Tehuantepec, there are metamorphic rocks of Mesozoic (?) age that are part of the possible suture zone between the Oaxaca and Maya blocks already proposed (López Ramos and Hernández Sánchez-Mejorada, 1979, Map).

In the continuation of the mountain range to the east, as the Central Cordillera of Guatemala, the geological situation is different in the sense that metasedimentary rocks prodominate in areal extent. This may be due

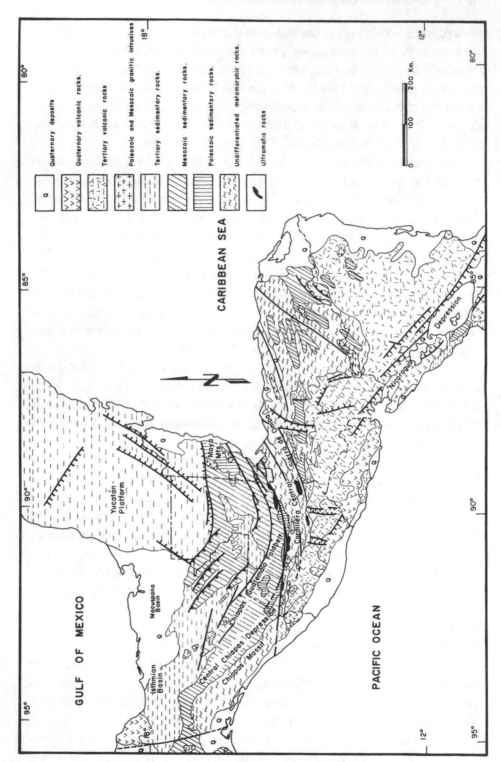

Fig. 6. Generalized geologic map, Maya and Chortis blocks.

to considerable offset along the Polochic Fault, which cuts obliquely across the range near the international border (Fig. 6).

The best-known part of the Cordillera is the central one, in the Department of Baja Verapaz, where the metasedimentary rocks are known as the Chuacús series (now Group), as designated by McBirney (1963). This sequence consists of garnet-biotite schists and gneisses with layers of amphibolite and marble and, to a lesser degree, staurolite-kyanite gneisses. In the western part of the Cordillera, near the Mexican border, Kesler *et al.* (1970) identified two main units in the Chuacus Group, one that is muscovite-rich and the other a banded gneiss, in addition to other minor ones including a marble layer. Because of the high quartz and potassium content in these metasediments the source for the original sedimentary rocks was mostly granitic. It is logical to correlate the Chuacús Group with the early Paleozoic metasedimentary rocks of the Chiapas Massif and to regard the Precambrian granites of the massif as a possible source for these. A different alternative, presented under the discussion of the regional tectonic history, is that during the Paleozoic the Maya block was to the east of the Oaxaca block, and that the Precambrian source was the Oaxaca Complex. Radiometric ages determined from detritic zircons taken from the Chuacus rocks (Gomberg *et al.*, 1968) also indicate a Precambrian source area.

The extension of the Precambrian–early Paleozoic rocks toward the north of the Maya block under the Chiapas–Guatemala Foldbelt and the Yucatán Platform can be inferred from a radiometric age of 420 m. y. from a basement oil-well core in the Yucatán Península (Bass and Zartman, 1969), and from a basement drill core of mylonitic gneiss and amphibolite from the Catoche Knoll, radiometrically dated as 500 m. y. (Dalimeyer, 1982).

Structurally, the Chiapas Massif–Central Cordillera forms an arc concave to the north. The trend of this arc parallels fold trends in the younger sedimentary rocks in the Chiapas–Guatemala Foldbelt. Analysis of the structural and lithological trends within the basement (Kesler, 1971) indicate that this ancestral orogenic zone also forms a regional arcuate pattern.

Middle Paleozoic intrusive rocks (McBirney and Bass, 1967) with radiometric ages of 375 m. y. (Devonian) have been dated in the Central Cordillera of Guatemala and granitic rocks, whose apparent ages range from Cambrian to possible Devonian, have been dated in the Chiapas Massif (Pantoja Alor *et al.*, 1971).

A thick sequence of sedimentary Mississippian–middle Permian rocks, the Santa Rosa Group, was deposited uncomformably over the basement rocks in the Chiapas–Guatemala Paleozoic basin (López Ramos, 1969; Dengo 1975). The basin is known from west-central Chiapas, in Mexico, across Guatemala, to the Maya Mountains of Belize (Kesler *et al.*, 1971).

Its possible continuation into the Caribbean is unknown. The lower part of the sequence, previously not well dated, is now known to be of late Mississippian age in Chiapas (Hernández-García, 1973), and consists of clastic, partially metamorphosed, continental and lagoonal sedimentary rocks. A possible unconformity separates them from another clastic sequence of which the upper part has late Pennsylvanian–early Permian marine shales and thin limestones and is conformably overlain by thick middle Permian fossiliferous carbonates (Chochal Formation). The entire sequence is over 4,500 m thick.

The late Paleozoic sedimentary sequence is intruded by granites of Permian age in the Maya Mountains of Belize and possibly in other parts where they are not dated, as well as in the Chiapas Massif–Central Cordillera mountain ranges. A regional unconformity separates the Paleozoic and Mesozoic sequences, corresponding to the late Permian–Triassic time interval.

The Mesozoic sedimentary rocks within the Maya block belong to the Mexican Geosyncline and now form the Chiapas–Guatemala Foldbelt (Benavides 1956, Sánchez-Montes de Oca, 1979). They extend at the subsurface under the Gulf of Mexico and the Yucatan Peninsula. On the southern slopes of the Chiapas Massif–Central Cordillera of Guatemala, marine Mesozoic (Cretaceous) rocks occur in small areas lying unconformably on the basement.

The basal part of the sequence consists of red-bed molasse deposits (Todos Santos Formation) of late Jurassic–early Cretaceous age. Although part of these rocks were considered earlier to be Triassic, sedimentary rocks of this age thus far are not known in the Maya block. Toward the north, in the subsurface, the Jurassic grades into a limestone–evaporite sequence.

The Lower Cretaceous is characterized by limestones and dolomites overlain upward by thick salt–anhydrite deposits and by thick limestones and dolomites of Aptian–Albian age (Sierra Madre Group and Cobán Formation). Conformably over these, there are Upper Cretaceous limestones, the upper portion of which is well dated as ranging from Coniacian to Campanian, although rocks of the latter age are missing over a large part of central Chiapas.

Along the central part of the foldbelt a thick sequence of flysch deposits of Maastrichtian to early Eocene age (Ocozocuautla Group in Chiapas and Sepur Group in Guatemala) mark a sharp change in sedimentation that resulted from tectonic activity. These rocks grade northward into platform carbonate deposits.

Other changes in Tertiary sedimentation that also reflect changing tectonic environments resulted in the deposition of a continental and shallow marine molasse sequence in the central part of the Maya block during the middle Eocene (El Bosque Formation).

Oligocene to Pliocene marine sediments are restricted to small areas in central Chiapas, the edges of the Yucatán Peninsula and western Guatemala, where they present a post-Miocene unconformity. In the Gulf of Mexico part of the block (Macuspana and Isthmian basins) sediments of these ages are not only extensive but also of great thickness.

Late Tertiary and Quaternary andesitic volcanic rocks in the Maya block are restricted to small areas to the east and north of Tuxtla Gutiérrez City, and more extensive pumiceous deposits occur in the upper Grijalva River Valley along a large syncline known as the Central Chiapas Depression (Fig. 6 and 9). At least one active volcano is still active, El Chichón, which had a large explosive eruption on March 28, 1982.

(b) Tectonic History. Whether the Precambrian (?) rocks of the Maya block are the eastern extension of the Oaxaca Complex is problematic because of lithological differences, particularly the predominance of granites in the Chiapas Massif. Possibly the two blocks may have been in relatively different positions during late Precambrian (?)–early Paleozoic time. The Precambrian (?) metasedimentary rocks in the Chiapas Massif are not well dated, but their contact with the granites could be an unconformity rather than an intrusive one.

If the metasedimentary rocks are of Grenville age, this possible unconformity would correspond to a previous tectonic event. This reasoning, which is not well supported by facts, is mentioned with the specific purpose of pointing out a problem that needs to be investigated in order to understand the early history of the region.

Based on radiometric age determinations McBirney and Bass (1967) proposed that the original sedimentary rocks of the Chuacus Group were highly deformed and metamorphosed during the Devonian as part of the Acadian orogeny of North America. This tectonic event was common to both the Oaxaca and Maya blocks, and was accompanied by granitic intrusions. Ortega-Gutiérrez (1978) has suggested that the Chuacus Group and the Acatlán Complex are correlatives.

After the late Mississippian–mid-Permian sedimentation in the Chiapas–Guatemala Paleozoic basin, perhaps during late Permian, another severe orogeny, corresponding to the Appalachian, affected the region. This orogeny was probably accompanied by a few granitic intrusions, such as that mentioned in the Maya Mountains of Belize.

Based on paleomagnetic interpretation Sánchez-Barreda (1981, p. 65) reached the following conclusions:

> The area of Oaxaca has remained without considerable rotations (less than 6°) and latitudinal displacements with respect to stable North America since the late Paleozoic . . . The areas of Oaxaca and Chiapas were located at different latitudes during the Permian. Chiapas were located to the north with respect to Oaxaca

. . . The area of Chiapas was displaced to the south and rotated close to 30° in a
clockwise direction with respect to stable North American and close to 24° in a
clockwise direction with respect to Oaxaca.

In the interpretation presented here in the regional tectonic history discussion, the original position of the Maya block with respect to the Oaxaca
block requires a counterclockwise motion of nearly 40°.

It is possible then to infer that at least until the end of the Permian, the
Maya block occupied a large part of the Gulf of Mexico. Its present position
relative to the Oaxaca block is presumably the result of southward migration
and rotation after the Appalachian orogeny, during the Triassic–early Jurassic. The Mesozoic metamorphic rocks in the Isthmus of Tehuantepec already mentioned, and considered to be a possible suture zone, would have
to be either Triassic or early Jurassic since the late Jurassic–Cretaceous
sedimentation in the Mexican Geosyncline was a common feature to both
the Oaxaca and Maya blocks.

During the Triassic practically all of the Maya block was uplifted, as
was also the case for the Oaxaca block. It remained emergent and was
disturbed mainly by taphrogenic movements, followed by molasse sedimentation (Todos Santos Formation) during the late Jurassic–early Cretaceous previous to subsidence and the extensive marine sedimentation and
transgression from the Gulf of Mexico (Mexican Geosyncline).

Widespread regional deformation occurred during late Cretaceous–
early Eocene time and corresponds to the Laramide orogeny. This orogeny
probably started as uplift of the southern margin of the geosyncline as indicated by the absence of Campanian rocks in central Chiapas, and was
stronger during the Maastrichtian–early Eocene, accompanied by concomitant flysch deposition (Ocozocuautla and Sepur groups) and by strong folding and thrusting. The folding was also produced by *decóllement* of carbonate
rocks sliding over evaporitic deposits or as result of evaporite diapirism.

The Laramide Orogeny undoubtedly established the tectonic pattern of
the Chiapas–Guatemala Foldbelt that resulted from the interaction of the
Chiapas Massif–Central Cordillera as the hinterland and the Yucatán Platform as the foreland.

Later tectonic activity was characterized by a short taphrogenic phase
during the middle Eocene evidenced by the molasse deposits of El Bosque
Formation; by renewed partial subsidence; and by a folding episode at the
end of the Miocene. Transcurrent faulting within the block, and along its
southern edge (Polochic and Motagua fault zones) are also part of the Miocene event.

Late Tertiary–Recent volcanic activity was only a localized phenomenon and not as widespread as in the neighboring Oaxaca and Chortis blocks.

The Maya–Chortis limits will be discussed latter as the boundary zone between the North America and Caribbean plates.

B. Caribbean Plate Segment

The Central American or western part of the Caribbean Plate extends from the Motagua Fault Zone in Central Guatemala to a poorly defined boundary in southern Nicaragua near the Costa Rican border. Present understanding of the tectonics of this area indicate that it is formed by a single crustal block, the Chortis, that contains the oldest rocks within the Caribbean plate. Rocks of older ages are found in the Mid-American–Caribbean region in the land masses bordering the plate on the northwest (Maya block) and the southeast (South America).

1. Chortis Block

Geographically the Chortis block includes southern Guatemala, El Salvador, Honduras, most of Nicaragua, and extends to the east under the Caribbean Sea as the Nicaragua Bank. In some of the proposed reconstructions (i.e., Carey, 1958; Freeland and Dietz, 1971, 1972; Malfait and Dinkelman, Helwig, 1975, and others) parts of the Chortis were called the Honduras or Nicaragua block.

Crustal thicknesses have been estimated as 35–40 km in its terrestrial part, indicating a continental type of crust, while in the Nicaragua Bank they vary 25–30 km (Arden, 1975), and the seismic velocities indicate an oceanic to intermediate crust. The change from one to another is still an unresolved problem. The following geological summary deals with the land portion of the block (Figs. 4 and 6).

(a) Stratigraphic Sequence. Low-grade metasedimentary basement rocks crop out extensively in the block and are mostly phyllites. These are difficult to correlate with metamorphic sequences in other parts of the region. In southern Guatemala they have been tentatively correlated with the Santa Rose Group (Clemons, 1966), but today they are distinguished as a different unit, called the San Diego Phyllite (Donnelly *et al.*, 1975, Map). A summary of the recent investigations of the basement rocks by Horne *et al.* (1976) shows that they are not so homogeneous. Some of the rocks described by them, in particular the El Tambor Group, are now interpreted as part of the suture zone between the Maya and Chortis blocks and will be discussed later, not as part of the Chortis basement.

Between the Motagua Fault Zone and the Chamelecón Fault (Fig. 6), a sequence of high-grade metasedimentary and meta-igneous rocks of garnet-staurolite and sillimanite schists and gneisses, amphibolite, migmatites, and marble, called the Las Ovejas Group, was described by Schwartz (1972). It was tentatively correlated with the Chuacús Group and Acatlán Complex by Ortega-Gutiérrez (1978). In the regional tectonic interpretation presented here this correlation, as well as others presented earlier by the author (Dengo, 1968), will be questioned. Because Las Ovejas is in fault contact with other metamorphic rocks in the Motagua Fault Zone and with the phyllitic sequence, it is not possible to establish clear stratigraphic relations. Based on structural and metamorphic differences, Donnelly *et al.* (1968) concluded that the Las Ovejas Group is older than the others and separated by a major unconformity.

In the Sierra de Omoa (northeastern Honduras) Horne *et al.* (1976) have identified metamorphic rocks that may be divided into several stratigraphic assemblages. In the northernmost part a low-grade metavolcanic assemblage includes metadacite, calcite-chlorite-albite schist (greenstone) and andesine-quartz-biotite metagraywacke, which may represent the upper part of an ophiolite assemblage. These rocks could also represent part of an oceanic suture zone now accreted to the Chortis block.

In the central part of the Omoa Mountains there are rocks of the almandine-amphibolite metamorphic facies that locally have staurolite and kyanite. The southern flank of the range consists dominantly of low greenschist facies metasedimentary strata.

The metasedimentary sequences are intruded by rocks of different ages. Of particular interest is that the basement that forms the core of the Sierra de Omoa at Bañaderos is a meta-igneous complex consisting of granodiorite gneiss, metatonalite, and metadiorite. A tentative date of 460 to 980 m.y. was determined and although it is not conclusive, it suggests a late Precambrian–early Paleozoic age. Horne *et al.* (1976 p. 578–579) who described these rocks, stated that "before attempts are made to correlate them with known Grenville-age rocks in southern Mexico, the age should be studied further".

Another meta-igneous complex occurs along the east flank of the Sierra de Omoa and intrudes metasedimentary rocks of the amphibolite facies. Because an isochron age of 305 ± 12 m.y. was established for some of these rocks, Horne *et al.* (1976 p. 578) state that "it appears that the amphibolite facies strata that are intruded by these rocks are of pre-Pennsylvanian age, and probably were metamorphosed regionally and intruded in the early Pennsylvanian."

Finally, on the basis of a granitic batholith (San Isidro) that intrudes the San Diego phyllitic metasediments and provides an estimated age of 180 m.y., Horne *et al.* (1976) suggest that the age of this younger metamorphic sequence is probably post-Mississippian and pre-Jurassic.

In summary, the Chortis basement includes at least the following simplified sequence of major units: (i) a possible late Precambrian–early Paleozoic meta-igneous complex (Bañaderos); (ii) a high-grade metasedimentary assemblage of the Las Ovejas Group and part of the Sierra de Omoa, of unknown age but possibly of early Paleozoic; and (iii) one, or more, low-grade assemblages which some authors (Clemons, 1966; Horne *et al.*, 1976) have proposed to be post-Mississippian or equivalent in part to the Santa Rosa Group of the Maya block. However, the possibility that it is older cannot be discarded. This generalized sequence, plus the occurrence of a metamorphosed ophiolitic assemblage in the Sierra de Omoa, bear more similarity with the Oaxaca Block.

Thus far, no unmetamorphosed Mississippian–middle Permian rocks, similar to those that occur extensively in the Maya block, have been identified in the Chortis, with one possible exception in southerwestern Guatemala, east of Lake Atitlán (Williams, 1960).

Overlaying the Chortis metamorphic basement there is a thick sequence of continental and marine Mesozoic sedimentary rocks described in Mills *et al.* (1967), Fakundiny and Everett (1976), and Finch (1972). These are considerably different from the Mesozoic sequence of the Maya Block.

The older Mesozoic rocks are plant-bearing shallow- and brackish-water sandstones and shales of late Triassic–early Jurassic age that occur in Central Honduras (El Plan Formation). Uncomformably over these, or directly over the basement, there are undated continental sandstones of variable thickness that, because of their stratigraphic position and lithology, have been correlated with the Todos Santos Formation of the Maya block. These rocks are probably late Jurassic or early Cretaceous because they are overlain by a thick early to "middle" Cretaceous limestone sequence of which the Aptian–Albian part (Atima Formation) is the thickest, topographically most conspicuous, and widespread from southern Guatemala across Honduras to northern Nicaragua.

The post-Albian rocks, which are also very thick and extensive, are prevailing continental red beds with local intercalations of limestone and gypsum (Valle de Angeles Group). Mesozoic and somewhat younger intrusive rocks are known in different parts of the Chortis block but they occur mainly along its northern edge, south of the Motagua Fault Zone (Fig. 6). Radiometric age determinations indicate that the intrusions range in age from middle Cretaceous to early Tertiary.

The Tertiary is characterized by possible Eocene, or even older, redbed deposits, restricted marine sedimentary rocks on the Caribbean side of the block, and very extensive and thick lava flows. Basalts and andesites of possible Oligocene age occur at the base, and Miocene to Pliocene rhyodacitic ignimbrites in the middle and upper parts.

The Quaternary volcanic rocks within the block fall into two groups, those that parallel the Pacific coast and that are part of the Central America

Volcanic Chain, and those associated with N–S trending faults and grabens in western El Salvador–southeastern Guatemala and Central Honduras (Williams *et al.*, 1964, Williams and McBirney, 1969).

(b) Tectonic History. From the description of the basement rocks of the Chortis block it is evident that a sound interpretation of the pre-Mesozoic tectonic history is difficult. In some aspects, there seem to be similarities with the basement of the Oaxaca block, and in others with the Maya. The meta-igneous complex exposed at Bañaderos could be related in its history to the Oaxaca Complex, or to the Precambrian (?) of the Chiapas Massif. Rocks of similar age and lithology thus far have not been identified in the Central Cordillera of Guatemala. In spite of the problems encountered by Horne *et al.* (1976) the Bañaderos rocks could be part of the Grenville Fold-belt and are separated by an unconformity from the overlying metasedimentary rocks. Whether this unconformity represents a late Precambrian or an early Paleozoic tectonic episode is not known. Nevertheless it is evidence of the oldest deformational activity with the Chortis block.

The lower metasedimentary sequence, namely, the Las Ovejas Complex and part of the Sierra de Omoa, are possibly Lower Paleozoic. In attempting a correlation with the basement of the other blocks, they may be compared, and perhaps are age-equivalent to part of the Acatlán Complex of the Oaxaca block, a correlation suggested by Ortega-Gutiérrez (1978). Because the unconformity that separates these rocks from the overlying phyllitic sequence is not well dated, it is problematic again to assign an age to the pre-unconformity tectonic events and regional metamorphism. It is possible that these events of large regional extent, correspond to the Taconic Orogeny suggested for the Oaxaca block, or they may be younger, as part of the Acadian orogeny, for which there is certain evidence in both the Oaxaca and Maya blocks.

The regional correlation of the upper metasedimentary sequence again poses problems in trying to compare it with the tectonic history of the other blocks. If they are post-Mississippian and pre-Jurassic as indicated by Horne *et al.* (1976), they could be age correlatives of the Santa Rosa Group of the Maya block. Other ideas to be discussed later, indicate that the Chortis and Maya blocks were in a different relative geographic position in pre-Jurassic times and, therefore, the San Diego Phyllite sequence should not be considered Santa Rosa, even if the two could be age-equivalent in part.

The stratigraphic gap between the basement and the Mesozoic sedimentary sequence indicates an episode of emergence and uplift at least from the late Permian to the late Triassic, which was also the case for the Oaxaca and Maya blocks.

The Mesozoic sedimentary events from late Triassic to early Jurassic time (El Plan Formation) bear more similarity with those of the State of

Guerrero in the Oaxaca block, both in the ages and types of sedimentary rocks, than with the late Jurassic–Cretaceous rocks of the Mexican Geosyncline that form the Chiapas-Guatemala Foldbelt of the Maya block.

The Laramide Orogeny events are evident in the Chortis block as folds and granitic intrusives. Folding, however, was not as intense as in the Maya block, except for an area in the Mosquitia region of eastern Honduras and northeastern Nicaragua (G. Escalante, personal communication).

The late Tertiary tectonic history was characterized by large volcanic flows–mainly ignimbritic, and by normal faulting perhaps due to the eastern migration of the block as part of the Caribbean plate—and to left-lateral displacement along its northern boundary (Polochic, Motagua, and Chamelecón fault zones). The ignimbrite flows, both in age and composition, indicate a similar and coeval tectonic episode in the Oaxaca and Chortis blocks not represented in the intervening Maya block. The normal faulting, which has a dominant N–S trend, is an episode of extensional deformation not encountered in the other blocks. Quaternary volcanism along the Pacific side of the block is of calc-alkaline type, while some of the volcanic rocks associated with the normal faluting belong to an alkaline suite (Williams and McBirney, 1969).

C. South America Plate Segment

The part regarded here as belonging to the South America plate extends approximately from the Nicaragua–Costa Rica border to northwestern Colombia (Fig. 3). This entire area is similar in its crustal structure and tectonic history and is also similar in many aspects to the Cordillera Occidental of Colombia and Coast Ranges of Colombia and Ecuador. Its boundaries will be discussed separately due to problems in defining certain parts.

The geological differences between northern and southern Central America have been known since the early work by Sapper (1937). Schuchert (1935) stressed the known differences both in types of basement and geological history and called the northern part "Nuclear Central America," and the southwestern portion (Costa Rica–Panama) the "Isthmian Link." Later Dengo (1969) subdivided Nuclear Central America into the Maya and Chortis blocks, and Dengo (1962) and Lloyd (1963) applied the name Southern Central American Orogen to the southern part. These name designations have been used extensively in the recent literature. The similarlity of southern Central America with northwestern South America was pointed out by Dengo (1962a) and later supported by other studies, in particular those of Case et al. (1971), Goosens et al. (1973), and Pichler et al. (1974).

Because a tectonic discontinuity exists at the position of the Panama Canal Zone, a new subdivision is presented in this paper where the name

Chorotega block (after the Chorotega Indian tribes of Costa Rica) is assigned to the part between the Nicaragua–Costa Rica border and the Canal Zone, and Chocó block (after Chocó Indian tribes in Colombia) to the portion that extends to the east of the Canal Zone and continues southward along the Serranía de Baudó in north-western Colombia, parallel to the Pacific Ocean.

1. *Chorotega Block*

The Chorotega block has the shape of an arc concave to the northeast, between the Pacific Ocean and the Caribbean Sea. In general, the major tectonic lines determine its shape, It can be divided into three more or less parallel units, the external or forearc on the Pacific side, an internal arc that forms the backbone of the block, and a back-arc basin (Limon) on the Caribbean side that extends as a submarine feature into the North Panama Deformed Belt (Case and Holcombe, 1980, Map). In the fore-arc in the peninsulas of Nicoya, Osa, Burica, Soná, and Azuero (Figs. 1 and 7) ophiolitic rocks of Mesozoic–early Tertiary age that form the basement of the block and are regarded as oceanic crust crop out extensively. Gravity and seismic data indicate a crustal thickness of 17 to 25 km (Case *et al.*, 1983; Matumoto *et al.*, 1977) which indicates that even if the crust is oceanic, it has been thickened probably as result of accretion or superposition of crustal slabs of different ages.

(a) *Stratigrahic Sequence.* In northwestern Costa Rica (Fig. 7) the basement rocks were designated as the Nicoya Complex (Dengo, 1962a) and described as consisting of basalts (including pillow lavas and agglomerates), diabase and gabbro intrusives associated with dark graywackes, cherts and siliceous limestones. A large serpentinized peridotite that forms most of the Santa Elena Peninsula was regarded as a separate unit. Based on scant eivdence, these rocks were regarded as ranging in age from possible late Jurassic to late Cretaceous (Campanian) time. Later studies have shown that the complex can be subdivided in units of different ages and with some lithological variations but at the same time with many common features characteristic of ophiolitic sequences. The name Nicoya, therefore, has become generic for the basement of the Chorotega block. The stratigraphic and structural problems have not yet been solved, but it is now known that mafic tholeiitic volcanism existed from late Jurassic to Paleocene or early Eocene time (Galli Olivier, 1977, 1979; Schmidt-Effing, 1979, 1980; Stibane *et al.*, 1977; Azema and Tournon, 1980; Kuijpers, 1980; de Boer, 1979; Weyl, 1966). In the Osa, Burica, and Soná peninsulas the exposed parts of the complex are composed largely of pillow basalts and radiolarites, while in the Azuero Peninsula minor greenschist metamorphic rocks and patches of mafic and ultramafic rocks are found along shear zones (Del Guidice, 1978).

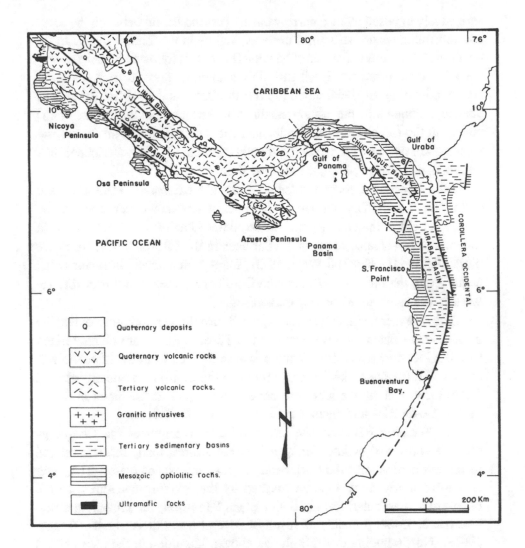

Fig. 7. Generalized geologic map, Chorotega and Chocó blocks.

In the Nicoya Peninsula the Campanian and older parts of the complex are overlain by a thick sequence of turbidites with thin micritic limestones that probably represent part of an earlier fore-arc accretionary wedge. These in turn, are followed by Paleocene reefal limestones and turbiditic calcarenites and sandstones. In other parts of Costa Rica (Quepos Point), there are Paleocene pillow basalts and a fairly thick sequence of siliceous radiolarian limestones, that must represent a younger slab of upper oceanic crust accreted to the present fore-arc.

Late Paleocene–late Eocene volcaniclastic sediments with lenticular Upper Eocene reefal limestones overlie the rocks described above and are extensively exposed along a narrow basin (Térraba Basin) between the fore-arc and the inner arc structural units as well as in the Limón Basin. In its lower part it has intercalations of basaltic flows. It is followed by Oligocene–early Miocene proximal flysch turbiditic sediments derived from a Pacific source (Henningsen, 1966) along the térraba Basin and by distal flysch and pelagic sediments in the Limón Basin. Miocene sedimentary rocks occur mainly in the Limón Basin as a volcaniclastic molasse, and in the inner- and fore-arc in restricted areas, which together indicate Pacific–Caribbean Sea connections (Fischer, 1980) during that time.

Late Miocene–Pliocene volcanic rocks are extensive in the inner arc and more so in the Panama part of the block where they cover, and may be somewhat older, than a large part of the Azuero Peninsula. In contrast with the extensive volcanic rocks of similar ages in the Chortis block, these are more alkaline (Pichler and Weyl, 1973). Large granodioritic intrusions also occur along the inner arc (Talamanca Cordillera). These were probably related to the same series of volcanic episodes.

The younger sedimentary rocks, which are thicker and more extensive in the Limón Basin but also occur in the Térraba Basin, are characterized by a boulder conglomerate volcanic molasse derived from the Miocene volcanic and intrusive rocks. Volcanic rocks of the Central American Volcanic Chain located along the inner arc cover a large part of the block in north-central Costa Rica and smaller areas to the east in Panama.

(b) Tectonic History. The structural configuration of the Chorotega block is controlled by late Tertiary–Recent features, many of them related to the interaction with the Caribbean, Cocos, and Nazca plates. The older structures are to a large extent masked by these younger features. In the older parts, particularly in the Santa Elena Peninsula, the prevailing structure trends E–W, probably as result of several transcurrent faults (Dengo, 1962a). North-dipping thrust faults have been identified in the Santa Elena Peninsula, indicating overriding of the peridotite body on complex basalts (Azema and Tournon, 1980), and also in the central part of the Nicoya Peninsula. Interpretation of magnetic anomalies by de Boer (1979) also indicates a structural grain of the Nicoya Complex with a WNW trend. These structural trends probably represent tectonic interaction between the ancestral Chorotega and Chortis blocks and correspond to late Cretaceous–early Tertiary events. Because the Nicoya complex includes rocks of late Jurassic (radiolarite xenoliths), and middle Cretaceous as well as well-defined Campanian and Paleocene assemblages, the earlier history is difficult to understand, but the present information points to several oceanic crustal slabs that were accreted in a single oceanic arc since the late Cretaceous, represented by a series of islands and not as a continuous landmass, (Lloyd,

1963; Dengo, 1962b). This arc provided the source for the late Cretaceous–Eocene volcaniclastic sediments. A volcanic chain along the inner arc must have existed during Eocene. Uplift of the fore-arc and submergence of the inner arc controlled the deposition history from Eocene to early Miocene, defining the Térraba fore-arc basin and the Limón back-arc basin, leaving marine connections between them.

Tectonic activity during middle Miocene–Pliocene time (Andean orogeny) was characterized by uplift, intrusions along the inner arc, extensive volcanism, and folding with reverse and thrust faulting in the Térraba and Limón basins, with the faults dipping in both cases toward the inner arc. This structural pattern is similar to the eastern Panama part of the Chocó block, and has been interpreted by Case (1974) and Lundberg (1983), as an indication of an uplifted oceanic block or an oceanic *zwischengebirge*.

2. *Chocó Block*

Northwestern South America, including the Cordillera Occidental of Colombia and the coast ranges of Colombia and Ecuador, present geological and geophysical characteristics of typical oceanic crustal basement (Goossens *et al.*, 1977; Case *et al.*, 1973; Case, 1974; Duque-Caro, 1980; Campbell, 1974; Stibane, 1967). This basement is separated from the continental crust to the east by the Romeral Fault, or Suture Zone.

An elongated Tertiary sedimentary basin (the Urabá Basin, previously called the Bolivar Geosyncline) separates the Cordillera Occidental and the coast ranges (Fig. 7). Along the foothills of the Cordillera, an east-dipping reverse fault zone (Atrato Fault) marks the boundary between the basin and the ranges. It extends from the Urabá Gulf on the Caribbean south to near the mouth of the Baudó River on the Pacific. The area that extends westward of this fault and continues from the northwest in Panama i.e., the Panama Spur (Lloyd, 1963), to the Canal Zone is the one called the Chocó block. The crustal thickness is 15–30 km (Case *et al.*, 1973; Case, 1974; Case *et al.*, 1983).

Although the Chocó and Chorotega blocks share characteristics in crustal composition and geologic history, a tectonic break separates them along the Panama Canal Zone, which is marked by a negative Bouguer gravity anomaly between large positive anomalies to the east and west (Case, 1974) and which corresponds to a Tertiary sedimentary basin between the Pacific and the Caribbean. Case and Holcombe (1980, Map) interpret this feature as a northwest-trending shear zone with left-lateral displacement. Other differences are that the late Tertiary continental volcanic rocks that occur extensively in the Chorotega block are only found sparsely in the Chocó.

(a) Stratigraphic Sequence. The basement rocks of the Cordillera Occidental and Colombia coast ranges were originally called the Grupo Dia-
básico and were regarded to be Cretaceous and possibly Paleocene (Julivert,
1968). More recent studies (Case *et al.*, 1971; Case *et al.*, 1973; Toussaint
and Restrepo, 1976; Toussaint, 1978; Barrero, 1979), indicate that, like the
basement of the Chorotega block, it ranges from Cretaceous to Paleocene
and is formed by magmatic arc rocks accreted to the continent by subduction–obduction processes.

In the Panama part of the block, two major areas of basement are known,
one that parallels the Caribbean coast and is made up of tholeiitic basalts,
basaltic andesite, and pelagic marine sedimentary rocks, and the other along
the Pacific coast where similar rocks are found interlayered with well-dated
Campanian radiolarites (Case, 1974). Between them, the low area of the
Chucunaque Basin is filled with a thick sequence of Eocene to Miocene
volcaniclastic sediments and some reefal limestones that, like the Térraba
Basin in the Chorotega block, originated as a forearc feature. The Chucunaque Basin joins to the south with the much larger Urabá Basin that has
sedimentary rocks of the same type and age but much thicker (>4,000 m).
Sedimentation in the Urabá Basin began in the Eocene and was characterized
by marine volcaniclastic rocks and reefal limestones, (late Eocene) followed
by early Oligocene black shales, conglomerates, and limestones, Oligocene—early Miocene shallow-water clastics, and, finally, by shallow-water
to continental facies and Pliocene continental deposits. The latter indicate
uplift of the bordering areas.

Late Cretaceous–Paleogene granodioritic intrusive rocks are known in
the Sierra de San Blas (Panama) and the Cordillera Occidental (Colombia),
which suggests that these mountain ranges were once continuous but are
now separated by the Atrato Fault and the Urabá Basin.

Quaternary volcanic rocks are uncommon in this block. Some small
volcanic cones are known on the Caribbean side of Panama near the Colombia border (Fig. 9). A chain of Quaternary volcanoes, related to the interaction of the Nazca and South America plates is well known south of the
Chocó block southern limit.

(b) Tectonic History. The structure of the Chocó block developed as
part of the tectonic histories of the Cordillera Occidental and, in many respects, the Chorotega block. Differences resulted from the fact that a large
part of the block and the Cordillera were accreted to a large continental
mass. On the other hand, the Panama spur part of the block (Lloyd, 1963)
and the Chorotega were formed by intraoceanic crust deformation, with only
one localized collision area with a small cratonic mass, between the Chorotega and the Chortis blocks.

A pre-Eocene unconformity and the more complicated structure of the basement indicate a late Cretaceous–early Tertiary strong tectonic series of events. Older unconformities are yet difficult to establish but they are not discarded. During the Oligocene and Miocene there were minor disturbances indicated by local unconformities.

The Andean orogeny (late Miocene–Pleistocene) produced folding, changes in sedimentation, and a general uplift such as in the Chorotega block. The Atrato Fault was very likely formed during this deformation. The present configuration of both the Chocó and Chorotega blocks probably started to take place at the beginning of the Andean orogeny.

D. Tectonic Plate Boundary Zones

The plate boundary zones are outlined under the regional tectonic framework in order to define the lithospheric blocks described (Fig. 3). These boundaries and those that separate blocks within one plate are mentioned in the description. The intraplate ones (Oaxaca–Chortis and Chorotega–Chocó) are better documented to indicate the reasoning behind the separation of the region into blocks. The aim of the following discussion is to present the geological features along the plate boundaries necessary to complete the block description already presented and to understand the regional tectonic history.

1. North America–Caribbean Boundary Zone

The boundary in many of the models presented (Figs. 2 and 3) is placed usually along the Motagua Fault Zone or along the Polochic Fault Zone which is the one that more closely cuts the Central American isthmus from the Pacific to the Caribbean. Its continuation in the Caribbean along the Cayman Ternch has been recognized by many authors.

What is referred to here as the boundary zone is not a single fault but an area of complex geology exposed mainly along the Motagua River Valley and the mountains north and south of it. Also within this zone is the Polochic Fault to the north of the Motagua and the Chamelecón Fault to the south of it. A large amount of field mapping has been done along the Motagua Valley area by Donnelly and his students, but most is still unpublished. The results have been summarized and compiled in one single map (Donnelly *et al.*, 1968; Donnelly *et al.*, 1975, Map).

The presence of several serpentinite bodies along the Motagua Valley and in the mountains to the north and south of it, has been known for a long time, and their age of emplacement has been a problem of much discussion (Dengo, 1968, 1972; Case, 1981). A sequence of metavolcanic rocks and phyllites was identified by Williams *et al.* (1964) and named El Tambor

Formation (now Group) by McBirney and Bass (1969), who in comparing these rocks with the Chuacús Group of the Maya block suggested that they could be younger. The metavolcanic rocks are associated with metasedimentary rocks, phyllites, and partially recrystallized limestones (Lawrence, 1976) that, in turn, in one locality to the south of the Motagua (Sansare), are mixed with pillow basalts and radiolarian cherts. Further studies in the area by Wilson (1974) established a late Cretaceous age for some of the limestones, and a similar age has been suggested for the cherts. Radiometric ages of some of the ultramafic rocks by Bertrand *et al.* (1978) indicate an age of 58.5 m.y. ($^{40}Ar/^{36}K$). The entire sequence including the serpentinite bodies is now regarded to be Mesozoic, possibly as young as Campanian, although the lower limit has not been established. On this basis as well as the structural features that indicate thrusting to the north over the Chuacús rocks and to the south over the Chortis block basement, and on the basis of the overall lithology, the sequence is now regarded as a metamorphosed ophiolitic assemblage representing an oceanic crustal segment between the Maya and Chortis blocks. This assemblage deformed due to collision of the two blocks during or after Campanian times, possibly during the Laramide Orogeny. Donnelly (personal communication) dates the collision at about 70 m.y. ago. To the west, rocks similar to those of El Tambor occur in a southeastern part of Mexico close to the Guatemala border, just south of the Polochic Fault, and have been interpreted by Carfantan (1977a, 1977b, 1979) and Gutiérrez-Coutiño and Carfantan (1979) as a continuation of the collision zone.

Further west, Mesozoic-metamorphosed rocks were already mentioned in the Isthmus of Tehuantepec. Whether or not these represent a continuation of the El Tambor Group is not yet known. To the east, on the Caribbean side, the metamorphosed ophiolitic rocks described by Horne *et al.* (1976) in the eastern part of the Sierra de Omoa, as well as similar rocks in Roatán Island (McBirney and Bass, 1967), could also represent the continuation of the El Tambor Group.

Several large serpentinized peridotite masses both to the north and south of the Motagua Fault Zone, and in one case north of the Polochic Fault, are considered to be allochtonous bodies thrust in both directions over the Maya and Chortis blocks. At least one case is well documented for this interpretation—the Sierra de Santa Cruz ultramafic body north of Lake Izabal (Williams, 1975; Rosenfeld, 1980). The peridotite and associated gabbroic rocks reached their present position by gravity-sliding over sedimentary rocks of the Sepur Group.

The other major structural feature along the plate boundary zone is the presence of large transcurrent faults (Fig. 8), namely, the Polochic, Motagua,

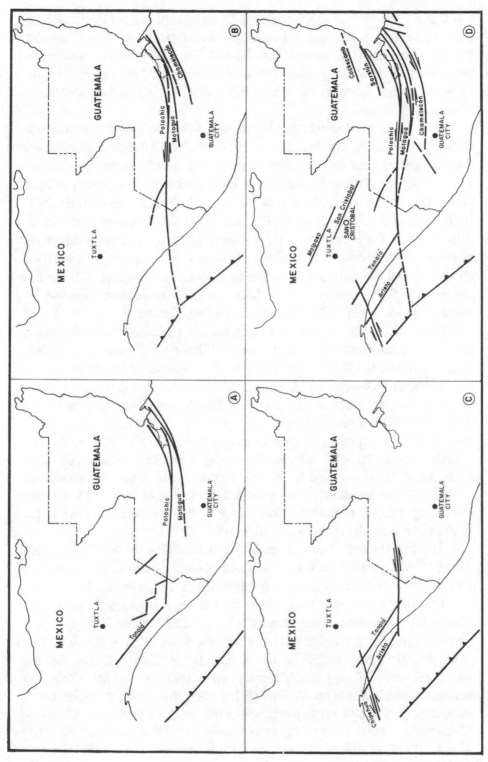

Fig. 8. Interpretations, North America–Caribbean–Cocos triple junction: A, after Muehlberger and Ritchie (1975); B, after Burkart (1978); C, after Sánchez-Barreda (1981); D, this paper.

and Chamelecón. Since the earliest studies of Caribbean geology some of these faults have been regarded as a continental extention of the Cayman Trough, a subject that is now well established. There have been considerable differences of opinion regarding the timing and type of faulting and in the amount of lateral displacement, but not in the direction of relative movement, which is left-lateral.

Before the plate-tectonic theory was fully developed, Hess and Maxwell (1953), in explaining the regional tectonics of the Caribbean, proposed a displacement of over 800 km along this major series of parallel fault zones. Even though it is difficult to establish offset of rock units along the individual faults, the cumulative movement of the major faults must be considerable. In the case of the Polochic Fault, Burkart (1978) has proposed a left-lateral displacement of approximately 125 km since Miocene, based on the offset of several large geomorphic features. This view is supported by a detailed structural analysis along a part of the fault by C. A. Dengo (1982). On the other hand, Erdlac and Anderson (1982) consider the amount of post-Cretaceous displacement to be no more than a few kilometers.

Some of the recent models to explain the Caribbean tectonics again propose displacements of several hundred kilometers (Sykes *et al.*, 1982; Kellog and Bonini, 1982). This problem has been discussed mostly in relation to Caribbean tectonics, but it is also of fundamental importance for the Pacific tectonic aspects of Mid-America. The Polochic Fault can be traced from the Caribbean to the Pacific, where near the Pacific coast it is covered by coastal plain deposits. Its extention over the continental shelf and to the Middle America Trench is inferred from bathymetry. The Motagua and Chamelecón faults are covered in western Guatemala by Quaternary volcanic deposits. There is a slight topographic indication that in western Guatemala the Motagua curves to the ENE, joins the Polochic, and continues as a single fault before it reaches the Pacific coast (Fig. 8D).

The continuation of one or more of these faults to the Middle America Trench forms a triple junction of the Cocos, North America, and Caribbean plates. Some of the proposed interpretations are shown in Fig. 8.

One final aspect relative to these fault zones is their age of inception. Earlier interpretations of the ultramafic belt, the El Tambor Group, as well as the Cayman Trough, considered a possible Paleozoic age for these features. Present understanding of the Motagua Suture Zone indicates that the age of the transcurrent faults is much younger (Schwartz *et al.*, 1979). The regional considerations by Burkart (1978, 1983), the detailed studies by C. A. Dengo (1982), plus the deformation of the southern part in the Chiapas–Guatemala Foldbelt of the Maya block point to a Miocene age. Seismicity along the faults (Molnar and Sykes, 1969; Sykes and Ewing, 1965; Spence and Person, 1976) indicates that they are presently active.

2. *Caribbean–South America Boundary Zone*

This boundary zone is very complex due to the variety of tectonic features in the northern termination of the Andes (Shagam, 1975; Campbell, 1974). The purpose here is to outline the present boundary in order to support the view that the Chorotega and Chocó blocks are now a part of the South America plate.

In its northern part it was proposed that the boundary zone is located approximately along the Nicaragua–Costa Rica border and that the Santa Elena Peninsula is part of it. It was also pointed out that the E–W structural trend in the peninsula and the structure of the Nicoya Complex south of it, may be interpreted as a collision zone between the oceanic crust to the south and the cratonic Chortis block (Caribbean plate) to the north (de Boer, 1979). This collision must have occurred during the late Cretaceous–early Tertiary.

The continuation of the structures of the Santa Elena Peninsula to the west, under the Pacific shelf, is not evident and has not been located by marine geophysics. It is only slightly indicated by some bathymetric changes of the 200-m contour (Case and Holcombe, 1980, Map). This may be due to superposition of younger structures related to the Middle America Trench. The proposed location of the plate boundary implies a triple junction, Cocos–Caribbean–South America, to the west of Santa Elena (Fig. 3).

To the east of Santa Elena, the structures are covered by Quaternary volcanic rocks and cut by superposed NW–SE faults that are features of the Central America Quaternary volcanic chain.

If the trend of the Hess Escarpment in the Caribbean (Case and Holcomb, 1980, Map) is projected westward on land, it practically coincides with the Santa Elena trend. This, however, is an intra-Caribbean plate feature that only on land may be related to the boundary zone with the South America plate.

The boundary of the two plates on the Caribbean side of southeastern Costa Rica and Panama is represented by a submarine thrust fault (Case and Holcombe, 1980, Map) approximately 100 km from the Panamanian coast (Fig. 3). This fault marks the limit of a highly folded area of Cenozoic sediments more than 1 km thick called the North Panama Deformed Belt by Case and Holcombe (1980, Map), and the less deformed sediments of the Colombia Basin.

On the land, to the west, this thrust zone seems to extend into Costa Rica, but does not cross the isthmus to the Pacific side. Because of Tertiary and Quaternary volcanic rocks and extensive alluvial cover of the older rocks in northeastern Costa Rica, it is not possible to locate the plate boundary in that part.

The thrust zone curves southwestward toward the Gulf of Urabá in Colombia (Fig. 3) where it converges with a similar southeast-trending fault zone known as the South Caribbean Marginal Fault (Kellog and Bonini, 1982) which bounds the South Caribbean Deformed Belt (Case and Holcombe, 1980, Map). Based on the structural evidence on land in the Limón and Urabá basins, both of these submarine thrust fault zones are young, probably not older than Miocene.

3. *Cocos–Mid-America Landmass Boundary Zone*

The boundary zone of Mid-America with the Pacific, Cocos, and Nazca plates is well defined by subduction and related features along the Middle America Trench, the northernmost part of the Peru–Chile Trench in front of Colombia, and a sediment-filled trench that indicates the original continuity of the former, on the north part of the Panama Basin (Fig. 3). The Cocos plate segment bounds the Mid-America landmass that is formed by parts of three other plates, namely, North America (Oaxaca block), Caribbean (Chortis block), and South America (part of the Chorotega block).

(a) Middle America Trench and Continental Slope. The middle America trench extends parallel to the Pacific Coast of southern Mexico and Central America from Cabo Corrientes in the State of Jalisco to the Nicoya Peninsula in Costa Rica, for a distance of 2600 km. Its maximum depth is 6662 ± 10 m, and it is the shallowest of the Pacific Ocean trenches.

Its northwestern end terminates at the intersection of the East Pacific Rise and the Rivera Fracture Zone, which are also part of the boundary between the Cocos and Pacific plates (Plate Tectonic Map of the Circum-Pacific Region). To the southeast it is not well defined between the Nicoya Peninsula and the Panama Fracture Zone because of the topographically high area of the Cocos Ridge as result of consumption of this aseismic ridge (Vogt *et al.*, 1976; Nur and Ben-Avraham, 1981). The N–S trending Panama fracture zone forms the limit of the Cocos and Nazca plates.

The trench is formed by two major parts, somewhat different in their topography and structure, whose separation is indicated by its intersection with the Tehuantepec Ridge. The southeastern part of the trench, in contrast to the northwestern part, is separated from the mainland by a broader continental shelf, has a crust of greater thickness, contains a lesser volume of sediments and, parallel to it, there extends the Central American Quaternary Volcanic Chain. Whether the Trans-Mexican Volcanic Chain is tectonically related to the Middle America trench or to a major fracture zone independent from the trench is a subject on which there is no concensus.

The tectonically active zone that includes the trench, the Quaternary volcanic chain (in the southeast portion) and the intermediate area, is regarded as the boundary between the Cocos plate and the Mid-America land-mass. The tectonic features along it are superposed on the older structures of the Oaxaca, Maya, Chortis, and part of the Chorotega blocks. It is along this zone that most of the earthquakes in the region originate.

The Middle America Trench has been known in its general features since the publications of Heacock and Worzel (1955), Fisher (1961), Shor and Fisher (1961), and Fisher and Hess (1963). Later geophysical investigations (Shor, 1974; Couch and Woodcock, 1981) and drilling (DSDP Legs 66, 67, and 84) have provided detailed information and a better understanding of the stratigraphy and tectonics of the trench (summarized by Karig *et al.*, 1978; Lundberg and Moore, 1982; Watkins *et al.*, 1982; and von Heune and Auboin, 1982). The following is a summary of the most relevant conclusions regarding the tectonics of the region.

Couch and Woodcock (1981) conclude that "geophysical measurements over the eastern end of the Tehuantepec Ridge and adjacent continental margins of southern Mexico and northern Guatemala indicate that the ridge is a fracture zone and that it marks the boundary between two different subduction provinces." They also state that "the continental margin of southern Mexico, north of the Tehuantepec Ridge, shows a markedly different margin structure with a relatively small amount of continental accretion and a continental crustal block extending to within approximately 25 km of the trench axis."

From the drilling results of DSDP Leg 66, Watkins *et al.* (1982) found no direct evidence of tectonic erosion in the trench northwest of the Tehuantepec Ridge. They inferred the following history from this part:

> 23–20 Ma, reorganization of plate motion: relative movement of Pacific and North American plate changes from oblique normal (subduction) to parallel (sinistral strike-slip) with margin subsidence and marine transgression. 19–17 Ma—margin sinks rapidly to, or just below CCD, then begins slow rise. Sinistral strike-slip motion continues. 10 m.y.—plate motion changes from strike-slip to normal—oblique accretion begins and continues to present.

In the portion of the trench to the southeast of the Tehuantepec Ridge the widest part of the continental shelf along the entire region forms the Gulf of Tehuantepec. According to Sánchez-Barreda (1981) this gulf contains a forearc basin which resulted from the eastward motion of the Cocos plate and subduction beneath the North America plate. Oil exploration drilling and seismic data indicate that the basin existed since late Cretaceous time. Paleocene uplift produced a landward migration of the basin axis and the sediments were exposed to subaerial erosion during the Oligocene and middle Miocene. The basin was folded and faulted during the Miocene.

Sánchez-Barreda (1981) proposed that the triple junction of the Cocos, Caribbean, and North America plates developed into a system of conjugate strike-slip faults during the middle Cenozoic (Fig. 8C). Farther to the southeast, drilling in the trench slope and bottom along the San José Canyon (in front of the port of San José, Guatemala) during DSDP Legs 66 and 84, indicates a different tectonic situation. The purpose of drilling this convergent margin was to investigate the continuous accretion and imbrication indicated in the earlier geophysical studies. The results, however, showed a different condition and a strong lithological contrast on both sides of the subduction zone. On the continental slope an Upper Cretaceous–Quaternary sequence was found, while in the opposite side a Miocene–Quaternary sequence overlays the oceanic crust. Blocks of oceanic basalt and serpentinite were found within the sediments.

In reference to the tectonic history von Heune and Auboin (1982; p. 789–790), conclude that

> The tectonic history of the Guatemalan margin began with deposition of deep-water oceanic late Cretaceous section . . . (Campanian–Maastrichtian) . . . This Upper Campanian–Maastrichtian deposition may have followed a strong tectonic episode seen on the Santa Elena Peninsula involving slabs of oceanic crust and the Nicoya Complex.
>
> In the Paleocene the present area of the outer fore-arc basin and certainly the upper slope were uplifted to about sea level; the present fore-arc basin has existed since the Eocene. During erosion in the Oligocene the outer shelf and much of the present upper slope were emergent to form a sharp unconformity at the shelf edge. During the early Miocene [they] received sediment that was rapidly deposited . . .
>
> . . . There are no products of Neogene accretion at the front of the margin, although some may have accumulated and have then been subducted.

A positive free-air gravity anomaly extends along the outer continental shelf of Guatemala, and to the northwest curves abruptly landward in the Gulf of Tehuantepec. It is on trend with the positive anomaly of the Nicoya Peninsula in Costa Rica. For this reason it has been pointed out that "rocks genetically related to Cretaceous rocks of the Nicoya Complex extend northwestward along the continental shelf to the Gulf of Tehuantepec" (Couch and Woodcock, 1981).

If the interpretation of the North America–Caribbean and Caribbean–South America plate boundaries presented here is correct, the crustal feature represented by high gravity should be younger than the collision zones along those boundaries. If that is the case, it does not represent the extension of the Nicoya Complex.

The extreme southeastern part of the trench in front of the Chorotega Fault is not only less defined topographically but also less known. This part differs from the others in having a narrow shelf and in that an earlier fore-arc crops out intermittently (Nicoya to Azuero peninsulas).

The interpretation of the depth and dips of the Benioff zone permits establishing that the subduction slab consists of different segments, each one with different inclination and which probably move independently (Stoiber and Carr, 1971; Carr *et al.*, 1974). The limits between these segments are also in part evident along the Central American Quaternary volcanic chain.

The Middle America Trench, on the whole, is now one continuous tectonic feature that developed as result of northwestward motion of the Cocos plate and subduction along the convergent margin with the mid-America landmass. Differences along the trench are due to variations in its development and interaction with blocks of different crustal thickness. The subduction and accretion processes must have been different since Cretaceous time (Couch and Woodcock, 1981).

(b) Trans-Mexican and Central America Quaternary Volcanic Chains. In the region under discussion, major Quaternary volcanic chains present different regional situations relative to the Middle America Trench. The Trans-Mexican Volcanic Chain trend forms an acute angle with respect to the trench axis. In its westernmost part the chain is 250 km from the trench axis, and its eastern portion their separation varies from 300 to 350 km. The Central America Chain, in general, is parallel to the trench axis at an average distance of 125 km in the Chortis block part and somewhat less along the Chorotega block (Fig. 9).

As stated earlier, the Trans-Mexican Volcanic Chain has been considered to have developed along a major fracture zone. A different interpretation proposed by Couch and Woodcock (1981) and Nixon (1982) considers it related to the subduction process of the Cocos plate, the differences in distance noted due to shallow dip of the oceanic crust beneath the continental margin of southern Mexico and a steep dip beneath the margin of Guatemala.

A somewhat similar opinion has been expressed by Demant (1978), who considers that the different spatial relations between the Mexican and Central American volcanic chains and the trench are due to the eastward motion of the Caribbean plate in relation to the North America plate since the early Oligocene and to the change of rotation pole of the Cocos plate during the late Miocene. Therefore, the subduction under the Oaxaca block must have occurred under an already fractured continental crust.

From the description of Mooser *et al.* (1974), Demant (1978), and Negendank (1980), it is known that the Trans-Mexican Volcanic Chain is formed by at least five major units or clusters of different structural and petrological characteristics. Most of the rocks belong to the calc-alkaline suite, but to the east become more alkaline, and near Jalapa (Cofre de Perote; Fig. 9) they are alkaline (Pichler and Weyl, 1973; Pichler, 1976). The San Andrés Tuxtla volcanic cluster, at the edge of the Gulf of Mexico and separated from the Trans-Mexican Belt, is predominantly alkaline in composition and,

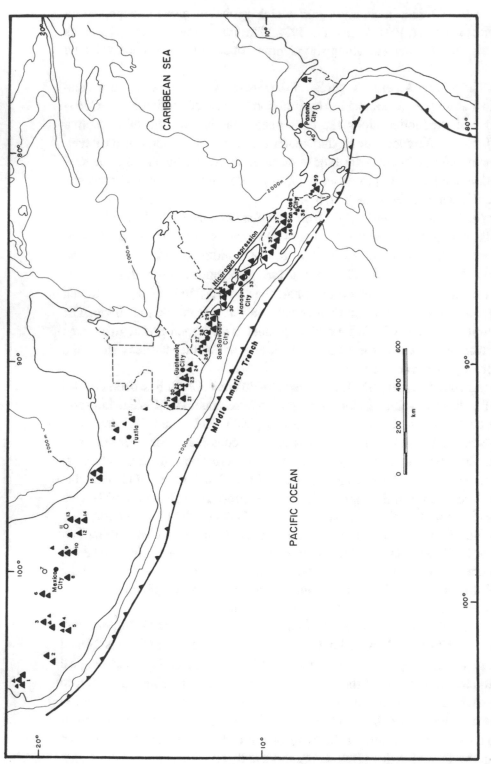

Fig. 9. Position of Quaternary volcanic chains relative to the Middle America Trench. *North boundary of Oaxaca block:* Trans-Mexican Volcanic Chain; (1) Ceboruco; (2) Colima; (3) Cerro Grande; (4) Tancítaro; (5) Buena Vista; (6) La Gavía; (7) Ixtacchuatl; (8) Toluca; (9) Ixtaccihuatl; (10) Popocatepetl; (11) Humeros Caldera; (12) La Malinche; (13) Cofre de Perote; (14) Orizaba. *Oaxaca–Maya block boundary:* (15) San Andrés Tuxtla. *Maya block:* (16) El Chichón; (17) Cerro San Cristóbal. *Chortís and Chortega blocks:* Central America Volcanic Chain; (18) Tacaná; (19) Tajumulco; (20) Santa María; (21) Santiaguito; (22) Atitlán Cauldrón; (23) Fuego; (24) Pacaya; (26) Santa Ana; (27) San Salvador; (29) San Miguel; (30) Cosigüina; (31) Momotombo; (32) Santiago; (33) Masaya Caldera; (34) Rincón de la Vieja; (35) Arenal; (36) Poás; (37) Irazú; (38) unnamed small centers; (39) Barú; (40) El Valle caldera. *Chocó block:* Río Pito cones.

as expressed previously, could be related to the Salina Cruz Fault Zone or to its intersection with the fracture zone along which the Trans-Mexican belt volcanoes were emplaced.

Two different structural settings are known along the belt, namely, N–S trending lines of large stratovolcanoes and NE–SE large groups of small cinder and lava cones (Demant, 1978). Evidently the volcanic centers in both cases are located along the axes of prevolcanic folds or along extension fractures perpendicular or oblique to the main trend of the belt.

Quaternary volcanic centers in the Maya block are few and far apart and do not clearly link the Trans-Mexican and Central American volcanic chains. The only active volcano in this part, El Chichón, is at a distance of 350 km from the Middle America Trench axis. Inactive eroded cones near San Cristóbal de las Casas and in the intervening area between them and El Chichón, follow a general WNW trend, nearly parallel to the trench axis. If these volcanic centers are related to the plate interaction process along the Pacific margin, they are an indication of a low-dipping subduction slab.

The Central America Quaternary volcanic chain forms a continuous range from the Tacaná volcano at the Mexico–Guatemala border southeastward to central Costa Rica, and includes several active centers (Fig. 9). Tectonically most of the volcanic clusters are associated with a large graben (Nicaragua Depression) or with other faults parallel to the Middle America Trench. Individual groups of cones are located along smaller N–S trending faults (Dengo *et al.*, 1970).

The Nicaragua Depression crosses the Central American isthmus in northcentral Costa Rica and becomes undefined where it enters the Limón backarc basin (Fig. 9). From central Costa Rica to the Panama Canal Zone Quaternary volcanic centers are fewer and discontinuous. Offsets along the alignement of the entire volcanic chain, according to Carr and Soiber (1977), are related to segmentation and dip changes of the Middle America Trench subduction zone.

The Central America Volcanic Chain is a common feature of the Chortis and Chorotega blocks, superposed on older structures and formed after the two blocks had been sutured. The Chortis block also presents the characteristic of volcanism associated with normal faulting due to an extention as a result of the eastward motion of the Caribbean plate.

The rocks of the Central America Volcanic Chain are of calc-alkaline composition (Pichler and Weyl, 1973). On the Chortis block part of the chain, one feature that is strikingly different from the Chorotega part is the abundance and large volumes of rhyodacitic pumiceous deposits and ignimbrites (Williams, 1960). Alkaline rocks occur in the Chortis block in some of the centers associated with extension fractures (Williams and McBirney, 1969) and in the Chorotega block in a small area in northeastern Costa Rica, on the Caribbean side of the Nicaragua Depression (Pichler, 1976).

4. *Nazca–South America Plate Boundary*

The northern Nazca plate, in the Panama Basin area, is separated from the Cocos plate by the Panama Fracture Zone on the west, and from the South America plate by a buried trench on the north and the Colombia Trench (northern end of the Peru–Chile Trench) on the east (Hey, 1977).

The Panama Fracture Zone, which is well defined by its high seismicity, consists of a series of N–S trending right-lateral transcurrent faults (Molnar and Sykes, 1969; van Andel *et al.*, 1971). Its landward extention can be identified only by a N–S fault on the west side of the Burica Peninsula between Costa Rica and Panama, which terminates against a major NE–SW trending fault that runs along the foothills of the Pacific Coast Ranges of Costa Rica (Río Esquinas Fault; Fig. 3). There is no evidence that any of the faults of the Panama Fracture Zone crosses the isthmus to the Caribbean. The northward motion of the Cocos plate on the west side of the fracture zone probalby results in right-lateral displacement along the Río Esquinas Fault and of faults along the Middle America Trench axis in front of Costa Rica (Fig. 3). In the interpretation of present plate boundaries this area represents the triple junction of the Cocos–Nazca–South America plates, and not of the Cocos–Nazca–Caribbean plates as interpreted by Londsdale and Kiltgord (1978).

The Panama Fracture Zone has been interpreted by Lowrie *et al.* (1979) as a faulted fossil spreading center. The northern part of the Panama Basin, along parallel 7° N, is now known to be formed by a buried trench with a sediment accumulation on the order of 2.5 km thick (Lowrie, 1978). This feature, which also forms the southern edge of the Gulf of Panama, had been previously interpreted as a possible E–W trending fault (Molnar and Sykes, 1969) and is represented as such in the Plate Tectonic Map of the Circum-Pacific Region (Addicott and Richards, 1981). According to Lowrie (1978), the thickness and undisturbed nature of the sediments in the trench are indications of slow subduction.

The tectonic history of the Panama Basin is complex and rather young since the basin seafloor is underlain by Miocene oceanic crust. Van Andel *et al.* (1971) have shown a westward migration of the Panama Fracture Zone and relate it to the formation of the Panama isthmus. This interpretation, however, does not fit with the on-land geology. The northern termination of the migrating fracture zone could have been limited by the buried trench which originally must have been an extension of the Middle America Trench. Londsdale and Kiltgord (1978) consider that the eastern part of the basin was formed by "highly asymmetric seafloor spreading along the Nazca and Cocos plates and that it contains a magnetic anomaly pattern of fossil spreading centers."

The Nazca-South America plate boundary on the east side of the Panama Basin is simpler in the sense that it is well defined by the seismically active Colombia Trench (Hayes, 1974; Meissner *et al.*, 1980). The continental margin along the trench is very narrow, and there are no Quaternary volcanoes associated with the Chocó block in front of the trench. The trench is somewhat over 4250 m deep in front of the Baudó Range (Londsdale and Kiltgord, 1978). Seismic activity along the trench not only is related to a subduction process but to right-lateral strike-slip motion (Meissner *et al.*, 1980).

IV. REGIONAL TECTONIC HISTORY: INTERPRETATIONS AND PROBLEMS

In the preceding descriptions the most important aspects of the tectonic history of each block were outlined, some comparisons were made, and some problems of correlating stratigraphic units and tectonic events were stated. The basement characteristics were emphasized because, in order to arrive at a clear understanding of the tectonic history of the region from late Precambrian to Holocene, the knowledge of the different types and ages of the basement is a key aspect.

Several tectonic models proposed for the region, or parts of it, deal with the origin and evolution of the Gulf of Mexico, the Caribbean, and the Panama Basin. It is not the purpose of this synthesis to review the numerous models that have been presented, but to try to integrate the geological facts in a way that will contribute to a full understanding of the tectonic history of the land area that links North and South America. Some of the major unsolved problems in the region have been posed recently by Ortega-Gutiérrez and Guerrero-García (1982) for the Mexico part and by Case and Dengo (1982) for the Central America and northwest Colombia area.

Without question, a Precambrian fit of North and South America is difficult, not only because of the problem of matching geological or tectonic units, but because there are extensive areas of possible Precambrian covered by younger rocks, or because part of the record was lost as result of tectonic transport and erosion.

The oldest rocks known in the region are those of the Oaxaca Complex, and the oldest orogeny thus far well established is the Grenville in the Oaxaca block. Kesler and Heath (1970) have proposed a possible continuation of the Grenville belt as far as the Chiapas Massif in the Maya block, but this is a problem that requires more research, in particular more accurate dating and more geological mapping in the Sierra Madre del Sur in the State of Chiapas. The structural discontinuity along the Salina Cruz fracture zone in the Isthmus of Tehuantepec suggests an originally different relative position of the Maya block with respect to the Oaxaca block. Several proposed predrift reconstructions (Carey, 1958; Dietz and Holden, 1970; Freeland and Dietz, 1971, 1972; Pindel and Dewey, 1982) place the Maya (or Yucatan) block in the area of the Gulf of Mexico. If this is the case the Oaxaca block should have been connected, or close, to South America during late Precambrian. The Grenville Tectonic Belt then must have been continuous at least for some distance in South America, bordering the older Guiana Shield either along its northern or its western side. At present the Precambrian rocks in northern South America that are geographically closer to the Oaxaca Complex are those of the Cordillera Central and the Sierra de Santa Marta in Colombia (Banks, 1975; Shagam, 1975; Bellizzia *et al.*, 1981 Campbell, 1974). It is not possible to suggest a correlation between those Precambrian areas, not only because of differences in ages but because, in the case of the Sierra de Santa Marta, it must have been subject to considerable change in its position due to transcurrent faulting.

The possible continuation of the Grenville belt along the northern part of the Guiana Shield is difficult to infer due to the thick cover of younger rocks and of the structural complexities of the northern border of South America. In relation to the possible extent of the Precambrian of northwestern South America at the subsurface, Shagam (1975; p. 409) states that "as early as 1300 m. y. ago there appears to have been thick continental crust extending as far west as the Cauca–Romeral–Sevilla fracture lineament." This lineament is considered today (Duque-Caro, 1980) as the boundary zone between continental crust to the east and younger oceanic crust to the west (Cordillera Occidental of Colombia). If the Grenville belt continued into South America it could well have extended along the edge of that major tectonic break or it could have been tectonically eroded as result of subduction along that zone during late Paleozoic or Mesozoic time. Although this is speculative, it is presented as one of the major outstanding problems in understanding the early tectonic history of the region, if the reconstructions are to be based on geological features and not on geometry.

Early Paleozoic correlations necessary to support plate reconstructions are also difficult because they involve metamorphic terranes. From the sedimentary record of the overlying rocks, both the Acatlán Complex and the Chuacús Group may be considered as pre-Mississippian; their correlation was suggested by Ortega Gutiérrez (1978). A correlation along the strike of these rocks would place the Oaxaca and Maya blocks essentially in the same relative position as they are today. Paleozoic reconstructions such as those presented by van der Voo *et al.* (1976), Keppie (1977), and Ross (1979) place the area covered by both blocks along the western margin of northwestern South America. Another interpretation by Pindell and Dewey (1982) locates the Maya block in the Gulf of Mexico and the Oaxaca and Chortis to the west of South America. The alternative of placing the Maya block in the Gulf of Mexico area, next to the Oaxaca block, during late Precambrian–early Paleozoic time implies a continental fit similar to the one suggested by Bullard *et al.*, (1965) and presents the problem of identifying the correlative terrane of the Chuacus Group in northern South America where early Paleozoic metamorphic rocks are known inthe Cordillera Oriental of Colombia and the Sierra de Perijá (Colombia–Venezuela).

The problem becomes larger when a correlation of the Chortis block basement rocks is attempted with other areas. Several models place this block in the Gulf of Mexico area (Carey, 1958; Freeland and Dietz, 1971, 1972; Helwig, 1975), although there are differences of opinion regarding its possible position. Other models (Malfait and Dinkelman, 1972; Gose and Swartz, 1977) suggest a position of the Chortis block on the Pacific side for Mesozoic reconstructions.

Some of the geological aspects previously described present indications that this latter situation is more possible and that it could have been the case in pre-Mesozoic time. To support a late Paleozoic position of the Chortis block to the southeast of the Oaxaca block requires a close comparison of the Acatlán Complex with the Chortis basement that is not possible at this time, but even so it should be considered as a possibility.

One important aspect of the early Paleozoic history of the region is the presence of Cambro–Ordovician and Mississippian rocks in the Tlaxiaco Basin of the Oaxaca block. This basin is considered as an isolated feature that during the Pennsylvanian and the Permian became integrated into a geosyncline (Dengo 1975). Considering again that the Oaxaca block was adjacent to northern South America, it is possible to find similar situations in northern Colombia, for instance in the Sierra de Macarena (Irving, 1971). The post-Ordovician or Silurian unconformity that was mentioned in the description of the Oaxaca block as a probably representing the Taconic orogeny has also been identified in Colombia.

The late Mississippian–middle Permian sedimentary record is well documented in the Chiapas–Guatemala Paleozoic basin in the Maya block as part of a geosyncline (López Ramos, 1969; Dengo, 1975). As was pointed out previously, this rock sequence terminates abruptly in Belize and cannot be traced along its trend into the Caribbean. A strong lithological similarity exists between this sequence and rocks of the same ages in the Mérida Andes of Venezuela where the Mucuchachi, Sabaneta, and Palmarito formations can almost be matched with the Santa Rosa Group and the Chóchal Formation of Guatemala. This possible correlation was already suggested by the author (Dengo, 1969) and has also been considered by Olson and Leyden (1971) who proposed a connection between both areas around an ancestral Caribbean Sea. It may be easier to consider again the Maya block to the east of the Oaxaca block and both joining the northern part of South America. This would require, however, a position for the Maya block different from its present one. Because the Chochal and Palmarito formations are fossiliferous, a faunal comparison of the middle Permian could support or discard this correlation. Such a comparison was made for the early Permian fusulinicean faunas by Ross (1979) who also presented a connection of the Mexico–Guatemala and the Andean Paleozoic geosynclines and placed them between the North and South American cratons. Similar models of Paleozoic connections between North and South America have been presented by Walper and Rowett (1972) and Helwig (1975).

Figure 10 shows a possible spatial relation between North and South America during the late Paleozoic, and the original position of the Maya and Chortis blocks that seems to fit better the known Precambrian and Paleozoic geology. The late Permian–Triassic history is more difficult to unravel or to present alternatives for because of the scanty, or absent, stratigraphic record not only in the region described but also in northern Mexico and northern South America. Some of the Mesozoic continental red-bed sequences such as the Todos Santos Formation in Mexico and Guatemala, the Girón in Colombia and La Quinta in Venezuela, originally considered to range from the late Permian–Jurassic, are now known to be of late Jurassic age.

In describing the Oaxaca and Maya blocks it has been indicated that the Appalachian orogeny can be well established in those areas and that it was followed by regional uplift during the late Permian–Triassic, a fact that is also evident in the Chortis block, regardless of its relative position with respect to the others. It has also been stated that the Maya block was displaced to the south and rotated counterclockwise with respect to the Oaxaca block at the end of the Permian or perhaps during the late Triassic–Early Jurassic (Fig. 10B).

Several worldwide drift models consider major continental breakup and separation during this long time span. Based on the differences of the Mesozoic stratigraphic sequences in Mexico–northern Central America and northern South America, it seems logical to consider a continental separation during the Triassic–early Jurassic (Fig. 10B). The lithological similarity and age of the Todos Santos, Girón, and La Quinta formations, pointed out earlier by the author (Dengo, 1969b) and by Olson and Leyden (1973), does not imply areal continuity of deposition but similarity of tectonic environments along rifts already produced as result of uplift and continental separation.

The alternative possibility for the Paleozoic position of the Chortis block next to the Pacific side of the Oaxaca block has been indicated as being more likely. The scant Triassic–early Jurassic stratigraphic record in the region is found in the Chapolopa and Consuelo formations in the Morelos–Guerrero Basin of the Oaxaca block and in the El Plan Formation, in Honduras, of the Chortis block. The ages and lithologies of these formations suggest a possible original continuity or, at least, similarity of tectonic and depositional environments. Furthermore, the overlying Cretaceous stratigraphic sequence is also similar.

Based on paleomagnetic results Gose and Swartz (1977) point out that the Chortis block was originally in the Pacific and not in the Gulf of Mexico area. This interpretation was refuted on other grounds by Wilson *et al.* (1978). Following the reasoning presented along this discussion the author considers the former a more acceptable one (Fig. 10A, 10B).

If the Chortis block was next to the Oaxaca block during the Paleozoic and during at least part of the Mesozoic, another problem is trying to establish the time of their separation. In the southwestern United States and in northcentral Mexico, Silver and Anderson (1974) have identified a megashear that extends from the Inyo Mountains in California to northeast of Hermosillo, Mexico, having an arcuate pattern, and have reported a left-lateral displacement of 700 to 800 km. Further work (Silver, personal communication) has established two parallel megashears with a total left-lateral displacement on the order of 1400 km. Numerous radiometric dates on both sides of the megashears indicate that their age is Jurassic. Is it possible that the present Pacific edge of the Oaxaca block follows another similar megashear zone that separated the Chortis block, and that a similar motion caused its migration southeastward? If so, this could also explain the present tectonic situation in the convergence zone between the Oaxaca block and the Cocos plate (Fig. 10C).

The separation of North and South America during the Triassic–early Jurassic must have resulted in the formation of oceanic crust in the intervening area. Remnants of Jurassic oceanic crust are known in the Nicoya Complex and in Costa Rica (Schmidt-Effing, 1979), and are also known in other places around the Caribbean. Likewise, the separation of the Chortis block from the Oaxaca block should have resulted in the formation of oceanic crust between the two blocks. In the southern part of the region the oceanic crust must have extended along the South America craton in the present position of the Cordillera Occidental of Colombia (Fig. 10).

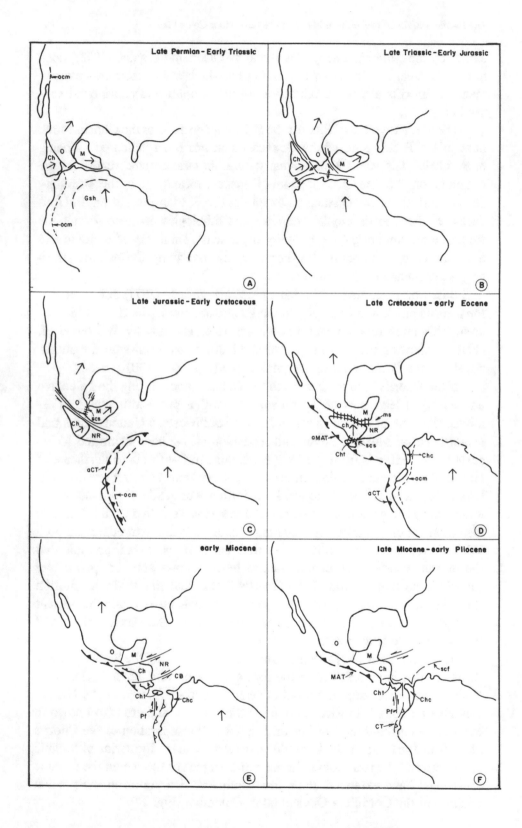

Several interpretations have been proposed for the Mesozoic (particularly the Cretaceous) and the Cenozoic history of the region, and are mostly related to the evolution of the Caribbean (Freeland and Dietz, 1972; Walper and Rowett, 1972; Malfait and Dinkelman, 1972; Donnelly, 1975; Ladd, 1976; Anderson and Schmidt, 1983; Tarling, 1981; Kellog and Bonini, 1982; and Sykes *et al.*, 1982). These present different opinions on certain aspects and are based on different types of evidence. Entirely different interpretations that question several of the plate-tectonic reconstructions have also been presented, particularly by Meyerhoff and Meyerhoff (1972).

The following simplified interpretations are based on the geology and tectonic history of the different parts previously described. Based on the analysis of South Atlantic magnetic anomalies Ladd (1976; p. 969) calculated the relative motion of South America with respect to North America for late Mesozoic and Cenozoic time and concluded that, "From Triassic to early Cretaceous time, South America moved southeast away from North America. From early Cretaceous to late Cretaceous, South America moved eastward with respect to North America. Southeastward motion of South America occurred again from late Cretaceous to early Tertiary time followed by northward motion from early to late Tertiary." This framework is useful in trying to understand the Mesozoic–Cenozoic history of the Central American isthmus.

The basic problems regarding the Mesozoic–Cenozoic plate history of the region are (1) the time of motion of the Chortis block to its present position and the evolution of the Motagua Suture Zone; (2) the time of accretion of oceanic crust of the Chorotega block to the cratonic Chortis block along the Santa Elena Suture Zone and the following displacements of this zone to its position; (3) evolution of the Chorotega–Chocó blocks; and (4) development of the Middle America–Colombia Trench in relation to the history of the land area.

If, as has been proposed, the Chortis block was separated from the Oaxaca block during Jurassic time and was displaced southeastward, it must have occupied a position in front of the Chiapas Massif of the Maya block, possibly during the early Cretaceous, with a strip of oceanic crust between both and oceanic crust also to the east of the cratonic part of the block (Fig. 10C). During the late Cretaceous–early Tertiary (Laramide orogeny) both

---

Fig. 10.   Sketches of relative positions of crustal blocks during their tectonic history: O, Oaxaca block; M, Maya block; CH, Chortis block; Cht, Chorotega block; Chc, Chocó block. Isolated arrows: present north directions. Gsh, Guiana Shield; MAT, Middle America Trench; aMAT, ancestral Middle America Trench; CT, Colombia Trench; aCT, ancestral Colombia Trench; NR, Nicaragua Rise; CB, Colombia Basin; SCS, Salina Cruz suture zone; ms, Motagua suture zone; ses, Santa Elena suture zone; PF, Panama fracture zone; scf, South Caribbean fault; ocm, old continental margin.

blocks collided by northeastward motion of the Chortis against the Maya, resulting in the development of the Motagua Suture Zone (Fig. 10D). The inception of the transcurrent faulting along this zone has been established by Burkart (1978, 1983) as Miocene, and for one of the faults (Polochic) he presents evidence of a left-lateral offset in the order of 130 km. However, to account for the present position of the Chortis block and move it from the position suggested during late Cretaceous time, the total displacement along the different faults of the suture zone (Polochic, Motagua, Chamele-cón) must have been much larger, perhaps in the order of the 800 km suggested by Hess and Maxwell (1953). If this interpretation is correct, the part of the Nicaragua Rise next to the land area of the Chortis block must have originated in the Pacific and migrated to its present Caribbean position (Fig. 10C, 10D). The extension-faulting within the Chortis block as well as earlier extensional fractures that acted as the locus for the widespread ignimbrite flows during the late Tertiary must have also taken place during the eastward migration of the block.

The Santa Elena Suture Zone also developed during the Laramide orogeny as collision of Pacific oceanic crust with the cratonic area of the Chortis block took place (Fig. 10D). Again, if the Chortis was in front of the Chiapas Massif during the late Cretaceous, the area represented today by the Santa Elena Peninsula must have been at least 500 km to the WNW, probably as part of a fore-arc in front of the Chortis block. The eastward motion of the Chortis block could have brought the Santa Elena Peninsula to its present position if there were no additional lateral displacements along the Santa Elena Suture Zone. There is evidence, however, of left-lateral strike-slip faulting in the Santa Elena Peninsula (Dengo, 1962a) but it is not possible to establish the amount of offset. It is possible then that the Chorotega block has been displaced eastward in relation to the Chortis block after the suture zone was formed, sometime during the Tertiary (Fig. 10E). In that case, and if the Hess Escarpment is tectonically continuous with Santa Elena, does this feature represent an original strike-slip fault now inactive? This poses a further problem in the long-standing controversy over whether the Caribbean oceanic crust is a displaced part of the Pacific or if it had an indigenous origin. From the preceding discussion it seems possible that part of the Nicaraguan Rise is of Pacific origin, while the crust of the Colombia Basin was originally formed within the Caribbean. As such, the Hess Escarpment–Santa Elena represent an old transcurrent fault zone with considerable left-lateral displacement (Fig. 10E).

The development of the Chorotega block and the Panama Spur part of the Chocó block has also been a subject of different interpretations (Dengo,

1962b, 1968; Lloyd, 1963; Weyl, 1966; Henningsen, 1966; de Boer, 1979; Case, 1974; Schmidt-Effing, 1979, 1980; Kuijpers, 1980; van Andel *et al.*, 1971; Lundberg, 1983). However, there are points of agreement in that it was formed as an intraoceanic feature, either an arc or an uplifted block. This arc developed from the late Mesozoic to the present, originally as a series of islands later joined as a continuous narrow isthmus but leaving Pacific–Caribbean marine connections until Pliocene time. Its history is related to that of northwestern Colombia.

The late Mesozoic separation of North and South America as well as the restriction of marine connections between the Atlantic and the Pacific until only narrow-sea areas were left during late Miocene–early Pliocene time is well documented by several studies of different paleobiological aspects (Woodring, 1966, 1978; Alencáster, 1978; Ferrusquía-Villafranca, 1978; Graham, 1978; Keigwin, 1978; Raven, 1979; Sloan, 1980; Webb, 1980; Marshall, 1980). The main points of disagreement are with respect to the original position of the islands and the mechanisms that produced the present geographical configuration. This last aspect is also closely related to the history of the Panama Basin and the interrelations of the Cocos–Nazca plates along the Panama Fracture Zone.

Sykes *et al.* (1982) have presented a quite different interpretation and consider that in the late Eocene (38 Ma) the southern Central America isthmus was already attached to northern Central America, separated by about 800 km from South America, and that the oceanic crust between the two areas was subducted under South America since that time. Because the interpretations presented here are based mostly on on-land geology, the author favors earlier ones of a closer relationship between southern Central America and northwestern Colombia, and that the northwestward motion of South America with respect to North America during Oligocene–Miocene time (Ladd, 1976, Fig. 2) reshaped an earlier island arc to conform the present shape of the isthmus.

The Middle America Trench and associated features, as described earlier, do not have a uniform history throughout their length (Fig. 10D–10F). The present trench probably contains parts of other ancestral trenches as indicated by Malfait and Dinkelman (1972). Recently Aubouin *et al.* (1982) concluded that

> Presently the Middle America Trench is a neotectonic feature. Its northern part, the Acapulco Trench, could have initiated by a large sinistral strike-slip motion giving to the Honduras Platform (*Chortis block*) an offset of about 800 km with regard to the Oaxaca Platform and cutting the Mexican structure off. Whereas its southern part, the Guatemala Trench, could have resulted from a long-lived subduction zone, perhaps related to the Franciscan subduction zone of North America.

The sediment-filled part of the original trench in front of Panama, indicates that the Middle America Trench and the Colombia Trench were continuous probably before the Miocene (Fig. 10D).

The latest major tectonic events in the whole region discussed are dated as post-middle Miocene, as the Cascadian orogeny in the northern part, and as the Andean orogeny in the Chorotega–Chocó blocks. Present seismic activity and volcanism associated with the Middle America Trench, high seismicity along some of the major strike-slip faults (Polochic–Motagua and Panama fracture zones), and fairly rapid uplift of parts of the region, are evidence that the Casacadian–Andean Orogeny is still in progress.

The regional history presented here is an effort to arrive at a congruent understanding of the complex geology of the area. It lacks integration of more geophysical aspects as well as radiometric and paleontological data to support or discard some of the long-range correlations proposed. In contrast with other interpretations proposed, even earlier ones by the author based on classical geological concepts or those supported mainly by geophysical interpretations or by paleogeographical reconstructions based on faunal distributions—it has the approach of a field geologist who believes that putting together isolated facts from a large region is not different from making a geological map of a large tropical jungle area, where only a few and weathered outcrops can be found. It is hoped that the author's view of this geological jungle was not obstructed by the trees.

ACKNOWLEDGMENTS

The part of this chapter dealing with Mexico was reviewed by Dr. Leon T. Silver and discussed with him and Dr. José Guerrero-García. The entire manuscript was critically reviewed by Drs. Thomas W. Donnelly, James E. Case, and Carlos A. Dengo. Some of the ideas presented were developed from discussions during a symposium on the tectonics of Mexico (November, 1979) and presented there orally at the invitation of Dr. Diego Córdoba.

The possible correlations with South America resulted from discussions with Dr. Alirio Bellizzia and field trips to study the Paleozoic of the Mérida Andes and Sierra de Perijá (Venezuela) with Drs. Raúl García-Jarpa, Oscar Odreman, and Gustavo Canelón. The author also had the benefit of attending the scientific panel for discussion of the results of DSDP Leg 84 at the invitation of Dr. Yves Lancelot. To these colleagues the author is greatly indebted.

Besides many geologists who have been with the author in several parts of the region, whose names would be lengthy to list, one person deserves a special gratitude, my wife Norma, whose encouragement and patience for being alone for long and frequent periods, has made possible the field activities of the author for over thirty years.

REFERENCES

Alencáster, G., 1978, Distribución de las faunas marinas del sur del México y Norte de América Central durante el Cretácico in: *Conexiones Terrestres entre Norte y Sudamérica, Univ. Nac. Auton. México, Inst. Geol., Bol. V.* 101 p. 47–63.

Alvarez Jr., M., 1958, Tectónica profunda de México. *Asoc. Mexicana Geol. Petrol. Bol.*, v. 10, p. 163–182.

Anderson, T. A. and Schmidt, V. A., 1983, The evolution of Middle America and the Gulf of Mexico—Caribbean Sea during Mesozoic time *Geol. Soc. Am. Bull.* v. 9, p. 941–966.

Arden, Jr., D. D., 1975, Geology of Jamaica and the Nicaragua Rise in: *The Ocean Basins and Margins*, v. 3, Nairn, A. E. M., and Stehli, F. G., eds., New York: Plenum Press, p. 617–661.

Auboin, J., Azéma, J., Carfantan, C. Demant, A., Rangin, C., Tardy, M., and Tournon, J., 1982, *Initial Reports of the Deep Sea Drilling Project*, v. 67, Washington, D.C.: U.S. Government Printing Office, p. 747–755.

Azéma, J. and Tournon, J., 1980, La Péninsule de Santa Elena, Costa Rica: un massif ultrabasique charrié en marge pacifique de l'Amerique Central *C.R. Acad. Sci. Paris, Ser. D.* v. 290, p. 9–12.

Banks, P. O., 1975, Basement rocks bordering the Gulf of Mexico and the Caribbean Sea, *The Ocean Basins and Margins*, v. 3, Nairn, A. E. M., and Stehli, F. G. eds., New York: Plenum, p. 181–199.

Barrero, C., 1979, Geology of the Central Western Cordillera, west of Buga and Roldanillo, Colombia, *Pub. Geol. Esp. INGEOMINAS* (Bogotá), no. 4, p. 1–75.

Bass, M. N. and Zartman, R. E., 1969, The basement of the Yucatan Peninsula (abstract), *Trans. Am. Geophys. Union*, v. 50, p. 313.

Bateson, J. H., 1972, New interpretation of the geology of the Maya Mountains, British Honduras, *Am. Assoc. Pet. Geol. Bull.*, v. 56, p. 956–963.

Bellizzia, A., Pimentel de Bellizzia, N., and Muñoz, M. T., 1981, Geology and tectonics of northern South America, *Dir. Geologia, Caracas, Bol. Geol. Pub. Esp.* v. 9, 140 p.

Benavides, L., 1956, Notas sobre la geología petrolera de México, *Twentieth Congr. Geol. Intern., México, Symposium Yacimientos Petróleo y Gas*, v. 3, p. 351–362.

Bertrand, J., Delaloye, M., Fontignie, D., and Vuagnat, M., 1978, Ages K/Ar sur divers ophiolites et roches associeés de la Cordillére Central du Guatemala, *Schweiz. Min. Petrogr. Mitteilungen*, v. 58, p. 130–136.

Bowin, C., 1976, Caribbean gravity field and plate tectonics, *Geol. Soc. Am. Spec. Pap.*, no. 169, 79 p.

Bullard, E. C., Everett, J. and Smith, A. G., 1965, A Symposium on continental drift, *Royal Soc. London Philos. Trans., Ser. A.*, v. 258, p. 41–51.

Burkart, B., 1978, Offset across the Polochic fault of Guatemala and Chiapas, Mexico, *Geology*, v. 6, p. 328–332.

Burkart, B., 1980, Plate-tectonic significance of the Polochic fault and reconstruction of northern Central America prior to measured displacement, *Geol. Soc. Am. Abstr. Progr.*, v. 10, p. 395.

Burkart, B., 1983, Neogene North American—Caribbean plate boundary across northern Central America: offset along the Polochic fault, *Tectonophysics* (in press).

Butterlin, J., 1977, *Géologie Structurale de la région des Caraibes (Mexique–Amérique Centrale–Antilles–Cordillére Caraibe*, Paris: Masson, 259 p.

Campbell, C. J., 1974, Colombian Andes, in: *Mesozoic—Cenozoic Orogenic Belts, Geol. Soc., London, Spec. Publ.*, no. 4, p. 705–724.

Carey, S. W., 1958, The tectonic approach to continental drift, in: *Continental Drift—A Symposium*, Hobart, Tasmania: University of Tasmania, p. 177–355.

Carfantan, J. C., 1977a, El Prolongamiento del sistema de fallas Polochic–Motagua en el S. E. de México, *Univ. Nac. Autón., México, Inst. Geol. Rev.*, v. 1, p. 138–141.

Carfantan, J. C., 1977b, La Cobijadura de Motozintla, un paleoarco volcánico en Chiapas, *Univ. Nac. Auton. México, Inst. Geol. Rev.*, v. 1, p. 138–141.

Carfantan, J. E., 1979, Evolución estructural del sureste de México: Paleografía e historia tectónica de las zonas internas mesozoicas (Resumen), *Univ. Nac. Auton. Méx. Inst. Geol.; Symposium sobre Evolución Tectónica de México*, p. 10–11.

Carr, M. J., and Stoiber, R. E., 1977, Geologic setting of some destructive earthquakes in Central America, *Geol. Soc. Am. Bull.*, v. 88, p. 151–156.

Carr, M. J., Stoiber, R. E. and Drake, C. L., 1974, The segmented nature of some continental margins, *The Geology of Continental Margins*, Burk, C. A., and Drake, C. L. eds, New York: Springer-Verlag, p. 105–114.

Case, J. E., 1974, Oceanic crust forms basement of Eastern Panama, *Geol. Soc. Am. Bull.*, v. 85, p. 645–652.

Case, J. E., 1981, Crustal setting of mafic and ultramafic rocks and associated ore deposits of the Caribbean Region, *U.S. Geol. Sur.*, *Open-File Rep.* v. 95, 80–304.

Case, J. E., and MacDonald, W. D., 1973, Regional gravity anomalies and crustal structures in northern Colombia, *Geol. Soc. Am. Bull.*, v. 84, p. 2905.

Case, J. E., Holcombe, T. L. and Martin, R. G., 1983, Geologic provinces in the Caribbean in: *Caribbean–South America Plate, Geol. Soc. Am. Mem.* 162 Bonini, W., and Hargraves, R. eds. p. 1–30.

Case, J. E., Barnes, J., González, G. P. Q., and Viña, I. H., 1973, Trans-Andean geophysical profile, southern Colombia, *Geol. Soc. Am. Bull.*, v. 76, p. 567–590.

Case, J. E., Durán, L. G., López, A. and Moore, W. R., 1971, Tectonic investigations in western Colombia and eastern Panama, *Geol. Soc. Am. Bull.*, v. 82, p. 2685–2712.

Clemons, R. E., 1966, *Geology of the Chiquimula quadrangle, southeastern Guatemala*, Ph.D. Dissertation, University of Texas, Austin, 156 p.

Couch, R., and Woodcock, S., 1981, Gravity and structure of continental margins of southwestern Mexico and northwestern Guatemala, *Geoph. Res.*, v. 86, p. 1829–1840.

Dalimeyer, R. D., 1982, Pre-Mesozoic basement of the southeastern Gulf of Mexico, *Geol. Soc. Am. Abstr. Progr.* v. 14, p. 471.

de Boer, J., 1979, The outer arc of the Costa Rican orogen (oceanic basement complexes of the Nicoya and Santa Elena peninsulas), *Tectonophysics*, v. 56, p. 221–259.

de Cserna, Z., 1960, Orogenesis in time and space in Mexico, *Geol. Rudsch.* v. 50, p. 595–604.

de Cserna, Z., 1971, Precambrian sedimentation, tectonics and magmatism in Mexico, *Geol. Rundsch.*, v. 60, p. 1488–1513.

de Cserna, Z., 1975, Mexico, *The Encyclopedia of World Regional Geology*, Fairbridge, R. W., ed., Stroudsburg, PA: Dowden, Hutchinson, and Ross, p. 348–360.

del Giudice, D., 1978, Características geológicas de la República de Panamá, in: *Conexiones Terrestres entre Norte y Sudamérica, Univ. Nac. Auton. México, Inst. Geol. Bol.* v. 100, p. 4–25.

Demant, A., 1978, Características del Eje Neovolcánico Transmexicano y sus problemas de interpretación, *Univ. Nac. Auton. México, Inst. Geol., Rev.* v. 2, p. 172–187.

Dengo, C. A., 1982, Structural analysis of the Polochic fault zone in western Guatemala: Ph.D. Dissertation, Texas A & M University, 293 p.

Dengo, G., 1962a, Estudio geológico de la región de Guanacaste, Costa Rica, *Inst. Geogr. Costa Rica*, 67 p.

Dengo, G., 1962b, Tectonic-igneous sequence in Costa Rica, *Buddington Vol., Geol. Soc. Am.*, p. 133–161.

Dengo, G., 1968, *Estructura Geológica, Historia Tectónica y Morfología de América Central: Centro Regional de Ayuda Técnica México*, Buenos Aires, 52 p.

Dengo, G., 1972, Review of Caribbean serpentinites and their tectonic implications, *Geol. Soc. Am. Mem.*, v. 132, p. 303–312.

Dengo, G., 1975, Paleozoic and Mesozoic tectonic belts in Mexico and Central America in: *The Ocean Basins and Margins*, v. 3, Nairn, A. E. M., and Stehli, F. G., eds., New York: Plenum, p. 283–323.

Dengo, G., Bohnenberger, O. and Bonis, S., 1970, Tectonics and volcanism along the Pacific marginal zone of Central America, *Geol. Rundsch.*, v. 59, p. 1215–1232.

Dietz, R. S., and Holden, J. C., 1970, Reconstruction of Pangea: break-up and dispersion of continents, Permian to present, *J. Geophys. Res.*, v. 75, p. 4939–4956.

Donnelly, T. W., Crane, D. and Burkart, B., 1968, Geologic history of the landward extention of the Bartlett Trough, *Trans. 4th. Caribbean Geol. Conf., Trinidad and Tobago*, p. 225–228.

Donelly, W. T., 1975, The geological evolution of the Caribbean and Gulf of Mexico—some critical problems and areas in: *The Ocean Basins and Margins*, v. 3, Nairn, A. E. M., and Stehli, F. G., eds., New York: Plenum, p. 663–689.

Duque-Caro, H., 1980, Geotectónica y evolución de la región noroccidental de Colombia, *Inst. Nac. Inv. Geol. Min., Colombia, Bol. Geol.* v. XIX, p. 4–37.

Durham, J. W., Applegate, S. H. and Espinoza-Arrubarrena, L., 1981, Onshore marine Cenozoic along southwest Pacific coast of Mexico, *Geol. Soc. Am. Bull.* v. 92, p. 384–394.

Erdlac, Jr., R. J., and Anderson, T. H., 1982, The Chixoy–Polochic fault and its associated fractures in western Guatemala, *Geol. Soc., Am. Bull.*, v. 93, p. 57–67.

Fakundiniy, R. H., and Everett, J. R., 1976, Re-examination of Mesozoic stratigraphy of El Rosario and Comayagua quadrangles, Central Honduras, *Pub. Geol. ICAITI*, v. 5, p. 5–17.

Ferrusguía-Villafranca, I., 1978, Distribution of Cenozoic vertebrate faunas in Middle America and problems of migration between North and South America, in: *Conexiones Terrestres entre Norte y Sur America, Univ. Nac. Autón., México, Inst. Geol. Bol.* v. 101, p. 193–321.

Fischer, R., 1980, Recent tectonic movements of the Costa Rican Pacific coast, *Tectonophysics*, v. 57, p. 25–33.

Fischer, R. L., 1961, Middle America Trench: topography and structure, *Geol. Soc. Am. Bull.*, v. 72, p. 703–720.

Fisher, R. L., and Hess, H. H., 1963, Trenches, in: *The Sea*, Hill, M. N. ed., New York: Interscience, v. 3, p. 411–436.

Finch, R. C., 1972, *Geology of the San Pedro Zacapa Quadrangle, Honduras*, Ph.D. Dissertation, University of Texas, Austin, 238 p.

Freeland, G. L., and Dietz, R. S., 1971, Plate-tectonic evolution of the Caribbean–Gulf of Mexico region, *Nature*, v. 232, p. 20–23.

Freeland, G., and Dietz, R., 1972, Plate-tectonic evolution of the Caribbean–Gulf of Mexico region, *6th Caribbean Geol. Conf., Mem.*, Margarita, Venezuela, p. 259–264.

Galli-Olivier, C., 1979, Ophiolite and island arc volcanism in Costa Rica, *Geol. Soc. Am. Bull.*, v. 90, p. 444–452.

Gomberg, D. N., Banks, P. D. and McBirney, A. R., 1968, Guatemala: preliminary zircon ages from the central Cordillera, *Science*, v. 162, p. 121–122.

Goossens, P. J., Rose, Jr., W. I. and Flores, D., 1977, Geochemistry of tholeiites of the basic igneous complex of northwestern South America, *Geol. Soc. Am. Bull.*, v. 88, p. 1711–1720.

Gose, W. A. and Swartz, D. K., 1977, Paleomagnetic results from Cretaceous sediments in Honduras: tectonic implications, *Geology*, v. 5; p. 505–508.

Graham, A., 1978, Distribution and migration of Cenozoic floras in Mesoamerica, in: *Conexiones Terrestres entre Norte y Sudamérica, Univ.Nac. Auton., Mexico, Inst. Geol. Bo.* v. 101, p. 166–180.

Gutiérrez-Coutiño, R., and Carfantan, J. C., 1979, Evolución estructural del sureste de México, zona externa (Resumen), *Symposium Sobre Evolución Tectónica de México, Univ. Nac. Auton., México, Inst. Geol.*, p. 20–21.

Guzmán, E. J., and de Cserna, Z., 1963, Tectonic history of Mexico: backbone of the Americas, *Am. Assoc. Pet. Geol. Mem.* v. 2., p. 113–129.

Hayes, D. E., 1974, Continental margin of western South America, in: *The Geology of Continental Margins*, Burk, C. A., and Drake, C. L., eds., New York: Springer–Verlag, p. 581–590.

Heacock, J. G., and Worzel, J. L., 1955, Submarine topography west of Mexico and Central America, *Geol. Soc. Am. Bull.*, v. 66, p. 773–776.

Helwig, J., 1975, Tectonic evolution of the southern continental margin of North America from a Paleozoic perspective in: *The Ocean Basins and Margins*, v. 3, Nairn, A. E. M., and Stehli, F. G., eds., New York: Plenum, p. 243–255.

Henningsen, D., 1966, Die Pazifische Küstenkordillere Costa Ricas und ihre Stellung innerhalb des súdzentral-amerikanischen Gebirges, *Geotekt. Forsch.*, v. 23, p. 3–66.

Henningsen, D., and Weyl, R., 1967, Ozeanische Kruste in Nicoya-Komplex von Costa Rica (Mittelamerika), *Geol. Rundsch.*, v. 57, p. 33–47.

Hernández García, R., 1973, Paleogeografía del Paleozoico de Chiapas, México, *Asoc. Mex. Geo. Petrol., Bol.*, v. XXV, p. 77–134.

Hess, H. H., and Maxwell, J. C., 1953, Caribbean research project, *Geol. Soc. Am. Bull.*, v. 64, p. 1–6.

Hey, R., 1977, Tectonic evolution of the Cocos–Nazca spreading center, *Geol. Soc. Am. Bul.*, v. 88, p. 1404–1420.

Horne, G. S., Clark, G. S. and Pushkar, P., 1976, Pre-Cretaceous rocks in Northwestern Honduras: basement terrane in Sierra de Omoa, *Am. Assoc. Pet. Geol. Bull.*, v. 60, p. 566–583.

Irving, E. M., 1971, La evolución estructural de los Andes más septentrionales de Colombia, *Inst. Nac. Invest. Geol. Min., Colombia, Bol. Geol.* v. XIX, p. 89.

Julivert, M., 1968, Colombie, *Lexique Stratigraphique International, V. V., Amérique Latine*, Paris: Centre National de la Recherche Scientifique, 648 p.

Karig, D. E., Cardwell, R. K., Moore, G. F. and Moore, D. G., 1978, Late Cenozoic subduction and continental margin along the northern Middle America Trench, *Geol. Soc. Am. Bull.* v. 89, p. 265–276.

Keigwin, Jr., L. D., 1978, Pliocene closing of the Isthmus of Panama, based on biostratigraphic evidence from nearby Pacific Ocean and Caribbean Sea cores, *Geology*, v. 6, p. 630–634.

Kellogg, J. N., and Bonini, W. E., 1982, Subduction of the Caribbean plate and basement uplifts in the overriding South America plate, *Tectonics*, v. 1, p. 251–276.

Keppie, J. D., 1977, Plate-tectonic interpretation of Paleozoic world maps, *Nova Scotia Dept. Mines, Paper* 77–3, 30 p.

Kesler, S. E., 1971, Nature of ancestral orogenic zone in nuclear Central America, *Am. Assoc. Pet. Geol. Bull.*, v. 55, p. 2116–2129.

Kesler, S. E., and Heath, S. A., 1970, Structural trends in southernmost North American Precambrian, Oaxaca, Mexico, *Geol. Soc. Am. Bull.*, v. 81, p. 2471–2476.

Kesler, S. E., Josey, W. L. and Collins, E. M., 1970, Basement rocks of western nuclear Central America: the western Chuacús Group, Guatemala, *Geol. Soc. Am.*, v. 81, p. 3307–3322.

Kesler, S. E., Bateson, H. J., Josey, W. L., Cramer, G. H. and Simmons, W. A., 1971, Mesoscopic structural homogeneity of Maya Series and Macal Series, Mountain Pine Ridge, British Honduras, *Am. Assoc. Pet. Geol. Bull.*, v. 55, p. 97–103.

Kuijpers, E. P., 1980, The geologic history of the Nicoya ophiolite complex, Costa Rica, and its geotectonic significance, *Tectonophysics*, v. 68, p. 233–255.

Ladd, J. W., 1976, Relative motion of South America with respect to North America and Caribbean tectonics, *Geol. Soc. Am. Bull.*, v. 87, p. 969–976.

Lawrence, D. P., 1976, Tectonic implications of the geochemistry and petrology of El Tambor Formation: probable oceanic crust in Central Guatemala, *Geol. Soc. Am., Abstr. Progr.*, v. 8, p. 973–974.

Lloyd, J. L., 1963, Tectonic history of the south Central American orogen, Backbone of the Americas, *Am. Assoc. Pet. Geol. Mem.*, v. 2, p. 88–100.

Londsdale, P., and Kiltgord, K. D., 1978, Structure and tectonic history of the eastern Panama Basin, *Geol. Soc. Am. Bull.*, v. 89, 981–999.

López Ramos, E., 1969, Marine Paleozoic rocks of Mexico, *Am. Assoc. Pet. Geol. Bull.*, v. 53, no. 12, p. 2399–2417.

López Ramos, E., 1973, Estudio Geológico de la Península de Yucatán, *Asoc. Mex. Geol. Petrol., Bol.*, v. XXV, p. 23–76.

López Ramos, E., 1979, *Geología de México*, México, D. F., t III, p. 445.

Lowrie, A., 1978, Buried trench south of the Gulf of Panama, *Geology*, v. 6, p. 434–436.

Lowrie, A., Aitken, T., Grim, P. and McRaney, L., 1979, Fossil spreading center and faults within the Panama fracture zone, *Mar. Geoph. Res.*, v. 4, p. 153–166.

Lundberg, N., 1983, Development of forearcs of intraoceanic subduction zones, *Tectonics*, v. 2, p. 51–61.

Lundberg, N., and Moore, J. C., 1982, Structural features of the Middle America Trench slope off southern Mexico, Deep Sea drilling Project Leg 66 in: *Initial Reports of the Deep Sea Drilling Project*, v. 66, Washington, D.C.: U.S. Government Printing Office, p. 793–805.

Malfait, B. T., and Dinkelman, M. G., 1972, Circum-Caribbean tectonic and igneous activity and the evolution of the Caribbean plate, *Geol. Soc. Am. Bull.*, v. 83, p. 251–272.

Marshall, L. G., 1980, South American mammalian chronology and inter-American interchange, *Geol. Soc. Am., Abst. Progr.* v. 12, p. 476.

Matumoto, T., Ohtake, M., Latham, G. and Umaña, J., 1977, Crustal structure in southern Central America, *Seis. Soc. Am. Bull.*, v. 67, p. 121–134.

McBirney, A. R., 1963, Geology of a part of the central Guatemala Cordillera, *Univ. California Publ. Geol. Sci.*, v. 38, p. 172–242.

McBirney, A. R., and Bass, M. N., 1967, Structural relations of pre-Mesozoic rocks in Central America, *Am. Assoc. Pet. Geol., Mem.* 11, p. 229–243.

Meissner, R., Flüh, E. R. and Muckelmann, R., 1980, Sobre la estructura de los Andes Septentrionales-Resultados de investigaciones geofísicas in: *Nuevos Resultados de la Investigación Geocientífica en Latinoamérica*, Bonn: Deutsche Forschungsgemeinschaft, p. 79–90.

Meissner, R. O., Flüh, E. R., Stibane, F. and Berg, E., 1976, Dynamics of the active plate boundary in southwest Colombia according to recent geophysical measurements, *Tectonophysics*, v. 35, p. 115–136.

Meyerhoff, A. A., and Meyerhoff, H. A., 1972, Continental drift; IV: The Caribbean "plate", *J. Geol.*, v. 80, p. 34–40.

Mills, R. A., Hugh, K. E., Feray, D. E. and Swolfs, H. C., 1967, Mesozoic stratigraphy of Honduras: *Am. Assoc. Pet. Geol. Bull.*, v. 51, p. 1711–1786.

Molnar, P., and Sykes, L. R., 1969, Tectonics of the Caribbean and Middle America regions from focal mechanisms and seismicity, *Geol. Soc. Am. Bull.* v. 80, p. 1639–1684.

Moore, G. W., and Del Castillo, L., 1974, Tectonic evolution of the southern Gulf of Mexico: *Geol. Soc. Am. Bull.*, v. 85, p. 607–618.

Muehlberger, W. R., and Ritchie, A. W., 1975, Caribbean–American plate boundary in Guatemala and southern Mexico as seen on Skylab IV orbital photography, *Geology*, v. 3, p. 232–235.

Mooser, F., Nairn, A. E. M. and Negendank, J. F. W., 1974, Paleomagnetic investigations of the Tertiary and Quaternary igneous rocks: VIII: a paleomagnetic and petrologic study of volcanics of the Valley of Mexico, *Geol. Rundsch.*, v. 63, p. 451–483.

Negendank, J., 1980, Investigaciones en el Cinturón Volcánico Trans-Mexicano in: *Resultados de la Investigación Geocientífica en Latinoamérica*, Bonn: Deutsche Forschungsgemeinschaft, p. 19–20.

Nixon, G. T., 1982, The relationship between Quaternary volcanism in central Mexico and the seismicity and structure of subducted ocean lithosphere, *Geol. Soc. Am. Bull.*, v. 93, p. 514–523.

Nur, A., and Ben-Avraham, Z., 1981, Volcanic gaps and consuption of a seismic ridges in South America, *Geol. Soc. Am. Mem.* 154, p. 729–740.

Olson, W. S., and Leyden, R. J., 1971, North Atlantic rifting in relation to Permian–Triassic salt deposition, *Int. Permian–Triassic. Conf.* University of Calgary, Canada, p. 720–732.

Ortega-Gutiérrez, F., 1978, Estratigrafía del Complejo de Acatlán en la Mixteca Baja, Estados de Puebla y Oaxaca, *Univ. Nac. Auton., México, Inst. Geol. Rev.*, v. 2, p. 112–131.

Ortega-Gutiérrez, F., 1979, La evolución tectónica Premisisípica del sur de México, *Symposium, Evolución Tectónica de México*, Univ. Nac. Auton., México, Inst. Geol. p. 27–28.

Ortega Gutiérrez, F., 1980, Algunas rocas miloníticas de México y su significado tectónico, *Soc. Geol. México, Resúmenes*, V. Convención Geol. Nac., November 9–16, 1980, p. 99–100.

Ortega-Gutiérrez, F., and Anderson, T. H., 1977, Lithologies and geochronology of the Precambrian craton of southern Mexico, *Geol. Soc. Am., Abstr. Progr.*, v. 9, no. 7, p. 1121–1122.

Pantoja Alor, J., and Robinson, R., 1967, Paleozoic sedimentary rocks in Oaxaca, Mexico, *Science*, v. 100, p. 47–58.

Pantoja Alor, J., Rincón Orta, C., Fries, Jr., C., Silver, L. T. and Solorio-Munguía, J., 1971, Contribución a la geocronología del Estado de Chiapas, *Univ. Nac. Auton., México, Inst. Geol. Bol.*, v. 100, p. 47–58.

Pichler, H., 1976, Quaternary alkaline volcanic rocks in eastern Mexico and Central America, *Múnster Forsch. Geol. Paleontol.*, v. 38, p. 159–178.

Pichler, H., and Weyl, R., 1973, Petrochemical aspects of Central American magmatism, *Geol. Rundsch.* v. 62, p. 357–396.

Pichler, H., Stibane, F. R., und Weyl, R., 1974, Basischer Magmatismus und krustenbau im südlichen Mittelamerika, Kolumbien und Ecuador, *Neues Jahr. Geol. Paläontol. Monatsh.*, v. 1974, p. 102–126.

Pindell, J., and Dewey, J. F., 1982, Permo–Triassic reconstruction of western Pangea and the evolution of the Gulf of Mexico–Caribbean region, *Tectonics*, v. 1, p. 179–211.

Raven, P. H., 1979, Plate tectonics and southern hemisphere biogeography in: *Tropical Botany*, Larsen, K., and Holm-Nielsen, L. B. eds., London: Academic Press, p. 3–24.

Rodríguez, R. T., 1970, Geología metamórfica del área de Acatlán, Estado de Puebla, *Soc. Geol. Mex.*, Libro Guía, Excursión México–Oaxaca.

Rosenfeld, J. H., 1980, The Santa Cruz ophiolite, Guatemala, Central America, *Trans. 9th Caribbean Geol. Conf.*, Dominican Republic, p. 451–452.

Ross, C. A., 1979, Late Paleozoic collision of North and South America, *Geology*, v. 7, p. 41–44.

Salinas, J. C., and Ramírez, J., 1980, Las capas rojas del Cretácico en la región de la Montaña de Guerrero, México, *Soc. Geol. Mex., Resúmenes, Conv. Geol. Nac.* November 9–16, 1980, p. 112–113.

Sánchez-Barreda, L. A., 1981, *Geologic evolution of the continental margin of the Gulf of Tehuantepec in southwestern Mexico*, Ph.D. Dissertation, University of Texas, Austin, 191 p.

Sánchez Montes de Oca, R., 1979, Geología Petrolera de la Sierra de Chiapas, *Asociación Mex. Geol. Petrol., Bol.*, v. XXXI, p. 67–97.

Sapper, K., 1937, *Mittelamerika: Handbuch der regionalen Geologie*, Heidelberg: Steinman und Wilckens, 160 p.

Schmidt-Effing, R., 1979, Alter und Genese des Nicoya-Komplexes, einer ozeanischen Paläokruste (Oberjura bis Eozän) im südlichen Zentralamerika, *Geol. Rundsch.*, v. 68, p. 457–494.

Schmidt-Effing, R., 1980, El origen del istmo centroamericano como vínculo de dos continentes, in: *Nuevos Resultados de la Investigación Geocientífica en Latinoamérica*, Bonn: Deutsche Forschungsgemeinschaft, p. 21–29.

Schwartz, D. P., 1972, Petrology and structural geology along the Motagua fault zone, Guatemala (abstract), *Trans. 7th Caribbean Geol. Conf. Guadaloupe*, p. 299.

Schwartz, D. P., Cluff, L. S. and Donnelly, T. W., 1979, Quaternary faulting along the Caribbean–North America plate boundary in Central America, *Tectonophysics*, v. 52, p. 431–445.

Schuchert, C., 1935, *Historical Geology of the Antillean–Caribbean Region*, New York: Wiley, 811 p.

Shagam, R., 1975, The northern termination of the Andes, in: *The Ocean Basins and Margins*, v. 3, Nairn, A. E. M., and Stehli, F. G., eds., New York: Plenum, p. 325–420.

Shor Jr., G. G., 1974, Continental margin of Middle America, in: *The Geology of Continental Margins*, Burk, C. A., and Drak, C. L., eds., New York: Springer-Verlag, p. 599–602.

Shor, G. C., and Fisher, R. L., 1961, Middle America Trench: seismic refraction studies, *Geol. Soc. Am. Bull.*, v. 72, p. 721–730.

Silver, L. T., and Anderson, T. H., 1974, Possible left-lateral early to middle Mesozoic disruption of the southwestern North American craton margin, *Geol. Soc. Am., Abstr. Progr.*, v. 6, p. 955–956.

Sloan, R. E., 1980, The late Cenozoic Caribbean bridge and barrier, *Geol. Soc. Am. Abstr. Progr.* v. 12, p. 523.

Spence, W., and Person, W., 1976, The Guatemala earthquake of February 4, 1976; Tectonic setting and seismicity, *U. S. Geol. Surv. Prof. Paper*, v. 1002, p. 4–11.

Stibane, F. R., 1967, Paläogeographie und Tektogenese der Kolumbianischen-Anden, *Geol. Rundsch*, v. 56, p. 629–642.

Stibane, F. R., Schmidt-Effing, R. and Madrigal, R., 1977, Zur stratigraphischtektonischen Entwicklung der Halbinsel Nicoya (Costa Rica) in der Zeit von Ober-Kreide bis Unter-Tertiär, *Gissen, Geol. Schrift.*, v. 12, p. 315–358.

Stoiber, R. E., and Carr, M. J., 1971, Lithospheric plates, Benioff zones and volcanoes, *Geol. Soc. Am. Bull.*, v. 82, p. 515–522.

Sykes, R. L., and Weing, M., 1965, The seismicity of the Caribbean region, *J. Geophys. Res.*, v. 70, p. 5065–5074.

Sykes, L. R., McCann, W. R. and Kafka, A. L., 1982, Motion of the Caribbean plate during the last 7 million years and implications for earlier Cenozoic movements, *J. Geophys. Res.*, v. 87, p. 10656–10676.

Tarling, D. H., 1981, The geologic evolution of South America with special reference to the last 200 million years, in: *Evolutionary Biology of the New World Monkeys and Continental Drift*, Chiochon, R. L., and Chiarelli, A. B., eds., New York: Plenum, p. 1–40.

Toussaint, J. F., 1978, Grandes rasgos tectónicos de la parte septentrional del occidente colombiano, *Bol. Ciencia de la Tierra, Medellín, Colombia*, no. 3, p. 331.

Toussaint, J. F., and Restrepo, J. J., 1976, Modelos Orogénicos de tectónica de placas en los andes colombianos, *Bol. Ciencias de la Tierra, Medellín, Colombia*, no. 1, p. 1–48.

Van Andel, T. H., Heath, G. R., Malfait, B. T., Henrichs, D. F. and Ewing, J. I., 1971, Tectonics of the Panama Basin, eastern equatorial Pacific, *Geol. Soc. Am. Bull.*, v. 82, p. 1489–1580.

Van der Voo, R., Mauk, F. J. and French, R. B., 1976, Permian–Triassic continental configurations and the origin of the Gulf of Mexico, *Geology*, v. 4, p. 177–180.

Viniegra, F., 1971, Age and evolution of the salt basins in southeastern Mexico, *Am. Assoc. Pet. Geol., Bull.*, v. 53, p. 478–494.

Vogat, P. R., Lowrie, A., Bracey, D. R. and Hey, R. N., 1976, Subduction of a seismic ocean ridges: effects on shape, seismicity, and other characteristics of consuming-plate boundaries, *Geol. Soc. Am. Spec. Paper* 172, p. 1–59.

von Heune, R., and Auboin, J., 1982, Summary—Leg 67, Middle America Trench off Guatemala in: *Initial Reports of the Deep Sea Drilling Project*, v. 67, Washington, D.C.: U.S. Government Printing Office, p. 775–793.

Walper, J. L., and Rowett, C. L., 1972, Plate tectonics and the origin of the Caribbean and the Gulf of Mexico, *Trans. Gulf Coast Assoc. Geol. Soc.*, v. 22, p. 105–116.

Watkins, J. S., McMillen, K. J., Bachman, S. B., Shipley, T. H., Moore, J. C. and Angevine, C., 1982, Tectonic synthesis, Leg. 66: transect and vicinity, in: *Initial Reports of the Deep Sea Drilling Project*, v. 66, Washington, D.C.: U.S. Government Printing Office, p. 837–755.

Webb, S. D., 1980, North American mammalian chronology and the inter-American interchange, *Geol. Soc. Am. Abstr. Progr.* v. 12, p. 546.

Weyl, R., 1966, Ozeanisch Kruste im südlichen Mittelamerica, *Neues Jahrb. Geol. Paläont. Monatsh.*, v. 1966, p. 275–281.

Weyl, R., 1980, *Geology of Central America*, Berlin: Geb. Borntraeger, 371 p.

Williams, H., 1960, Volcanic history of the Guatemalan highlands, *Univ. California Pub. Geol. Sci.*, v. 32, p. 1–64.

Williams, H., McBirney, A. R. and Dengo, G., 1964, Geologic reconnaissance of southeastern Guatemala, *Univ. California Pub. Geol. Sci.*, v. 50 1–56.

Williams, H., and McBirney, A. R., 1969, Volcanic history of Honduras, *Univ. California Pub. Geol. Sci.*, v. 85, p. 1–101.

Williams, M. D., 1975, Emplacement of the Sierra de Santa Cruz, eastern Guatemala, *Am. Assoc. Pet. Geol. Bull.*, v. 59, p. 1211–1216.

Wilson, H. H., 1974, Cretaceous sedimentation and orogeny in nuclear Central America, *Am. Assoc. Pet. Geol. Bull.*, v. 58, p. 1348–1396.

Wilson, H. J., Meyerhoff, A. A. and MacDonald, W. D., 1978, Paleomagnetic results from Cretaceous sediments in Honduras, comments, *Geology*, v. 6, p. 440–444.

Woodring, W. P., 1966, The Panama landbridge as a sea barrier, *Am. Philos. Soc. Proc.*, v. 110, p. 425–433.

Woodring, W. P., 1978, Distribution of Tertiary marine molluscan faunas in southern Central America and northern South America, in: *Conexiones Tererestres entre Norte y Sudamérica, Univ. Nac., Mexico, Inst. Geol., Bol.*, v. 101, p. 153–165.

Woollard, G. P., and Monges-Caldera, J., 1956, Gravedad, geología regional, y estructura cortical de México, *Anal. Inst. Geof. México*, v. II, p. 60–96.

MAPS

Addicott, W. O., and Richards, P. W. (compilation coordinators), 1981, Plate tectonic map of the Circum-Pacific region, Northeast Quadrant, 1:10,000,000: *Am. Assoc. Pet. Geol.*

Arango-Cálad, J. L., Kassem-Bustamante, T., and Duque-Caro, H. (compilers), 1976, Mapa geológico de Colombia, 1:1,500,000: Inst. Nac. Invest. Geol. Mineras, Colombia.

Case, J. E., and Holcombe, T. L., 1980, Geologic-tectonic map of the Caribbean region, 1:1,250,000: U.S. Geol. Survey.

Dengo, G., Levy, E., Bohnenberger, O. H. and Caballeros, R., 1969, Mapa metalogenético de América Central, 1:2,000,000: Pub. Geol. ICAITI 3. Inst. Centroamericano de Tecnologia Industrial.

Donnelly, T. W. (compiler), 1975, Geological map of the Central Motagua river, Guatemala, 1:125,000: Ozalid reproduction.

López-Ramos, E., and Hernández Sánchez-Mejorada, S., 1976, Carta geológica de la República Mexicana, 1:2,000,000: Comité de la Carta Geológica de México.

Martín, C., 1978, Mapa tectónico, norte de América del Sur, 1:1,250,000: Minist. Energía y Minas, Venezuela.

Chapter 5

THE NORTHERN ANDES: A REVIEW OF THE ECUADORIAN PACIFIC MARGIN

John W. Baldock

British Geological Survey
Keyworth, Nottingham NG12 5GG, England

I. INTRODUCTION

The territory of Ecuador includes an active oceanic spreading zone, the Galapagos or Cocos–Nazca Rift (Hey, 1977; Hey *et al.*, 1977), and an active subduction zone beneath the continental margin (Lonsdale, 1978; Feininger and Bristow, 1980). The mainland is dominated by the Andes, which attain altitudes of around 4000 m with isolated volcanoes up to a height of 6000 m, and which divide the country into three natural, quite distinct, regions. To the east is the low-lying Oriente, part of the upper Amazon Basin. In the middle the comparatively narrow (100–150 km wide) Andean Mountain Belt, or Sierra, comprises the eastern Cordillera Real and Western Cordillera, separated by the inter-Andean valley and other discontinuous intermontane basins. The coastal zone, or Costa, includes various hill ranges close to the Pacific littoral as well as the extensive, low-lying, internal Guayas Basin, west of the Andes.

Offshore, a narrow continental platform and slope are terminated by the present-day submarine trench, which here barely reaches depths of 3000 m. Beyond the trench the ocean floor is more uniform, but the transverse Carnegie Ridge is a prominent feature, especially where it joins the Cocos Ridge forming the Galapagos Platform (Holden and Dietz, 1972), on which the Galapagos Islands have been built up, constituting the fourth separate region of Ecuador, some 1000 km west of the mainland. South of the Carnegie Ridge lies the Carnegie Platform and Peru Basin; to the north is the complex Panama Basin (Lonsdale and Klitgord, 1978).

Systematic mapping of much of the Costa and Sierra of Ecuador during the last decade has led to a better understanding of the onshore geology and the publication of over 60 geological maps by the Dirección General de Geología y Minas, Quito, with the cooperation of the Institute of Geological Sciences, United Kingdom. The terminology, lithology, paleontology, age, thickness, and distribution of Ecuadorian stratal units, where known, has been comprehensively revised by Bristow and Hoffstetter (1977). Other publications arising from recent work include an outline of the geology of Ecuador (Kennerley, 1980), several detailed, specific studies such as Bristow (1973), Henderson (1979, 1981), Evans and Whittaker (1981), and Wilkinson (1982), as well as an annotated bibliography of Ecuadorian geology (Bristow, 1981) and a new edition (with summary explanation) of the national Geological Map of Ecuador (Baldock and Longo, 1982).

II. GEOLOGICAL FRAMEWORK

Ecuador is part of a type area, the whole western coast of South America, for active convergence of an oceanic plate with a continental lithosphere (Mitchell and Reading, 1969). Consequently, it is reasonable to suggest that a plate-tectonic regime was the dominant force in the development of the geological structure of the region, at least since Cretaceous times (Lonsdale, 1978; Feininger and Bristow, 1980), but little is known concerning the pre-Cretaceous history of the Costa and Sierra. The Proterozoic and Paleozoic (and early Mesozoic) geology of the Oriente (Tschopp, 1953) and of the proto-Cordillera (Kennerley, 1980) could have been related to the development of a foldbelt on the margin of the Archean Guyana Craton (Engel et al., 1974; Fig. 12) followed by diastrophism as a result of polyphase (late Mesozoic–Pliocene) Andean Orogeny. Precambrian and Paleozoic tectonism may or may not have depended on subduction processes for its driving force (Engel et al., 1974), whereas Andean tectonism almost certainly was the result of, or response to, such forces.

The transverse tectonic segmentation of the Andean chain into regions with differing characteristics is particularly marked between the Central and Northern Andes (Gansser, 1973; Sillitoe, 1974). Here, at the diffuse Huancabamba deflection (Ham and Herrera, 1963), the Andean strike alters from NW in northern Peru, to NNE in Ecuador (see Fig. 2). This is reflected in the change at the Guayaquil–Dolores Suture, or Megashear (Case et al., 1971; Shepherd and Moberly, 1975), from an ensialic basement in the south, with felsic metamorphic and abundant plutonic rocks close to the Pacific margin, to a northern ensimatic province, comprising Cretaceous basalts with no sialic basement near the coast (Feininger, 1977; Henderson, 1979; Feininger and Bristow, 1980; Shepherd and Moberly, 1981). Another segment boundary may occur in northern Ecuador, offsetting the Cretaceous volcanics of the Western Cordillera (see Fig. 2).

A. Geology of the Oriente

The Oriente represents part of the pericratonic (foreland) platform, or laterally a back-arc basin, developed between the Guyana Craton and the Cordilleran Orogenic Belt, a tectono-sedimentary environment stretching from Venezuela to Bolivia on the east of the Andes (Tschopp, 1956; Harrington, 1962). The Oriente Basin of Ecuador was divided into two structural provinces in the late Tertiary when the sub-Andean zone (see Fig. 4), which partly corresponds to the Eastern Cordillera of Colombia, was subjected to Andean tectonism. The whole region is probably underlain at depth by crystalline rocks of the Guyana Shield (Campbell, 1970), which is exposed in southeastern Colombia (INGEOMINAS, 1976), but is covered by a thick epicontinental sedimentary sequence (see Fig. 1) in Ecuador (Tschopp, 1953).

Unmetamorphosed Paleozoic rocks crop out only in an isolated antiformal structure in the Cordillera de Cutucu in the southern sub-Andean region (Tschopp, 1953). The succession is at least 2500 m thick and comprises black argillites and sandstones (Pumbuiza Formation, probably of Devonian age) overlain by limestones and shales (Macuma Formation) of Carboniferous age (Sigal, 1968). Their extent beneath the thick Mesozoic–Tertiary cover is unknown, but they may occur as widespread but thin ($\sim$ 1000 m) platform deposits directly overlying the ancient Precambrian basement in some areas. Metamorphosed (and deeper-water?) equivalents of these Paleozoic formations may form parts of the Cordillera Real. Tschopp (1948) and more recently Mortimer (1980a) have correlated the weakly metamorphosed black slates and phyllites of the Margajitas Formation occurring in a narrow belt along the eastern flank of the Cordillera Real with the Pumbuiza. However, Bristow and Hoffstetter (1977) have pointed out that the Margajitas might equally well be correlated with the Napo Formation (see p. 185). It apparently passes southward into unmetamorphosed, undifferentiated, Cretaceous sediments (Limon Group) (Baldock and Longo, 1982).

A sequence of Mesozoic–Tertiary sediments, some 8–12 km thick, was deposited in the Oriente Basin by a succession of sedimentary cycles (Campbell, 1974). Marine conditions prevailed during deposition of the Lower Jurassic (Sinemurian) Santiago Formation, represented by 2500 m of shelf lime-

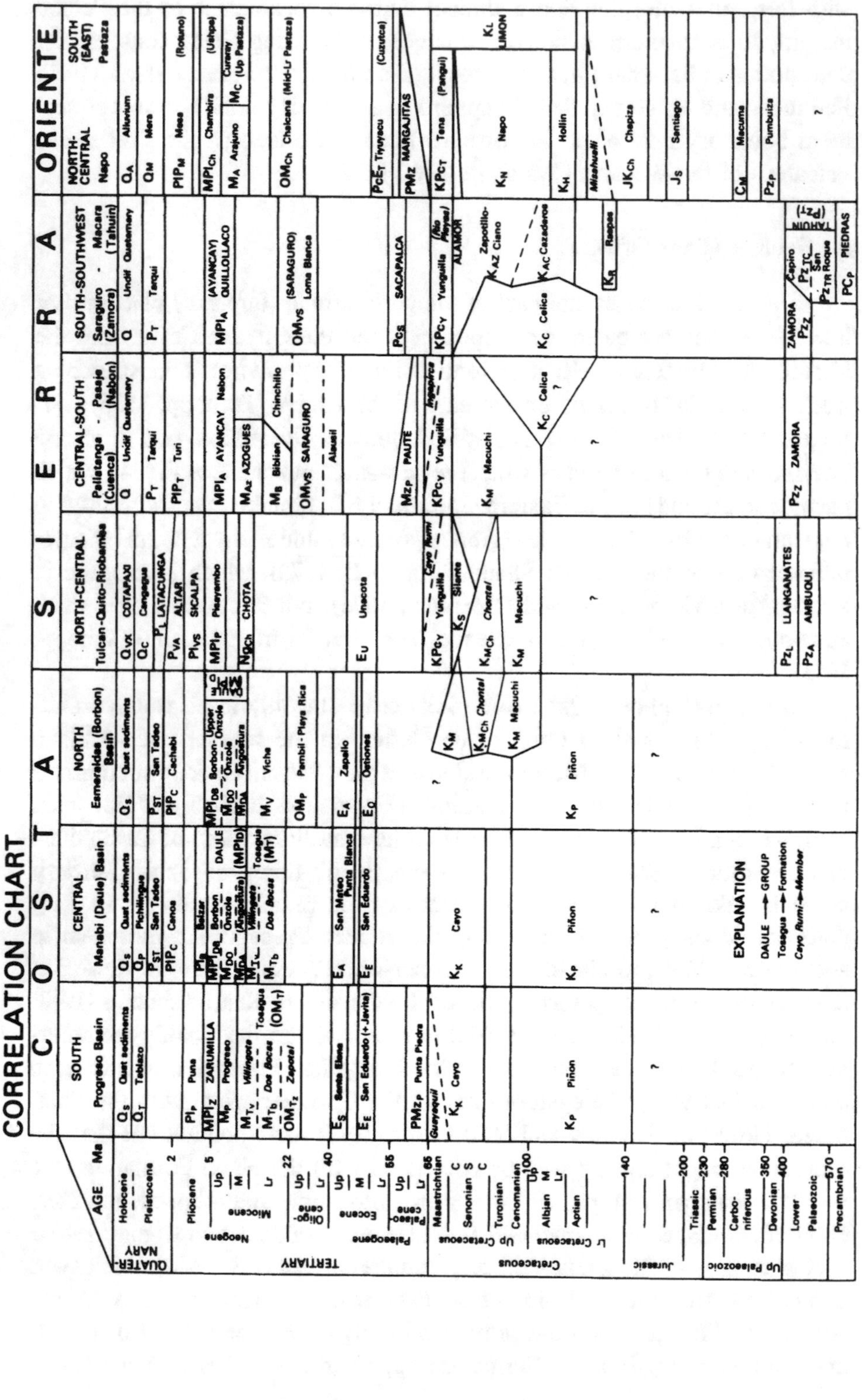

Fig. 1.   Correlation chart. K_p: symbol composed of system letter (U) and stratigraphic name (p). See text.

stones, sandstones, and shales. This sequence is restricted to the southern half of the Ecuadorian Oriente (Cutucu Uplift), although southward it can probably be correlated with part of the much more extensive Pucara Group in Peru (Cobbing *et al.*, 1981). The unconformably overlying Upper Jurassic to lowermost Cretaceous Chapiza Formation, which varies from 1000 to 4500 m in thickness, comprises red-bed sandstones and shales with intercalated pyroclastics. Andesitic lavas with subordinate pyroclastics (Misahualli Member) become dominant in the upper part of the sequence in the northern Oriente (Tschopp, 1956). Three major fault-bounded batholiths (Zamoro, Abitagua, and Cuchilla) probably of middle Jurassic age (Hall and Calle, 1982) were emplaced along the extreme western margin of the belt, at its thrust-faulted contact with the metamorphic rocks of the Cordillera Real (Fig. 3). Subvolcanic rocks of the Guacamayos Series (Tschopp, 1953) may be related to this magnetic episode.

A widespread Cretaceous marine transgression was heralded by deposition of thin epicontinental sandstones (Hollin Formation), and succeeded by the thick limestone–shale sequence of the (miogeosynclinal) Napo Formation, which is now thought to be of Albian to Campanian or even middle Maastrichtian age (Whittaker and Hodgkinson, 1979; Wilkinson, 1982). Clastic material was derived almost entirely from denudation of the Guyana Craton to the east. Wilkinson (1982) has summarized the evidence for postulating a proto-Andean (Cordillera Real), but not necessarily emergent, positive feature as proposed by Sauer (1965) and Campbell (1970), dividing the miogeosynclinal Napo Basin from the continental margin basin, and island arc, to the west. His palaeogeographic/lithofacies map (Wilkinson, 1982; Fig. 2 therein) reproduced here as Fig. 8, is based on a detailed analysis of the facies variations, including much unpublished data from oil exploration (Kehrer, 1975), within the Napo (see also Wilkinson, 1982, Figs. 3, 4, 5 therein): It supersedes Kennerley's (1980; Fig. 3 therein) earlier interpretation (reproduced here as Fig. 9), which suggested the presence of a hinge line to deeper open-water conditions, as also proposed by Feininger (1975).

Red-bed shales, sandstones, and marls of Maastrichtian–Lower Paleocene age (Tena Formation) unconformably overlie the Napo, testifying to a marine regression and showing an increasing derivation of clastic material from the west. By the end of the Cretaceous the proto-Cordillera Real had become an emergent feature providing a source of sediment for both the Oriente Basin and the continental margin basin to the west (Kennerley, 1980, Fig. 4 therein, reproduced as Fig. 10 in this chapter)—as it continued to do for much of the Tertiary. The Tena thins from 600 m in the west to 270 m in the eastern Oriente (Faucher and Savoyat, 1973; Feininger and Bristow,

1980). In the southern part of the Oriente Basin equivalent red-bed sediments were originally denominated as the Pangui Formation (Tschopp, 1953; Bristow and Hoffstetter, 1977), but recent mapping and compilation (Mortimer, 1980b; Baldock and Longo, 1982) has extended the northern terminology to the whole basin.

The thick Tertiary sequence comprises a series of mainly continental clastic units that show a slight facies variation between the northern (Napo) basin and the southern (Pastaza) basin (Tschopp, 1953). The Tena Formation is unconformably overlain by Paleocene–Lower Eocene conglomerates, sandstones, and shales of the Tiyuyacu Formation (Cuzutca in the SE Oriente), but no deposits of middle Eocene–Upper Oligocene age have been documented. The uppermost Oligocene–Lower Miocene Chalcana Formation consists of red-bed shales with gypsum deposits (Tschopp, 1953). This in turn is overlain by Miocene brackish-water clays, sandstones, and lignites of the Arajuno Formation, with the continental Curaray facies (red clays, tuffaceous shales, and gypsum) in the south and east of the Oriente Basin. Mio–Pliocene sandstones, shales, and tuffs (Chambira Formation), were tilted and warped before deposition of the Plio–Quaternary piedmont/mesa conglomerates, sandstones, and volcaniclastics derived from the uplifted Sierra.

The Oriente Basin of Ecuador deepens northward and more evidently southward from the central E–W trending Cononaco Arch (Fig. 4) (Wilkinson, 1982; Fig. 3 therein, reproduced as Fig. 7 in this chapter). It is principally along the main N–S axis of this back-arc basin that the oil fields of northeastern Ecuador have been discovered (Kehrer, 1975; Zuñiga y Rivero et al., 1976). The trough is bounded to the west by the fold/thrustbelt of the sub-Andean zone (Fig. 2), which was subjected to Tertiary deformation, and which is characterized by high-angle thrust faults and steep to open folds (Fig. 5; sections 6 and 7). Subsequent uplift and erosion has led to exposure of folded Upper Paleozoic–Tertiary (Miocene) rocks, originally deposited on the western edge of the miogeosynclinal Oriente Basin, which are partly covered by Plio–Quaternary piedmont deposits from the uplifted Cordillera.

B. Geology of the Sierra

The Sierra of Ecuador is divided into two parallel, geologically distinct, zones trending NNE. Metamorphic rocks of various ages (see p. 188) underlie the Cordillera Real, whereas mainly marine Cretaceous to Paleogene volcanic and subsidiary sedimentary rocks comprise the bulk of the Western Cordillera (Fig. 2). Neogene volcanics cover parts of both ranges and infill the inter-Andean valley, which separates the two Cordillera, although in the

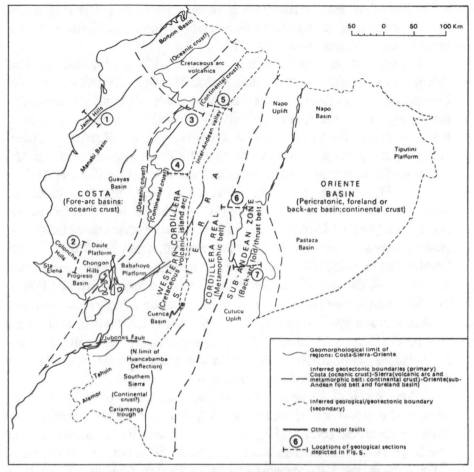

Fig. 2. The geomorphological and geological framework of Ecuador.

south the distinction is morphologically less obvious. The southwestern part of the country (south of the Jubones Fault, see Fig. 2) is quite different, having a pre-Mesozoic metamorphic basement (Tahuin block) overlain by Cretaceous volcanics and sediments (Kennerley, 1973).

1. Cordillera Real

A major zone of high-angle thrust faulting, running the entire length of Ecuador, separates the sub-Andean back-arc foldbelt of the Oriente from the metamorphic rocks of the Cordillera Real (SNGM, 1969; Baldock and Longo, 1982). Parts of the flanks of this belt are evidently underlain by Mesozoic phyllites, schists, and metavolcanics: the Margajitas Group on the eastern flank may represent metamorphosed Cretaceous shales (p. 183) and a transition from fossiliferous Maastrichtian shales (Yunguilla Formation) and volcanics into schists and metavolcanics of the Paute Group has been reported east of Cuenca (Bristow, 1973).

The core of the Cordillera Real, however, is probably composed of Paleozoic (and older?) metamorphic rocks. Mortimer (1980a) mapped the Llanganates Group (schists and gneisses) as being of (undifferentiated) Paleozoic age, following Kennerley's (1971) demonstration of two phases of folding in the central Cordillera Real. Pre-Mesozoic, possibly Lower Paleozoic, metamorphic rocks, mainly schists of the Zamora Group, with a northerly or northeasterly strike are exposed in the southern Cordillera Real (Kennerley, 1973; Kennerley, 1980). East of Cuenca, the late Mesozoic metamorphics (Paute Group) can be distinguished from the older (Paleozoic?) core (Zamora Group) on the basis of tectonic style and metamorphic grade. Although the contact has not been mapped, it may be (thrust?) faulted (Baldock and Longo, 1982). Northward the metamorphic rocks (Llanganates and Ambuqui Groups) of the Cordillera Real, described by Herbert (1977), may link with those that have been considered to be Paleozoic and Precambrian in southernmost Colombia (INGEOMINAS, 1976), although again some may be Cretaceous (Feininger, 1974). Recent attempts to resolve the metamorphic history by Rb/Sr age determinations along sections across the Cordillera Real have produced confusing results.

Along much of its length the Cordillera Real is capped by Neogene andesitic lavas and acid pyroclastics and ignimbrites, usually directly overlying metamorphic rocks. It is bounded to the west by the depressions or faulted grabens of the inter-Andean valley (Colombian border to Riobamba), Cuenca Basin, and Cariamanga–Gonzamana Trough, filled with Cretaceous shales/volcaniclastics (Alamor Group), and flysch-facies shales and turbidites (Yunguilla Formation, see p. 192), both partially overlain by Mio–Pliocene freshwater sedimentary and volcaniclastic deposits.

2. *Western Cordillera*

The Western Cordillera is mainly composed of an 8 km thick sequence of Cretaceous to Paleogene or Eocene andesite and basaltic andesite lavas and volcaniclastics, now distinguished as the Macuchi Formation, (Henderson, 1979, 1981). The lavas have spilitic affinities and exhibit pillow structures. Isotopic studies show that the parental calc-alkali magma was mantle-derived (Francis *et al.*, 1977). Intercalated marine (turbiditic) sediments are present (Chontal Member, where differentiated) especially in northern Ecuador (Guzman, 1980). The succession represents a (eugeosynclinal) volcanic island arc assemblage, mainly of relatively deep-water origin (Hen-

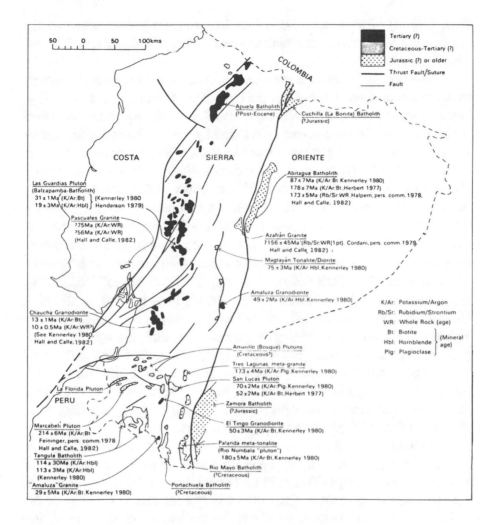

Fig. 3. The distribution and inferred ages of plutonic rocks in Ecuador.

derson, 1979). It is presumed that the sequence is now floored at depth by continental lithosphere (metamorphic rocks) or transitional crust, but this is nowhere exposed, except perhaps in isolated schistose outcrops (Punta Piedra Formation) of dubious origin in the Guayaquil region (Fig. 1). In the extreme north there is gravity evidence of a change to an ensimatic arc (Feininger, 1977). The Macuchi Formation extends with little lithological change from immediately north of the Jubones Fault in southern Ecuador (Zuñiga, 1980) to the border with Colombia and beyond, where it is known as the Diabase Group. Recent mapping indicates that there is no major lithological change across the Pallatanga (Puna–Pallatanga–Riobamba) Fault

Zone (Fig. 4) (Guevarra, 1979), which has previously been taken as the tectonic contact between the Macuchi and Celica (Feininger and Bristow, 1980; Hall and Calle, 1982).

However, the Macuchi is not a simple, continuous, north–northeasterly trending volcanic arc assemblage: the bulk of the formation is (Upper) Cretaceous, and, in places, it is overlain by fossiliferous Maastrichtian sediments of the Yunguilla Formation (see p. 192) and by volcaniclastics of the Silante Formation, which may perhaps be pre-Maastrichtian (Henderson, 1979). In other localities Macuchi-type volcaniclastics contain Eocene faunas associated with the local development of limestones (Echeverría, 1977; Henderson, 1979; Longo, 1980). The volcanic arc sequence, therefore, was evidently formed during the Upper Cretaceous–Eocene period—perhaps in two separate phases of activity. It may have subsequently (in the late Eocene or early Oligocene) been emplaced tectonically (Feininger and Bristow, 1980), probably in a series of northeasterly trending "slices" bounded by discontinuous, partly en echelon faults, which branch off from the main Guayaquil–Dolores suture (Fig. 4). Such faults may have brought volcano-sedimentary successions of similar lithology but differing age into juxtaposition in adjacent blocks (Fig. 4 and Fig. 5; sections 3, 4, and 5) (Baldock and Longo, 1982), but more detailed work is needed to verify this interpretation. The overall, slightly arcuate, northerly trend suggests that the older, stabilized metamorphic belt of the proto-Cordillera Real played an important role in the final configuration of the volcanic arc.

In southern Ecuador Cretaceous volcanism appears to have mainly subaerial, with deposition partly on a positive Paleozoic microcratonic block—the transverse Amotape–Tahuin Ridge (Kennerley, 1980)—and partly in the marine Cariamanga Trough (part of the NW Peru Basin, Cobbing *et al.*, 1981). The volcanic rocks of the Celica Formation, which show much less evidence of submarine eruption, are limited to the area south of the Jubones Fault system (Figs. 2 and 4), which is a major structure (possibly also a branch of the Guayaquil–Dolores Suture). The Jubones Fault appears to form the northern limit of the Huancabamba deflection, and delimit the northern margin of the Precambrian to Paleozoic Tahuin metamorphic block, composed of the Piedras and Tahuin Groups (Feininger, 1978; Zuñiga, 1980). The Cretaceous volcanic arc is associated with the marine sediments and volcaniclastics of the Alamor Group (Cazaderos and Zapotillo–Ciano Formations) in the extreme SW of Ecudaor (Kennerley, 1973). It apparently continues southward and may once have been linked to the andesitic Casma Arc in Peru (Hall and Calle, 1982), which also exhibits both submarine and subaerial facies (Cobbing et al., 1981). The Casma Group forms the envelope of the great coastal batholith of Peru (Cobbing *et al.*, 1981; and in Chapter 6 of this volume), which continues at least into southern Ecuador where it

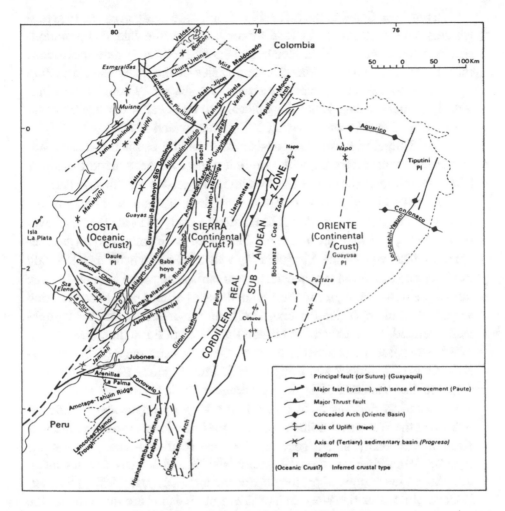

Fig. 4.   The main structural features of Ecuador.

is present as the middle Cretaceous Tangula Batholith and other related
intrusive bodies (Fig. 3; Kennerley, 1973, 1980). Paleocene volcanics (Sacapalca Formation) in the Cariamanga Graben of southern Ecuador also
appear to correlate southward with the Older Calipuy volcanic cycle of
northern Peru. The overall geographical continuity of Cretaceous–Paleocene
volcanism from Peru to Colombia suggests that the Ecuadorian section of
the arc also built up as a result of subduction over a more or less continuous
earlier trench system (from Peru to Colombia) located a little (50–100 km)
to the west of the present position of the Cretaceous arc volcanics in the
Western Cordillera. A more complex interpretation, involving three successive subduction zones and the creation and destruction of a hypothetical
oceanic plate, has been proposed by Feininger and Bristow (1980).

In northern Ecuador the Macuchi volcanics are overlain by red to green-ish-grey volcaniclastic rocks (and minor lavas) of the Silante Formation, which is some 5400 m thick. Henderson (1979) reported (way-up) evidence that the Silante underlies Maastrichtian shales of the Yunguilla Formation (Fig. 5; section 3) and is therefore Upper Cretaceous, although it was pre-viously thought to overlie the Yunguilla and to be of Paleocene age (Savoyat et al., 1970; Faucher and Savoyat, 1973; Bristow and Hoffstetter, 1977).

The Yunguilla Formation consists of turbiditic flyschlike shales, silt-stones, and volcaniclastics, with rare limestones. A conglomeratic member (Cayo Rumi) near the top of the formation suggests a shallower-water en-vironment and rapid erosion of proto-Cordillera metamorphic rocks to the east. The Yunguilla Formation is Maastrichtian in age (Sigal, 1968; Savoyat et al., 1970), but may include some Paleogene sediments. Although Hen-derson (1979) viewed the Yunguilla (as well as the Macuchi) as a strongly diachronous formation, volcaniclastic and carbonate rocks with an Eocene fauna, previously assigned to the Yunguilla, are perhaps better denominated separately (Feininger and Bristow, 1980). Although not yet lithostratigraph-ically defined, these have been termed the Unacota Formation (see Longo, 1980); such Eocene sediments have so far only been mapped in very re-stricted areas (Baldock and Longo, 1982). Further study and remapping of the flyschlike Yunguilla of the Sierra is required to clarify these relationships.

Numerous plutonic bodies intrude the Upper Cretaceous–Eocene se-quence of the Western Cordillera. Most of the intrusions north of the Jubones Fault are probably of Neogene age and are relatively small, except the Apuela Batholith in northern Ecuador (Fig. 3), which has contact meta-morphosed limestones of presumed Eocene age (Guzman, 1980). The post-Eocene plutons of the Western Cordillera of Ecuador are quite unlike the composite coastal batholith of Peru (Cobbing and Pitcher, 1972; Cobbing et al., 1981; and Cobbing, Chapter 6 of this volume).

Widespread Oligo–Miocene volcanic deposits (andesitic to rhyolitic lavas and pyroclastics) of the Saraguro Group (Alausi, Saraguro, Loma Blanca, Chinchillo formations) cover the southern sector of the Ecuadorian Andes. Some of the pyroclastics were deposited in the freshwater sedimen-tary basins developing within the proto-Cordillera (Kennerley, 1980). In the Cuenca Basin (Liddle and Palmer, 1941; Bristow, 1973) shales and arena-ceous sediments of the Biblian Formation and Azogues and Ayancay Groups include subsidiary volcaniclastics; coals were developed in the Mangan For-mation of the Ayancay Group (UNDP, 1969). Further south, similar (mainly lacustrine) sequences (Nabon Formation and Quillollaco Group) occur within the Nabon, Loja, and Malacatus basins (Kennerley, 1973; Bristow,

1976b). In northern Ecuador conglomerates and tuffaceous sediments (Chota Group) within the Chota Basin (Hall *et al.*, in press) may be of a similar age.

Late Neogene and Quaternary volcanic deposits (for example, the andesitic lavas and pyroclastics of the Pisayambo Formation in the central Sierra, and the widespread Pleistocene rhyolites and ignimbrites of the Tarqui Formation in the south) cover much of the Cordillera. The numerous volcanic centers and major stratovolcanoes of Pliocene–Recent age that dominate the Ecuadorian Andes are confined to the central and northern parts of the country, but continue into southern Colombia (Hall, 1977). Plio-Quaternary pyroclastic and ash deposits (Cangagua Formation) infill the down-faulted inter-Andean valley (Fig. 5; section 5), in places to a depth of more than 1000 m.

C. Geology of the Costa

The Costa represents a late Cretaceous–Tertiary fore-arc basin (or series of basins, formed in the arc–trench gap) underlain by Cretaceous basaltic volcanics of the Piñon Formation, which is exposed in the Chongón–Colonche and Jama–Mache Hills (Fig. 2). The term "Piñon" (Tschopp, 1948) has previously had a varied usage (see Bristow and Hoffstetter, 1977), but it is now more strictly applied to the sequence of basalts, basaltic pyroclastics, and minor associated sediments and volcaniclastics in the Ecuadorian coastal region (Feininger and Bristow, 1980; Baldock and Longo, 1982). The Piñon basement of the Costa is tilted and block-faulted (Fig. 5; sections 1 and 2), but not substantially metamorphosed. Gravity data (Feininger, 1977) indicate that the Piñon is characterized by extremely strong positive Bouguer anomalies and cannot be underlain by felsic continental crust. Chemical analyses show that the basalts are tholeiitic (Goosens and Rose, 1973; Goosens et al., 1977); they could be of true oceanic (Feininger and Bristow, 1980) or of primitive island arc origin (Henderson, 1979). The Piñon Basalt, therefore, must either represent ocean floor material or directly overlie Mesozoic oceanic crust, accreted into the continental South American plate.

In much of the Costa the Piñon is covered by a thick Upper Cretaceous and Tertiary sedimentary succession that was principally deposited in several troughs within the NNE-trending Bolivar Geosyncline (Nygren, 1950) or in the fore-arc basin extending from NW Peru to Panama: the Progreso Basin to the south of the Chongón–Colonche Hills; the elongate Manabi Basin of central coastal Ecuador, divided into southern and northern depocenters (Fig. 4); and the Esmeraldas–Borbon Basin in the north, which continues as the Tumaco Basin of southwestern Colombia. The Piñon Basalts are overlain by Upper Cretaceous (Senonian–Maastrichtian) volcani-

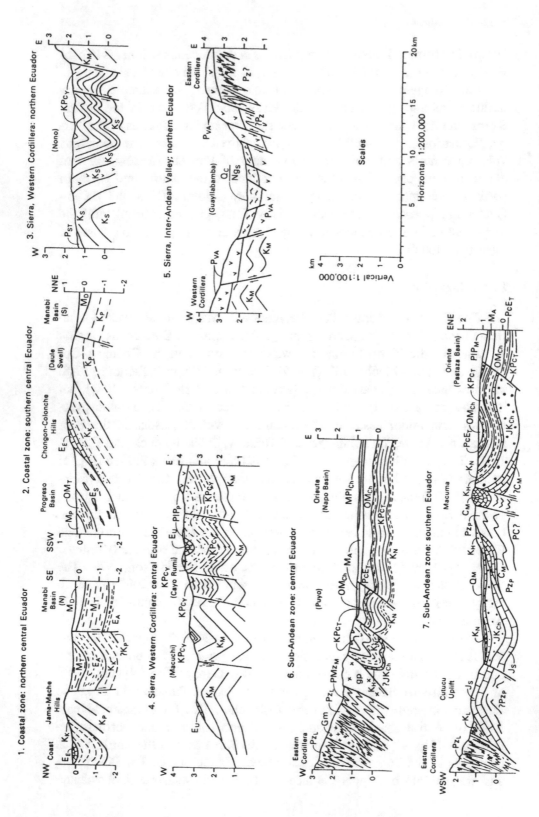

Fig. 5. Diagrammatic sections to illustrate the differences in structural styles. See Fig. 1 for stratigraphic symbols and Fig. 2 for locations.

clastic and sedimentary rocks (silicified tuffs and shales, with sandstones and agglomerates towards the base) of the Cayo Formation (Bristow, 1976a; Bristow and Hoffstetter, 1977), which correlates directly with the Macuchi Island Arc Volcanics of the Sierra (Feininger and Bristow, 1980), but were deposited to the west of the Guayaquil–Dolores Suture in the ensimatic fore-arc basin.

Tertiary sedimentation in the Costa commenced in the middle Eocene with the San Eduardo (and Ostiones) Limestones (Mills, 1968), which are considered to be of redeposited, turbiditic origin (Feininger and Bristow, 1980; Evans and Whittaker, 1981). After a period of little or no deposition in the Oligocene, sedimentation resumed in the late Oligocene–early Miocene, resulting in sequences up to 9 km thick in the Progreso Basin of southern Ecuador (Colman, 1966, 1970) and up to 4 km thick (Manabi and Esmeraldas–Borbon basins; see below) in central and northern Ecuador (Canfield, 1966; Faucher et al., 1971; Evans and Whittaker, 1981).

Submarine slumping on a gigantic scale into the Progreso Basin in the late Eocene led to the emplacement of the allochthonous Paleocene–Eocene Santa Elena Olistostromic Complex in the Santa Elena Peninsula (Azad, 1964; Colman, 1970). A condensed description of the complex, and one interpretation of its origin and derivation, has been given by Feininger and Bristow (1980). In the Manabi and Esmeraldas–Borbon regions, more normal Eocene sedimentation deposited the shale, sandstone, and mudstone sequences of the San Mateo–Punta Blanca and Zapallo Formations (Mills, 1967; Evans and Whittaker, 1981). In the Borbon Basin the Zapallo is overlain (apparently conformably) by Oligocene mudstones and sandstones of the Pambil Formation (Evans and Whittaker, 1981).

Following the postulated widespread middle-Oligocene unconformity, late Oligocene and Miocene sediments of the Tosagua Formation were deposited in the Progreso and Manabi basins: the lowest conglomeratic/arenaceous member (Zapotal) is restricted to the region south of the Chongón–Colonche Hills (Bristow, 1975). The Dos Bocas Member (siltstones and mudstones) spans the Oligocene–Miocene boundary and is overlain by the Villingota Member (Bristow and Hoffstetter, 1977). The equivalent Lower Miocene Viche Formation (mudstones with thin siltstones) was deposited in the Esmeraldas–Borbon Basin. The middle to late Miocene Daule Group (arenaceous and argillaceous sediments of the Angostura, Onzole, and Borbon Formations) occurs in both the Esmeraldas–Borbon and Manabi regions (Bristow, 1976c). South of the Chongón–Colonche Ridge, a thick Miocene sequence of sands, silts, and clays (Progreso Formation) was deposited in the subsiding Progreso–Jambeli Basin, which was formed by a pull-apart mechanism, and is still active (see Shepherd and Moberly, 1981). Sedimentation continued during the Plio–Quatenary in parts of the present coastal region (Puna, Tablazo, and Cachabi formations). In the central and southern Costa, Neogene uplift of the marginal hills, which represents the rise of the mid-slope high (Evans and Whittaker, 1981) caused an eastward migration of the axis of the sedimentary trough: at least 1500 m of piedmont, laharitic, and fluviatile material (Balzar, San Tadeo, and Pichilingue formations) was deposited closer to the uplifted Andes in the Guayas Basin–Gulf of Guayaquil.

In northern Ecuador the Tertiary sequence on the southeast of the Borbon Basin apparently overlies island arc andesites (Macuchi); it is not known whether the Andean foothills in this region partly comprise Piñon, but positive gravity anomalies (Feininger, 1977) suggest that oceanic crustal material may underlie the Macuchi in that area, which would be typical for a residual fore-arc basin developed "on oceanic or transitional crust trapped between the arc massif and the subduction zone" (Dickinson and Sealy, 1979; see also Evans and Whittaker, 1981). In the central Costa (Guayas Basin) the thick Plio–Quatenary piedmont deposits conceal the supposed more abrupt change from the felsic lithosphere beneath the Western Cordillera, to the oceanic crustal environment of the Costa (Figs. 2 and 4; see also the gravity data in Feininger, 1977).

Apart from the very small Paleocene–Eocene Pascuales Granite near Guayaquil (Fig. 3), no plutonic rocks are known in coastal Ecuador—which highlights the difference between this sector of the eastern Pacific margin, compared with the Peru region to the south.

D. Geology of the Offshore and Galapagos Islands

1. Continental Shelf and Trench

Part of the continental shelf between the present coastline of Ecuador and the offshore trench has been investigated in outline (Aldrich, 1977; Lonsdale, 1978; Shepherd and Moberly, 1981). Piñon Basalts crop out on Isla La Plata (Sheppard, 1927) off the coast of Manabi (Fig. 4). Geophysical and borehole evidence (unpublished data; Shepherd and Moberly, 1981) suggests that thick Upper Cretaceous and Cenozoic sequences, similar to those onshore, were deposited in various (partly fault-bounded) troughs within fore-arc and trench–slope basins. In the Gulf of Guayaquil, the NE–SW-trending sector of the Progreso Basin continues to the south as the Jambeli–Tumbes Basin, in which more than 10,000 m of Tertiary sediments have been deposited (Olsson, 1932). The Jambeli (Gulf of Guayaquil) Basin (see Fig. 4) obscures the important Guayaquil–Babahoyo–Santa Domingo Suture, which was probably a major transform fault (also called the Guayaquil–

Romeral Fault; Fig. 6) separating the continental crust (South American plate) to the east from the old (Mesozoic) oceanic crust (proto-Nazca plate) (Feininger and Bristow, 1980). This isolated and partially preserved piece of old oceanic crust, accreted to the continental plate, is limited to the west by the present trench system. Structurally the boundary may be a zone of imbricate thrust faulting (Lonsdale, 1978; Lonsdale and Klitgord, 1978) on the inner slope of the trench, which reflects the junction between the inactive Mesozoic oceanic crust and the spreading/subducting Neogene oceanic crust of the Nazca plate (Fig. 6) (Shepherd and Moberly, 1981).

2. *Pacific Margin*

The margin of the Pacific Basin west of the Ecudaor Trench is composed of Miocene and younger oceanic lithosphere (Nazca plate), formed at the Cocos–Nazca (or Galapagos) Rift Zone, and overlain by pelagic sediments. To the west of the Gulf of Guayaquil and southeast of the Grijalva Fracture Zone, the northernmost part of the Chile–Peru Trench is subducting Upper Eocene and Oligocene crust (Fig. 6) generated at the older Galapagos Rise spreading system (Menard *et al.*, 1964; Lonsdale, 1978).

West of the trench, off central Ecuador, lies the submarine Carnegie Ridge, which is interpreted as an aseismic volcanic feature formed either as the Nazca plate moved eastward over the Galapagos hot spot (Hey *et al.*, 1977) or by volcanic thickening of the oceanic crust during slow spreading from the E–W Galapagos Rift Zone. The relatively shallow nature of much of the Ecuador Trench results from subduction of this thickened oceanic lithosphere at the eastern end of the ridge (Lonsdale, 1978). Southward, between the Carnegie Ridge and the Grijalva Fracture Zone, lies an area of older, smoother-surfaced and thinner crust, presumably formed before development of the hot spot or during more rapid spreading during the middle Miocene (Macdonald and Mudie, 1974). North of the Carnegie Ridge (off northern Ecudaor and southern Colombia) there is a complex zone of late Miocene–Pliocene oceanic crust that was generated by spreading from the reactivated E–W-trending spreading axis and displaced along the Panama Fracture Zone and other transform faults, in the southern sector of the Panama Basin (Fig. 6) (Lonsdale and Klitgord, 1978).

The Carnegie Ridge continues (and gets younger) westward, joining the Cocos Ridge some 1000 km west of the mainland to form the submarine Galapagos Platform (Holden and Dietz, 1972) on which the Galapagos Islands have been built up to above sea level by Plio–Quaternary volcanism over a zone of crustal weakness (fossil transform fault) or a hot spot (Rea and Malfait, 1974; Anderson *et al.*, 1975).

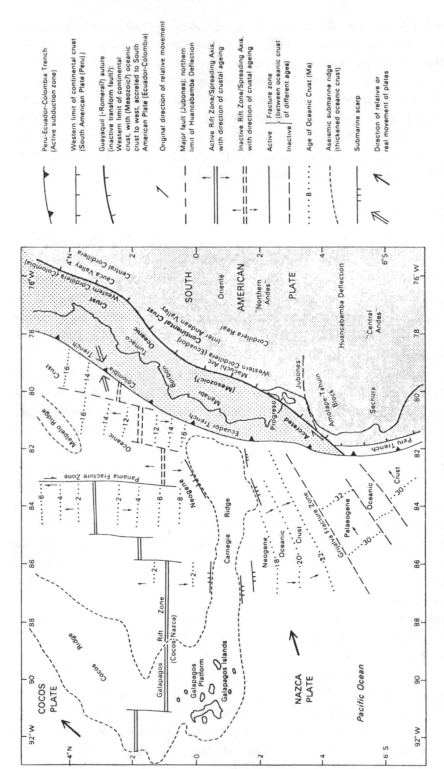

Fig. 6. Plate-tectonic framework of Ecuador (mainland, ocean floor, and Galapagos Islands).

3. *Galapagos Islands*

The islands of the Galapagos Archipelago thus occur on the northern edge of the Nazca plate, at the intersection of the Carnegie and Cocos submarine ridges (Cox, 1975), roughly midway between the South American coast and the East Pacific Rise. The Galapagos comprise some 14 main islands and numerous islets, all composed principally of Plio–Quaternary volcanic rocks, (McBirney and Williams, 1969; Nordlie, 1973) largely derived from the 15 major volcanoes that rise some 4500 m from the seafloor or Galapagos Platform. The islands have been classified into five groups (Hall, 1977) on the basis of their age and petrology, mainly comprising typical shield volcanoes (Banfield *et al.*, 1956; Nordlie, 1973) or their partially dissected remnants, and composed of olivine-poor tholeiitic basalt or of alkaline olivine–basalts, although a distinct group comprise uplifted blocks of older submarine basalts.

III. CORRELATION AND GEOLOGICAL HISTORY

Having reviewed the geological framework of the separate regions of Ecuador, an attempt can now be made briefly to correlate these distinct environments and present a tentative history of this sector of the Pacific Margin.

A. Precambrian and Paleozoic

Archean cratonic rocks of the Guyana Craton are nowhere exposed in Ecuador, but underlie the sedimentary epiplatform cover in the Oriente Basin. Precambrian rocks (Piedras Group) crop out in a small area of the Tahuin block, perhaps representing a relic of greenstone terrain.

Metamorphic rocks (schists and phyllites) of the Excelsior Group in the Olmos Arch region in the (western) Cordillera of northern Peru, which forms a direct continuation of the Cordillera Real of southern Ecuador, are clearly overlain by fossiliferous Triassic, and probably also Devonian, sediments (Cobbing *et al.*, 1981), while further south in central Peru the Excelsior Schists have been shown to the pre-Ordovician (Megard *et al.*, 1971). Hence it is supposed that the equivalent metamorphic rocks of the Cordillera Real, at least those of the (southernmost) Zamora Group (Kennerley, 1973) perhaps also those of the Llanganates (Mortimer, 1980a) and Ambuqui Groups, may also be substantially of Lower Paleozoic age. Metamorphic rocks of probable Paleozoic age (Capiro Formation of the Tahuin Group) are also exposed in the Tahuin block, but their relationship with the Cordillera Real is uncertain.

Upper Paleozoic sediments (Pumbuiza and Macuma formations) were deposited in the Oriente on the western margin of the Guyana Craton; unmetamorphosed remnants are only exposed in the Cutucu Uplift and perhaps along the flank of the sub-Andean zone, but may occur beneath the Mesozoic cover in the Oriente. Coeval, but perhaps deeper-water, sediments (e.g., Margajitas Formation) further west may have subsequently been deformed together with the Lower Paleozoic(?) core of the Cordillera Real.

B. Mesozoic: Pre-Cretaceous

Confused (unpublished) radiometric data for rocks collected from the central (Baños) and northern (Papallacta) Cordillera Real could suggest a period of Triassic metamorphism. Between the supposed proto-Cordillera Real positive axis and the margin of the craton, Mesozoic deposition is first recorded in the Oriente Basin by the dominantly continental/red-bed sediments of Upper Jurassic age (Chapiza Formation). In Ecuador it was accompanied, but chiefly followed, by widespread Jurassic–Cretaceous igneous activity (Misahualli volcanics and Guacamayos subvolcanics). Original intrusion of the three large batholiths along the edge of the Cordillera Real may have been earlier (middle Jurassic) or contemporaneous (Hall and Calle, 1982).

C. Mesozoic: Cretaceous

The oldest rocks exposed in the Costa and Western Cordillera (excepting the metamorphics of the Tahuin block) are Cretaceous (Goossens and Rose, 1973; Bristow and Hoffstetter, 1977). Consequently, virtually nothing is known of the pre-Cretaceous geological environment of Ecuador west of the Cordillera Real. The occurrence of ultrabasic rocks and high-pressure metamorphics (Feininger, 1980), perhaps of Cretaceous age, to the north of, or thrust into, the Tahuin block suggest that the Jubones and Arenillas–La Palma faults (Fig. 4) might be surficial expressions of Cretaceous or earlier subduction. The earliest Piñon Basalts of the Costa are probably of later Lower Cretaceous and early Upper Cretaceous age (Bristow and Hoffstetter, 1977), and probably represent an upper layer of Cretaceous ocean floor material (Feininger and Bristow, 1980; Hall and Calle, 1982).

Aptian and younger Cretaceous rocks are considered to have been deposited in a paired geosynclinal couple, separated by the positive Olmos Arch–proto-Cordillera Real axis. The miogeosynclinal, foreland (or back-arc) basin of the Oriente received a mainly nonvolcanic clastic/carbonate sequence: the Hollin and Napo formations (Fig. 7; reproduced from Wilkinson, 1982), which span the Aptian to mid-Maastrichtian period (Wilkinson,

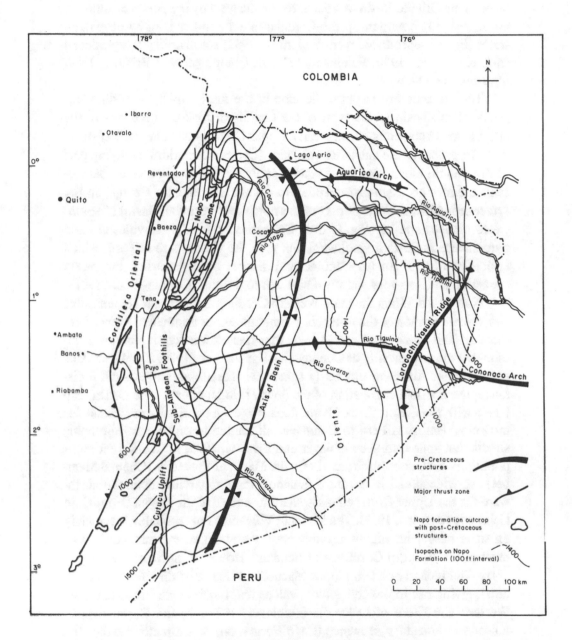

Fig. 7. Napo Formation thickness and outcrop areas, showing pre-Cretaceous structural elements and post-Cretaceous thrust zones. Reproduced from Wilkinson (1982).

1982; Whittaker and Hodgkinson, 1979). Compelling evidence, summarized by Wilkinson (1982), does indicate that even during the early Upper Cretaceous the Oriente Basin was delimited on the west by the positive, although not necessarily emergent, proto-Cordilleran axis, rather than by deep open water (Fig. 8; reproduced from Wilkinson, 1982; see also Fig. 9 reproduced from Kennerley, 1980; Feininger, 1975; and Feininger and Bristow, 1980, for a previous view).

The volcanic arc began to develop in the more rapidly subsiding (eugeosynclinal) basin to the west of the Cordillera Real positive axis in the late Lower Cretaceous or very early Upper Cretaceous. The earliest products are probably the subaerial volcanics (Celica Formation) overlying part of the Tahuin block, and their marine equivalents deposited as a volcano-sedimentary succession (the Aptian to Campanian Alamor Group), in the Lancones–Alamor Trough (part of the northwest Peruvian Basin) of southwesternmost Ecuador (Fig. 9; from Kennerley, 1980). These volcanics and pyroclastics indicate a southward link with the Casma Arc in Peru, which is dominantly Albian (to Cenomanian) in age, mainly marine, but partly subaerial, and also very sharply delineated to the east (Cobbing et al., 1981).

Northward, along the main Western Cordillera, the island arc andesites and volcaniclastics of the Macuchi Formation are of undoubted Upper Cretaceous (and Paleogene) age at their top; their lower age is unknown, although it may be as early as Cenomanian (Henderson, 1979, Fig. 4), perhaps even Albian. Although obviously somewhat diachronous over such a distance, the possible correlation of the (lower?) Macuchi with the Celica, and hence with the Casma Arc of Peru, does suggest a continuity of Cretaceous arc volcanism from Peru to Colombia, all related to one eastward-dipping subduction zone (but see Feininger and Bristow, 1980, for a different interpretation). Sediments associated with the Macuchi Volcanics (Chontal Member) are widespread in northern Ecuador: they apparently continue northward (as the Upper Cretaceous Espinal Formation of the Dagua Group) into Colombia (Barrero, 1979). The Macuchi Volcanics now probably rest mainly on sialic basement: highly negative gravity anomalies are characteristic of most of the Western Cordillera of Ecudaor. However, north of the Esmeraldas–Pichincha Fault (Fig. 4) the Macuchi arc may rest directly on oceanic crust, giving rise to positive gravity values that continue northward into the Western Cordillera of Colombia (Feininger, 1977). In the Costa, a thick volcano-sedimentary sequence (Cayo Formation) that directly overlies the Piñon Basalts, was deposited in a developing, ensimatic, fore-arc basin (Fig. 9) in the arc–trench gap and is the lateral equivalent of the Macuchi Formation (Feininger and Bristow, 1980).

By the end of the Cretaceous and in early Tertiary time, much of proto-Cordillera Real was emergent and undergoing erosion to provide the source

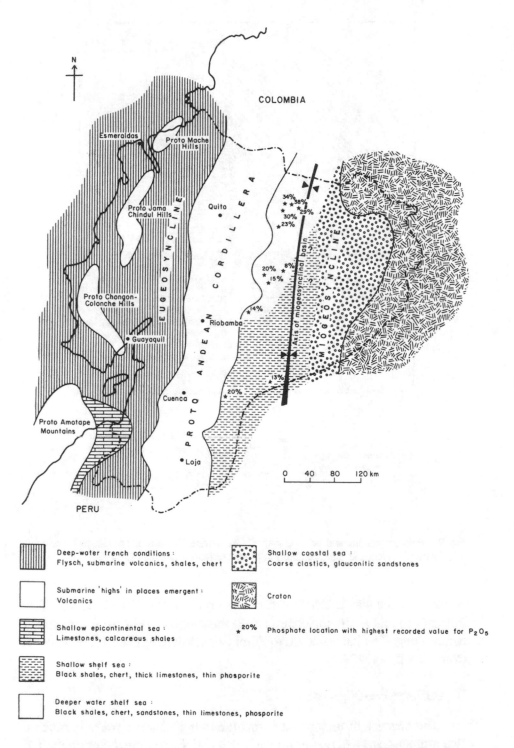

Fig. 8. Paleogeographic and lithofacies map (nonpalinspastic) of Ecuadorian Upper Cretaceous, showing the supposed relationship between depositional basins and the proto-Cordillera.
Reproduced from Wilkinson (1982).

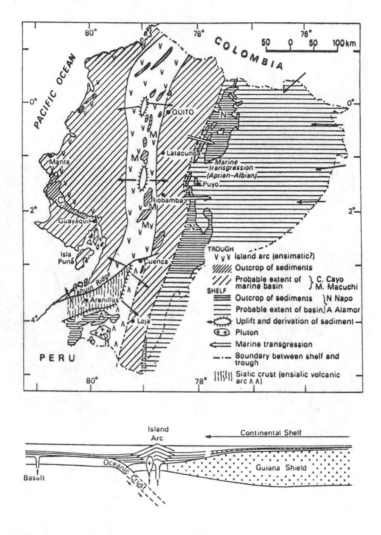

Fig. 9. Paleogeographic and outcrop map of Ecuadorian Cretaceous (Aptian to Campanian) rocks with plate-tectonic interpretation. Reproduced from Kennerley (1980).

of some of the clastic material deposited both in the Oriente Basin (Tena Formation) and in the (discontinuous?) interarc basins of the Sierra (flysh-facies Yunguilla Shales and Cayo Rumi Conglomerates (Fig. 10; reproduced from Kennerley, 1980).

D. Tertiary: Pre-Oligocene

The absence of undoubted autochthonous Paleocene rocks in much of Ecuador suggests a period of Lower Tertiary deformation and/or uplift. This Incaic or Peruvian phase of the Andean Orogeny was developed strongly to

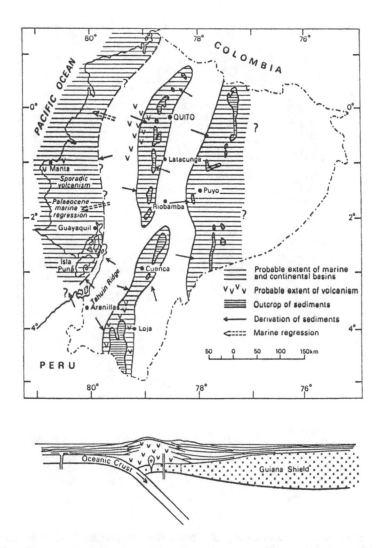

Fig. 10. Paleogeographic and outcrop map of Maastrichtian–Paleocene rocks in Ecuador. Reproduced from Kennerley (1980).

the south, in Peru (Cobbing *et al.*, 1981). It was accompanied or succeeded by plutonism and volcanism (Sacapalca Formation) in the southern Sierra, and may have affected the Ecuadorian Cordillera belt (for example the metamorphism of the Paute Group Phyllites and Metavolcanics).

The Lower Tertiary sedimentary fore-arc basins of the Costa continued to be separated from the back-arc Oriente Basin not only by the periodically inactive volcanic arc itself, but also by the emergent massif of the proto-Cordillera Real. Red-bed sedimentation (Tiyuyacu Formation) remained dominant in the Oriente; reef limestones, sometimes interbedded with vol-

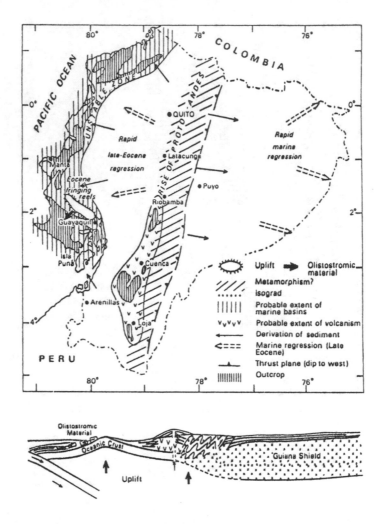

Fig. 11. Distribution of Eocene–Oligocene rocks and paleogeographic reconstruction of the principal features of the late Eocene orogeny in Ecuador, with an interpreted westward "jump" in the position of the subduction zone. Reproduced from Kennerley (1980).

canics, were developed in interarc basins in the middle Eocene (Unacota Formation), while turbiditic calcareous flysch derived from these fringing reefs was deposited in deeper-water fore-arc basins in the present coastal region (San Eduardo, Ostiones Limestones). During the Upper Eocene the coastal fore-arc basin was an unstable zone that developed three major (and several minor) depocenters (Fig. 11; reproduced from Kennerley, 1980). Allochthonous rocks of Cretaceous to Eocene age (Santa Elena Olistostromic Complex) were deposited chaotically in the deep Progreso Basin (Colman, 1970; Feininger and Bristow, 1980), while more normal turbiditic sedimentation continued in the Manabi Basin (San Mateo Formation) and in the Esmeraldas/Borbon Basin (Zapallo Formation).

Late Eocene–early Oligocene diastrophism, correlated with the Quechua 1 phase of the Andean orogenic cycle in Peru (Cobbing *et al.*, 1981), principally affected the Sierra. Folding, perhaps accompanied by low-grade metamorphism, was important locally, and plutonism was widespread. Folding and fracturing of the Cretaceous–Lower Tertiary sequence occurred mainly during tectonic emplacement of the Macuchi Volcanic Arc against the proto-Cordillera. Clogging of the late Cretaceous–Paleogene subduction system led to the establishment of a new eastward-dipping Benioff zone further to the west (Fig. 11; reproduced from Kennerley, 1980)—the so-called "jump" of the trench. Uplift and minor overthrusting commenced along the sub-Andean fault zone, contributing to the rise of the proto-Andes and the formation of freshwater intermontane, mainly Miocene basins, principally in southern Ecuador (Figs. 11 and 12; reproduced from Kennerley, 1980).

E. Tertiary: Post-Eocene Orogeny

Volcanism recommenced in the southern and central Sierra during the late Oligocene and continued in Miocene times (Saraguro Group, including Alausi, Saraguro, Loma Blanca, and Chinchillo formations), but in the north there is no evidence for volcanic activity until the late Miocene–early Pliocene (Pisayambo Formation). Sedimentation occurred from the late Oligocene, throughout the Miocene and into the Pliocene, in the Progreso, Manabi, and Esmeraldas–Borbon depocenters, within the fore-arc basin of the Costa (Oligo–Miocene Tosagua and Viche formations; middle to late Miocene Progreso Formation and Daule Group), as well as in the continental back-arc basin of the Oriente (Chalcana, Arajuno–Curaray, and Chambira formations) (Fig. 1 and Fig. 12; reproduced from Kennerley, 1980). Miocene sediments were also deposited within intermontane basins in the progressively uplifted Cordillera (Bristow, 1973; Kennerley, 1973) and plutonism was evidently widespread along the Western Cordillera (Fig. 3). Much of the massive uplift of the Ecuadorian Andes is of late Neogene age, perhaps a final response to the Cretaceous to Miocene period of crustal thickening resulting from volcano-sedimentary deposition and associated underplating. In the sub-Andean back-arc foldbelt, late-stage deformation caused folding of Miocene sediments and also led to the Cutucu and Napo uplifts. Arching of the uplifted Sierra reintroduced a tensional regime in the late Miocene, that caused the development of the inter-Andean valley, bounded by reactivated normal faults (Figs. 4 and 5).

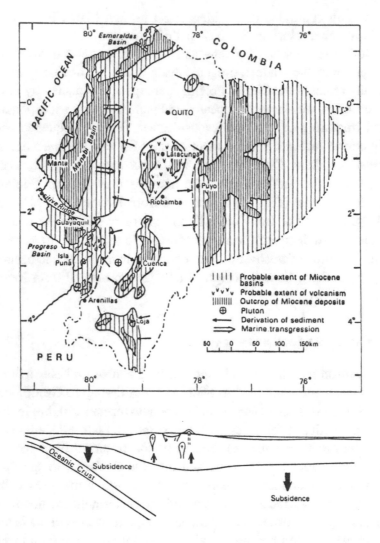

Fig. 12. Outcrop map and paleogeographic reconstruction of the Ecuadorian Miocene. Reproduced from Kennerley (1980).

F. Plio–Quaternary

Continuing Plio–Pleistocene volcanism produced thick lava and pyroclastic cover squences (Pisayambo and Tarqui formations). Erosion in the Sierra led to the accumulation of thick laharitic, piedmont, and terrace deposits over the sub-Andean and western Oriente regions (Mera Formation and Mesa deposits). Uplift of the coastal hills, or mid-slope high (Evans and Whittaker, 1981), caused the depositional axis in the Costa to move eastward, closer to the Sierra, so that a thick Quaternary sequence (~ 1500 m)

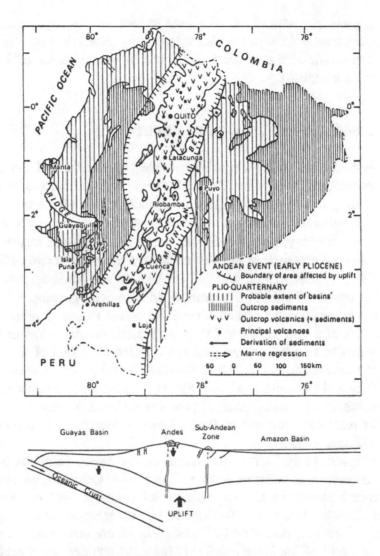

Fig. 13. Outcrop of Plio–Quaternary rocks and principal features of early Pliocene Andean uplift in Ecuador. Reproduced from Kennerley (1980).

occupies the Guayas Basin (Fig. 13; reproduced from Kennerley, 1980) and continues southward into the pull-apart Gulf of Guayaquil Basin, which is still undergoing active sedimentation.

Offshore, slow spreading of Neogene oceanic crust from the E–W (Galapagos or Cocos–Nazca) spreading axis built the thickened Carnegie Ridge: renewed, mainly submarine Plio–Quaternary volcanism in a region of crustal weakness led to the formation of the numerous shield volcanoes, whose summits are the Galapagos Islands. Subduction beneath the existing trench system is reflected by continuing volcanic activity in the Ecuadorian Andes

to the present day with the development of two belts of major strato-volcanoes, the most currently active being to the east, along the Cordillera Real
and in the sub-Andean zone (Hall, 1977), and by the deposition of thick ($\sim$
50–80 m) ash deposits.

IV. SUMMARIZED TECTONIC INTERPRETATION

The geology of Ecuador exhibits both clear similarities and distinct differences compared with that of Peru to the south and Colombia to the north.
In presenting a summarized tectonic interpretation it is necessary to draw
a balance between the essential continuity of the Andean range with its
flanking sedimentary basins or platforms, and the transverse segmentation
that is also characteristic of the western margin of South America (Sillitoe,
1974; Cobbing *et al.*, 1981). Much of the Central and Northern Andes has
a core of presumed ancient (Precambrian to Paleozoic) metamorphic rocks
marginal to the Archean Craton. Precambrian and Paleozoic relics on the
coastal side of the orogenic belt (Arequipa Massif, Amotape–Tahuin block)
vary widely in age, but suggest that the Western Cordillera of the Andes,
at least as far north as the Gulf of Guayaquil, and probably even further
north (Esmeraldas–Pichincha Fault; Fig. 4), is underlain by a thickened continental lithosphere. Relationships in the Tahuin block suggest that parts of
it might represent remnants of Precambrian greenstone terrain typical of
other cratonic nucleii.

Northwest of the Guayaquil–Dolores Suture, isolated and inactive Mesozoic oceanic crust underlies western Ecuador and Colombia, an important
difference between the Central Andes (Peru–Ecuador) and the Northern
Andes (Ecuador–Colombia). The transverse Tahuin block and the major E–
W faults (Jubones, Arenillas–La Palma; Fig. 4) are considered to be the
northern limits of the loosely defined Huancabamba deflection separating
these two major Andean segments, but it is the NNE–trending Guayaquil–
Dolores suture, which is believed to follow the base (or western flank) of
the Andes in most of Ecuador, that separates continental lithosphere to the
east from old oceanic crust to the west (see Fig. 6).

The Guayaquil–Dolores Suture may represent an old dextral transform
fault, probably linking with the Romeral Fault system in Colombia (Feininger
and Bristow, 1980). It was evidently a complex crustal feature with numerous
subsidiary (horsetail) faults (Jubones, Jambeli–Naranjal, Puna–Pallantanga,
Milagro–Guaranda, Alluriquin–Mindo, and other faults) that sliced up the
volcanic arc (Fig. 4). In northern Ecuador–southern Colombia the Cretaceous (Macuchi–Diabase Group) arc changes from ensialic to ensimatic,
partly at another major tectonic break that could be superficially expressed

by the transverse (NW–SE) Esmeraldas–Pichincha Fault (Fig. 4). Small, tectonically emplaced, ultrabasic bodies associated with a few of the NE-trending major crustal fault structures may be slivers of ocean floor material (Hall and Calle, 1982) perhaps representing remnants of dismembered ophiolite (Juteau *et al.*, 1977). Gravity data suggest that in the Gulf of Guayaquil the total dextral displacement across the southernmost end of the Guayaquil–Dolores Suture is some 80–100 km (Shepherd and Moberly, 1981).

In the Paleozoic and early Mesozoic there was probably little to distinguish the Ecuadorian section of the proto-Andean belt from regions to the north and south. By middle or late Mesozoic time a miogeosynclinal basin or foreland platform was still undergoing periodic sedimentation on the west of the Guyana Shield. It was separated from a developing volcanic arc and deep (eugeosynclinal) fore-arc basin further to the west by structurally positive axes with cores of older metamorphic rocks, such as the Marañon Geanticline and Olmos Arch in northern Peru (Cobbing *et al.*, 1981; Fig. 6 therein), the proto-Cordillera Real in Ecuador (Fig. 8; reproduced from Wilkinson, 1982), the Garsón massif and ancestral Central Cordillera in Colombia. The change in the trend of these old positive axes from northwesterly in Peru to north–northeasterly in Ecuador and Colombia is believed to have resulted from uplift of Paleozoic (or older) sublinear to arcuate mobile belts.

By the Cretaceous, the western edge of the continental South American plate was bounded by Mesozoic oceanic crust that was being subducted northeastward beneath it. Along the Ecuadorian section of that trench system, the angle of convergence was highly oblique to the NNE-trending continental margin, resulting in the formation of a major transform fault (ancestral Guayaquil–Dolores Suture) at the plate boundary along and down which oceanic crust was "transducted".

Paleocene volcanism (Sacapala Group) was restricted to the Huancabamba–Cariamanga Graben between the emergent positive highs of the Tahuin block and the proto-Cordillera in southern Ecuador, but was widespread in Peru (Older Calipuy Group, Cobbing *et al.*, 1981). North of the Gulf of Guayaquil convergence at the mid-Tertiary trench had virtually ceased, but volcanism did reoccur in the Eocene. The final phase of transduction of Mesozoic oceanic crust, which may have been partly overlain by Cretaceous (and Eocene?) volcanics, along the continental margin, perhaps led to the development of a series of dextral faults that sliced up the volcanic arc sequences and emplaced them during the late Eocene or early Oligocene against the deep-rooted metamorphic belt of the proto-Cordillera Real.

The late Eocene–early Oligocene phase of tectonism (Quechua 1 phase in Peru; see Cobbing *et al.*, 1981) was caused by the development of a new subduction system that partly resulted from clogging of the older subduction system and partly from a reorganization of crustal plates in the Pacific at this time (Herron, 1972). A more oceanward trench was developed, isolating the region of inactive (Mesozoic) oceanic crust that thus became accreted to the continental plate. Subduction at the new trench led to subaerial Oligo–Miocene volcanism (Saraguro Group) in the proto-Andean belt; contemporary erosion permitted thick clastic sedimentation to continue in the fore- and back-arc basins as well as intermontane troughs.

Further reorientation of the Pacific plate system occurred in the late Neogene (Rea and Malfait, 1974; Lonsdale and Klitgord, 1978), indirectly resulting in Andean uplift (Quechua 2 phase in Peru), with a brief compressional phase causing thrusting to the east, especially in the back-arc fold/thrust belt. However, deformation and uplift by vertical tectonics in a tensional regime (arching of the Andean orogen) was dominant, and many of the old dextral shear zones were reactivated, mainly as faults with a vertical (normal) sense of movement, especially along the Inter-Andean valley between the two Cordillera.

In conclusion, it might be suggested that several important geotectonic differences between the Central Andes (Peru to southern Ecuador) and the Northern Andes (Ecuador–Colombia) can be ascribed, at least partly, to the change from normal convergent subduction of both the proto-Nazca and Nazca oceanic plates beneath the Central Andean sector, compared with transduction of the Mesozoic ocean floor (proto-Nazca plate) along the continental margin of the Northern Andean sector. Renewed convergence, originating from a trench further to the west, was therefore between Neogene oceanic crust and old, inactive, oceanic crust accreted to the continental South American plate.

The reason for the highly oblique convergence to the north of the principal bend (Huancabamba deflection) in the Andes may perhaps be found in the trend of Paleozoic (and older), as yet poorly defined, sublinear fold-belts around the southwestern and northwestern edges of the Precambrian cratonic nucleus of northern South America. This interpretation ultimately ascribes to the configuration of Paleozoic and Precambrian features such differences between the Central and Northern Andes as older metamorphic rocks and abundant Cretaceous/Paleogene plutons in the south compared with oceanic crust (ophiolite?) and no granitoid plutons to the north; destructive convergence in Peru compared with "accretion" in Ecuador–Colombia; eastward younging of magmatism, as a result of eastward migration of the Benioff zone, beneath the Central Andes, compared with possible westward younging of plutonism resulting from the development of a new subduction zone, further to the west, beneath the old oceanic plate and the Northern Andes.

ACKNOWLEDGMENTS

This chapter is published with the permission of the Director, British Geological Survey, NERC. Grateful acknowledgement is made of helpful discussions with colleagues in the DGGM, Quito and BGS, United Kingdom, and of critical reviews by Drs. Cobbing, Bristow, and Evans (BGS).

REFERENCES

Adrich, L. T., 1977, An onshore–offshore geophysical study of southern Colombia and Ecuador, in: *Narino; Proyecto Cooperativo Internacional, 1973; La Transicion Oceano-Continente en el Suroeste de Colombia*, Ramirez, J. E., and Aldrich, L. T. ed., Bogotá: Instituto Geofisico, Universidad Javeriana, p. 9–15.

Anderson, R. N., Clague, D. A., Klitgord, K. D., Marshall, M., and Nishimori, R. K., 1975, Magnetic and petrologic variation along the Galapagos spreading center and their relation to the Galapagos melting anomaly, *Geol. Soc. Am. Bull.*, v. 86, p. 683–694.

Azad, J., 1964, The Santa Elena Peninsula (Ecuador): a review of the geology and prospects. *Rep. Anglo Ecuadorian Oilfields Ltd.*, no. J.A.7. (unpublished).

Baldock, J. W., and Longo, R., 1982, Geology of Ecuador (Explanation of the 1:1,000,000 Geological Map Quito: Dirección General de Geología y Minas, Ministerio de Recursos Naturales y Energéticos.

Banfield, A. F., Behre, Jr., C. H., and St. Clair, D., 1956, Geology of Isabela (Albemarle) Island, Archipiélago de Colón (Galápagos), *Geol. Soc. Am Bull.*, v. 67, p. 215–234.

Baranzangi, M., and Isacks, B. L., 1976, Spatial distribution of earthquakes and subduction of the Nazca plate beneath South America, *Geology*, v. 4, p. 686–692.

Barrero, D., 1979, *Geology of the central Western Cordillera west of Buga and Roldanillo, Colombia*, Bogotá: Pub. Geol. Esp. 4, Ingeominas, Ministerio de Minas y Energía.

Bristow, C. R., 1973, *Guide to the Geology of the Cuenca Basin, Southern Ecuador*, Quito: Ecuadorian Geological and Geophysical Society.

Bristow, C. R., 1975, On the age of the Zapotal sands of southwest Ecuador, *Newsl. Stratigr.*, v. 4(2), p. 119–134.

Bristow, C. R., 1976a, The age of the Cayo Formation, Ecuador, *Newsl. Stratigr.*, v. 4(3), p. 169–173.

Bristow, C. R., 1976b, On the age of the Nabón Formation, Ecuador, *Newsl. Stratigr.*, v. 5(2/3), p. 104–107.

Bristow, C. R., 1976c, The Daule Group, Ecuador, *Newsl. Stratigr.*, v. 5(2/3), p. 190–200.

Bristow, C. R., 1981, An annotated bibliography of Ecuadorian geology, *Overseas Geol. Miner. Resour.*, no. 58.

Bristow, C. R., and Hoffstetter, R., 1977, *Lexique Stratigraphique International.* 2nd Ed., Paris: Centre National de la Recherche Scientifique.

Campbell, C. J., 1970, *Guidebook to the Puerto Napo Area, Eastern Ecuador, with Notes on the Regional Geology of the Oriente Basin*, Quito: Ecuadorian Geological and Geophysical Society.

Campbell, C. J., 1974, Ecuadorian Andes. in: *Mesozoic–Cenozoic Orogenic Belts. Data for Orogenic Studies. Circum-Pacific and Caribbean Orogens. Geol. Soc. London, Spec. Publ.* no. 4, p. 725–732.

Canfield, R. W., 1966, *Geological Report of the Coast of Ecuador.* Quito: Ministerio de Industrias y Commercio.

Case, J. E., Durán, L. G., López, A., and Moore, W. R., 1971, Tectonic investigations in western Colombia and eastern Panamá. *Geol. Soc. Am. Bull.*, v. 82, p. 2685–2711.

Cobbing, E. J., and Pitcher, W. S., 1972, The coastal batholith of central Peru. *J. Geol. Soc. London*, v. 128, p. 421–460.

Cobbing, E. J., Wilson, J. J., Baldock, J. W., Taylor, W. P., McCourt, W., and Snelling, N. J., 1981, The geology of the Western Cordillera of northern Peru, *Inst. Geol. Sci. London Overseas Mem. no. 5*.

Colman, J. A. R., 1966, Summary of information relating to test wells and coreholes of the Progreso Basin, with notes on the stratigraphic units of the Guayas region, *Rep. Anglo Ecuadorian Oilfields Ltd.*, no. J.A.C.12. (unpublished).

Colman, J. A. R., 1970, *Guidebook to the geology of the Santa Elena Peninsula*, Quito: Ecuadorian Geological and Geophysical Society.

Cox, A., 1975, The Galapagos Islands: a migrating volcanic chain near a triple junction, *Newsl. Geol. Soc. London*, v. 4(1), p. 11.

Dickinson, W. R., and Seely, D. R., 1979, Structure and stratigraphy of fore-arc regions. *Am. Ass. Pet. Geol. Bull.* 63, p. 2–31.

Echeverría, J. M., 1977, *Geología del cuerpo calcáreo de Unacota, Pilaló, Cotopaxi*, Unpublished Thesis, Esc. Politéc. Nal., Quito, 123 p.

Engel, A. E. J., Itson, S. P., Engel, C. J., Stickney, D. M., and Cray, E. J., 1974, Crustal evolution and global tectonics: a petrogenetic view, *Geol. Soc. Am. Bull.*, v. 85, p. 843–858.

Evans, C. D. R., and Whittaker, J. E., 1981, The geology of the western part of the Borbon Basin, Northwest Ecuador, in: *Trench and Fore-Arc Sedimentation and Tectonics in Modern and Ancient Subduction Zones*, Leggett, J. K., ed., *Geol. Soc. London, Spec. Publ. no. 10*, p. 191–198.

Faucher, E., and Savoyat, E., 1973, Esquisse géologique des Andes de l'Equateru. *Rev. Géogr. Phys. Géol. Dyn. Ser. 2.* v. 15, p. 115–142.

Faucher, B. Vernet, R., Bizon, G., Bizon, J. J., Grekoff, N., Lys, M., and Signal, J., 1971, *Sedimentary Formations in Ecuador: A Stratigraphic and Micropaleontological Survey*, Bureau d'Études Industrielles et de Cooperation de l'Institut Français de Pétrole (BEICIP).

Feininger, T., 1974, Andean metamorphic rocks of Colombia and Ecuador. *Geol. Soc. Am. Abstr. Progr.* v. 6(7).

Feiningen, T., 1975, Origin of petroleum in the Oriente of Ecudaor. *Am. Assoc. Pet. Geol. Bull.*, v. 59(7), p. 1166–1175.

Feininger, T., 1977, *Mapa gravimétrico Bouguer del Ecuador (1:1,000,000)*, Instituto Geográfico Militar, Quito.

Feininger, T., 1978, *Mapa geológico de la parte occidental de la Provincia de El Oro (1:50,000)*, Quito: Escuela Politécnica Nacional.

Feininger, T., 1980, Ecologite and related high-pressure regional metamorphic rocks from the Andes of Ecuador, *J. Petrol.*, v. 21(1).

Feininger, T., and Bristow, C. R., 1980, Cretaceous and Paleogene geologic history of coastal Ecuador, *Geol. Rundsch.*, v. 69, p. 849–874.

Francis, P. W., Moorbath, S., and Thorpe, R. S., 1977, Strontium isotope data for Recent andesites in Ecuador and north Chile, *Earth Planet. Sci. Lett.*, v. 37(2), p. 197–202.

Gansser, A., 1973, Facts and theories on the Andes, *J. Geol. Soc. London*, v. 129, p. 93–131.

Goossens, P. J., and Flores, D., 1977, Geochemistry of tholeiites of the Basic Igneous Complex of northwestern South America, *Geol. Soc. Am. Bull.*, v. 88, p. 1711–1720.

Goossens, P. I., and Rose, W. I., 1973, Chemical composition and age determination of tholeiitic rocks in the Basic Igneous Complex, Ecuador, *Geol. Soc. Am. Bull.*, v. 84, p. 1043–1052.

Guevarra, S., 1979, *Explicación Breve de la Geología de la Hoja de Bucay (51)*, Quito: Dirección General de Geología y Minas, Ministerio de Recursos Naturales y Energéticos.

Guzman, J., 1980, Explicación Breve de la Geología de la Hoja de Pacto (64), Quito: Dirección General de Geología y Minas, Ministerio de Recursos Naturales y Energéticos.

Hall, M. L., 1977, *El Volcanismo en el Ecuador*, Quito: Instituto Panamericano de Historia y Geografía.

Hall, M. L., and Calle, J., 1982, Geochronological control for the main tectonic–magmatic events of Ecuador, *Earth Sci. Rev.*, v. 18, p. 215–239.

Hall, M. L., *et al. Mapa Geológico de la Cuenca Terciaria del Río Chota, Provincias de Imbabura y Carchi (1:25,000).* Quito: Escuela Politécnica Nacional, in press.

Ham, C. K., and Herrera, L. J., 1963, Role of sub-Andean fault system in tectonics of eastern Peru and Ecuador, in: *Backbone of the Americas, Am. Assoc. Pet. Geol. Mem.*, v. 2. p. 47–61.

Harrington, H. J., 1962, Paleogeographical development of South America, *Am. Assoc. Pet. Geol. Bull.*, v. 46, p. 1773–1814.

Henderson, W. G., 1979, Cretaceous to Eocene volcanic arc activity in the Andes of northern Ecuador, *J. Geol. Soc. London*, v. 136, p. 367–378.

Henderson, W. G., 1981, The Volcanic Macuchi Formation, Andes of Northern Ecuador, *Newsl. Stratigr.*, v. 9(3), p. 157–168.

Herbert, H., 1977, Petrochemie und Ausgangmaterial von Grünschiefern aus der E-cordillere Ecuadors, *Fortschr. Mineral.*, v. 55(1), p. 45–46.

Herron, E. M., 1972, Seafloor spreading and the Cenozoic history of the east-central Pacific, *Geol. Soc. Am. Bull.*, v. 83, p. 1671–1692.

Hey, R. N., 1977, Tectonic evolution of the Cocos–Nazca spreading center, *Geol. Soc. Am. Bull.*, v. 88, p. 1414–1420.

Hey, R. N., Johnson, G. L., and Lowrie, A., 1977, Recent plate motions in the Galapagos area. *Geol. Soc. Am. Bull.*, v. 88, p. 1385–1403.

Holden, J. C., and Dietz, R. S., 1972, Galapagos Gore, NazCoPac triple junction and Carnegie–Cocos ridges, *Nature,(London)*, v. 235, p. 266–269.

Instituto Nacional de Investigaciones Geológico-Mineras, 1976, *Mapa Geológico de Colombia 1:1,500,000*, Bogotá Ministerio de Minas y Energía.

Juteau, T., Mégard, F., Raharison, L., and Whitechurch, H., 1977, Les assemblages ophiolitiques de l'occident équatorien: nature pétrographique et position structurale, *Bull. Géol. Soc. Fr.*, v. 19(5), p. 1127–1132.

Kehrer, W., 1975, *Reporte Final de Grupo de Asesores Tecnicos Allemanes, CEPE, 1974–5*, Unpublished report. Quito: Dirección de Hidrocarburos.

Kennerley, J. B., 1971, Geology of the Llanganates area, Ecudaor, *Inst. Geol. Sci. London, Overseas Div. Unpublished report,* no. 21.

Kennerley, J. B., 1973, Geology of Loja Province, *Inst. Geol. Sci. London, Overseas Div., Unpublished report,* no. 23.

Kennerley, J. B., 1980, Outline of the geology of Ecuador, *Overseas Geol. Miner. Resour.*, no. 55.

Liddle, R. A., and Palmer, K. V. W., 1941, The geology and paleontology of the Cuenca–Azogues–Biblian region, provinces of Canar and Azuay, Ecuador, *Bull. Am. Paleontol.*, v. 26(100), p. 357–418.

Longo, R., 1980, *Explicación Breve de la Geología de la Hoja de Latacunga (67)*, Quito: Dirección General de Geología y Minas, Ministerio de Recursos Naturales y Engergéticos.

Lonsdale, P., 1978, Ecuadorian subduction system, *Am. Assoc. Pet. Geol. Bull.*, v. 62, p. 2454–2477.

Lonsdale, P., and Klitgord, K. D., 1978, Structure and tectonic history of the eastern Panama basin, *Geol. Soc. Am. Bull.*, v. 89, p. 1–9.

Macdonald, K. C., and Mudie, J. D., 1974, Microearthquakes on the Galapagos spreading centre and the seismicity of fast-spreading ridges, *Geophys. J. R. Astron. Soc.*, v. 36, p. 245–257.

McBirney, A. R., and Williams, H., 1969, Geology and petrology of the Galápagos Islands, *Geol. Soc. Am. Mem.*, no. 118.

Megard, F., Dalmayrac, B., Laubacher, G., Marocco, R., Martinez, C. L., Paredes, J., and Tomasi, P., 1971, La chaîne hercynienne au Pérou et en Bolivie; premiers résultats, *Cah. O.R.S.T.O.M. Ser. Geol.*, v. 3, p. 5–43.

Menard, H. W., Chase, T. E., and Smith, S. M., 1964, Galapagos Rise in the southeastern Pacific, *Deep Sea Res.*, v. 11, p. 233–242.

Mills, S. J., 1967, Tertiary stratigraphy in coastal Ecuador. The stratigraphy of the Tertiary rocks of southern Manabi and Guayas provinces (excluding the Santa Elena Peninsula) with notes on Esmeraldas Province, Ecuador, *Rep. Anglo Ecuadorian Oilfields Ltd.*, no. S.J.M. 1 (unpublished).

Mills, S. J., 1968, The micropalaeontology of the San Eduardo limestone and associated stratigraphical units, *Palaeontol. Note Anglo Ecuadorian Oilfields Ltd., Ecuador*, no. 1 (unpublished).

Mitchell, A. H., and Reading, H. G., 1969, Continental margins, geosynclines, and ocean floor spreading, *J. Geoel.*, v. 77, p. 626–646.

Mortimer, C., 1980a, *Explicación Breve de la Geología de la Hoja de Baños (88)*, Quito: Dirección General de Geología y Minas, Ministerio de Recursos Naturales y Energéticos.

Mortimer, C., 1980b, *Explicación Breve de la Geología de la Hoja de Puyo (103)*, Quito: Dirección General de Geología y Minas, Ministerio de Recursos Naturales y Energéticos.

Nordlie, B. E., 1973, Morphology and structure of the western Galápagos volcanoes and a model for their origin, *Geol. Soc. Am. Bull.*, v. 84, p. 2931–2956.

Nygren, W. E., 1950, Bolivar Geosyncline of northwestern South America, *Am. Assoc. Pet. Geol. Bull.*, v. 34(10), p. 1998–2006.

Olsson, A. A., 1932, Contributions to the Tertiary paleontology of northern Peru. Part 5, The Peruvian Miocene, *Bull. Am. Paleontol.*, v. 19(68), p. 5–216.

Rea, D. K., and Malfait, B. T., 1974, Geologic evolution of the northern Nazca plate, *Geology*, v. 2(7), p. 317–320.

Sauer, W., 1965, *Geología del Ecuador*, Quito: Ministerio de Educación

Savoyat, E., Vernet, R., Signal, J., Mosquera, C., Granja, J., and Guevara, G., 1970, *Formaciones Sedimentarias de la Sierra Tectónica Andina en el Ecuador*, Quito: Instituto Francés del Petroleo; Servicio Nacional de Geología y Minería.

Shepherd, G. L., and Moberly, R., 1975, Southern extension of the Dolores–Guayaquil megashear across the continental margin of northwest Peru and the Gulf of Guayaquil (abstract), *Trans. Am. Geophy. Union*, v. 56, p. 442.

Shepherd, G. L., and Moberly, R., 1981. Coastal structure of the continental margin, northwest Peru and southwest Ecuador. *Geol. Soc. Am., Mem.*, v. 154, p. 351–391.

Sheppard, G., 1927, Geological observations on Isla de la Plata, Ecuador, South America, *Am. J. Sci.*, v. 13, p. 480–486.

Sigal, J., 1968, *Estratigrafía Micropaleontológica del Ecuador, Datos Anteriores y Nuevos.* Quito: Instituto Francés del Petroleo; Servicio Nacional de Geología y Minería.

Sillitoe, R. H., 1974, Tectonic segmentation of the Andes: implication for magmatism and metallogeny, *Nature (London)*, v. 250, p. 542–545.

Servicio Nacional de Geología y Minería, 1969, *Mapa Geológico de la República del Ecuador (1:1,000,000)*. Quito: Ministerio de Recursos Naturales y Energéticos.

Tschopp, H. J., 1948, Geologische Skizze von Ekuador, *Bull. Assoc. Suisse Geol. Ing. Pet.*, v. 15(48), p. 14–45.

Tschopp, H. J., 1953. Oil explorations in the Oriente of Ecuador, 1938–50. *Am. Assoc. Pet. Geol. Bull.*, v. 37(10), p. 2303–2347.

Tschopp, H. J., 1956, Upper Amazon Basin geological province. in: *Handbook of South American Geology. Geol. Soc. Am. Mem.*, v. 65, p. 253–267.

United Nations Development Programme, 1969, Survey of metallic and nonmetallic minerals. Coal investigations (Operation No. 1, Cuenca-Biblián and Loja), *Tech. Rep. U.N. Dev. Programme, Quito–New York*, no. 1, annex no. 1.

Whittaker, J. E., and Hodgkinson, R. L., 1979, Micropalaeontological report on the Napo Formation, Eastern Ecuador. *Br. Mus. (Nat. His.)* Rep. OGS, no. 1979/1, 4p.

Wilkinson, A. F., 1982, Exploration for phosphate in Ecuador, *Trans. Inst. Min. Metall. Sect. B*, v. 91, p. 130–145.

Zuñiga, A., 1980, *Explicación Breve de la Geología de la Hoja de Machala* (*36*), Quito: Dirección General de Geología y Minas, Ministerio de Recursos Naturales y Energéticos.

Zuñiga y Rivero, F., Prado, A., Valdivia, H., and Velvarde, P., 1976, Hydrocarbon potential of Amazon basins of Colombia, Ecuador and Peru, *Am. Assoc. Pet. Geol. Mem., no. 25,* p. 339–348.

Chapter 6

THE CENTRAL ANDES: PERU AND BOLIVIA

Edwin John Cobbing

British Geological Survey
Keyworth, Nottingham NG12 5GG, England

I. INTRODUCTION

The Andean mountain chain in Peru is divided into two physiographic belts, the Eastern and the Western Cordillera, which correspond to orogenic belts of different ages; Paleozoic to the east and Mesozoic–Tertiary to the west (Fig. 1). Between the two Cordilleras there is a somewhat poorly defined region that, though intermittent, is nevertheless real. It has been named by Megard (1978) the region of the High Plateau and it forms a kind of buffer zone between the East and West Cordilleras. Intermontane basins are sporadically developed along its length, and the largest of these in southern Peru, Bolivia, and northern Chile forms the Altiplano of those countries. This zone of the High Plateau and Altiplano is commonly bounded to the west by a zone of faulting and thrusting.

The chain is flanked to the east by broad Mesozoic and Tertiary sedimentary basins which are the site of hydrocarbon exploration and which rest on crystalline rocks of the Brazilian Shield. To the west crystalline metamorphic rocks crop out along the coast in southern Peru and in the Amotape Hills of northern Peru. In southern Peru these rocks extend inland to form a crystalline massif, the Arequipa Massif, which at this latitude provides the visible floor to the Mesozoic sedimentary and volcanic formations.

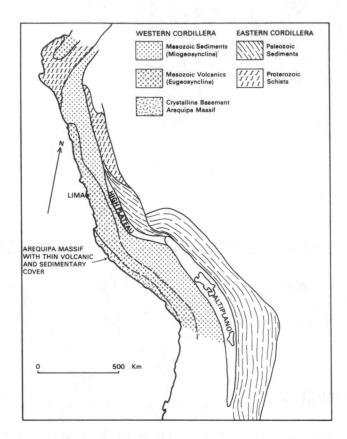

Fig. 1. Map of Peru and Bolivia showing the major morphostructural elements of the Andes.

II. CRYSTALLINE BASEMENT

A. The Arequipa Massif and Metamorphic Inliers

The Arequipa Massif underlies a large area in southern Peru, exposed in numerous windows through the Mesozoic and Tertiary cover, and rises to an altitude of about 4000 m. Over much of southern Peru the crystalline rocks, together with elements of the Mesozoic cover, provide an erosion surface upon which Tertiary–Recent volcanic deposits have been erupted, concealing large areas of the underlying basement.

The geology of the gneisses is very complex and there has never been any serious attempt to study them systematically. Reconnaissance studies of some of the structural elements have been made by Shackleton *et al.* (1979), while good descriptions of some of the lithologies present are given in Dalmayrac *et al.* (1980). Metasedimentary and metaigneous gneisses and

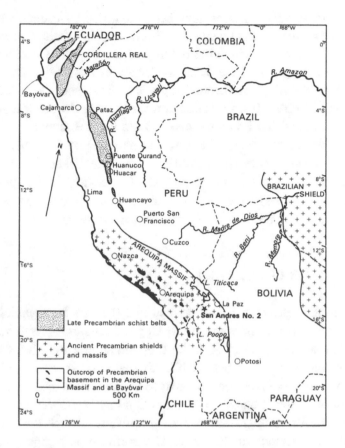

Fig. 2. Distribution of Proterozoic schist belts and ancient cratons in Peru and Bolivia.

schists are widespread, but it is at present impossible to give any estimate of their relative abundance, or their geological relationships. Reconnaissance geochronology (Cobbing *et al.*, 1977; Dalmayrac *et al.*, 1977; Dalmayrac *et al.*, 1980; Shackleton *et al.*, 1979) has shown that parts of the massif have ages of the order of 2000 Ma, an age similar to that for the Trans-Amazonian event of the Brazilian shield. However, correlation remains speculative and a variety of younger ages ranging from 720 Ma to about 600 Ma have also been recorded from the massif (Cobbing *et al.*, 1977; Lancelot *et al.*, 1977) that suggest a complex geological history. An exploration borehole for oil, San Andreas No. 2 (Fig. 2) sited in the Altiplano of Bolivia, encountered crystalline basement at 2744 m. The rock, which was a hornblende biotite meta granite, was dated at 1050 Ma (Lehmann, 1978). This discovery extends the area of the Arequipa Massif as far as the Altiplano, and it has very considerable implications for the interpretation of Paleozoic and Mesozoic geology.

Nuclei of high-grade gneisses have been identified by French workers in the region of Huanuco and Puente San Francisco, Apurimac (Dalmayrac *et al.*, 1980). These occur within the outcrop of the Eastern Cordillera Paleozoic belt and suggest that the belt is underlain by crystalline basement. Charnockitic, enderbitic, and khondalitic gneisses have been recorded from these areas and U/Pb analyses on zircons from these rocks suggest ages of 1140 Ma from Puente San Francisco and 600 Ma from Huanuco.

In the extreme northwest of Peru, recent mapping by geologists of the Peruvian Geological Survey (INGEMMET), has established the presence of banded crystalline gneisses in the district of Bayóvar (Caldas, personal communication). These gneisses resemble those of the Arequipa Massif, but at present there is no more information on this occurrence.

III. THE EASTERN CORDILLERA

This Cordillera provides a well-defined belt that has been geologically active from the late Precambrian to the Upper Permian, an interval that includes orogenies assigned to both the Caledonian and the Hercynian in other parts of the world. The belt itself is divided into two major portions, a northerly part that extends from Huancayo in central Peru to Ecuador and, thereafter, through the Huancabamba deflection into the Cordillera Real of Ecuador, and a southern part that extends southward through Peru and Bolivia into northern Argentina and Chile. The northern part consists principally of crystalline schists of late Precambrian age and the southern part of clastic sediments of Lower to Upper Paleozoic age.

A. The Precambrian Part of the Eastern Cordillera: The Marañon Complex

In general, the geology of the Eastern Cordillera is poorly known. The region is generally remote and inaccessible and the geology itself is difficult and resistant to investigation by reconnaissance methods. For the most part the rocks consist of low- to medium-grade metasediments, phyllites, and schists. They are mainly micaceous, and they are the metamorphosed representatives of a thick sequence of clastic shales and quartzites that were deposited in an elongate trough along the site of the present-day Eastern Cordillera. The formation of this trough probably represents the first event in the formation of the geological edifice that ultimately gave rise to the Andes, and it is unfortunate that so little is known about it. Myers (1974) has speculated that the formation of the ancestral Pacific may have been related to continental rifting in the late Precambrian, which established the outline of the South American continent. It is possible that the formation of

TABLE I
Structural Sequence in the Marañon Complex at Puente Durand

6   Low-grade schists and slates with thin bands of quartzite

5   A thick sequence of metatuffs and/or metagreywackes with ocellar albite

4   A thick sequence of pelites giving schists and mica schists with staurolite and garnet associated at some levels with basic tuffs and more rarely with impure carbonates

3   Feldspathic pelites locally transformed to migmatitic gneiss with garnet and sillimanite

2   A basic episode giving hornblendites and serpentinites

1   Granulite gneisses that may represent the nuclei of a more ancient orogenic edifice

the trough might have been related to such a process. The belt, however, is physically connected through Ecuador, Colombia, Venezuela, Bolivia, Chile, and Argentina with a network of foldbelts of late Precambrian age that surround cratonic nuclei, and are known as the Brazilides. It is now known that the Brazilides, at the time of their formation, were physically continuous with the similar Pan-African network of foldbelts in Africa and it now appears to be a real possibility that the Proterozoic belts of Peru and the other countries mentioned, might have begun their geological life as part of this Gondwanawide network which, instead of dying at the beginning of the Phanerozoic, has spawned daughter foldbelts that are active to the present day. Regrettably there are not many hard facts available at the moment to verify this hypothesis, which has been lucidly documented by Dalmayrac et al, (1980).

The main outcrop of the Marañon Complex (Wilson et al., 1967) consists chiefly of pelitic schists and feldspathic quartzites metamorphosed in the greenschist and lower amphibolite facies. Albite schists, which have been termed Prasinites, are also present (Dalmayrac et al., 1980) and are interpreted as metatuffs and greywackes. Ultramafic rocks occur in small discontinuous outcrops and high-grade metamorphic nuclei are present at Puente Durand, near Huanuco, and also along the Rio Marañon south of Pataz. The latter have only been identified from boulders, however. Dalmayrac et al. (1980) have interpreted the field relationships at Puente Durand as suggesting a more ancient basement, and have suggested a provisional sequence (Table I). It is likely that the Olmos Schists of northern Peru correspond to units 4, 5, and 6 of Table I.

The Marañon Complex is characterised by polyphase folding and according to Dalmayrac et al. (1980) four phases can be recognized (Table II). The author has observed that structures of the last category (Table II) are overlain unconformably by Mesozoic sedimentary sequences and are therefore of pre-Andean age. They have been affected by coaxial superimposed Andean folds.

TABLE II
Structural Phases in the Marañon Complex

---

1 Isoclinial with E–W axes, recumbent axial planes, northward facing with a flow cleavage
   or schistosity

2 Similar folds on N–W axes with inclined axial planes and crenulation cleavages, with no
   consistent facing direction

3 Chevron folds and kink bands in both N–S and E–W directions having vertical axial planes
   and a rough cleavage

4 Cylindrical folds of the Andean phase with NNW SSE axes and with no cleavage

---

1. Age of the Marañon Complex

Schists of the Marañon Complex are unconformably overlain by fossiliferous Ordovician sediments of Llanvirnian age at Pataz, and also at Huacar near Huanuco in central Peru. The upper limit for the metamorphism of these schists is thus Lower Ordovician. However, because unmetamorphosed fossiliferous Cambrian strata occur in Bolivia and Argentina within the same Lower Paleozoic belt, there is a strong probability that the schists and gneisses of the Marañon Complex are actually of Precambrian age.

The granulite gneiss from Puente Durand 90 km to the north of Huanuco has been dated by the U/Pb method on zircons (Dalmayrac *et al.*, 1980). The analyzed points clustered in a group at about 600 Ma on the Concordia curve, which rendered the interpretation of the results uncertain. However, these authors, combining this information with the evidence for a resetting at 600 Ma in the Arequipa Massif (Cobbing *et al.*, 1977; Shackleton *et al.*, 1979; Dalmayrac *et al.*, 1980), have suggested that the age of 600 Ma represents a culminative metamorphic event in this sector of the Brazilide foldbelt network. The suggestion, therefore, is that this belt experienced the complete geotectonic cycle from sedimentation, through to metamorphism, orogeny, and uplift by the end of the Precambrian.

B. The Lower Paleozoic Part of the Eastern Cordillera

During the Cambrian the whole of the Eastern Cordillera was probably emergent, but sediments of Cambrian age were deposited in Bolivia and northwest Argentina. This marked the first phase of the development of an enormous sedimentary basin that, in Peru and Bolivia, attained its maximum during Ordovician to Devonian times. Dalmayrac *et al.* (1980) suggest that the Peru–Bolivia basin continued southward through Argentina and Patagonia to connect via a system of linked basins with the Samfrau Geosyncline of Du Toit (1937), which formed a marginal geosyncline to Gondwanaland from Australia, through Antarctica to the Andes. Be that as it may, the basin

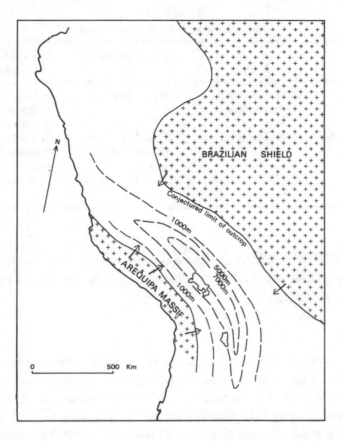

Fig. 3. Distribution of Ordovician strata in Peru and Bolivia with approximate isopachytes indicated (after Harrington, 1962, Turner, 1972; Dalmayrac *et al.*, 1980).

in Peru and Bolivia was directly superimposed on the former Proterozoic crystalline belt, and although it was in a sense marginal to the continent, it was actually formed at some distance within it and was bounded to the west by the minicraton of the Arequipa Massif and to the east by the Brazilian shield (Fig. 3). It is a notable feature of this basin that the deposits of the axial zone are much thicker than their stratigraphic equivalents to the SW and NE and there is consequently no doubt that the basin formed as a relatively narrow, elongate trough contained within cratonic crystalline basement. The sedimentary fill reflects this tectonic situation, for it consists almost entirely of shales and sandstones that have been derived from the adjacent forelands.

1. Ordovician

The Ordovician is readily divided into three lithostratigraphic divisions (Table III). The sequence is entirely clastic and attains its maximum thick-

TABLE III
Lithostratigraphy of the Ordovician

Age	Lithology	Thickness in axial zone	Thickness in marginal zone
Caradocian	Quartzites with intercalations of shale	3500 m S. Peru	500–1500 m (Carcelpunco Sub-Andean zone)
Arenig–Lower	Graptolitic black shales with pyrite	2500 m N. of Cuzco; 3500 m SE Peru and Bolivia	4–700 m Contaya Huanuco Maray Pataz
Arenig	Conglomerates and quartzites	300 m Ollantaytambo near Cuzco	1–20 m Huacar

ness of 7000 + m in southeast Peru and Bolivia, while the quartzites of the Upper Ordovician are confined to that area and do not extend northward much beyond Cuzco.

2. *Silurian*

The transition from Ordovician is marked by a regional discordance whereby the basal Silurian rests discordantly on different parts of the Ordovician sequence. The discordance, although regional, is not noticeably angular, and no deformation is assigned to this interval. Silurian strata occur in two main zones, the marginal zone of the Altiplano and the axial zone of the Eastern Cordillera (Table IV).

In the Altiplano marginal zone a series of marine shales and sandstones 600 m thick concordantly overlie Caradocian quartzites. These rocks, which consist of thin beds of sandstone regularly alternating with shales and siltstones, also contain dolomitic concretions. They are of Lower Llandoverian age and although there is faunal evidence of their deposition in cold-water conditions (Boucot, 1972), e.g., in the Malvinokaffric province, there is no indication of the widespread Lower Silurian tillite, that occurs in the axial zone.

The Lower Llandoverian strata are overlain by fine sandstones and siltstones that pass gradually upward to fine micaceous violet sandstones of Llandoverian to Ludlovian age, which in turn are transitional to the Devonian.

The basal unit of the Silurian in the axial zone is the widespread tillite or diamictite of possibly glacial origin (Turneaure, 1960). It occurs through northern Argentina and Bolivia to Peru and ranges in thickness from 30 to 40 m in northern Argentina to 600 m at Potosi in Bolivia, reducing to 200 m through-out most of Bolivia and southern Peru; this area is known as the

TABLE IV
Lithostratigraphy of the Silurian and Devonian

Age	Thickness and lithology in the axial zone	Thickness and lithology in the marginal zone altiplano
Upper Devonian		
Givetian–Frasnian		
Middle Devonian	4000 m	200 m shales
Eifelian	Black shale	200 m white quartzites
Lower Devonian		600 m shales and sandstones
Gedinnian–Emsian		500 m sandstones and microconglomerates
Siluro–Devonian passage		100 m Taya Taya quartzite
Upper Silurian	1000 m	100 m sandstones and siltstones
Wenlockian–Ludlovian	Black shale	
Lower Silurian	Zapla–Cancarini marine tillite	600 m sandstone and shale
Llandoverian	40 m in N. Argentina to 600 m at Potosi Bolivia	

Zapla Formation, or Cancarini Formation (Turneaure, 1960). The formation discordantly overlies various units of the underlying Ordovician and is in turn overlain concordantly by a thick sequence of black shales. The thickness of these shales is not properly known, but they are estimated as being about 1000 m thick, and they pass upward into the Devonian.

The lateral variation within the lithofacies of the Silurian suggests the existence of a source area to the southwest of the Altiplano (Berry and Boucot, 1972). It is most likely that the Arequipa Massif was that source area (Fig. 4).

3. *Devonian*

Devonian deposits in southern and central Peru occur in three distinct zones—on the Arequipa Massif near the present-day coast; in the area of the Altiplano; and in the axial zone of the Eastern Cordillera.

Devonian sediments have been recorded from two localities on the Arequipa Massif, Toran, and Cocachacra (Fig. 5) (Boucot *et al.*, 1980), and in both areas the sediments rest directly on crystalline basement. The section at Toran is 400 m thick and consists of sandstones and greenish-grey shales overlying a basal conglomerate 40 m thick. It contains a poor fauna of Malvinokaffric affinities. The sequence at Cocachacra is about 100 m thick and consists of a basal conglomerate 30 m thick overlain by quartzites and indurated lutites. The fauna from this area suggests a warm-water realm in contrast to the cold-water Malvinokaffric realm of Toran, and all other Devonian and Silurian faunas in southern Peru and Bolivia (Boucot *et al.*, 1980).

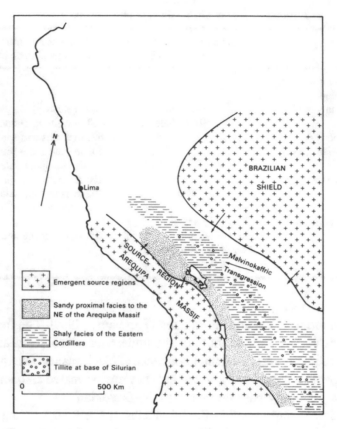

Fig. 4.   Distribution of Silurian strata in Peru and Bolivia (after Berry and Boucot, 1972).

In the Altiplano marginal zone the passage from Silurian to Devonian is represented by the quartzites of Taya Taya (Laubacher, 1977), which are overlain by about 500 m of sublittoral microconglomerates followed by 600 m of shales and fine sandstones containing a fauna of Emsian age (Newell, 1949; Boucot in Laubacher, 1977). These Lower Devonian strata are overlain by a middle Devonian sequence that comprises 200 m of white quartzites of Eifelian age (Wolfart and Voges, 1968) and 200 meters of shales with do-lomitic concretions.

In the axial zone the Devonian follows the Silurian concordantly, and because of the similar lithologies and paucity of fauna, it is not possible to distinguish the precise junction between the two systems. However, it is known that the Devonian sequence is thick throughout Bolivia and southern Peru, reaching about 4000 m in the region of Quillabamba. In this region a fauna of Chitinozoans suggests a Givetian to Frasnian age (Doubinger in Marocco, 1977). Toward the east of the axial zone the sequence is somewhat

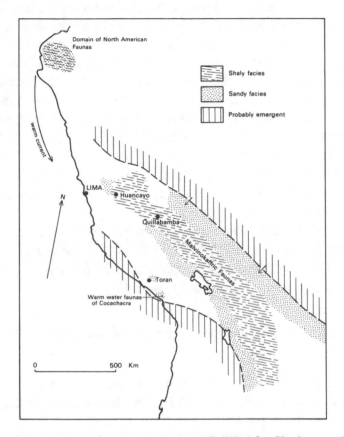

Fig. 5.   Distribution of Devonian strata in Peru and Bolivia (after Harrington, 1962; Turner 1972; Dalmayrac *et al.*, 1980).

reduced in thickness to about 200 m and becomes more arenaceous in character, suggesting a measure of supply from the Brazilian shield (Dalmayrac *et al.*, 1980). A somewhat more proximal sequence of early to middle Devonian age, which is about 3000 m thick and which has Malvinokaffric fauna, has been recognized in central Peru to the east of Huancayo (Paredes, 1972). This most probably represents a northerly extension of the axial zone of southern Peru and Bolivia. In the Amotape hills of northwest Peru about 1300 m of shales and sandstones rest discordantly upon the Olmos Schists. These rocks contain a warm-water fauna similar to that at Cocachacra.

The division of the Devonian outcrop into marginal and axial zones seems to be well established (Fig 4). Isaacson (1975) showed that the sedimentary facies were consistently more proximal toward the Arequipa Massif, and more distal towards the axial zone and the Brazilian Shield. He argued from these data that the principal source of supply for Devonian sediment was the Arequipa Massif, and because the outcrop area of the

Massif was inadequate to supply the calculated volume of sediment, he proposed an extra continental source region, in effect a much enlarged Arequipa Massif extending into what is now the Pacific Ocean. However, Dalmayrac *et al.* (1980) suggest that the Brazilian shield probably also supplied sediment to the axial zone, and it is also known (Lehmann, 1978) that crystallines of the Arequipa Massif underlie the Altiplano. Consequently, the known extent of the massif at the present time is much greater than was formerly considered to be the case. These factors, while affecting the argument of Isaacson, do not nullify it, and if in addition to the Devonian, allowances are made for the Ordovician and Silurian deposits in the axial zone, for which no volumetric calculations have been made, it may be that an extra continental source area would be necessary.

The Malvinokaffric cold-water domain was constant in the axial zone throughout the Silurian and Devonian. Devonian warm-water faunas from northwest Peru and Cocachacra are attributed by Boucot *et al.* (1980) to the affect of cross-latitudinal warm-water currents.

C. Eohercynian Folding and Upper Paleozoic Sedimentation in the Eastern Cordillera

At the end of the Devonian or the beginning of the Mississipian the Lower Paleozoic deposits were folded. The Eohercynian folding is polyphase and up to five phases, that range in intensity from isoclinal folds to kink bands, have been recognized (Dalmayrac *et al.*, 1980). The most important and widesread phase is F2 (Dalmayrac *et al.*, 1980), which dominates the foldbelt from Bolivia to Central Peru. Fold styles range from isoclinal to concentric and are developed at all scales from tens of meters to several kilometers. A vertical flow cleavage is normally developed in the axial zone, but toward the eastern and western borders, is recumbent. The overall structure of the belt is that of an anticlinorium with the older rocks exposed in the center of the axial zone and younger rocks on the flanks. The fold episode was followed by uplift and erosion that was particularly well marked in the axial zone. In this region Upper Paleozoic strata rest most discordantly upon those of Lower Paleozoic age.

The Eohercynian folding resulted in a complete change in the character of sedimentation. Whereas the Lower Paleozoic was a uniformly thick sequence of marine shales and sandstones, the Upper Paleozoic varied from continental to neritic deposits with rapid lateral and vertical variation in both facies and thickness. The main basin of deposition was superimposed on the former Lower Paleozoic axial zone and also on the schists of the late Proterozoic foldbelt.

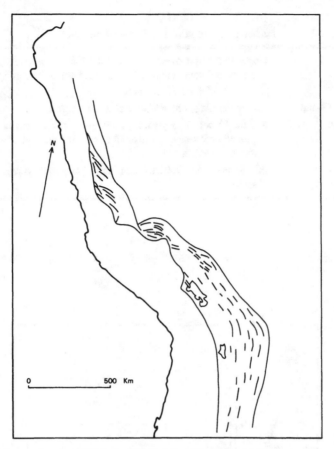

Fig. 6.   Fold axes of major folds formed during the Eohercynian deformation (Upper Devonian/Lower Mississipian) in the Eastern Cordillera of Peru and Bolivia.

Newell *et al.* (1953) have defined four major stratigraphic groups (Table V). A major phase of uplift with minor folding, the Tardihercynian episode (Mégard *et al.*, 1971), occurred in the late Lower Permian during the interval between deposition of the Copacabana Group and that of the Mitu Group.

1. *The Ambo Group*

Continental and marine deposits attributed to the Ambo Group occur along the western border of the Eastern Cordillera and also in isolated outcrops in the coastal areas of northwest and southwest Peru (Martinez, 1970). The Group was considered by Steinmann (1929) and by Newell (1953) to be entirely continental, but Mégard *et al.* (1971) have described fossiliferous marine sediments intercalated within the continental deposits. The continental deposits are sandstones and shales containing plant fragments, while the marine sediments are also comprised of sandstones and shales but con-

TABLE V
Lithostratigraphy of the Upper Paleozoic

Mitu Group	Upper Permian–Lower Triassic, 50–2000 m, andesitic plateau volcanics and continental red beds, sandstones, and conglomerates
	Major unconformity
Copacabana Group	Lower Permian, 400–600 m, neritic limestones
Tarma Group	Middle–Upper Pennsylvania, 2000–3000 m, dark shales with intercalated limestones in the axial zone; epicontinental limestones in the Sub-Andean range
Ambo Group	Mississippian, 100–1000 m, continental and marine sandstones and shales

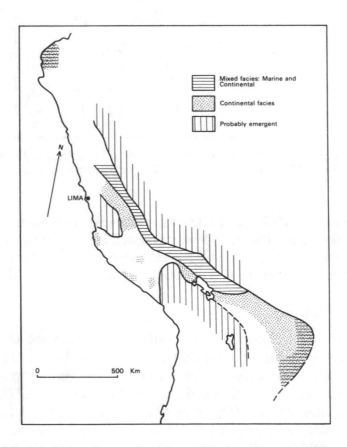

Fig. 7.   Distribution of Mississipian strata in Peru and Bolivia (after Harrington, 1962; Ahlfeld and Branisa 1960; Dalmayrac *et al.*, 1980).

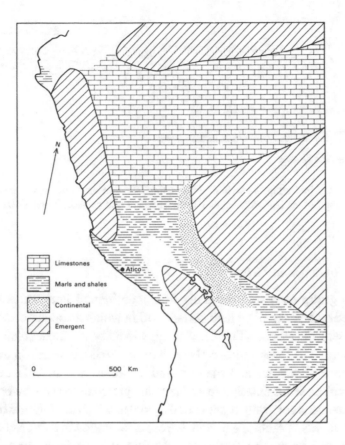

Fig. 8. Distribution of Pennsylvanian and Lower Permian strata in Peru and Bolivia (after Newell *et al.*, 1953; Mégard, 1973; Dalmayrac *et al.*, 1980).

tain brachiopods and marine lamellibranchs. The thickness of sediments over the length of the outcrop, which extends southward into the Cordillera Real of Bolivia, ranges from 100 to 1000 m.

2. *The Tarma Group*

The sediments of this group consist of epicontinental limestones and shales. Limestones occur in the sub-Andean region but the main basin of deposition coincides with the axial zone of the Lower Paleozoic where 2000 m to 3000 m of dark shales with intercalated limestones were deposited (Fig. 8). This facies overlaps onto the Arequipa Massif where about 1500 m of sandstones, shales, and limestones occur in the region of Atico. In Bolivia the Pennsylvanian is represented by the Gondwana tillite (Ahlfeld and Branisa, 1960). In northern Bolivia a deltaic facies shows that the Pennsylvanian Sea did not extend south of the Peru–Bolivia border (Helwig, 1972).

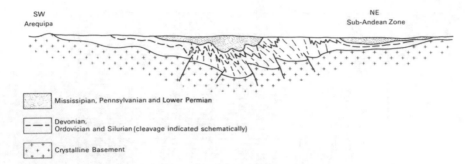

Fig. 9. Sketch indicating the style of deformation affecting the Paleozoic strata of southern Peru as a result of the Eohercynian and Tardihercynian fold episodes (after Dalmayrac *et al.*, 1980).

3. *The Copacabana Group*

The group consists essentially of epicontinental limestones that were deposited on a peneplaned land surface within which no specific sedimentary basins were developed (Fig. 8). The type locality is Copacabana, Bolivia, on the shores of Lake Titicaca (Newell *et al.*, 1953). The group outcrops in the Eastern Cordillera and the sub-Andean zone, and in both areas there was unbroken sedimentation from the Pennsylvanian to the Lower Permian. The sequence is generally represented by 400 m to 600 m of neritic limestones with fusulinids, brachiopods and bryozoans. Following deposition of the Copacabana Group, the Tardihercynian fold episode took place, which affected strata of Upper and Lower Paleozoic age (Fig. 9).

4. *The Mitu Group: A Post-Tectonic Molasse*

The Tardihercynian uplift gave rise to a general emergence, and the resulting land surface was subjected to very rapid erosion. Continental deposits were laid down in a series of elongate troughs, which were formed along the entire length of the Eastern Cordillera by a strongly developed system of horst and graben tectonics related to the regional uplift (Fig. 10). A massive andesitic vulcanicity was generated in the same zone and contributed to the basin fill, which consequently consists of highly variable quantities of continental sandstones and volcanics.

In spite of the local character of sedimentation the typical detrital facies of the Mitu is relatively uniform. It consists of breccias, conglomerates, arkosic sandstones, and shales of a red or violet color. The nature of the deposits suggests rapid erosion and transport over short distances. The volcaniclastic portion is derived directly from the Mitu volcanoes and the proportion of volcanic material diminishes with distance from the volcanic centers. This is illustrated by the fact that deposits of the Mitu on the Arequipa

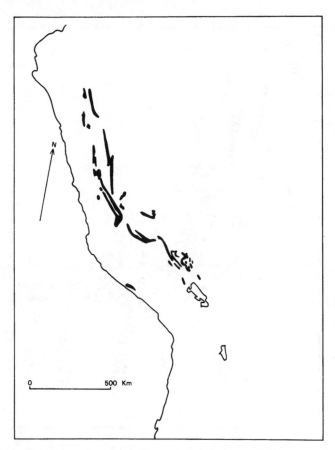

Fig. 10.   Outcrop of the Mitu Formation, Upper Permian plateau volcanics, and red beds (molasse) in the Eastern Cordillera of Peru (after Dalmayrac, 1980).

Massif in the region of Atico consist entirely of continental sediments. Evaporite deposits are locally developed and, toward the north of the Peru salt domes in the sub-Andean zone, penetrate younger rocks.

The vulcanicity consists of breccias, ignimbrites, tuffs, and flows of andesitic to rhyolitic composition, while basalts are locally present at the base of the volcanic sequence. The volcanics are chemically calcalkali to alkali while some of the more acid lavas are peralkaline. Noble *et al.* (1978) have suggested that the combination of peralkaline and calcalkaline chemical characteristics is consistent with a hypothesis of continental rifting as the most appropriate tectonic setting for the generation of the Mitu Volcanics.

C. Permo–Triassic Plutonism in the Eastern Cordillera

A chain of large granite plutons are emplaced along the length of the Eastern Cordillera from northern Bolivia to central Peru (Fig. 11) The plutons intrude both the late Precambrian schists and the Paleozoic sediments

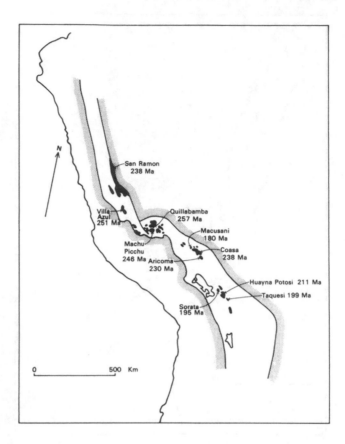

Fig. 11.   Distribution of Permo–Triassic granite plutons in the Eastern Cordillera of Peru and Bolivia (after Lancelot *et al.*, 1978; Mégard *et al.*, 1971). Radiometric ages quoted: *San Ramon* 238 ± 10 Ma Rb/Sr, Capdevila *et al.*, 1977; *Villa Azual* 251 Ma K/Ar, Stewart *et al.*, 1974; *Quillabama* 257 ± 5 Ma U/Pb, Lancelot *et al.*; Machu Picchu 246 ± 10 Ma Rb/Sr, Priem *et al.*, unpublished, quoted in Lancelot *et al.*, 1978; *Macusani* 180 Ma, Stewart *et al.*, 1974; *Coasa* 238 ± 11 U/Pb, Lancelot *et al.*, 1978; *Aricoma* 230 ± 10 Ma U/Pb, Dalmayrac *et al.*, 1980; *Sorata* 195 Ma K/Ar, *Huayna Potosi* 211 Ma, *Taquesi* 199 Ma, Clarke and Farrar, 1972.

that form the two elements of the cordillera, and the plutonism together with the Mitu vulcanicity provided a final magmatic term that stabilized geological activity in the Eastern Cordillera and unified it into a single structural whole.

These intrusions are mainly granites and granodiorites with a range of SiO_2 from 66 to 74% They are normally biotite granites, commonly megacrystic with pink or grey orthoclase. Mafic dykes are present, and the granites are not normally xenolithic except in the marginal zones. The plutons intrude rocks that range from late Precambrian to late Permian in age and they are surrounded by well-developed thermal aureoles in which andalusite, cordierite, and biotite are common. Some of the plutons cut the Mitu Group and isotopic dating by Rb/Sr, U/Pb, and K/Ar indicates a range of ages from

TABLE VI
Lithostratigraphy of the Marcona Area

Phyllites and limestones	1000 m	Marcona Formation
Dolomites and schists	1000 m	San Juan Formation
Diamictite or tillite	600 m	Chiquerio Formation

260 Ma to 180 Ma (Capdevila *et al.*, 1977; Lancelot *et al.* 1978; Dalmayrac *et al.*, 1980).

The granites are chemically calc-alkaline to sub-alkaline with a K_2O/Na_2O ration greater than one, and thus differ from the later Andean granitoids. Similarly, the $^{87}Sr/^{86}Sr$ initial ratio of 0.7075 (Dalmayrac *et al.*, 1980) suggests an anatectic origin, or a mixed crustal/mantle source area for these granites. The granites of this belt in Bolivia are of particular interest since they are foliated two-mica granites and carry cassiterite. They form an essential component of the Bolivian tin province.

IV. PALEOZOIC SEDIMENTS AND PLUTONISM IN THE AREQUIPA MASSIF

There are sediments and granitoid plutons that occur along the western side of the Massif, that cannot be readily correlated with any episode in the Peruvian or Bolivian Andes and seem to form a separate and quite distinctive province of their own. For this reason they may be conveniently considered apart.

1. *The Marcona Formation*

The Marcona formation is the name that has been traditionally applied by geologists of the Marcona Mining Company to a sedimentary sequence containing the stratiform iron deposits worked by the company, now known as Hierro Peru. Marcona itself is situated about 50 km south of Nazca and the sediments occupy a faulted outcrop of about 40 km strike length. The sequence, which is folded and metamorphosed in the chlorite to biotite grade, rests unconformably on Precambrian gneisses and is divided into three units that have been renamed by Caldas (1978) (Table VI). There seems to be good evidence that the Chiquerio Formation is of glacial origin, and it has been correlated with the Zapla–Cancarini Formation of Bolivia and Peru by Shackleton *et al.* (1979). The evidence for this correlation, however, is not at present compelling and depends on the isotopic dating of two Paleozoic batholiths that occur in the area, which are now briefly described.

2. *The Lower Paleozoic Plutonism*

The San Nicolás Batholith, which was formerly considered to be of Jurassic age (Pitcher, 1974), was dated by the K/Ar method (Wilson, 1975) and was shown to have an age of 440 Ma. The batholith consists of hornblendic diorites, tonalites, and granodiorites, and it cuts the regionally metamorphosed sediments of the Marcona Formation. The batholith was dated again by Shackleton *et al.* (1979) who obtained an age of 400 Ma by the Rb/Sr method. The Marcona Formation is therefore either pre-Upper Ordovician (Wilson, 1975) or pre-Upper Silurian (Shackleton *et al.*, 1979). The latter figure would permit correlation with the Zapla–Cancarini Formation, but the former would not.

Shackleton *et al.* also dated a suite of plutonic rocks, the Atico Igneous Complex, and the Camana Atico Batholith, and obtained an age of 440 Ma by Rb/Sr. Although they state that the Marcona Formation is bracketed by the ages 400 and 440 Ma, their map shows that the Atico Igneous Complex does not underlie the formation. There is at present no good indication of the lower age limit for the formation.

The initial $^{87}Sr/^{86}Sr$ ratios of 0.7097 for the Atico Complex and 0.7087 for the San Nicolás Batholith are interesting since they suggest a crustal source area for these rocks. This plutonism remains a somewhat enigmatic factor in the geology of Peru, for although only locally developed, both plutonic sequences represent considerable plutonic episodes that cannot presently be satisfactorily related to other geological events of similar age in the Andean region.

V. DEVELOPMENT OF THE WEST PERUVIAN TROUGH

Once the orogenic cycle in the Eastern Cordillera had been completed by Permian vulcanicity and plutonism a new cycle was initiated on the western flank of the earlier foldbelt. It would appear that this new cycle was provoked by subduction of oceanic crust at the continental margin, which ushered in a tectonic regime that has dominated the region up to the present.

The new belt was formed in crust that was marginal to the earlier belt. In southern Peru this is represented by the Arequipa Massif, but over the rest of the Western Cordillera the basement is concealed by younger rocks. However, the presence of epicontinental deposits of Upper Paleozoic age in the coastal areas of northwest Peru and on the Arequipa Massif (Dalmayrac *et al.*, 1980) suggests that all of Peru was a stable continental area at this period. If this supposition is correct the Mesozoic formations are probably floored by crystalline rocks similar to the Arequipa Massif, with thin marginal Lower Paleozoic sediments and a patchy distribution of Upper Paleozoic epicontinental sediments. The new orogenic cycle was initiated by a widespread horst and graben development essentially similar to that begun during the terminal phase of the earlier cycle, which controlled the vulcanicity and sedimentation of the Permian molasse. However, granite emplacement in the Eastern Cordillera had thickened and stabilized the crust in that area, with the result that the former western foreland became relatively thinner and more subject to graben formation under the new system of tensional tectonics which now become dominant, and which was to continue throughout the Mesozoic. The former marginal zones to the older belt naturally became reactivated under such a regime and the sedimentary basins that developed to the west received sediments from the now emergent Eastern Cordillera. The High Plateau area also was established at this time and continued to develop throughout the Mesozoic as an intermediate zone where relatively thin shelf deposits accumulated in contrast to the thicker sequences to the west.

From this time onward the axial zone of the Eastern Cordillera acted as a positive element, the Marañon Geanticline, which separated Mesozoic sedimentation into two separate areas—the West Peruvian Trough and the East Peruvian Trough (Wilson, 1963). It was the cumulative geological activity within the West Peruvian Trough that ultimately gave rise to the Western Cordillera as a physical structure.

A. The West Peruvian Trough: A Continental Margin Back-Arc Basin

Throughout the history of this depositional structure a chain of volcanoes that contributed material to the back-arc basin on the landward side of the chain was present along the continental margin. At the same time clastic sedimentary material, derived from the Marañon Geanticline, was fed into the basin on the other side. In this way a polarity from eastern sediments to western volcanics was established early on, and continued throughout the life of the structure. The trough itself was a very large structure that was not simple but divided into internal horsts and grabens each of which had a distinctive history of its own (Cobbing, 1978). These internal basins were all interconnected and the result of this complexity is that whereas in some regions the volcanic and sedimentary components are neatly separated into well-defined belts that can be labeled miogeosynclines and eugeosynclines in the classical sense, in other parts there is a poorly defined transition. This is the general case in southern Peru, whereas in north and central Peru the belts are very clearly separate.

1. *Triassic to Middle Jurassic*

Throughout the Triassic and Lower Jurassic massive carbonates were deposited in the region of the High Plateau and on the Marañon Geanticline. The thickness of these limestones is very variable since they were controlled by the horst and graben structure established during the Permian. For example, the sequence to the west of the Cerro de Pasco Fault is 600 m thick, whereas on the east side it is 3000 m thick (Jenks, 1951). In general, the Triassic–Jurassic carbonate sequence follows the Mitu Group concordantly and inherits the same sedimentary and structural basins. This rather complex limestone sequence is collectively known as the Pucará Group (Jenks, 1951).

Andesitic flows and pyroclastic deposits occur on the coast in southern Peru where they are known as the Chocolate Formation, and also in the coastal areas in northern Peru where they consist of about 1500 m of breccias and pyroclastics known as the Zaña Formation (Fischer, 1956). In this area the volcanics are transitional to Pucará limestones to the north of Cajamarca (Cobbing *et al.*, 1981). There are no known outcrops of Triassic or Lower Jurassic strata over the greater part of the area covered by the West Peruvian Trough. Consequently, it is not known whether deposits of this age were ever laid down; whether they are simply concealed; or whether they have been removed.

There is no record of strata of middle Jurassic age anywhere in Peru. It is probable that this time span was marked by a general emergence accompanied by lack of activity in the volcanic arc. This episode has been termed the Nevadan epeirogeny (Ruegg, 1962).

2. *Upper Jurassic to Upper Cretaceous*

The volcanic arc was reestablished in the Upper Jurassic, and volcanic sequences were deposited in the coastal regions from southern Peru to north of Lima. Further north volcanics of this age are concealed by Cretaceous formations. These Jurassic volcanics are known by different names in different places—Puente Piedra near Lima, Rio Grande at Nazca, and Guaneros at Chala. On the eastern side of the trough clastic sediments were contributed from the Marañon Geanticline to form the 2000+ m Chicama Formation in northern Peru and the Yura Formation of similar thickness in southern Peru (Fig. 12). In both cases shales predominate over sandstones, though sandstones are more important in the upper part of the Chicama Formation. The Marañon Geanticline and the High Plateau were mainly emergent at this time.

During the Neocomian the High Plateau and the Marañon Geanticline became progressively submerged and were the repository of 500 m of orthoquartzites—the Goyllarisquizga Formation (McLaughlin, 1924)—while

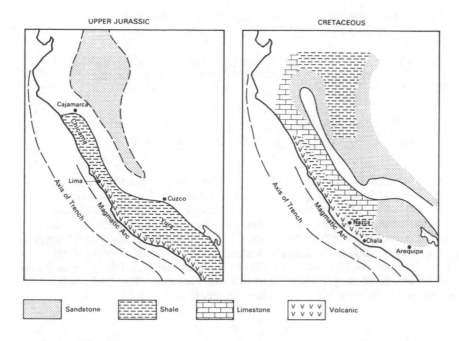

Fig. 12.   Distribution of lithofacies in the West and East Peruvian troughs during Upper Jurassic and Cretaceous times (after Mégard, 1978).

within the trough the sequence was expanded to give a mainly shaley section 1500–2500 m thick (Fig. 13). Toward the coast these shales are replaced by volcanics, and in the Lima area and over a substantial distance to the south, a thick shaley sequence is present. It was during the Upper Jurassic and Neocomian that individual basins within the West Peruvian Trough became well established, and they persisted to the end of the Cretaceous (Cobbing, 1978).

During the Albian, carbonate sequences were deposited that were thicker in the trough (400–500 m) than on the High Plateau (100 m). In the middle Albian a marine transgression took place, which finally submerged the Marañon Geanticline (Wilson, 1963), and a sequence of about 100 m of bituminous limestones—the Pariatambo Formation—was deposited over the Geanticline, the High Plateau, and in the eastern basins of the West Peruvian Trough. This period marked a resurgence of activity in the volcanic arc and great sequences of volcanics which are referred to as the Casma Group, were deposited in the western basins of the trough at this time (Cobbing, 1978). The thickness of this group is not known, but it is probably in excess of 5000 m, and in the region of Chimbote it has been estimated as being more than 9000 m (Bussell, 1975). Volcanics of this age extend from Trujillo in northern Peru to Chala in southern Peru, and comprise pillow lavas, flows,

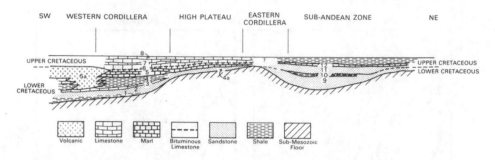

Fig. 13. Stratigraphic correlation chart for the Upper Jurassic and Cretaceous in a section across Northern Peru. (1) Chicama Formation: Tithonian. (2) Chimu Formation: Berriasian–Lower Valanginian. (3) Santa Formation: Upper Valanginian. (4) Carhuaz Formation: Hauterivian–Barremian. (4a) Goyllarisquizga Formation: Neocomian. (5) Pariahuanca, Chulec Formations: Aptian–Lower Albian. (5a) Casma Group: Neocomian–Albian. (6) Pariatambo Formation: Middle Albian. (7) Jumasha Formation: Upper Albian–Coniacian. (8) Celedin Formation: Santonian. (9) Lower Agua Caliente Formation: Berriasian. (10) Esperanza member, base of: Middle Albian. (11) Upper Agua Caliembe Formation: Middle Albian. (12) Chonta Formation: Upper Albian–Coniacian. (After Mégard, 1978.)

and volcaniclastic deposits of basaltic and andesitic composition with sporadic shaley intercalations (Myers, 1974).

The Upper Cretaceous consists entirely of carbonates. In central Peru these are comprised of two formations: the Jumasha Formation of Upper Albian–Turonian age, and the Celendin Formation of Coniacian–Santonian age, which have a combined thickness of 1500 m in the trough and about 1000 m on the High Plateau (Benavides, 1956; Wilson, 1963). Northward along the trough these carbonates merge into an expanded sequence that includes intercalated marls and have a combined thickness of 2500 m. The Pulluicana Group, the Quilquiñan Group, and the Otuzco Group are equivalent to the Jumasha and Celendin formations of central Peru (Benavides, 1956; Wilson, 1963; Cobbing *et al.*, 1980).

In Peru south of 13°S, or about the latitude of Cuzco, the Cretaceous sequence is much reduced and is of continental type both in the trough and in the High Plateau (Marocco, 1971). Two marine incursions are present: the Albian Ayavaca Limestone (Newell 1949) and the Vilquechico Formation of Campanian age (Newell, 1949).

B. Uplift and Molasse Formation

At some rather indeterminate time toward the end of the Cretaceous or the beginning of the Tertiary the West Peruvian Trough became emergent, and clastic sediments were shed eastward onto the High Plateau and Marañon Geanticline. This episode was named the Peruvian Phase by Steinmann

TABLE VII
Distribution of Lithologies in the Coastal Batholith

Lithology	Southern Peru[a]	Central Peru[b]	Northern Peru[c]
Gabbro–Meladiorite	7.0	15.9	11.5
Tonalite	55.0	57.9	46.3
Granodiorite			20.0
Monzogranite		25.6	20.2
Syenogranite	4.0	0.6	2.0

[a] Jenks, 1948.
[b] Cobbing and Pitcher, 1972.
[c] Child, 1976.

(1929). These red beds of the Chota and Casapalca formations are commonly concordant with upper Cretaceous strata, but they overstep eastward onto progressively older formations. The sediments consist of rhythmic molasse deposits of conglomerates, sandstones, and shales which are predominantly red; lacustrine limestones, evaporites, and andesitic volcanics also occur locally. Distinct basins of red-bed deposition were formed in the High Plateau (Fig. 1), and the thickness of the deposits is somewhat variable, ranging from 1000 to 4000 m (Mégard, 1978). The largest of these basins survived subsequent deformation and continued as intermontane basins of inland drainage—the Altiplano of the present day. Many of these have now been breached and drain into the Atlantic, but the largest of them, the Altiplano of southern Peru, Bolivia, and Chile, continues as a basin of inland drainage to the present time.

C. The Coastal Batholith: A Lineament of Cretaceous Plutonism

The batholith is a composite plutonic structure about 60 km wide that has a marked linearity and occurs in the coastal region along the length of Peru. Batholithic rocks of similar type and dimensions occur throughout the Andes, North America, and the whole circum-Pacific region, forming one of its most distinctive features. In Peru the batholith is a composite structure, built up of about 1000 plutons that range in size from 30 km^2 for the monzogranites to 800 km^2 for the tonalites (Pitcher, 1978). From the figures in Table VII it will be seen that the most important rocks are tonalites and granodiorites.

1. *Plutons, Superunits, and Segments*

The Coastal Batholith has been found to be organized in a systematic way (Cobbing et al., 1977), and its constituent components are the pluton,

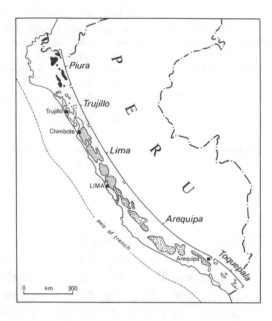

Fig. 14   The segmented Coastal Batholith of Peru.

the superunit, and the batholithic segment (Fig. 14). The primary element is the pluton, which may itself be simple or complex and internally zoned by pulses and surges of related magma, yielding a compositional variation from a relatively basic margin to a more acid core.

Chains of plutons of identical lithology have been recognized over distances of several hundred kilometers. These chains, although large, are of finite length and always occupy one particular area of the batholith. It is considered that they are lithological magmatic elements each of which was produced by a particular fusion event in a source area that was of the same linear extent in the lower crust or upper mantle as the outcrop at the surface. These chains of plutons can in practice be mapped as lithostratigraphic formations and have been termed "superunits" (Cobbing *et al.*, 1977). Superunits usually show a compositional range from diorite to monzogranite, but some are more basic, or more acid, at the outset, than the norm, and consequently give a particular compositional bias to the sector of the batholith in which they occur.

It has been found that each segment of the Coastal Batholith is characterized by a particular assemblage of superunits that are specific to that segment (Fig. 15). Five of these segments have been recognized to date and three have been mapped in reasonable detail—the Lima segment, the Arequipa segment, and the Toquepala segment. The superunits in each segment are characterized by overall geochemical characteristics that distinguish

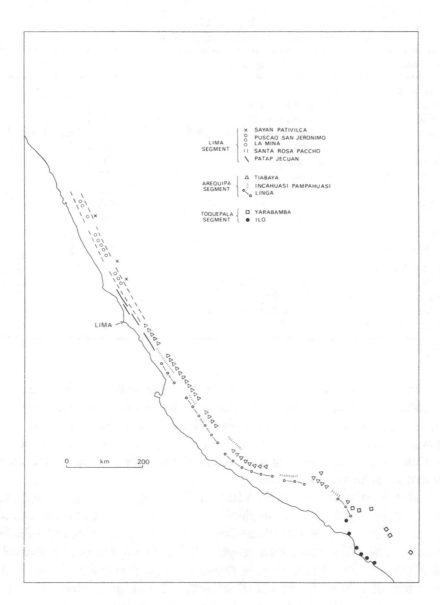

Fig. 15. Distribution of superunits within the batholithic segments.

them from superunits in adjacent segments (Atherton *et al.*, 1979). For example, the K_2O/SiO_2 ration is higher for those of the Arequipa segment than for the Lima segment. The Arequipa segment is more heavily mineralized than the Lima segment, which is relatively barren. Porphyry copper deposits occur in the Toquepala segment and in the Piura segment (Cobbing *et al.*, 1981).

TABLE VIII
Superunits in the Lima and Arequipa Batholithic Segments

Superunit	Composition	Age (Ma)	
Lima Segment			
Cañas	Monzogranite	58	
Sayán	Monzogranite	61	
Puscao	Grandiorite to Monzogranite	62–65	Ring complexes
San Jeronimo	Monzogranite	61	
La Mina	Tonalite	66	
Dyke swarm	Andesite	70	
Humaya	Granodiorite	72	
Santa Rosa	Tonalite to Granodiorite	80–90	
Paccho	Diorite to Tonalite	95	
Jecuan	Granodiorite	102	
Patap (Precursor)	Gabbro	102?	
Arequipa Segment			
Tiabaya	Tonalite to Granodiorite	80	
Incahuasi	Monzodiorite	82	
Linga	Monzodiorite to Granodiorite	97	
Patap (Precursor)	Gabbro	102?	

In comparing the Lima and the Arequipa segments it is clear that the tonalites that comprise the main volume of the batholith in both segments were emplaced fairly early on between 95 and 75 Ma. Activity in the Arequipa segment then stopped but in the Lima segment there was a renewal at about 60 Ma, and monzogranitic superunits that form the ring complexes in that segment were generated.

It will be seen from Table VIII that the Patap Gabbro is common to both segments and is not geographically limited in the way that the granitoid superunits are. It is best considered as a basic precursor to the batholith as a whole. The gabbros are rather variable olivine, two-pyroxene, hornblende, plagioclase rocks. They have been shown to be geochemically distinct from the granitoid units of the batholith (McCourt, 1981) and are of olivine tholeiite composition, geochemically similar to the more basic volcanics of the Casma Group (McCourt, 1981). The gabbros were emplaced as large plutons, up to 50 km − 10 km on the west side of the batholith, and also as somewhat smaller bodies, across the width. These latter have been somewhat dissected by later, more acid, arrivals, but they occur throughout the length and breadth of the batholith.

The 60-Ma granites of the ring complexes occur in a well-defined line down the center of the batholith. This emphasizes the structural control of the locus of generation of the later, more-acid magmas, which are not located at a greater distance from the subduction zone than earlier magmas but were generated within precisely the same zone. The initial $^{87}Sr/^{86}Sr$ ratios for all plutons of the batholith are 0.7042 (McCourt, 1981), which suggests a primitive origin for all batholithic magmas. A crustal underplate has been suggested as a likely source area (McCourt, 1981).

Plutons of the batholith were typically emplaced into volcanic country rocks of Cretaceous age; however, sedimentary rocks of Cretaceous and Upper Jurassic age form part of the envelope in northern and southern Peru, and in southern Peru metamorphic rocks of the Arequipa Massif form the envelope rocks at outcrop. In these different situations the granitoids were emplaced as epizonal plutons having steep walls and flat roofs and with very restricted thermal aureoles commonly only tens of meters, even in situations where limestones and shales are the country rock.

The plutons appear to have been emplaced by stoping and the evidence for this in the case of the monzogranites is good. Swarms of accidental xenoliths in the roof and marginal zones can be successfully related to stoping mechanisms. The tonalites and granodiorites, on the other hand, carry ubiquitous microdiorite xenoliths that are not accidentally derived since they are equally characteristic of plutons where the country rock is gneiss, volcanic, or sedimentary rock. Xenoliths of this type are considered by Chappell and White (1974) to represent "restite," i.e., refractory material from a source area that has been incorporated into the magma as a remnant and carried into the upper crust during emplacement of the pluton.

In the Lima segment plutons are typically not foliated. In the Arequipa segment Tiabaya and Linga plutons are not foliated, but Incahuasi plutons very often are. This foliation may be quite intense and xenoliths severely flattened. The local country rocks, however, are not comparably affected by penetrative cleavages, even though they are folded and the folds are cut by the foliated plutons. It is evident that the foliation is a result of emplacement tectonics, which must reflect the relative ductility of the plutons and the host rocks (Berger and Pitcher, 1970). The foliations follow Andean and Andean normal directions, which also commonly define pluton contacts. Resurgent movement along fractures in these directions during emplacement both facilitated stoping in the brittle country rock and foliation in the more ductile pluton. Most plutons have a suite of mafic dykes that are invariably of andesitic composition attached to them. In some cases these form major dyke swarms.

All of the plutons were emplaced at epizonal levels with a cover of probably not more than 2 or 3 km. In the case of the ring complexes it was probably less, and there is evidence that the ring dykes vented to the surface. It has been argued that these complexes contributed substantially to the Tertiary volcanic cover (Pitcher, 1972; Cobbing and Pitcher, 1972a). However, it has now been shown that the volcanics are geochemically distinct

(McCourt, 1981); that the volcanics are not of the same age as the plutons (Atherton *et al.*, 1979); and that the ring complexes and Tertiary volcanics are not geographically coextensive (Atherton *et al.*, 1979). Accordingly, it would seem that if these complexes did vent to the surface, which they seem to have done, their contribution to the volcanic cover was slight.

In both the Arequipa and the Lima segments large plutons that cut folded country rocks of Lower Cretaceous age have been dated at between 90 and 100 Ma. These ages require that folding, which is coaxial with the later Paleocene Incaic folding, must have taken place in the Lower Cretaceous (Cobbing, 1978).

A well-defined metallogenic belt is associated with the Coastal Batholith, and the degree of mineralization within the belt closely reflects the segmentation of the batholith. The belt is a copper–molybdenum–gold province, but tungsten and base metals are locally important. Porphyry copper deposits are confined to the Toquepala and Piura segments (Cobbing *et al.*, 1981).

The great bulk of the granitoids were emplaced in the already folded eugeosynclinal zone during the Upper Cretaceous, and it is known that molasse deposition began in red-bed basins on the High Plateau at the end of the Cretaceous and continued into the Paleocene. It is most likely that the crustal thickening caused by batholith emplacement induced this uplift, and it is possible that by the time of the emplacement of the ring complexes there may have already been an emergent land surface in the eugeosynclinal zone that was providing a source region for the molasse deposits.

D. The Incaic Folding and Formation of the Eocene Erosion Surface

At some time during the Eocene the Cretaceous–Tertiary sedimentary pile within the miogeosynclinal basins and the High Plateau was intensely folded. Within the miogeosynclinal area huge concentric folds were developed, the dimensions of which were controlled by the competent Chimu Quartzites and Upper Cretaceous limestones, the quartzites usually forming the cores of the anticlines and the limestones the synclines. The whole sedimentary sequence deformed as a packet, detached from any underlying basement by the insulating medium of the thick, plastic Upper Jurassic shales that were folded disharmonically with respect to the concentric structures in the Cretaceous sequence. Incompetent shaley formations within the Cretaceous sequence were also deformed disharmonically but to a lesser degree than within the thicker Chicama Formation. Faults were commonly developed along the flanks and crests of structures, and while some of these were superficial and confined to the sedimentary pile, others were more deeply rooted and represented the reactivation of earlier horst-and-graben structures. The most important of these resurgent faults was at the juntion of the miogeosyncline with the High Plateau where the earlier hinge line that controlled the Cretaceous stratigraphy was reactivated to provide a zone of intense high-angle and low-angle thrusting—the imbricate zone of Wilson *et al.* (1967).

Within the High Plateau the tectonic style was different, and this reflected the thinner Cretaceous sequence, the absence of thick incompetent shaley formations, and the presence of the existing red-bed basins. A system of domes and basins was developed in which the structural basins inherited the older red-bed basins. Toward the junction with the miogeosyncline, deformation was more intense. Also, in particular areas where thick Triassic and Jurassic carbonates were present, concentric folds of similar dimensions to those in the miogeosyncline were developed. With the completion of the folding a surface of erosion was developed on the emergent Andes that was of regional extent and covered the High Plateau, the miogeosyncline and eugeosyncline, where the surface was cut into the exposed tops of the batholithic plutons.

1. *Tertiary Plateau Volcanics*

A pile of mainly fragmental andesites and dacites about 2000–3000 m thick, was erupted onto the subaerial Incaic erosion surface over a long period between the Eocene and upper Miocene (Cobbing *et al.*, 1981) (Fig. 16). The sequence is known as the Calipuy Group in central and northern Peru and as the Tacaza Formation in southern Peru, where it is overlain by a very widespread outcrop of middle to late Miocene dacitic tuffs (Noble *et al.*, 1979).

The axis of this great volcanic field is located about 70 km inland, well to the east of the batholith and approximately along the junction between the Cretaceous miogeosynclinal and eugeosynclinal zones. This volcanic sequence rests on the Incaic structures with great unconformity, but is itself folded on coaxial Andean trends.

2. *Tertiary Plutonism*

A new axis of plutonism was established during the Tertiary that was different from that of the Cretaceous, being much further inland (Fig 16), and corresponding approximately to the outcrop area of the Tertiary Plateau volcanic field, although some plutons, and among them the most important, do extend beyond the eastern limit of the field. These plutons are usually small, four to ten km^2 in size, and are biotite, hornblende, plagioclase porphyries of granodioritic composition (Cobbing *et al.*, 1981). They occupy a time span of 30 to 10 Ma and they extend from the east side of the Coastal

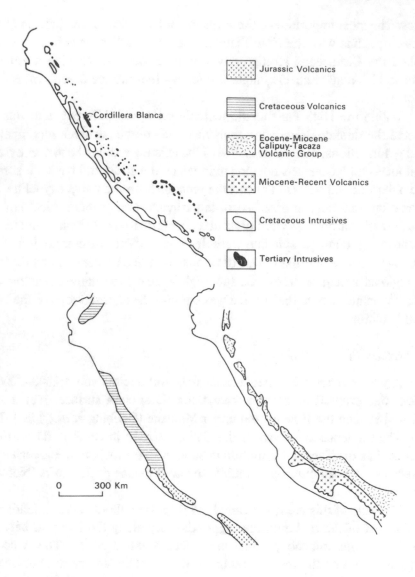

Jurassic Volcanics

Cretaceous Volcanics

Eocene-Miocene
Calipuy-Tacaza
Volcanic Group

Miocene-Recent Volcanics

Cretaceous Intrusives

Tertiary Intrusives

Cordillera Blanca

0   300 Km

Fig. 16. Distribution of Cretaceous and Tertiary intrusives, and Cretaceous marine volcanics and Tertiary plateau volcanics.

Batholith to the zone of the miogeosyncline and the High Plateau. They are emplaced as small stocks within the Tertiary volcanic field and also within the folded miogeosynclinal sediments and in the High Plateau where they are a constant element in the base metal metallogenesis of the high Andes. Some larger batholiths do occur, for example the Cordillera Blanca Batholith near Huaraz (Egeler and De Booey, 1956; Cobbing *et al.*, 1981), and the Abancay Batholith near Cuzco (Marocco, 1978). These batholiths are associated with important tungsten and base metal deposits in the Cordillera

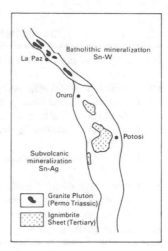

Fig. 17. The tin belt in Bolivia, divided into a batholithic zone and a subvolcanic zone (after Sillitoe *et al.*, 1975).

Blanca and skarn ironstones at Abancay. In southern Bolivia stocks of rhyodacite (Grant *et al.*, 1977) are associated with the major tin field of the Altiplano centered on Potosi (Fig. 17).

3. *Present-Day Vulcanicity*

In southern Peru a chain of andesitic volcanoes sits astride the older Tertiary field and extends southward into Chile. These have been the object of much recent work, and James *et al.* (1976) have shown that each volcano is petrologically distinct from its neighbor, and that the magmas may have been contaminated by crustal material (Tilton and Barreiro, 1980; James, 1981).

VI. DISCUSSION: COMPARISON OF GEOTECTONIC CYCLES IN THE EASTERN AND WESTERN CORDILLERAS

From the foregoing account it is clear that each of these orogenic belts had a unique and well-developed geotectonic cycle. The implication is that the different cycles reflect different tectonic settings.

A. The Western Cordillera

This is really the story of a magmatic arc superimposed on a continental margin. The geotectonic cycle here is essentially one of crustal thickening provoked and maintained by oceanic subduction which by initially generating the back-arc basin of the West Peruvian Trough, produced a crustal

TABLE IX
The Geotectonic Cycle in the Western Cordillera

(a) Formation of basin by crustal necking, horst- and-graben tectonics	Triassic
(b) Establishment of a magmatic arc	Triassic
(c) Preorogenic marine vulcanicity; Eugeosyncline/Miogeosyncline polarity in back-arc basin	Triassic–Upper Cretaceous
(d) Preorogenic plutonism in eugeosynclinal zone; Coastal Batholith	Upper Cretaceous
(e) Uplift and molasse formation	Upper Cretaceous Paleocene
(f) Incaic deformation, decollement tectonics, no crustal shortening: development of regional erosion surface	Paleocene
(g) Post-orogenic, plateau volcanics	Eocene–Miocene
(h) Post-orogenic plutonism	Miocene

anomaly where geological processes were most actively brought into play. The features of the cycle can be conveniently labeled, and it will be noted that different magmatic products are characteristic of different stages of the cycle (Table IX). The outstanding feature of this cycle is the complexity of the magmatic products produced at different stages of the cycle. It would seem that the underlying control of oceanic subduction is in some way perturbed at different growth stages of the belt so that successive products are different. Each of these four magmatic episodes can be correlated with different rates of seafloor spreading in the ocean (Larsen and Pitman, 1972; Charrier, 1973; Levi and Aguirre, 1981). The Western Cordillera orogenic belt can therefore be successfully linked to a tectonic regime of ocean floor spreading and subduction of oceanic crust beneath the continental margin. The case with the Eastern Cordillera is far different.

B. The Eastern Cordillera

The main difficulty in reconstructing the tectonic regimes of Paleozoic and earlier foldbelts is the absence of oceanic crust of that age. Consequently resort must be made to drawing analogies with the features of Mesozoic foldbelts or by identifying remnants of earlier ocean basins. In the case of the Eastern Cordillera these analogies and remnants are few.

A major difference between the two belts, as indicated in Tables IX and X, is their length of life, 400–500 Ma in the Eastern Cordillera as opposed to 150–200 Ma for the Western, but like that of the Western Cordillera this is a cycle of crustal thickening which is, however, achieved by an entirely clastic sedimentary process and by deformation with crustal shortening.

TABLE X
The Geotectonic Cycle in the Eastern Cordillera

(a) Formation of encratonic trough	Late Proterozoic
(b) Filling of trough by clastic sedimentation	Late Proterozoic
(c) Regional metamorphism, polyphase folding, crustal shortening	Late Proterozoic; 600 Ma
(d) Formation of Paleozoic trough	Cambrian, Ordovician
(e) Filling of trough by clastic sedimentation	Ordovician to Devonian
(f) Regional folding with penetrative cleavages, crustal shortening	Upper Devonian
(g) Uplift, Upper Paleozoic protomolasse	Mississipian to Lower Permian
(h) Major uplift minor folding	Middle Permian
(i) Volcanic sedimentary molasse, rift-and-graben tectonics	Upper Permian
(j) Granite plutonism	Upper Permian

There is a total lack of preorogenic magmatism, and the only magmatic products are the uplift associated Upper Permian plateau volcanics and granite plutonism that terminated the cycle. It is clear that the geotectonic cycle of the Eastern Cordillera was controlled by a different tectonic regime than that of the Western Cordillera, but it is not easy to suggest what that regime may have been. However, there is such a contrast in the volume and type of magmatic product within the two cycles that a comparison may shed some light on the mechanisms that generated them. The products that can at present be most directly compared are the granites, although it is to be hoped that more work will be undertaken with the volcanics, which will enable a proper comparison in that area.

C. The Granites Compared

It has been suggested by Chappell and White (1974), and by Pitcher (1979), that different kinds of granitoids characterize different tectonic settings. The idea behind this is that under different tectonic regimes different source areas become mobilized, and the resulting granites display differences that reflect those sources. Thus, subduction-related regimes mobilize source areas in the mantle and produce *I*-type granites, whereas collision related regimes result in crustal anatexis and the *S*-type granites of Chappell and White (1974). Each tectonic setting is distinguished by a particular suite of granites and a whole array of tectonic, volcanic, and sedimentary features. However, of all the sciences, geology is most characterized by its infinite variety, and things are not often as simple as they seem. In the case of the two belts under consideration a wide range of differences has been documented that, taken together, suggest that they developed under different tectonic regimes. The granites of the two belts will now be briefly considered to see whether they are sufficiently different to support the hypothesis that different source areas have been mobilized in the two belts.

The main difference between the granitoids of the Western and Eastern belts is that the former are typically tonalites and granodiorites, whereas the latter are chiefly granites. The compositionally expanded suite is characteristically Andino type whereas the contracted suite is Hercynotype in the scheme of Pitcher (1979). Moreover the close association of the latter with a postorogenic sedimentary volcanic molasse is a feature that has been documented from other foldbelts, for example the Caledonides and Hercynides of Europe. Equally, the association of a compositionally expanded granitoid suite with thick marine volcanics is a characteristically circum-Pacific phenomenon.

D. Major Element Analyses

.   In Tables XI and XII sets of major element analyses for the Coastal Batholith and the Eastern Cordillera are set out. Only granites of more than 70% SiO_2 have been listed in order to ensure that the comparisons made are valid. Simple inspection of these tables reveals that in the case of the Coastal Batholith Na_2O is invariably greater than K_2O, whereas with the Permian granites the reverse is true and alumina is also somewhat higher. Other differences are that CaO is higher in most analyses of the Coastal Batholith than in those of the Eastern Cordillera and that Fe_2O_3 is lower. Mineralogically and chemically the granites from both belts exhibit the main features of *I*-type granites (Chappell and White, 1974). Nevertheless there are also chemical differences between them as noted above. More crucially the initial $^{87}Sr/^{86}Sr$ ratios are different, 0.7042 for the Coastal Batholith (McCourt, 1981), and 0.7075 for the Eastern Cordillera (Dalmayrac, 1981). These ratios suggest a mantle source for the Coastal Batholith and a crustal source for the Eastern Cordillera granites. In both cases the source region would have been of igneous material thus satisfying the criteria of Chappell and White (1974), but in the case of the Eastern Cordillera the igneous rocks of the source area would have had a long residence time in the crust before being mobilized by the orogenic process. There would thus seem to be both similarities with and differences between the granites of the two belts that support in a general way the hypothesis of a different source region for the granites. In order to consider this matter further the granites from the two belts will be compared with granites from other orogenic belts of contrasting type.

TABLE XI

Peru Coastal Batholith Ring Complex Granites, ca. 60 Ma[a]

Sample / Composition	Cañas Pluton		Sayan Pluton		Puscao Pluton				San Jeronimo Pluton				Mean	Range
	A21	A150	A172	B5	P13	P17	A27	A38	A84	A86	A152	A105		
SiO_2	72.30	74.10	72.60	74.00	72.90	73.14	74.50	75.00	71.00	73.90	70.28	72.31	73.00	70.28–75.00
TiO_2	0.29	0.20	0.28	0.20	0.31	0.31	0.20	0.29	0.31	0.21	0.33	0.28	0.29	0.20–0.33
Al_2O_3	13.77	13.90	13.62	13.68	14.13	14.19	13.94	13.61	14.30	12.95	13.46	13.70	13.77	12.95–14.30
Fe_2O_3	0.87	0.47	0.80	0.44	1.21	1.31	0.49	0.60	0.88	0.82	0.66	1.05	0.80	0.44–1.31
FeO	1.00	0.52	1.12	0.38	1.13	1.22	0.60	0.80	1.44	0.93	1.87	1.00	1.00	0.38–1.87
MnO	0.06	0.08	0.07	0.08	0.06	0.06	0.10	0.05	0.06	0.07	0.08	0.08	0.07	0.05–0.10
MgO	0.57	0.67	0.51	0.31	0.62	0.66	0.30	0.31	0.70	0.32	0.78	0.62	0.53	0.30–0.78
CaO	2.18	2.00	1.60	1.67	2.49	2.50	1.66	1.65	2.14	0.64	2.10	1.76	1.86	0.64–2.50
Na_2O	3.80	3.90	3.84	4.00	3.82	3.75	3.82	3.57	4.60	4.43	4.22	4.08	3.98	3.57–4.60
K_2O	3.54	4.03	3.80	4.00	3.29	3.49	4.09	4.03	3.30	4.62	3.20	3.67	3.75	3.29–4.62
H_2O														
P_2O_5	0.06	0.09	0.06	0.82	0.08	0.12	0.03	0.04	0.05	0.03	0.09	0.04	0.12	0.03–0.82

[a] From McCourt (1981).

TABLE XII
Peru Permian Granites Eastern Cordillera[a]

Composition	Machu Picchu Pluton P537	Panta Pluton FW151	Coasa Pluton P381	Coasa Pluton P386	Aricoma Pluton P393	Mean	Range	San Ramon Ta Mn 319B	San Ramon JP71B	San Ramon P170	San Ramon P169
SiO_2	74.13	73.00	73.79	73.98	70.41	73.06	70.41–74.13	70.31	71.38	72.07	74.22
TiO_2	0.18	0.08	0.31	0.23	0.57	0.27	0.08–0.31	0.38	0.32	0.41	0.18
Al_2O_3	13.68	13.80	14.28	13.45	14.95	14.03	13.45–14.95	13.66	13.66	13.51	12.87
Fe_2O_3	1.50	0.72	1.50	1.50	1.50	1.34	0.72–1.50	} 2.91	2.61	2.97	1.96
FeO	0.21	0.07	0.77	0.65	1.35	0.61	0.21–1.35				
MnO					0.07			0.07	0.07	0.06	0.03
MgO	0.32	0.20	0.48	0.10	0.98	0.41	0.10–0.48	0.64	0.53	0.63	0.08
CaO	1.29	0.80	1.56	0.95	2.15	1.35	0.80–2.15	1.44	1.00	1.12	0.21
Na_2O	3.80	4.10	3.51	3.67	3.47	3.71	3.51–4.10	4.08	3.62	3.82	3.83
K_2O	3.99	5.50	4.26	4.64	3.82	4.44	3.82–5.50	4.60	4.52	4.68	5.01
H_2O	0.40	1.40	0.55	0.26	0.59	0.64	0.26–1.40	0.71	1.29	0.89	0.57
P_2O_5								0.11	0.10		

[a] From Dalmayrac et al., 1980.

TABLE XIII
Characteristic Features of *I*- and *S*-Type Granites[a]

I-Types	*S*-Types
Relatively high sodium content	Relatively low sodium content <3.2% Na_2O in rocks with 5% K_2O
Mol $Al_2O_3/(Na_2O_3 + K_2O + CaO)$ <1.0	Mol $Al_2O_3/(Na_2O_3 + K_2O + CaO)$ >1.0
Low initial $^{87}Sr/^{86}Sr$ ratios <0.708	High initial $^{87}Sr/^{86}Sr$ ratios >0.708
Magmas with relatively high oxygen fugacity; relatively high ferric/ferrous ratios; characterized by magnetite	Magmas with relatively low oxygen fugacity; relatively low ferric/ferrous ratios; characterized by ilmenite
Expanded compositional range from diorite to granite	Contracted compositional range from granodiorite to granite
Hornblende and sphene commonly present	Muscovite commonly present

[a] Chappell and White, 1974; Beckinsale, 1979.

The two orogenic belts that have been chosen for comparison with the Peruvian belts are the Caledonides of Europe and the Kimmerian orogenic belt of Southeast Asia, which constitutes the Southeast Asian tin belt, and within which the Main Range Batholith of the Malay Peninsula is located. These two belts have been chosen because the Caledonian granites resemble the Eastern Cordillera granites, whereas the Kimmerian granites, which are crustal, anatectic, tin-bearing *S*-type granites (Chappell and White, 1974), contrast both with the Peruvian granites and with the Caledonides, and thus provide a wider and more comprehensive background to the discussion.

The granites of the Caledonides have been most carefully studied in Donegal, western Ireland (Pitcher and Berger, 1972). These granites are all highly silicic and have a restricted compositional range, and in this sense they correspond to the Hercynotype granites of Pitcher (1979). Nevertheless, mineralogically and chemically they have features in common with the *I*-type granites of Chappell and White (1974) (Table XIII), and on this basis they are similar to the granites of the Peruvian Coastal Batholith and also to the granites of the Eastern Belt both of which are *I*-type in the scheme of Chappell and White (1974). It is thus clear that Chappell and White's scheme includes granites from belts of markedly different character, and it is the Cordilleran or circum-Pacific *I*-type granites that are compositionally expanded. The Caledonian *I*-types, on the other hand, are compositionally contracted even though chemically and mineralogically they may be similar to the cirum-Pacific granites. It is critical to distinguish Caledonide *I*-type granites from circum-Pacific *I*-type granites because the two varieties belong

TABLE XIV
Peru Cordillera Blanca Batholith Quebrada de Llanganuco *ca.* **10 Ma**[a]

Composition \ Sample no.	CB1	CB4	CB5	CB6	CB8	CB9	Mean	Range
SiO_2	72.63	71.49	73.91	73.89	71.70	71.70	72.55	71.49–73.91
TiO_2	0.23	0.19	0.19	0.18	0.26	0.24	0.21	0.18–0.26
Al_2O_3	15.41	14.93	15.29	14.93	15.68	16.03	15.37	14.93–16.03
Fe_2O_3	0.42	0.42	0.50	0.21	0.62	0.60	0.46	0.21–0.62
FeO	1.03	0.70	0.59	0.08	0.78	0.75	0.65	0.08–1.03
MnO	0.06	0.04	0.03	0.05	0.07	0.04	0.05	0.03–0.07
MgO	0.33	0.45	0.19	0.38	0.42	0.30	0.34	0.19–0.45
CaO	1.75	1.54	1.30	1.28	0.84	1.90	1.43	0.84–1.90
Na_2O	4.03	4.30	4.00	4.23	4.60	4.42	4.26	4.00–4.60
K_2O	3.95	3.85	4.42	3.88	3.50	3.74	3.89	3.50–4.42
H_2O								
P_2O_5	0.04	0.04	0.05	0.04	0.08	0.06	0.05	0.04–0.08

[a] From McCourt, 1981.

to foldbelts that have undergone different geological histories, and the composition of the granites probably reflects this.

The granites of the Main Range Batholith in the Kimmerian orogenic belt of Southeast Asia are highly stanniferous and are true *S*-type granites in the scheme of Chappell and White (1974), being derived from a source region that has undergone a weathering cycle. Although the geological history of this foldbelt is not well understood, the granites, which provide the last major episode in the history of the belt, are very distinctive, and it has been suggested by Beckinsale (1979) and Mitchell (1977) that these granites are characteristic of a collisional tectonic setting.

The chemical analyses from these two belts are of sufficient contrast to enable a reasonable comparison to be made between them and the Peruvian granites of the Eastern and Western belts. In making this comparison analyses from the Cordillera Blanca have been included as well as those from the Coastal Batholith. This has been done because the granites of this batholith are the most differentiated in Peru and in many respects resemble *S*-type granites, which is rather unusual for a major granite batholith in a Cordilleran situation (Cobbing *et al.*, 1981). It is of considerable interest to judge whether this batholith, which occurs in a truly circum-Pacific setting, really does have *S*-type characteristics or whether those features are super-ficial and could be explained in an alternative way. At present it is only possible to compare major element analyses from the different belts.

TABLE XV
Donegal European Caledonides[a]

Sample / Composition	Rosses Pluton			Trawenagh Bay Pluton		Main Donegal Pluton		Mean	Range
	G1	G2	G3	Normal	Marginal	NW	SE		
SiO_2	73.10	74.00	74.40	72.90	75.30	71.90	71.60	73.30	71.60–75.30
TiO_2	0.21	0.19	0.20	0.23	0.80	0.27	0.32	0.31	0.19–0.80
Al_2O_3	14.20	13.80	13.60	14.70	13.80	15.00	15.00	14.30	13.60–15.00
Fe_2O_3	0.50	0.56	0.50	0.57	0.14	0.14	0.52	0.46	0.14–0.56
FeO	0.93	0.79	0.70	1.02	0.41	1.27	1.37	0.92	0.41–1.37
MnO	0.03	0.04	0.03	0.03	0.01	0.03	0.03	0.03	0.01–0.04
MgO	0.62	0.57	0.52	0.56		0.61	0.64	0.58	0.52–0.64
CaO	1.24	0.97	0.91	1.52	0.76	1.95	2.12	1.35	0.76–2.12
Na_2O	4.05	3.91	4.04	3.85	3.60	3.74	3.87	3.86	3.60–4.05
K_2O	4.60	4.66	4.62	4.10	5.00	4.31	3.80	4.44	3.80–5.00
H_2O	0.73	0.67	0.68	0.67	0.02	0.53	0.57	0.55	0.02–0.73
P_2O_5	0.08	0.06	0.05	0.07	0.27	0.09	0.10	0.10	0.05–0.27

[a] From Pitcher and Berger, 1972.

From the analyses given in Tables XI–XVI, and summarized in Table XVII, the S-type granites of the Main Range Batholith in Malaysia are distinctive on two counts; the first is their very low sodium content, and the second is the very low ferric/ferrous ratio for iron. Both these features are

TABLE XVI
Main Range Malaysia[a]

Composition / Sample no.	012002	012005	012003	012007	012008	012011	012012	012014	Mean	Range
SiO_2	71.64	70.24	71.58	70.02	73.10	71.67	70.71	70.61	71.19	70.02–73.10
TiO_2	0.41	0.34	0.66	0.50	0.39	0.40	0.44	0.40	0.44	0.34–0.66
Al_2O_3	14.09	17.43	12.96	14.47	13.44	13.59	13.29	14.01	14.16	12.96–17.43
Fe_2O_3	0.52	0.33	0.61	0.48	0.31	0.43	0.44	0.45	0.44	0.33–0.61
FeO	2.26	2.11	3.12	2.88	2.03	3.01	3.38	2.75	2.69	2.03–3.38
MnO	0.09	0.03	0.15	0.01	0.02	0.01	0.03	0.07	0.05	0.01–0.09
MgO	0.04	0.99	0.35	0.72	0.80	0.76	0.77	0.90	0.66	0.04–0.99
CaO	1.92	2.08	1.78	2.15	2.10	0.81	1.52	1.85	1.77	0.81–2.15
Na_2O	3.83	2.10	3.39	2.92	2.89	2.92	2.77	2.75	2.94	2.10–3.83
K_2O	4.68	3.39	5.03	4.27	4.37	4.43	4.83	4.93	4.49	3.39–5.03
H_2O	0.57	0.73	0.48	1.46	0.89	1.86	1.74	1.02	1.09	0.48–1.86
P_2O_5	0.07	0.48	0.11	0.17	0.04	0.14	0.15	0.17	0.16	0.04–0.48

[a] Alexander et al., 1964.

TABLE XVII
Comparison of Averaged Granite Analyses from Different Orogenic Belts

Composition	Peru				
	Coastal Batholith	Cordillera Blanca	Permian E. Cordillera	Donegal	Malaysia Main Range
SiO_2	73.00	72.50	73.06	73.30	71.90
TiO_2	0.29	0.21	0.27	0.31	0.44
Al_2O_3	13.72	15.37	14.03	14.30	14.16
Fe_2O_3	0.80	0.46	1.34	0.46	0.44
FeO	1.00	0.65	0.61	0.92	2.69
MnO	0.07	0.05		0.03	0.05
MgO	0.53	0.34	0.41	0.58	0.66
CaO	1.86	1.43	1.35	1.35	1.77
Na_2O	3.98	4.26	3.71	3.86	2.94
K_2O	3.75	3.89	4.44	4.44	4.49
H_2O			0.64	0.55	1.09
P_2O_5	0.12	0.05		0.10	0.16
K_2O/Na_2O	0.94	0.91	1.19	1.15	1.52
Fe_2O_3/FeO	0.80	0.70	2.19	0.50	0.16
$MgO/Mgo + FeO^+$	0.23	0.24	0.18	0.30	0.17
Mol $Al_2O_3/(Na_2O + K_2O + CaO)$	0.98	1.11	1.048	1.049	1.094

characteristic of S-type granites according to Chappell and White (1974). Conversely all the granites in Table XV are peraluminous with the exception of the Coastal Batholith, and the Cordillera Blanca is the most peraluminous of all. Thus the granites of the Cordillera Blanca, the Eastern Cordillera, and Donegal are S-type with respect to alumina, but I-type with respect to sodium. It is evident that the analyses have to be interpreted with care, and in conjunction with the field characteristics and petrographic considerations indicated in Table XIII.

It is certain from the analyses alone that the granites of the Coastal Batholith are clearly I-type, whereas those of the Malaysian Main Range Batholith are equally clearly S-type. In the case of the Cordillera Blanca Batholith the very high soda content contrasts strongly with all the other rocks and most strongly of all with the Malaysian Main Range Batholith. Since a low soda content is a specified character for S-type granites, this very marked deviation seems to push the Cordillera Blanca Batholith toward the field of I-type granites. Other features such as the K_2O/Na_2O ratio and the Fe_2O_3/FeO ratio are concordant with this interpretation and, accordingly, the provisional view is taken that the Cordillera Blanca Batholith is a very highly differentiated example of a Cordilleran I-type has entered the field of peraluminous granites.

TABLE XVIII
Characteristics Shared by the Permian Granites of Peru and the Caledonides of Donegal

(a) They are chemically similar in major elements
(b) They are classified as I-type granites in the scheme of Chappell and White (1974), but they have a restricted compositional range and in this respect contrast with circum-Pacific I-types.
(c) They are petrographically similar and carry biotite, hornblende, and sphene.
(d) They have similar $^{87}Sr/^{86}Sr$ initial ratios, 0.7075 for the Eastern Belt (Dalmayrac et al., 1980) and 0.708 for Donegal (Pitcher and Berger, 1972).
(e) They are both attached to orogenic belts with a long and complex history in which they are the final term.
(f) They are both associated with plateau vulcanicity and molasse deposition.

According to Tables XV, XVI, and XVII the granites that are most similar on major elements are those from the Eastern Cordillera of Peru and the Caledonides of Donegal. The analyses of the granites from these two belts are strikingly similar for every major element except iron, and they are distinguished from the Main Range Batholith by their higher soda and lower ferrous iron content, and from the Cordillera Blanca Batholith by their lower alumina, lower soda, and higher potash. Geochemically they are considered as I-types although they are slightly peraluminous. Petrographically they are similar in that examples from both areas contain hornblende and sphene, and the opaque mineral is characteristically magnetite; these features distinguish them from the S-types of the Main Range Batholith. The granites of the Eastern Belt of Peru and the Caledonides of Donegal appear to have a number of features in common (Table XVIII).

VII. CONCLUSION

Enough has been written to show that the geology of the Eastern and Western Cordilleras is different and that each developed under a different set of geotectonic conditions that produced granite magmas of different composition. Whereas the Western Cordillera may be firmly related to subduction of oceanic crust beneath a continental margin, the tectonic setting for the Eastern Cordillera is more problematical. However, it can at least be said that the Eastern Cordillera was not related to subduction. A cause has to be sought in some other mechanism, and it may well be that continental rifting as proposed by Noble et al. (1978), or mantle plume tectonics suggested by Schermerhorn (1981), may be appropriate. Attention has been drawn to the similarities of the Eastern Belt granites to those of the European Caledonides, and without pressing the analogy too far, it seems logical that the belt should have some affinities with precontinental-drift orogenic belts. The question of an appropriate mechanism is now a matter of research and debate, but it is at least clear that the tectonic setting should be compatible with a predrift configuration for Gondwanaland.

ACKNOWLEDGMENTS

I wish to acknowledge my debt to my French colleagues F. Mégard, B. Dalmayrac, R. Marocco, and B. Laubacher, whose pioneering work in the Eastern Cordillera I have drawn on heavily in preparing this manuscript. I also wish to thank my British colleagues from Liverpool University and the Insitute of Geological Sciences and my many Peruvian colleagues with whom I have worked for many years.

This paper is published with the permission of the Director of the British Geological Survey, Natural Environment Research Council.

REFERENCES

Ahlfeld, F., and Branisa, L., 1960, *Geología de Bolivia*, La Paz: Institute Boliviano de Petroleol, 245p.

Alexander, J. B., Harral, S. M., and Flinter, B. H., 1964, Chemical Analyses of Malayan Rocks, *Professional Paper E-64 1-C*, Geological Survey of Malaya IPOH.

Atherton, M. P., McCourt, W. J., Sanderson, L. M., and Taylor, W. P., 1979, The geochemical character of the segmented Peruvian Coastal Batholith and associated volcanics, in: *Origin of Granite Batholiths: Geochemical Evidence*, Atherton, M. P., and Tarney, J., eds., Orpington: Shiva Press, p. 45–65.

Beckinsale, R. D., 1979, Granite magmatism in the tin belt of Southeast Asia, in: *Origin of Granite Batholiths: Geochemical Evidence*. Atherton, M. P., and Tarney, J., eds, Orpington: Shiva, p. 34–44.

Benavides, V., 1956, Cretaceous system in northern Peru, *Bull. Am. Mus. Nat. Hist.* v. 108, p. 355–493.

Berger, A. R., and Pitcher, W. S., 1970, Structures in granitic rocks: a commentary and a critique on granite tectonics, *Proc. Geol. Assoc.* v. 81, p. 441–461.

Berry W. B. N., and Boucot, A. J., 1972, Correlation of the South American Silurian rocks, *Geol. Soc. Am. Spec. Paper* no. 133, p. 1–59.

Boucot, A. J., Isaacson, P. E., and Laubacher, G., 1980, An early Devonian, eastern Americas realm faunule from the coast of southern Peru, *J. Paleontol.*, v. 54, p. 359–365.

Bussell, A., 1975, *The Structural Evolution of the Coastal Batholith in the Provinces of Ancash and Lima*, Ph.D. Dissertation, University of Liverpool (unpublished).

Caldas, J., 1978, Geologia de los Cuadrangolos de San Juan, Acari y Yauca, *Inst. Geol. Min., Bol.* 30, Lima, Peru, p. 1–78.

Capdevila, R., Megard, F., and Vidal, P. L., 1977, Le Batholite de San Ramou, Cordillere Oriontal du Perou central, *Geol. Rundsch.* v. 67, p. 434–446.

Chappell, B.W., and White, A. J. R., 1974, Two contrasting granite types, *Pac. Geol.*, v. 8, p. 173–174.

Charrier, R., 1973, Interruption of spreading and the compressive tectonic phases of the meridional Andes, *Earth and Planet. Sci. Lett.*, v. 20, p. 242–249.

Child, R., 1976, *The Coastal Batholith and its Envelope in the Casma Region of Peru*, Ph.D. Dissertation, University of Liverpool, (unpublished).

Cobbing, E. J., 1978, The Andean geosyncline in Peru and its distinction from Alpine geosynclines, *J. Geol. Soc. London*, v. 135, p. 207–218.

Cobbing, E. J., and Pitcher, W. S., The Coastal Batholith of Central Peru, *J. Geol. Soc. London*, v. 128 p. 421–460.

Cobbing, E. J., Ozard, J. M., and Snelling, N. J., 1977, Reconnaissance geochronology of the crystalline basement rocks of the Coastal Cordillera of Southern Peru, *Geol. Soc. Am., Bull.* v. 88, p. 241–246.

Cobbing, E. J., Pitcher, W. S., and Taylor, W. P., 1977, Segments and superunits in the coastal batholith of Peru, *J. Geol.*, v. 85, p. 625–631.

Cobbing E. J., Pitcher, W. S., Wilson, J. J., Baldock, J. W., Taylor, W. P., McCourt, W. J., and Snelling, N. J., 1981. The geology of the Western Cordillera of northern Peru, *Inst. Geol. Sci. Overseas Mem.* no. 5, p. 1–143.

Dalmayrac, B., Laubacher, G., and Marocco, R., 1980, Geologie des Andes Peruviennes, Travaux et Documents de L'ORSTOM, no. 122, Paris, p. 1–501.

Du Toit, A. L., 1937, *Our Wandering Continents*. Edinburgh: Oliver and Boyd, 266p.

Grant, J. N., Halls, C., Avila, W., and Snelling, N. J., 1977, Edades potasio--argon de las rocas igneas y la mineralizacion de parte de la Cordillera Oriental, Bolivia, *Bol. Serv. Geol. de Bolivia Serie A*, v. 1 (1), p. 33–60.

Helwig, J., 1972, Stratigraphy; sedimentation, palaeogeography, and palaeoclimate of Carboniferous (Gondwana) and Peruvian of Bolivia, *Am. Assoc. Pet. Geol. Bull.*, v. 56, p. 1008–1033.

Isaacson, J., 1974, Evidence for a western extracontinental land source during the Devonian period in the Central Andes, *Geol. Soc. Am.* Bull, v. 81, p. 39–46.

Ishihara, S., 1977, The magnetite series and the ilmenite series granite rocks, *Min. Geol. Tokyo*, v. 27, p. 293–305.

James, D., 1981, A crustal contamination model for central Andean Lavas, *Ann. Rep. Dept. Terr. Magn. Carnegie Inst., Washington, D.C.*, p. 479–488.

James, D., Brooks, C., and Cuyubamba, A., 1976, Andean Cenozoic volcanism: magma genesis in the light of strontium isotopic composition and trace-element geochemistry, *Geol. Soc. Am. Bull.*, v. 87, p. 592–600.

Jenks, W. F., 1951, Triassic to Tertiary stratigraphy near Cerro de Pasco Peru, *Geol. Soc. Am. Bull.*, v. 62, p. 203–220.

Lancelot, J. R., Laubacher, G., Marocco, R., and Renaud, U., 1978, U/Pb radiochronology of two granitic plutons from the Eastern Cordillera (Peru). Extent of Permian magmatic activity and consequences, *Geol. Rundsch.*, v. 67, p. 236–243.

Larson, R. L., and Pitman, W. C., 1972, Worldwide correlation of Mesozoic magnetic anomalies and its implications, *Geol. Soc. Am. Bull.*, v. 83, p. 3624–3662.

Laubacher, G., 1977, *Geologie de l'Altiplano de la Cordillere Orientale au Nord et au Nord Quest du lac Titicaca (Peru)*, Ph.D. Dissertation, Laboratoire de Géologie Structurale Université des Sciences et Techniques du Languedoc, Montpellier.

Lehmahn., 1978, A Precambrian core sample from the Altiplano, Bolivia, *Geol. Rundsch.*, v. 67, p. 270–278.

Levi, B., and Aguirre, L., 1981, Ensialic spreading-subsidence in the Mesozoic and Paleogene Andes of Central Chile, *J. Geol. Soc. London*, v. 138, p. 75–81.

Marocco, R., 1971, Etude geologique de la chaine andine au niveau de la deflexion d'Abancay, *Cah. ORSTOM*, v. 3 p. 45–57.

Marocco, R., 1978, Estudio Geologico de la Cordillera de Vilcabama, *Inst. Geol. Min. Bol. 4 Estudios Especiales*. Lima Peru, p. 1–157.

McCourt, W. J., 1981, The geochemistry and petrography of the Coastal Batholith of Peru, Lima segment, *J. Geol. Soc. London*, v. 138, p. 407–420.

McLaughlin, D. H., 1924, Geology and physiography of the Peruvian Cordillera, departments of Junin and Lima, *Geol. Soc. Am. Bull*, v. 35, p. 591–632.

Megard, F., 1978, Etude Geologique des Andes du Perou Central, *Cah. ORSTOM Mem. No. 86* Paris, 310p.

Megard, F., Dalmayrac, B., Laubacher, G., Marocco, R., Martinez, C. L., Paredes, P. J., and Tomasi, P., 1971, La chaine hereynienne au Perou et en Bolivie. Premiers resultats, *Cah. O.R.S.T.O.M.* v. III, pl. 5–43.

Myers, J. S., 1974, Cretaceous stratigraphy and structure; western Andes of Peru between latitudes 10° and 10°30'S, *Bull. Am. Assoc. Pet. Geol.*, v. 58, p. 474–487.

Newell, N. D., 1949, Geology of the Lake Titicaca region, Peru, and Bolivia, *Geol. Soc. Am. Mem.*, v. 36, 111p.

Newell, N. D., Chronic, J., and Roberts, T., 1953, Upper Paleozoic of Peru, *Geol. Soc. Am. Mem.*, v. 58, 276p.

Noble, D. C., Silbermann, M. L., Mégard, F., and Bowman, H. R., 1978, Comendite (peralkaline rhyolite) in the Mitu Group Central Peru: evidence of Permian-Triassic crustal extension in the Central *Andes, J. Res. U.S. Geol. Surv.*, v. 6, p. 453–457.

Paredes, P. J., 1972, *Etude geologique de la feuille de Jauja (Andes du Perou Central)*, Ph.D. Dissertation, Laboratoire de Geologie Structurale Université des Sciences et Techniques du Languedoc, Montpellier.

Pitcher, W. S., 1972, The Coastal Batholith of Peru, some structural aspects, *Rep. 24th Sess. Int. Geol. Congr. Section 2*, no. 24, p. 156–163.

Pitcher, W. S., 1974, The Mesozoic and Cenozoic batholiths of Peru, *Pac. Geol.*, v. 8, p. 51–62.

Pitcher, W. S., 1978, The anatomy of a batholith, *J. Geol. Soc. London*, v. 135, p. 157–182.

Pitcher, W. S., 1979, The nature, ascent, and emplacement of granitic magmas, *J. Geol. Soc. London*, v. 136, p. 627–662.

Pitcher, W. S., and Berger, A. R., 1972, *The Geology of Donegal: A Study of Granite Emplacement and Unroofing*, Regional Geology Series, London: Wiley-Interscience, 435 p.

Ruegg, W., 1962, Rasgos morfologicos intramarinos y sus contrapartes en el suelo continental peruano, *Bol. Soc. Geol. Peru*, v. 38, p. 97–142.

Schermerhorn, L. J. G., 1981, The West Congo orogen: a key to Pan-African tectonism, *Bull. Geol. Rundsch.*, v. 70 p. 850–867.

Steinmann, S., 1929, *Geologia von Peru*, Heidelberg: Karl Winter, 488p.

Tilton, G. R., and Barreiro, B. A., 1980, Origin of lead in Andean calcalkaline lavas, southern Peru, *Science*, v. 210, p. 1245–1247.

Wilson, J. J., 1963, Cretaceous stratigraphy of central Andes of Peru, *Bull. Am. Assoc. Pet. Geol.*, v. 47, p. 1–34.

Wilson, J. J., Reyes, L., and Garayar, J., 1967, Geología de los cuadrangulos de Mollebamba, Tayabamba, Huaylas, Pomabamba, Carhuas y Huari, *Bol. Serv. Geol. Min. Peru*, v. 16, p. 1–95 p.

Wilson, P. A., 1975, *K-Ar Studies in Peru with Special Reference to the Emplacement of the Coastal Batholith*, University of Liverpool, Ph.D. Dissertation (unpublished).

Wolfart, R., and Voges, A., 1968, Beitrage zur Kenntnis des Devon von Bolivien, *Geol. Jahrb Beih*, v. 74, 214p.

Chapter 7

THE SOUTHERN ANDES

Luis Aguirre*

Department of Geology
University of Liverpool
Liverpool, L69 3BX England

I. INTRODUCTION

The term "Southern Andes" as used in this paper refers to the *ca.* 4000 km long segment of the Andean Orogen between the Arica deflection (18°S latitude) and Cape Horn (56°S latitude).

A. Morphological Units and Geographical Divisions

Several morphological units are recognized in the Southern Andes (Fig. 1); reference is made to this framework throughout the text. These units are:

(*i*) the *Coastal Range*;

(*ii*) the *Andean Range* (Cordillera Principal of some authors), or simply Andes. It is mainly used here to refer to the High Cordillera region between 18° and 41°S. The term Patagonian Cordillera would be its equivalent south of 41°S and it includes the Cordillera Darwin, the name for the main range between 54° and 55°S;

(*iii*) a *Central Depression* separating the Coastal and Andean Ranges and known as Pampa del Tamarugal in the north and Central Valley Graben in the center;

* Present address: Laboratoire de Petrologie, Faculté des Sciences et Techniques de St. Jerôme, Universitié d'Aix Marseille .III`, 13397 Marseille, France.

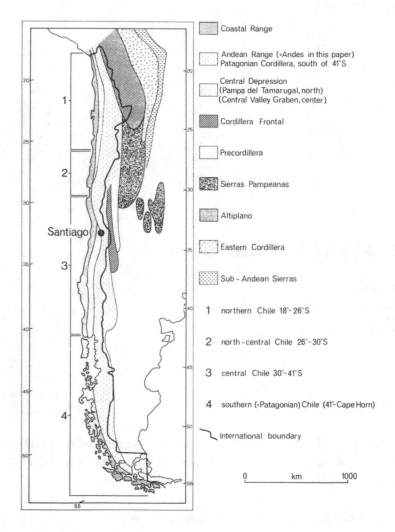

Fig. 1. Morphological units of the Southern Andes and geographical divisions as used in the text. Morphology is partly based on Aubouin *et al.* (1973).

  (iv) the *Cordillera Frontal*;
   (v) the *Precordillera*;
  (vi) the *Sierras Pampeanas*, three units that are present in Argentina and flank the Southern Andes along their middle section;
 (vii) the *Altiplano*, a high plateau region developed in Bolivia, Chile, Peru and Argentina;
(viii) the *Eastern Cordillera*;
  (ix) the *Sub-Andean Sierras*, two units that are developed toward the northeast, beyond Chilean territory.

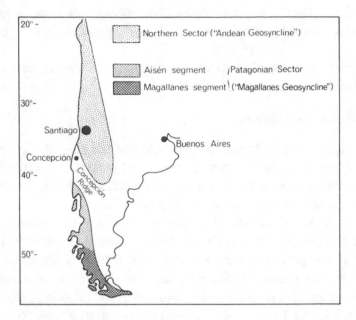

Fig. 2. Geotectonic division of the Southern Andean edifice. Sector boundaries modified after Aubouin *et al.* (1973).

Geographical divisions of the country (Fig. 1) have been adopted for the purposes of presentation and discussion as follows: northern Chile (18°–26°S); north-central Chile (26°–30°S); central Chile (30°–41°S) and southern (Patagonian) Chile (41°–Cape Horn).

B. Geotectonical Divisions

A primary geotectonical division is made between the *pre-Andean basement* and the *Andean edifice*.

The pre-Andean basement comprises metamorphic, sedimentary and igneous rocks from the Precambrian to the Upper Paleozoic–Lower Triassic(?). Unconformable relations reflecting the presence of a late Hercynian orogeny are observed between the basement and the Andean edifice. The term "pre-middle Jurassic basement" is used by some authors as synonymous with pre-Andean basement in southern Chile.

The southern Andean edifice is the outcome of the Andean orogenic evolution, a process extending from the beginning of the Mesozoic to the Present. In this paper, the Andean edifice is divided into two sectors (Fig. 2): the *Northern Sector*, between 18° and 41°S, that corresponds to the "Andean Geosyncline" in most of Chilean geological literature, and the *Patagonian Sector*, between 41° and Cape Horn, corresponding to the "Magallanes Geosyncline''. In the Patagonian Sector two segments are distinguished: the *Magallanes segment*, from Cape Horn to 49°S, and the *Aisén segment* from 49°S to 41°S.

A positive area, the *Concepción Ridge*, separated the northern and Patagonian sectors during their evolution.

C. Methodology and Outline

The methodology adopted for discussion of the geotectonic divisions consists in a presentation of factual information (geological record) followed by an attempt at interpretation (geological evolution). This approach is reflected in the outline of the corresponding sections.

The treatment of the Andean edifice is based on a scheme considering its cyclic evolutionary nature. Cyclic development is characterized by alternation of long periods of extension and short periods of compression and is punctuated by successive folding phases expressed as unconformities. These unconformities separate *stratigraphical-structural units*, a term which refers to the rock piles originated in each cycle and to the events involved in their generation. The chronological sequence of these events is followed to describe the various stratigraphical-structural units throughout the Andean orogenic evolution: (i) depositional processes (volcanic, sedimentary); (ii) burial (or ocean floor) low-grade metamorphism; (iii) folding; (iv) granitoid emplacement; (v) uplift, erosion. A variable degree of overlapping of these events exists in each cycle.

A closing section discusses major contrasts and analogies between the Northern and Patagonian sectors of the Andean edifice.

No attempt was made to cover the field of mineral economic resources. However, major metallogenic features have been briefly referred to in the framework of the overall geotectonic evolution.

D. Main Conclusions

A long-standing plate interaction between a paleo-Pacific plate and the South American continent is evidenced in the geological record. The beginning of such interaction can be dated at least as middle to late Paleozoic and it accounts for considerable oceanic accretion along most of Chile during the Paleozoic. A paired metamorphic belt bears witness of this process. Hercynian events closed the Paleozoic evolution and changes in the geotectonic style of the continental margin took place.

During the Mesozoic and Cenozoic a process of cyclic development of island arcs and marginal (back-arc) basins occurred under ensialic conditions. This process is thought to have been controlled by the complementary and mainly alternate action of two major mechanisms: *continental spreading-subsidence* and *oceanic plate subduction*. The first is attributed to mantle diapirism related to hot spots, hot lines, or induced flow while the second is characterized by its different spreading rates. An interplay of several parameters determined the stage of development reached by the marginal basins in each cycle. Among these parameters upwelling rates of diapiric mantle material, thickness of the continental crust, ocean spreading rates and dip of the subducting slab are of particular significance.

In the southernmost Andes, continental spreading-subsidence achieved splitting of the crust and generation of ocean floor (ophiolitic complexes) during late Jurassic–early Cretaceous time. However, further north, no oceanic crust was ever exposed although great volumes of mantle-derived flood basalts were extruded in different epochs.

Cycles of marginal basin generation along the Chilean continental margin seem to have occurred with an average duration of *ca.* 40 Ma from the early Mesozoic to the Present. Extensional régimes largely prevailed over compressional ones during most of the evolution of the Andean edifice.

II. THE PRE-ANDEAN BASEMENT

The pre-Andean basement in Chile is represented by metamorphic, sedimentary and igneous rocks of Precambrian to late Paleozoic/early Triassic age. Outcrops of this basement are found all along the extent of the Chilean territory (Fig. 3). The metamorphic rocks are mainly exposed in the Coastal Range of northern and central Chile and in the Patagonian archipelagos and are characterized by facies ranging from greenschist to granulite. Radiometric ages indicate that metamorphic events took place from late Proterozoic to late Paleozoic time with most dates concentrated in the Carboniferous. Sedimentary rocks with no (or slight) metamorphic effects are represented by relatively small and scattered exposures mainly found in the Altiplano and the central coast line and by extensive areas in the Patagonian archipelagos. Among them, both marine and continental facies containing Ordovician–Permian fossil fauna and flora are known. Unmetamorphosed volcanic rocks, or those affected by low-grade burial metamorphism, are restricted to a thick series of acid volcanoclastic units of Permo–Triassic age that crop out along the Andean Range between latitudes 21° and 31°S. The plutonic rocks are granitoids and ultramafics. The granitoids extend from northernmost Chile to latitude 39°S where they are shifted sharply eastward into Argentina. The most continuous exposures occur along the Andes (28°–31°S) and the Costal Range (33°–39°S) and vary from gabbro to monzogranite with tonalite and granodiorite being predominant. Radiometric ages of granitoids range from Cambro–Ordovician in the Altiplano to Permo–Triassic in the Andean Range between 28° and 31°S. Granitoids of the Coastal Range in central Chile are predominantly Carboniferous. Ultramafic rocks are

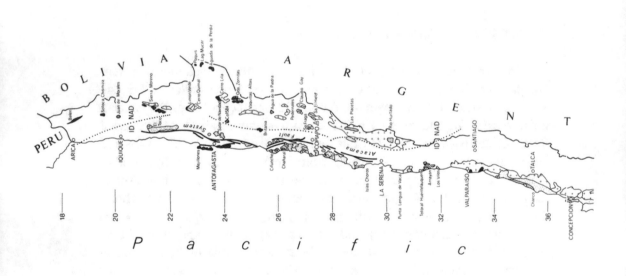

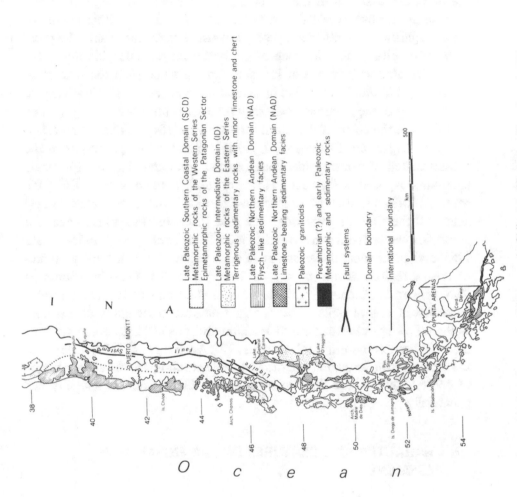

Fig. 3.　Distribution of the pre-Andean basement in Chile. Modified after Hervé *et al.* (1981a).

mainly restricted to the Coastal Range between 38° and 44°S and consist of peridotite and serpentinite bodies closely associated in space with metamorphic rocks.

Knowledge of the pre-Andean basement is largely the result of research done since radiometric dating became possible. Previously, a Precambrian age had been assigned to the metamorphic basement (see Aguirre, 1965) because the exposures of Paleozoic sedimentary rocks are either devoid of metamorphism or are affected by very low-grade metamorphism. The presence of itabiritic schists in some of the basement metamorphic units also was interpreted as evidence of Precambrian age since such rocks seem to be restricted to the Precambrian in several parts of the world. Stratigraphic relationships of the granitoids had early revealed a pre-Liassic age for some batholiths in the north of the country (Domeyko, 1845; Willis, 1929) and it had been suggested (Steinmann, 1929) that some of the granitoids in the coastal region of central Chile could actually be older than the Mesozoic batholiths exposed farther east. However, until the end of the 1950's, the names "Andean Batholith" or "Andean Diorite" were used widely to designate all the Chilean granitoids. The first Paleozoic radiometric ages for granitoids were obtained in the early 1960's (Ruiz et al., 1960; Muñoz Cristi, 1962) and, since then, the existence of large volumes of Paleozoic plutonics in several regions of the country has been evidenced. González-Bonorino (1967) reported Paleozoic radiometric ages for rocks of the metamorphic basement in central Chile as part of the first systematic study of this unit. In the following years the term "Crystalline Basement" has commonly been used (González-Bonorino and Aguirre, 1970; Aguirre et al., 1972) to embrace the metamorphic basement and the Paleozoic granitoids in the coastal region of central Chile where the two units appear closely related spatially and genetically.

III. CHARACTERISTIC FEATURES OF THE PRE-ANDEAN BASEMENT

A. Metamorphic Rocks

1. *Precambrian and Early Paleozoic (Cambrian to Devonian) Metamorphic Rocks*

Several isolated outcrops of metamorphic rocks are present in the Andes of northern Chile. One of them, the Esquistos de Belén Formation (Montecinos, 1963) of gneisses and schists is exposed as a NNW, discontinuous strip 30 km long and 0.7 km wide (Fig. 3). Samples of biotitic gneiss, mica schists, and amphibolic and chloritic schists from 3 km north of the village of Belén gave a whole-rock Rb-Sr isochron of 1,000 Ma with a $^{87}Sr/^{86}Sr$ initial ratio [Sr(i) in the following] of 0.718 (Pacci *et al.*, 1981). This date is interpreted by Pacci *et al.* (1981) as the probable imprint of a regional metamorphic event of that age. The 1,000 Ma date is the first geochronological evidence for the existence of Precambrian rocks in Chile and confirms the age that Montecinos (1963) had assigned to the Esquistos de Belén Formation on the basis of indirect geological evidence.

Further south, along the Andes, metamorphic rocks of uncertain age crop out in few other places: (a) At Chismisa (Aguirre, 1965) quartz-sericitic schists with strong microfolding are exposed in a 3 km long and 1.5 km wide outcrop; a Precambrian age was suggested for these rocks. (b) Between 21° and 22°S, in the Quehuita–Sierra Moreno area, quartz-feldspar-mica schists and migmatites (injection gneisses) crop out as a NNE elongated, discontinuous band (Vergara, 1978a; Maksaev, 1978; Venegas and Niemeyer, 1982). The metamorphic rocks are intruded by pre-Jurassic granitoid bodies with K-Ar ages ranging from 431 ± 10 Ma to 271 ± 8 Ma (Huete *et al.*, 1977). Based on this relationship, Vergara (1978a) has assigned a minimum pre-Silurian age to the regional metamorphic event that generated the schists and an Ordovician–Silurian age (intrusion of a 431 Ma old muscovite granite) to the formation of the migmatites (injection gneisses). (c) At the Limón Verde–Cerro Quimal area (*ca.* 23°S) scattered, small outcrops of micaceous schists and gneisses were described by Harrington (1961) who considered them as probably late Precambrian by comparison with unmetamorphosed sedimentary rocks of Cambrian and Ordovician age that crop out further east at the Puna de Atacama and Argentina.

Along the coastal region, metamorphic rocks partly attributed to the Precambrian are exposed on the Mejillones peninsula (*ca.* 23°30'S) and have been described by Ferraris and Di Biase (1978) and Venegas (1979). They are represented by garnet-bearing biotite gneisses, amphibolites and metaplutonics closely associated with ultramafic to felsic plutonic rocks. According to Venegas (1979) this complex is polymetamorphic and retrogressive in character. NE to EW foliation and microfold axes have been interpreted by him as indicative of late Precambrian age for the first tectonometamorphic event since similar orientations have been recorded for structures in rocks assigned to the Precambrian both in Chile and Argentina. An early Precambrian age is attributed by Damm and Pichowiak (1982) to the metamorphism of the highest grade in the Mejillones peninsula. Geochronological studies in progress (Hervé, personal communication) have failed to obtain Precambrian ages for these rocks; preliminary values would rather suggest they were metamorphosed in early Paleozoic time.

In the coastal region of central Chile, between latitudes 33°S and 33°35'S, the metamorphic Quintay Formation of gneisses, amphibolites, quartzites and schists has been described by Corvalán and Dávila (1964). Two Pb-U ages of 383 Ma were obtained for samples of amphibolite and gneiss from the northern exposures (Corvalán and Munizaga, 1972). Broadly similar Pb-U values, ranging from 386 to 405 Ma, came from granitoids that intrude the metamorphic unit. On this evidence Corvalán and Munizaga (1972) concluded that at least part of the batholithic and metamorphic complex recorded development of an early Devonian orogenic event, the Quintay phase. Hervé (1976) described charnockitic rocks (opdalites), pyroxene gneisses, amphibolites and migmatites plus minor paragneisses from the northern outcrops of the Quintay Formation. According to him, the metamorphic event associated to the Quintay phase partially transformed the opdalite into gneiss with relict pyroxene or into hornblende-biotite augen gneiss in the zones of higher deformation. He concludes that the opdalite is older than early Devonian and suggests a possible Precambrian age.

2. *Late Paleozoic (Carboniferous and Permian) Metamorphic Rocks*

Outcrops of these rocks are present in most of Chile from latitude 29°S (Isla and Punta Choros) down to the Strait of Magellan (Fig. 3). They are mainly located along the western, coastal, border of the territory where they are best and most continually exposed as a belt between latitudes 34°S (Pichilemu) and 47°S (Taitao Peninsula). Metamorphic facies and facies series in the central belt (34° to 44°S) were first identified by González-Bonorino (1970, 1971) who described an intermediate high pressure, an intermediate low pressure and a low pressure series. It was later assumed (González-Bonorino and Aguirre, 1970) that the first two constituted a paired belt. Modifications to González-Bonorino's scheme led to a general division of the metamorphic rocks of central Chile into two major units, the Western and Eastern series (Aguirre *et al.*, 1972), a scheme later extended by different authors to other areas of the country where similar rocks have been studied, i.e. the Patagonian archipelagos.

The Western series occupies the Pacific side of the belt and includes micaschists, greenschists, blueschists, metacherts and some ultramafic bodies. Ocean floor tholeiitic basalts have been geochemically identified among metabasites that in several places present a relict pillow structure (Hervé *et al.*, 1976a; Godoy, 1979). The unit shows polyphase metamorphism and deformation. Tight folds with low-angle axial planes overturned to the west dominate the structure, which in places consists of disrupted beds in a "mélange" type association (Hervé *et al.*, 1981a). Mineral assemblages correspond largely to the greenschist facies but the presence of lawsonite (Saliot,

1968) and crossite in some local associations would indicate high P/T gradients.

The Eastern series occupies the continental side of the metamorphic belt between the Western series and the Coastal Range Batholith of late Paleozoic age. It corresponds mainly to metamorphic equivalents of a shale-graywacke alternation with minor amounts of calc-silicate rocks. Metas-andstones, slates, phyllites, schists and scarce gneisses are represented in the series, which is considered to be the result of polyphase deformation and metamorphism from greenschist to amphibolite–granulite facies grade under intermediate low P/T gradients (Aguirre et al., 1972; Hervé, 1977; Hervé et al., 1981a). Together with the Western series of central Chile, these rocks form a paired metamorphic belt. The contact between the two series was referred to as the Coast Range Suture by Ernst (1975). Although in places it has the characteristics of a fault, a transitional nature is commonly observed. Transitions are expressed by a progressively stronger development toward the west of an S_2 crenulation foliation that tends to transpose and obliterate the S_1 and S_0 structures predominant in the rocks of the Eastern series (Hervé et al., 1982).

Geochronological data concerning the two series described above are mainly found in Munizaga et al. (1973); Hervé et al. (1974a); Hervé et al. (1976b); Cordani et al. (1976) and Hervé et al. (1982). Rb-Sr whole-rock limiting isochrons covering the span 342 to 273 Ma with a Sr(i) ratio of 0.711 ± 0.001 were obtained by Munizaga et al. (1973) for rocks belonging to both series and from widely separated areas in central Chile (34° to 40°S). Hervé et al. (1976b) reported an Rb-Sr whole-rock isochron of 316 Ma with a Sr(i) ratio of 0.711 for metamorphic rocks of the Eastern Series. Hervé et al. (1982, 1984) have obtained Rb-Sr whole-rock isochrons for the Western and Eastern series from two localities 15 km apart near the northernmost contact between the series (ca. 34°30'S). The Eastern series metamorphics gave older whole-rock isochron ages (347 ± 32 Ma) than the Western series metamorphics (311 ± 10 Ma) with Sr(i) ratios of 0.7124 ± 0.0007 and 0.7060 ± 0.0005 respectively.

According to the data above, a Carboniferous age is indicated for the main metamorphic episode in central Chile. Both series must have been uplifted shortly after metamorphism as suggested by some K-Ar muscovite ages in the range of 270 to 300 Ma obtained for rocks of both series (Dávila et al., 1979; Hervé et al., 1982).

Metamorphic rocks correlative with those of the Western and Eastern series of central Chile have been identified north of latitude 34°S and south of latitude 42°S. To the north of 34°S sequences of actinolitic and quartzitic schists exposed at Punta Choros and adjacent islands (Aguirre, 1967) have been considered by Hervé *et al.* (1981a) to be the northernmost equivalents of the Western Series. Rocks akin to both series crop out sporadically along the coast between 30°10' and 31°30'S (Godoy, 1975; Hervé *et al.*, 1981a) where the nature of their contact with unmetamorphosed sedimentary sequences and their age have been a matter of controversy in the last years. Hervé *et al.* (1976a) and Mundaca *et al.* (1979) have pointed out that a gradual transition exists between the metamorphic rocks and Paleozoic sedimentary units while Maass and Roeschmann (1971) mapped tectonic contacts between them. If a transition exists, the age of the regional metamorphic events should be extended to a time between the Permian age of the youngest transitional unit involved, and the middle Triassic, the age of the sedimentary units unconformably lying over them (Hervé *et al.*, 1981a). However, Miller (1970) assigns a pre-Devonian age to the metamorphic rocks of this region based on the presence of albite grains with helicitic inclusions, supposedly derived from those metamorphics, in sandstones of the contiguous Arrayán Formation of Devonian age. The easternmost equivalents of the Eastern series in this sector are found at 29°S in the upper course of the Tránsito River where garnet-bearing, biotite-muscovite schists and migmatitic granites (Aguirre, 1965) are exposed in a N–S elongated outcrop *ca*. 25 km long. Recent Rb-Sr whole-rock isochrons obtained by Hervé (personal communication) indicate ages of 261 and 232 Ma (Permian) for rocks of this area; these dates could, however, correspond to superimposed dynamic metamorphic events in the region (Hervé, personal communication).

South of 42°S, metamorphic rocks of late Paleozoic age have been studied in the Chiloé-Aisén region, the Patagonian Archipelagos, the Patagonian mainland and Tierra del Fuego. However, use of the term "basement" to signify a Paleozoic age for metamorphic rocks of this region is not free from ambiguity. In most places metamorphic ages are unknown while arguments have been presented in favor of the existence of Mesozoic metamorphic events.

(a) Chiloé-Aisén Region. Levi *et al.* (1966) established the presence of several areas of outcrop of metamorphic rocks between Puerto Montt and Bajo Palena (41°30'–43°) (Fig. 3) where the most representative rock type is a quartz-muscovite-biotite schist. These rocks have been considered as equivalent with the Eastern series metamorphics by Hervé *et al.* (1981a). Hervé *et al.* (1976c) and Hervé *et al.* (1981b) have described the metamorphic rocks exposed at the Guaitecas Islands and the central part of the Chonos Archipelago, respectively. The Chonos Archipelago has also been studied by Miller (1979a, 1979b) and Miller Sprechmann (1978). The conclusions reached by Miller and Hervé and coworkers conflict with each other. Miller (1979a) distinguished three formations of different ages in the Chonos Ar-

chipelago: Canal King Formation of metasedimentary and metavolcanic rocks of pre-Devonian, probably Cambro–Ordovician age; Potranca Formation on Isla Potranca, consisting of weakly metamorphosed sedimentary rocks with Devonian fauna (Miller and Sprechmann, 1978) and; Canal Pérez Sur Formation with sedimentary rocks of possible Permo–Carboniferous age. According to Miller (1979a) folding would have taken place in the Silurian (?), Carboniferous, and late Permian–early Triassic(?). Except for the fossil fauna of Isla Potranca, the ages assigned by Miller to the formations above and to folding events are based on structural considerations. According to Hervé et al. (1981b), on the other hand, two structural zones are present in the metamorphic rocks of the central Chonos Archipelago; an eastern zone of black shales, slates, sandstones and radiolarian cherts with well preserved primary sedimentary structures, and a western phyllite, "quartzite" and greenschist-bearing zone where primary structures have been completely transposed. Two deformational events took place. Isoclinal and tight chevron folds with a variously striking axial plane cleavage (S_1) of the western zone that developed at an early stage were progressively refolded toward the west while a steeply dipping crenulation cleavage (S_2) was developed. Transition from one zone to the other is gradual and the fact that S_0 is restricted to the eastern areas is due to its obliteration toward the west by fabrics of D_2 type. No structural or lithological breaks allowing a formational division of these metamorphic rocks was recognized. Hervé et al. (1981b) point out that similar structural and lithological conditions have been observed in the Guaitecas Archipelago and in the metamorphic belt of central Chile. This transitional nature of the metamorphic rocks of the Guaitecas and Chonos islands would question Miller's chrono-stratigraphic scheme.

In mainland Aisén, from Lake General Carrera to Lake O'Higgins, low-grade metamorphic rocks are widely represented (Lagally, 1975; Skarmeta, 1978). They have been considered as partly correlative with the Eastern series of central Chile and partly with types akin to the Western series (Hervé et al., 1981a; Lagally, 1975). Metabasites from Lake General Carrera have alkaline affinities (Godoy, 1979).

(b) Patagonian Archipelagos. Units lithologically similar to those of the Western series, although with a weaker metamorphism, are found at the Patagonian Archipelagos between 50° and 53°S (Madre de Dios Archipelago, Nelson Strait) where they are represented by abyssal tholeiitic pillow basalts, ferruginous cherts, thick flyschoid sequences and massive fusulinid-bearing limestones of late Carboniferous to early Permian age (Cecioni, 1955, 1956; Douglas and Nestell, 1972; Forsythe and Mpodozis, 1979).

(c) Patagonian Mainland and Tierra del Fuego. Greenschists and metacherts are present at Staines Peninsula (Forsythe and Allen, 1980) although

in extended areas such as the Cordillera Darwin the protolith seems to be restricted to pelitic and pelito–arenaceous facies (Kranck, 1932; Dalziel and Cortés, 1972). Hervé *et al.* (1979) have reported a Rb-Sr whole-rock isochron of 236 ± 9 Ma for the schists of Bahia Plüschow (*ca.* 54°30′S). This age is interpreted by them as evidence of a metamorphic event of late Permian age in the Cordillera Darwin.

3. *Pre-Andean Metamorphism in Adjacent Argentina*

The Paleozoic evolution in Chilean territory cannot be properly evaluated without a brief consideration of the main characteristics of the pre-Andean basement in Argentina.

K-Ar dates (whole-rock and micas) for metamorphic rocks of the Sierras Pampeanas, Precordillera and Cordillera Frontal between latitudes 32° and 34°S were reported by González-Bonorino and Aguirre (1970). The oldest age (591 Ma) is for a rock from the easternmost range, the Sierras Pampeanas, while toward the west two metamorphic rocks from the Precordillera gave values of 406 and 365 Ma and two from the Cordillera Frontal of 263 and 251 Ma.

McBride *et al.* (1976) reported K-Ar age determinations for mineral separates and whole-rock samples from igneous and metamorphic rocks of northwest Argentina between latitudes 25° and 30°S. Values obtained according to these authors would define three thermal events: Ordovician–Silurian (400–450 Ma); middle Carboniferous (310–340 Ma); and Permian (225–270 Ma).

Recent studies by Caminos *et al.* (1979, 1982) show that metamorphic rocks with ages from Precambrian to early Paleozoic (900 Ma and 650–400 Ma) are present in the Sierras Pampeanas while a Rb-Sr isochron age of 500 ± 50 Ma with a Sr(i) ratio of 0.7076 was obtained for metamorphic rocks from the Precordillera and Cordillera Frontal. Younger ages obtained by Caminos *et al.* (1982) for this same area were interpreted as being the result of granitoid emplacement, a point that should be kept in mind when considering the also younger dates reported by González-Bonorino and Aguirre (1970) from similar localities.

4. *Alternative Schemes for a Treatment of the Chilean Paleozoic Basement Rocks*

Miller (1970, 1973, 1979a) has criticized the series' scheme for the Paleozoic metamorphic rocks. Based on work carried out mainly in the coastal region north of 34°S and in the Chonos Archipelago, he has distinguished two major groups for the rocks of the Paleozoic basement. These groups, known as series "A" and "B" correspond essentially with the metamorphic rocks of the Western and Eastern series of Aguirre *et al.* (1972) respectively. However, different arguments based on structural features and petrography added to local paleontological evidence (see above pp. 276, 277) have led Miller to conclude a pre-Devonian age for the sedimentation, tectonism and first metamorphism of his series "A" (Canal King Formation of the coastal Aisén region, Miller, 1976). Rocks of the series "B" experienced their main metamorphism during the late Paleozoic Variscan events when the pre-Devonian age of rocks of the series "A" was radiometrically reset. Miller's chrono-stratigraphic scheme, however, conflicts with evidence found in the field both in northern Chile and the insular part of Aisén by Hervé and coworkers.

A comprehensive framework that includes both sedimentary and metamorphic sequences of late Paleozoic age has been recently proposed by Hervé *et al.* (1981a). It consists of a three-fold geographic division of the country based on the lithology, fossil content, structure and metamorphism of the rocks in their principal areas of outcrop. Three "domains" have been distinguished: (a) the northern Andean domain (NAD); (b) the southern coastal domain (SCD) and, (c) the intermediate domain (ID). It is further suggested that each domain had a particular geotectonic significance during late Paleozoic evolution. The metamorphic rocks of the SCD and the ID correspond to the Western and Eastern series of central Chile respectively. Although the scheme of series was retained for the treatment of metamorphic rocks in the present paper, the division into domains by Hervé *et al.* (1981a) is adopted for presentation of the late Paleozoic sedimentary rocks and for geotectonic interpretation of Paleozoic evolution.

B. Sedimentary Rocks

Terrigenous and marine sedimentary rocks of the pre-Andean basement are present as scattered outcrops from 19°30'S down to the Strait of Magellan. Paleontological evidence shows the presence of Ordovician, Devonian, Carboniferous and Permian rocks. Most of these Paleozoic sequences are affected by epimetamorphism and, in a few places, they present transitional boundaries with rocks belonging to the metamorphic basement. The three depositional domains defined by Hervé *et al.* (1981a) are used to divide the pre-Andean sedimentary rocks (Fig. 3).

1. *The Northern Andean Domain (NAD)*

Discontinuous outcrops of terrigenous detritic sequences and marine fossiliferous limestones of Devonian to Permian age characterize the NAD, which occupies the eastern half of the country from approximately 31°S to the boundary with Peru. These rocks were deposited over a Precambrian and/or early Paleozoic basement to which the metamorphic rocks described in A.1 belong. The sedimentary rocks of that basement will be dealt with now, before proceeding to the description of the sequences of the NAD proper.

(a) *Early Paleozoic Sedimentary "Basement" of the NAD.* The sedimentary basement rocks are Ordovician graptolite-bearing sequences known at: (i) Sotoca where *Dictyonema flabelliforme* (Lower Tremadocian) is found in quartzites, quartzfeldspar sandstones and limolites affected by low-grade metamorphism (Cecioni, 1979); (ii) Poquis (Marinovic, 1979); (iii) Laguna Mucar (Hoffstetter *et al.*, 1957) and (iv) Aguada de la Perdiz, where early Ordovician graptolites are found in shales intercalated in a 2 km thick sequence of quartzitic sandstones and chert (García *et al.*, 1962).

(b) *Late Paleozoic Sedimentary Sequences of the NAD.* Exposures are found between 20° and 30°S west of the Precambrian-early Paleozoic metamorphic and sedimentary rocks already described (Fig. 3). A sandstone–limestone–shale facies characterizes the northern outcrops while a flyschlike unit of graywacke and shale with subordinate limestone predominates in the more southern areas. Hervé *et al.* (1981b) interpreted this lithological change in terms of two types of environmental conditions replacing each other from north to south.

The limestone-bearing facies are known from the Cerro Juan de Morales area where the upper portion of a 150 m thick sequence of limestone, sandstone and shale contains brachiopods, pelecypods, bryozoa and corals (Galli, 1956; 1957, 1968). A late Carboniferous age was assigned to this fauna by Corvalán (*in* Galli, 1968) while Barthel (*in* Zeil, 1964) assigned an early to middle Permian age to fossils from the same locality. A similar rock sequence, 400 m thick, is exposed at 24°S and contains a fauna of pelecypods, brachiopods and crinoid stems covering the age interval early Carboniferous–late Permian (Chong and Cecioni, 1976). At latitude 25°45'S, pelecypod fragments of late Paleozoic age were collected from sandstone–limestone outcrops (Hervé *et al.*, 1981a). Further south, at *ca.* 27°S, at the eastern flank of Sierra de Fraga, a 200 m thick neritic sequence of limestone, marls and sandstone has been described by Hillebrandt and Davidson (1979). The sequence is transgressive over granitoids and is unconformably overlain by continental sedimentary rocks with a Triassic paleoflora. Fossil fauna from the marine sequence indicate a late Carboniferous to early Permian age.

The flyschlike facies, apparently contemporaneous with the limestone-bearing facies, is exposed at the Cordillera Claudio Gay, Pedernales (Kubanek and Zeil, 1971; Cisternas, 1977) and at the upper reaches of Quebrada Paipote (Davidson and Mpodozis, 1978). Further south, at 29°S, similar units are present in the Andes of Vallenar where a flora-bearing formation (Las

Placetas Fm.) was described by Reutter (1974). The southernmost Paleozoic sedimentary rocks so far recognized in the Chilean Andes are found at the upper valley of Rio Hurtado (30°30'S) and correspond to a 1500 m thick sequence of epimetamorphic graywackes and black shales. The sequence, of possible Carboniferous age, is covered by Permo–Triassic acid volcanoclastic deposits (Cornejo and Mpodozis, 1979).

2. *The Southern Coastal Domain (SCD)*

Sedimentary rocks belonging to this domain are found in the Patagonian Archipelagos between 50° and 53°S. Due to the fact that they are affected by low-grade regional metamorphism, they have been mostly described in the paragraph concerning late Paleozoic metamorphic rocks (see p. 277).

Forsythe and Mpodozis (1979) have described three "lithologic complexes" in the Madre de Dios Archipelago: (i) Tarlton Limestones, a sequence over 500 m thick of fusulinid-bearing limestones of late Carboniferous to early Permian age (Cecioni, 1955, 1956; Douglas and Nestell, 1972); (ii) Denaro Complex, a unit less than 200 m in thickness composed of cherts with an upper section of black siliceous shales and calcarenites containing reworked fossils and shell fragments. These rocks overlie a basal section of tholeiitic pillow basalt; (iii) Duque de York Complex, constituted by a flyschoid sequence, several thousands meters thick, containing rhythmically alternating graywackes with graded bedding, black shales and conglomerates. This late complex is considered to be of post early Permian age since it overlies Tarlton and Denaro rocks.

The metamorphic equivalents of the SCD's sedimentary rocks are represented, according to Hervé *et al.*'s scheme (1981a), by the Western series of Aguirre *et al.* (1972).

3. The Intermediate Domain (ID)

Paleozoic sedimentary rocks belonging to this domain appear as discontinuous outcrops all along the country between the NAD and the SCD (Hervé *et al.*, 1981a). They are mainly of terrigenous origin and include minor amounts of limestone and chert. Deformational and regional metamorphism are stronger than in rocks of the NAD and, unlike the case of the SCD, no oceanic rock types are represented. Their metamorphic equivalents are to be found in the Eastern series of the basement.

In northern Chile, a sequence of sandstone and shale with minor conglomerate and limestone, the El Toco Formation (Wetzel, 1927; Harrington, 1961; García, 1967; Maksaev and Marinovic, 1980), crops out along the eastern slope of the Coastal Range between 20° and 23°S. It has been assigned a Carboniferous–Permian age based on the presence of *Dadoxylon sp.*

(Gothan *in* Wetzel, 1927) although recently a fossil known from the middle/
late Devonian of Bolivia (*Haplostigma furquei* Frenguelli) has been found
in rocks of this belt (Covacevich and Troncoso (1980). A Devonian age for
a similar thick sequence of marine sandstone located further south, east of
Antofagasta (24°S) also has been suggested by paleontological evidence (Ferraris and Di Biase, 1978; Covacevich, 1977). Metasedimentary and metavolcanic rocks exposed along the Coastal Range between 25°30′ and 27°S
have been described by Bell (1982). The sediments are probably Ordovician
to Devonian in age and their deposition was accompanied by eruption of
alkali basalts partly as pillow lavas. According to Bell (1982) these rocks
were deformed and experienced greenschist facies metamorphism prior to
the intrusion of Lower Permian granitoids. A tectonic mélange (Chañaral
mélange) is also present. Although correlated with the El Toco Formation
by Bell (1982), the characteristics of these rocks seem closer to those belonging to the SCD and their inclusion in the ID (Hervé *et al.*, 1981a) should
be revised.

   According to Hervé *et al.* (1981a) this northernmost part of the ID would
be characterized by a largely homogeneous epimetamorphic flyschlike sequence of probable Devonian age. The rocks involved would have been
deposited on a basement possibly represented by the crystalline complex of
Mejillones Peninsula.

   Between latitudes 29° and 32°S ("Norte Chico") several formations of
Paleozoic age crop out in the Coastal Range (Muñoz Cristi, 1942, 1973;
Cecioni, 1974; Charrier, 1977; Mundaca *et al.*, 1979 among others). According to Hervé *et al.* (1981a) they correspond to two major lithostratigraphic units. The apparently older one is a flyschoid, plant-bearing marine
sequence of graywacke and shale with recumbent tight folds, incipient cleavage and local low-grade metamorphic effects. The base and thickness of the
unit are unknown but it is unconformably covered by Triassic sedimentary
rocks in one locality (Muñoz Cristi, 1942). Different fossil findings would
suggest that the older unit was deposited during a long span ranging from
the Devonian to the Carboniferous and possibly Permian (Cecioni, 1962,
1974; Cecioni and Westermann, 1968; Tavera *in* Paredes *et al.*, 1977; Mundaca *et al.*, 1979). The younger unit, mainly identified with the Huentelauquén and Totoral formations (Muñoz Cristi, 1942, 1973; Charrier, 1977),
consists of *ca.* 500 m of marine black shales, sandstones, minor limestones
and conglomerates. It contains a fossil fauna of Permian age, probably early
to middle Permian, according to Minato and Tazawa (1977). As already
mentioned, transitions between sedimentary rocks belonging to both, the
older and younger units, and metamorphic rocks have been reported for this
region.

In central Chile, south of 32°S, sedimentary rocks of the ID are present 50 km NE of Valparaiso (*ca.* 33°S) where, according to Thomas (1958) a fauna of *Productus* contained in sandstones would suggest a Carboniferous–Permian age for the outcrops. Corvalán and Dávila (1964), based on poorly preserved fragments of *Productus*, assigned a similar age to phyllites exposed at a locality (Colliguay) close to the previous one.

East from Valdivia, at latitude 40°S, a sequence of metasandstones, slates, phyllites, quartzites and minor quartz conglomerate beds crops out in several places from Lake Calafquén to Lake Ranco. This unit was named Panguipulli Formation by Aguirre and Levi (1964) who, in the absence of paleontological evidence assigned it early Paleozoic–Permian age. Subsequent finding of fossil flora has permitted a late Carboniferous–Permian age assignment for this formation (Tavera, 1971; Minato and Tanai, 1977). According to Parada (1975) clasts in the conglomerates include metamorphic rocks of the Western and Eastern series of the basement.

In the Chiloé–Aisén region (42°–50°S), shales containing trilobites of Devonian age have been found in beach boulders at Caleta Buill located at *ca.* 42°45′S (Biese, 1953; Levi *et al.*, 1966).

Based on a brachiopod fauna a Devonian age has been assigned by Miller and Sprechmann (1978) to an epimetamorphic arenaceous and pelitic rock sequence, the Potranca Formation (Miller, 1976), exposed in the Chonos Archipelago area.

A sequence of unmetamorphosed feldspathic sandstone, pelitic beds and bedded radiolarite also present at the Chonos Archipelago has been named Canal Pérez Sur by Miller (1976). Based on structural considerations, a Permo–Carboniferous age for this formation has been suggested by Miller (1976, 1979a).

C. Igneous Rocks

1. *The Granitoids*

No Precambrian igneous rocks are known to occur either among the granitoids or the volcanic series of the Chilean pre-Andean basement. Paleozoic batholiths are found distributed in two discontinuous belts along the Andean and the Coastal ranges respectively (Fig. 3 and Fig. 4). The Andean Range granitoidal belt extends from 21°S to 32°S; between 28°30′ and 32°S it constitutes most of the Andean Cordillera where it is associated with large volumes of volcanoclastic silicic rocks. The Coastal granitoidal belt extends between 25° and 38°30′S to reappear in Argentina after a sharp deflection near latitude 38°S. No volcanic rocks are associated with this belt.

The main rock types represented in the Paleozoic batholiths are granodiorite, quartz diorite, granite and tonalite, with subordinate diorite and

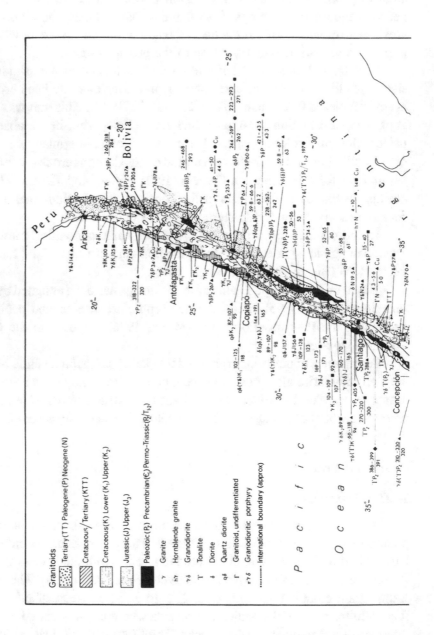

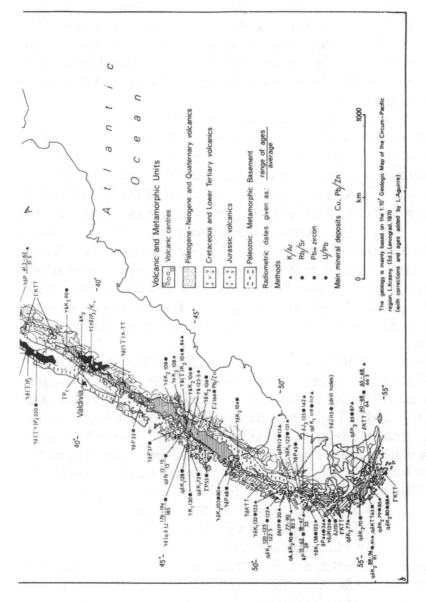

Fig. 4. Distribution of granitoids in Chile. Modified after Aguirre (1983).

gabbro and minor porphyries. In some instances the presence of several of these petrographic phases has been interpreted as representing a comagmatic suite (i.e. El Salvador, Tobar 1977; Andes of Coquimbo, Mpodozis *et al.*, 1976).

A calc-alkaline nature characterizes the Chilean Paleozoic batholiths with a tendency toward alkalinity at the end of the cycle in the case of the Andean belt of Coquimbo as indicated by the presence of hypersolvus granites (Mpodozis *et al.*, 1976). According to Parada (1981) original peralkalinity of these granites was depleted due to loss of alkalies produced by groundwater interaction.

The Paleozoic granitoids cut across stratified sequences ranging in age from late Ordovician (Caradoc) (McBride *et al.*, 1976) to latest Permian (Mpodozis *et al.*, 1976) including the metamorphic basement of central Chile whose main metamorphic event took place during the Carboniferous. They are, in turn, unconformably covered in several localities by sedimentary marine rocks of early Jurassic (Liassic) and also probable Triassic age and locally, as in the Andes of Coquimbo (Mpodozis *et al.*, 1976), by continental rocks of late Triassic (Keuper) age. Volcanic and sedimentary rocks of early Cretaceous (Neocomian) and late Cretaceous age unconformably overlie the Paleozoic granitoids west of Santiago (Corvalán and Dávila, 1964) and in the lake region of Valdivia (Hervé *et al.*, 1974b; Parada, 1975), respectively. Cretaceous and Tertiary granitoids intrude the Paleozoic batholiths in several areas.

Contacts with the country rock can be sharply discordant, devoid of migmatites and dike swarms and with thermal aureoles narrow or absent as in the Andes of Coquimbo (Mpodozis *et al.*, 1976) or in the Coastal Range between 34° and 38°S (González-Bonorino, 1970). On the other extreme, broad migmatitic zones associated with high-grade facies (amphibolite–granulite transition) of the metamorphic basement and with abundant pegmatitic veins are characteristically found in the coastal area west of Santiago (Muñoz Cristi, 1964) and in the Nahuelbuta Mountains (38°S, Hervé, 1977). Contamination of the Paleozoic granitoids with the metamorphic country rocks has been observed by Hervé (1977) and Tobar (1977) for the Nahuelbuta and Salvador–Potrerillos areas respectively.

Radiometric ages for the Chilean Paleozoic granitoids (Fig. 4) have been obtained by the Rb-Sr, Pb-U, K-Ar and Pb-alpha methods (see Aguirre, 1983 Appendix). The oldest plutonic ages have been recorded at the northernmost part of the Andean granitoidal belt where Rb-Sr and K-Ar dates between 468 ± 100 Ma and 431 Ma have been interpreted as representing a late Ordovician–early Silurian activity (Huete *et al.*, 1977; Halpern, 1978). Younger plutonic events of Carboniferous and Permian age at the same latitude and further south (27°30'S) have also been recorded in the Andean Range granitoidal belt (Huete *et al.*, 1977; McBride *et al.*, 1976). Whole-rock Rb-Sr reference isochrons of 328 ± 21 Ma with Sr(i) ratio = 0.7058 (tonalite–granodiorite) and 197 ± 5 Ma with Sr(i) ratio = 0.7073 (leuco-granite) have been reported for the Elqui–Limarí composite batholith (Parada *et al.*, 1981). These dates are consistent with the stratigraphic framework and with Pb-alpha ages previously obtained by Dediós (1967) for similar rocks of the area that ranged for 306 to 373 Ma for the tonalite–granodiorite unit and gave a value of 200 ± 20 Ma for the leucogranites. South of latitude 31°S the Elqui–Limarí Batholith extends into Argentina where it is known as the Composite Batholith of the Cordillera Frontal (Polanski, 1958). Rb-Sr isochrons and K-Ar dates would indicate the existence of three plutonic cycles in the Cordillera Frontal (32° to 34°S app.): Silurian to early Carbon-iferous (400–330 Ma); early Permian (275 ± 30 Ma) and late Permian to early Triassic (225 ± 20 Ma) (Caminos *et al.*, 1979; Caminos *et al.*, 1982).

In the Coastal granitoidal belt the oldest ages recorded correspond to the region west of Santiago where Pb-U dates between 386 and 405 Ma suggest plutonic activity in the early Devonian related to the orogenic Quin-tay phase (Corvalán and Munizaga, 1972). A Rb-Sr isochron of 320 Ma (late Carboniferous) was obtained by Hervé *et al.* (1976b) for the Nahuelbuta Batholith (*ca.* 38°S). Finally, most of the ages compiled in Aguirre (1983) fall in the interval Carboniferous–Permian.

The Paleozoic granitoids in Chile appear to be almost unrelated to the genesis of ore deposits; one exception would be the copper mineralization in the Imilac area, Andes of Antofagasta, which Halpern (1978), on the basis of geochronology, reported as Paleozoic.

2. *The Volcanics*

A widespread volcanic activity of calc-alkaline character and silicic composition accompanied plutonism in the Andean granitoidal belt. The rocks are rhyolitic ignimbrites and tuffs, dacitic flows and pyroclastics, an-desitic flows and terrigenous sedimentary rocks with fossil flora. This vol-canic sequence known as Choiyoi Formation in Argentina (Rolleri and Criado Roque, 1969) and first recognized in Chile by Dediós in the Andes of Coquimbo (Matahuaico Formation, Dediós 1967) is known to occur in the Chilean Andean slope from *ca.* 30°30'S (upper Hurtado River, Cornejo and Mpodozis, 1979) to 20°S (Juan de Morales, Galli, 1968). In Argentina these volcanic rocks overlie sedimentary units of Carboniferous to Lower Permian age and are unconformably covered by continental sediments of middle to late Triassic age (Caminos *et al.*, 1982). According to Mpodozis and Davidson (1979) the silicic volcanic unit is widely represented in the pre-Cordillera of Copiapó where it has a thickness of *ca*. 2000 m and un-conformably rests on epimetamorphic sedimentary rocks of possible Devonian–Carboniferous age.

A close spatial and chronological relationship exists between the Choiyoi–Matahuaico Formation and the granitoids belonging to the Andean belt; in several places granitoids are observed intruding the silicic volcanic rocks while in others the plutonic units are cut by rhyolitic dikes associated to the volcanic complex (Cornejo and Mpodozis, 1979).

Concerning the age of the volcanic rocks, Letelier (1977) has reported the finding of a probable Permian flora in shales interbedded in the silicic volcanoclastics of the Matahuaico Formation. Geochronological determinations of rocks from the Choiyoi Formation indicate a Permian to early Triassic age (Caminos *et al*., 1979; Caminos *et al*., 1982).

Rocks equivalent to the Choiyoi–Matahuaico unit are known further south between latitudes 39° and 41°30'S in Argentina near the boundary with Chile where they have been named Aluminé Formation (Turner, 1965). According to Moreno and Parada (1976) they are volcanic breccias and tuffs of andesitic–rhyolitic composition.

3. *The Ultramafics*

Small but numerous bodies of ultramafic rocks are found closely associated in space with the metamorphic rocks of the Western series. Outcrops are known in the Coastal Range between 35° and 36°S (Hervé, personal communication) and near Valdivia (*ca*. 40°S) where strongly sheared serpentinites probably originated by alteration of peridotite and pyroxenite (Flores, 1947; Illies, 1960; Zamarski *et al*., 1972). The age of these bodies is unknown but they are presently interpreted as being linked with the processes that generated the Western series.

IV. GEOTECTONIC INTERPRETATION OF THE PALEOZOIC EVOLUTION

Present knowledge concerning the pre-Andean basement in Chile permits us to sketch a broad but coherent evolutionary model. Central to this model is the existence, during the Paleozoic, of a plate interaction mechanism between Pacific oceanic crust and the continental mass of Gondwana. Evidence for such a mechanism is provided by geological knowledge of the region and by strong analogy with subduction patterns in the circum-Pacific realm during the Mesozoic and Cenozoic.

The abyssal tholeiitic affinities of the basaltic lavas of the Western series (p. 274), the structural features of this unit which include mélange type deposition and strong penetrative deformation (p. 274), the presence in it of blueschist assemblages and ultramafic rocks (p. 274, p. 288), the isotopic and geochronological data (p. 275) all strongly suggest that the Western series (SCD) was generated as a subduction complex accreted to the Pacific margin of Gondwana during the late Paleozoic (Hervé et al., 1976a; Kato, 1976; Forsythe and Mpodozis, 1979; Hervé et al., 1981a; Levi and Aguirre, 1981; Forsythe, 1982; Hervé et al., 1982). The broadly contemporaneous Eastern series (ID), for which isotopic ratios (p. 275) would indicate an origin from recycled crustal material (Hervé et al., 1982), is interpreted as a unit probably deposited in a fore-arc basin type of environment the nature of whose basement, either continental or oceanic, is yet unknown (Hervé et al., 1981a). The sedimentary units of the NAD are considered by Hervé et al. (1981a) as intracontinental deposits overlying an arc massif represented by the early Paleozoic and Precambrian rocks sporadically present along the NAD. The existence of contemporaneous mobile intracontinental basins controlled by deep faults is suggested by Godoy (1979) based on the alkaline affinities of metabasites from Lake General Carrera (p. 277). these rocks differ from the predominant tholeiitic metabasites of the Western series scraped off and welded to the continent.

Forsythe (1982) has interpreted the late Paleozoic to early Mesozoic evolution of southern South America in terms of the existence of three tectono-stratigraphic zones of the ancestral Pacific margin of Gondwana. They correspond to: (i) a fore-arc region consisting in part of "oceanic" components; (ii) a magmatic arc that is coincident with the Carboniferous to Triassic continental sedimentation; and (iii) a back-arc region comprising the epicratonic sequences of the "Samfrau Geosyncline". These three zones can be traced from 29° to 56°S and would closely correspond to Hervé's SCD (Western series), ID (Eastern series) and NAD. According to Forsythe (1982) the zones defined document a semicontinuous subduction process from the middle Devonian to the Triassic.

Rather continuous plutonic activity throughout the Paleozoic is recorded by the chronological data for the Coastal and Andean granitoidal belts (see Forsythe 1982, Fig. 2; Aguirre, 1983). Possible culminations might have taken place in the late Silurian–early Devonian, middle Carboniferous, early Permian and Permo–Triassic. McBride et al. (1976) point out that, by late Ordovician, the focus of major orogenic activity in South America was located along the present western and southern margins of the craton and tended to migrate westward during the Paleozoic. Such an interpretation is supported by the radiometric ages (p. 278) reported in González-Bonorino and Aguirre (1970).

The late Paleozoic magmatic belts can be interpreted as representing the magmatic arc activity resulting from the subduction process that accompanied accretion of oceanic material at the SCD. Sr(i) ratios for granitoids of the Elqui–Limarí composite batholith have been interpreted by Parada *et al.* (1981) as indicative of an upper mantle source for the magmas accompanied by late crustal contamination. A crustal origin for the late peralkaline plutonism in this area is also considered by them and has been taken as indirect evidence of crustal extension, probably associated with rifting, during the Lower Triassic (Parada, 1981). The coastal Carboniferous granitoids are known to have been closely related with the thermal régimes that resulted in the generation of the high T/P Eastern series (González-Bonorino, 1970; González-Bonorino and Aguirre, 1970; Hervé, 1977). On the other hand, batholithic emplacement in the Andean Range during Permian and early Triassic time was intimately linked with volcanic activity of predominantly silicic composition (p. 288). These contrasting features are interpreted by Hervé *et al.* (1981a) as possibly representing different levels of exposure of the same magmatic belt.

Late Paleozoic evolution from early Carboniferous (late Devonian?) to the end of the Permian and slightly beyond, comprises several plutonic and metamorphic events spread along the entire duration of the "Hercynian orogeny". Earlier Paleozoic events seem to have occurred during late Silurian–early Ordovician and in early Ordovician times suggesting a probable "Caledonian" imprint in this margin of South America. The only known Precambrian age of 1000 Ma could indicate a late Algonkian event.

An "orthodox" type of plate interaction, ultimately one typified by accretion, has not taken place in this segment of the continent since the late Paleozoic.

Finally, the plutonic activity linked with widespread silicic volcanism of ignimbritic type that characterizes the transition from the Paleozoic to the Triassic in the Andean belt is seen here as a prelude to the action of the major mechanism that will command the Andean evolution of this segment.

V. THE ANDEAN OROGENIC EVOLUTION

The geological development of Chile during the Mesozoic and Cenozoic has been mainly interpreted as the evolutionary process of ensialic geosynclinal basins (Ruiz *et al.*, 1965; Aubouin *et al.*, 1973; Aguirre *et al.*, 1974; Aguirre and Levi, 1977; Zeil, 1979, among several others). Two principal sectors have been classically distinguished: (i) north from 41°S a pericontinental, geoliminal sector, commonly referred to as "Andean Geosyncline", and (ii) from 41°S to Cape Horn, a geosynclinal sector proper usually known as "Magallanes Geosyncline" (Aubouin and Borrello, 1966; Ruiz *et al.*, 1965; Vicente, 1970, 1972; Aubouin *et al.*, 1973; Aguirre *et al.*, 1974). In this paper

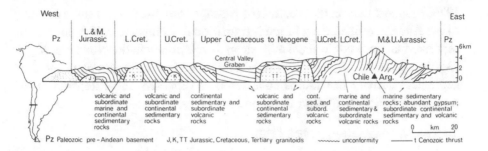

Fig. 5 Schematic cross section through the ensialic synclinorium in central Chile at *ca.* 33°S. Modified from Åberg *et al.* (1984).

they will be referred to as the Northern Sector and the Patagonian Sector respectively (Fig. 2). The Concepción Ridge, a positive area, separated the two sectors during their evolution.

In the Northern Sector the Mesozoic–Cenozoic edifice consists of several, partly overlapping, stratigraphical-structural units separated by unconformities. Each of these units would be the result of a sequence of events forming a cycle (Charrier, 1973; Aguirre *et al.*, 1974) in which volcanicity, sedimentation, subsidence, burial metamorphism, folding, granitoid intrusion, contact metamorphism and uplift closely followed each other and partly overlapped in time (Aguirre *et al.*, 1978). Along most of the extent of the Northern Sector, the Mesozoic–Cenozoic stratigraphical-structural units conform a broad synclinorium (Fig. 5) in which flat-lying young volcanic units and their feeder dikes are located in the axial part while increasingly steeper, older units, occur toward the flanks.

In the Patagonian Sector a marginal basin developed in the late Jurassic and was typified by generation of oceanic crust (Katz, 1972; Dalziel *et al.*, 1974; Dalziel, 1981, among others). Several regional deformational episodes, interpreted as representing orogenic phases, are also known in this region (Klohn, 1965; Aubouin *et al.*, 1973; Katz, 1973; Natland *et al.*, 1974; Skarmeta and Charrier, 1976). They can be considered as boundaries of stratigraphical-structural units of the type present in the Northern Sector. However, the two sectors show different organizations and structures, both in space and time, that require separate description and discussion.

Crucial to the understanding of the Andean orogenic evolution is the question of its inception in time. It has been generally agreed that, in the Northern Sector, the formation of marine basins at the beginning of the Jurassic would mark the onset of "Andean" events (Ruiz *et al.*, 1965). However, a better knowledge of the evolution that followed the climax of the Hercynian orogeny, notably the Permo–Triassic magmatism (Mpodozis *et al.*, 1976) and recently documented Triassic orogenic phases (Hervé *et al.*,

1976d) suggest that, in that sector, the passage from the Gondwanide orogeny to the Andean orogenic evolution is transitional. In this view, Triassic events represent an embryonic stage of Andean development in the Northern Sector (Aguirre *et al.*, 1974), a line of thought followed in the present paper. Recent geological activity in Chile is considered to represent the extension of the Andean orogenic evolution to the present.

A. The Northern Sector (18°–41°S)

1. *Geological Record*

The following summary of Mesozoic and Cenozoic geology is based on numerous papers published since the middle of the 1960's (see References). The most comprehensively studied areas are between 26° and 29°S (i.e. Zentilli, 1974 and Clark *et al.*, 1976) and between 30° and 35°S (i.e. Aguirre *et al.*, 1974, Charrier, 1981; Drake *et al.*, 1982a) and they will be referred to frequently.

The presentation of data will consider the division in the following unconformity bounded stratigraphical-structural units: (a) middle to Upper Triassic; (b) Lower to Upper Jurassic; (c) Tithonian to middle Cretaceous; (d) Upper Cretaceous to Paleocene; (e) Palocene to Upper Miocene, and (f) Pliocene to Recent.

During the Triassic, continental sediments and acid calc-alkaline ignimbrites were deposited on Paleozoic basement partly as the extension of volcanic activity started in the Permian. Marine sediments were laid down along the westernmost areas.

Marine depositional basins of "geosynclinal" type in which a western "eugeosynclinal" and an eastern "miogeosynclinal" side are coupled, can be recognized during the Jurassic and the early Cretaceous. Shallow marine to continental sediments and large volumes of calc-alkaline ignimbrites and flood basalts were deposited within N–S elongated basins bordered by Paleozoic basement, both at the eastern and western flanks. Units (b) and (c) represent this stage of the Andean orogenic evolution. Continental conditions were established during the middle Cretaceous due to uplifting, infilling by flood basalts and broad folding produced by the Subhercynian orogenic phase. Except for local and narrow transgressions of the Pacific along the western border, the continental character of the region persisted during the late Cretaceous and Paleogene, which were characterized by the ejection of huge quantities of calc-alkaline, mainly fissure-erupted volcanic material, and by block tectonics. Units (d) to (f) are included in this continental stage of the Andean orogenic evolution.

The absence of marine fossils together with the predominantly volcanic character of all the units older than middle Cretaceous constitute a major limitation in trying to establish the ages involved. Correlation of observed unconformities with regional and world-wide orogenic phases has been the main criterion on which ages have been assigned to the several major formations. In recent times, however, radiometric determinations in volcanic rocks have provided the bases for drastic changes in the chronology of these volcanic units. A systematic younging of the accepted ages of most of the formations has been suggested as a result. Nevertheless, the fact that all the rocks concerned are affected by low-grade burial metamorphism introduces an important factor of uncertainty for the dates obtained since the younger K-Ar ages may be due to Ar loss. Younging of ages caused by mild metamorphic alteration is known to have occurred in other metamorphic terranes, e.g. Iceland (Wood *et al.*, 1976; McDougall *et al.*, 1977).

(*a*) *Middle to Upper Triassic* Outcrops of Triassic rocks in Chile (Fig. 6) are randomly distributed, both along the Andean and Coastal ranges, from the northernmost extreme to the region of Valdivia (40°S). An angular unconformity between Triassic rocks and late Paleozoic strata is observed at several localities along the Coastal Range between 31° and 32°S (Los Vilos, Talinai). South of 32°S Triassic rocks unconformably cover pre-Andean crystalline basement.

Two different sedimentary environments are recognized, one predominantly marine that prevailed along the present Coastal Range (west), the other exclusively continental developed further east, mainly along the present Andean Range. Volcanism in the form of lava flows and ignimbrites took place in both environments. Silicic rocks predominate but basalts and andesites are also represented (Vicente, 1976). Basalts and tholeiitic andesites are high in TiO_2 and total FeO and low in Al_2O_3. K_2O is low in basalts. Iron tends to be high in dacites and rhyolites (Table I, p. 328, analyses 1–5).

Predominantly marine sedimentary rocks of the western side form series of cyclic character that consist, from bottom to top, of: (i) conglomerate and sandstone; (ii) a rythmic sequence of graywacke and shale; (iii) shale; (iv) another rythmic sequence, as before and (v) sandstone and conglomerate (Charrier, 1979). Continental sedimentary rocks are often intercalated and in the upper levels they are represented by plant-bearing shales and, in places, by coal seams (Charrier, 1979 and references therein; Hervé *et al.*, 1976d). Thick units of silicic volcanics are present in the middle to upper part of this generalized column. The sedimentological data suggest transgression-regression of the sea (Charrier, 1979; Hervé *et al.*, 1976d). The continental rocks on the east consist of thick detritic sequences of fluvial and limnic character with intercalated coal seams. Silicic volcanics commonly appear in the upper levels. Further east, in Argentina, Triassic rocks are

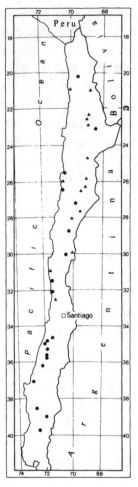

Fig. 6.   Distribution of Triassic deposits in Chile. (●) Paleontologically dated deposits; (▲) deposits assigned to the Triassic. Modified after Charrier (1979).

also continental and include thick detrital members, tuffs, coal seams and evaporites; some of these units are bituminous. The sequences are also cyclic as shown by grain size and color (see Charrier, 1979).

Although the Triassic sequences in Chile are mainly present as isolated, remnant outcrops, their thickness may reach up to 3500 m, especially where volcanics are predominant. The age of fossiliferous deposits is from Anisian to Rhaetian. No rocks of early Triassic age are known in Chile although the uppermost volcanics of the Choiyoi–Matahuaico unit of the Andean Range could be of that age.

The Triassic volcanic rocks have undergone low-grade burial metamorphism with conversion of rhyolites into keratophyres and basalts into spilites. Palynological studies of Triassic sedimentary rocks show that burial metamorphism occurred extensively; in most areas of central Chile it was equivalent to the anthracite/meta-anthracite coal rank (Askin *et al.*, 1981).

Subsidence, accompanied by marine transgression and intense silicic volcanism took place during the early(?) and late Triassic and were followed by uplift that, together with sedimentation and erosion, led to the establishment of continental conditions during the Rhaetic (Corvalán, 1965; Charrier, 1979). Hervé *et al.* (1976d) suggest that vertical movements recorded in the lithology can be associated to epeirogenic events only for the first transgressive stage. The strong subsidence and volcanism which followed are due, according to the same authors to orogenic movements which led to folding of Triassic strata and to their intrusion by granitoids.

Widespread Triassic plutonism is not yet sufficiently documented. However, plutonic bodies of Triassic age could be represented among the youngest members of the Paleozoic Andean Range granitoidal belt in Chile since several Triassic ages have been obtained from the extension of this same belt in the Cordillera Frontal of Argentina (Caminos *et al.*, 1979). Dávila *et al.* (1979) have interpreted some Triassic ages from the Coastal Range (*ca.* 34°S) to be Paleozoic intrusives rejuvenated during a regional tectonic event in early to middle Triassic. This event would have caused mylonitization and large scale faulting. However, the same authors point out that the existence of Triassic granitoid bodies in the area cannot be ruled out. Hervé *et al.* (1982) have reported a K-Ar (biotite) date of 208 Ma for the granitoid that crops out in Constitución (*ca.* 35°20'S) and Hervé and Munizaga (1978) have obtained geochronological evidence in favour of late Triassic–Jurassic magmatism in the Coastal Range between 35°30' and 36°30'S. Drake *et al.* (1982a) have compiled and contributed geochronological data on magmatism in central Chile (31°–36°S); according to these authors an age group of 230–170 Ma (early Triassic to middle Jurassic) is apparent for the data. The granitoids included are plutons lying in a zone south and east of the Paleozoic metamorphics and granitoids of the pre-Andean basement.

A great number of dikes cutting across Paleozoic and Triassic units in the Coastal Range at approximately 31°30'S have been described by Muñoz Cristi (1973); they are diabases and granophyres with alternate crosscutting relations.

(b) Lower to Upper Jurassic The Jurassic system is extensively developed in the northern sector between the Peruvian boundary and 39°S (Fig. 7). It outcrops in two fairly continuous belts with a N–S trend along the Coastal Range (west) and the Andes (east) respectively. Predominance of volcanic material in the western domain and of sedimentary rocks in the eastern one, has been interpreted as representing deposition in a geosynclinal basin with well-defined eugeosynclinal and miogeosynclinal troughs (Ruiz

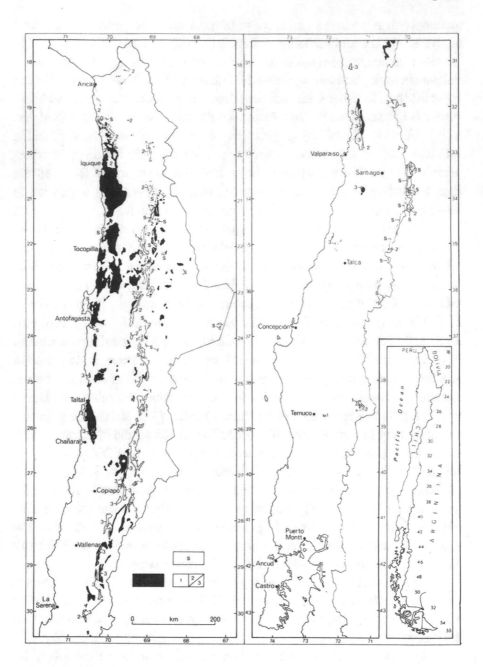

Fig. 7.   Distribution of Jurassic deposits in the Northern Sector. Black, volcanic rocks with
minor intercalations of marine and/or continental sedimentary rocks in places; 1, marine sedimentary rocks and subordinate volcanics of general Jurassic age; 2, marine sedimentary rocks
and subordinate volcanics of the Dogger–Malm; 3, marine sedimentary rocks and subordinate
volcanics of the Lias–Dogger; *S*, continental sedimentary rocks (partly red beds) and subordinate volcanics of the Upper Jurassic. (Geology from *Mapa Geológico de Chile*, 1:1,000,000,
1982).

et al., 1965; Corvalán, 1965; Aubouin *et al.*, 1973; Aguirre *et al.*, 1974). More recent information concerning the longitudinal (N–S) changes in volcanic distribution and composition, the configuration of the predominantly sedimentary basins to the east and their facies changes toward the west with increasing participation of volcanic material, all have led to modified views about the Jurassic evolution of the territory (see Zentilli, 1974; Rivano and Mpodozis, 1976; Jensen *et al.*, 1976, among others).

In the Coastal Range the southernmost Jurassic exposures are found east of Temuco (*ca.* 38°40'S) and consist of shales and subordinate andesitic volcanic rocks. Fossils indicate a middle Jurassic (Dogger) age (García, Tavera *in* Ruiz and Corvalán, 1966). Toward the north small and restricted basins formed *ca.* 35°S near Vichuquén–Tilicura and Hualañé. Stratigraphic discontinuities expressed either as angular unconformities or as abrupt lithological changes exist in those areas at the boundary between Upper Triassic (Norian) and Lower Jurassic (Hettangian) sedimentary units. Local stratigraphic continuity has also been documented in localities of the same region (Corvalán, 1976, 1982). Further north, between 32° and 34°S, the Jurassic is represented by a predominantly volcanic section up to 8000 m thick composed of acid ignimbrites (dacitic and rhyolitic) and lavas, subordinate basic flow breccias and lahars and marine, nearshore, clastic rocks and limestones. Palentological evidence indicates that this section is Lower and middle Jurassic (Sinemurian to Upper Bajocian) (Thomas, 1958; Levi, 1970; Vergara, 1972; Piracés, 1976; Covacevich and Piracés, 1976). The most continuous outcrops of Jurassic rocks along the Coastal Range are found between 27°S and the Peruvian border where great volumes of lava flows of intermediate and basic composition (La Negra Formation) occur reaching thicknesses in excess of 10,000 m. From 26°S to 20°30'S the lavas of the La Negra extend for nearly 700 km along the Pacific border as a continuous sequence of flows with a few intercalations of cross-bedded, fluviatile, red sandstones. From 20°30'S to the Peruvian boundary a Jurassic sequence composed of La Negra type volcanics and marine sediments makes up most of the Coastal Range. Its average thickness is 3000 m and it is characterized by graded and rhythmic graywackes, limestones and pillow lavas. Paleontological evidence indicates that this sequence includes units of Bajocian, Bathonian, Callovian, Oxfordian and possibly Lower Kimmeridgian age (Cecioni and García, 1960; Thomas, 1970). Overlying Oxfordian sediments is a 20 m thick unit of gypsum.

From the preceding description it can be appreciated that the "eugeosynclinal" Jurassic trough is dominated by the presence of volcanic material that, from south to north, changes drastically both in terms of thickness and composition. While in central Chile the volcanics are largely silicic and were deposited in a predominantly shallow marine environment, in the region between 26° and 20°30'S the lavas are intermediate and basic, reach thicknesses of more than 10,000 m and were deposited in a continental environment. Volcanism under marine conditions is again observed north of 20°30'S where the intermediate–basic composition of the lavas persists although a decrease in their thickness is observed.

In the Andean Range, where sedimentary rocks predominate, all stages of the Jurassic are represented. Several detailed sections have been described between 22° and 23°30'S and between 24° and 30°S, this last interval containing the longest continuous outcrop of marine rocks in Chile. Sections at Cerritos Bayos (Biese, 1957), Caracoles (Harrington, 1961), Quebrada Asientos (Pérez, 1959, 1978), Juntas–Vegas de San Andrés (Segerstrom, 1959; Cisternas and Vicente, 1976), 25° to 32°30'S (Hillebrandt, 1971) and 30°30' to 31°20'S (Rivano and Mpodozis, 1976), can be mentioned. Together with the more meridional Andean Jurassic outcrops of the Northern Sector these sections can be generalized as follows:

(i) A *ca.* 1700 m thick marine sequence mainly of sandstones, shales, marls and limestones lies at the bottom of the section and represents the period from early Lias to middle Oxfordian (Rauracian). Volcanic material is quantitatively important in most of the sections with andesitic lavas and volcanic breccias being represented (Davidson *et al.*, 1976; Rivano and Mpodozis, 1976).

(ii) A level of evaporites (gypsum, anhydrite) up to 200 m thick is found on top of the previous sequence.

(iii) A sequence of continental red conglomerates, sandstones, shales and intermediate to basic volcanic rocks represents the Kimmeridgian and probably part of the Tithonian.

Volcanism during the Jurassic is characterized by the intermediate and basic La Negra Formation in the north and by the predominantly rhyolitic and dacitic sequences of the central region (Ajial and Melón formations in particular). The La Negra volcanics have been studied by Losert (1974a, 1974b) immediately south of Tocopilla (*ca.* 22°S). They consist of alternating sequences of clinopyroxene andesites, low-silica andesites and olivine basalts with a few intercalations of volcanic breccias, tuffs and volcanogenic sandstones and agglomerates. According to Losert (1974a) the series is about 2800 m thick and bears evidence of rhythmic volcanic activity and deposition in a subaereal or shallow subaqueous continental environment. The thickness of the flows is constant and does not exceed 10 m, individual flows can be followed for several kilometers along the strike morphologically resembling flood basalts. Subvolcanic plugs and dikes of andesitic to basaltic composition are common in the series. Dike swarms with a N–S orientation abundantly present along the Coastal Range of the Chañaral region (*ca.* 26°20'S) have been described as possible La Negra feeders by Zentilli (1974). Volcanics in the Tocopilla area have calc-alkaline affinities, are low in Al_2O_3 and Sr and high in Rb (Palacios, 1976). According to Rogers (1983) they are high-potassium basaltic andesites with Sr/Nd < 18 and their parental liquids were enriched in LREE and trace elements relative to MORB (Table I, p. 328, analyses 10–12 for major element composition of the Tocopilla Volcanics and 13–17 for other areas). In the northernmost outcrops, at the Morro de Arica, pillow lavas of La Negra are characterized by the presence of enstatite, augite and labradorite in a glassy groundmass (Cofré, 1955).

The Jurassic volcanics of central Chile (32°–34°S) are mainly ignimbrites and lavas of rhyolitic and dacitic composition; subordinate basalts and andesites are intercalated (Table I, p. 328, analyses 6–9). According to Levi (1970) the silicic rocks have a glassy to crystalline groundmass enclosing crystals of high oligoclase to andesine, quartz, sanidine and biotite and/or hornblende. Levi and Nyström (1982) have shown that the Zr–Ti–Y ratio and the K_2O content of basic rocks from this coastal sequence are similar to those obtained from coeval volcanics present as minor intercalations in the sedimentary sequences of the Andean Range at this latitude (Fig. 5).

A bimodal magmatism in which considerable volumes of acid rocks (keratophyres) are accompanied by basic products has been described by Thiele (1964) from the Andes of Coquimbo (Cordillera de la Punilla, 29°40'S) between the La Negra volcanics of the north and the silicic volcanism of central Chile and could represent a transitional kind of magmatic activity.

The Jurassic age interval for volcanism of the La Negra is a matter of argument. In the coastal region the type section is without fossils. Considerable volumes of volcanics correlative with La Negra appear, however, in several fossiliferous marine sections, both in the Coastal Range north of 20°30'S and in the Andean Range. Based either on the paleontological record or on petrographic and /or stratigraphic correlations, different ages ranging from Sinemurian to Upper Dogger-Malm, have been assigned to La Negra type of volcanics in several areas where they originated in marine and/or continental environments (Dediós, 1967, 1978; Davidson *et al.*, 1976; Cecioni and García, 1960; Vila, 1976; Zentilli, 1974; Tobar, 1977; Ruiz *et al.*, 1965; García, 1967; Rivano and Mpodozis, 1976; Cisternas and Vicente, 1976; Skarmeta and Marinovic, 1981). The length of the time interval involved would suggest that the Jurassic intermediate–basic volcanism developed in several stages, probably related to the regional geotectonic evolution of the continental margin. The age of the silicic volcanism of central Chile extends from the Lias to the Upper Bajocian as indicated by the paleontological record from interbedded marine strata (Levi, 1960, 1970; Thomas, 1958; Piracés, 1976).

Jurassic deposits, both in the Coastal and Andean belts, are affected by nondeformational, low-grade regional metamorphism of the burial type and by alteration on a local scale (Levi, 1970; Losert, 1974a; Vila, 1976; Campano *et al.*, 1976). In the coastal region of central Chile, the ignimbrites and lavas of original rhyolitic and dacitic composition have been changed into quartz keratophyres while the subordinate basic and intermediate flows have undergone spilitization. The metamorphic assemblages thus generated correspond to the prehnite–pumpellyite facies in the upper levels and to the greenschist facies in the lowermost levels; isograds conform to the regional structures. In the basic and intermediate rocks of the La Negra Formation (Tocopilla region) two types of alteration have been described by Losert (1974a). The first type is regional and has produced mineralogical associations similar to those formed in burial metamorphism under zeolite and prehnite–pumpellyite facies. A widespread epidotization, selectively related to some basaltic levels, took place as part of the regional alteration resulting in K metasomatism, selective redistribution of Si, Na, Ca and Sr and concentration of Cu in a well defined level in metasomatic high-potassium andesites underlying the sequence of epidotized basalts (Losert, 1974a). The second type of alteration is local and superimposed on the first and is associated with economic copper mineralization of stratiform and vein type.

No major Jurassic folding is known and diastrophism is characterized by block faulting. Unconformities or pseudoconformities have been observed between the Triassic and Lower Jurassic (Muñoz Cristi, 1950, 1973; Corvalán, 1976, 1982; Zentilli, 1974; Thiele and Morel, 1981) and may reflect late Triassic orogenic movements as suggested by Hervé *et al.*, 1976d) and/ or epeirogenic tectonism of the same age as indicated by Gutiérrez (*in* Corvalán, 1982). An anomalous contact between Lower and middle Jurassic strata is present in the coastal area of Chañaral–Taltal where the volcanic La Negra Formation of Bajocian (Callovian?) age unconformably overlies the Sinemurian Tres Hidalgos Formation. According to Zentilli (1974) a *lacuna* exists in the Andes of Copiapó at the level of the late Toarcian–early Bajocian. Rocks of this age rest with clear erosional unconformity on older Jurassic rocks at Pedernales (Hillebrandt, 1970). An erosional unconformity (disconformity?) between the Lower and the middle Upper Jurassic in the area of Hualañé (35°S) has also been reported by Thiele and Morel (1981). The existence of a widespread hiatus in marine Bathonian sedimentation marked by unconformity between the Bajocian and the Callovian is a matter of discussion. Muñoz Cristi (1973) has attributed to that event the unconformity observed on the top of middle Upper Bajocian sediments (Caleta Lígate Formation) in the coastal region of Iquique (20°S) and Zentilli (1974) points out that the Bathonian unconformity is clearly defined in the Copiapó region (*ca.* 27°S). However, the presence of a 100 m thick sequence of Bathonian sedimentary rocks has been documented by Thomas (1970) from the Iquique area (21°30'S).

During middle late Oxfordian time the first major Andean diastrophism, the Araucanian orogenic phase, took place (Muñoz Cristi, 1942; Stipanicic and Rodrigo, 1970; Charrier and Vicente, 1972; Charrier, 1973; Aubouin *et al.*, 1973; Aguirre *et al.*, 1974). In the coastal belt its effects are reflected in an upper Jurassic hiatus (Muñoz Cristi, 1956; Thomas, 1958), a regional block fault pattern (Ruiz *et al.*, 1965) and an angular unconformity between the upper Bajocian and Neocomian units (Carter, 1963). Along the Andean belt almost no direct tectogenic expression of this phase is observed but emergence of the trough, evaporite deposition (gypsum and anhydrite) and continental red bed sedimentation are indirect effects. All these features consitute clear evidence for a considerable uplift of the "miogeosynclinal" zone during the Upper Jurassic. However, no angular unconformity is present.

Granitoid emplacement occurred during the Jurassic (Fig. 4). Outcrops are found along the Coastal Range from 18°30' to 27°S where these rocks appear as extensive multiple batholiths, massifs and minor bodies with mesozonal and postkinematic characteristics. The Jurassic granitoids cut across the pre-Andean basement in several areas between 27° and 29°S and are, in turn, intruded by Cretaceous granitoids along most of their eastern boundary south of 26°S. A tectonic contact with the pre-Andean basement exists along most of the western boundary of the Jurassic intrusives south of latitude 30°S. Rock types range from gabbro to granite with predominant granodiorite; the rocks are oversaturated and calc-alkaline (Aguirre *et al.*, 1974; López-Escobar *et al.*, 1979; Montecinos, 1979; Aguirre, 1983). Radiometric data indicate that plutonic activity was continuous during the Jurassic (Drake *et al.*, 1982a) with most ages concentrated in the intervals 190–170 Ma and 160–140 Ma (Zentilli, 1974; Clark *et al.*, 1976; Hervé and Munizaga, 1978; Montecinos, 1979; Drake *et al.*, 1982a; Aguirre, 1983). Damm *et al.* (1981) report Sr(i) ratios "markedly above 0.7100" for granitoids in the age interval 190–130 Ma cropping out along the coast between 25°30' and 26°35'S. According to these authors the intrusives would represent *S*-type granitoids. In strong contrast, Sr(i) ratios for Jurassic (and partly Cretaceous) coastal granitoids between 26° and 29°S with K-Ar ages in the interval 195–125 Ma are lower and vary from 0.7043 to 0.7059 (McNutt *et al.*, 1975).

Copper veins accompanied by small amounts of cobalt, molybdenum and uranium have been genetically linked with Jurassic plutonism (Ruiz *et al.*, 1965).

(c) Tithonian to Middle Cretaceous At the close of the Jurassic, following the Araucanian orogenic uplift, subsidence caused a marine transgression that in several areas started during the Tithonian. The early Cretaceous is characterized by development of a volcanic arc and depositional trough. The arc was formed within the continental margin on eroded Jurassic rocks and pre-Andean basement and its volcanic products were deposited in the trough interfingered with sediments from the east. This situation is well exemplified in central Chile (25°–36°S) where outcrops of Lower Cretaceous rocks form a N–S band 100 to 150 km wide, narrower than the preceding Jurassic depositional basin in the same area (Fig. 8). Predominantly volcanic sequences occur along the western border of the band (Coastal Range) while sedimentary rocks with subordinate volcanics appear to the east (Andean Range). This distribution has been commonly interpreted in terms of a "eu/mio" geosynclinal couple (Ruiz *et al.*, 1965; Corvalán, 1965; Levi, 1970; Vergara, 1972; Aguirre *et al.*, 1974; Charrier, 1981; Levi and Aguirre, 1981).

Along the Coastal Range considerable volumes of dacitic and rhyolitic ignimbrites and minor basaltic flows are found together with subordinate turbidites, marine limestones and terrestrial sediments of Neocomian age (Fig. 9). This lithology reflects changes in the depositional "geosynclinal" environment with time from shallow-marine to alternating shallow-marine and continental conditions. Between 31°15′ and 33°30′S these rocks are represented, in decreasing order of age, by the Patagua, Pachacama and Lo Prado Formations with a total thickness of 2000 to 4000 m. The section contains fossils of Berriasian to Hauterivian and possibly Barremian age (Thomas, 1958; Aliste *et al.*, 1960; Corvalán and Dávila, 1964; Herm, 1965, 1967; Levi, 1968, 1970). South of 33°30′S Valanginian marine sedimentary rocks are intercalated with rhyolitic volcanics and minor andesitic lavas and reach a total thickness of 13,000 m (La Lajuela Formation, Vergara, 1969). The southernmost exposures of marine Neocomian are found in the Santa Cruz area at *ca.* 34°45′S (Muñoz Cristi, 1960). North of 31°15′S the coastal Neocomian sequences can be followed through the localities of Punitaqui, Ovalle, La Serena (Cruz de Caña–La Cruz limestone) to the north of the Elqui River (30°S) where 1250 m of basaltic lava flows and marine limestones, the Arqueros Formation occurs. The Arqueros is devoid of silicic volcanics and corresponds to the Hauterivian–Barremian (Aguirre and Egert, 1962, 1965). From the Elqui area northward to 25°20′S, an almost continuous strip 500 km long and 30–40 km wide of Neocomian strata is exposed. The best known sections are present south of Copiapó where a shallow marine unit of limestones, shales, marls and minor spilitic pillowed andesites has been defined and described as the Chañarcillo Group (Biese, 1942; Corvalán, 1955;

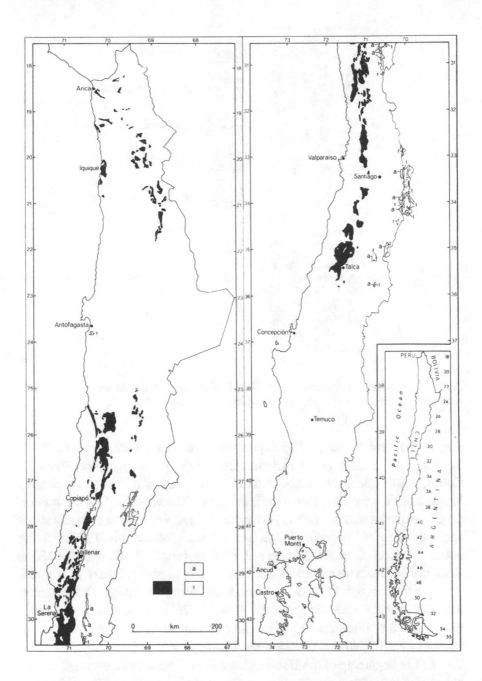

Fig. 8.   Distribution of Lower Cretaceous deposits in the Northern Sector. Black, volcanic
rocks with minor intercalations of marine and/or continental sedimentary rocks in places; 1,
marine sedimentary rocks and subordinate volcanics (including the Tithonian south of 33°S);
a, continental sedimentary rocks (partly red beds) and subordinate volcanics of Aptian–Albian
age. (Geology from *Mapa Geológico de Chile*, 1:1,000,000, 1982.)

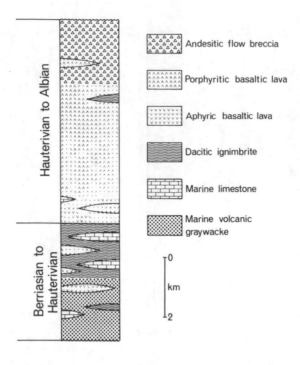

Fig. 9. Generalized stratigraphic column of the Lower Cretaceous in the western flank of the synclinorium, central Chile at *ca*. 33°S. Modified after Åberg *et al*. (1984).

Segerstrom and Parker, 1959; Segerstrom and Ruiz, 1962; Zentilli, 1974). The *ca*. 2500 m thick group includes, from older to younger, the Punta del Cobre, Abundancia, Nantoco, Totoralillo and Pabellón formations ranging in age from Upper Valanginian to Barremian. The Chañarcillo Group inter-fingers with the partly marine, partly continental volcanic and clastic strata of the Bandurrias Formation (Segerstrom, 1960, 1968; Zentilli, 1974). Further north, Neocomian strata reappear at El Way (*ca*. 23°50'S) where a 350 m thick unit of limestones, marls and sandstones contains fossils of Hauteri-vian–Barremian age (Tavera *in* Alarcón and Vergara, 1964). The northern-most Neocomian rocks are exposed at *ca*. 19°20'S in the region of Atajaña where a 400 m thick section of calcareous sandstones and shales contains fossils of Berriasian age (Cecioni and García, 1960).

At the beginning of the Hauterivian volcanic activity in the western side of the "geosynclinal" basin changed markedly in character. The silicic vol-canism of the early Neocomian gave way to eruption of basic lavas along deep and extended north-trending faults. A thick volcanic pile was built up along the western basinal margin: the lower and middle levels consist of flow basalts, whereas the upper levels are dominated by andesitic flow breccias (Fig. 9). Volcanism continued up to Albian time. It was less intense at the northern and southern ends of the basin where deposition of marine sediments predominated. The maximum thickness of the pile (5 to 8 km) is found in the south central section (between 30° and 34°S) where the interbedded sedimentary rocks are largely continental. In this section, the Veta Negra Formation (Levi, 1968) characterizes this evolution. North of 30°S basic to intermediate volcanism of Hauterivian and younger age is represented by Arqueros (Hauterivian–Barremian) and Quebrada Marquesa (Aptian–Albian?) formations of the Coquimbo region (Aguirre and Egert, 1962, 1965; Moscoso, 1976) and by the upper levels of the Bandurrias Formation (Segerstrom, 1960, 1968; Zentilli, 1974; Tobar, 1977).

In the Andean ("miogeosynclinal") flank, marine rocks of early Cretaceous age are known to occur from *ca.* 25°S (Sierra Candeleros, Chong, 1976; Naranjo and Covacevich, 1979) to 36°S. North of 22°S extensive volcanic sequences deposited under continental conditions have been assigned to the Lower Cretaceous based on geochronological data and field relations (Maksaev, 1978; Vergara, 1978a). South of 36°S marine Cretaceous strata are found in Argentina.

Sedimentary rocks of Tithonian age present in the Andean regions of Atacama (Pedernales Formation, Harrington, 1961; García, 1967; Tobar, 1977; Pérez, 1978) and Santiago–Curicó (Lo Valdés Formation and Baños del Flaco Formation; Biro, 1964; González, 1963; Covacevich *et al.*, 1976) indicate that the transgressive phase that characterizes the Neocomian had started in the late Jurassic, a situation also implied by the conformably overlying Lower Cretaceous in those areas.

Neocomian marine strata along the Andean Range are mainly represented by clastic rocks, limestone, gypsum and intercalated basaltic and andesitic flows reaching thicknesses between 300 m and 2000 m and of Berriasian to Barremian age. In most areas, however, regression of the sea seems to have taken place during the Hauterivian. Continental sedimentary red beds conformably cover the Neocomian and in central Chile they are typified by the Colimapu and Pucalume formations (Klohn, 1960; Aguirre, 1960; Dediós, 1967, 1978) which are lacustrine and deltaic deposits, subordinate fluviatile sediments, basic lavas (partly brecciated), pyroclastics and subordinate gypsum layers with a total average thickness of *ca.* 2000 m. An Aptian–Albian age is assigned to these rocks which thus correlate with the predominantly volcanic Quebrada Marquesa and Bandurrias (upper) formations of the coastal areas.

Widespread continental volcanism of early Cretaceous age in the Andes of Tarapacá (*ca.* 21°S) has been described by Vergara (1978a, 1978b) and Maksaev (1978). At the Quehuita quadrangle a 1700 m to 3200 m thick sequence of andesites, andesitic flow breccias, conglomeratic breccias, coarse sandstones and minor dacitic tuffs, the Macata Formation, unconformably overlies a Jurassic sequence of Callovian—Kimmeridgian(?) age. The Macata Formation (Arca Formation of Maksaev, 1978) is intruded by rhyolitic porphyries for which a Barremian–Albian age is suggested by K-Ar dating (Vergara, 1978a). Based on these relationships, a Tithonian–Hauterivian age is inferred for the Macata Formation. Toward the northeast, in the Ujina quadrangle (Vergara, 1978b), the Macata Formation presents a lower section of continental clastic rocks and porphyritic andesites ("ocoites") and an upper section of more than 200 m thick made up of rhyolitic tuffs. An igneous complex of rhyolitic porphyries, rhyolites and rhyolitic ignimbrites (Collahuasi Formation, Vergara, 1978a, 1978b and Peña Morada Formation, Maksaev, 1978) overlies the Macata–Arca Formation in the same area. K-Ar determinations on hypabyssal and effusive units of the complex suggest a Barremian–Albian age.

The volcanic-sedimentary units of the Tarapacá region are lithologically similar to coeval ones in coastal central Chile. However, they were deposited in a continental environment implying that the ocean had withdrawn during the Neocomian while marine units were still being formed in central Chile.

Lower to middle Cretaceous volcanism is markedly bimodal in character. The silicic rocks are dacites, rhyolites and trachyandesites (Table I, p. 329, analysis 19) with a glassy to crystalline trachytic groundmass and phenocrysts of high oligoclase to andesine, quartz, sanidine, and biotite and/ or hornblende. Flood basalts with a remarkable uniform primary mineralogical and chemical composition predominate in the late Neocomian (Chávez and Nisterenko, 1974; Levi et al., 1982; Oyarzún and Frutos, 1982; Åberg et al., 1984). Their spatial and temporal variations are extremely limited considering the huge volumes involved (exceeding 6 x 10^4 km^3) and the temporal persistence of these lavas. The commonest type is porphyritic with unzoned phenocrysts of labradorite (0.5–3.0 cm), clinopyroxene, magnetite and olivine (altered) in a groundmass that includes these same minerals and contains minor K-feldspar/quartz intergrowths (Levi, 1969). The name "ocoite" (from Ocoa, the type locality west of Santiago) is used by Chilean geologists to refer to this type of highly porphyritic flow. Numerous feeder dikes of similar composition, locally present as dike swarms parallel to the trend of the volcanic belt, are observed and bear witness to deep persistent regional faults.

The lavas are K-rich calc-alkaline basalts according to the classification of Irvine and Baragar (1971), though transitional to andesites (Table I, p. 329, analyses 18 and 20). A calc-alkaline affinity is also shown by their Zr–Ti–Y ratio. They are Fe-rich with a low Mg/Fe ratio and with a REE pattern showing enrichment in LREE indicative of an evolved basaltic type (Fig.

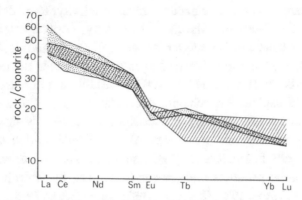

Fig. 10.   Chondrite normalized REE pattern for basalts of early Cretaceous age, Northern Sector. Dots, range in porphyritic flood basalts (ocoites) from central Chile (Åberg *et al.*, 1984); hatching, range in basalts from north central Chile (Dostal *et al.*, 1977).

10). The Sr(i) ratio is low; 0.7038 ± 0.0004 for "unaltered" samples of porphyritic basalt (ocoite) from central Chile (Åberg *et al.*, 1984). In north central Chile, between 26° and 29°S, Sr(i) ratios of 0.7035 and 0.7047 have been reported by McNutt *et al.* (1975) for early Cretaceous basalts (Table I, analyses 21, 22 and 23). In this same region, the REE pattern for early Cretaceous basalts shows a moderate enrichment in LREE and only small fractionation of HREE (Dostal *et al.*, 1977) and contrasts with the steeper slopes observed in the patterns for the flood basalts (ocoites) of coastal central Chile referred above (Fig. 10). The Zr–Ti–Y ratio and K_2O content in ocoites from the Coastal Range in central Chile are similar to the ones found in coeval, subordinate, basalts that appear in the sedimentary sequences of the Andean Range at the same latitude (Levi and Nyström, 1982) (Fig. 5). A widespread change to lavas of intermediate composition took place during late Aptian to Albian time in most of Chile.

The age of the predominantly silicic volcanic activity of coastal central Chile is indicated by fossils in the interbedded shallow-marine strata to range from Berriasian to Hauterivian (partly Barremian?). The porphyritic basalts of the overlying flood lava sequence (Fig. 9) were extruded sometime during the Hauterivian to Aptian time span (126–108 Ma, van Hinte, 1976) according to paleontological data (Aguirre and Egert, 1965). This age has been recently confirmed by Rb-Sr geochronology on "unaltered" samples of these basalts (Åberg *et al.*, 1984). A K-Ar age of 105 Ma for a "fresh" ocoite from the type locality was reported by Vergara and Drake (1979). K-Ar ages of 110 and 117 Ma for basalts from two different Lower Cretaceous localities between 26° and 29°S (McNutt *et al.*, 1975) corroborate this age assignment. Further north, in Tarapacá (21°S), K-Ar dates of 109 ± 2 Ma to 114 ± 2 Ma obtained in silicic rocks of the Peña Morada Formation (Maksaev, 1978) confirm widespread early Cretaceous volcanic activity. K-Ar determinations on rocks of the Quebrada Marquesa Formation (Aptian–Albian?) of Coquimbo show considerable scatter between 91.9 Ma and 61.3 Ma with a general tendency for the calculated ages to be younger than expected on stratigraphic bases (Palmer *et al.*, 1980). Variable argon losses due to alteration would explain this behavior according to those authors.

Iron, manganese and copper deposits are genetically linked with the early Cretaceous volcanic processes. A belt of apatitie-rich iron deposits several hundreds kilometers long is associated with the flow basalts and their comagmatic products and appear to be largely controlled by rifting (Oyarzún and Frutos, 1982; Oyarzún, 1982; Oyarzún and Frutos, 1984). Sedimentary manganese and copper deposits generated in paralic basins by metal enrichment due to volcanism are present in a large area of Atacama and Coquimbo (28° to 30°30′ app.) (Aguirre and Mehech, 1964; Klohn and Aguirre, 1965). Manto-type copper deposits related to early Cretaceous volcanism are known in the coastal region of Aconcagua, mainly between 32° and 33°S (Ruiz *et al.*, 1965).

The subsiding volcanics and sediments deposited in the early Cretaceous trough were widely affected by nondeformational burial metamorphism, folded without penetrative deformation and intruded by granitoids.

Burial metamorphism followed closely after extrusion in the case of the lavas; a Rb-Sr isochron of 102 ± 3 Ma was obtained from samples of strongly altered porphyritic flood basalts (spilites) and amygdules representing metadomains in the same lava flows from which "unaltered" domains were dated (see p. 307). No change in the Sr(i) ratio with metamorphism was recorded, a value of 0.07040 was obtained for the spilitized basalts (Åberg *et al.*, 1984). According to these values, the Rb-Sr whole-rock systems would have closed about 10 to 20 Ma after extrusion of the flows. At the Coast Range side of the basin mineral assemblages of zeolite to greenschist facies were produced from top to bottom in the *ca.* 10,000 m thick early Cretaceous sequence while in the *ca.* 4000 m thick, Andean Range pile, the assemblages belong to zeolite and prehnite–pumpellyite facies (Levi, 1968, 1969, 1970; Levi *et al.*, 1982). Metamorphic isograds are parallel to the overall fold structure; unconformably overlying late Cretaceous conglomerates contain metamorphosed pebbles from the Lower Cretaceous volcanic pile covering the entire spectrum of metamorphic facies represented in it (Levi, 1970). These relationships demonstrate that burial metamorphism of the Lower Cretaceous rocks preceded folding and uplift.

The regression that started in the Hauterivian–Barremian heralded the orogenic events of the second major Andean diastrophism, the Meso–Cretaceous or Subhercynian orogenic phase (Steinmann, 1929; Charrier and

Vicente, 1972; Aguirre *et al.*, 1974). These movements are reflected in the angular and erosional unconformity separating the Lower Cretaceous series from the molassic-type deposits that represent the basal units of the Upper Cretaceous in most of central and north central Chile. This orogenic phase, at the close of the Albian, marks a definite regression of the sea in the whole Northern Sector.

Batholithic intrusion occurred during the Cretaceous (Aguirre, 1983); outcrops are distributed in a N–S band to the east of the older, Paleozoic, Triassic and Jurassic, batholithic belts (Fig. 4). Cretaceous plutons are generally in close spatial relationship with the Lower Cretaceous volcano-sedimentary units. The granitoids are calc-alkaline and oversaturated; ultrabasic and alkaline types are unknown. According to radiometric dating and geochronological compilations by various authors (Zentilli, 1974; Clark *et al.*, 1976; Montecinos, 1979; Drake *et al.*, 1982a; Aguirre, 1983) plutonic activity during the Cretaceous appears as a fairly continuous process with some culminating intervals notably between 80 and 110 Ma BP. The postkinematic features of most Cretaceous granitoids together with their main interval of incidence would suggest that an important fraction of them might have been emplaced after the Meso–Cretaceous orogenic phase, at the opening of the late Cretaceous.

Four Sr(i) ratios from early Cretaceous granitoids between 26°30′ and 27°30′S reported by McNutt *et al.* (1975) range from 0.7050 to 0.7022, suggesting a mantle origin. The two lowest values, 0.7023 and 0.7022, correspond to granitoids of 117 and 107 Ma respectively, coeval with flood basalt effusion.

Mineral deposits genetically related to the Cretaceous granitoids include iron, copper, gold and silver (Ruiz *et al.*, 1965; Oyarzún, 1971; Oyarzún and Frutos, 1982; Oyarzún, 1982).

(d) Upper Cretaceous to Paleocene. Following the Meso-Cretaceous phase sustained volcanic activity took place along a rapidly subsiding, tectonically controlled, N–S elongated basin that represents the persistence of the Mesozoic mobile zone under continental conditons. Lava flows and pyroclastics of basic and intermediate composition, silicic ignimbrites and continental sedimentary rocks characterize sequences that crop out from northernmost Chile to 40°S and which are located immediately east of the Lower Cretaceous belt of Fig. 8. Between 26° and 35°S, where they are best exposed, they unconformably overlie Lower Cretaceous and older units; in several of these areas the basal section consists of coarse conglomerates, several thousand meters thick, representing the rapid destruction of relief originated by the Meso–Cretaceous folding phase. A facies change, from more sedimentary material in the western side of the basin to predominantly

volcanic rocks in the eastern flank exists. Broadly correlative units representing the products of this volcano-sedimentary activity have been formationally defined in various latitudes. Among the best known, from north to south, are: Cerrillos Formation (Segerstrom and Parker, 1959) in Atacama (26°–29°S); Viñita Formation (Aguirre and Egert, 1965) in Coquimbo (29° – 32°S); Abanico Formation (Aguirre, 1960) and Coya–Machalí Formation (Klohn, 1960) in the Andean Range between 32° and 35°S, and Las Chilcas Formation (Thomas, 1958) in the Coastal Range west of Santiago.

The Cerrillos Formation unconformably overlies strata of the Lower Cretaceous Chañarcillo Group and Bandurrias Formation. Its thickness reaches 4500 m in the type locality but lateral variations and lenticularity are strong and the formation wedges out to a few meters in places. It consists of a conglomeratic lower member (2300 m thick) with intercalations of tuffs, freshwater limestone and siltstone, and a predominantly volcanic upper member of basic lava flows and pyroclastics partly with flow units of rhyolitic to dacitic ignimbrites near the top. Reworked marine fossils of Hauterivian–Barremian age are found as pebbles in the conglomerate of the lower member (Segerstrom and Parker, 1959; Segerstrom, 1968; Zentilli, 1974).

The Viñita Formation rests unconformably on the Lower Cretaceous Quebrada Marquesa Formation. Due to lateral changes in facies its thickness may vary from 1000 m to 4500 m. Its basal member (520 m) is represented by coarse conglomerates and subordinate sandstone and is followed by a middle, sedimentary unit, *ca.* 300 m thick of sandstone, freshwater limestone, limy shales, shales, gypsum lenses and minor andesitic flows. The uppermost member, *ca.* 650 m thick, consists of andesitic flows and volcaniclastic rocks. Neocomian fossils and ocoites are common among the pebbles of the basal conglomerate (Aguirre and Egert, 1965; Aguirre and Thomas, 1964).

The Abanico (Coya Machalí) Formation rests unconformably in most places on early Cretaceous red continental sediments of the Colimapu (Cristo Redentor) Formation or on Upper Jurassic units. Its thickness is close to 6000 m and it is composed of rhyolitic ignimbrites, tuffs, volcanic breccias and conglomerates that predominate in the lower half of the section and finer continental sediments, basic to intermediate lava flows and silicic flows preferently in the upper levels (Aguirre, 1960; Klohn, 1960; Thiele, 1980; Charrier, 1981).

The Las Chilcas Formation, located in the Coastal Range between 32°30′ and 33°30′S, unconformably covers the Lower Cretaceous Veta Negra Formation (Levi, 1968). It is predominantly sedimentary and reaches a maximum thickness of 7000 m although it is characteristically lenticular. It is represented by thick conglomerates alternating with thin beds of red sandstone and shales to make up 1500 m in the lower part of the formation. These sediments laterally interfinger with andesitic breccias and lavas, dacitic ignimbrites and limestones, marls and limy shales (Thomas, 1958; Godoy, 1982). Fossil algae, brackish-water pelecypods and gastropods (Levi, 1968) and lagoonal marine fossils found in the limestone units (Corvalán and Vergara, 1980; Osorio *in* Godoy, 1982) suggest proximity to the open sea during deposition.

The formations described above are unconformably covered by Tertiary volcanics known as the Hornitos, Los Elquinos, Farellones and Lo Valle formations in Atacama, Coquimbo, Aconcagua and Santiago respectively. The volcanic products are calc-alkaline and mainly correspond to basalt, aluminous andesite and rhyolite (Table I, analyses 24, 25 and 26). Sr(i) ratios for two intermediates and one basic volcanic of late Cretaceous age of the Atacama region (Cerrillos Formation) are 0.7055, 0.7037 and 0.7036 respectively (McNutt *et al.*, 1975).

The volcano-sedimentary rocks belonging to this stratigraphical-structural unit are strongly affected by burial metamorphism with facies ranging from prehnite–pumpellyite at the top to greenschist at the bottom in some places, and from zeolite to prehnite–pumpellyite in others (Levi, 1970).

Brackish water and lagoonal fossils from these rocks do not permit an age assignment. The presence of a *Nothofagus* flora east of Santiago could indicate the Upper Cretaceous without excluding an early Tertiary are (Kräusel, *in* Klohn, 1960). Late Maastrichtian dinosaur remains in the upper Viñita Formation of Coquimbo (Casamiquela *et al.*, 1969) are the best biostratigraphic control available. Combining paleontological and radiometric data, Zentilli (1974) has suggested that the Cerrillos Formation of Atacama is pre-early Paleocene and post-Barremian. Vergara and Drake (1979) postulated that the Coastal Range's Las Chilcas Formation was deposited after 105 Ma BP, the age of a "fresh" ocoite from the unconformably underlying Veta Negra Formation. K-Ar dates for rocks of the Abanico (Coya Machalí) Formation between 33° and 34°S cover the range 62.3 to 16.4 Ma with most values around 23 Ma (see Drake *et al.*, 1982b). This spread of ages may primarily reflect pervasive alteration by burial or may result partly from insufficient stratigraphic control of some of the samples analysed which, in cases, have been indistinctly assigned to the Abanico and to the younger Farellones Formation (i.e. Drake *et al.*, 1982b). Until further clarification these younger dates will not be considered here as crystallization ages of the Abanicos Formation.

During the Maastrichtian, marine transgression from the Pacific and Atlantic sides took place (Charrier, 1981; Charrier and Malumián, 1975); in Chile this episode is best represented by narrow strips of marine sedimentary rocks along the coast between 36°30′ and 38°S. Regression seems to have preceded the third Andean diastrophism at the opening of the Tertiary, currently referred to as the Laramian orogenic phase (Aguirre *et al.*, 1974; Frutos, 1981). The angular unconformity separating the Abanico (Coya Machalí) and the Farellones formations and the Las Chilcas and Lo Valle formations in central Chile would reflect this Paleocene orogenic event which also may be represented in north central Chile by the unconformity between the Viñita and Los Elquinos and the Cerrillos and Hornitos formations. On the coast of central Chile this diastrophic event is marked by the unconformity between Senonian and Eocene marine strata at the Algarrobo Beach, due west of Santiago (Brüggen, 1950).

Granitoids intruded during the 80–110 Ma interval partly coincided with the volcanic activity of late Cretaceous age. Radiometric ages suggest that plutonism reached another culmination between 60 and 65 Ma BP, immediately after the Laramian phase (Fig. 4). This last age interval is well represented among granitoids of northern Chile between 26° and 29°S while only a few dates in that interval are known in central Chile (Zentilli, 1974; Drake *et al.*, 1982a; Aguirre, 1983). In northern Chile, an increase in the Sr(i) ratio with a decrease in age is observed for granitoids of early Cretaceous, early late Cretaceous and Cretaceous–Tertiary age for which the average values are 0.7031, 0.7043 and 0.7051 respectively (McNutt, 1975).

(e) Paleocene to Upper Miocene. Almost continuous volcanic activity in several N–S elongated, intracontinental basins, took place during this interval. This activity was accompanied by widespread tectonism that accounts for most of the present Andean morphology. Aguirre *et al.* (1974) defined two stratigraphical-structural units representative of the geological evolution of central Chile during this period, namely the Farellones Formation and the Coast Range Miocene volcanism. They are separated by an angular unconformity of early Oligocene age reflecting an orogenic episode correlated with the Incaic phase of Steinmann (1929) in Peru. In the last decade, however, a considerable number of radiometric age determinations and new field information have led to a global reappraisal of events that occurred during this interval and, as a result, to a redefinition of the structural units involved. The pattern obtained from the radiometric ages, as opposed to the one for the Upper Cretaceous rocks, is fairly consistent, probably due to the fact that the rocks concerned have suffered less metamorphic alteration.

In this section we will geographically summarize the information available rather than trying to set correlations of structural units throughout the whole territory. Common features for the different areas discussed will then be analyzed with the intention of sketching a global trend of evolution for this interval.

Central Chile (30°–41°S). A *ca*. 2000 m thick continental volcano-sedimentary sequence consisting of rhyolitic ignimbrites, andesitic lavas and pyroclastics, lacustrine sedimentary rocks and basaltic flows unconformably covers the Upper Cretaceous of the region (Abanico = Coya Machalí). In the Andean Range this sequence is classically known as the Farellones Formation (Klohn, 1960; Aguirre, 1960); its outcrops are located immediately east of the Upper Cretaceous volcanic belt and form a N–S band along the whole segment. Toward the west, particularly between 32° and 33°S, rocks lithologically similar known as the Lo Valle Formation (Thomas, 1958) occupy the eastern slopes of the Coastal Range where they unconformably overlie the Upper Cretaceous Las Chilcas Formation (Thomas, 1958) and crop out in a number of hills and ridges emerging at the northern end of the Central Valley Graben around Santiago.

The Farellones rocks are calc-alkaline and a bimodal tendency is shown by the predominance of silicic ignimbrites in the lower members and basaltic lavas in the upper ones (Aguirre, 1960). Representative analyses of basic, intermediate and silicic rocks are given in Table I, analyses 34, 35 and 36.

The Farellones and the Lo Valle strata have been affected by low-grade metamorphism and mineral assemblages of the zeolite and prehnite–pumpellyite facies are present (Levi, 1970). The intensity of the alteration is, however, less than in the underlying, Upper Cretaceous, sequences.

Based purely on stratigraphic and relative age criteria an early Paleogene age, probably extending to the late Eocene, was assigned to the Farellones formation (i.e. Aguirre *et al.*, 1974). K-Ar dates on the Lo Valle and Farellones formations (see Drake *et al.*, 1976, 1982b) partially conflict with this assignment. Three analyses in dacitic ignimbrites and one in andesite belonging to the Lo Valle Formation gave values between 77.8 and 64.5 Ma, a Campanian to Cretaceous–Tertiary interval (Fig. 11). Godoy (1981, 1982) doubts an unconformable relationship between the Lo Valle and Las Chilcas formations and has demonstrated lateral merging between Lo Valle welded tuffs and Las Chilcas andesites. Toward the east, the Lo Valle Formation has been considered as a partial equivalent of the Abanico Formation (Thiele, 1980; Thiele *et al.*, 1980; Godoy, 1981, 1982). Although new data and interpretations do not substantially modify the suggested age (Thomas, 1958) for the Lo Valle Formation, its stratigraphic relations with the Upper Cretaceous are being drastically revised. In the Andean Range, rocks belonging to the type section of the Farellones Formation east of Santiago give early Miocene K-Ar ages of 17.3 and 18.5 Ma respectively (Vergara and Drake, 1978), considerably younger than currently accepted values. This tendency is confirmed by 21 K-Ar dates listed in Drake *et al.* (1982b) corresponding to analyses of dacitic ignimbrites and andesitic flows of the Farellones Formation exposed between 31°40' and 34°25'S that range from 25.2 ± 0.5 Ma to 7.8 ± 0.3 Ma covering most of the Miocene and uppermost Oligocene (Fig. 11). Further north, in the Andes of Coquimbo (30°45' to 31°30'S), a sequence of andesitic lavas and breccias, the Estero Cenicero Formation (Rivano, 1980), has yielded four K-Ar ages in the range 17.4 ± 0.6 Ma and 6.1 ± 0.5 Ma (Quirt *et al.*, 1971; Munizaga and Vicente, 1978) that cover most of the Miocene (Fig. 11). This formation rests unconformably on the Upper Cretaceous Viñita Formation and has been correlated by Rivano (1980) with the Farellones Formation. At *ca.* 36°S, in the Andean Range east of Linares, a unit lithologically similar to the Farellones Formation was identified and named Campanario Formation by Drake (1976a, 1976b). Six K-Ar dates for rocks of this unit (Drake *et al.*, 1982b) range from 15.3 ± 0.8 Ma to 6.4 ± 1.0 Ma or from middle to late Miocene (Fig. 11). In the Central Valley Graben, between 37°30' and 41°30'S, scattered remnants of an eroded coastal volcanic belt (Vergara and Munizaga, 1974) crop out. Characteristic rocks are two-pyroxene andesitic lava flows (Table I, analysis 37), andesites, dacites and corresponding pyroclastics. These rocks are geochemically similar to the calc-alkaline suite of circum-Pacific island arcs (López-Escobar *et al.*, 1976). They are subhorizontal beds and crosscutting necks. In the coastal areas of Temuco and Chiloé they interfinger with marine sedimentary rocks of Miocene age (García, 1968). Four K-Ar ages (20.4, 20.7, 22.0 and 27.7 Ma) reported by Vergara and Munizaga (1974) indicate that this unit formed in late Oligocene–early Miocene time and is thus partly contemporaneous with the lower Farellones Formation.

Considering the consistency of the dates obtained and the fact that they come from samples with only minor metamorphic alteration, a Miocene age seems now well established for the Farellones and correlative formations of the Northern Sector.

The unconformable relation between marine-continental strata of Eocene age and marine Miocene beds along the coastal line between 36°30' and 38°S indicates a deformational event (Ruiz *et al.*, 1965; Charrier, 1981). In this coastal margin transgressive-regressive movements that started during the Senonian continued during the Tertiary over a narrow strip of the continent. According to Ruiz *et al.* (1965) Oligocene tectonism, preceded by epeirogenesis, ended Eocene sedimentation and produced block faulting and extensive erosion. This deformational phase probably can be considered as a pulse of the Incaic phase of Oligocene age in the region although Charrier (1981) assigns an early Miocene age to the folding event. No indication of corresponding deformation is found in the Andean Range where no ages between 62.3 Ma and 25.2 Ma are known (Drake *et al.*, 1982a, 1982b). However, if the youngest K-Ar dates for the Abanico Formation are interpreted

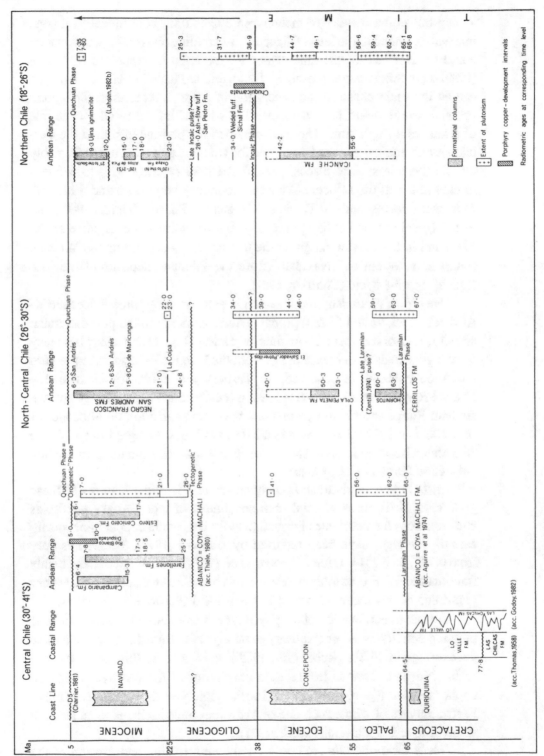

Fig. 11. Correlation chart of events. Paleocene to Upper Miocene, Northern Sector.

as crystallization instead of metamorphic ages and, consequently, its age interval extended to the latest Oligocene, the folding phase affecting Abanico could be ascribed to this time. This interpretation is supported by Thiele (1980) who defines a compressive "Tectogenetic Phase" of latest Oligocene age as the main phase in the generation of Andean structures. This compressive event would have deformed the whole Jurassic to Oligocene pile of strata as a single unit. The ages obtained for plutons intruding the Farellones and equivalent formations, and for volcanic units that unconformably overlie them in several places, permitted Drake *et al.* (1982b) to define a folding phase at the Miocene–Pliocene boundary between 6 and 5 Ma BP. This phase corresponds to Thiele's "Orogenetic Phase" (Thiele, 1980), and to the Quechuan or Pontian phase of previous authors (i.e. Aguirre *et al.*, 1974) and in the coastal margin to the unconformity separating the Miocene sedimentary strata of "Navidad" from the Pliocene deposits of the "La Cueva" marine cycle (Charrier, 1981).

The oldest plutonism in this region is evidenced by three Paleocene K-Ar dates of 65, 62 and 56 Ma, probably corresponding to the post-Laramian activity, and one Eocene K-Ar date of 41 Ma (Fig. 11). A clear tendency for the granitoids to decrease in age from the Central Valley Graben toward the Andean Range is observed; no Tertiary granitoids are found in the Coastal Range (Fig. 4). Intrusive ages in the Central Valley Graben and the Andean Range spread in a continuous way from early to late Miocene (21 ± 0.2 to 7 ± 0.5 Ma) as shown in a list of twelve K-Ar ages from 32°30' to 36°S that also includes one date of 26 ± 0.4 Ma corresponding to the late Oligocene (Drake *et al.*, 1982b).

Little is known about the geochemistry and isotopic features of these granitoids. Sr(i) ratios of total rock specimens of late Tertiary granitoids cutting across the Farellones Formation at La Disputada copper mine, northeast of Santiago, have been reported by Halpern (1979). The values vary from 0.7037 to 0.7044. López-Escobar *et al.* (1979) have noted that the highly fractionated REE distribution relative to chondrites exhibited by Tertiary granitoids of this region resemble those of local Pleistocene andesites.

Copper mineralization of porphyry type is associated with late Miocene plutonic activity. A K-Ar (biotite) age on a granite from La Disputada mine gave an age of 10 Ma (Ruiz *et al.*, 1965)) while a K-Ar (biotite) value of 5 ± 0.5 Ma was obtained from quartz porphyries of the nearby Rio Blanco copper deposit (Quirt *et al.*, 1971). Further south, at the El Teniente mine, K-Ar analyses of whole rock and of hydrothermal sericite gave an average value of 5 ± 0.1 Ma (Quirt *et al.*, 1971).

North Central Chile (26°–30°S). Following folding and erosion of the Upper Cretaceous Cerrillos Formation, a *ca.* 2400 m thick continental volcano-sedimentary sequence, the Hornitos Formation (Segerstrom, 1959; Segerstrom and Parker, 1959), was deposited. It consists of silicic ignimbrites, conglomerates, volcanic wackes with intercalated gypsum beds, basic volcanic flows and pyroclastic and lacustrine sedimentary rocks; lateral facies changes are numerous and abrupt. Based on lithologic criteria and relative age relations and in spite of distance, a correlation between the Hornitos and the Farellones formations has been currently assumed. The Hornitos volcanics are calc-alkaline and range in composition from high-alumina basalts to rhyolites (Zentilli, 1974) with a neat bimodal character (Table I, analyses 27, 28, 29 and 30). The rocks of Hornitos have been affected by low-grade burial metamorphism of zeolite and prehnite–pumpellyite facies.

According to Zentilli (1974), geochronological K-Ar evidence indicates that much of the Hornitos sequence was deposited between about 63 and 60 Ma BP and folded, faulted and intruded by early Paleocene plutons (Fig. 11). A correlation with the Farellones Formation, as newly timed, is no longer justified. Erosion following deformation of Hornitos modelled a landscape upon which a new volcanic unit, the Cerro de la Peineta Formation (Mortimer, 1969), was deposited. This unit caps several mountains of the Precordillera lying unconformably on folded Hornitos and Cerrillos formations and reaching a maximum thickness of *ca.* 1000 m at its type locality (Segerstrom, 1959). The rocks are predominantly agglomeratic with a high proportion of partially absorbed plutonic rock fragments set in a rhyolitic mass consisting of numerous horizontal flows (Zentilli, 1974) (Table I, analysis 31). K-Ar ages of 53 $\pm$ 2.4 Ma, 53.7 $\pm$ 1.8 Ma, 52.5 $\pm$ 2 Ma, 53.2 $\pm$ 2 Ma have been obtained by different authors (Zentilli, 1974) and a Rb-Sr isochron of 50.3 $\pm$ 3.2 Ma is known from the locality of Indio Muerto (Gustafson and Hunt, 1975). Based on Rb-Sr dates an approximate age of 40 to 50 Ma can be assigned to the Cerro de la Peineta type volcanics in the El Salvador area (Fig. 11). McNutt *et al.* (1975) have reported Sr(i) ratios of 0.7043 and 0.7057 for these early Eocene volcanic rocks.

No major magmatism is recorded during most of the Oligocene. However, volcanic activity was sporadic throughout the Miocene. It is documented along the peaks of the western flank of the "Alta Cordillera" where K-Ar dates of 24.8 to 21.0 Ma (La Coipa), 15.8 $\pm$ 0.6 Ma (Ojo de Maricunga volcano) have been obtained in andesites and porphyritic dacites (Zentilli, 1974). The last author assigns these volcanics to the Negro Francisco Formation (Segerstrom, 1967) and points out that they interfinger, toward the west and north, with the Miocene rhyolitic ignimbrites of the San Andrés Formation dated by K-Ar as between 12.6 and 6.0 Ma (Clark *et al.*, 1976; Mortimer, 1969). Chemical analyses of two of these rocks are shown in Table I, analyses 32 and 33. Sr(i) ratios of 0.7075 and 0.7060 have been reported by McNutt *et al.* (1975) for a dacite (10.8 Ma) and an andesite (7.7 Ma) of this series in Argentina *ca.* 27°20'S. The Negro Francisco–San Andrés volcanism correlates with the Altos de Pica (Liparítica) Formation of northern Chile (Fig. 11).

A diastrophic event during the early Paleogene is recorded in the unconformity separating the Hornitos and the Cerro de la Peineta Formations (Fig. 11). Considering the ages of these two units that event might have occurred sometime between 60 and 40 Ma, probably during the late Paleocene as a pulse of the Incaic phase (i.e. Incaic phase I of Frutos, 1981).

Since the early Eocene the modelling of the Andes has progressively taken place (see Mortimer, 1969; Clark *et al.*, 1976; Mortimer *et al.*, 1974; Zentilli, 1974; Tobar, 1977, among others). Following deposition of the Cerro de la Peineta volcanics a regional topographic surface was generated and incised by deep valleys during the Oligocene that was a time of strong uplift in the north central Chilean Andes. The erosional régime abruptly changed at the end of the Paleogene probably due to uplift resulting in the burial of the post-Peineta erosional surface under the Atacama Gravel and the development of the Atacama pedi-plain (Sillitoe *et al.*, 1968). This epeirogenic episode could reflect a second pulse of the Incaic phase (Incaic II of Frutos, 1981?). Uplift and erosion were renewed at the close of the Miocene.

Granitoid intrusion in north central Chile for the interval considered culminated between 67–59 Ma (latest Cretaceous–early Paleogene), 44–34 Ma (late Eocene–early Oligocene) and 23–22 Ma (Oligocene–Miocene boundary) (Fig. 11 and also Fig. 4). Sr(i) ratios of 0.7055 and 0.7047 are known for the Cretaceous–Tertiary granitoids, of 0.7044 for a 42.5 Ma old granodiorite and of 0.7061 and 0.7052 for dacite porphyries of 23.1 and 22.3 Ma respectively, all K-Ar ages (McNutt *et al.*, 1975).

Large porphyry copper deposits are related to middle Paleogene plutonic activity. K-Ar and Rb-Sr ages for the different intrusive units involved in the mineralization at El Salvador cover the interval between 46 and 39 Ma (middle to late Eocene) (Fig. 11). Sr(i) ratios range between 0.7040 and 0.7042 (Gustafson and Hunt, 1975). K-Ar ages of 37.7 ± 1.2 Ma and 34.1 ± 1.0 Ma, corresponding to the early Oligocene, have been reported by Quirt *et al.* (1971) for feldspar porphyries associated with the copper mineralization in the Potrerillos District. Finally, at Agua Amarga, 40 km due north of Potrerillos, a K-Ar age of 36.2 ± 1.0 Ma was obtained by Tobar (1977) for a feldspar porphyry.

Northern Chile (18°–26°S). The description of this segment is largely based on publications by Maksaev, (1978); Huete *et al.* (1977); Clark *et al.* (1976); Mortimer *et al.* (1974); Baker and Francis, (1978); Noble *et al.* (1974); Roobol *et al.* (1976); Francis and Rundle, (1976); Baker, (1977); Lahsen, (1982a, 1982b); Thorpe *et al.* (1982) and others.

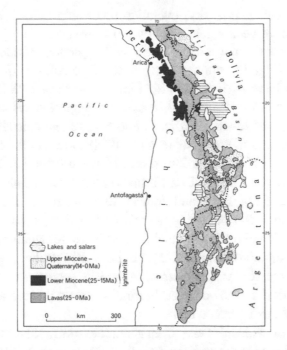

Fig. 12. Distribution of Miocene–Recent volcanic deposits in the Northern Province (18°–28°S) of the Northern Sector. Modified after Thorpe *et al.* (1982).

Rock units of Paleogene age have been defined in the Andean Range of Antofagasta by Maksaev (1978) who named and described the Icanche Formation (Fig. 11), a 1500 m thick sequence of andesitic rocks yielding K-Ar whole-rock ages of 55.6 ± 6.9 Ma to 42.2 ± 2.3 Ma or latest Paleocene to late Eocene (Huete *et al.*, 1977). These rocks are considered to be products of subaerial volcanism in a continental environment. The Icanche Formation is unconformably covered by a 1300 m thick continental sequence, predominantly conglomeratic, that carries a 30 m thick intercalation of breccia-bearing welded tuff of dacitic composition. The sequence is known as the Sichal Formation (Maksaev, 1978) and a K-Ar (biotite) date of 34 ± 1.0 Ma (early Oligocene) was obtained from the welded tuff. Similar sequences with intercalations of evaporites and scarce volcanics were deposited in different places during the Oligocene and, locally up to the early Miocene.

Starting about 25 Ma BP, in the late Oligocene, and continuing until the Pleistocene, volcanism developed in the area. Two main episodes of Neogene volcanism are recognized (Fig. 12), one represented by sheets of folded rhyolitic to dacitic ignimbrites with intercalations of andesitic lavas in the Miocene and one including ignimbrites, dacitic to rhyolitic lava domes and andesitic to dacitic strato-volcanoes in the Plio–Pleistocene (Lahsen, 1969).

Variations in timing, intensity, evolution and composition along the N–S volcanic belt have been emphasized by most authors.

The Miocene event was inaugurated by the ejection of widespread silicic ash-flow tuffs (Fig. 12) and moved from north to south with time (Baker, 1977; see also Lahsen, 1982a and his Fig. 3). North of 20°S ignimbrites of the Oxaya Formation (Henríquez, 1963), contemporaneous with andesitic flows, yield ages of 23–18 Ma (Fig. 11 and Fig. 12), an episode also known in central and southern Peru (Noble *et al.*, 1974) and in southwestern Bolivia (Kussmaul *et al.*, 1975). Toward the south, between 20° and 21°S, the earliest ash-flow tuffs of the Altos de Pica Formation (Galli and Dingman, 1962) are between 17 and 15 Ma (Baker and Francis, 1978). South of 21°S considerable ignimbrite ejection started during the middle Miocene, about 12 Ma BP. A similar age was found by Schawb and Lippolt (1976) for ignimbritic eruptions in northwest Argentina between 22° and 23°S. Still further south, between 26° and 27°S, the Negro Francisco–San Andrés sequences, considered as correlative with the Altos de Pica Formation, have yielded ages between 12.6 and 6.0 Ma as indicated earlier.

The Miocene volcanics vary from alkali feldspar rhyolites to quartz lati-andesites with a concentration in the alkali feldspar rhyolite, rhyolite and rhyodacite fields of the Streckeisen triangle (Pichler and Zeil, 1972; Klerkx *et al.*, 1977) (Table I, analysis 38). The ignimbrites of northern Chile are characterized by high Rb/Sr ratios and intermediate Sr(i) ratios ranging from 0.705 to 0.709 (Klerkx *et al.*, 1977).

Baker and Francis (1978) have estimated the volume of Upper Cenozoic lava and ignimbrite for a segment (19°30' to 22°30'S) of this volcanic province arriving at values of 10×10^3 km^3 and 2×10^3 km^3. They have also established that the proportion of lava to ignimbrite varies along the length of the Central Andes. Thus, between 22°15' and 23°15'S, volumes of both types are approximately equal while, further north, in the Arequipa region of Peru, ignimbrites are less than 1% of the total volcanic rocks. If the volume of volcanics is plotted against time for the same segment (19°30' to 22°30'S) two dissimilar patterns are obtained; in the northern subsegment (19°30' to 21°S) the largest volumes were erupted at 12 to 9 Ma BP and 6 to 3 Ma BP while in the southern subsegment (21° to 22°30'S) a large proportion of the material was erupted between 6 and 0 Ma BP (Baker and Francis, 1978). These examples illustrate the regional variations commonly found in volcanic provinces of this nature and which should be kept in mind when attempting correlations.

Various regional unconformities have been radiometrically bracketed in time. Upper Cenozoic volcanic unconformably overlie folded Mesozoic to Paleogene rocks in several localities. K-Ar ages of 42 Ma and 34 Ma from rocks below and above the unconformity in the upper Loa Valley have been interpreted to mean that an orogenic event took place in the late Eocene–early Oligocene (Maksaev, 1979; Lahsen, 1982a, 1982b) immediately followed by extension and development of intramontane basins (Maksaev, 1979; Lahsen, 1982b). Another deformational event has been fixed as between 28 and 17 Ma BP or late Oligocene early Miocene (Lahsen, 1982a). The beginning of the Neogene volcanic episode would have closely followed this folding event while block faulting was already active in the Precordillera of Arica reflecting the uplift of the Coastal Range and Precordillera relative to the Central Depression (Mortimer *et al.*, 1974; Lahsen, 1982a). Another tectonic phase followed by uplift and erosion occurred during the middle Miocene and is probably linked to the reactivation of magmatism throughout the region (Lahsen, 1982b).

A deformational period between late Miocene and Pliocene has been dated between 12 and 4 Ma BP in the Andes at *ca.* 19°30'S (Puchuldiza area) by Lahsen and Munizaga (1979). To the south, at El Tatio and San Bartolo areas (between 22° and 23°S, *ca.* 68°30'W), this event can be placed between 7 and 4.2 Ma BP (Rutland *et al.*, 1965; Lahsen, 1976) while in the Argentinian Altiplano according to Schwap and Lippolt (1976) these movements occurred between 7.2 and 4.8 Ma BP.

Granitoids in the area (Fig. 4) vary from diorites to monzogranites and porphyries and are characteristically calc-alkaline. They are found as stocks or minor subvolcanic bodies along the easternmost part of the territory. As compared with the Mesozoic granitoids they show higher contents of SiO_2, FeO (total), Al_2O_3 and MgO and lower contents of alkalies and TiO_2; no variation in CaO is apparent (Montecinos, 1979). K-Ar dates for 14 granitoids between 19°50' and 22°17'S (Huete *et al.*, 1977; Maksaev, 1978; Montecinos, 1979 and references therein) suggest that several plutonic events took place during the Tertiary. Five of these dates ranging from 56.6 ± 2.6 Ma to 65.8 ± 1.7 Ma (Fig. 11) show that plutonism was active during the Paleocene and probably during the latest Cretaceous. Two values of 44.7 ± 2.5 Ma and 49.1 ± 1.2 Ma indicate plutonism during the middle Eocene while five ages between 36.9 ± 0.6 Ma and 25.3 ± 0.5 Ma suggest conspicuous early Oligocene activity. This last intrusive phase is connected with the development of hydrothermal alteration processes and tourmaline brecciation to which the porphyry copper mineralization in Chuquicamata appears to be related. Finally, two ages corresponding to the late Miocene (7.6 ± 0.2 Ma and 7.35 ± 0.95 Ma) have been listed by Montecinos (1979).

Main Characteristics of the Paleocene–Late Miocene Interval in the Northern Sector. From the record presented for each of three latitudinal segments (Fig. 11), the following major features emerge:

Volcanism:

(*i*) Volcanism was intense during the Paleocene and Eocene in north central and northern Chile. Unless the wide range of radiometric ages obtained for the Abanico (Coya Machalí) Formation in central Chile is interpreted as indicating crystallization events (Thiele, 1980), no volcanism of those ages has yet been shown to have occurred there.

(*ii*) Volcanic activity is conspicuously absent during most of the Oligocene in the three segments considered (but see Thiele, 1980).

(*iii*) Widespread volcanism, commencing in latest Oligocene time, characterizes all three segments during the whole extent of the Miocene.

(*iv*) An eastward shift of the volcanic foci by about 100 km took place between the Miocene and the middle Pleistocene (?) (Vergara and Munizaga, 1974). This feature is particularly evident in central Chile.

(*v*) A bimodal type of volcanism represented by mixed effusion of silicic ignimbrites and basic and intermediate flows and pyroclastics is typical in the Paleogene and Neogene sequences.

Orogeny:

(*i*) A Laramian phase at the Cretaceous–Tertiary boundary or earliest Paleocene, is recognized in central (coastal margin and Andes) and north central Chile. The position of such a phase in the Coastal Range of central Chile should be revised (cf. Godoy, 1982; Drake *et al.*, 1982b).

(*ii*) A folding episode in the late Paleocene only recorded in north central Chile (Zentilli, 1974), might be considered as a late pulse of the Laramian phase.

(*iii*) Two Oligocene folding phases, one close to the earliest and the other to the late part of this epoch, are recorded in northern Chile. The second seems to have shortly preceded the onset of the Neogene magmatism. Similar events during the Oligocene can be inferred, but not precisely dated, in north central Chile and at the coastal margin of central Chile. The oldest Oligocene event might be correlated with the main Incaic phase while the younger could represent its late pulse. Thiele's "Tectogenetic Phase" (Thiele, 1980) in central Chile would coincide in time with this last event.

(*iv*) A middle Miocene tectonic phase affected northern Chile approximately 12 Ma BP (Lahsen, 1982b).

(*v*) A late Miocene folding phase is recorded in all three segments and is known as the Quechuan phase. The middle Miocene event could represent an early Quechuan pulse.

Plutonism:

(*i*) Significant granitoid emplacement took place during the Paleocene in northern and north central Chile where it closely followed the Laramian phase and widely overlapped the volcanic activity. In central Chile Paleocene plutonism seems to have been restricted.

(*ii*) Late Eocene to early Oligocene plutonism is recorded in north central and northern Chile where it is related to porphyry copper mineralization (El Salvador, Potrerillos, Chuquicamata). Ages known for the mineralization are close to the inferred age of the Incaic folding phase.

(*iii*) Plutonism was active in central Chile since the latest Oligocene and extended throughout the whole Miocene. Although the geochronological record is far less complete in the two other segments, a similar situation is suggested.

The scarce information availabe to support any comprehensive model of Tertiary evolution is made apparent in Fig. 11, which also illustrates the difficulties arising from application of long-distance correlations to continental volcanic sequences.

(f) Pliocene to Recent. Intense volcanic activity initiated *ca.* 25 Ma BP continued during the Pliocene and extended to the present as demonstrated by the considerable number of active volcanoes found along the Chilean Andes. Two provinces are clearly distinguishable, one located between 16° and 28°S and identified with the "Liparitic" or "Rhyolitic" formations of northern Chile and southern Peru (Fig. 12), the other between 31° and 42°S and corresponding to a southern "Andesitic" province. Quaternary volcanism is conspicuously absent from the Andes between 27° and 33°S (Pérez and Aguirre, 1969).

Northern Province (18° –28°S). Two main units are distinguished here: the ignimbrite plateau at *ca.* 3500 m and the andesitic strato-volcanoes rising up to about 6000 m (Katsui and González, 1968; Pichler and Zeil, 1972; Roobol *et al.*, 1976). The traditional idea that andesitic volcanoes sitting on the ignimbritic platform indicate that andesitic volcanism postdated ignimbritic activity has been largely disproved; ignimbrites and andesitic lavas have erupted contemporaneously throughout most of the volcanic history of the region (Baker, 1977). Interdigitation of ignimbrite flow units and andesitic products of strato-volcanoes has been evidenced in several places (Roobol *et al.*, 1976).

Approximately 500 major volcanic structures are recognizable in this region; their basement is Cenozoic and Mesozoic igneous rocks, minor sedimentary units of similar age and crystalline Paleozoic and Precambrian rocks. According to their morphology, they correspond to the following types defined by Thorpe *et al.* (1982): *composite cones* representing the classic andesite volcanoes with symmetrical profiles and heights of *ca.* 6000 m ; *monogenetic lava extrusions* typified by dacite–rhyolite domes; *compound volcanoes* corresponding to structures in which two or more morphological units can be recognized and which lack radial symmetry; *ignimbrite centers* whose products cover about 200,000 km² of the region and which in many cases correspond to ignimbrite shields with a maximum thickness of 100 m at the center of the structure. Some of these shields are collapse, caldera type, centers (Baker, 1981).

The most common rocks are andesites, rhyolites, dacites and dacitic-rhyolitic ignimbrites (Table I, analyses 40, 41, 42 and 43). Basalts are practically absent. Andesites and dacites are usually present as lava flows although domes of both compositions are known. The andesites are pyroxene and hornblende types with the boundary at about 60% SiO_2 (Roobol *et al.*, 1976). Pyroxene andesites contain phenocrysts of labradorite, clino and orthopyroxenes, green olivine and iron-titanium oxides in a glassy groundmass. Hornblende andesites are characterized by phenocrysts of andesine, hornblende, biotite and minor clinopyroxene and/or orthopyroxene. Dacites have a similar mineralogy but quartz, as embayed crystals, is a conspicuous phase. Biotite predominates in the ignimbrites.

Francis and Rundle (1976) have estimated the volumes and rates of production of andesitic rocks and ignimbrites younger than 10 Ma for the one degree latitude strip between 21° and 22°S. They conclude that andesitic rocks amount to 2×10^3 km³ with an average rate of production varying between a minimum of 2×10^{-4} km³/yr and a maximum of 3.3×10^{-4} km³/yr while ignimbrites total 1.5×10^3 km³ erupted at a maximum average rate of 1.5×10^{-4} km³/yr. According to the ages reported by Francis and Rundle (1976) most of the rock units included in these calculations have Pliocene to Pleistocene ages. These authors conclude that the total volume of ignimbrite is less than is suggested by the huge area blanketed and that the rate of extrusive volcanism is about two orders of magnitude less than that at the Icelandic constructive plate margin and several times less than the rate of intrusion of batholithic material.

The rocks described belong to a calc-alkaline suite; shoshonitic associations have been described for a few localities east of the calc-alkaline belt between 18° and 26°S (i.e. Déruelle, 1978; Munizaga and Marinovic, 1979).

Andesites from the calc-alkaline suite have SiO_2 ranging from 53.5% to 66%; they are high in Al_2O_3 (>16%), high in K_2O (2–3%), Sr and Ba but relatively poor in TiO_2 (<1.2%) and Sc (3.6–9.3 ppm) and have low $Mg/(Mg + Fe_{tot})$ ratios (0.22–0.29). They fall in the high-potassium andesite field of Taylor (1969) and their Sr(i) ratios range from 0.7051 to 0.7072. They are enriched in LREE and show a positive correlation between Ce_N and Ce_N/Yb_N with SiO_2 (Palacios and López-Escobar, 1979; Roobol et al., 1976; Pichler and Zeil, 1972; Francis et al., 1977; Thorpe et al., 1976; Thorpe and Francis, 1979; Thorpe et al., 1982).

Dacites have SiO_2 contents higher than 63% and their overall geochemistry is continuous with the andesitic lavas. The ignimbrite magmas have SiO_2 ranging between 64%–75% and show continuous major-element chemical variation with respect to the andesite–dacite lavas. However, they have lower Sr and higher Rb contents than the lavas, while their Sr(i) ratios (0.705–0.710) overlap with those of andesites and dacites. A well-defined eastward increase in K_2O, Rb, Ba, Th and Y accompanied by a decrease in Sr and Ni has been detected (Roobol et al., 1976; Thorpe and Francis, 1979); shoshonites would appear to the east as a culmination of K_2O increase.

Gravimetric and seismic data for this Andean segment indicate that the volcanic belt is located approximately 140 km above the present Benioff zone and overlying a ca. 70 km thick continental crust (Barazangi and Isacks, 1976; James, 1971; Dragićević, 1970). These particular conditions seem to determine the geochemical nature of the volcanism.

Radiometric ages for andesites and ignimbrites cover the span between Pliocene and Recent showing that volcanism in northern Chile has been particularly intense in the period 6 to 0 Ma BP (Baker and Francis, 1978). A migration of the younger volcanic centers from NE to SW is observed (Vergara and González, 1972).

The Andean Cordillera, mainly the result of the orogenic events of the Eocene–late Miocene interval was subjected to block faulting and vertical oscillatory movements during the Pliocene and Quaternary. In the coastal margin, marine terraces and marine deposits formed at various levels while block faulting resulted in a steep coastline (Mortimer, 1969; Herm, 1969; Stiefel, 1972).

Southern Province (31°–42°S). The volcanic rocks of southern Chile sharply differ in composition from those of northern Chile. They correspond to a basalt–andesite association where acid rocks are absent or scarcely represented. Two main units have been proposed by Vergara and González (1972) in order to group the Andean volcanics in southern central Chile: the Andesitic Plateau Series and the Strato-volcanoes of Central Type.

The Andesitic Plateau Series crop out between 33° and 42°S, both in Chile and Argentina and consists of subhorizontal andesite flows (Table I, analysis 39) and pyroclastics, amphibole trachyandesites and lesser amounts of olivine-bearing basaltic andesites. Acidic pyroclastic flows are minor components in the lowermost part of this unit. The series rests unconformably on Mesozoic and Paleogene basement and has a thickness of between 100 m and 800 m. Andesites have phenocrysts of andesine–labradorite, hypersthene, augite and subordinate amphibole; olivine is scarce (Vergara and González, 1972). Affinities between this series, the calc-alkaline hypersthenic series of Japan and the Cascade Range volcanics have been pointed out by Vergara and González (1972). Transitions of the Plateau Series along and across the Andes are suggested. Thus, the predominantly silicic Pliocene volcanics of the Colorado–La Parva Formation (Thiele, 1980) that unconformably cover the Farellones strata in the Andes east of Santiago (33°S), could represent a transitions from the Plateau Series toward the compositions observed in coeval units of the north. Transitions from the Plateau Series toward the alkali olivine basalt series across the Andes are also suggested by the geochemical characteristics of rocks from the easternmost exposures (i.e. Pino Hachado, 38°30'S and 70°55'W; López-Escobar et al., 1976).

The Andesitic Plateau Series is equivalent to the volcanic formation known as Cola de Zorro (González and Vergara, 1962; Vergara, 1978) present in the Andean Range from Santiago to the latitude of Puerto Montt. Pliocene K-Ar ages have been obtained for rocks of this formation in several localities (Vergara and Munizaga, 1974; Drake, 1976b).

The strata of the Plateau Series are gently folded and affected by intense block faulting and deep glacial erosion.

The Strato-volcanoes of Central Type are numerous in south central Chile; Katsui (1972) mentions about sixty of such structures with more than twenty still active (Casertano, 1963). Between 33° and 34°S the Andean strato-volcanoes constitute a single belt, south of 34°S this belt splits into the subparallel Western and Eastern belts whose axes are about 30–50 km apart (Moreno, 1976; López-Escobar et al., 1977). The volcanism belongs to the high-alumina basalt series; most representative rocks in the Western belt are olivine-bearing basalts, olivine-pyroxene mafic andesites, pyroxene andesites and Fe olivine-bearing dacites while olivine basalts, hornblende pyroxene andesites and hornblende dacites are characteristically found in the Eastern belt (Moreno, 1976). In the single belt segment (33°–34°S) amphibole is a typical phase as in the Marmolejo and Tupungato volcanoes where andesites are similar to those from northern Chile.

Lavas of the strato-volcanoes between 37°30'S and 41°S are high in Al_2O_3 (19%–22% in the high-alumina basalts) and Na_2O, low in TiO_2, K_2O and $MgO/(MgO + FeO_{tot.})$ (López-Escobar et al., 1977; Moreno, 1976) (Table I, anal. 46 and 47). According to López-Escobar et al. (1977) these features characterize basaltic rocks of island arcs and continental margin environments. Sr(i) ratios are low and range between 0.7040 and 0.7045, much lower than those from the andesites in northern Chile (Klerkx et al.,

1977). The La/Yb ratio is about 4 and 7 for the Western and Eastern belts respectively, much lower than in the northern volcanics. Geochemical features of the andesites from the Marmolejo and Tupungato strato-volcanoes located between 33° and 34°S largely contrast with the ones just described for the high-alumina basalts further south (Table I, analyses 44 and 45). They have significant amounts of normative quartz (10.2 to 12.7 wt%), are enriched in alkalies and present a highly fractionated REE pattern with a La/Yb ratio of 18 for the Marmolejo andesites and 27 for those of Tupungato (López-Escobar et al., 1977). Similar features are found in the andesites of northern Chile.

Gravimetric and seismic data for this southern province indicate that the volcanic Andean belts are located ca. 90 km above the present Benioff zone and overlie a 30–40 km thick continental crust (Barazangi and Isacks, 1976; James, 1971; Dragićević, 1970). This setting sharply differs from the one inferred for the Northern Province.

Radiometric dates and the relationship with glacial phenomena support that strato-volcanoes in the region have been active since the late Pliocene. Historic eruptions have occurred in several centers of the chain, most recently in the Villarrica, Llaima, Calbuco, Osorno and Puyehue–Carrán. In some of them, solfataric, fumarolic and geyser activity is taking place. During the late Pleistocene to Holocene, migration of the volcanic centers toward the west has occurred (Vergara and Munizaga, 1974). Regional lineaments with directions N130°, N60° and N10° corresponding to major fractures have controlled volcanism in the province since late Pliocene (Moreno, 1976). The main trend, N10°, coincides with that of the Liquiñe Fault system, one of the outstanding tectonic features in the development of the Andes of southern Chile.

Transgressive episodes over a narrow coastal strip continued during this period as a result of block faulting, oscillatory movements and glacio-eustatic effects (Stiefel, 1971, 1972; Herm and Paskoff, 1966 and Paskoff, 1967, 1970).

2. Geotectonic Interpretation of the Mesozoic–Cenozoic Evolution of the Northern Sector

(a) The General Mechanisms. The Gondwanide Orogeny, represented by Hercynian deformation and subsequent volcanism, brought about a definite change in the geotectonic evolution of the South American Pacific margin. A link between the contrasting régimes characterizing the pre-Andean and the Andean may be found, however, in the events that took place during the Triassic, a period in which silicic volcanism initiated in the Permian persisted in large areas probably related to tectonic basins (Hervé et al., 1976d; Charrier, 1979) of embryonic Andean style.

TABLE I
Northern Sector: Volcanic Rocks, Selected Analyses

	Middle to Upper Triassic					Lower to Middle Jurassic							Lower to Upper Jurassic				
	1	2	3	4	5	6	7	8	9	10	11	12	13	14	15	16	17
SiO_2	50.74	55.22	57.75	61.35	72.98	65.76	72.46	70.17	59.21	53.38	55.74	56.80	52.35	51.80	51.40	53.29	55.50
TiO_2	2.99	2.08	0.89	0.86	0.16	1.06	0.46	0.69	0.81	1.21	1.25	1.23	1.43	1.27	1.39	1.36	1.26
Al_2O_3	13.82	16.16	17.23	17.44	15.39	16.51	14.57	15.43	17.47	14.16	13.59	14.14	15.93	17.72	17.43	18.59	18.18
Fe_2O_3	3.66	3.54	2.27	0.98	0.21	5.22	1.93	3.09	1.60	2.15	1.94	7.58	5.32	1.76	8.30	2.86	5.72
FeO	9.34	6.12	5.86	6.24	2.62	0.15	0.20	0.42	5.38	10.07	9.19	3.27	3.95	5.60	0.61	5.37	2.42
MnO	0.22	0.12	0.15	0.06	0.04	0.00		0.02	0.16	0.19	0.21	0.19	0.12	0.12	0.22	0.24	0.08
MgO	5.97	2.53	3.61	3.11	0.79	0.05		0.04	3.24	2.97	2.79	3.06	4.15	6.95	2.17	6.50	4.12
CaO	8.42	4.11	6.26	4.38	0.26	0.43	0.33	0.36	5.61	5.29	4.97	4.30	7.64	6.60	9.65	4.53	6.70
Na_2O	3.72	4.46	4.27	3.67	5.47	8.38	4.44	6.01	4.73	3.50	2.62	3.98	3.78	4.90	2.81	5.61	3.52
K_2O	0.66	3.22	1.35	1.68	1.97	1.48	4.63	3.71	1.53	2.40	3.04	2.80	1.42	0.52	1.15	1.34	2.04
P_2O_5	0.29	0.41	0.24	0.18	0.02	0.09	0.02	0.04	0.25	0.45	0.46	0.17	0.31	0.33	0.38	0.33	0.41
H_2O						1.34	1.04			2.43	2.78	1.32					
CO_2										1.17	1.06	1.90					
TOTAL	99.83	98.27	99.88	99.95	99.91	100.47	100.08	99.98	99.99	99.37	99.64	100.74	96.40	97.57	95.51	100.02	99.95
FeO^*/MgO	2.12	3.80	2.19	2.29	3.56	96.95	80.02	80.02	2.11	4.04	3.92	3.30	1.15	1.03	3.72	1.22	1.84

	Lower Cretaceous						Upper Cretaceous				Paleocene			Eocene	Miocene		
	18	19	20	21	22	23	24	25	26	27	28	29	30	31	32	33	34
SiO_2	54.19	68.61	48.5–53.5	49.80	50.40	53.32	51.25	56.78	71.22	49.03	52.70	74.47	75.10	69.10	58.60	72.43	49.32
TiO_2	0.99	0.53	0.8–1.1	1.36	1.46	1.19	1.62	0.95	0.29	1.03	0.95	0.15	0.20	0.47	0.80	0.14	1.05
Al_2O_3	17.90	15.41	15.9–17.7	17.15	16.48	18.18	14.65	17.89	15.15	18.47	17.80	11.85	12.68	15.70	17.91	13.20	17.63
Fe_2O_3	5.04	3.10	8.1[a]–11.9[a]	7.75	8.51	9.06[a]	9.30	5.14	2.23	7.71	7.05	0.59	0.99	1.98	3.37	0.72	6.02
FeO	4.51	2.54		2.39	1.43		2.63	3.92	1.67	1.74	2.28	0.73	0.44	0.29	1.45	0.34	6.08
MnO	0.24	0.08	0.2–0.3	0.09	0.25		0.16	0.19	0.07	0.14	0.22	0.02	0.01	0.03	0.07	0.08	0.23
MgO	3.66	0.67	2.0–4.6	5.20	5.80	4.32	1.90	2.89	0.64	3.69	1.67	0.37	0.25	0.50	2.13	0.41	3.71
CaO	6.10	1.13	6.7–9.6	3.21	6.40	7.87	8.10	5.28	2.13	8.35	8.72	2.72	0.99	0.66	6.25	1.00	9.28
Na_2O	4.22	3.37	1.8–3.6	3.98	3.33	4.18	3.40	4.76	3.19	4.07	4.77	0.83	3.04	3.35	3.86	3.07	3.71
K_2O	2.56	3.89	1.4–3.4	2.47	1.48	1.12	0.94	1.15	3.43	0.95	1.17	4.71	4.54	4.87	1.88	4.76	0.30
P_2O_5	0.34	0.21	0.1–0.4	0.22	0.27		0.34	0.27	0.10	0.26	0.27	0.08	0.03	0.08	0.27	0.09	0.46
H_2O																	2.01
CO_2																	
TOTAL	99.75	99.54		93.62	95.81	99.44	94.29	98.27	100.12	95.44	97.60	96.52	98.27	97.03	96.59	96.24	100.58
FeO^d	2.47	7.96	4.1–2.6	1.80	1.57	2.10	5.79	2.96	5.75	2.35	5.16	3.41	5.32	4.14	2.11	2.41	3.10
MgO																	

(continued)

TABLE I. (continued)

	Miocene			Miocene–Pliocene	Pliocene	Pliocene to Recent								
	35	36	37	38	39	40	41	42	43	44	45	46	47
SiO_2	58.16	70.11	58.37	71.10	57.64	61.01	67.90	68.22	74.61	61.23	60.82	59.70	51.89
TiO_2	0.79	0.59	0.79	0.40	0.83	0.72	0.40	0.45	0.19	0.79	0.65	0.90	1.01
Al_2O_3	18.15	15.34	17.15	14.50	17.98	16.50	15.30	15.13	12.79	16.61	18.04	17.50	20.36
Fe_2O_3	3.84	2.50	2.68		2.74	1.99	1.60	2.80	0.99		4.54		
FeO	3.60	1.00	3.60	2.70[a]	3.03	3.35	0.90	0.28	0.22	4.71[a]	0.84	6.70[a]	7.82[a]
MnO	0.14	0.09	0.14		0.13	0.09	0.05	0.05	0.05	0.09	0.11		0.18
MgO	3.02	0.82	4.73	0.90	2.28	3.06	1.00	0.98	0.36	3.25	1.84	3.20	4.41
CaO	6.08	2.36	6.87	2.40	7.52	5.33	2.60	3.13	1.06	4.48	5.19	5.60	9.41
Na_2O	4.27	3.54	3.73	3.50	4.95	4.09	4.40	3.50	3.56	5.05	4.27	4.20	3.50
K_2O	1.54	3.53	1.42	4.20	1.94	2.43	3.30	3.25	4.34	2.11	2.45	1.90	0.54
P_2O_5	0.34	0.13	0.15	0.10	0.13	0.21	0.18	0.12	0.05	0.21	0.32	0.30	0.21
H_2O			0.84		1.24	1.13	1.90	1.60	1.35		1.32		0.71
CO_2								0.13					
TOTAL	99.93	100.01	100.47	99.80	100.41	99.91	99.53	99.64	99.57	98.53	100.39	100.00	100.14
FeO^a/MgO	2.34	3.96	1.27	3.00	2.41	1.68	2.34	2.86	3.09	1.45	2.68	2.09	1.77

ᵃ Total iron as FeO

KEY: *1*, Basalt (average of 2 rocks). Pichidangui Formation. Coastal Range, central Chile (*ca.* 32°S). Vicente (1976, Samples F229 and A3508). *2*, Tholeiitic andesite (average of 3 rocks). Pichidangui Formation. Coastal Range, central Chile (*ca.* 32°S). Vicente (1976, Samples F236, A3644 and A3645). *3*, Andesite (average of 3 rocks). Pichidangui Formation. Coastal Range, central Chile (*ca.* 32°S). Vicente (1976, Samples F252, A2 and A3670). *4*, Dacite. Pichidangui Formation. Coastal Range, central Chile (*ca.* 32°S). Vicente (1976, Sample A3518). *5*, Rhyolite. Pichidangui Formation. Coastal Range, central Chile (*ca.* 32°S). Vicente (1976, Sample 3641). *6*, Dacite (keratophyre). Melón Formation. Coastal Range, central Chile (32° 45' S). Levi (1960, Sample B1-3). *7*, Rhyolite. Melón Formation. Coastal Range, central Chile (32° 45' S). Levi (1960, Sample B1-1a). *8*, Rhyolite (average of 3 rocks). Coastal Range, central Chile (32° 40'-33° 40' S). Oyarzún & Villalobos (1969, p. 10-12); Vergara (1972, Sample 1B). *9*, Andesite (average of 2 rocks). Coastal Range, central Chile (32° 40'-33° 40' S). Oyarzún & Villalobos (1969, p. 10-12); Vergara (1972, Sample 1A). *10*, Low-silica andesite. La Negra Formation. Coastal Range, northern Chile, Tocopilla (*ca.* 22°S). Losert (1974b, Sample 146/B). *11*, High-K andesite. La Negra Formation. Coastal Range, northern Chile, Tocopilla (*ca.* 22°S). Losert (1974b, Sample 146). *12*, Andesite (with native copper). La Negra Formation. Coastal Range, northern Chile, Tocopilla (*ca.* 22°S). Losert (1974b, Sample 136/B). *13*, Andesite (average of 2 rocks). North central Chile, Atacama, Qda. Vicuñita (*ca.* 27°S). Zentilli (1974, Samples 52 and 53). *14*, Andesite. North central Chile, Atacama, Manflas (*ca.* 28°S). Zentilli (1974, Sample 51). *15*, Basalt. North central Chile, Atacama, Amolanas (*ca.* 28°S). Zentilli (1974, Sample 54). *16*, Andesite. Aalenian unit. Andean Range, central Chile (30° 30'-31° 20' S). Rivano and Mpodozis (1976, Sample 11). *17*, Andesite. Kimmeridgian unit. Andean Range, central Chile (30° 30'-31° 20' S). Rivano and Mpodozis (1976, Sample 11). *18*, Andesite (average of 83 rocks). Coastal Range, central Chile. Oyarzún and Villalobos (1969, p. 6-8, 10-12 and 16-23); Vergara (1972, Sample 2A). *19*, Dacite (average of 8 rocks). Coastal Range, central Chile. Oyarzún and Villalobos, ibid., Vergara (1972, Sample 2B). *20*, Basalt (range in 8 rocks). Veta Negra Formation, Ocoa member. Coastal Range, central Chile, sections at latitudes 30° and 33° S. Åberg *et al.* (1984). *21*, Basalt. North central Chile, Atacama, Qda. Puquios (*ca.* 28°S). Zentilli (1974, Sample 49). *22*, Basalt. North central Chile, Atacama, Qda. Puquios (*ca.* 28°S). Zentilli (1974, Sample 50). *23*, Basalt. North central Chile, Atacama, transect between 26°-28°S. Dostal *et al.* (1977, Sample DZ-1). *24*, Basalt. Cerrillos Formation. Andean Range, north central Chile, Atacama. Los Loros (*ca.* 27° 50'S) Zentilli (1974, Sample 46). *25*, Aluminous andesite (average of 12 rocks). Abanico (Coya–Machalí) Formation. Andean Range, central Chile (32°S–35°S). Vergara (1972, Sample 3A). *26*, Rhyolite (average of 5 rocks). Abanico (Coya–Machalí) Formation. Andean Range, central Chile (32°-35°S). Oyarzún and Villalobos (1969, p. 10, 11, 18, 21, 27, 28); Vergara (1972, Sample 3B). *27*, Basalt (average of 5 rocks). Hornitos Formation. Andean Range, north central Chile, Atacama. Zentilli (1974, Samples 29, 30, 41, 42 & 43). *28*, Andesite. Hornitos Formation. Andean Range, north central Chile, Atacama. Zentilli (1974, Sample 44). *29*, Rhyolite (average of 9 rocks). Hornitos Formation. Andean Range, north central Chile, Atacama. Zentilli (1974, Samples 27, 28 and 31-37). *30*, Rhyolitic ignimbrite (average of 2 rocks). Hornitos Formation. Andean Range, north central Chile, Atacama. Zentilli (1974, Samples 38 & 39). *31*, Rhyolite. Cerro de la Peineta Formation. Andean Range, north central Chile, Atacama. Zentilli (1974, Sample 23). *32*, Andesite (average of 3 rocks). Negro Francisco Formation. Andean Range, north central Chile, Atacama. Zentilli (1974, Sample 12). *34*, Basalt. Farellones Formation. Andean Range, central Chile. Santiago. Oyarzún and Villalobos (1969, p. 24). *35*, Aluminous andesite (average of 19 rocks). Farellones Formation. Andean Range, central Chile. Oyarzún and Villalobos (1969, p. 10-11, 24-25, 27); Vergara (1972, Sample 4A). *36*, Rhyolite (average of 9 rocks). Farellones Formation. Andean Range, central Chile. Oyarzún and Villalobos (1969, ibid.); Vergara (1972, Sample 4B). *37*, Andesite. Coast Range Miocene belt. Central Valley Graben, central Chile (37° 30'-41°S). Aguirre *et al.* (1974, Sample 5). *38*, Rhyolitic ignimbrite (average of 60 rocks). Rhyolite Formation, Andean Range, northern Chile. Klerkx *et al.* (1977, p. 53). *39*, High-Al andesite. Andesitic Plateau series (Cola de Zorro Formation). Andean Range, central Chile (33°-41°S). Aguirre *et al.* (1974, Sample 6). *40*, High K-andesite (average of 21 rocks). Pliocene to Recent volcanism. Northern province (18°-28°S). Andean Range, northern Chile. Thorpe *et al.* (1976). *41*, Rhyodacite. Northern province (18°-28°S). Andean Range, northern Chile. Zeil and Pichler (1967, sample 12). *42*, Rhyodacitic ignimbrite (average of 4 rocks). Northern province (18°-28°S). Andean Range, northern Chile. Zeil and Pichler (1967). *43*, Rhyolitic ignimbrite (average of 14 rocks). Northern province (18°-28°S). Andean Range, northern Chile. Zeil & Pichler (1967). *44*, Andesite. Stratovolcanoes of Central type. Tupungato. Andean Range, central Chile (33° 15'S). López-Escobar *et al.* (1977, Sample 797). *45*, Andesite. Stratovolcanoes of Central type. Marmolejo. Andean Range, central Chile (33° 45'S). López-Escobar *et al.* (1977, Sample 803-Y31). *46*, High-Al andesite (average of 97 rocks). Southern province (31°-42°S). Andean Range, central Chile. Klerkx *et al.* (1977). *47*, High-Al basalt (average of 8 rocks). Stratovolcanoes of Central type. Andean Range, central Chile. Recent volcanism (37° 30'-40° 45'S). López-Escobar *et al.* (1977, Samples 798, 799, 800, 801, 802, GV164a, GV170 and GV177).

The edifice generated during the Mesozoic–Cenozoic differs from the pre-Andean one in several basic aspects: (*1*) its ensialic nature; (*2*) the absence of oceanic accretion; (*3*) the predominance of a tensional régime without signs of significant shortening of the continental crust; (*4*) the bimodal and calc-alkaline nature of its volcanics and (*5*) the absence of regional metamorphism of deformational type. The events leading to the building of the Mesozoic–Cenozoic edifice have an episodic character and are arranged in cycles (Charrier, 1973; Aguirre *et al.*, 1974; Aguirre *et al.*, 1978) comprising: (a) volcanism and sedimentation (depositional stage) accompanied by subsidence and vertical oscillatory movements; (b) burial metamorphism and chemical redistribution taking place as a continuous process from deposition onwards; (c) folding; (d) granitoid emplacement accompanied by development of narrow contact aureoles and occasional mineralization, and (e) uplift and erosion. Events (a) to (e) followed each other and partly overlapped in time during a cycle. Each cycle generated a stratigraphic-structural unit separated by unconformities from those of the preceding and next cycle. Seven such major units are recognized from the Triassic to the Recent with an average cycle duration of 40 million years (Fig. 13).

The loci of these cyclic processes were N–S elongated basins largely dominated by extensional forces as shown by megalineament structural control, persistence of volcanic activity, abundance of dikes parallel to basin elongation (pp. 299, 306, 308), broad and gentle folding of the sequences, absence of regional deformational metamorphism and subdued expression of most unconformities. From the early Jurassic to the end of the Neogene, the axes of the successive basins migrated eastward. Following the middle Cretaceous or Subhercynian orogenic phase and the marked regression of the sea, basinal mechanisms persisted under continental conditions.

The present structural expression of this longstanding cyclic activity is a N–S trending synclinorium composed of successive, partially superimposed, stratigraphic-structural units extending along most of the Northern Sector (Fig. 5). An overall symmetry (pp. 299, 300, 307 and Fig. 5), combined with secondary asymmetric features characterizes this synclinorium. The processes involved in its construction have been explained on the basis of geosynclinal theory, particularly for its marine development during the Jurassic and early Cretaceous when the existence of a "eu/mio" couple is suggested by the record (pp. 295–298, 302–305, Figs. 7 and 8). Since the advent of plate tectonics, a reappraisal of the geosynclinal model has taken place with strong emphasis on the extrapolation of the Recent "Andean Model" to account for the Mesozoic and Cenozoic history of the continental margin.

The interpretation of some Mesozoic "eugeosynclinal" troughs in the Peruvian and Chilean Andes in terms of marginal basins has been suggested by some authors (Chotin, 1976; Mégard, 1978; Dalziel, 1981). Recently, the evolution of the Northern Sector has been explained as reflecting a cyclic process of generation of ensialic marginal basins controlled by a coupled action of *plate subduction* and *intracontinental spreading-subsidence* (Levi and Aguirre, 1981). The basins thus resulted from a thinning of the continental crust through uplift (doming), produced by upwelling of mantle material, and from rifting, spreading and subsidence on the continental block. Eruption of mantle-derived flood basalts took place but no oceanic crust was generated; in this respect the marginal basins of the Northern Sector can be classified as aborted. The events of the early (to middle) Cretaceous would typically illustrate the evolution of one of these basins (Åberg *et al.*, 1984; geological record in the present paper; Coira *et al.*, 1982).

The spreading-subsidence mechanism offers an explanation for the global transverse (E–W) symmetry while plate subduction would account for some of the secondary asymmetries present (i.e. granitoid age polarity, predominance of volcanics in the western flank of the Jurassic and early Cretaceous basins). The contrasting nature of contemporaneous volcanism along (N–S) the sector during some periods (i.e. Jurassic, early Cretaceous, Eocene, Pliocene–Pleistocene; Fig. 13) could be interpreted as portraying different stages of basin construction, which would be determined by different rates of upwelling of mantle material. The regional thickness and composition of the continental crust. the segmentation of a subducting oceanic plate or combinations of these two or more variables could condition upwelling rates.

(b) Identification and Interpretation of Events. Reference is made to Fig. 13 where the main magmatic, metamorphic and orogenic events described in the section titled "Geological Record" have been represented.

Volcanism. During the Triassic, the inaugural stage of Andean development, volcanism was mainly silicic and ignimbritic and extended until the middle Jurassic in central Chile. In the early stages of the Jurassic, early Cretaceous, Paleocene, Miocene and Plio–Pleistocene cycles, particularly in central Chile, a marked tendency for a volcanism of silicic type is observed. This tendency is interpreted as the outcome of regional crustal uplift (updoming?) and melting produced by slow upwelling of mantle material in the *early stages* of basin building and commonly following orogenic events. During this phase volcanic arcs formed within the continental margin and contributed materials to a contiguous ensialic trough located to the east; subsidence was initiated due to rifting and deposition. Volcanic products predominate toward the arc side (west) but the extension of their interfingering with sediments toward the east varied widely (e.g. Jurassic record in the Andean flank, pp. 295–297, p. 298). This phase is best exemplified by

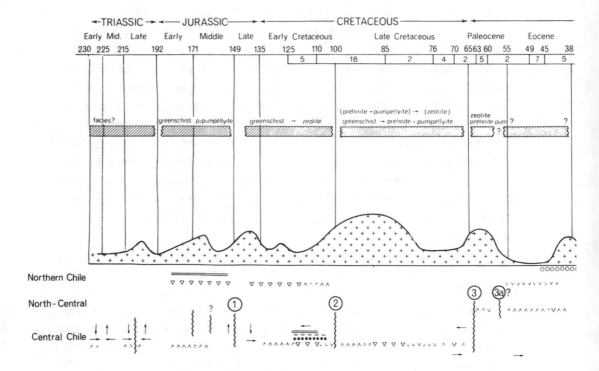

Fig. 13.   Schematic chart of events for the Andean orogenic evolution in the Northern Sector.

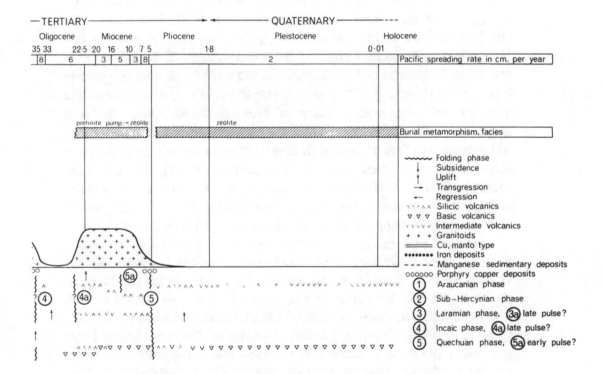

Fig. 13.   (*continued*)

the early Cretaceous cycle of central Chile. The next, *mature stage*, in the cycles under analysis corresponds to the eruption of basic to intermediate lavas, partly flood basalts, that took place along deep north-trending faults and resulted in the buildup of thick volcanic piles fed by dikes parallel to the basin trend. Flood basalts are characteristically present in the unusually thick units of the Jurassic (La Negra in northern Chile) and of the early Cretaceous (Veta Negra in central Chile) and their Sr(i) ratios are indicative of a mantle origin. This effusive stage would be linked to spreading and rapid subsidence of the basin and would mark the farthest step reached toward the production of a lithospheric break. In the case of the early Cretaceous basin, a variation in upwelling rates of mantle material along its extension is suggested by the comparative characteristics of the REE patterns of basalts from northern and central locations (p. 307 and Fig. 10) with possible higher rates in the north. In most of the cycles considered the *late stage* of volcanism is characterized by andesitic, or more silicic, products suggesting a slowing of upwelling rates probably due to heat dissipation, increase in viscosity and, consequently, longer residence of the ascending magma at crustal levels. This is best observed in the early Cretaceous, late Cretaceous and Paleocene cycles. Moreover, from the Paleocene onwards, a progressive increase in the comparative amount of silicic volcanics (Fig. 13) and in the values of Sr(i) ratios notably in northern and central Chile could reflect a progressive thickening of the crust or a mechanism of recycling of "aged" sialic material through subduction as the one suggested by James (1981).

Marginal basin generation seems to have proceeded at a different pace along the Northern Sector as evidenced by the Jurassic and early Cretaceous volcanic records of northern and central Chile. Thus, during the Jurassic, the La Negra basic to intermediate lava flows built up a *ca.* 10,000 m thick pile during a time in which, more than 1000 km further south in central Chile, rhyolitic–dacitic ignimbrites and silicic lavas up to 8000 m thick were being generated. The reverse situation is observed during the early Cretaceous with huge volumes of basic effusives in central Chile (the ocoites of the Veta Negra Formation) largely contemporaneous with silicic volcanism in northern Chile represented by the Collahuasi–Peña Morada formations (p. 306). If the chemistry of volcanics can be taken as an indicator of a particular stage in the evolution of a marginal basin, then mature and early stages would seem to have alternated in time between northern and central Chile during most of the Mesozoic. The contrast observed between the volcanic types of the northern and southern provinces from the Pliocene onwards may be tentatively explained in similar terms.

Burial Metamorphism. Metamorphic reactions took place in the basins from the very beginning of deposition and progressed as subsidence proceeded. A metamorphic facies series in which grade increases with depth was generated in each stratigraphic-structural unit. Lateral motion due to intracontinental spreading, on the other hand, protected the subsiding deposits from long periods of exposure at high temperatures and pressures. This last interpretation is supported by the prevalence of metastable mineral assemblages and by the appearance of greenschist facies assemblages at stratigraphical depths of several kilometers in the absence of pumpellyite–actinolite facies assemblages (Levi et al., 1982: Åberg et al., 1984). The fact that burial metamorphism predates the folding events closing each cycle is firmly substantiated by the geological record (i.e. isograds parallel to regional structures, metamorphosed pebbles in basal conglomerates, p. 300, p. 308). Moreover, the coincidence of mineralogical and structural breaks between adjacent stratigraphic-structural units involving a reverse in metamorphic grade when unconformities are crossed (Fig. 13) constitutes one of the most conclusive arguments in favor of an episodic character of Andean development (Levi, 1970; Aguirre et al., 1978). The time elapsed between effusion and the closing of metamorphism in the volcanic materials would represent between a quarter and a half of the duration of a total cycle as suggested by Rb-Sr geochronology for the early Cretaceous one (p. 308).

Geothermal gradient values for each metamorphic event are thought to be reflected in the extension of the facies series involved in relation to the thickness of the corresponding stratigraphic-structural unit. Spreading of facies appears to be largest in the early Cretaceous cycle with a progressive decrease in variety and grade starting in the Tertiary cycles (Fig. 13). Considering that the intensity of a geothermal gradient is directly related to the rate of upwelling of mantle material and to its volume, it again appears that the most favorable conditions for the generation of a marginal basin were reached during the Cretaceous. It is noteworthy that, in spite of the sharp change in environment brought about by the Meso–Cretaceous Orogeny, conditions for basin development as judged by the volcanic and burial metamorphic patterns of the late Cretaceous cycle (and its implied geothermal gradient) were basically similar to those of the preceding cycle. Geothermal gradients seem to have decreased in intensity during the Cenozoic, a situation that could be related to crustal thickening determining the slowing down in the ascent of mantle material.

Orogenesis. Orogenic phases occurred at the closing stage of cycles as defined here. Folding, when it took place, produced broad structures; penetrative deformation is absent and crustal shortening is minimal. More generally epeirogenic events occurred in which block faulting and vertical oscillatory movements played the leading role (i.e. Triassic, p. 295; late Jurassic in the Andean Range, p. 301; Oligocene, p. 314). Hiatuses are evidenced in several periods, notably during the Jurassic (Bathonian, Upper Jurassic in the Coastal Range of central Chile, p. 300) and the Tertiary (Paleogene in central Chile, p. 314–316). All these features reinforce the hypothesis of an Andean edifice largely controlled by vertical forces (mantle diapirism?) giving rise to updoming, crustal thinning, block faulting, rifting and continental spreading during long standing tensional conditions. Horizontal compressional movements are subordinate and short-lived as indicated by the complete lack of regional deformational metamorphism and collision-accretion events.

The main orogenic episodes described in the record (Fig. 13) have a regional expression and can be correlated beyond the Chilean territory with similar contemporaneous events in Argentina, Peru, Ecuador and Colombia (Aguirre, 1976). This is specially valid for the Meso–Cretaceous (Subhercynian), Incaic and Quechuan phases although their relative importance in the Andean building process varies from place to place.

Plutonisn. Large volumes of granitoids have been contributed to the making of the Andean edifice throughout its history. They are represented by composite batholiths of calc-alkaline oversaturated nature. Their features suggest that the majority are postorogenic and postburial metamorphic events in each cycle. A close spatial relationship with coeval volcanics, particularly evident for the Cretaceous and Tertiary granitoids, is often explained in terms of consanguinity but geochemical data supporting that assertion are lacking except for the late Tertiary granitoids in central Chile where genetic links with Pleistocene andesites have been pointed out (p. 316).

The emplacement of granitoids in successive N–S trending belts with ages decreasing from west to east introduces an asymmetric feature in the overall symmetry of the Andean edifice and constitutes a strong argument for the presence of an oceanic plate being subducted below the continental margin. According to Drake *et al.* (1982a) migration of plutonic lines during the Mesozoic and Cenozoic took place at an average speed of about 1 mm/yr (1 km per million years). The same authors argue that, in central Chile, the successive emplacement of these plutonic belts has rifted the pre-Mesozoic continental margin in a way analogous to mid-ocean ridge spreading. This assertion focuses our attention on the time of granitoid intrusion inside each cycle. Based on scarce geochronological data Aguirre *et al.* (1974) and Aguirre (1983) pointed out that, in spite of a rather continuous process of granitoid production and emplacement during the whole of the Andean evolution, a close relationship between orogenic phases and main plutonic culminations could be observed. This same relationship is apparent from Fig. 13 where culminations commonly follow orogenic phases suggesting that granitoid emplacement took place under conditions of fading compression to beginning of extension. Thus, it largely coincided with updoming and silicic volcanism marking the early stage of marginal basin generation. However, no common magmatic origin would exist between the granitoids, whose Sr(i) ratios have consistent mantle values, and the silicic volcanics presumably derived from crustal melting. Generation of granitoidal melts at depth during compressive phases and their passive emplacement after the fading of compression have been postulated by Damm *et al.* (1981) for the Coastal Range in northern Chile. Moreover, some of their granitoids would have *S*-type features (p. 301). The possibility of granitoids being intruded contemporaneously with basic to intermediate volcanics during the spreading stage of a marginal basin seems to be supported by the development of the early Cretaceous in central Chile (Åberg *et al.*, 1984) and by the contemporaneous production of flood basalts and granitoids with extremely low Sr(i) ratios in north central Chile during the same period (p. 307, p. 309).

The progressive increase in Sr(i) ratios observed in granitoids with ages ranging from middle Cretaceous to late Tertiary has been interpreted by McNutt *et al.* (1975) as reflecting a migration of the locus of melting from along or close to the upper surface of the subducting slab into hanging wall mantle peridotite while subduction proceeded. This feature could also be related to the process of progressive crustal thickening implied in the evolution of volcanism and burial metamorphism as pointed out earlier. Similar trends with younging ages expressed in the increasing values of the LREE/HREE ratio, Sr and alkali content of granitoids would agree with the interpretations given above.

(c) Ocean-Floor Spreading Rates, Plate Interaction and Geotectonic Evolution. A relationship between ocean floor spreading rates in the Pacific and Andean geotectonic evolution seems to exist. Fig. 13 shows values of spreading rates in cm/yr for the early Cretaceous–Recent span as compiled by Frutos (1981) from publications by various authors.

Periods of mantle upwelling in the continental margin causing continental spreading would slow the movement of the oceanic plate and would considerably increase its angle of dip. Extrusion of basic and intermediate volcanics typifies this stage, which may also include emplacement of granitoids. This relationship seems well illustrated in different parts of the Northern Sector for the early Cretaceous (125–110 Ma), the late Cretaceous (85–63 Ma), the Paleocene–Eocene (60–49 Ma) and the Pliocene–Recent.

Slowing in the upwelling rates of mantle material in the continent would result in the attenuation or interruption of continental spreading. Due to the almost vertical position reached by the oceanic slab during the preceding phase, the contact surface between plates is also near vertical and the reinitiation of subduction is confronted by maximum opposition at the start. These short "triggering" periods would correspond with the short-lived compressional régimes reflected as orogenic phases in the continental margin (see Luyendick, 1970; Mégard, 1978; Frutos, 1981). In Fig. 13 the main orogenic episodes, with minor exceptions, fall inside periods of fast ocean spreading rates (relative to preceding and subsequent ones).

Periods of fast ocean spreading, commonly coinciding with a low dip of the subducting plate, seem to be mainly related to a high rate in granitoid emplacement and, probably, to silicic volcanic activity. The batholithic culmination of the middle to late Cretaceous, which coincides with the fastest recorded period of ocean spreading (18 cm/yr for the span 110–85 Ma) represents the best example and has been identified worldwide (Larson and Pitman, 1972). The conspicuous absence of volcanism during the Oligocene could be tentatively explained by the prevalence of similar plate interrelations, notably the low angle of dip of the oceanic slab.

The remarkable and widespread reactivation of magmatism started at the end of the Oligocene in the Northern Sector, and in different circum-Pacific regions, is seen as the effect of an increase in the rotation of the Pacific plate about 20–25 Ma BP (Clague and Jarrard, 1973). In Chile this event could also account for an accelerated eastward migration of volcanic foci.

Geophysical data, particularly seismological, magnetic and oceanographic studies, are providing a more tangible picture of the Nazca plate (e.g. Kulm et al., 1981). Its segmentation, the distribution of sediments along trenches, together with hypothetical processes of consumption of aseismic ridges such as the Nazca and Juan Fernández (Ben-Avraham et al., 1981; Nur and Ben-Avraham, 1981) will contribute to the understanding of many of the features characterizing the Quaternary in the Chilean Andes. Among them the distribution and nature of the volcanics belonging to the rhyolitic and andesitic provinces (p. 323, p. 325), the puzzling absence of volcanism between 27° and 33°S (p. 323), the regression of volcanic foci toward the west started in the late Pleistocene (p. 325, p. 327) and others.

B. The Patagonian Sector (41°S to Cape Horn)

1. Geological Record

(a) Early Stage of Development. The Patagonian Sector ("Magallanes Geosyncline") extends from approximately 41°S to Cape Horn. It is separated from the Northern Sector by the Concepción Ridge (Fig. 2) and, toward the south, its geological connections with Antarctica through the Scotia Arc seem well established.

The beginning of the Andean orogenic evolution in this sector can be traced back to the middle Jurassic when widespread, mainly subaerial, volcanic activity controlled by faults, developed. Its products, predominantly silicic in composition, are rhyolites, andesites, welded and nonwelded acidic tuffs and breccias and were unconformably deposited on pre-Andean basement. They cover an approximate surface of 1,000,000 km^2 (Feruglio, 1949 *in* Suárez, 1976) all along the sector and are known as "Serie Tobífera" in the Magallanes region and Elizalde Formation (Skarmeta and Charrier, 1976) or Ibáñez Formation (Niemeyer, 1975) in the Aisén region. K-Ar ages of 166 ± 6 Ma (Creer *et al.*, 1972) are known for rocks in Argentina suggesting that volcanism was contemporaneous with initiation of granitoidal plutonism along the present continental margin (see Patagonian Batholith, p. 351). Activity ceased during the late Jurassic as shown by the occurrence of sedimentary units of this age concordantly overlying the silicic volcanics. The chemical characteristics of these rocks in terms of major and trace elements are presented in Table II, p. 354 analyses 1, 2 and 3. The "Tobífera" volcanics have suffered metamorphism of prehnite–pumpellyite and greenschist facies (Bruhn *et al.*, 1978). No signs of a late Jurassic (Araucanian) folding phase are recorded.

(*b*) *Segmentation of the Patagonian Sector.* After the close of the "Tobífera" activity, probably during late Jurassic–early Cretaceous time, the geological evolution of the Patagonian Sector followed different paths along its extent suggesting a process of segmentation of the continental margin. A division of the Patagonian Sector into two geotectonic domains, here called the Aisén and the Magallanes segments, is well defined. The Aisén segment, to the north, extends from *ca.* 41°S to a latitude between 47°30'S and 49°S that corresponds to the present intersection of the Chile Rise with the continent and the northernmost exposures of the Ophiolitic Complex of Magallanes (see below) respectively (Suárez, 1976 and references therein). The Magallanes segment covers from 49°S to Cape Horn. The geological record of these two segments will be presented from south to north.

The Magallanes Segment. The Magallanes segment is characterized by the development of an island arc and a marginal basin along the Pacific side (eugeosynclinal = internal zone) from late Jurassic to middle Cretaceous and by the existence of sedimentary basins toward the continental side (miogeosynclinal = external zone) from late Oxfordian to Pliocene (Fig. 14, Fig. 15).

The eugeosynclinal zone. The most conspicuous unit is the Ophiolitic Complex or "rocas verdes" and is present as discontinuous exposures along the entire 1000 km length of the Magallanes segment (Fig. 14). Two areas of continuous outcrop of these rocks located at *ca.* 51°30'S and 55°S are

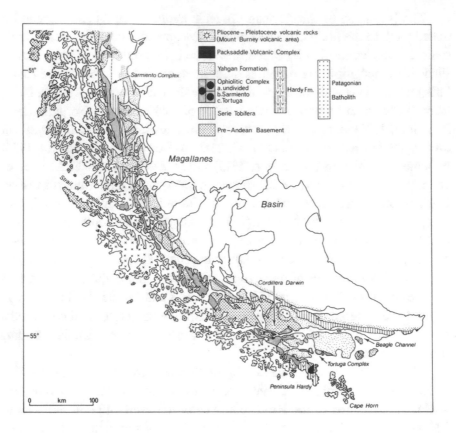

Fig. 14. Geologic map of the eugeosynclinal zone, Magallanes segment, Patagonian Sector (from Dalziel, 1981 with addenda by L. Aguirre, from *Mapa Geológico de Chile*, 1:1,000,000, 1982).

known as Sarmiento and Tortuga complexes respectively (Dalziel, 1981 and references therein).

The Ophiolitic Complex consists of an intrusive unit of gabbro with local layering and sheeted dikes and an extrusive unit of pillow lava, pillow breccia and tuff (Fig. 15); plagiogranites and associated siliceous dikes are present in subordinate volume (Dalziel *et al.*, 1974; Saunders *et al.*, 1979). Toward the west the Ophiolitic Complex is flanked by the Patagonian Batholith which intrudes pre-Andean basement and "Tobífera" while, toward the east, it is bounded by the "Tobífera" unconformably deposited on basement. In some areas of the Sarmiento Complex silicic rocks of the "Tobífera" flank mafic rocks on both sides. The width of the Ophiolitic Complex seems to increase substantially from north to south.

Stratigraphic relations of the Complex are commonly obscured by fault tectonics. Thus, the Sarmiento rocks are in fault contact with pre-Andean

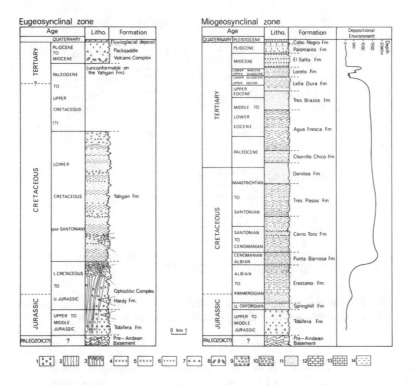

Fig. 15. Generalized stratigraphic columns for the eugeosynclinal and miogeosynclinal zones, Magallanes segment, Patagonian Sector. Mainly based on Aubouin *et al.* (1973); Natland *et al.* (1974), and Winn, (1978). KEY: 1, gabbro; 2, sheeted dikes; 3, pillow lavas; 4, intermediate lava; 5, silicic lava; 6, rhyolitic and dacitic ignimbrite; 7, volcanic breccia; 8, agglomerate; 9, conglomerate; 10, sedimentary breccia; 11, sandstone; 12, limestone; 13, marl; 14, shale.

basement and "Tobífera" strata while the Tortuga Complex is upfaulted into Lower Cretaceous volcanoclastic turbidites. However, in places such as the southernmost archipelagos (Hoste Island, Hardy Peninsula) pillow lavas of the upper levels of the Ophiolite Complex overlie silicic volcanics that are also cut by dolerite feeder dikes of the pillow lavas (Suárez, 1979). The silicic volcanics there are part of the Hardy Formation (Suárez and Pettigrew, 1976), a 600 m thick sequence of pyroclastics and lava flows ranging in composition from rhyolite to basalt. According to Suárez (1979) this calc-alkaline unit with island arc features would have interfingered toward the Atlantic with silicic volcanics of the "Tobífera" prior to the generation of the Ophiolitic Complex. Still in other areas, interbedding of mafic extrusives and Upper Jurassic silicic to intermediate volcanic rocks suggests (Dalziel, 1981), that the ophiolitic rocks were emplaced late during the development of the "Tobífera". The Ophiolitic Complex is conformably covered by lower Cretaceous turbidites. Interdigitation observed in Hoste Island between

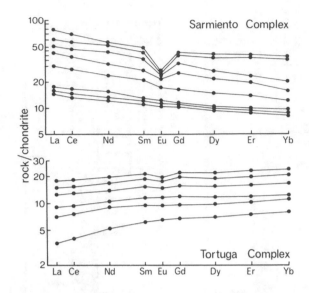

Fig. 16.   Chondrite-normalized REE pattern for lavas and dikes from the Sarmiento and Tortuga complexes, Magallanes segment, Patagonian Sector (modified after Stern, 1980).

these turbidites and andesitic rocks presumably belonging to the Hardy Formation indicates that island arc activity persisted in postophiolitic time (Suárez, 1979).

Thicknesses of 3000 m to 3500 m, from sea level to the top, for the Tortuga and Sarmiento complexes have been recorded (Stern and Elthon, 1979). Suárez and Pettigrew (1976) indicate that the thickness of the Tortuga Complex in the southwestern area of Navarino Island amounts to 820 m.

A late Jurassic–early Cretaceous age is assigned to the Ophiolitic Complex mainly on the basis of paleontological evidence from the overlying turbidites.

Chemical characteristics of the mafic products are shown in Table II, analyses 4 to 9. The rocks present a tholeiitic differentiation trend (Stern, 1980; Saunders et al., 1979); the extent of differentiation is much greater in the Sarmiento Complex where silicic dikes and lavas similar in chemistry to oceanic plagiogranites have been formed. The least differentiated basalts belong to the Tortuga Complex and are olivine normative (Suárez, 1977; Stern, 1980). Affinities with ocean ridge basalts rather than with island arc tholeiites are apparent (Suárez, 1977; Saunders et al., 1979; Stern, 1980). Lavas and dikes from the Sarmiento Complex have higher $(Ce/Yb)_N$ than those from the Tortuga Complex (Fig. 16) for rocks of similar FeO^*/MgO, Zr and Yb. Finally, a wide range of K/Rb, Rb/Sr and $(Ce/Yb)_N$ is shown by the basalts of the Ophiolitic Complex, a feature also observed in basalts from present-day back-arc basins (Stern, 1980).

The rocks of the Ophiolitic Complex have experienced nondeformative metamorphism characterized by extremely steep vertical metamorphic gradient passing downward from zeolite to upper actinolite facies over 2000 m, followed by transition to fresh gabbro (Elthon and Stern, 1978; Stern and Elthon, 1979). These authors describe such alteration as hydrothermal "ocean floor metamorphism," a phenomenon closely associated with circulation of sea water related to igneous and tectonic activity at a spreading center. The metamorphic conditions reached during this ocean floor metamorphism are mainly a function of temperature and fluid phases (Stern *et al.*, 1976).

A unit of volcanoclastic turbidites, the Yahgan Formation (Katz and Watters, 1966) crops out in the southernmost area of Magallanes. Deposited as a marginal basin infilling, it rests directly upon basic lavas of the Ophiolitic Complex (Tortuga Complex) in several localities (Fig. 14, Fig. 15); no top of the Yahgan Formation is exposed in those places. Graywackes, argillites and cherts are predominant and a high content of basic and intermediate volcanic débris, as well as fragments of volcanic glass, is characteristically present. A number of thick quartz dolerite sills are part of this unit. The formation reaches 4000 m to 5000 m in thickness in southern Tierra del Fuego (Winn, 1978).

Based on paleontological evidence, notably the presence of *Inoceramus, Favrella americana* or *F. steinmanni* and *Belemnopsis* (Hoffstetter, 1957; Halpern and Rex, 1972; Harrington, 1943), an early Cretaceous age is assigned to the Yahgan Formation without excluding the possibility that its deposition may have started in latest Jurassic time. Granitoids of 79 ± 5 Ma intrude Yahgan strata in Navarino and Hoste islands (Halpern and Rex, 1972) setting a minimum age for the formation.

The Yahgan rocks show low-grade metamorphism with most mineral assemblages corresponding to prehnite–pumpellyite facies; prehnite is abundant (Watters, 1965) and pumpellyite scarce. Zeolites (heulandite, analcite, laumontite) accompany prehnite in southwest Navarino Island and southern Hoste Island (Winn, 1978). The presence of penetrative deformation in the Yahgan rocks distinguishes this low-grade metamorphism from strict burial types suggesting that it may be related to the middle Cretaceous orogenic phase that followed depositon of the Yahgan Formation.

The age of the youngest rocks deposited as marginal basin infilling is considered to be late Lower Cretaceous to early Upper Cretaceous (Dott *et al.*, 1977). These strata were deformed during middle Cretaceous time, probably during the Cenomanian, before the intrusion of 80–90 Ma granitoids (Nelson *et al.*, 1980). This first orogenic phase in the Andean history of the Patagonian Sector was confined to the eugeosynclinal, internal domain, and marks the passage from a longstanding extensional régime to a short period of compression.

The middle Cretaceous orogenic phase involved vertical uplift, strike-slip displacements and crustal shortening at right angle to the Pacific margin in the eugeosynclinal trough (Dalziel *et al.*, 1975; Bruhn and Dalziel, 1977; Bruhn, 1979; Dalziel and Palmer, 1979; Nelson *et al.*, 1980; Dalziel, 1981). Structures generated in rocks of the eugeosynclinal terrane are comparatively simple; volcanics are gently folded and local complications are observed in shear zones (Dalziel, 1981). Considerable shortening of the sedimentary pile is suggested by the presence in the Yahgan strata of tight asymmetric to isoclinal folding with folds overturned toward the foreland (Katz, 1973; Dalziel, 1981). The low-grade, penetrative metamorphism may be related to this event. Locally, high-grade penetrative metamorphism took place (Dalziel and Cortés, 1972; Katz, 1973; Nelson *et al.*, 1980) as in the Cordillera Darwin where high amphibolite facies conditions were reached in pre-Andean reactivated rocks as well as in the Andean eugeosynclinal cover (Nelson *et al.*, 1980). The deformed Yahgan Formation contrasts with the Tortuga Complex whose mafic components appear undeformed (Dalziel, 1981).

The middle Cretaceous orogenic phase resulted in development of a foredeep, the Magallanes Basin, between the uplifted Patagonian Cordillera and the stable interior toward the Atlantic. The destruction of the relief so generated subsequently exerted a major control on deposition in the miogeosynclinal basin to the east.

Tertiary volcanic activity in the internal belt is recorded from several archipelagos south of the Beagle Channel. The Packsaddle Volcanic Complex (Suárez, 1978) consisting of columnar olivine basalts, pillow lavas, pyroclastic breccias, agglomerates, tuffs and laharic deposits unconformably overlies the Yahgan Formation. According to Suárez (1978) its age is uncertain and it could be synchronous with the Mio–Pliocene plateau basalts of Patagonia and Antarctica.

The miogeosynclinal zone. East of the eugeosynclinal belt, marine sedimentary deposition took place from the late Oxfordian to the Pliocene (Fig. 15) resulting in the accumulation of *ca.* 13,000 m of strata. Three sections can be distinguished in the vicinity of latitude 51°S (Ultima Esperanza region):

(*i*) A late Oxfordian–Aptian section, transgressive on the "Tobífera", composed of basal sandstone and black calcareous sandstone with turbiditic facies. It includes the Springhill and Erezcano formations.

(*ii*) An Albian–Maastrichtian section represented by a rhythmic alternation of black sandstone and calcareous sandstone known as Magellan Flysch (Cecioni, 1957; Scott, 1966). This mainly Cenomanian flysch is considered to relate to the middle Cretaceous Orogeny that occurred in the eugeosynclinal belt to the west. The section includes the Punta Barrosa, Cerro Toro, Tres Pasos and most of the Dorotea formations. In the Lago Sofía conglomerate (Cerro Toro) pebbles of plutonic, metamorphic and ophiolitic rocks derived from the uplifted area to the west are abundant (Katz, 1973; Zeil, 1958; Scott, 1966). During this period the miogeosynclinal basin, north and south of the Strait of Magellan, collected sediments at an accelerating rate; in Ultima Esperanza the Campanian–Maastrichtian alone measures 5000 m (Katz, 1973). No folding due to the middle Cretaceous orogenic phase is observed in this basin.

Between the late Cretaceous and the Paleogene a folding phase (Laramian) affected the most internal (foothill) areas of the miogeosynclinal belt (Katz, 1961, 1962, 1973; Aubouin *et al.*, 1973). Its effects are illustrated by (*1*) an angular unconformity between isolated remnants of Lower Tertiary and Senonian strata (Katz, 1973); (*2*) an hiatus between the Cretaceous and the Tertiary (Klohn, 1965; Katz, 1973); (*3*) the presence of Neocomian redeposited microflora in Eocene strata (Klohn, 1965). Toward the most external parts of the miogeosyncline no breaks are observed in the marine succession, e.g. near the Strait of Magellan the sedimentary sequence between the Cretaceous and the Tertiary is continuous and completely gradational (Charrier and Lahsen, 1969). At the beginning of the Tertiary the sedimentary basin was displaced to the easternmost, or most external area, of the preceding trough where a new unit started to be deposited.

(*iii*) A Paleogene unit in which green sandstone with massive stratification predominates; it corresponds to the Magellan Molasse (of various authors), which resulted from the Laramian orogenic phase. The Paleogene deposits reach up to 5500 m in thickness (Thomas, 1949 *in* Katz, 1973) and are followed by late Oligocene to Miocene nonmarine beds measuring up to 2000 m. This unit comprises from the upper Dorotea to the Loreto Formation (Fig. 15).

During the late Miocene compression and folding of the Paleogene basin took place in its most external border and was strongly pronounced in places. This event is marked by sharp angular unconformity between Lower Tertiary beds and the Miocene–Pliocene Palomares Formation exposed at Cordillera Vidal north of Seno Skyring (52°–52°30′S, 72°W) (Katz, 1973).

An episode of gentle folding (Pliocene?) recorded in strata of the Palomares Formation is the youngest manifestation of Andean folding. Basaltic effusions in the extra-Andean foreland in the region north of the Strait of Magellan and throughout Patagonia, followed. Progressive uplift of the continental block, particularly at its Atlantic side, has taken place during the Quaternary (Feruglio, 1950 *in* Katz, 1973).

The Aisén Segment. The Aisén segment represents a transitional geotectonic domain between the Magallanes segment and the Northern Sector. Extending from 49° to 41°S approximately, it can be divided into three subsegments with different geological characteristics which, in terms of latitude, are: (1) 49°–47°S; (2) 47°–43°S; (3) 43°–41°S. They are described below from south to north and this account largely relies on recent publications such as Skarmeta and Charrier, (1976); Skarmeta (1976, 1978); González–Bonorino, (1979); Haller and Lapido, (1982) and Ramos *et al.* (1982).

Subsegment 49°–47°S. In contrast with the immediately adjacent Magallanes segment, no lower Cretaceous volcanic rocks are known in the Patagonian Cordillera here. Magmatism is represented by dikes and sills of a basic alkaline nature, constituting a series of complexes located in the lower half of a sedimentary unit of black shales of Neocomian age (Ramos *et al.*, 1982). According to these authors these complexes extend for approximately 200 km following the position of the former shale basin and they decrease in number from south to north.

Volcanism of Upper Cretaceous age is sparsely represented, but one example occurs in the Lake San Martín area, in Argentina, where flows of a calc-alkaline nature unconformably overlie folded Neocomian black shales. K-Ar dates from 76.7 ± 5 Ma to 84 ± 5 Ma have been obtained from the flows (Riccardi, 1971). On this evidence it is possible to infer the occurrence of a folding episode between the Neocomian and pre-Senonian time, an event that probably reflects the middle Cretaceous orogenic phase in the eugeosynclinal belt of the Magallanes segment further south.

Volcanism in the Cordillera reappears in the Upper Cenozoic (Recent). A remarkable discontinuity exists at 47°S; between this latitude and 49°S a spatial gap in the volcanicity is apparent with wide scattering of the volcanic centers from 49°S to the south. This break is also chemically expressed; north to 47°S high-alumina olivine basalts similar to those in the southern province of the Northern Sector (p. 326) are present but are replaced to the south by olivine-free types in which hornblende accompanies pyroxene as in Mount Burney (Stern *et al.*, 1976).

In the extra-Andean domain of this subsegment, deposition of plateau basalts (p. 351) was widespread.

Subsegment 47°–43°S. Volcanism was intense in this area during the Cretaceous. Lower Cretaceous strata are found along two main belts on

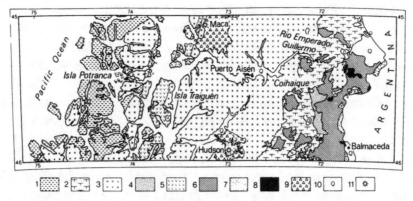

Fig. 17. Geologic map of the Aisén Segment, Patagonian Sector, between 45° and 46°S (based on *Mapa Geológico de Chile*, 1:1,000,000, 1982, and Skarmeta, 1976). Key: 1, pre-Andean basement; 2, Serie Tobífera; 3, Patagonian Batholith; 4, Coihaique Formation; 5, basic and intermediate volcanic rocks of the Coihaique Formation flanking the Patagonian Batholith; 6, Divisadero Formation; 7, continental sedimentary rocks, Ñirehuao and Galera Formations; 8, Patagonian plateau basalts; 9, Quaternary volcanics; 10, Quaternary sediments; 11, active volcanoes.

both sides of the Patagonian Batholith; an eastern belt in the continental region of Aisén and a western belt along the fjord area (Fig. 17). The rocks, grouped in the Coihaique Formation (Lahsen, 1966), correspond to marine shales and fossiliferous sandstones that interfinger with basaltic, andesitic and rhyolitic lavas in the proximity to both flanks of the batholith (Skarmeta, 1976). The Coihaique Formation, with a maximum thickness of 400 m (Skarmeta and Charrier, 1976), concordantly overlies the "Tobífera" (Ibáñez Formation) and gradually passes upward into the Divisadero Formation. Paleontological evidence indicates a Valanginian–Barremian age for the unit suggesting that the post-"Tobífera" transgression was later at this latitude as compared with the Magallanes segment. Chemical analyses of basic and intermediate lavas from the Coihaique Formation are presented in Table II, analyses 10 to 13 and a calc-alkaline nature is apparent.

The Divisadero Formation (Heim, 1940) a *ca.* 650 m thick sequence of subaerial tuffaceous sandstones, rhyolitic and dacitic tuffs that toward the west becomes enriched in porphyric andesitic components, concordantly overlies the Coihaique Formation (Skarmeta, 1978). Exposures of this unit are confined to the eastern flank of the Patagonian Batholith. A late Hauterivian to Cenomanian age is assigned to the Divisadero Formation by Skarmeta (1978) while Haller and Lapido (1982) consider it as Aptian. A minimum Santonian age is indicated by the fact that the Formation is intruded by 80 Ma old granitoids (Skarmeta, 1978). The Divisadero Formation marks a definite regression of the sea in this latitude during the middle Cretaceous followed by a pronounced depositional hiatus.

During the Paleogene a sequence of andesitic flows, volcanic breccias and plant-bearing shales known as the Ñirehuao Formation (Espinoza and Fuenzalida, 1971) or Chile Chico Formation (Niemeyer, 1975) was deposited between 45° and 46°S and south of Lake General Carrera (*ca.* 46°45'S) along the boundary with Argentina. This unit, 350 m thick in average, would appear to unconformably overlie the Divisadero Formation, a stratigraphic relation mainly based on contrasting folding styles, since their actual contact, with the exception of the Lake General Carrera area (Niemeyer, 1975 *in* Skarmeta, 1978), has not been observed. In the boundary region *ca.* 45°25'S the Ñirehuao Formation is concordantly covered by the Punta del Monte Formation, a 150 m thick pile of andesitic flows of supposed Oligocene age. In nearby areas, e.g. 45°20'S, the Ñirehuao Formation is unconformably overlain by the Galera Formation, a sequence of conglomerates, sandstones, shales and tuffites up to 600 m thick to which a Miocene–Lower Pliocene age has been assigned on the basis of paleontological correlation with units in Argentina (Skarmeta and Charrier, 1976). Upper Cenozoic plateau basalts (p. 351) crop out near the boundary and extends toward the east.

According to Skarmeta and Charrier (1976) the geological record of the region (45°–46°S) as presented above evidences the existence of at least three compressive (folding) phases: (f_1) folding of the Coihaique and Divisadero formations in the middle to late Cretaceous; (f_2) folding of the Ñirehuao Formation in the early Tertiary, and (f_3) folding of the Galera Formation in the late Miocene–early Pliocene.

Subsegment 43°–41°S. The geological evolution of this subsegment strongly contrasts with those to the south. González-Bonorino (1979) has studied the boundary region between Argentina and Chile where the oldest Andean strata correspond to Liassic shallow marine clastic rocks unconformably deposited on metamorphic pre-Andean basement. No record is known from the Dogger to the Paleocene, probably due to nondeposition. According to González-Bonorino (1979) the scanty Mesozoic record is a characteristic feature of the North Patagonian Cordillera suggesting that the region constituted an emergent area at the time.

During the Paleogene volcanic and marine sedimentary rocks, the Nahuel Huapi Group, were unconformably deposited on pre-Andean basement and on Liassic strata. Fossils found in rocks of this group only permit to assign it an early to middle Tertiary age while radiometric dates between 60 and 42 Ma from volcanic rocks (González-Bonorino, 1979) would indicate an early Paleocene to late Eocene age. Unconformably upon the Nahuel Huapi Group rocks of the Miocene volcano-sedimentary Collón Curá Formation are found (González-Bonorino, 1979). Volcanism continued during the Plio–Pleistocene to the present.

(c) The Patagonian Batholith. Granitoids belonging to the Patagonian Batholith are aligned in a narrow belt 1500 km long and 40–100 km wide forming the axis of the Patagonian Andes from 39° to 56°S (Fig. 4). Rock types range from gabbro to granite with tonalite and granodiorite being the most abundant. Their chemical trend is calc-alkaline. In relation to their SiO_2 content, a negative correlation for FeO*, MgO, CaO and Sr and a positive one for K_2O, Rb, Th, Ce and Y has been reported (Wells, 1979 *in* Ramos *et al.*, 1982).

The age range for the rocks of the Patagonian Batholith extends from Jurassic to Tertiary with oldest dates about 155 Ma and youngest ones *ca.* 10 Ma (Fig. 4). The oldest group varies from 155 to 141 Ma (Halpern, 1973; Hervé *et al.*, 1979) and partly coincides with the beginning of middle Jurassic volcanic silicic activity in the Patagonian Sector. According to Halpern (1973) three igneous episodes can be distinguished south of 50°S: (*1*) 150–120 Ma (late Jurassic–early Cretaceous); (*2*) 100–75 Ma (late Cretaceous); (*3*) 50–10 Ma (middle to late Tertiary). In granitoids exposed south of the Beagle Channel five determinations (four Rb-Sr in hornblende and one K-Ar in biotite) give values from 88 ± 5 Ma to 77 ± 5 Ma representing the interval late Turonian–early Campanian (Halpern and Rex, 1972; Halpern, 1973).

According to Ramos *et al.* (1982) ages of granitoids exposed between 45° and 52°S can be grouped as follows: (*1*) 125–100 Ma (late Neocomian–middle Cretaceous); (*2*) 85–70 Ma (late Cretaceous); (*3*) 50–10 Ma (Eocene–Miocene) and, (*4*) 5–3 Ma (Pliocene), probably related to thermal events coinciding with the late plateau magmatism in extra-Andean Patagonia. The same authors point out that, although the magmatism persisted during most of the Cretaceous, a maximum activity can be located at 98 ± 4 Ma for the whole Patagonian Batholith (Ramos and Ramos, 1978).

Most authors conclude that there is no evidence of the plutonic bodies becoming younger or older toward the east.

Few Sr(i) ratios are known for Patagonian granitoids; values of 0.7049 and 0.7070 have been reported by Halpern (1973) for rocks of Jurassic–Cretaceous and Tertiary age respectively and, more recently, Hervé *et al.* (1979) have determined a value of 0.7092 ± 0.001 for rocks of 154 ± 5 Ma from Bahía Pia in the Cordillera Darwin.

(d) The Patagonian Plateau Basalts. The Patagonian plateau basalts cover an approximate surface of 120,000 km^2 in the extra-Andean region between 40° and 52°S and unconformably overlie Mesozoic and Cenozoic volcanic and sedimentary strata. In some areas, as in Lake Buenos Aires, *ca.* 46°30′S near the boundary with Chile, they form a pile approximately 700 m thick (Baker *et al.*, 1981).

The lavas are alkaline with average of 47% SiO_2 and 5% alkali (Ramos *et al.*, 1982). Mafic and ultramafic nodules are present (Niemeyer, 1978;

Skewes and Stern, 1979). Age determinations have shown that these basalts are products of longstanding volcanic activity extending from late Cretaceous to latest Cenozoic with oldest ages about 81–77 Ma (see Baker *et al.*, 1981; Charrier *et al.*, 1979). Four main age and compositional groups are distinguished (Baker *et al.*, 1981): (*1*) *ca.* 80 Ma, tholeiitic, showing some calc-alkaline affinities and resembling basalts from marginal basins. One Sr(i) ratio of 0.70443 (Hawkesworth *et al.*, 1979) is known for lavas of this group (Table II, analysis 14); (*2*) 57–43 Ma ranging from olivine tholeiites through alkali basalts to basanites (Table II, analyses 15 to 18); (*3*) 25–9 Ma with dominantly alkali basalt composition (Table II, analysis 19); (*4*) 4–0.2 Ma, postplateau flows from small cinder cones on the surface of the plateau range, corresponding to undersaturated basanites and minor leucite basanites (Table II, analyses 20 and 21).

2. *Geotectonic Interpretation of Mesozoic–Cenozoic Evolution of the Patagonian Sector*

The *early stage* in the evolution of the Patagonian Sector, initially characterized by predominantly silicic volcanism, followed the Gondwanide Orogeny and persisted until latest Jurassic time. The volcanic activity was principally related to a process of regional updoming, extension and rifting of the continental crust (Fig. 18A) and its products have been attributed to crustal anatexis induced by a rising mantle diapir below a broad tectonic rift zone (Bruhn *et al.*, 1978; Suárez, 1979). The same anatectic processes would account for the generation of contemporaneous granitoids with high Sr(i) ratios as pointed out by Hervé *et al.* (1979).

During its *mature stage*, from late Jurassic to middle Cretaceous (Fig. 18, B1 and B2), the Patagonian edifice is typified by the presence, from west to east, of: (*1*) a volcanic island arc that is part of a broader tectono-magmatic belt including granitoidal plutonism; (*2*) a marginal basin in different stages of evolution and; (*3*) a sedimentary depositional basin (external zone).

In the Magallanes segment, where the eu/mio geosynclinal couple is best defined at this stage, the presence of an island arc, partly contemporaneous with the latest manifestations of the silicic "Tobífera" and active until middle Cretaceous, seems well established (Dalziel *et al.*, 1974; Suárez and Pettigrew, 1976; Suárez, 1979). According to Suárez (1979) the arc and its calc-alkaline volcanic products can be attributed to plate subduction. The continuity of this island arc is interrupted between 49° and 47°S, in the Aisén segment, only to reappear further north (46°–45°S) where calc-alkaline volcanics, which Skarmeta (1976) considers as genetically linked with the granitoids from the Patagonian Batholith, interfinger with the sedimentary Coihaique Formation (Fig. 18C).

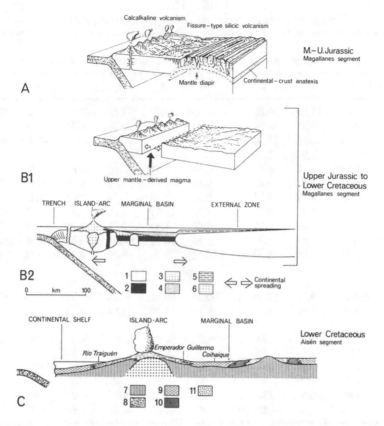

Fig. 18. Evolution of the Patagonian Sector during the Jurassic–Lower Cretaceous (based on interpretations by Skarmeta, 1976; Suárez and Pettigrew, 1976; and Suárez, 1979).

An extensional period, a further expression of mantle diapirism (Bruhn *et al.*, 1978), developed during late Jurassic–early Cretaceous time and originated a marginal basin floored by oceanic crust. Crustal spreading, illustrated by the presence of a 100% sheeted dike unit, resulted in the splitting apart of the Jurassic tectono-magmatic belt leaving an active island arc flanking the basin to the west and a remnant, inactive volcanic terrane, along its eastern border (Suárez, 1979)(Fig. 18, B1 and B2). The ophiolitic suite present is incomplete since no ultramafic units are exposed. Conditions of generation of the basaltic magmas in the suite seem to have varied along the extent of the belt as indicated by an increase in width of the Ophiolitic Complex from north to south and by concomitant geochemical changes in the basalts. These chemical differences have been explained as reflecting compositional differences in the mantle source region (Stern, 1980). They could also be determined by the interplay of parameters such as rate of upwelling of mantle material and its volume, and thickness of the continental

TABLE II
Patagonian Sector: Volcanic Rocks, Selected Analyses

	Middle to Upper Jurassic			Upper Jurassic–Lower Cretaceous					
	1	2	3	4	5	6	7	8	9
SiO_2	77.40	71.50	56.60	52.99	50.97	75.35	49.68	50.00	50.19
TiO_2	0.13	0.21	0.84	0.87	1.04	0.24	1.33	1.26	1.46
Al_2O_3	12.13	14.80	16.90	16.16	16.67	11.21	15.45	16.55	16.17
Fe_2O_3		1.97	2.51						
FeO	1.37^a	1.32	4.62	8.50^a	8.51^a	2.86^a	10.82^a	10.90^a	10.47^a
MnO	0.03	n.d.	n.d.	0.15	0.15	0.02	0.16	0.14	0.18
MgO	1.12	0.88	3.23	8.96	7.33	0.24	9.63	8.22	7.05
CaO	0.67	1.97	5.35	5.77	12.99	7.18	10.51	10.57	10.94
Na_2O	2.69	3.60	5.29	2.89	1.49	1.84	1.76	2.09	2.44
K_2O	3.53	3.82	1.14	3.65	0.01	0.01	0.03	0.56	0.71
P_2O_5	0.03	0.05	0.30	0.11	0.25	0.05	0.15	0.16	0.23
H_2O									
CO_2									
TOTAL	99.10	100.12	96.78	100.05	99.41	99.00	99.52	100.45	99.84
$FeO^a/$ MgO	1.22	3.52	2.13	0.95	1.16	11.91	1.12	1.33	1.49
Ba	872	617	190	n.d.	n.d.	n.d.	n.d.	n.d.	n.d.
Rb	117	195	44	95	<1	<1	1	14	22
Sr	126	247	623	157	245	245	40	116	119
Y	n.d.	39	32	24	23	102	30	27	34
Zr	159	205	154	54	75	481	85	65	92
Nb	12	15	12	n.d.	n.d.	n.d.	n.d.	n.d.	n.d.

[a] Total iron as FeO
[b] Total iron as Fe_2O_3

KEY: *1*, Silicic volcanics (average of 13 rocks). Serie Tobífera, Magallanes segment. Bruhn *et al.* (1978). *2*, Rhyolite (average of 3 rocks). Serie Tobífera (Ibáñez Formation), Aisén segment. Baker *et al.* (1981). *3*, Andesite (average of 3 rocks). Serie Tobífera (Ibáñez Formation), Aisén segment. Baker *et al.* (1981). *4*, Basalt, vesicular pillow lava. Ophiolitic Complex, Sarmiento. Magallanes segment. Stern (1980, Sample FL70A). *5*, Basic dike, sheeted dike unit. Ophiolitic Complex, Sarmiento. Magallanes segment. Stern (1980, Sample PA28B). *6*, Silicic dike, sheeted dike unit. Ophiolitic Complex, Sarmiento. Magallanes segment. Stern (1980, Sample FL43E). *7*, Basic dikes cross cutting gabbro (average of 2 rocks). Ophiolitic Complex, Tortuga. Magallanes segment. Stern (1980, Samples P2 and NB 86-1). *8*, Basalt, pillow lava (average of 2 rocks). Ophiolitic Complex, Tortuga. Magallanes segment. Stern (1980, Samples H60A, and NB93-2). *9*, Basic dikes cross cutting pillow lavas (average of 2 rocks). Ophiolitic Complex, Tortuga. Magallanes segment. Stern (1980, Samples NB93-1 and NB24-1). *10*, Basalt. Coihaique Formation. Rio Emperador Guillermo. Aisén segment. Skarmeta (1976, Sample 1). *11*, Andesite. Coihaique Formation. Rio Emperador Guillermo. Aisén segment. Skarmeta (1976, Sample 2). *12*, Basalt, Coihaique Formation. Isla Traiguén. Aisén segment. Skarmeta (1976, Sample 1). *13*, Basalt. Coihaique Formation. Isla Traiguén. Aisén segment. Skarmeta (1976, Sample 2). *14*, Quartz tholeiite (average of 5 rocks). Patagonian Plateau Basalts. Baker *et al.* (1981, Samples P117, P119, P120, P121 and P124). *15*, Olivine tholeiite (average of 3 rocks). Patagonian plateau basalts. Baker *et al.* (1981, Samples P17, P18 and P20). *16*, Olivine tholeiite (average of 3 rocks). Patagonian plateau basalts. Baker *et al.* (1981, Samples P68, P69 and P71). *17*, Alkali basalt (average of 3 rocks). Patagonian plateau basalts. Baker *et al.* (1981, Samples P40, P66 and P70). *18*, Basanite. Patagonian plateau basalts. Baker *et al.* (1981, Sample P67). *19*, Alkali basalt (average of 4 rocks). Patagonian plateau basalts. Baker *et al.* (1981, Samples P110, P111, P127, and P131). *20*, Basanite (average of 2 rocks). Patagonian plateau basalts. Baker *et al.* (1981, Samples P115 and P133). *21*, Leucite basanite (average of 2 rocks). Patagonian plateau basalts. Baker *et al.* (1981, Samples P88 and P90).

The Southern Andes 355

Lower Cretaceous				Upper Cretaceous		Eocene			Miocene	Pliocene–Pleistocene	
10	11	12	13	14	15	16	17	18	19	20	21
52.37	54.07	50.09	47.64	51.89	50.65	48.32	45.13	44.35	48.28	46.13	42.77
1.05	0.81	1.56	1.80	1.69	1.50	2.45	2.93	3.11	2.38	2.74	2.90
17.66	15.44	17.85	17.66	16.92	17.64	16.00	14.96	14.98	15.99	15.32	13.94
2.62	2.64	1.65	0.82	9.89^b	9.23^b	12.01^b	12.95^b	12.71^b	11.66^b	11.55^b	11.65^b
6.84	5.58	8.55	8.46								
0.17	0.17	0.23	0.45	0.18	0.15	0.17	0.19	0.19	0.15	0.16	0.18
4.56	6.10	4.16	7.94	4.35	5.33	7.15	8.94	9.43	6.97	8.60	10.81
4.40	7.12	6.84	6.64	7.57	9.26	9.11	9.37	10.30	8.85	9.18	10.38
3.69	3.60	3.84	2.53	3.90	3.84	3.20	3.66	3.86	4.03	4.21	3.81
2.00	1.30	1.60	1.74	1.50	0.68	0.95	0.83	1.13	1.46	1.40	2.44
0.28	0.22	0.20	0.22	0.92	0.44	0.52	0.70	0.90	0.53	0.76	1.13
4.02	2.62	3.30	3.78								
0.04		0.22	0.05								
99.70	99.67	100.09	99.73	98.81	98.72	99.88	99.66	100.96	100.30	100.05	100.01
2.02	1.30	2.41	1.16	2.05	1.56	1.51	1.30	1.21	1.51	1.21	0.97
				596	184	216	346	470	318	572	774
				28	14	13	13	23	23	38	40
				545	424	465	432	881	591	867	1116
				44	30	25	21	26	22	20	28
				351	200	164	214	248	165	245	316
				19	8	23	50	74	31	60	80

crust among others. Thus, the greater width of the Tortuga Complex in the south may reflect a more permissive emplacement régime that facilitated upwelling of mantle material, probably in greater volumes than in the north (Sarmiento Complex). The primitive character of the basalts, suggesting ocean ridge affinities, would further support such an interpretation (see Fig. 16). Hydrothermal, ocean floor metamorphism, characterized by high thermal gradients and lack of deformation affected the ophiolitic rocks during their evolution.

Along the length of the Patagonian Sector, the early Cretaceous marginal basin evolved from a "proper" type to an aborted one. The quasi-ophiolitic complex of the Magallanes segment disappears in the southernmost part of the Aisén segment (49°–47°S) together with its western island arc; its last traces being probably represented by complexes of alkaline dikes and sills emplaced in Neocomian shales (p. 348). North of 46°S the marginal basin is definitely aborted in nature; no oceanic components are found and its basic volcanics are chemically evolved with a calc-alkaline trend (Bartholomew and Tarney, 1984). Subaerially deposited volcanics of andesitic and silicic composition (Divisadero Formation) precede the middle Cretaceous folding and their composition may be interpreted as the outcome of extreme sluggishness in the ascent of mantle material with consequent widespread contamination and partial anatexis. Slowing of the upwelling rate of magma would suggest the approaching compressional regime. In this respect, the Aisén segment of the Patagonian Sector represents a distinctive transition toward the marginal basins of the Northern Sector.

The Magallanes sedimentary depositional basin was formed by downwarping during the late Jurassic (Natland *et al.*, 1974) and received continuous marine deposition from the Oxfordian to the Pliocene. Contemporaneously with the Jurassic–early Cretaceous volcanic activity to the west, sandstones and black calcareous sandstones were deposited; no deformational effects related to the middle Cretaceous orogeny are observed in these deposits. However, the uplift and subsequent destruction of the eugeosynclinal belt is reflected here by the accumulation, during the Cenomanian, of flysch and the presence of abundant volcanic, plutonic and metamorphic debris.

The middle Cretaceous orogenic phase terminated the eugeosynclinal evolution in the Magallanes segment and brought about the destruction of the marginal basin; volcanism in this internal belt only reappeared in the late Tertiary. A significant amount of crustal shortening related to this deformational event is evidenced by the folding style of the sedimentary units (Yahgan Formation) and by the presence of regional penetrative metamorphism reaching the high amphibolite facies. The middle Cretaceous orogeny closely corresponds in time with the Subhercynian phase of the Northern Sector.

Paleogene molasse deposits in the Magallanes segment are currently interpreted as a result of a possible Laramian orogenic phase that affected the westernmost border of the miogeosynclinal basin. The presence of a late Miocene folding phase is evidenced by an angular unconformity; this event could be related in time with the Quechuan Orogeny of the Northern Sector.

In the Aisén segment orogenic phases during the middle Cretaceous, the early Tertiary and the late Miocene–early Pliocene have been identified by Skarmeta and Charrier (1976) who correlate them with the Subhercynian, Incaic and Quechuan phases of the Northern Sector respectively. The absence of Upper Cretaceous deposits in the area of Aisén studied by those authors precludes the direct identification of a Laramian phase.

Apart from the Patagonian plateau basalts, post-middle Cretaceous volcanism in the Aisén segment took place mainly during the Paleogene. The activity, under an extensional régime, occurred along a belt lying to the east of the earlier eugeosyncline (Skarmeta and Charrier, 1976).

The Patagonian Batholith stretches along the Magallanes and Aisén segments and comprises rocks that range in age from at least late Jurassic to Neogene. The granitoids are mainly post-tectonic but the number of ages determined is insufficient to establish a clear relationship in time between orogenic phases and plutonic episodes. An origin related to plate subduction processes is accepted for most of the granitoids although crustal anatexis has also been invoked to explain the genesis of some suites. Consanguinity of granitoidal phases and volcanic products has been suggested on the bases of chemical composition and isotopic relations. A distinctive feature of the Patagonian Batholith consists in the spatial permanence of the plutonic foci for a span covering *ca.* 150 million years; only the youngest intrusives of the Miocene Paine Pluton (12 Ma) differ from the main intrusive cordillera in having been emplaced approximately 70 km to the east in the sub-Andean belt.

C. The Northern and Patagonian Sectors: Contrasts and Analogies

Contrasting features in the evolution of these two sectors were pointed out in early studies of Chilean geology. The distinction between an Andean Geosyncline and a Magellan Geosyncline is well established and has been further elaborated by the introduction of concepts such as geoliminal and Alpine geosynclinal belts (Aubouin and co-workers). The actual boundary between both domains and the characteristics of their coupling remains a major problem. However, recent studies have revealed the existence of a transitional terrane, here named the Aisén segment, which would represent a bridge between the sectors.

The contrasts and analogies discussed are based on a comparison between the Northern Sector and the Magallanes segment of the Patagonian Sector. Transitional features inherent to the Aisén segment are mentioned at the end.

Fundamental differences between the two sectors refer to their time of inception and to the structure of their edifices. The oldest activity in the Patagonian domain is dated as middle Jurassic while a Triassic–early Jurassic age is assigned to the inaugural events in the Northern Sector. The cyclic building up of stratigraphic-structural units separated by unconformities, which characterizes the development of the Northern Sector until the Neogene is not identifiable in the Patagonian Sector with the exception of the Jurassic–middle Cretaceous interval.

Albeit at different times in each sector (Triassic and locally reaching to the middle Jurassic in the north, middle to late Jurassic in the south), widespread silicic volcanism typifies as *early stage* of updoming followed by rifting and extension which is attributed to mantle diapirism. Island arcs were established on the continent along the Pacific border in both cases, probably in response to the presence of a subducting plate. Immediately east, toward the continent, progress in the ascent of mantle diapirs enhanced the tensional régime and resulted in crustal thinning, spreading and subsidence in both domains. From this stage onwards (Hauterivian in the Northern Sector, latest Jurassic in the Magallanes segment) a *mature phase* takes place at both latitudes comprising the generation of ensialic marginal basins of contrasting evolution. In Magallanes rupture of the continental crust and generation of oceanic floor took place leading to the formation of a marginal basin "proper". In the Northern Sector mantle-derived basalts were extruded in large volumes but no oceanic crust was formed; an aborted marginal basin resulted. Differences in the geochemical features of the magmatic products from both types of basins suggest that in the north the upwelling rates of mantle material were considerably slower than those in Magallanes. Differences in behavior of the associated island arcs also exist; island arc activity contemporaneous with basin development is recorded in Magallanes but is absent in the Northern Sector.

Metamorphism affecting the products generated and deposited in the respective marginal basins is of low-grade burial type with low thermal gradient in the Northern Sector and of hydrothermal ocean floor type with steep thermal gradient in Magallanes.

A middle Cretaceous orogenic phase (Subhercynian) is recorded in both domains with remarkably different characteristics. In the Northern Sector it affected the "eugeosynclinal" and "miogeosynclinal" belts alike with no evidence of crustal shortening; a definite regression of the sea occurred. In the Magallanes segment its effects are restricted to the eugeosynclinal domain where strong crustal shortening is recorded in the sedimentary infilling of the marginal basin accompanied by thrusting of the internal zone over the external zone in places. Regression of the sea from the eugeosynclinal belt occurred but marine conditions persisted in the external zone where flysch sequences reflect the orogenic disturbance experienced by its adjacent western flank.

Penetrative regional metamorphism associated with the Andean orogeneses is unknown in the Northern Sector. In Magallanes the middle Cretaceous event is responsible for widespread deformational metamorphism reaching the amphibolite facies locally and affecting reactivated pre-Andean basement and Andean eugeosynclinal cover alike. Postkinematic granitoid emplacement culminating between 80–100 Ma BP is recorded in both sectors.

Following the middle Cretaceous orogenic phase the trends in the development of both domains diverge further. In the Northern Sector the process of aborted marginal basin generation, which continued under continental conditions during the Upper Cretaceous–Paleocene, experienced an hiatus lasting until the late Oligocene in many areas. Thereafter it resumed from the Neogene to the Present. In the Magallanes segment the eugeosynclinal activity faded out and the evolution is mainly recorded in the continuous sedimentary sequences accumulated in the external domain, the Magallanes Basin.

The sequence of orogenic phases present in the Northern Sector from the late Jurassic until the close of the Tertiary is also recorded in the Magallanes segment with the possible exceptions of the Araucanian and Incaic phases (this last present in the transitional Aisén segment). Although broadly coincident in time, their characteristics and effects show important contrasts. While in the north postorogenic sedimentation is commonly represented by continental red beds and, at certain times by molasse *s.l.* (p. 308, p. 309), in the Magallanes Basin the same processes gave rise to typical sequences of flysch and molasse *s.s.* Also, an orogenic polarity is well displayed in the Magallanes segment with the Subhercynian, Laramian and Quechuan phases progressively moving toward the Atlantic. This feature is only poorly reflected in the Northern Sector. On the other hand, polarity of granitoid emplacement as observed in this last sector, with ages younging toward the east from the Triassic to the late Tertiary is not found in the Patagonian Batholith. The alkaline nature of the Patagonian plateau basalts together with their volume and persistence in time is a unique feature of the Patagonian Sector.

Of all the characteristics outlined above, the most significant is considered to be the difference in the type of marginal basin generated in each sector. Any explanation for this has to be sought among the fundamental mechanisms governing the interaction between crust and mantle and risks entry into the realm of mere speculation. However, the author believes that the combined action of plate subduction and spreading-subsidence within the continental crust along the western edge of the present Chilean territory since the Mesozoic can account for such major difference. Spreading-subsidence is seen as the direct outcome of mantle diapirism related to hot spots or hot lines or resulting from secondary flow. Plate subduction, in turn, will depend on the global plate configuration at a given time and will be commonly reflected in segmentation as changes in parameters such as spreading rate, dip, thickness and age of the subducting slab, and others take place. In the Northern Sector an alternation of continental crust spreading and variation in Pacific spreading rates permitted to maintain a "rebound" interaction leading to the cyclic construction of the Andean edifice. A thick continental crust and a slow upwelling rate of mantle material in the continental border prevented the exposure of oceanic floor. Contrasting conditions, probably represented by thin continental crust, slow rates of subduction and large volume of rapidly ascending mantle diapirs led to splitting and oceanic crust generation in the Magallanes segment.

The passage from a Magallanes to a Northern Sector type basin is made at the connection between the Aisén and the Magallanes segment. Here all effusive Cretaceous magmatism disappears and only the presence of scattered dikes and sills reminds us of the Ophiolitic Complex southward or of the flood basalts of Cretaceous age in the north. Everything happens as if a "stitch" had been applied there to a continental crust in the process of splitting. The close proximity of this transitional zone to the present location of the intersection of the Chile Rise and the continent is indeed remarkable, and can scarcely be fortuitous.

ACKNOWLEDGMENTS

The author is grateful to R. Charrier, F. Hervé and J. Skarmeta for constructive suggestions and criticism of parts of the manuscript. B. Levi read and commented on a complete version and contributed illuminating ideas. F. Hervé kindly allowed me to include personal communications on unpublished matters concerning the pre-Andean basement. J. Lynch patiently and skillfully prepared the illustrations.

A. E. M. Nairn and F. G. Stehli, editors, are thanked for their careful correction of the text. Research grants from the University of Liverpool and the Guggenheim Foundation helped to materialize this work.

REFERENCES

Åberg, G., Aguirre, L., Levi, B., and Nyström, J. O., 1984, Spreading subsidence and generation of ensialic marginal basins: an example from the early Cretaceous of central Chile, in: *Marginal Basin Geology: Volcanic and Associated Sedimentary and Tectonic Processes in Modern and Ancient Marginal Basins,* Kokelaar, B. P. and Howells, M. F., eds., *Geol. Soc. London. Spec. Publ.* 16 p. 185–193.

Aguirre, L., 1960, Geología de los Andes de Chile Central (Provincia de Aconcagua), *Inst. Invest. Geol. Bol.,* v. 9, p. 1–68.

Aguirre, L., 1965, Basamento Cristalino Precámbrico, in: *Geología y Yacimientos Metalíferos de Chile,* Ruiz, C. ed., Santiago: Instituto de Investigaciones Geologicas, p. 6–18.

Aguirre, L., 1967, Geología de las islas Choros y Damas y de Punta Choros, Provincia de Coquimbo, *Rev. Min.,* v. 96/97 p. 73–83

Aguirre, L., 1976, Essay review: structural evolution of the northernmost Andes, Colombia, by E. M. Irving, *Geol. Mag..* v. 113 p. 475–485.

Aguirre, L., 1983, Granitoids in Chile, in: *Circum-Pacific Plutonic Terranes, Geol. Soc. Am. Mem. 159* (Roddick, J. A. ed.), Boulder: Geological Society of America, p. 293–316.

Aguirre, L., and Egert, E., 1962, Las formaciones manganesíferas de la región de Quebrada Marquesa, Provincia de Coquimbo, *Rev. Min.*, v. 76, p. 25–37.

Aguirre, L., and Egert, E., 1965, Cuadrángulo Quebrada Marquesa, Provincia de Coquimbo, *Carta Geol Chile Inst. Invest. Geol.*, v. 15, p. 1–92.

Aguirre, L., and Levi, B., 1964, Geología de la Cordillera de los Andes de las provincias de Cautín, Valdivia, Osorno y Llanquihue, *Inst. Invest. Geol. Bol.*, v. 17, p. 1–37.

Aguirre, L., and Levi, B., 1977, Relation between metamorphism, plutonism and geotectonics in the Paleozoic and Mesozoic edifices of the northern segment of the meridional Andes (extended abstract) in: *Plutonism in Relation to Volcanism and Metamorphism* 7th CPPP-IGCP, Toyama, Japan, p. 75–76.

Aguirre, L., and Mehech, S., 1964, Stratigraphy and mineralogy of the manganese sedimentary deposits of the Coquimbo province, Chile, *Econ. Geol.* v. 59, p. 428–442.

Aguirre, L., and Thomas, H., 1964, El contacto discordante entre las formaciones cretácicas Quebrada Marquesa y Viñita en la Provincia de Coquimbo, *Rev. Min.*, v. 84, p. 30–37.

Aguirre, L., Hervé, F., and Godoy, E., 1972, Distribution of metamorphic facies in Chile, an outline, *Krystalinikum* v. 9, p. 7–19.

Aguirre, L., Charrier, R., Davidson, J., Mpodozis, A., Rivano, S., Thiele, R., Tidy, E., Vergara, M., and Vicente, J.-C., 1974, Andean magmatism: its paleogeographical and structural setting in the central part (30°–35°S) of the Southern Andes, *Pac. Geol.* v. 8, p. 1–38.

Aguirre, L., Levi, B., and Offler, R., 1978, Unconformities as mineralogical breaks in the burial metamorphism of the Andes, *Contrib. Mineral. Petrol.*, v. 66, p. 361–366.

Alarcón, B., and Vergara, M., 1964, Nuevos antecedentes sobre la geología de la quebrada El Way, *Dept. Geol. Univ. Chile Publ.* v. 26, p. 1–28.

Aliste, N., Pérez, E., and Carter, W. D., 1960, Definición y edad de la Formación Patagua, Provincia de Aconcagua, Chile, *Rev. Min.* v. 71, p. 40–50.

Askin, R. A., Charrier, R., Hervé, F., Thiele, R., and Frutos, J., 1981, Palynological investigations of Paleozoic and lower Mesozoic sedimentary rocks of central Chile, *Dept. Geol. Univ. Chile Communicaciones* v. 32, p. 10–25.

Aubouin, J., and Borrello, A. V., 1966, Chaînes andines et chaînes alpines: regard sur la géologie de la Cordillère des Andes au parallèle de l'Argentine moyenne, *Bull. Soc. Geol. Fr.* (7), v. 8, p. 1050–1070.

Aubouin, J., Borrello, A. V., Cecioni, G., Charrier, R., Chotin, P., Frutos, J., Thiele, R., and Vicente, J-C., 1973, Esquisse paléogéographique et structurale des Andes Méridionales, *Rev. Geogr. Phys. Geol. Dyn.*, v. 15, p. 11–72.

Baker, M. C. W., 1977, Geochronology of upper Tertiary volcanic activity in the Andes of north Chile, *Geol. Rundsch.* v. 66, p. 455–465.

Baker, M. C. W. 1981, The nature and distribution of upper Cenozoic ignimbrite centres in the Central Andes, *J. Volcanol. Geotherm. Res.* v. 11, p. 293–315.

Baker, M. C. W., and Francis, P. W., 1978, Upper Cenozoic volcanism in the Central Andes—ages and volumes, *Earth Planet. Sci. Lett.*, v. 41, p. 175–187.

Baker, P. E., Rea, W. J., Skarmeta, J., Caminos, R., and Rex, D. C., 1981, Igneous history of the Andean cordillera and Patagonian plateau around latitude 46°S, *Phil. Trans. R. Soc. London*, v. A303, p. 105–149.

Barazangi, M., and Isacks, B. L., 1976, Spatial distribution of earthquakes and subduction of the Nazca plate beneath South America, *Geology*, v. 4, p. 686–692.

Bartholomew, D., and Tarney, J., 1984, Ensialic marginal basin in the Southern Andes, in: *Marginal Basin Geology: Volcanic and Associated Sedimentary and Tectonic Processes in Modern and Ancient Marginal Basins*, Kokelaar, B. P., and Howells, M. F. eds. *Geol. Soc. London Spec. Publ.* 16 p. 195–205.

Bell, C. M., 1982, The lower Paleozoic metasedimentary basement of the Coastal Ranges of Chile between 25°30' and 27°S, *Rev. Geol. Chile*, v. 17, p. 21–29.

Ben-Avraham, Z., Nur, A., Jones, D., and Cox, A., 1981, Continental accretion: from oceanic plateaus to allochthonous terranes, *Science*, v. 213, p. 47–54.

Biese, W., 1942, La distribución del Cretáceo Inferior al sur de Copiapó, *An. 1er Congr. Panam. Ing. Minas. Geol.*, v. 2, p. 429–466.

Biese, W., 1953, Chile, *Zentralbl. Geol. Palaeontol.* v. 1, p. 555–563.

Biese, W., 1957, Der Jura von Cerritos Bayos—Calama, República de Chile, Provinz Antofagasta, *Geol. Jahrb.*, v. 72, p. 439–494.

Biro, L., 1964, *El límite Titoniano–Neocomiano de Lo Valdés*, thesis Universidad de Chile, Santiago.

Brüggen, J., 1950, *Fundamentos de la Geología de Chile*, Santiago: Instituto Geográfico Militar.

Bruhn, R. L., 1979, Rock structures formed during back-arc basin deformation in the Andes of Tierra del Fuego, *Geol. Soc. Am. Bull.* v. 90, p. 998–1012.

Bruhn, R. L., and Dalziel, I. W. D., 1977, Destruction of the early Cretaceous marginal basin in the Andes of Tierra del Fuego in: *Island Arcs, Deep Sea Trenches, and Back-Arc Basins* Talwani, M. and Pitman, W. C. eds., Washington D.C.: American Geophysical Union, p. 395–405.

Bruhn, R. L., Stern, C. R., and de Witt, M. J., 1978, Field and geochemical data bearing on the development of a Mesozoic volcano-tectonic rift zone and back-arc basin in southernmost South America, *Earth Planet. Sci. Lett.* v. 41, p. 32–46.

Caminos, R., Cordani, U., and Linares, E., 1979, Geología y geocronología de las rocas metamórficas y eruptivas de la Precordillera y Cordillera Frontal de Mendoza, Rep. Argentina, *Actas II Congr. Geol. Chileno*, v. 1, p. F43–F61.

Caminos, R., Cingolani, C. A., Hervé, F., and Linares, E., 1982, Geochronology of the pre-Andean metamorphism and magmatism in the Andean cordillera between latitudes 30° and 36° S, *Earth Sci Rev.* v. 18, p. 333–352.

Campano, P., Guerra, N., and Oyarzún, J., 1976, Contenido de Cu, Zn, Pb, Ni, y Co en rocas extrusivas, intrusivas y sedimentarias del norte de Chile, *Actas I Congr. Geol. Chileno*, v. II p. F43–F58.

Carter, W. D., 1963, Unconformity marking the Jurassic–Cretaceous boundary in the La Ligua area, Aconcagua Province, Chile, *U.S. Geol. Surv. Prof. Pap.* 450E, p. E61–E63.

Casamiquela, R., Corvalán, J., and Franqueza, F., 1969, Hallazgo de dinosaurios en el Cretácico Superior de Chile: su importancia cronológico-estratigráfica, *Inst. Invest. Geol. Bol.* v. 25, p. 1–31.

Casertano, L., 1963, General characteristics of the active Andean volcanoes and summary of their activities during recent centuries, *Bull. Seismol. Soc. Am.*, v. 53, p. 1415–1433.

Cecioni, A., 1979, El Tremadociano de Sotoca, I Región, norte de Chile, *Actas II Congr. Geol. Chileno* v. 3, p. H159–H164.

Cecioni, G., 1955, Noticias preliminares sobre el hallazgo del Paleozoico Superior en el Archipiélago Patagónico, *Univ. Chile, Fac. Cienc. Fís. Mat. An.*, v. 12, p. 258–259.

Cecioni, G., 1956, Primeras noticias sobre la existencia del Paleozoico Superior en el Archipiélago Patagónico entre los paralelos 50° y 52° S, *Univ. Chile, Fac. Cienc. Fís. Mat. An.* v. 13, p. 184–202.

Cecioni, G., 1957, Cretaceous flysch and molasse in Departamento Ultima Esperanza, Magallanes Province, Chile, *Am. Assoc. Pet. Geol. Bull.*, v. 41, p. 538–564.

Cecioni, G., 1962, La Formación Arrayán, Devónica, en la Provincia de Coquimbo, *Univ. Chile Bol.*, v. 34, p. 40–47.

Cecioni, G., 1974, Flysch devónico y orogénesis Bretónica en Chile, *Dept. Geol. Univ. Chile, Publ.*, v. 42, p. 19–21.

Cecioni, G., and García, F., 1960, Observaciones geológicas en la Cordillera de la Costa de Tarapacá, *Inst. Invest. Geol. Bol.*, v. 6, p. 1–28.

Cecioni, G., and Westermann, G., 1968, The Triassic/Jurassic marine transition of coastal central Chile, *Pac. Geol.* v. 1, p. 41–75.

Charrier, R., 1973, Interruptions of spreading and the compressive tectonic phases of the meridional Andes, *Earth Planet. Sci. Lett.* v. 20, p. 242–249.

Charrier, R., 1977, Geology of the region of Huentelauquén, Coquimbo Province, Chile, in: *Comparative studies of the Circum-Pacific Orogenic Belt in Japan and Chile* Ishikawa, T., and Aguirre, L., eds., Tokyo: Japan Society for the Promotion of Science, p. 81–94.

Charrier, R., 1979, El Triásico en Chile y regiones adyacentes de Argentina: Una reconstrucción paleogeográfica y paleoclimática, *Dept. Geol. Univ. Chile Comunicaciones*, v. 26, p. 1–37.

Charrier, R., 1981, Mesozoic and Cenozoic stratigraphy of the central Argentinian–Chilean Andes (32°–35° S) and chronology of their tectonic evolution, *Zentralbl. Geol. Palaeontol., Teil I*, v. H3/4, p. 344–355.

Charrier, R., and Lahsen, A., 1969, Stratigraphy of late Cretaceous–early Eocene, Seno Skyring–Strait of Magellan area, Magallanes province, Chile, *Am. Assoc. Pet. Geol. Bull.*, v. 53, p. 568–590.

Charrier, R., and Malumián, N., 1975, Orogénesis y epeirogénesis en la región austral de América del Sur durante el Mesozoico y el Cenozoico, *Rev. Asoc. Geolog. Argent.*, v. 20, p. 193–207.

Charrier, R., and Vicente, J-C., 1972, Liminary and geosynclinal Andes: major orogenic phases and synchronical evolutions of the central and Magellan sectors of the Argentine–Chilean Andes, *Solid Earth Probl. Conf. Upper Mantle Project* v. 2, p. 451–470. Buenos Aires.

Charrier, R., Linares, E., Niemeyer, H., and Skarmeta, J., 1979, K-Ar ages of basalt flows of the Meseta Buenos Aires in southern Chile and their relation to the southeast Pacific junction, *Geology*, v. 7, p. 436–439.

Chávez, L., and Nisterenko, G., 1974, Algunos aspectos de la geoquímica de las andesitas de Chile, *Dept. Geol. Univ. Chile Publ.*, v. 41, p. 97–127.

Chong, G., 1976, Las relaciones de los sistemas Jurásico y Cretácico en la zona preandina del norte de Chile, *Actas I Congr. Geol. Chileno* v. I p. A21–A42.

Chong, G., and Cecioni, A., 1976, Presencia de una secuencia marina de probable edad Paleozoica Superior en la Provincia de Antofagasta, Actas I Congr. Geol. Chileno. v. I p. A11–A20.

Chotin, P., 1976, Essai d'interprétation du Bassin Andin Chiléno–Argentin Mesozoïque en tant que bassin marginal, *Ann. Soc. Géol. Nord.*, v. 96, p. 177–184.

Cisternas, M. E., 1977, *Estudio Geológico del Flanco Occidental de la Cordillera Claudio Gay: El Sector de La Ola, al sur de Pedernales (26° 30' S), III Región, Chile, Thesis*, Universidad de Chile, Santiago.

Cisternas, M. E., and Vicente, J-C., 1976, Estudio geológico del sector de las Vegas de San Andrés (Provincia de Atacama, Chile), *Actas I Congr. Geol. Chileno* v. I p. A227–A252.

Clague, D. A., and Jarrard, R. D., 1973, Tertiary Pacific plate motion deduced from the Hawaiian–Emperor chain, *Geol. Soc. Am. Bull.*, v. 84, p. 1135–1154.

Clark, A. H., Farrar, E., Caelles, J. C., Haynes, S. J., Lortie, R. B., McBride, S. L., Quirt, G. S., Robertson, R. C. R., and Zentilli, M., 1976, Longitudinal variations in the metallogenetic evolution of the Central Andes: a progress report, *Geol. Assoc. Can. Spec. Pap.* 14, p. 23–58.

Cofré, C., 1955, *Informe Geológico sobre Reconocimiento de las Canteras Ubicadas en Arica*, unpublished report, Instituto Geológico Universidad de Chile, Santiago.

Coira, B., Davidson, J., Mpodozis, C., and Ramos, V., 1982, Tectonic and magmatic evolution of the Andes of northern Argentina and Chile, *Earth Sci. Rev.* v. 18, p. 303–332.

Cordani, U., Munizaga, F., Hervé, F., and Hervé, M., 1976, Edades radiométricas provenientes del basamento cristalino de la Cordillera de la Costa de las provincias de Valparaiso y Santiago, Chile. *Actas I Congr. Geol. Chileno*, v. II, p. F213–F222.

Cornejo, P., and Mpodozis, A., 1979, Las sedimentitas del Paleozoico Superior del Alto Valle del Río Hurtado, Coquimbo, IV Región, Chile, *Actas II Congr. Geol. Chileno*, v. 1 p. A87–A101.

Corvalán, J., 1955, *El Cretácico Inferior en la Provincia de Atacama*, unpublished report, Corp. Fomento Produc. (CORFO), Santiago.

Corvalán, J., 1965, Geología General, in: *Geografía Económica de Chile*, Santiago: Corp. Fomento Produc., p. 35–82.

Corvalán, J., 1976, El Triásico y Jurásico de Vichuquén–Tilicura y de Hualañé, Provincia de Curicó. Implicaciones paleogeográficas. *Actas I Congr. Geol. Chileno*, v. I, p. A137–A154.

Corvalán, J., 1982, El límite Triásico–Jurásico en la Cordillera de la Costa en las provincias de Curicó y Talca, *Actas III Congr. Geol. Chileno* v. III, p. F63–F65.

Corvalán, J., and Dávila, A., 1964, Observaciones geológicas en la Cordillera de la Costa entre los ríos Aconcagua y Mataquito, *Soc. Geol. Chile, Resúmenes* v. 9, p. 1–4.

Corvalán, J., and Munizaga, F., 1972, Edades radiométricas de rocas intrusivas y metamórficas de la Hoja Valparaiso–San Antonio, *Inst. Invest. Geol. Bol.*, v. 28, p. 1–40.

Corvalán, J., and Vergara, M., 1980, Presencia de fósiles marinos en las calizas de Polpaico, implicaciones paleoecológicas y paleogeográficas, *Rev. Geol. Chile* v., 10, p. 75–83.

Covacevich, V., 1977, *Faunas Fósiles del Noroeste de la Hoja Palestina*, unpublished report, Instituto de Investigaciones Geológicas, Santiago,

Covacevich, V., and Piracés, R., 1976, Hallazgo de ammonites del Bajociano Superior en la Cordillera de la Costa de Chile Central entre la Cuesta Melón y Limache, *Actas I Congr. Geol. Chileno*, v. I, p. C67–C86.

Covacevich, V., and Troncoso, A., 1980, Presencia de *Haplostigma furquei* Frenguelli (Pteridophyta, Lycopsida *Incertae Sedis*) en la formación El Toco, in: *Cuadrángulo Cerro de la Mica, Quillagua, Cerro Posada y Oficina Prosperidad* (by Maksaev, V., and Marinovic, N.), *Carta Geol Chile* v. 45/48, p. 1–63.

Covacevich, V., Varela, J., and Vergara, M., 1976, Estratigrafía y sedimentación de la Formación Baños del Flaco al sur del río Tinguiririca, Cordillera de los Andes, Provincia de Curicó, Chile, *Actas I Congr. Geol. Chileno*, v. I, p. A191–A212.

Creer, K., Mitchel, J., and Aboudeeb, J., 1972, Paleomagnetism and radiometric ages of the Jurassic Chon Aike Formation from Santa Cruz Province, Argentina: implications for the opening of the South Atlantic ocean, *Earth Planet. Sci. Lett.* v. 15, p. 131–138.

Dalziel, I. W. D., 1981, Back-arc extension in the Southern Andes: a review and critical reappraisal, *Philos. Trans. R. Soc. London*, v. A300, p. 319–335.

Dalziel, I. W. D., and Cortés, R., 1972, Tectonic style of the southernmost Andes and the Antarctandes, *Proc. 24th Int. Geol. Congr.* v. 3, p. 316–327, Montreal.

Dalziel, I. W. D., and Palmer, K. F., 1979, Progressive deformation and orogenic uplift at the southern extremity of the Andes, *Geol. Soc. Am. Bull.*, v. 90, p. 259–280.

Dalziel, I. W. D., de Witt, M. J., and Palmer, K. F., 1974, Fossil marginal basin in the Southern Andes, *Nature*, v. 250, p. 291–294.

Dalziel, I. W. D., Dott, R. H., Winn, R. D., and Bruhn, R. L., 1975, Tectonic relations of South Georgia Island to the southernmost Andes, *Geol. Soc. Am. Bull.*, v. 86, p. 1034–1040.

Damm, W. K., and Pichowiak, S., 1982, Nuevos resultados sobre el basamento cristalino chileno en las provincias Antofagasta y Tarapacá, *Actas III Congr. Geol. Chileno*, v. III, p. F155–F159.

Damm, W. K., Pichowiak, S., and Zeil, W., 1981, The plutonism in the north Chilean Coast Range and its geodynamic significance, *Geol. Rundsch.* v. 70, p. 1054–1076.

Davidson, J., and Mpodozis, A., 1978, Geología de la precordillera de Copiapó: las nacientes de la quebrada Paipote al oeste del Salar de Maricunga, *Dept. Geol. Univ. Chile Comunicaciones* v. 24, p. 1–34.

Davidson, J., Godoy, E., and Covacevich, V., 1976, El Bajociano marino de Sierra Minillas (70° 30′L.O.–26° L.S.) y Sierra Fraga (69° 50′L.O.–27°L.S.), Provincia de Atacama, Chile: edad y marco geotectónico de la Formación La Negra en esta latitud, *Actas I Congr. Geol. Chileno*, v. I, v. A255–A272.

Dávila, A., Hervé, F., and Munizaga, F., 1979, Edades K/Ar en granitoides de la Cordillera de la Costa de la Provincia de Colchagua, VI Region, Chile Central, *Actas II Congr. Geol. Chileno*, V. 1, p. F107–F120.

Dediós, P., 1967, Cuadrángulo Vicuña, Provincia de Coquimbo, *Carta Geol. Chile Inst. Invest. Geol.*, v. 16, p. 1–65.

Dediós, P. 1978, Cuadrángulo Rivadavia, Región de Coquimbo, *Carta Geológica de Chile Inst. Invest. Geol.* v. 28, p. 1–20.

Déruelle, B., 1978, Calc-alkaline and shoshonitic lavas from five Andean volcanoes (between latitudes 21° 45′ and 24° 30′ S) and the distribution of the Plio–Quaternary volcanism of the south central and Southern Andes, *J. Volcanol. Geotherm. Res.*, v. 3, p. 281–298.

Domeyko, I., 1845, Memoria sobre la constitución jeolójica de Chile. Constitución jeolójica del sistema andino i de los terrenos que atraviesan en la latitud de Coquimbo, *Mineralojía (jeolojía)*, v. 5 p. 173–294.

Dostal, J., Zentilli, M., Caelles, J. C., and Clark, A. H., 1977, Geochemistry and origin of volcanic rocks of the Andes (26°–28°S), *Contrib. Mineral. Petrol.*, v. 63, p. 113–128.

Dott, R. H., Jr., Winn, R. D., Jr., de Witt, M. J., and Bruhn, R. L., 1977, Tectonic and sedimentary significance of Cretaceous Tekenika Beds of Tierra del Fuego, *Nature*, v. 266, p. 620–622.

Douglas, R., and Nestell, M., 1972, Late Paleozoic Foraminifera from southern Chile, *U.S. Geol. Surv. Prof. Pap.* 858, p. 1–49.

Dragićević, M., 1970, Carta gravimétrica de los Andes meridionales e interpretación de las anomalías de gravedad de Chile central, *Dept. Geofis. y Geodes. Univ. Chile Publ.*, v. 93, p. 1–42.

Drake, R. E., 1976a, The chronology of Cenozoic igneous and tectonic events in the central Chilean Andes, in: *Proceedings of the Symposium on Andean and Antarctic volcanology problems, IAVCEI* González, O. ed. Napoli: Giannini & Figli, p. 670–697.

Drake, R. E., 1976b, Chronology of Cenozoic igneous and tectonic events in the central Chilean Andes—latitudes 35° 30′ to 36° S, *J. Volcanol. Geotherm. Res.*, v. 1, p. 265–284.

Drake, R. E., Curtis, G., and Vergara, M., 1976, Potassium-Argon dating of igneous activity in the central Chilean Andes–latitude 33° S, *J. Volcanol. Geotherm. Res.*, v. 1, p. 285–295.

Drake, R. E., Vergara, M., Munizaga, F., and Vicente, J-C., 1982a, Geochronology of Mesozoic–Cenozoic magmatism in central Chile, lat. 31°–36° S, *Earth Sci. Rev.*, v. 18, p. 353–363.

Drake, R. E., Charrier, R., Thiele, R., Munizaga, F., Padilla, H., and Vergara, M., 1982b, Distribución y edades K/Ar de volcanitas post-Neocomianas en la Cordillera Principal entre 32° y 36° L.S.: Implicaciones estratigráficas y tectónicas para el Meso–Cenozoico de Chile central, *Actas III Congr. Geol. Chileno*, v. II, p. D41–D78.

Elthon, D., and Stern, C. R., 1978, Metamorphic petrology of the Sarmiento ophiolite complex, Chile, *Geology*, v. 6, p. 464–468.

Ernst, W. G., 1975, Systematics of large-scale tectonics and age progressions in Alpine and Circum-Pacific blueschist belts, *Tectonophysics*, v. 26, p. 229–246.

Espinoza, W., and Fuenzalida, R., 1971, *Geología del territorio de Aysén entre los 45° y 46° Lat. Sur.*, Unpublished report, Instituto de Investigaciones Geológicas, Santiago.

Ferraris, F., and Di Biase, F., 1978, Hoja Antofagasta, Región de Antofagasta, *Carta Geol. Chile Inst. Invest. Geol.*, v. 30, p. 1–48.

Flores, H., 1947, *Informe geológico sobre los yacimientos de asbesto de Morro Bonifacio (Valdivia)*, unpublished report, Santiago, Chile.

Forsythe, R., 1982, The late Paleozoic to early Mesozoic evolution of southern South America: a plate tectonic interpretation, *J. Geol. Soc. London.*, v. 139, p. 671–682.

Forsythe, R., and Allen, R. B., 1980, The basement rocks of Peninsula Staines, Region XII, Province of Ultima Esperanza, Chile, *Rev. Geol. Chile*, v. 10, p. 3–15.

Forsythe, R., and Mpodozis, C., 1979, El Archipiélago Madre de Dios, Patagonia occidental, Magallanes: rasgos generales de la estratigrafía y estructura del "basamento" prejurásico, *Rev. Geol. Chile*, v. 7, p. 13–29.

Francis, P. W., and Rundle, C. C., 1976, Rates of production of the main magma types in the Central Andes, *Geol. Soc. Am. Bull.*, v. 87, p. 474–480.

Francis, P. W., Moorbath, S., and Thorpe, R. S., 1977, Strontium isotope data for recent andesites in Ecuador and north Chile, *Earth Planet. Sci. Lett.*, v. 37, p. 197–202.

Frutos, J., 1981, Andean tectonics as a consequence of seafloor spreading, *Tectonophysics*, v. 72, p. T21–T32.

Galli, C., 1956, Nota sobre el hallazgo del Paleozoico Superior en la Provincia de Taracapá, *Rev. Mineral.* v. 53/54, p. 30–31.

Galli, C., 1957, Las formaciones geológicas en el borde occidental de la Puna de Atacama, sector de Pica, Tarapacá, *Rev. Mineral.*, v. 56, p. 14–26.

Galli, C., 1968, Cuadrángulo Juan de Morales, Provincia de Tarapacá, *Carta Geol. Chile Inst. Invest. Geol.* v. 18, p. 1–53.

Galli, C., and Dingman, R., 1962, Cuadrángulos Pica, Alca, Matilla y Chacarilla: con un estudio sobre los recursos de agua subterránea; provincia de Tarapacá, *Carta Geol. Chile Inst. Invest. Geol.*, v. 3, p. 1–125.

García, F., 1967, Geología del Norte Grande de Chile, Simp. Geosincl. Andino (1962), *Soc. Geol. Chile Publ.*, v. 3, p. 1–138.

García, F., 1968, Estratigrafía del Terciario de Chile central, in: *El Terciario de Chile—Zona central* (Cecioni, G. ed.) pp. 25–57, Santiago: Andrés Bello, p. 25–57.

García, F., Pérez, E., and Ceballos, E., 1962, El Ordovícico de Aguada de la Perdiz, Puna de Atacama, Provincia de Antofagasta, *Rev. Mineral.* v. 77, p. 52–61.

Godoy, E., 1975, Geología del basamento cristalino de Punta Lengua de Vaca, Provincia de Coquimbo, Chile, *Actas VI Congr. Geol. Argentino* v. 1, p. 89–99.

Godoy, E., 1979, Metabasitas del basamento metamórfico Chileno. Nuevos datos geoquímicos, *Actas II Congr. Geol. Chileno*, v. 3, p. E133–E148.

Godoy, E., 1981, Sobre la discordancia intrasenoniana y el origen de los depósitos de caolín de Montenegro, Región Metropolitana, Chile, *Actas VIII Congr. Geol. Argentino*, v. III, p. 733–741.

Godoy, E., 1982, Geología del área Montenegro–Cuesta de Chacabuco, Región Metropolitana: el "problema" de la Formación Lo Valle, *Actas III Congr. Geol. Chileno*, v. I, p. A124–A146.

González-Bonorino, F., 1967, Nuevos datos de edad absoluta del Basamento Cristalino de la Cordillera de la Costa de Chile central, *Dept. Geol. Univ. Chile, Notas y Comunicaciones*, v. 1, p. 1–7.

González-Bonorino, F., 1970, Series metamórficas del Basamento Cristalino de la Cordillera de la Costa, Chile central, *Dept. Geol. Univ. Chile Publ.*, v. 37, p. 1–68.

González-Bonorino, F., 1971, Metamorphism of the crystalline basement of central Chile, *J. Petrol.*, v. 12, p. 149–175.

González-Bonorino, F., 1979, Esquema de la evolución geológica de la Cordillera Norpatagónica, *Rev. Asoc. Geol. Argent.* v. 34, p. 184–202.

González-Bonorino, F., and Aguirre, L., 1970, Metamorphic facies series of the crystalline basement of Chile, *Geol. Rundsch.* v. 59, p. 979–994.

González, O., 1963, Observaciones geológicas en el valle del Río Volcán, *Rev. Mineral.* v. 81, p. 20–54.

González, O., and Vergara, M., 1962, Reconocimiento geológico de la Cordillera de los Andes entre los paralelos 35° y 38° latitud sur, *Dept. Geol. Univ. Chile Publ.* v. 24, p. 1–121.

Gustafson, L. B., and Hunt, J. P., 1975, The porphyry copper deposit at El Salvador, Chile, *Econ. Geol.*, v. 70, p. 857–912.

Haller, M. J., and Lapido, O. R., 1982, The Jurassic–Cretaceous volcanism in the septentrional Patagonian Andes, *Earth Sci. Rev.*, v. 18, p. 394–410.

Halpern, M., 1973, Regional geochronology of Chile south of 50° latitude, *Geol. Soc. Am. Bull.*, v. 84, p. 2407–2422.

Halpern, M., 1978, Geological significance of Rb-Sr isotopic data of northern Chile crystalline rocks of the Andean orogen between latitudes 23° and 27° south, *Geol. Soc. Am. Bull.*, v. 89, p. 522–532.

Halpern, M., 1979, Strontium isotope composition of rocks from the Disputada copper mine, Chile, *Econ. Geol.*, v. 74, p. 129–130.

Halpern, M., and Rex, D. C., 1972, Time of folding of the Yahgan Formation and age of the Tekenika Beds, southern Chile, South America, *Geol. Soc. Am. Bull.*, v. 83, p. 1881–1886.

Harrington, H. H., 1943, Observaciones geológicas en la Isla de los Estados, *An. Mus. Argent. Cienc. Nat.* v. 41, p. 29–52.

Harrington, H. H., 1961, Geology of parts of Antofagasta and Atacama provinces, northern Chile, *Am. Assoc. Pet. Geol. Bull.*, v. 45, p. 169–197.

Hawkesworth, C. J., Norry, M. J., Roddick, J. C., Baker, P. E., Francis, P. W., and Thorpe, R. S., 1979, $^{143}Nd/^{144}Nd$ and $^{87}Sr/^{86}Sr$ variations in calc-alkaline andesites and plateau lavas from South America, *Earth Planet. Sci. Lett.* v. 42, p. 45–57.

Heim, A., 1940, Geological observations in the Patagonian Cordillera (Preliminary Report), *Eclogae Geol. Helv.*, v. 33, p. 25–51.

Henríquez, H., 1963, *Reconocimiento Geológico del extremo Norte del Departmento de Arica*, thesis, Universidad de Chile, Santiago.

Herm, D., 1965, Microfacies de algunos sedimentos calcáreos del Jurásico y Cretácico Inferior de Chile Central (Cordillera de la Costa y Cordillera de los Andes) Abstract, *Soc. Geol. Chile, Resúmenes* v. 11, p. 13–17.

Herm, D., 1967, Zur Mikrofazies Kalkiger Sedimenteinschaltungen in Vulkaniten der andinen Geosynklinale Mittel-chiles, *Geol. Rundsch.* v. 56, p. 657–669.

Herm, D., 1969, Marines Pliozän und Pleistozän in Nord-und Mittel-Chile unter besonderer Berücksichtigung der Entwicklung der Mollusken-Faunen, *Zitteliana*, v. 2, p. 1–59.

Herm, D., and Paskoff, R., 1966, Note préliminaire sur le Tertiaire supérieur du Chili centre-nord, *Bull. Soc. Geol. Fr.* v. 8, p. 760–765.

Hervé, F., 1976, Petrografía del basamento cristalino en el área Laguna Verde–Quintay, Provincia de Valparaiso, Chile, *Actas I Congr. Geol. Chileno*, v. II, p. F125–F143.

Hervé, F., 1977, Petrology of the crystalline basement of the Nahuelbuta Mountains, south central Chile, in: *Comparative Studies of the Circum-Pacific Orogenic Belt in Japan and Chile* Ishikawa, T., and Aguirre, L., eds., Tokyo. Japan Society for the Promotion of Science. p. 1–51.

Hervé, F., and Munizaga, F., 1978, Evidencias geocronológicas de un magmatismo intrusivo Triásico Superior–Jurásico en la Cordillera de la Costa de Chile entre los 35° 30′ y 36°30′ S, *Actas VII Congr. Geol. Argentino* v. II, p. 43–52.

Hervé, F., Munizaga, F., Godoy, E., and Aguirre, L., 1974a, Late Paleozoic K/Ar ages of blueschists from Pichilemu, central Chile, *Earth Planet. Sci. Lett.* v. 23, p. 261–264.

Hervé, F., Moreno, H., Parada, M. A., 1974b, Granitoids of the Andean Range of Valdivia Province, Chile, *Pac. Geol.* v. 8, p. 39–45.

Hervé, F., Godoy, E., Del Campo, M., and Ojeda, J. M., 1976a, Las metabasitas del basamento metamórfico de Chile central y austral, *Actas I Congr. Geol. Chileno*, v. II, p. F175–F187.

Hervé, F., Munizaga, F., Mantovani, M., and Hervé, M., 1976b, Edades Rb/Sr neopaleozoicas del basamento cristalino de la Cordillera de Nahuelbuta, *Actas I Congr. Geol. Chileno*, v. II, p. F19–F26.

Hervé, F., Thiele, R., and Parada, M. A., 1976c, El basamento metamórfico del archipiélago de Las Guaitecas, Aysén, Chile, *Actas I Congr. Geol. Chileno*, v. I, p. B73–B85.

Hervé, F., Thiele, R., and Parada, M. A., 1976d, Observaciones geológicas en el Triásico de Chile central entre las latitudes 35° 30′ y 40° 00′ S, *Actas I Congr. Geol. Chileno*, v. I, p. A297–A313.

Hervé, F., Nelson, E., and Suárez, M., 1979, Edades radiométricas de granitoides y metamorfitas provenientes de Cordillera Darwin, XII Región, Chile, *Rev. Geol. Chile*, v. 7, p. 31–40.

Hervé, F., Davidson, J., Godoy, E., Mpodozis, C., and Covacevich, V., 1981a, The late Paleozoic in Chile: stratigraphy, structure and possible tectonic framework, *An. Acad. Brasil. Cienc.*, v. 53, p. 361–373.

Hervé, F., Mpodozis, C., Davidson, J., and Godoy, E., 1981b, Observaciones estructurales y petrográficas en el basamento metamórfico del archipiélago de Los Chonos, entre el canal King y el canal Ninualac, Aisén, *Rev. Geol. Chile* v. 13/14, p. 3–16.

Hervé, F., Kawashita, K., Munizaga, F., and Bassei, M., 1982, Edades Rb-Sr de los cinturones metamórficos pareados de Chile central, *Actas III Congr. Geol. Chileno* v., II, p. D116–D135.

Hervé, F., Kawashita, K., Munizaga, F., and Bassei, M., 1984, Rb-Sr isotopic ages from late Palaezoic metamorphic rocks of central Chile, *J. Geol. Soc. London*, v. 141. p. 877–884.

Hillebrandt, A., 1970, Zur Biostratigraphie und Ammoniten-Fauna des südamerikanischen Jura (insbes. Chile), *Neues Jahrb. Geol. Paläontol. Abh*, v. 136, p. 166–211.

Hillebrandt, A., 1971, Der Jura in der chilenisch-argentinischen, Hochkordillere (25° bis 32°30'S), *Münstersch Forsch. Geol. Paläont.*, v., H20/21, p. 63–87.

Hillebrandt, A., and Davidson, J., 1979, Hallazgo de Paleozoico Superior marino en el flanco oriental de Sierra de Fraga, Región de Atacama, *Rev. Geol. Chile*, v. 8, p. 87–90.

Hinte, van, J. E., 1976, A Cretaceous time scale, *Am. Assoc. Pet. Geol. Bull.* 60, 498–516.

Hoffstetter, R., Fuenzalida, H., and Cecioni, G., 1957, *Léxique Stratigraphique International, Amérique Latine, Chili-Chile, Paris*: Centre National de la Recherche Scientifique, v. 7, p. 1–444.

Huete, C., Maksaev, V., Moscoso, R., Ulricksen, C., and Vergara, H., 1977, Antecedentes geocronológicos de rocas intrusivas y volcánicas en la Cordillera de los Andes comprendida entre la Sierra Moreno y el Rio Loa y los 21° y 22° lat. S., II Región, Chile, *Rev. Geol. Chile*, v. 4, p. 35–41.

Illies, H., 1960, Geologie der Gegend von Valdivia (Chile), *Neues Jahrb. Geol. Paläontol. Abh.* v. 111, p. 30–110.

Irvine, T. N., and Baragar, W. R. A., 1971, A guide to the chemical classification of common volcanic rocks, *Can. J. Earth Sci.*, v. 8, p. 523–548.

James, D. E., 1971, Plate tectonic model for the evolution of the Central Andes, *Geol. Soc. Am. Bull.*, v. 82, p. 3325–3346.

James, D. E., 1981, Role of subducted continental material in the genesis of calc-alkaline volcanics of the Central Andes, in: *Nazca Plate: Crustal Formation and Andean Convergence*, Kulm, L. D., Dymond, J., Dash, E. J., and Hussong, D. M., eds., *Geol. Soc. Am. Mem.* 154 p. 769–790.

Jensen, O., Vicente, J-C., Davidson, J., and Godoy, E., 1976, Etapas de la evolución marina Jurásica de la cuenca Andina externa (mioliminar) entre los paralelos 26° y 29° 30'S, *Actas I Congr. Geol. Chileno* v. I, p. A273–A296.

Kato, T., 1976, *The Relationship between Low-Grade Metamorphism and Tectonics in the Coast Ranges of Central Chile*, Ph.D. Dissertation, University of California, Los Angeles.

Katsui, Y., 1972, Late Cenozoic volcanism and petrographic provinces in the Andes and Antarctica, *J. Fac. Sci. Hokkaido Univ., Ser. 4*, v. 15, p. 27–39.

Katsui, Y., and González, O., 1968, Geología del area neovolcánica de los Nevados de Payachata. Consideraciones acerca del volcanismo Cenozoico Superior en los Andes Chilenos, *Dept. Geol. Univ. Chile Publ.* v. 29, p. 1–61.

Katz, H., 1961, Descubrimiento de una microflora neocomiana en la Formación Agua Fresca (eocena) de Magallanes y su significado con respecto a la evolución tectónica de la zona, *Univ. Chile Fac. Cienc. Fis. Mat. An.*, v. 18, p. 131–141.

Katz, H., 1962, Fracture patterns and structural history in the sub-Andean belt of southernmost Chile, *J. Geol.*, v. 70, p. 595–603.

Katz, H., 1972, Plate tectonics and orogenic belts in the southeast Pacific, *Nature*, v. 237, p. 331–332.

Katz, H., 1973, Contrasts in tectonic evolution of orogenic belts in the southeast Pacific, *J. R. Soc. N. Z.*, v. 3, p. 333–362.

Katz, H., and Watters, W. A., 1966, Geological investigation of the Yahgan Formation (Upper Mesozoic) and associated igneous rocks of Navarino Island, southern Chile, *N. Z. J. Geol. Geophys.*, v. 9, p. 323–359.

Klerkx, S., Deutsch, H., Pichler, H., and Zeil, W., 1977, Strontium isotopic composition and trace-element data bearing on the origin of Cenozoic volcanic rocks of the Central and Southern Andes, *J. Volcanol. Geotherm. Res.*, v. 2, p. 49–71.

Klohn, C., 1960, Geología de la Cordillera de los Andes de Chile central; provincias de Santiago, O'Higgins, Colchagua y Curicó, *Inst. Invest. Geol. Bol.*, v. 8, p. 1–95.

Klohn, C., 1965, Geosinclinal Magallánico, in: *Geología y Yacimientos Metalíferos de Chile* Ruiz, C., ed., Santiago: Instituto de Investigaciones Geológicas, p. 75–82.

Klohn, E., and Aguirre, L., 1965, Manganeso, in *Geología y Yacimientos Metalíferos de Chile* Ruiz, C., ed., Santiago: Instituto de Investigaciones Geológicas, p. 263–280.

Kranck, E. H., 1932, Geological investigations in the cordillera of Tierra del Fuego, *Acta Geogr.*, v. 4, p. 1–231.

Kubanek, F., and Zeil, W., 1971, Beitrag zur Kenntnis der Cordillera Claudio Gay (Nordchile), *Geol. Rundsch.*, v. 60, p. 1009–1024.

Kulm, L. D., Dymond, J., Dasch, E. J., and Hussong, D. M., eds., 1981, Nazca Plate: Crustal formation and Andean convergence, *Geol. Soc. Am. Mem.* 154, 824p.

Kussmaul, S., Jordan, L., and Ploskonka, E., 1975, Isotopic ages of Tertiary volcanic rocks of SW Bolivia, *Geol. Jahrb.*, v. 14, p. 111–120.

Lagally, U., 1975, *Geologische Untersuchungen im Gebiet Lago General Carrera–Lago Cochrane, Prov. Aysén, Chile, unter besonderer Berücksichtigung des Grundgebirges und seiner Tektonik*, Ph.D. Dissertation, Universität München.

Lahsen, A., 1966, *Geología de la Región Continental de Aisén*, Santiago: Instituto de Investigaciones Recursos Naturales, CORFO, Informe, 25 p.

Lahsen, A., 1969, *Geología del Área Comprendida entre El Tatio y los Cerros de Ayquina*, unpublished report, Corp. Fomento. Produc., Santiago.

Lahsen, A., 1976, Geothermal exploration in northern Chile. Summary, *Am. Assoc. Pet. Geol. Mem. 25*, p. 169–175.

Lahsen, A., 1982a, *Upper Cenozoic volcanism and tectonism in the Andes of northern Chile*, *Earth Sci. Rev.*, v. 18, p. 285–302.

Lahsen, A., 1982b, Evolución tectónica, solevantamiento y actividad volcánica de los Andes del norte de Chile durante el Cenozoico Superior, *Actas III Congr. Geol. Chileno*, v. I, p. B1–B27.

Lahsen, A., and Munizaga, F., 1979, Nuevos antecedentes cronológicos del volcanismo Cenozoico Superior de los Andes del norte de Chile, entre los 19° y los 22°30'S, *Actas II Congr. Geol. Chileno*, v. 1, p. F61–F82.

Larson, R. L., and Pitman, W. C., 1972, Worldwide correlation of Mesozoic magnetic anomalies and its implications, *Geol. Soc. Am. Bull.*, v. 83, p. 3645–3662.

Letelier, M., 1977, *Petrología, Ambiente de Depositación y Estructura de las Formaciones Matahuaico, Las Breas y Tres Cruces en el Área Rivadavia-Alcohuás, Elqui, IV Región, Chile*, Thesis, Universidad de Chile, Santiago.

Levi, B., 1960, Estratigrafía del Jurásico y Cretáceo Inferior de la Cordillera de la Costa entre las latitudes 32°40' y 33°40'S, *Univ. Chile Fac. Cienc. Fis. Mat. An.* v. 17, p. 219–271.

Levi, B., 1968, *Cretaceous Volcanic Rocks from a Part of the Coast Range West from Santiago, Chile Ph.D.*, Dissertation, University of California, Berkeley.

Levi, B., 1969, Burial metamorphism of a Cretaceous volcanic sequence west from Santiago, Chile, *Contrib. Mineral. Petrol.*, v. 24, p. 30–49.

Levi, B., 1970, Burial metamorphic episodes in the Andean geosyncline, central Chile, *Geol. Rundsch.*, v. 59, p. 994–1013.

Levi, B., and Aguirre, L., 1981, Ensialic spreading-subsidence in the Mesozoic and Paleogene Andes of central Chile, *J. Geol. Soc. London*, v. 138, p. 75–81.

Levi, B., and Nyström, J. O., 1982, Spreading-subsidence and subduction in central Chile: a preliminary geochemical test in Mesozoic–Paleogene volcanic rocks, *Actas III Congr. Geol. Chileno*, v. I, p. B28–B36.

Levi, B., Aguilar, A., and Fuenzalida, R., 1966, Reconocimiento geológico en las provincias de Llanquihue y Chiloé, *Inst. Invest. Geol. Bol.*, v. 19, p. 1–45.

Levi, B., Aguirre, L., and Nyström, J. O., 1982, Metamorphic gradients in burial metamorphosed vesicular lavas: comparison of basalt and spilite in Cretaceous basic flows from central Chile, *Contrib. Mineral. Petrol.*, v. 80, p. 49–58.

López-Escobar, L., Frey, F. A., and Vergara, M., 1976, Andesites from central-south Chile: trace element abundances and petrogenesis, in: *Proceedings of the Symposium on Andean and Antarctic volcanology problems, IAVCEI* González, O., ed., Napoli: Giannini & Figli, p. 725–761.

López-Escobar, L., Frey, F. A., and Vergara, M., 1977, Andesites and high-alumina basalts from the central-south Chile High Andes: geochemical evidence bearing on their petrogenesis, *Contrib. Mineral. Petrol.*, v. 63, p. 199–228.

López-Escobar, L., Frey, F. A., and Oyarzún, J., 1979, Geochemical characteristics of central Chile (33°–34°S) granitoids, *Contrib. Mineral. Petrol.*, v. 70, p. 439–450.

Losert, J., 1974a, Alteration and associated copper mineralizations in the Jurassic volcanic rocks of the Buena Esperanza mining area (Antofagasta Province, northern Chile), *Dept. Geol. Univ. Chile Publ.*, v. 41, p. 51–86.

Losert, J., 1974b, The formation of stratiform copper deposits in relation to alteration of volcanic series (on north Chilean examples), *Rezpravy Československé Akad. Věd. Ročnik*, v. 84, p. 1–77.

Luyendyck, B. P., 1970, Dips of downgoing lithospheric plates beneath island arcs, *Geol. Soc. Am. Bull.*, v. 81, p. 3411–3416.

Maass, R., and Roeschmann, C., 1971, Über die präandine Entwicklung am Beispiel der südlichen Provinz Coquimbo, *Münstersch Forsch. Geol. Paläont*, v. 20/21, p. 101–148.

Maksaev, V., 1978, Cuadrángulo Chitigua y sector occidental del Cuadrángulo Cerro Palpana, Región de Antofagasta, *Carta Geol. Chile Inst. Invest. Geol.* v. 31, p. 1–55.

Maksaev, V., 1979, Las fases tectónicas Incaica y Quechua en la Cordillera de los Andes del Norte Grande de Chile, *Actas II Congr. Geol. Chileno* v. 1, p. B63–B77.

Maksaev, V., and Marinovic, N., 1980, Cuadrángulos Cerro de la Mica, Quillagua, Cerro Posada y Oficina Prosperidad, Región de Antofagasta, *Carta Geol. Chile Inst. Invest. Geol. Santiago.* v. 45/48, p. 1–63.

Mapa Geológico de Chile, 1:1,000,000, 1982, *Servicio Nacional Geología y Minería*, Santiago, Chile.

Marinovic, N., 1979, Geología de los cuadrángulos Zapaleri y Nevados de Poquis, II Región, Chile, Thesis, Universidad de Chile, Santiago.

McBride, S. L., Caelles, J. C., Clark, A. H., and Farrar, E., 1976, Paleozoic radiometric age provinces in the Andean basement, latitudes 25°–30°S, *Earth Planet. Sci. Lett.*, v. 29, p. 373–383.

McDougall, I., Saemundsson, K., Johannesson, H., Watkins, N. D., and Kristjansson, L., 1977, Extension of the geomagnetic polarity time scale to 6.5 m.y.: K-Ar dating, geological and paleomagnetic study of a 3,500 m lava succession in western Iceland, *Geol. Soc. Am. Bull.*, v. 88, p. 1–15.

McNutt, R. H., Crocket, J. H., Clark, A. H., Caelles, J. C., Farrar, E., Haynes, S. J., and Zentilli, M., 1975, Initial $^{87}Sr/^{86}Sr$ ratios of plutonic and volcanic rocks of the Central Andes between latitudes 26° and 29°S, *Earth Planet. Sci. Lett.*, v. 27, p. 305–313.

Mégard, F., 1978, Étude géologique des Andes du Perou central, *Mem. ORSTOM* v. 86, p. 1–310, Paris.

Miller, H., 1970, Vergleichende Studien an prämesozoischen Gesteinen Chiles unter besonderer Berücksichsichtigung ihrer Kleintektonik, *Geotektonische Forsch.* v. 34, p. 1–64.

Miller, H., 1973, Características estructurales del basamento geológico chileno, *Actas V Congr. Geol. Argentino*, v. IV, p. 101–115.

Miller, H., 1976, El basamento de la Provincia de Aisén (Chile) y sus correlaciones con las rocas premesozoicas de la Patagonia Argentina, *Actas VI Congr. Geol. Argentino*, v. I, p. 125–141.

Miller, H., 1979a, Unidades estratigráficas y estructurales del basamento andino en el archipiélago de Los Chonos, Aisén, Chile, *Actas II Congr. Geol. Chileno*, v. I, p. A103–A120.

Miller, H., 1979b, Das Grundgebirge der Anden im Chonos-Archipel, Region Aisén, Chile, *Geol. Rundsch.*, v. 68, p. 428–456.

Miller, H., and Sprechmann, P., 1978, Eine devonische Faunula aus dem Chonos-Archipel, Region Aisén, Chile, und ihre stratigraphische Bedeutung, *Geol. Jahrb.*, v. B28, p. 37–45.

Minato, M., and Tanai, T., 1977, Carboniferous–Permian plant remains found at the border of Lake Panguipulli, Valdivia Province, Chile, in: *Comparative studies of the Circum-Pacific Orogenic Belt in Japan and Chile* Ishikawa T., and Aguirre, L., eds., Tokyo: Japan Society for the Promotion of Science, p. 69–80.

Minato, M., and Tazawa, J., 1977, Fossils of the Huentelauquén Formation at the locality F, Coquimbo Province, Chile, in: *Comparative studies of the Circum-Pacific Orogenic Belt in Japan and Chile* Ishikawa T., and Aguirre, L., eds., Tokyo: Japan Society for the Promotion of Science, p. 95–117.

Montecinos, F., 1963, *Observaciones de Geología en el Cuadrángulo Campanani, Departmento de Arica, Provincia de Tarapacá*, Thesis, Universidad de Chile, Santiago.

Montecinos, P., 1979, Plutonismo durante el Ciclo Tectónico Andino en el Norte de Chile entre los 18°–29° lat. Sur., *Actas II Congr. Geol. Chileno*, v. 3, p. E89–E108.

Moreno, H., 1976, The Upper Cenozoic volcanism in the Andes of Southern Chile (from 40° to 41°30′ S.L.), in: *Proceedings of the Symposium on Andean and Antarctic volcanology problems, IAVCEI*, González, O., ed., Napoli: Giannini & Figli, p. 143–171.

Moreno, H., and Parada, M. A., 1976, Esquema geológico de la Cordillera de los Andes entre los paralelos 39°00′ y 41°30′S, *Actas I Congr. Geol. Chileno*, v. I, p. A213–A225.

Mortimer, C., 1969, The geomorphological evolution of the southern Atacama Desert, Chile, Ph.D. Dissertation, University College, London.

Mortimer, C., Farrar, E., and Sarič, N., 1974, K-Ar ages from Tertiary lavas of the northernmost Chilean Andes, *Geol. Rundsch.*, v. 63, p. 484–491.

Moscoso, R., 1976, Antecedentes sobre un engranaje volcánico-sedimentario marino del Neo-comiano en el área de Tres Cruces, IV Región, Chile, *Actas I Congr. Geol. Chileno,*, v. I, p. A155–A168.

Mpodozis, C., and Davidson, J., 1979, Observaciones tectónicas en la precordillera de Copiapó: el sector de Puquios–Sierras La Ternera–Varillar, *Actas II Congr. Geol. Chileno*, v. 1, p. B111–B145.

Mpodozis, C., Parada, M. A., Rivano, S., and Vicente, J-C., 1976, Acerca del plutonismo tardi-Hercínico de la Cordillera Frontal entre los 30° y 33°S (provincias de Mendoza y San Juan-Argentina; Coquimbo-Chile), *Actas VI Congr. Geol. Argentino*, v. I, p. 143–171.

Mundaca, P., Padilla, H., and Charrier, R., 1979, Geología del área comprendida entre Quebrada Angostura, Cerro Talinai y Punta Claditas, Provincia de Choapa, IV Región, Chile, *Actas II Congr. Geol. Chileno*, v. 1, p. A121–A161.

Munizaga, F., Aguirre, L., and Hervé, F., 1973, Rb/Sr ages of rocks from the Chilean meta-morphic basement, *Earth Planet. Sci. Lett.*, v. 18, p. 87–92.

Munizaga, F., and Vicente, J-C., 1978, Zonación plutónica y volcanismo miocénico a la latitud 33° S de los Andes, *Actas VII Congr. Geol. Argentino* (resumé).

Munizaga, F., and Marinovic, N., 1979, Evidencias preliminares de un volcanismo Cenozoico Superior en el área del volcán Zapaleri, II Región, Chile, *Actas II Congr. Geol. Chileno*, v. 3, p. E237–E255.

Muñoz, Cristi, J., 1942, Rasgos generales de la constitución geológica de la Cordillera de la Costa, especialmente en la Provincia de Coquimbo, *An. 1 er Congr. Panamer. Ing. Minas Geol.*, v. 2, p. 285–318, Santiago, Chile.

Muñoz Cristi, J., 1950, Geología, in: *Geografía Económica de Chile*, Santiago: Corp. Fomento Produc., t. 1.

Muñoz, Cristi, J., 1956, Chile, *Geol. Soc. Am. Mem.*, v. 65, p. 187–214.

Muñoz Cristi, J., 1960, Contribución al conocimiento geológico de la Cordillera de la Costa de la zona central, *Rev. Mineral.*, v. 69, p. 28–47.

Muñoz Cristi, J., 1962, Comentarios sobre los granitos chilenos, *Rev. Mineral.*, v. 78, p. 15–19.

Muñoz Cristi, J., 1964, Estudios petrográficos y petrológicos sobre el Batolito de la Costa de las provincias de Santiago y Valparaiso, *Dept. Geol. Univ. Chile Publ.*, v. 25, p. 1–93.

Muñoz Cristi, J., 1973, *Geología de Chile, Prepaleozoico–Paleozoico y Mesozoico.* Santiago: Editorial Andrés Bello.

Naranjo, J. A., and Covacevich, V., 1979, Nuevos antecedentes sobre la geología de la Cordillera de Domeyko en el área de Sierra Vaquillas Altas, Región de Antofagasta, *Actas II Congr. Geol. Chileno*, v. 1, p. A45–A64.

Natland, M. L., González, E., Cañón, A., and Ernst, M., 1974, A system of stages for correlation of Magallanes Basin sediments, *Geol. Soc. Am. Mem.* 139, 126 p.

Nelson, E. P., Dalziel, I. W. D., and Milnes, A. G., 1980, Structural geology of the Cordillera Darwin—collisional style orogenesis in the southernmost Chilean Andes, *Eclogae Geol. Helv.*, v. 73/3, p. 727–751.

Niemeyer, H., 1975, Geología de la región entre el Lago General Carrera y el Rio Chacabuco, Provincia de Aysén, Chile, Thesis, Universidad de Chile, Santiago.

Niemeyer, H., 1978, Nódulos máficos y ultramáficos en basaltos alcalinos de la Meseta Buenos Aires, Lago General Carrera, Provincia de Aysén, Chile, *Rev. Asoc. Geol. Argent.*, v. 33, p. 63–75.

Noble, D. C., McKee, E. H., Farrar, E., and Petersen, U., 1974, Episodic Cenozoic volcanism and tectonism in the Andes of Peru, *Earth Planet. Sci. Lett.* v. 21, p. 213–220.

Nur, A., and Ben-Avraham, Z., 1981, Volcanic gaps and the consumption of aseismic ridges in South America, in: *Nazca Plate: Crustal Formation and Andean Convergence*, Kulm, L. D., Dymond, J., Dash, E. J., and Hussong, D. M. eds., *Geol. Soc. Am. Mem.* 154, p. 729–740.

Oyarzún, J., 1971, Contribution à l'étude geochimique des roches volcaniques et plutoniques du Chili, Doctoral Dissertation, Université de Paris Sud.

Oyarzún, J., 1982, El potencial ferrífero y cuprífero de los magmas en función de su hidratación inicial, evolución y condiciones de emplazamiento, *Actas III Congr. Geol. Chileno*, v. II, p. A349–A363.

Oyarzún, J., and Frutos, J., 1982, Proposición de un modelo para los depósitos cretácicos de magnetita del norte de Chile. Discusión de un esquema general para las mineralizaciones ferríferas asociadas al magmatismo calcoalcalino, *Actas V Congr. Latinoamericano Geología*, v. III, p. 25–39. Buenos Aires.

Oyarzún, J. and Frutos, J., 1984, Tectonic and petrological frame of the Cretaceous iron deposits of north Chile, *Min. Geol.*, v. 39, p. 21–31.

Oyarzún, J., and Villalobos, J., 1969, Recopilación de análisis químicos de rocas chilenas, *Dept. Geol. Univ. Chile Publ.*, v. 33, p. 1–47.

Pacci, D., Hervé, F., Munizaga, F., Kawashita, K., and Cordani, U., 1981, Acerca de la edad Rb-Sr Precámbrica de rocas de la Formación Esquistos de Belén, Departamento de Parinacota, Chile, *Rev. Geol. Chile*, v. 11, p. 43–50.

Palacios, C., 1976, Notes about the Jurassic paleovolcanism in northern Chile in *Proceedings of the Symposium on Andean and Antarctic volcanology problems, IAVCEI* González, O., ed., Napoli: Giannini & Figli, p. 238–248.

Palacios, C., and López-Escobar, L., 1979, Geoquímica y petrología de andesitas cuaternarias de los Andes centrales (18°57′–19°28′S), *Actas II Congr. Geol. Chileno*, v. 3, p. E73–E88.

Palmer, H. C., Hayatsu, A., and MacDonald, W. D., 1980, Paleomagnetic and K-Ar age studies of a 6 km-thick Cretaceous section from the Chilean Andes, *Geophys. J. R. Astr. Soc.*, v. 62, p. 133–153.

Parada, M. A., 1975, *Estudio Geológico de los Alrededores de los Lagos Calafquén, Panguipulli y Riñihue, Provincia de Valdivia*, Thesis, Universidad de Chile, Santiago.

Parada, M. A., 1981, Lower Triassic alkaline granites of central Chile (30°S) in the high-Andean Cordillera, *Geol. Rundsch.* v. 70, p. 1043–1053.

Parada, M. A., Munizaga, F., and Kawashita, K., 1981, Edades Rb-Sr roca total del batolito compuesto de los rios Elqui–Limarí a la latitud 30°S, *Rev. Geol. Chile*, v. 13/14, p. 87–93.

Paredes, Y., Mora, R., and Ilabaca, P., 1977, *Estudio Preliminar de los Estratos de Puerto Manso en su Localidad Típica*, unpublished report, Departamento Geología, Universidad de Chile, Santiago.

Paskoff, R., 1967, Antecedentes generales sobre la evolución del litoral de Chile del Norte durante el Plioceno y el Cuaternario, *Asoc. Geogr. Chile Bol.*, v. 3, p. 3–4, Santiago.

Paskoff, R., 1970, *Recherches Géomorphologiques dans le Chile Semi-Aride*, Bordeaux: Biscaye Frères.

Pérez, E., 1959, *Estratigrafía del Jurásico de Quebrada Asientos*, unpublished report, Instituto de Investigaciones Geológicas, Santiago.

Pérez, E., 1978, *Bioestratigrafía del Jurásico de Quebrada Asientos, Norte de Potrerillos, Tercera Región*, Thesis, Universidad de Chile, Santiago.

Pérez, E., and Aguirre, L., 1969, Relación entre estructura y volcanismo Cuaternario andino en Chile, *Pan Am. Symp. Upper Mantle, Petrol. Tectonics*, v. II, p. 39–46, Mexico.

Pichler, H., and Zeil, W., 1972, Chilean "Andesites"—crustal or mantle derivation?, *Int. Upper Mantle Proj. Conf. Sol. Earth Probl.*, v. 2, p. 361–371, Buenos Aires.

Piracés, R., 1976, Estratigrafía de la Cordillera de la Costa entre la Cuesta El Melón y Limache, Provincia de Valparaiso, Chile, *Actas I Congr. Geol. Chileno*, v. I, p. A65–A82.

Polanski, J., 1958, El bloque varíscico de la Cordillera Frontal de Mendoza, *Rev. Asoc. Geol. Argent.*, v. 12, p. 165–196.

Quirt, G. S., Stewart, J., Clark, A. H., and Farrar, E., 1971, Potassium-Argon ages of porphyry copper deposits in northern and central Chile, *Geol. Soc. Am. Abstr. Progr.*, v. 3, p. 676–677.

Ramos, E. D., and Ramos, V. A., 1978, Los ciclos magmáticos de la República Argentina, *Actas VII Congr. Geol. Argentino*, v. I, p. 771–786.

Ramos, V. A., Niemeyer, H., Skarmeta, J., and Muñoz, J., 1982, Magmatic evolution of the austral Patagonian Andes, *Earth Sci. Rev.* v. 18, p. 411–443.

Reutter, K., 1974, Entwicklung und Bauplan der Chilenischen Hochkordillere im Bereich 29° südlicher Breite, *Neues Jahrb. Geol. Paläontol. Abh.*, v. 146, p. 153–178.

Riccardi, A. C., 1971, Estratigrafía en el oriente de la Bahía de La Lancha, Lago San Martín–Santa Cruz–Argentina, *Rev. Mus. La Plata, Secc. Geol.*, v. 7, p. 245–318.

Rivano, S., 1980, Cuadrángulos D86, Las Ramadas, Carrizal y Paso Río Negro, Región de Coquimbo, *Carta Geol. Chile Inst. Invest. Geol.*, v. 41/44, p. 1–68.

Rivano, S., and Mpodozis, C., 1976, Note on the Jurassic (Dogger-Malm) paleovolcanism in the Main Range between 30°30′ and 31°20′ south latitude (Coquimbo Province, Chile), in *Proceedings of the Symposium on Andean and Antarctic Volcanology Problems, IAVCEI*, González, O., ed., Napoli: Giannini & Figli, p. 249–266.

Rogers, G., 1983, The petrogenesis of the "La Negra" Formation, N. Chile, Eos, v. 64, p. 329.

Rolleri, E., and Criado Roque, P., 1969, Geología de la Provincia de Mendoza, *Actas Jornadas Geol. Argent.*, v. 2, p. 1–60.

Roobol, M. J., Francis, P. W., Ridley, W. I., Rhodes, M., and Walker, G. P. L., 1976, Physicochemical characters of the Andean volcanic chain between latitudes 21° and 22° South, in: *Proceedings of the Symposium on Andean and Antarctic Volcanology Problems, IAVCEI* González, O., ed., Napoli: Giannini & Figli, p. 450–464.

Ruiz, C., and Corvalán, J., 1966, Geología, in: *Geografía Económica de Chile*, (primer apéndice) Santiago: Corp. Fomento Product. p. 18–30.

Ruiz, C., Segerstrom, K., Aguirre, L., Corvalán, J., Rose, H. J., and Stern, T. W., 1960, Edades plomo-alfa y marco estratigráfico de granitos chilenos; con una discusión acerca de su relación con la orogénesis, *Inst. Invest. Geol. Bol.*, v. 7, p. 1–26.

Ruiz, C., Aguirre, L., Corvalán, J., Klohn, C., Klohn, E., and Levi, B., 1965, *Geología y Yacimientos Metalíferos de Chile*, Santiago: Instituto de Investigaciones Geológicas.

Rutland, R. W. R., Guest, J. E., and Grasty, R. L., 1965, Isotopic ages and Andean uplift, *Nature*, v. 208, p. 677–678.

Saliot, P., 1968, Sur la présence et la signification de la lawsonite dans la cordillère cotière du Chili (Île de Chiloé), *C.12, As. Sc.*, v. 267, p. 1183–1185.

Saunders, A. D., Tarney, J., Stern, C. R., and Dalziel, I. W. D., 1979, Geochemistry of Mesozoic marginal basin floor igneous rocks from southern Chile, *Geol. Soc. Am. Bull.*, v. 90, p. 237–258.

Schawb, K., and Lippolt, H., 1976, K-Ar mineral ages and late Cenozoic history of the Salar de Cauchari area, Argentina–Puna, in: *Proceedings of the Symposium on Andean and Antarctic Volcanology Problems, IAVCEI*, González, O., ed., Napoli: Giannini & Figli, p. 698–714.

Scott, K. M., 1966, Sedimentology and dispersal pattern of a Cretaceous flysch sequence, Patagonian Andes, southern Chile, *Am. Assoc. Pet. Geol. Bull.* v. 50, p. 72–107.

Segerstrom, K., 1959, Cuadrángulo Los Loros; Provincia de Atacama, *Carta Geol. Chile Inst. Invest. Geol.*, v. 1 (1) p. 1–33.

Segerstrom, K., 1960, Cuadrángulo Quebrada Paipote; Provincia de Atacama, *Carta Geol. Chile Inst. Invest. Geol.*, v. 2, p. 1–35.

Segerstrom, K., 1967, Geology and ore deposits of central Atacama Province, Chile, *Geol. Soc. Am. Bull.*, v. 78, p. 305–318.

Segerstrom, K., 1968, Geología de las hojas Copiapó y Ojos del Salado, Provincia de Atacama, *Inst. Invest. Geol. Bol.*, v. 24, p. 1–58.

Segerstrom, K., and Parker, R. L., 1959, Cuadrángulo Cerrillos; Provincia de Atacama, *Carta Geol. Chile Inst. Invest. Geol.*, v. (2) p. 1–33.

Segerstrom, K., and Ruiz, C., 1962, Cuadrángulo Copiapó; Provincia de Atacama, *Carta Geol. Chile Inst. Invest. Geol.*, v. 3, p. 1–115.

Sillitoe, R. H., Mortimer, C., and Clark, A. H., 1968, A chronology from landform evolution and supergene mineral alteration, southern Atacama Desert, Chile, *Inst. Min. Metall. Trans.*, v. 77, p. 166–169.

Skarmeta, J., 1976, Evolución tectónica y paleogeográfica de los Andes Patagónicos de Aysén (Chile) durante el Neocomiano, *Actas I Congr. Geol. Chileno* v. I, p. B1–B15.

Skarmeta, J., 1978, Región Continental de Aysén entre el Lago General Carrera y la Cordillera Castillo, *Carta Geol. Chile Inst. Invest. Geol.*, v. 29, p. 1–53.

Skarmeta, J., and Charrier, R., 1976, Geología del sector fronterizo de Aysén entre los 45° y 46° lat. sur, Chile, *Actas VI Congr. Geol. Argentino*, v. I, p. 267–286.

Skarmeta, J., and Marinovic, N., 1981, Hoja Quillagua, Región de Antofagasta, *Carta Geol. Chile Inst. Invest. Geol.*, v. 51, p. 1–63.

Skewes, M. A., and Stern, C. R., 1979, Petrology and geochemistry of alkali basalts and ultramafic inclusions from the Pali–Aike volcanic field in southern Chile and the origin of the Patagonian plateau lavas, *J. Volcanol. Geotherm. Res.*, v. 6, p. 3–25.

Steinmann, G., 1929, *Geologie von Peru*, Heidelberg: C. Winters.

Stern, C. R., 1980, Geochemistry of Chilean ophiolites: evidence for the compositional evolution of the mantle source of back-arc basin basalts, *J. Geophys. Res.*, v. 85, p. 955–966.

Stern, C. R., and Elthon, D., 1979, Vertical variations in the effects of hydrothermal metamorphism in Chilean ophiolites: their implications for ocean floor metamorphism, *Tectonophysics*, v. 55, p. 179–213.

Stern, C. R., de Witt, M. J., and Lawrence, J. R., 1976, Igneous and metamorphic processes associated with the formation of Chilean ophiolites and their implication for ocean floor metamorphism, seismic layering, and magnetism, *J. Geophys. Res.*, v. 81, p. 4370–4380.

Stiefel, J., 1971, Kaenozoische Sedimentfolgen im Umkreis der Küstenkordillere Mittel-und Südchiles, *Münstersche Forsch. Geol. Paläontol*, v. 20/21, p. 277–291.

Stiefel, J., 1972, Zur tektonischen Entwicklung Chiles im Känozoikum, *Geol. Rundsch.* v. 61, p. 1109–1125.

Stipanicic, P., and Rodrigo, F., 1970, El diastrofismo jurásico en Argentina y Chile, *Actas Jornadas Geol. Argent.*, v. 2, p. 353–368.

Suárez, M., 1976, La Cordillera Patagónica: su división y relación con la Península Antártica, *An. Inst. Patagonia*, v. 7, p. 105–113.

Suárez, M., 1977, Aspectos geoquímicos del Complejo Ofiolítico Tortuga en la Cordillera Patagónica del Sur, Chile, *Rev. Geol. Chile*, v. 4, p. 3–14.

Suárez, M., 1978, Región al sur del Canal Beagle, Región Magallanes y Antártica Chilena, *Carta Geol. Chile Inst. Invest. Geol.*, v. 36, p. 1–48.

Suárez, M., 1979, A late Mesozoic island arc in the southern Andes, Chile, *Geol. Mag.*, v. 116, p. 191–201.

Suárez, M., and Pettigrew, T. H., 1976, An Upper Mesozoic island-arc–back-arc system in the southern Andes and South Georgia, *Geol. Mag.*, v. 113, p. 305–328.

Tavera, J., 1971, *Informe-estudio sobre material paleontológico florístico proveniente de la Formación Panguipulli y cerro Tralcán*, unpublished report, Departmento Geología, Universidad de Chile, Santiago.

Taylor, S. R., 1969, Trace element chemistry of andesites and associated calc-alkaline rocks, in: *Proceedings of the Andesite Conference*, McBirney, A. R. ed., Internation Upper Mantle Project, Scientific Report 16, State of Oregon Dept. of Geol. and Mineral. Indust. p. 43–64.

Thiele, R., 1964, Reconocimiento geológico de la Alta Cordillera de Elqui, *Dept. Geol. Univ. Chile, Publ.*, v. 27, p. 1–80.

Thiele, R., 1980, Hoja Santiago, Región Metropolitana, *Carta Geol. Chile Inst. Invest. Geol.*, v. 39, p. 1–51.

Thiele, R., and Morel, R., 1981, Tectónica triásico–jurásica en la Cordillera de la Costa, al norte y sur del Rio Mataquito (34°45′–35°15′ lat S), Chile, *Rev. Geol. Chile*, v. 13/14, p. 49–61.

Thiele, R., Bobenrieth, L., and Boric, R., 1980, Geología de los cerros Ruiz, Renca y Colorado (Santiago): contribución a la estratigrafía de Chile central, *Dept. Geol. Univ. Chile Communicaciones*, v. 30, p. 1–14.

Thomas, A., 1970, Cuadrángulo Iquique y Caleta Molle, Provincia de Tarapáca, *Carta Geol. Chile Inst. Invest. Geol.*, v. 21/22, p. 1–52.

Thomas, H., 1958, Geología de la Cordillera de la Costa entre el valle de La Ligua y la Cuesta de Barriga, *Inst. Invest. Geol. Bol.*, v. 2, p. 1–86.

Thorpe, R. S., and Francis, P. W., 1979, Variations in Andean andesite compositions and their petrogenetic significance, *Tectonophysics*, v. 57, p. 53–70.

Thorpe, R. S., Potts, P. J., and Francis, P. W., 1976, Rare Earth data and petrogenesis of andesite from the north Chilean Andes, *Contrib. Mineral. Petrol.*, v. 54, p. 65–78.

Thorpe, R. S., Francis, P. W., Hammill, M., and Baker, M. C. W., 1982, The Andes, in: *Andesites* Thorpe, R. S., ed., New York: John Wiley & Sons, p. 187–205.

Tobar, A., 1977, Stratigraphy and structure of the El Salvador–Potrerillos region, Atacama, Chile Ph.D. Dissertation, University of California, Berkeley.

Turner, J. C., 1965, Estratigrafía de Aluminé y adyacencias (Provincia de Neuquén), *Rev. Asoc. Geol. Argent.* v. 20, p. 153–184.

Venegas, R., 1979, Rocas metamórficas y plutónicas de la Península de Mejillones al sur de los 23°17′ lat. Sur y al oeste de los 70°30′ long. Oeste, II Región, Chile, *Actas II Congr. Geol. Chileno*, v. 3, p. E1–E20.

Venegas, R., and Niemeyer, H., 1982, Noticia sobre un probable Precámbrico sedimentario-metamórfico en el borde occidental de la Puna, al norte de Chuquicamata, *Actas III Congr. Geol. Chileno* v. III, p. F143–F154.

Vergara, H., 1978a, Cuadrángulo Quehuita y sector occidental del Cuadrángulo Volcán Miño, Región de Tarapacá, *Carta Geol. Chile* Inst. Invest. Geol., v. 32, p. 1–44.

Vergara, H., 1978b, Cuadrángulo Ujina, Región de Tarapacá, *Carta Geol. Chile Inst. Invest. Geol.*, v. 33, p. 1–63.

Vergara, M., 1969, Rocas volcánicas y sedimentario-volcánicas mesozoicas y cenozoicas en la latitud 34°30′S, Chile, *Dept. Geol. Univ. Chile Publ.*, v. 32, p. 1–36.

Vergara, M., 1972, Note on the paleovolcanism in the Andean Geosyncline from the central part of Chile, *24th Int. Geol. Congr.*, v. 2, p. 222–230. Montreal.

Vergara, M., 1978, Comentario sobre la edad de las formaciones Cola de Zorro y Farellones, Chile central, *Rev. Geol. Chile*, v. 5, p. 59–61.

Vergara, M., and Drake, R., 1978, Edades Potasio-Argón y su implicancia en la geología regional de Chile central, *Dept. Geol. Univ. Chile Communicaciones*, v. 23, p. 1–11.

Vergara, M., and Drake, R., 1979, Edades K/Ar en secuencias volcánicas continentales post-neocomianas de Chile central; su depositación en cuencas intermontanas restringidas, *Rev. Asoc. Geol. Argent.*, v. 34, p. 42–52.

Vergara, M., and González, O., 1972, Structural and petrological characteristics of the late Cenozoic volcanism from Chilean Andean region and west Antarctica, *Krystalinikum*, v. 9, p. 157–184.

Vergara, M., and Munizaga, F., 1974, Age and evolution of the Upper Cenozoic andesitic volcanism in central-south Chile, *Geol. Soc. Am. Bull.*, v. 85, p. 603–606.

Vicente, J-C., 1970, Reflexiones sobre la porción meridional del sistema peripacífico oriental, *Solid Earth Problem Conference Upper Mantle Project*, v. 1, p. 158–184, Buenos Aires.

Vicente, J-C., 1972, Aperçu sur l'organisation et l'evolution des Andes Argentino–Chiliennes centrales au parallèle de l'Aconcagua, *24th Int. Geol. Congr.* v. 3, p. 423–436, Montreal.

Vicente, J-C., 1976, Exemple de "volcanisme initial euliminaire": les complexes albitophyriques neo-triasiques et méso-jurassiques du secteur côtier des Andes Méridionales centrales (32° à 33°L. sud), in: *Proceedings of the Symposium on Andean and Antarctic Volcanology Problems, IAVCEI*, González, O., ed., Napoli: Giannini & Figli, p. 267–329.

Vila, T., 1976, Secuencia estratigráfica del Morro de Arica, Provincia de Tarapacá, *Actas I Congr. Geol. Chileno*, v. I, p. A1–A10.

Watters, W. A., 1965, Prehnitization in the Yahgan Formation of Navarino Island, southernmost Chile, *Mineral. Mag.*, v. 34, p. 517–527.

Wetzel, W., 1927, Beiträge zur Erdgeschichte der mittleren Atacama, *Neues Jahrb. Geol. Palaeontol.*, v. 58, p. 505–578.

Willis, B., 1929, Earthquake conditions in Chile, *Carnegie Inst. Washington Publ.*, v. 382, p. 1–178. Washington D.C.

Winn, R. D. Jr., 1978, Upper Mesozoic flysch of Tierra del Fuego and South Georgia Island: A sedimentologic approach to lithosphere plate restoration, *Geol. Soc. Am. Bull.*, v. 89, p. 533–547.

Wood, D. A., Gibson, I. L., and Thompson, R. N., 1976, Elemental mobility during zeolite facies metamorphism of the Tertiary basalts of Eastern Iceland, *Contrib. Mineral. Petrol.*, v. 55, p. 241–254.

Zamarsky, V., Conn, H., and Tabak, M., 1972, *Estudio Geoquímico de los Productos de Intemperismo de las Rocas Ultrabásicas (Serpentinitas) en la Provincia de Valdivia, Chile*, unpublished report, Instituto de Investigaciones Geologicas 67 p, Santiago, Chile.

Zeil, W., 1958, Sedimentation in der Magallanes-Geosynclinale mit besonderer Berücksichtigung des Flyschs, *Geol. Rundsch.* v. 47, p. 425–443.

Zeil, W., 1964, *Geologie von Chile*, Berlin: Gebrüder Borntraeger.

Zeil, W., 1979, *The Andes, a Geological Review*. Berlin: Gebrüder Borntraeger.

Zeil, W., and Pichler, H., 1967, Die Känozoische Rhyolith-Formations in mittleren Abschnitt der Andes, *Geol. Rundsch.* v. 57, p. 48–81.

Zentilli, M., 1974, Geological evolution and metallogenetic relationships in the Andes of Northern Chile between 26° and 29° south, Ph.D. Dissertation Queen's University Kingston, Canada.

Chapter 8

THE SEA OF OKHOTSK–KURIL ISLANDS RIDGE AND KURIL–KAMCHATKA TRENCH

Helios S. Gnibidenko

Institute of Marine Geology and Geophysics
Yuzhno-Sakhalinsk, Sakhalin 693002, USSR

I. INTRODUCTION

The area discussed in this chapter shows a considerable degree of heterogeneity. The northern and central parts of the Sea of Okhotsk form part of an epi-Mesozoic platform built up of deformed geosynclinal rocks ranging in age from Precambrian to Cretaceous covered by only slightly deformed Upper Paleogene and Neogene rocks (Fig. 1). The southern part of the Sea of Okhotsk, the Yuzno (South) Okhotsk basin is a fault-bounded back-arc basin. On the seaward side lies first the Kuril Islands Ridge and then the Kuril–Kamchatka Trench (fig. 22).

The structure of the continental slope down into the trench is made up of horst–anticlinorial uplifts of acoustic basement separated by a partly compensated grabenlike trough. The horsts appear to be composed of pre-Neogene (Upper Cretaceous and older) deformed volcanic and sedimentary deposits invaded by gabbroic, granodioritic and granitoid rocks. In the graben-synclinal troughs Neogene–Quaternary rocks, sometimes exceeding 3 km in thickness are found. The oceanic slope of the trench has a cover of 100–300 m of sediment lying upon layer 2 of the oceanic crust. Dredging of the acoustic basement on the Hokkaido Rise suggests it is composed of metamorphosed basalts with metasedimentary intercalations. Whole-rock K-Ar data indicate that the period of intensive volcanism extends from the Cre-

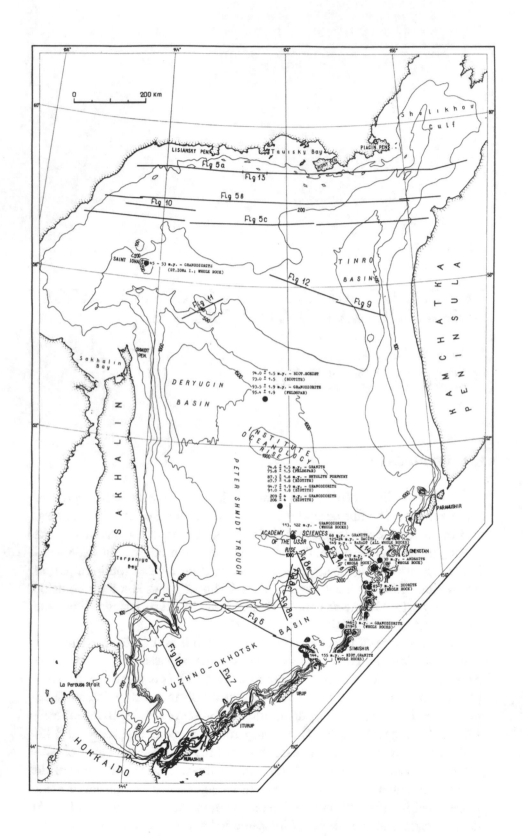

taceous to the Paleogene. The faults bounding the trough are normal, pointing to formation during a tensional regime. The vast region of oceanic plate adjacent to the central and northeastern parts of the Kuril–Kamchatka Trench is characterized by the absence of linear magnetic anomalies. Such anomalies are found, however, subparallel to the southern part of the trench.

In view of the major differences between the Sea of Okhotsk and the ridge and trench, it is convenient to deal with the two separately.

II. THE SEA OF OKHOTSK

A. Introduction

The principal geomorphic elements of the Sea of Okhotsk are the shelf in the north, the continental slope area of the central part of the sea, and the deep-sea basin in the south (Fig. 1). The Academy of Science of the USSR Rise, the Institute of Oceanology Rise, the Lebed, Kashevarov and Saint Iona banks all appear to have been subaerial at one time. East of Sakhalin lies the Deryugin Basin into which open the Saint Iona, Lebed, Makarov and Peter Schmidt troughs. The average depth of the Deryugin Basin is about 1600 m (maximum depth 1780 m), the Tinro Basin immediately west of Kamchatka with an average depth of only 850 m is much shallower.

The continental slope drops off 3000 m at 8°–10° along the mid-slope scarp between Sakhalin and Hokkaido and along the Kuril Shelf, into the Yuzhno Okhotsk Basin. Along some parts of the Greater Kuril Island Ridge the angle of slope may reach 20°–30°. The floor of the basin, at a depth of more than 3200 m, is a plain formed by the accumulation of recent sediments.

The Greater Kuril Island Ridge stretches for more than 1200 km, rising 3.5 to 4.0 km above the floor of the Yuzhno Okhotsk Basin. The two principal straits that cut through the ridge, the Bussol and Krusenstern straits have depths of 1000 m and 1400 m, respectively.

During periods of glacio-eustatic low sea level, the drop of 100 to 110 m exposed large areas of the North Okhotsk Shelf (Kulakov, 1973). However, some tectonic cause must be sought to explain the submergence of the formerly exposed surfaces of the Academy of Sciences and Institute of Oceanology rises, which now lie at depths of 1000 m.

The Sea of Okhotsk is crisscrossed by a dense network of seismic profiles (Fig. 2). Seismic profiles and deep seismic soundings have been carried

---

Fig. 1. Bathymetry of the Sea of Okhotsk: The locations of acoustic basin outcrops are indicated together with the K-Ar whole-rock age determinations (Burk and Gnibidenko, 1977; Gnibidenko, 1979; Kornev et al., 1982). The locations of seismic sections (Figs. 5–14, 18) are also shown.

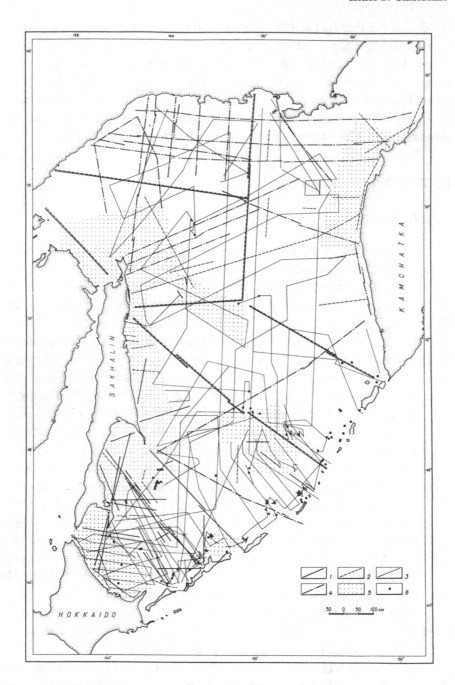

Fig. 2.   The location of seismic lines in the Sea of Okhotsk: *1*, deep seismic sounding (DSS) refraction lines; *2*, seismic sounding by reflection; *3*, single-channel continuous seismic profiling; *4*, multichannel (24 and 48) continuous seismic profiling; *5*, areas of detailed seismic profiling; *6*, the location of dredge haul sites. Data are from investigations up to 1975 (see Gnibidenko, 1979).

out, as have magnetic and gravimetric surveys, and numbers of heat flow measurements. Geological research has been restricted to sediment coring and to dredging within areas of submarine outcrop. In the surrounding land areas, and in near-shore regions there are more detailed geological and geophysical surveys and deep drilling has been carried out. This makes it possible to attempt to extend knowledge of the geology of the land area into the offshore areas.

The first dredge hauls in the Sea of Okhotsk were made during an expedition of the RV *Seifu-Maru* by the Maizuru Marine Metorological Observatory (Aoki, 1967). This was followed by many surveys carried out by the Sakhalin Complex Scientific Research Institute and the Institute of Oceanology of the USSR Academy of Sciences (Gnibidenko, 1979). The most recent dredging of outcrops of the acoustic basement were made during cruise N21 of RV *Pegas* in the South Okhotsk Basin in the late autumn, 1980 (Kornev *et al.*, 1982).

B. Basement and Cover

1. *Basement*

From seismic profiling, refraction and deep seismic sounding (Fig. 2) the basement is found to be characterized by seismic P-wave velocities in the 4.0–6.2 km/s range. It is overlain by a sedimentary cover in which average P-wave velocities are 2.0–2.5 km/s. In the basins two or three velocity subdivisions of the sediments is possible with individual layers having P-wave velocities in the 1.7 to 3.5–4.5 km/s range (Figs. 3–6).

In some parts of the Sea of Okhotsk acoustic basement crops out (Fig. 4). The examination of dredge hauls from these locations shows that acoustic basement is made up of an assemblage of lithologies characteristic of deformed geosynclinal beds. Among the rock types identified are phyllite, greenschist, graywacke, aleurolite and sandstone, tuff and tuff breccias made up of andesite, andesite–basalts and rhyolites, quartz diorites, granodiorites, granites, andesite, andesite basalts, quartz porphyries and dacite. The presence of metamorphic rocks with well developed cleavage (greenschist, phyllite and occasionally mica schist) indicates that the acoustic basement had been subjected to tectonic processes and provides the basis for the assumption that the basement, at least in the areas dredged, formed part of a folded complex.

Information on the age of this basement is inferred from radiometric ages determined from dredged samples (Fig. 1) and from attempts to project basement rock from a knowledge of the onshore geology into the offshore area. The K-Ar ages determined from the igneous rocks (granite, grano-

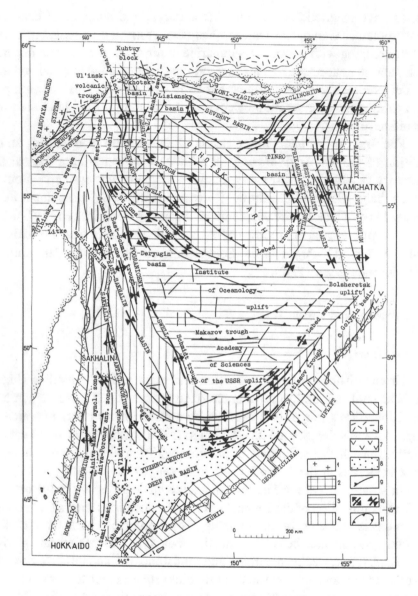

Fig. 3. A generalized tectonic map of the Sea of Okhotsk: *1,*, Precambrian deformed and metamorphosed complexes of the Stanovaya fold system and the Kuhtuy block; *2.-5*, Paleozoic, Mesozoic, and Cenozoic fold complexes; *2*, area deformed in the Paleozoic, sedimentation beginning in Precambrian and early Paleozoic; *3*, area deformed in the Mesozoic, sedimentation in the early Paleozoic and late Precambrian; *4*, area deformed in the late Mesozoic–early Paleogene, sedimentation beginning in early Paleozoic; *5*, area deformed in the late Neogene and continuing to the present, sedimentation in the late Paleozoic–early Mesozoic; *6*, Okhotsk postgeosynclinal Cretaceous and Cenozoic foldbelt; *7*, East Kamchatka Neogene–Quaternary postgeosynclinal volcanic complex; *8*, Neogene to Recent turbidite sequence of the Yuzhno (South) Okhotsk Basin; *9*, principal faults; *10*, horst-anticlinoria (a).

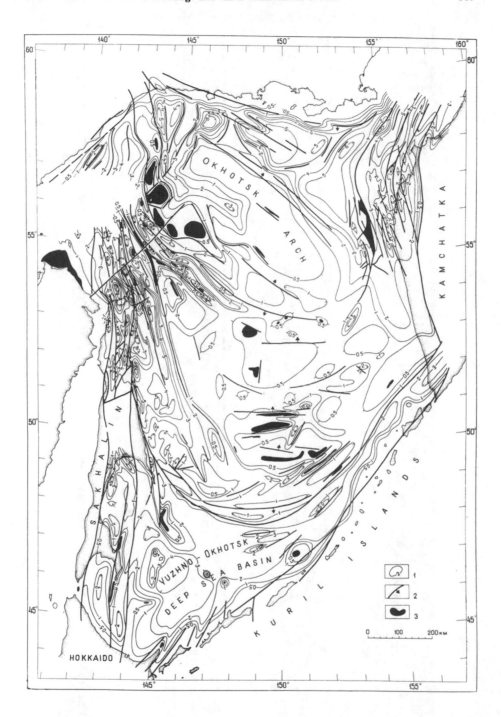

Fig. 4.   Isopach map of the Sea of Okhotsk: *1*, Isopachs in km of the Cenozoic sedimentary cover with seismic *P* velocities in the 1.7–4.8 km/s range based upon data from seismic refraction and reflection (see Fig. 2); *2*, main faults; *3*, Outcrops of acoustic basement (mainly pre-Cenozoic deformed geosynclinal sedimentary, igneous, and metamorphic rocks).

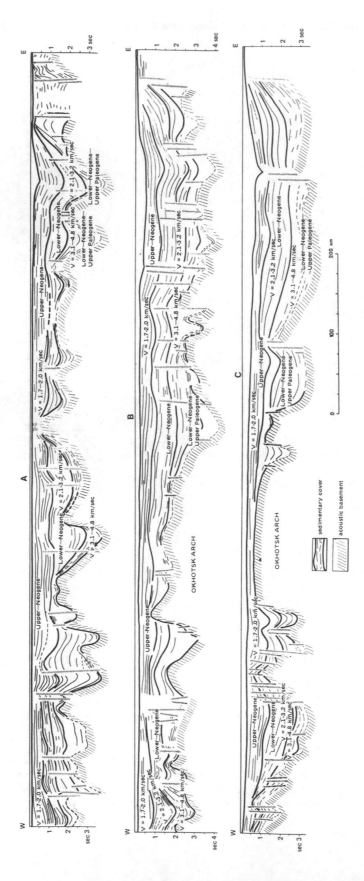

Fig. 5. Geological cross sections across the northern part of the Sea of Okhotsk interpreted from multichannel seismic profiles. Location of profiles are shown in Figure 1.

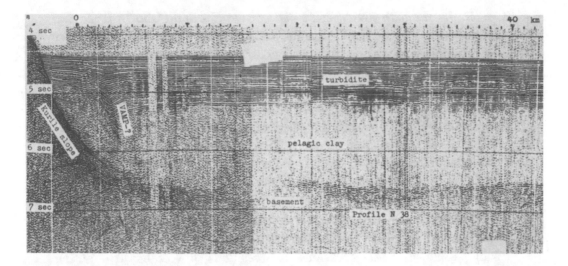

Fig. 7.   Single-channel seismic profile across the southeast slope of the Yuzhno (South) Okhotsk Basin. Location of profile is shown in Figure 1.

diorite, basalt, rhyolite, and dacite) range from 45 m.y. for the granodiorite of St. Iona Island to 209 m.y. calculated from the mica of a granite dredged from the Academy of Science Rise.

By comparing seismic velocities measured in dredged samples (Gnibidenko and Iliev, 1976) with velocities measured at the top of the acoustic basement (Galperin and Kosminskaya, 1964; Geodekyan *et al.*, 1978, 1980; Popov and Anosov, 1978) taken in conjunction with the radiometric data it is inferred that the basement of the Sea of Okhotsk Platform consists of deformed Cretaceous and other Mesozoic rocks and possibly Upper Paleozoic geosynclinal formations. The possibility that pre-Paleozoic rocks may also be present at some outcrops cannot be excluded.

2. *Cover*

Over most of the Sea of Okhotsk the sedimentary covers rests on a rough, irregular basement surface (Fig. 4). Around the rises the thickness of this cover ranges from 0.2 to 0.5 km but in the troughs this may rise to more than 5 to 6 km. There are, as a rule, good reflecting horizons within the sediment that may be traced many kilometers. Two, or more layers can be recognized with the velocity of the lowest layers reaching 3.3 km/s (Geodekyan *et al.*, 1978; Figs. 5–7). In some areas where the sedimentary cover can be traced into the onshore, as in Sakhalin, Kamchatka and northeastern USSR, it can be correlated with Upper Paleogene and Neogene–Quaternary sequences.

In the northern and western parts of the Yuzhno Okhotsk Basin the sedimentary cover on the continental slope has been intensively eroded and is dissected by submarine canyons. Landslides have occurred (Fig. 8a) and the slopes transect the layering of the sedimentary cover (Fig. 8b). It has also been subjected to folding with the formation of large brachyanticlines, and to faulting. Linear subsidence of basement blocks led, in some cases, to the formation of basins within which regularly folded structures (Fig. 8c) presumably resulted from the gravitational migration of the sediments towards the axial part of the basin.

Within the Yuzhno Okhotsk Basin the sedimentary cover is subdivided into an upper stratified unit overlying a lower acoustically transparent layer (Snegovskoy, 1974, Soloviev *et al.*, 1977, Figs. 6 and 7). The upper unit, up to 1.0–5.0 km thick, appears to consist of interbedded turbidites with pelagic ooze and occasional intercalations of volcanic ash. Within these layers seismic P velocities vary from 1.8 km/s near the seafloor to 2.0–2.5 km/s at depths of 1.5 km to 3.0 km/s in the lowest parts of the layer. The lower, acoustically transparent layer, with a thickness of up to 3.0 km seems to be composed of pelagic clays and argillites. Velocities within the transparent layer lie in the 2.5–3.0 km/s range. This layer rests upon layer 2 of volcanic and other sediments about 1.5 km thick, in which seismic P velocities range from 4.2–4.8 km/s (Snegovskoy 1974; Popov and Anosov, 1978).

C. Regional Descriptions

1. *West Kamchatka Region*

A region characterized by late Pliocene folding, lies in the eastern part of the Sea of Okhotsk (Figs. 3, 4) west of the Tigil–Malkin horst-anticlinorium (Gnibidenko *et al.*, 1974) while to the west of this is West Kamchatka Basin. In the basin there are some 4.0 km of sediment (Smirnov, 1971) that include the Kavransk Neogene terrigenous deposits and the Lower Miocene Voyampolsk series (Geolology of the U.S.S.R., v. 31, 1967). *P*-wave velocities up to 3.0–3.2 km/s have been recorded in these beds (Skorikova, 1972). The margin of the Kamchatka Basin is marked by the Pri–Kamchatsky swell and by the Bolsheretsk uplift in the south (Fig. 3). This Cretaceous uplift of basement, exposed in the axial part of the swell (Figs. 4 and 9), can be traced into the Kamchatka Peninsula. Near the village of Ust–Bolsheretsk amphibolized gabbroids with seismic velocities of 5.0–5.2 km/s (Vlasov, 1964; Suprunenko and Schwartz, 1967) were found at a depth of 534 m. Westwards, the Bolsheretsk uplift can be traced towards the Institute of Oceanology uplift from which it is separated by a basement saddle.

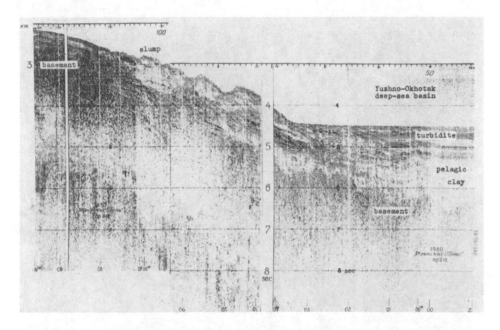

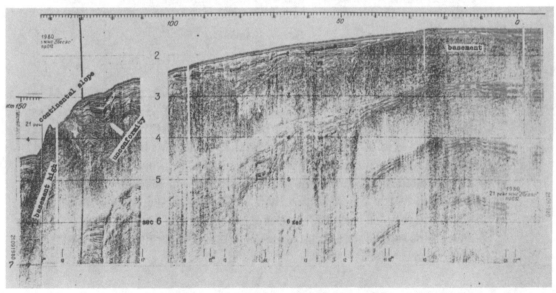

Fig. 8.   Single-channel seismic profiles across the north slope of the Yuzhno (South) Okhotsk Basin. For location of profiles see Figure 1.

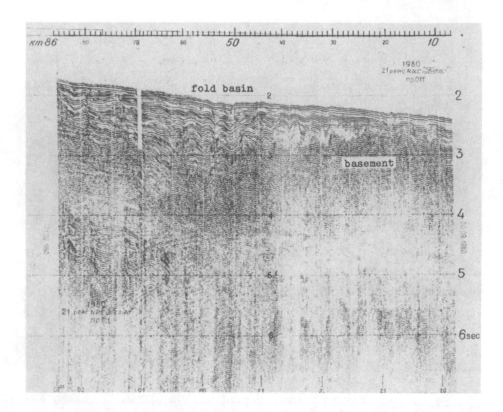

Fig. 8.   (*Continued*)

The southwestern branch of the Bolsheretsk uplift, the Lebed swell is clearly outlined by the 0.5 km isopach. In the axial part of the swell the acoustic basement crops out, here sandstone and pelitic schist are found (Geodekyan *et al.* 1974). Basement velocities of 5.5 km/s have been recorded (Galperin and Kosminskaya, 1964). The Lebed swell appears to be structurally related to the central block of the Academy of Sciences of the USSR uplift from which it is separated by a saddle with a sedimentary thickness of no more than 1.0 km. Southeast of the Lebed swell lies the Golygin Basin in which the sediment thickness exceeds 3.0 km (Galperin and Kosminskaya, 1964; Smirnov, 1971).

2. *North Sea of Okhotsk Region*

The Yurovsky block is made up of about 1 km of Upper Cambrian and Ordovician carbonate and terrigeneous rocks that overlie a pre-Upper Cambrian metamorphic complex (Drabkin, 1970; Korolkov, 1972). This block forms the northerly continuation of the Kashevarov swell stretching northwestwards towards the mainland (Fig. 3). The swell appears to form the eastern boundary of the Mongolo–Okhotsk fold system.

The Kashevarov swell (Figs. 3, 4) extends south-southeast a distance of more than 800 km from the Yurovsk block by way of the Saint Iona Islet and the Kashevarov Bank. The width of the swell is restricted by faulting to about 50 to 100 km. The acoustic basement is made up of deformed geosynclinal rocks intruded by granodiorites. The age of the rocks ranges from Precambrian through Paleozoic to Triassic in the Yurovsky block and to Upper Cretaceous–Lower Paleogene in the Saint Iona block where additionally, Eocene granodiorites dated by K/Ar as 45–53 m.y. old, are exposed (Fig. 1). The Kashevarov swell is connected with the Institute of Oceanology uplift by a basement welt in which granodiorite with a K/Ar age of 94 ± 1.5 m.y. and biotite schist (hornfels) with a K/Ar age of 73.5 m.y. (Burk and Gnibidenko, 1977) occur.

The base of the sedimentary overburden on the acoustic basement of the Kashevarov swell (Figs. 3, 4, 10 and 11) has an age older than Oligocene. It is suggested that the sedimentary cover in the adjacent trough accumulated during Oligocene–Pleistocene time. The thinness of sediments over the axial region of the swell (0.25–0.5 km) and their absence over considerable areas at depths of about 200 m (Saint Iona and Kashevarov banks) suggests that the swell was a large island as far back as late Pliocene–Pleistocene time.

The Kashevarov Trough (Fig. 3) more than 600 km long and 100 km wide, which separates the swell from the Okhotsk Arch, can be divided into two subordinate basins: the more northerly, Okhotsk Basin, filled by more than 5,000 m of Neogene terrigenous deposits and a southerly basin in which more than 3,000 m of sediments accumulated (Figs. 4, 10). The former lies close to land where the Okhotsk Graben (Chikov, 1970) is identified. It is filled with the Moransk coal-bearing series of late Pliocene beds about 700 m in thickness, followed by Quaternary deposits. They rest upon Cretaceous volcanic rocks.

The Okhotsk Arch, with a maximum width of 200 km, stretches southeastwards a distance of more than 600 km. To the northwest the Okhotsk Anticlinorium is continued through the Lisiansky swell (Fig. 3) as a basement arch. The axial zone of this anticlinorium, the Kuhtuy block, consists of metamorphosed eugeosynclinal Proterozoic sediments overlain by a Paleozoic to middle Jurassic miogeosynclinal sequence, covered in turn by Cretaceous to Paleogene volcanics of the Chukotka–Katasiatic volcanic belt (Grinberg, 1968; Veldyaksov *et al.*, 1970). It may be anticipated that the acoustic basement of the Okhotsk Arch will be formed by these geosynclinal sediments which dip southeastwards below younger deposits. It appears to lie at a depth of about 1 km below the seafloor beneath thin Cretaceous–

Paleogene volcanics in the northeast, and to have velocities of the order of 5.2 km/s (Galperin and Kosminskaya, 1964). The thickness of the Mesozoic layer increases towards the Severny Trough and the axial zone of the arch where it may reach a thickness of 3–4 km. Occasional outcrops (Fig. 12) suggest that the acoustic basement in the axial zone of the arch is probably formed by the Mesozoic geosynclinal complex with granitoid and volcanic bodies of Cretaceous–Paleogene age.

The Severny Trough northeast of the Okhotsk Arch (Figs. 3–5), has approximately the same dimensions, 600 km long by 200 km wide. It is divided by swells into three basins, from northwest to southeast these are the Lisiansky, Severny and Tinro basins. They are bounded by major northwest-trending faults. Trending towards the basin is the axial zone of the Iniisko–Kavinsk synclinorium formed of a Permian–Lower Mesozoic molasse complex overlain by Cretaceous–Paleogene volcanics and intruded by Cretaceous granitoid bodies. The thickness of continental Neogene deposits in the coastal region between the Lisiansk and Khmitevsk peninsulas does not exceed 100 m (Vereschagin, 1970; Chikov, 1970) but some 50 km to the south the sedimentary cover reaches thicknesses of 3–4 km (Fig. 4, 5, 13). The acoustic basement of the basin is presumed to be formed by the Permian and Lower Mesozoic geosynclinal deposits although in the northwestern extremity of the Severny Trough Cretaceous–Paleogene volcanics may occupy that role.

The sediments of the Severny Trough may locally exceed 6 km in thickness (Galperin and Kosminskaya, 1964; Figs. 4, 5). They are divisible into three or four seismic units and apparently form a continuous Upper Paleogene to Pliocene sequence. The Tinro Basin fill reaches 5 km in thickness (Geodekyan et al., 1978; Figs. 3, 4, 5, 9). Seismic profiling indicates a 2 km thickness for the layer with 2 km/s velocity. During the 1972 Cruise 53 of RV *Vityaz* it was shown that the southern part of the basin had a bottom relief of alternating hills and depressions trending in a submeridional direction. The deformation of the sedimentary cover is concordant with the relief (Udintsev et al., 1976).

Along the northern margin of the Severny Trough is the late Mesozoic Okhotsk fold system (Drabkin, 1970), the Koni Piyagin Anticlinorium (Fig. 3). Outcrops onshore indicate that the axial zone of this structure consists of Triassic–Jurassic andesites intruded by Lower Cretaceous granitoids. Rocks of this type and age seem to form the acoustic basement of the northern flank of the Severny Basin.

3. *Central Sea of Okhotsk Region*

In the Deryugin Basin, which has dimensions of 200 km by 100 km, a sedimentary complex buries a rough acoustic basement. In the smaller subbasins a two-layer sedimentary cover can be recognized as an Upper Pliocene–Quaternary layer with a maximum thickness of about 1.0 km resting upon a lower layer 1–2 km thick. The basement topography is very rough, particularly in the eastern part of the basin where it is found in outcrop. It is characterized by velocities of 4.1–5.3 km/s (Galperin and Kosminskaya, 1964). In the central and northern parts of the basin, basement rises 150–200 m above the floor of the basin. The sedimentary cover decreases to 250 m, and the acoustic basement is exposed along some parts of the Pogranichny swell (Fig. 3).

The nature of the basement of the Deryugin Basin is problematic. In the eastern part of the basin, structurally related to the Lebed Trough (Fig. 3) it may be presumed to have the same composition, a geosynclinal folded complex intruded by granodiorites with K/Ar ages of 73.5–95.4 m.y. (Fig. 1) such as have been determined in blocks dredged from basement uplift to the north on the Institute of Oceanology uplift (Burk and Gnibidenko, 1977).

As the seismic velocities found in the granodiorite samples dredged from the Lebed Trough are in the 4.3–5.5 km/s range and those in the phyllites are from 4.7–6.6 km/s it seems possible that the dredged samples are part of the sample complex in which velocities of 4.1–5.3 km/s were earlier recorded (Galperin and Kosminskaya, 1964). If this is the case, and the basement of the eastern part of the Deryugin Basin may be regarded as of Cretaceous–Lower Paleocene age and the lower layer of the sedimentary cover shoould be Eocene–Oligocene in age consistent with a Neogene–Quaternary age for the upper layer.

The basement surface of the Institute of Oceanology uplift is planar so that the sedimentary cover rests on an abraded flat acoustic basement in which seismic velocities of 6 km/s have been recorded (Galperin and Kosminskaya, 1964). There are large basement outcrops in the northwest and west, usually on the uplifted sides of fault blocks. The thin sedimentary cover appears to consist of only late Pliocene–Quaternary deposits.

The Makarov Trough (Figs. 3, 4) which separates the Institute of Oceanology uplift from the Academy of Sciences of the USSR uplift is filled by about 1 km of probable Pliocene–Quaternary sediments. The relief of the acoustic basement is rough and there are basement outcrops along the central zone of the trough.

The Academy of Sciences of the USSR uplift (Figs. 3, 4) trends east–west for a distance of about 400 km. It can be divided into northern, central and southern blocks. The first two of these are connected with the Pogranichny swell, the eastern boundary of the East Sakhalin Basin, while to the east they link through the Lebed swell and Bolsheretsk uplift with the Tigil–Malkinsky Anticlinorium (Fig. 3) in a broad arc. The southern block is connected en echelon via the west with the Sakhalin Anticlinorium, while in the east it is structurally related with the horst-anticlinal uplift or the Paramushir Island. The link in the east is via a basement arch (Figs. 1, 4, 14a) that separates the Atlasov Trough from the Yuzhno Okhotsk Basin.

The surface of the uplift is abraded and lies at an average depth of 1.0–1.1 km. The 0.5 km isopach, particularly in the northern and eastern parts of the uplift has a complicated configuration that seems to be due to relative motion of blocks across the faults (Fig. 14b). The acoustic basement crops out in the central and northern parts of the uplift or is covered by a thin veneer of sediments. Rocks dredged from those acoustic basement outcrops indicate that they consist of a Mesozoic and possible Upper Paleozoic assemblage of eugeosynclinal rocks (Gnibidenko and Iliev, 1976; Burk and Gnibidenko 1977; Korenbaum et al., 1977). K/Ar age determinations of the igneous rocks are in the 68 to 209 m.y. range (Fig. 1). Laboratory seismic velocity determinations on basement samples (Gnibidenko and Iliev, 1976) are in good agreement with velocities determined by deep seismic sounding techniques (Galperin and Kosminskaya, 1964) and range from 6.0–7.0 km/s.

4. *East Sakhalin Offshore Region*

The Pegas Trough, which opens into the Yuzhno Okhotsk Basin and lies east and south of the Sakhalin Anticlinorium (Fig. 3), is the southernmost structural element of the region. To the northwest the Pegas Trough is connected with the Pogranichny Basin. Within the latter basin the Upper Miocene–Pliocene and Quaternary sediments exceed 3 km in thickness (Soloviev et al., 1979). In the axial region of the trough siliceous argillites and aleurolites of early middle Miocene age seem to form acoustic basement (Pilgensk suite; Geology of the USSR, v. 33, 1970).

The East Sakhalin Basin (Figs. 3, 4) lies between the Pogranichny swell and the Sakhalin anticlinorium to the west. It is subdivided by the Schmidt anticlinal zone into the westerly Piltun–Chaivinsk Trough and the East Schmidt Trough east of the anticlinal zone. The Neogene section within the Piltun–Chaivinsk trough has a thickness of 5 to 9 km (Telegin 1969; Alperovich and Chernyavsky 1973, Fig. 4) over a metamorphosed Upper Cretaceous sedimentary-volcanic basement. The Neogene thickness in the East Schmidt Trough exceeds 7 km (Margulis et al., 1979; Fig. 4).

The Pogranichny swell which lies east of the East Sakhalin Basin (Figs. 3, 4) continues northward along the western slope of the Deryugin Basin as a series of en echelon uplifts of the acoustic basement. The margin follows approximately the line of the 1,000 m isobath along the Sakhalin Shelf.

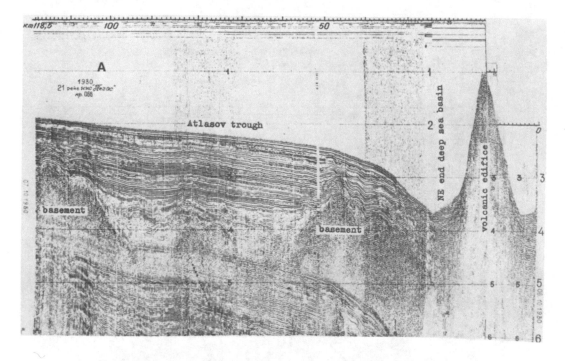

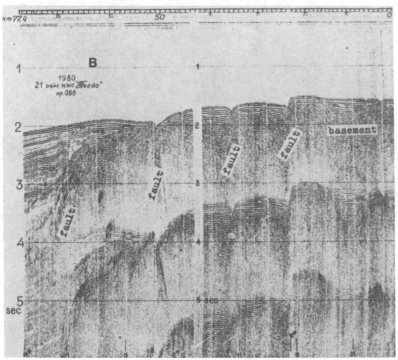

Fig. 14. Single-channel seismic profiles (A) and (B) of the eastern part of the northern slope of the Yuzno (South) Okhotsk deep basin. For location see Figure 1.

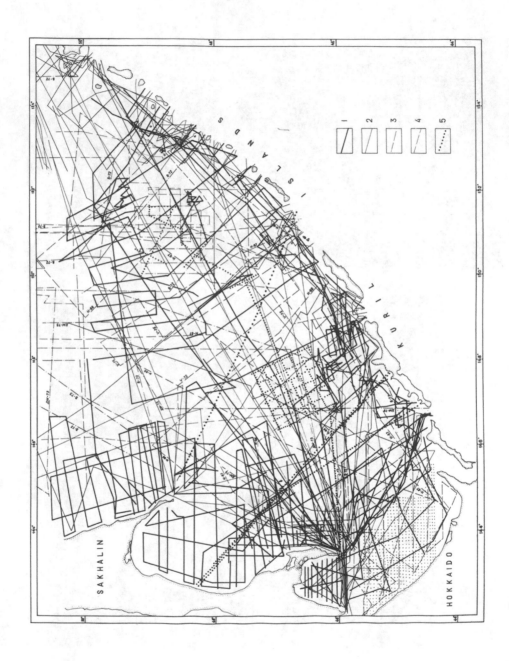

Thus in the northern, western, central and eastern parts of the Sea of Okhotsk the basement has been broken up by Cenozoic movements along deep faults with a predominantly northwesterly or sublatitudinal strike (Gnibidenko, 1976). The majority of the faults are normal or growth faults and have produced a taphrogenic pattern in the basement. This basement is now buried for the most part beneath an Upper Paleogene–Neogene sedimentary blanket. These basins have considerable hydrocarbon potential (Geodekyan *et al.*, 1977; Gnibidenko 1977).

5. *Yuzhno Okhotsk Basin*

The Yuzhno Okhotsk Basin is a classic back-arc basin (Fig. 1) behind the Kuril Island arc. It has a relatively thick sedimentary fill. The interpretation of its structural evolution is basic to the understanding of continental margin geodynamics. The basin, and other Far Eastern marginal basins like it are regarded by Karig (1971) and Matsuda and Uyeda (1971) as secondary spreading basins in front of subducting oceanic plate, or by Belousov (1968) as resulting from the basification of sialic crust. Vassilkovsky (1967) and Gnibidenko and Sychev (1972) have also considered the basins as relict elements of oceanic crust at a different stage in its transformation to continental crust. The detailed study, by regional geological and geophysical methods (Fig. 15) provides a basis for the discussion of the alternative models.

The principal morphological elements of the Yuzhno Okhotsk Basin are the continental slopes and the deep-sea basin relief. The basin has a level floor at a depth of 3,200–3,300 m the only mountains are some peaks rising 1,000–1,500 m above the abyssal plain at the foot of the main Kuril Ridge which is crowned by the Greater Kuril Island volcanic arc. Some of the peaks appear to be volcanic. Despite a relatively dense network of echo sounding and seismic profiling in the basin, elements indicative of active erosional relief such as scarps, hollows or channels have yet to be found.

←—————————————————————————————————————

Fig. 15. Seismic and echo-sounding lines in the Yuzno (South) Okhotsk Basin: *1*, continuous single-channel profiling by RV *Pegas* and *Morskoy Geofizik* (1973–1980 Sakhalin Complex Scientific Research Institute); *2*, echo-sounding by RV *Vityas* (1949–1955 Institute of Oceanology, USSR Academy of Science); *3*, continuous single-channel seismic profiling by American and Japanese research vessels (Yasui *et al.*, 1967, Verzhbitsky *et al.*, 1976, Honza, 1978); *4*, continuous single-channel seismic profiling by RV *Vityas* and *Dr. Mendeleev* (1973–1974, Institute of Oceanology, USSR Academy of Science); *5*, multichannel seismic profiling by RV *Morskoy Geofizik, Edward Toll*, and *Iskatel* (Sakhalin Complex Scientific Research Institute and Pacific Expedition Soyuzmorgeo). The stippled region is covered by a bathymetric survey and continuous seismic single-channel profiling carried out by the Japanese Hydrographic Survey (Nagano *et al.*, 1974).

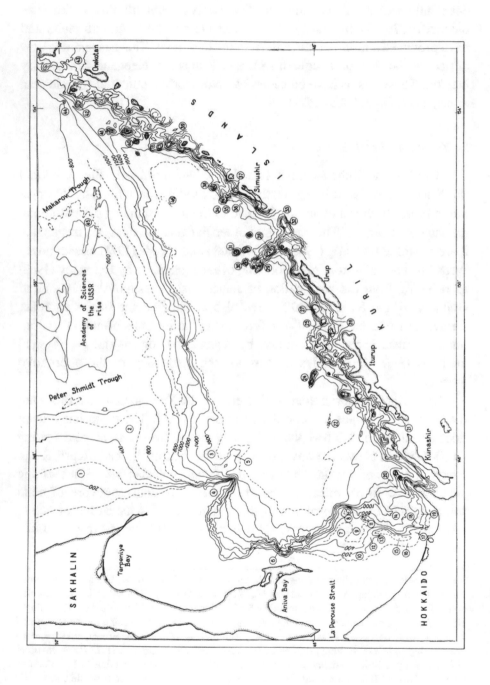

The Hokkaido–Sakhalin continental slope has an angle of about 10° reaching as much as 15° along southern Sakhalin. The continental shelf break with the scarp is at a depth of 145–160 m. The slope is considerably dissected by submarine canyons, and is marked by the presence of deep-sea terraces. The northern continental slope is less steep at 5° and rises smoothly 2,000 m to the Academy of Sciences of the USSR Rise. The slope is only moderately dissected by submarine canyons when compared to Hokkaido–Sakhalin slope. Four spurs dipping gently basinward can be recognized (Fig. 16) along the slope.

The Kuril slope of the basin dips at from 10° to 20–25° with a characteristic en echelon system of short ridges at an angle of about 25° to the main Kuril trend particularly in the southwestern part of the basin. Some of the high points on these ridges appear to be volcanic in origin but others appear to be basement uplifts.

The Yuzhno Okhotsk Basin is bounded on all sides by structural elements having a continental or subcontinental crust. These structural elements are assumed to have developed at different times (Gnibidenko, 1979). The margin of the basin is usually distinguished by deep faults (Fig. 3).

The basement of the folded system of the Academy of Sciences of the USSR Rise, which delimits the Yuzhno Okhotsk Basin to the north, is made up of Mesozoic and probably some Upper Paleozoic eugeosynclinal beds (Gnibidenko, 1979). This suggests that the geosynclinal system was established here as early as Upper Paleozoic. The Sakhalin–Hokkaido and Kuril uplifts are also regarded as formed by Mesozoic and probably, Upper Paleozoic rocks, and are considered to be structurally related (Gnibidenko and Snegovskoy 1975; Gnibidenko *et al.*, 1980). The fold system that thus bounds the basin to the west and southeast has an arclike form and probably began forming in the Upper Paleozoic and Mesozoic. Older structural elements can be recognized in this fold system along the boundary of the deep sea

---

Fig. 16. A bathymetric map of the South Okhotsk deep basin. The contours are isochronous lines *T* (where *T* = 1/400 of sound velocity in water). Elements of relief are: *1*, East Sakhalin Plain; *2*, Polevoy Plateau; *3*, Pegas Basin; *4*, Terpeniya Plateau; *5*, Pegas Rise; *6*, Levenorn Basin; *7*, Novoye Plateau; *8*, Vorobievy Gory Rise; *9*, North Hokkaido Plateau; *10*, Monbetsu Rise; *11*, Kitami Rise; *12*, Kitami Valley; *13*, Yubetsu Rise; *14*, Abashiri Basin; *15*, Kitami–Yamato Plateau; *16*, Notori Valley; *17*, Notori Rise; *18*, Abashiri Valley; *19*, Sayri Valley; *20*, Shiretoko Rise; *21*, Ekaterina Basin; *22*, Loskutov Mountain; *23*, Krylatka Mountain; *24*, Hydrographers Ridge; *25*, Berg Mountain; *26*, Friez Basin; *27*, Shokalsky Rise; *28*, Browton Mountain; *29*, Vavilov Mountain; *30*, Obrutchev Mountain; *31*, Mironow Mountain; *32*, Arkhangelsky Ridge; *33*, Bussol Basin; *34*, Otvazhny Mountain; *35*, Pegas Mountain; *36*, Srednyaya Mountain; *37*, Lisyansky Mountain; *38*, Sluchainaya Mountain; *39*, Kruzenstern Basin; *40*, Makarov Mountain; *41*, Edelstein Mountain; *42*, Orlik Mountain; *43*, Fourth Kuril Basin; *44*, Belyankin Mountain; *45*, Pegas Valley; *46*, Morskoy Geofizik Rise.

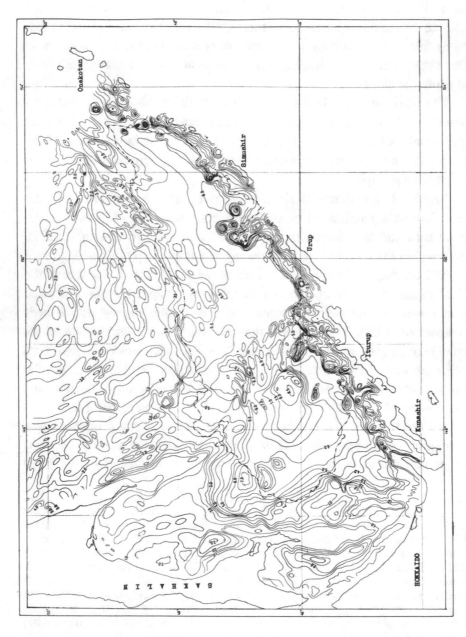

Fig. 17. The relief of the acoustic basement in the southern part of the Sea of Okhotsk (structural contour map). The dash and dot line marks the 3000 m isobath. Contour lines are in km from sea level.

basin. These elements are discordant with the basin contours and extend considerable distances into it.

Within the basin the basement is formed by the rough surface (Fig. 17) of layer 2 in which velocities of 4 km/s have been recorded. Neither the composition nor the age of the basement is known; however, results of deep seismic sounding (Galperin and Kosminskaya, 1964) indicates that the crust approximates to oceanic in type and differs from it only by the presence of a thick sedimentary layer (Figs. 4, 6). These investigations showed velocities of 6.5–7.0 km/s in the lowest crustal layer, presumed to be composed of gabbro to basaltic rocks resting on a periodotitic mantle characterized by minimum velocities of 8.0 km/s. Snegovskoy (1974), Popov et al. (1978), and Bikkenina and Argentov (personal communication) in later investigations found a velocity of 5 km/s for layer two which is probably composed of basic volcanogenic and sedimentary rocks. The overlying sediments of layer 1 have a thickness of 0.5–1.0 km. They also indicated that the upper part of basaltic layer 3 was 0.5–1.5 km thick.

As there are basement outcrops of igneous rocks along the basin slopes that have been dated as Lower Mesozoic (Krasny et al., 1981; Gnibidenko and Khvedchuk, 1982) it is concluded that rocks of this age should be traceable into the basement of the basin. A multichannel seismic profile (Fig. 18) across the continental slope near Sakhalin (Fig. 1) shows that the basement top can be traced into the Yuzhno Okhotsk Basin. Below that unit it seems probable that Lower Mesozoic and more ancient rocks may form the upper part of the basement in the basin.

The sedimentary cover in the basin is divided into an upper, well-stratified unit and a lower "transparent" sequence (Snegovsky, 1974; Figs. 6, 7, 18). The upper sequence, about 1 km thick, is apparently made up of an alternation of turbidites and pelagic oozes with intercalations of ash. Seismic P velocities increase from 1.8 km/s at the top of the unit to 2.5 km/s at the base. The 3 km thick acoustically transparent sequence below, is presumed to consist of pelagic clays and argillites in which seismic velocities vary with depth from 2.5 to 4.3 km/s.

Sedimentation of the upper turbidite sequence began in late Miocene early Pliocene time (Snegovsky, 1974). If a sedimentation rate of 2 mm/1,000 yrs can be assumed, even without taking sedimentary compaction into account, deposition of the acoustically transparent layer may have begun in pre-Upper Cretaceous time. If compaction has reduced the sediments to half their original thickness a Lower Mesozoic or even Upper Paleozoic age is not impossible. If this is true then the basin morphology in roughly its present form may extent back to pre-Cretaceous time. The margins of the basin seem to be defined by early Mesozoic folded complexes. The history of these systems seems to indicate that by Lower Paleogene the basin may have been partly closed by geanticlinal uplifts of island arcs. That there were substantial land areas in the Neogene seems predicated by the occurrence of turbidites, which suggests the island arcs were extant by the beginning of the Neogene.

The undeformed sediments with growth faults recognized in some parts of the basin suggest stable depositional conditions in relatively deep water, probably since pre-Cretaceous time. The sediments were laid down over the rough surface of layer 2. The absence of deformation in the sedimentary cover militates back-arc spreading in the basin during the period of its deposition.

Recent (Holocene) sedimentation in the Yuzhno Okhotsk Basin seems to follow an inherited pattern. It is characterized by a regular decrease in grain size of the terriginous sediments towards the center of the abyssal zone where only aleuritic oozes, probably a distal turbidite deposit, are recorded (Iliev *et al.*, 1979). The regime for the basin is miogeosynclinal. The heat flow, twice the average level (Ehara, 1978) indicates a relatively hot upper mantle below the basin. The alternation of layers of higher (8 km/s) and lower (7 km/s) seismic velocities down to depths of 30 km (Starshinova, 1980) seems to indicate partial melting. It is hard to account for the layered mantle structure below the basin if appeal is made to secondary convection flow subducting beneath the Kuril Arc.

D. Deep Structure of the Sea of Okhotsk

Three main crustal types are distinguished in the Sea of Okhotsk, each corresponding to one of the main morpho-structural elements. Continental crust typically occurs in Kamchatka, the northeast of the USSR, in the Sakhalin and in the Kuril geanticlinal uplift. The central part of the Okhotsk platform and the central part of the Kuril geanticlinal uplift have crust of subcontinental type. Suboceanic crust underlies the Yuzhno Okhotsk deep basin. Crustal thicknesses vary from 12 to 35 km (Figs. 19, 20). Deep seismic sounding indicates that the upper mantle below the Yuzhno Okhotsk Basin is characterized by anomalously low C values and high seismic attenuation to depths of 30 km below the crust (Starshinova, 1980). This anomalous upper mantle is related to layered structure with alternating P-wave velocities of 7.0 and 8.0 km/s, a difference regarded as due to composition.

The crust of the whole Okhotsk Sea consists of metabasalts, granite, metamorphic and sedimentary–volcanic layers with a total thickness of no more than 25–30 km. Subcrustal P velocities are about 8 km/s and below the K discontinuity range between 6.6 and 6.7 km/s. Granitic metamorphic crust, clearly distinguishable in the northern part of the Sea of Okhotsk, is

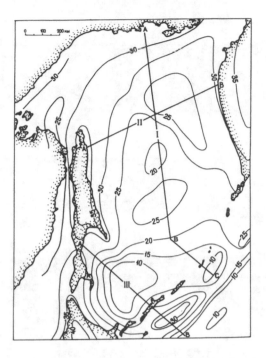

Fig. 19.   Crustal thickness in the Sea of Okhotsk. Data sources: Gilperin and Kosminskaya, 1964; Zverev and Tulina, 1971. The location of sections of Figure 20 are shown.

replaced south of the central part of the sea by a metadioritic crust. Below the sedimentary layer, seismic velocities in the granite and metamorphic crust of between 5.5 km/s and 6.0 km/s are more typical (Galperin and Kosminskaya, 1964; Tulina, 1969; Zverev and Tulina, 1971).

The heat flow values recorded by Watanabe *et al.* (1977), Ehara (1978) and shown in Fig. 21 point to different thermodynamic conditions at comparable depths (Tikhomirov, 1970). The considerable isostatic unbalance in the central part of the Sea of Okhotsk seems to require continued vertical movement (Giananov *et al.*, 1971) although recent seismic activity in the area is not great (Poplavskay and Oskorbin, 1977).

E. Geological History of the Sea of Okhotsk

The general conclusion from the data presented is that the basement of the Sea of Okhotsk Platform is made up of folded geosynclinal beds of various ages over which lies a sedimentary cover. The deposition of this cover began not earlier than Oligocene and it consists of rocks of Upper Paleogene and Neogene–Quaternary age.

The oldest structural element is probably the Okhotsk Arch made up of deformed eugeosynclinal early Proterozoic (Grinberg, 1968) or Archean

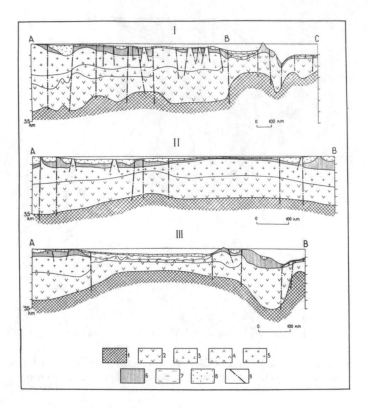

Fig. 20.   Sections across the Sea of Okhotsk (location shown in figure 19): *1*, Upper mantle; *2*, metabasalt layer; *3*, diorite (meta-andesitic) layer; *4*, second metabasalt layer in the South Okhotsk deep basin; *5*, granite–metamorphic layer; *6*, volcanic–sedimentary geosynclinal complex; *7*, probable pelagic clays and argillites; *8*, terrigenous sedimentary cover; *9*, faults.

sediments (about 4 b.y. according to Korolkov *et al.*, 1974) according to estimates based on studies of its exposed northern continuation, the Okhotsk continental block. The eugeosynclinal system was transformed to continental crust in Upper Proterozoic and since then has been involved in other orogenic regimes up to the Mesozoic, if a history similar to the exposed arc can be assumed (Gavrikov 1965; Veldyakov *et al.*, 1970; Chikov, 1970). From middle Paleogene time the Okhotsk Arch was probably subaerial forming a rigid massif until, in Pliocene times, it subsided and was covered by Pliocene–Quaternary deposits.

Available data (Drabkin, 1970; Bely and Kotlyar, 1975) indicate that the geosyncline that evolved into the Okhotsk fold system lying north of the Okhotsk Arch began forming in Precambrian time. In Upper Paleozoic a miogeosynclinal sequence can be identified. Uplift of the axial region dates from Triassic–Jurassic time with andesitic volcanism and with further Cre-

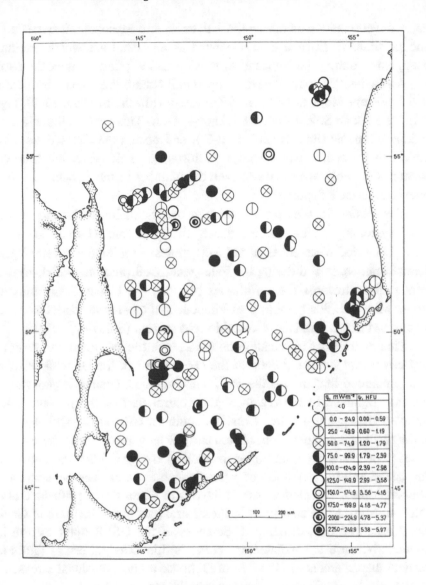

Fig. 21. Heat flow in the Sea of Okhotsk. Data compiled by O.V. Veselov from Watanabe *et al.* (1977), Ehara (1978), and data from the Sakhalin Complex Science Research Institute.

taceous–Paleogene volcanism (Osipov, 1975) in the Okhotsk branch of the volcanic belt. Taphrogenic activity at the end of Paleogene time resulted in subsidence of the Okhotsk Sea Platform affecting in particular the region that was formerly the geosynclinal basin.

The Yurovsky block (Korolkov, 1972) and by extension the Kasherov swell (Fig. 3) along the southwestern edge of the Okhotsk Arch developed as a miogeanticlinal ridge in Upper Cambrian–Ordovician time. While its major development was during the Paleozoic and Mesozoic, it persisted to Eocene time. In Eocene time it existed as an island arc which gradually disappeared during the Neogene as erosion and subsidence took their toll.

An ensimatic geosyncline developed in Precambrian time in the location of the present Mongol–Okhotsk fold system (Shashkin, 1969, 1970), separating the Aldan Shield from the Okhotsk Arch. This geosyncline closed in middle Mesozoic time (Nagibina, 1969), and during the Cretaceous a volcanic belt formed, it was a land area during the Paleogene, but following erosion and subsidence it was covered beginning in the Pliocene by Neogene–Quaternary deposits.

The ensimatic Western Kamchatka and Sakhalin–Hokkaido geosynclinal systems probably formed during the Paleozoic (Gnibidenko *et al.*, 1974, 1975) and were zones of volcanic molasse sedimentation until geanticlinal uplift occurred during the Paleogene. Deformation spread from the axial regions towards the peripheries between the Upper Cretaceous and Upper Miocene. Posthumous post-Pliocene and Quaternary movements are recognized in the Sakhalin–Hokkaido fold system (Soloviev, 1972).

Piecing together information on the ages of the basement and sedimentary cover and structural data on the central part of the Okhotsk Platform it is concluded that the middle or Upper Paleozoic ensimatic geosynclinal system originated in the region of the central part of the present Sea of Okhotsk occupying the area from the southern edge of the Okhotsk Arch to the recent continental slope. From early Paleogene onwards there existed a series of island arcs, and miogeosynclinal troughs. With the end of the Paleogene and the cessation of volcanism, the region was peneplaned by Miocene times. The central part of the Sea of Okhotsk then subsided about 1,000 m. Subsidence also affected the central part of the Institute of Oceanology Rise and the Academy of Sciences of the USSR uplift by the late Pliocene because the sedimentary cover is thin. The pattern of faults and vertical displacements in the Sea of Okhotsk require tensional stresses directed in a submeridional, northeast–southwest direction.

Volcanism and the formation and growth of the Kuril geanticline uplift seems to have begun in pre-Cretaceous time possibly as early as early Mesozoic. Beginning with the Paleogene eugeanticlinal uplift was located along the line of the Lesser Kuril ridge (Gavrilov and Solovieva, 1973), and it seems probable that the intensive development of the Greater Kuril Islands ridge dates from the same time. Since the Neogene it has behaved as a mature uplift with the development of andesites (Piskunov, 1975; Sergeev, 1976). The uplift separated the Yuzhno Okhotsk Basin from the Pacific Ocean. Pelagic sedimentation resulted in the formation of an acoustically transparent layer that seems to consist of clays and argillites. Only in Upper Miocene was it supplanted by the turbiditic-pelagic sedimentation that persisted until recent times in association with a thick accumulation of diatoms (Bezrukov, 1960).

The Okhotsk Platform and the Yuzhno Okhotsk Basin are both isostatically uncompensated which, in combination with anomalously high heat flow (Fig. 21), testifies to continued upper mantle differentiation. The positive isostatic anomalies under the Yuzhno Okhotsk Basin, and the particularly intense isostatic anomalies of the volcanic Greater Kuril Ridge require such differentiation to maintain their nonequilibrium state. The ridge rose 200–250 m during the Pleistocene (Grabkov and Pavlov, 1972; Fedorchenko and Piskunov, 1974) and this together with recent intensive seismic and volcanic activity and evidence of upper mantle density inhomogeneities (Vaschilov and Gainanov, 1970; Averianova, 1972; Tarakanov, 1972) is supporting evidence of continuing evolution of the basin and arc.

III. THE KURIL ISLANDS RIDGE AND KURIL–KAMCHATKA TRENCH

A. Introduction

The area covers both the Kuril Islands Ridge, the Kuril–Kamchatka Trench (Fig. 22) and its junction with the Aleutian and Japan trenches (Honza, 1977; Gnibidenko *et al.*, 1980, 1982). From the structure and geological history of the geanticlinal uplift of these island arcs and the associated trenches it is possible to evaluate the various geological models of crustal evolution, and in the particular case of the Kuril–Kamchatka Trench the geodynamic processes in the transition zone from the Asian continent to the Pacific Ocean.

B. Kuril Islands Ridge

The Kuril Islands, Ridge (Figs. 3. 22) is made up of an Upper Cretaceous–Cenozoic volcanic–sedimentary complex (Bevz *et al.*, 1971; Gavrilov and Solovieva, 1973; Sergeev, 1976) consisting of an andesitic sequence (Piskunov and Gavrilov, 1970; Piskunov, 1975) and a volcanic molasse. The internal structure of the ridge is one of alternating horsts and grabens complicated by brachyform folds and tectonic depressions associated with volcanic activity.

The volcanic complex provides evidence of the existence of the ridge since early Miocene time (Gavrilov and Solovieva, 1973). A Paleogene (and possibly Mesozoic) metamorphosed geosynclinal complex may exist below

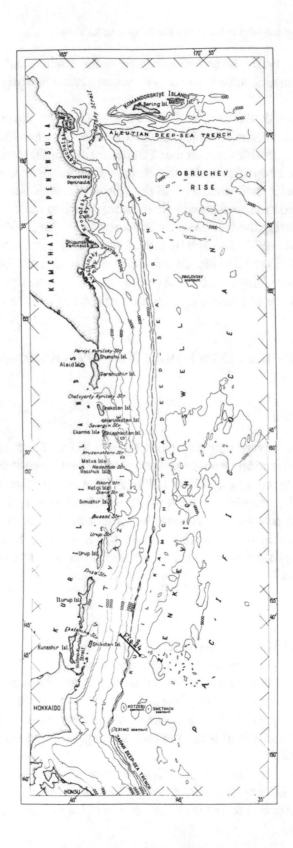

the Neogene. Within beds of the Neogene complex, xenoliths are found which, according to Fedorchenko and Rodionova (1975), represent the pre-Cenozoic ophiolitic basement of the ridge. They consist mostly of meta-morphosed basic rocks.

The anomalous upper mantle below the Kuril geanticlinal uplift is marked by an alternation of four asthenospheric layers within the seismo-focal zone associated with magma formation. This is confirmed by the at-tenuation of body waves at depths of 70–100 km below the volcanic arc (Fedotov and Boldyrev, 1969). Uncertainty in the recognition of the M dis-continuity seems to result from the passage of magma and fluids through it. Focal mechanism studies (Averianova, 1975) suggest that below the ridge, zones of compression and tension alternate. These seem to correspond with zones of mantle differentiation.

C. Tectonics of the Kuril–Kamchatka Trench

Seismic data and deep-sea drilling and dredge hauls in the Kuril–Kam-chatka, Aleutian and Japan trenches and in the adjacent areas of the north-west Pacific plate have provided information on the acoustic basement and its sedimentary cover (Fig. 23).

1. Basement

The acoustic basement of the trench continental slope consists of con-siderably deformed and metamorphosed igneous and sedimentary rocks in which seismic P waves exceed 3–4.0 km/s. In the axial zone of the Lesser Kuril Islands horst-anticlinorium metamorphic rocks with P-wave velocities of 5 km/s are found. The age of these deformed geosynclinal sediments is Paleogene–Mesozoic and possibly they may be as old as Upper Paleozoic (Gnibidenko et al., 1980). Lower Miocene deposits also seem to be incor-porated in the acoustic basement in the volcanic segment of the Kuril uplift.

The acoustic basement of the trench oceanic slope coincides with the top of oceanic layer two and consists of basalts in which some sedimentary layers appear to be incorporated. P-wave velocities range from about 4.3 to 5.6 km/s in the basalts. Siliceous horizons and porcellanites alternating with clays and nanno-oozes have been recorded on top of the basement at DSDP sites 303 and 304. Seismic P-wave velocities in these beds reach 3.0–5.0 km/s due to the presence of the siliceous interbeds. The thickness of the layers is about 50 m (Creager Scholl et al., 1973; Larson et al., 1975).

Fig. 22.   A Bathymatric map of the Kurile–Kamchatka Trench (from Gnibidenko et al., 1980). The location of the profile in Figure 24 is shown.

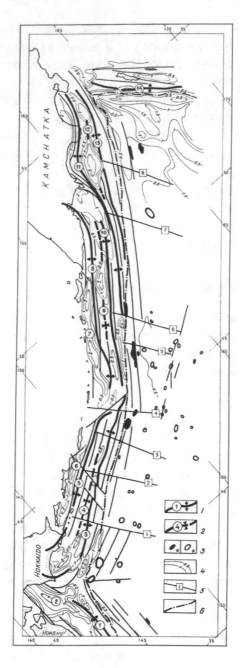

Fig. 23.   General pattern of the Kurile–Kamchatka Trench tectonics: *1*, horst (anticlinorial) uplifts; *2*, graben (synclinorial) troughs; *3*, acoustic basement outcrop on the seafloor (a) and (b) on the slopes of submarine volcanoes; *4*, isopachs in km of the sedimentary thickness in compensated and uncompensated basins; *5*, principal faults; *6*, thalwegs of deep-sea trenches. The main structural elements: ① Oyashio Horst (anticlinorial) uplift); ② Ishikari Graben (synclinorial troughs); ③ Hidaka–Frontal Horst (anticlinorial uplift); ④ Vneshny Graben (synclinorial trough); ⑤ Lesser Kuril Ridge Horst (anticlinorial uplift); ⑥–⑦ Central Kuril Graben

3. *Cover*

The sedimentary cover on the continental slope fills a system of graben-synclinal troughs (Fig. 23). The sediments are considerably less deformed than the acoustic basement underlying them. Seismic *P*-wave velocities vary from 1.6 to 2.2 km/s near the surface to 3.0 to 3.2 km/s at the base of the cover in the troughs.

On the continental slope two or three layers, sometimes separated by angular unconformities can be recognized. The internal structure of the sediments in the trenches usually shows good bedding reflecting the synsedimentational character of turbidite deposits. There is also evidence of slump structures in the sediments of the continental slope.

Over the oceanic plate flat-lying sediments averaging 250–300 m in thickness occur. In some places the sedimentary thickness increases to 400–600 m or even to as much as 1000 m at the northern end of the Emperor Seamounts (Obruchev Rise). The sediments are generally well stratified and conformably overlie basement. From continuous seismic profiling and deep-sea drilling data they can generally be divided into an upper and lower sequence.

In some parts of the thalwegs of the Kuril–Kamchatka, Aleutian and Japan trenches the lower sequence of sediments is overlain by a well-stratified turbidite sequence that may be 1 km thick. Within the trench thalweg these turbidites are flatlying and without perceptible deformation but in some parts of the Kuril–Kamchatka Trench up to 1,000 m of slump beds may also be distinguished (Gnibidenko *et al.*, 1980). These slump beds are characterized by poorly correlated reflections or by the absence of reflections and by a rough, poorly reflecting lower surface.

D. Geological Development

1. *Paleogeography*

The information of the rock associations of the acoustic basement on the trench continental slope suggest an advanced stage in the tectonic development of the Hokkaido–Kamchatka horst-mega-anticlinorium. The Hidaka–Frontal, Avacha, Kronotsk and, probably, Pri–Kamchatka horst-an-

---

(synclinorial trough) with the South Kuril ⑥ and North Kuril ⑦ basins; ⑧ Pegas Horst (anticlinorial uplift); ⑨ Avacha Graben (synclinorial trough); ⑩ Avacha Horst (anticlinorial uplift); ⑪ Kronotsk Horst (anticlinorial uplift), ⑫ Kronotsk Graben, (synclinoral trough), ⑬ Near Kamchatka Horst (anticlinorial uplift); ⑭ South Komandor Horst (anticlinorial uplift). The principal faults: ① Tuscarora; ② Iturup; ③ Urup; ④ Bussol; ⑤ Onekotan; ⑥ Paramushir; ⑦ Avacha; ⑦ Kronotsk.

ticlinorial uplifts (Fig. 23) are characterized by approximately normal continental crustal thicknesses (25–45 km). The structural relations between the Lesser Kuril Islands and the Hidaka horst-anticlinorial uplifts suggest that the Paleogene–Miocene structural complex is superposed on the north-western extension of the pre-Paleogene complex of the Hidaka–Frontal horst-anticlinorial uplift. That is, that the late geosynclinal formations of the Lesser Kuril Islands developed in pre-Paleogene time on the sialic basement of the Hidaka–Frontal zone. This latter zone appears to have developed as a result Upper Cretaceous–Paleogene deformation (Tanaka and Nozawa, eds., 1977) of rocks that began to form in Lower Mesozoic–Upper Paleozoic time. K/Ar ages of diorite dredged southeast of the Hidaka–Frontal uplift (Site 1113 Gnibidenko et al., 1980) yielded an apparent Upper Carboniferous age (327 m.y.).

Prior to the Upper Cretaceous–Lower Paleogene deformation of the Hidaka–Frontal, Avacha, Kronotsk and Pri–Kamchatka structural zones, mature geanticlinal uplift had already occurred in the positions now occupied by the Pegas and Lesser Kuril Islands. The axial regions of the latter two were uplifted and thus the Paleogene–Miocene deformation came to be imposed on the sialic basement of the Hidaka–Frontal, Avacha, Kronotsk and Near Kamchatka horst-anticlinorial uplifts which had gradually subsided from the beginning of the Paleogene. As these zones are at the edge of the northwest Pacific plate, clearly the Paleogene was the time when the trench morphology began to develop.

The principal deformation of the Pegas and Lesser Kuril Islands horst-anticlinorial uplift seem to have occurred in Upper Miocene–Lower Pliocene time. In conjunction with these movements not only did folding of the horst-anticlinorium take place, but also bevelling of the axial zone and the development of the platform (shelf) occurred.

The association of Miocene–Pliocene volcanics and deposits of the Greater Kuril Islands uplift and southern Kamchatka giving the andesitic association referred to earlier (p. 405), characterize a mature stage of geosynclinal development. Thus clearly the beginnings of the geosynclinal system must be assigned to the Paleogene or earlier. From late Miocene to Holocene the Greater Kuril Islands uplift was upwarped and the formation of intrageanticlinal South Kuril and North Kuril troughs as a single sedimentary basin dates from this time.

2. Recent Dynamic Regime

A large part of the transition zone between the Asiatic Craton and the Pacific, involving deep-sea basins, island arc systems and oceanic trenches can be ascribed to a geosynclinal system. The Kuril–Kamchatka trench and

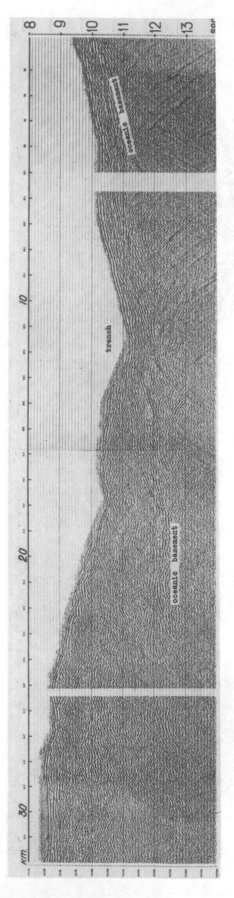

Fig. 24. A multichannel (24) seismic reflection profile across the southern part of the Kuril–Kamchatka Trench recorded on **RV** *Morskoy Geofizik*, 1979. For location see Figure 22.

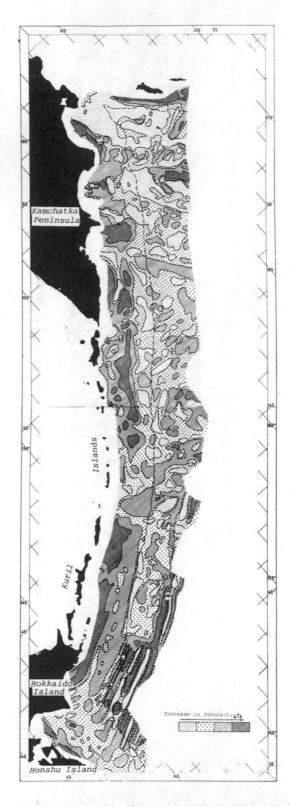

Fig. 25.   Magnetic anomaly map of the Kuril–Kamchatka Trench (from Gnibidenko 1980).

its junctions with the Aleutian and Japan trenches are elements of this system in different stages of development. In this terminology the Greater Kuril Islands geanticlinal uplift is in a eugeanticlinal state whereas the continental slope and deep-sea trench are in a miogeanticlinal regime with turbidite sedimentation. The Obruchev and Komandor rises also are in the latter regime. The adjacent northwestern Pacific plate is, in the same terms, in a pregeosynclinal thalassocratonic regime.

If the attempt is made to relate the structural pattern of the Sea of Okhotsk to the northwesterly motion of the Pacific plate since Miocene times as postulated by Hilde *et al.*, (1976, 1977) it is difficult to develop an internally consistent scheme of geodynamics for the continent-ocean transition. The plate tectonics model to explain the crustal structure of the trenches requires, simultaneously, different directions of movement of the northwest Pacific plate in neighboring areas. Thus to explain the evolution of the active Asiatic continental margin in terms of plate tectonics, a model of differently oriented plate motions with different rotations is required, which in turn implies an improbable pattern of mantle movement.

The seismicity recorded along the Kuril–Kamchatka continental slope indicates vertical block movements along steep and subvertical faults that penetrate the crust and the Benioff zone and are oriented parallel to and transverse to the trench (Balakina, 1979, 1981). These data, together with the indications that the Pacific plate dips at a shallow angle (10°, Fig. 24) beneath the Kuril–Kamchatka trench as it does beneath the Japan trench (6–7° over a 50 km distance; Matsuzawa *et al.*, 1980) and the generally nonlinear pattern of magnetic anomalies (Fig. 25) make it difficult to interpret the Kuril–Kamchatka trench geodynamics in terms of plate tectonics only.

REFERENCES

Aoki, S., 1967, Report of the *Hecauteus* Expedition-1 in the Okhotsk Sea, 1966, *Oceanogr. Mag.* v. 19, p. 57–59.

Alperovich, I. M., and Chernyavsky, G. A., 1973, Thickness of the northern Sakhalin sedimentary deposits from the magnetotelluric sounding data. *Oil Gas Geol.*, v. 6, p. 55–59 (in Russian).

Averianova, V. N., 1972, Seismological description of processes occurring within upper mantle of the northwestern Pacific, in: *Deep Structure of the Far East Marginal Seas and Island Arcs*, Touezov, I. K. and Asano, S., eds., *Proc. Sakhalin Complex Sci. Res. Inst.*, v. 33, p. 229–244 (in Russian).

Averianova, V. N., 1975, *Deep seismotectonics of Island Arcs (Northwestern Pacific)*, Moscow: Nauka, 219 p (in Russian).

Balakina, L. M., 1979, Orientation and ruptures and movements in the sources of strong earthquakes of the north and northwest parts of the Pacific Ocean, *Phys. Earth*, v. 4, p. 43–52 (in Russian).

Balakina, L. M., 1981, Source mechanism of intermediate earthquakes in the Kuril–Kamchatka focal zone. *Phys. Earth*, v. 8, p. 3–24 (in Russian).

Beloussov, V. V., 1968, *The Earth's Crust and Upper Mantle of Oceans*. Moscow: Nauka, 256 p, (in Russian).

Bely, V. F., and Kotlyar, I. N. 1975, New data on geology of the Piyagin Peninsula western part, in: *The Data on Geology and Mineral Resources of the Northeast of the USSR*, Shilo, N. A., ed *Proc. Northeast Complex Sci. Res. Inst.*, v. 22, p. 74–85 (in Russian).

Bevz, V. E., Smirnov, I. G. and Korolkova, T. P. 1971, On the geological structure of Big Kuril ridge islands, *Proc. Sakhalin Geog. Soc. USSR*, v. 2, p. 33–101 (in Russian).

Bezrukov, P. L., 1960, The Okhotsk Sea bottom deposits, *Proc. Oceanol. Inst. USSR*, v. 32, p. 15–95 (in Russian).

Burk, C. A., and Gnibidenko, H. S. 1977, The structure and age of acoustic basement in the Okhotsk Sea, in; *Island Arcs, Deep Sea Trenches and Back-Arc Basins*. Talwani, M. and Pitman III, W. C., eds. *Maurice Ewing Ser.* v. 1, Washington, D.C.: American Geophysical Union, p. 451–466.

Chikov, B. M., 1970, *Tectonics of the Okhotsk Medial Massif*, Moscow: Nauka, 152 p. (in Russian).

Creager, J. S., and Scholl D. W. *et al.*, 1973, *Initial Reports of the Deep Sea Drilling Project*, v. 19, Washington D. C.: United States Government Printing Office, 980 p.

Drabkin, I. E., ed., 1970, *Northeast of the USSR. Books 1 and 2*, v. 30, Moscow: Nedra, 548 p. and 536 p. (in Russian).

Ehara, S., 1978, The heat flow in Hokkaido–Okhotsk sea region and its tectonics implications, in: *Structure and Geodynamics of Northwest Pacific Lithosphere According the Geophysical Data*. Soloviev, S. L., ed., Vladivostok: Far East Central Academy of Sciences of the USSR, p. 86–98 (in Russian).

Fedorchenko, V. I., and Piskunov, B. N. 1974, On recent vertical tectonic movements of the Kuril Island arc, in: *Problem of Geology of Kuril and Sakhalin Islands*, Zhidkova L. S. ed., *Proc. Sakhalin Complex Sci. Res. Inst.*, v. 31 p. 158–163 (in Russian).

Fedorchenko, V. I., and Rodionova, R. I. 1975, *Xenoliths from the Lavas of Kuril Islands*, Novosibirsk: 139 p. (in Russian).

Fedotov, S. A., and Boldyrev, S. A. 1969, On absorption of body waves in the Earth's crust and upper mantle of the Kuril Island arc, *Phys. Earth*, v. 9, p. 17–33 (in Russian).

Gainanov, A. G., Isaev, E. N., Stroev, P. A., and Ushakov, S. A. 1971, Isostasy and crustal structure of the Okhotsk Sea region. *Geophys. Bull. Intergovernmental Geoph. Commission Acad. Sci. USSR*, v. 2, p. 37–43 (in Russian).

Gainanov, A. G., Pavlov, Yu. A., Stroev, P. A., Sychev, P. M., and Touezov, L. K. 1974, *Anomalous Gravity Fields of the Far East Marginal Seas and the Adjacent Part of the Pacific Ocean,* Novosibirsk: Nauka, 108 p. (in Russian).

Galperin, E. G. and Kosminskaya, I. P., eds., 1964, *Crustal Structure in the Asia-to-Pacific Transition Zone*, Moscow: Nauka, 307 p. (in Russian).

Gavrikov, S. I., 1965. On the history of geological development of the Okhotsk massif, *Sov. Geol.* v. 2, p. 71–81 (in Russian).

Gavrilov, V. K., and Solovieva, N. A., 1973, *Volcanogeneous-Sedimentary Formations of Geoanticlinal Uplifts of the Greater and Lesser Kuril Islands*, Novosibirsk: Nauka, 152 p. (in Russian).

Geodekyan, A. A., Neprochnov, Yu. P., Yelnikov, I. N. and Yaroshevskaya, G. A. 1980, The results of the seismic studies on the Academy of Sciences of the USSR Rise in Okhotsk Sea, *Proc. Acad. Sci. USSR*, v. 250, p. 919–923 (in Russian).

Geodekyan, A. A., Neprochnov, Yu. P., Yelnikov, I. N., Yaroshevskaya, G. A. Trotsyuk, V. Ya., Pokryshkin, A. A. and Sheina, L. P. 1978, New data on deep structure of the TINRO Basin in the Okhotsk Sea, *Proc. Acad. Sci. USSR*, v. 243, p. 449–452 (in Russian).

Geodekyan, A. A., Trotsyuk, V. Ya., Ulmishek, G. F. and Pilyak, V. L. 1977, About katagenetic transformation of the organic matter in Okhotsk Sea, *Sov. Geol.* v. 1 p. 99–108 (in Russian).

Geodekyan, A. A., Udintsev, G. B., Beresnev, A. F., and Trotsyuk, V. Ya., 1974, Geological, geophysical and geochemical investigation in the Okhotsk Sea. *Sov. Geol.* v. 1 p. 43–52 (in Russian).

Gnibidenko, H. S., 1969, *The Metamorphic Complexes in the Northwest Sector of the Pacific Belt Structures*, Moscow: Nauka, 136 p. (in Russian).

Gnibidenko, H. S., 1976. On the rift system of the Okhotsk sea floor. *Proc. Acad. Sci. USSR*, v. 229, p. 163–165 (in Russian).

Gnibidenko, H. S., 1977, The hydrocarbons potential of the Okhotsk Sea. in: *Geology and Hydrocarbons in Sakhalin*, B. K. Ostisty ed., *Geo. Soc. USSR* Leningrad, p. 7–14 (in Russian).

Gnibidenko, H. S., 1979, *The Tectonics of the Far East Marginal Seas*, Moscow: Nauka, 161 p. (in Russian).

Gnibidenko, H. S., Bykova, T. G. Veselov, O. V. Vorobiev, V. M. Un, K. C., and Tarakanov, R. Z., 1980. *The Tectonics of the Kuril–Kamchatka Deep-Sea Trench*, Moscow: Nauka, 180 p. (in Russian).

Gnibidenko, H. S., Bykova T. G., Veselov O. V., Vorobiev V. M., and Svarichevsky, A. S., 1982, The tectonics of the Kuril–Kamchatka deep-sea trench, in: *Geodynamics Programme Final Report v. 1: Western Pacific and Indonesian Region*, Hilde, T. W. C., ed., Washington, D.C.: America Geophysical Union.

Gnibidenko, H. S., Gorbachev, S. Z., Lebedev, M. M. and Marakhanov, V. I. 1974. Geology and deep structure of Kamchatka Peninsula, *Pac. Geol.*, v. 7 p. 1–32.

Gnibidenko, H. S., and Iliev, A. Ya. 1976, On composition, age and seismic wave velocities of the acoustic basement of the Okhotsk Sea central part, *Proc. Acad. Sci. USSR*, v. 229 p. 431–434 (in Russian).

Gnibidenko, H. S., and Khvedchuk, I. I., 1982, The main features of the Okhotsk Sea tectonics, in: *Geological Structure of the Okhotsk Sea Region*. Korney, O. S. ed., Vladivostok: Far East Center Academy of Sciences USSR, p. 3–25 (in Russian).

Gnibidenko, H. S., Saito K., Zassu S., and Ozima M., 1975, On the age and genesis of the rocks from the basement of the Sakhalin Island. in: *Natural Resources of the Sakhalin, Its Development and Using*. Gnibidenko, H. S., ed., Yuzhno–Sakhalinsk: *Sakhalin Branch* of the Geographical Society of the USSR, p. 93–103 (in Russian).

Gnibidenko, H. S., and Snegovskoy, S. S., 1975, Structural relation between Sakhalin and Hokkaido, *Proc. Acad. Sci. USSR*, v. 224 p. 1391–1394 (in Russian).

Gnibidenko, H. S., and Sychev, P. M. 1972, Crustal structure and evolution of the Bering, Okhotsk and Japan Seas. Montreal, *Proc. 24th Int. Geophys. Conf.* p. 43–46.

Grabkov, V. K., and Pavlov, Yu. A., 1972, Crustal recent movements and isostatic conditions within the Kuril Island arc region, *Proc. Acad. Sci. USSR*, v. 203 p. 650–653 (in Russian).

Grinberg, G. A., 1968, *The Okhotsk Precambrian*. Moscow: Nauka, 187 p. (in Russian).

Hilde, T. W. C., Isesaki N., and Wageman, J. M. 1976, Mesozoic seafloor spreading in the North Pacific. in: *Geophysics of the Pacific Oceanic Basin and Margin, Geophysics Monograph 19*, Washington, D.C.: American Geophysical Union, p. 205–226.

Hilde, T. W. C., Uyeda S., and Kroenke, L., 1977, Evolution of the western Pacific and its margin, *Tectonophysics*, v. 38, p. 145–165.

Honza, E., ed., 1977, *Geological Investigations of the Southern Kuril Trench and Slope Areas, GH76-2 Cruise April–June, 1976*, Kawasaki-shi: Geological Survey of Japan, 127 p.

Honza, E., ed., 1978, Geological Investigations of the Okhotsk and Japan seas off Hokkaido, GH77-3 Cruise, June–July, 1977, *Geol. Surv. Japan Cruise Pap.* 11, 72 p.

Iliev, A. Ya., Voronova, V. A. Zakharova, M. A. Nesterova, O. N., Tarakanova, L. I., Sheremetieva, G. N. and Shustov, L. N. 1979. *Bottom Sediments of the Okhotsk Sea Southern Part*, Moscow: Nauka, 148 p. (in Russian).

Karig, D. C. E., 1971, Origin and developments of marginal basins in the western Pacific, *J. Geophys. Res.*, v. 76, p. 2542–2561.

Korenbaum, S. A., Mishkin, M. A., Gnibidenko, H. S., Valui, G. A., and Kurentsova, N. A., 1977, Magmatic complexes and processes of metamorphism in the Okhotsk seafloor rocks. in: *Mineralogy and Petrography of Metamorphic and Metasomatic Rocks of the Far East Center*, Korenbaum, S. A., and Avchenko, O. V., eds., Vladivostok: Far East Center Academy of Sciences USSR, p. 51–79 (in Russian).

Kornev, O. S., Neverov, Yu. L. Ostapenko, V. F., Zhigulev, V. V., Kichina, E. N. Krasny, M. L. and Khomyakov, V. D., 1982, Geological dredging in Okhotsk Sea during RV *Pegas* cruise N 21, in: *Geological Structure of the Okhotsk Sea Region*, Kornev, O. S., ed., Vladivostok: Far East Center Academy of Sciences USSR, p. 36–52 (in Russian).

Korolkov, V. G., 1972, On stratigraphy of the Okhotsk region Lower Ordovician deposits, in: *Data on Geology and Mineral Resources of the Northeast of the USSR*, Drabkin, N. E., ed., *Proc. Northeast Geol. Surv. USSR*, v. 20, p. 31–33 (in Russian).

Korolkov, V. G., Rudnik, V. A. and Sobotovich, E. A., 1974, On late Paleozoic-early Archean age of the most ancient rocks of the Okhotsk medial massif, *Proc. Acad. Sci. USSR.* v. 219, p. 1441–1444 (in Russian).

Krasny, M. L., Neverov, Yu. L. and Kornev, O. S., Ostapenko, V. F. Baranov, V. V. Zhigulev, V. V., 1981, *Geological Structure of the Basement of the South Okhotsk Deep-Sea Basin Frame from the Results of RV Pegas N 21 Cruise*. Novoalexandrovsk: Sakhalin Complex Scientific Research Institute, 20 p. (in Russian).

Kulakov, A. P., 1973, *Quaternary Coastal Lines of the Okhotsk and Japan Seas*, Novosibirsk: Nauka, 188 p. (in Russian).

Larson, R. L., and Moberly R., *et al.*, 1975, *Initial Reports of the Deep Sea Drilling Project*, v. 32, Washington D. C.: United States Government Printing Office, 980 p.

Margulis, L. S., Mudretzov, V. B., Sapojnikov, B. G., Fedotov, G. P., and Khvedchuk, I. I. 1979, Geology of the northwest part of the Okhotsk Sea. *Sov. Geol.* v. 7, p. 61–71 (in Russian).

Matsuda, T., and Uyeda, S., 1971, On the Pacific-type orogeny and its model-extension of the paired belts concept and possible origin of marginal seas, *Tectonophysics*, v. 11, p. 5–27.

Matsuzawa, A., Tamano, T., Aoki, Y., and Ikawa, I. 1980. Structure of the Japan Trench subduction zone, from multichannel seismic reflection records. *Mar. Geol.* v. 35, p. 171–182.

Nagano, M., Sakurai, M., Uchida, M., Ikeda, K., Taguchi, H., and Omori, T., 1974. Submarine geology off northeast coast of Hokkaido district, *Rep. Hydrogr. Res.* v. 9, p. 1–22.

Nagibina, M. S., 1969, *Stratigraphy and Formations of the Mongol–Okhotsk Belt*. Moscow: Nauka, 400 p. (in Russian).

Osipov, A. P., 1975, Magmatic formations of the Okhotsk–Chukotsk volcanic belt, *Geol. Explor.* v. 3 p. 16–22 (in Russian).

Piskunov, B. N., 1975, *Volcanism of the Greater Kuril Ridge and Petrology of Rocks of High-Alumineferous Series the Urup and Simushir Islands Taken as an Example*, Novosibirsk: Nauka, 187 p. (in Russian).

Piskunov, B. N., and Gavrilov, V. K., 1970, Neogene volcanogeneous-sedimentary formations of the Kuril Islands, *Proc. Acad. Sci. USSR*, v. 192, p. 1111–1113 (in Russian).

Poplavskaya, L. N., and Oskorbin, L. S., 1977, Crustal earthquakes activity of the shelf areas of the Okhotsk Sea. in: *Earthquakes of the Kuril Islands, Primorie and Priamurie*. Soloviev, S. L., ed., Vladivostok: Far East Center Academy of Sciences USSR, p. 120–127 (in Russian).

Popov, A. A., and Anosov, G. I., 1978, New data on crustal structure of the Kuril Basin, *Proc. Acad. Sci. USSR*, v. 240 p. 166–168 (in Russian).

Popov, A. A., Anosov, G. A., Argentov, V. V., Bikkenina, S. K., and Sivkova, I. G. 1978, Seismic refraction measurements carried out on the Far East sea polygons. *Geol. Geophys.* v. 10, p. 109–118 (in Russian).

Sergeev, K. F., 1976, *The Kuril Island Arc System Tectonics*. Moscow: Nauka, 239 p. (in Russian).

Shashkin, K. S., 1969, *Peculiarities of Geological Development of Mongol–Okhotsk Region*, Synopsis thesis, Vladivostok, 23 p. (in Russian).

Shashkin, K. S., 1970, Structure-formational zones of Mongol–Okhotsk belt and some features of their development, in: *Some Problems Related with Geochemistry and Metallogeny of Northwestern Pacific Belt*, Govorov, I. N., ed., Vladivostok: Far East Center Academy of Sciences USSR, p. 31–33 (in Russian).

Skorikova, M. F., 1972, On possible utilizing of elastic and density characteristics of rocks for stratification of the Kamchatka deposits, *Proc. Sakhalin Complex Sci. Res. Inst.*, v. 26, p. 54–65 (in Russian).

Smirnov, L. M., 1971, Tectonics of western Kamchatka, *Geotectonics*, v. 3 p. 104–117 (in Russian).

Snegovskoy, S. S., 1974, *Seismic Reflection Studies and Tectonics of the Okhotsk Sea Southern Part and the Adjacent Pacific Margin*, Novosibirsk: Nauka, 88 p. (in Russian).

Soloviev, V. V., 1972, Recent tectonics of the Sakhalin Island, in: *Problems of Study of the Quaternary Period*, Khomentovsky, A. S. and Tseitlin, S. M. eds., Moscow: Nauka, p. 322–326 (in Russian).

Soloviev, S. L., Touezov, I. K. and, Vasiliev, B. I., 1977, The structure and origin of the Okhotsk and Japan Sea abyssal depressions according to new geophysical and geological data, *Tectonophysics*, v. 37 p. 153–166.

Soloviev, S. L., Touezov, I. K., Snegovskoy, S. S., Krasny, M. L., Iliev, A. Ya., Svarichevsky, A. S., Taboyakov, A. Ya. and, Tyutrin, I. I. 1979, The deep structure of the Okhotsk shelf of the central Sakhalin. *Geol. Geophys.* v. 5 p. 104–116 (in Russian).

Starshinova, E. A., 1980, Crustal and upper mantle heterogeneity of the Okhotsk Sea, *Proc. Acad. Sci. USSR*, v. 255 p. 1339–1343 (in Russian).

Suprunenko, O. I., and Schwartz, Ya. V., 1967, Results of the seismic refraction survey in the Bolsheretsk Basin (western Kamchatka), *Sov. Geol.*, v. 3 p. 122–124 (in Russian).

Tanaka, K., and Nozawa, T., eds., *Geology and Mineral Resources of Japan*, v. 1. Kawasaki-shi: Geological Survey of Japan.

Tarakanov, R. Z., 1972, Upper mantle structure peculiarities of the Kuril–Japanese region, in: *Deep Structure of the Far East Marginal Seas and Island Arcs*, Touezov, I. K., and Asano, S., eds., *Proc. Sakhalin Complex Sci. Res. Inst.*, v. 33 p. 220–228 (in Russian).

Telegin, A. N., 1969, New data on geological structure of the Baikal depression, in: *Results of Investigations on Geology and Oil and Gas Prospecting in Sakhalin*. Abramov, A. L., ed., *Proc. All-Union Sci. Res. Inst. Geol.*, v. 255, p. 78–81 (in Russian).

Tikhomirov, V. M., 1970, Thermodynamic conditions within the Earth's crust and upper mantle of the Okhotsk Sea, Kuril Islands and Near-Kuril parts of the Pacific Ocean, in: *Geology and Geophysics of the Pacific Belt*. Gnibidenko, H. S., ed., *Proc. Sakhalin Complex Sci. Res. Inst.*, v. 25, p. 23–33 (in Russian).

Tulina, Yu. V., 1969, Detail seismic investigations of the Earth's crust near the Kuril Islands, in: *Structure and Development of the Earth's Crust in the Soviet Far East*, Fotiadi, E. E., and Touezov, I. K., eds., Moscow: Nauka, p. 90–96 (in Russian).

Udintsev, G. B., Beresnev, A. F., Geodekyan, A. A., Mirlin, E. G. Savostin, L. A., Schreider, A. A. Baranov, V. V. and Belyaev, A. V. 1976, Prelininary data of geological and geophysical investigations carried out in the Okhotsk Sea and in the Northwest Pacific on board RV *Vityaz*, in: *Geological and Geophysical Investigations of the Asia-to-Pacific Transition Zone*, Volvovsky, B. S., and Rodnikov, A. G., eds., *Sov. Radio* (Moscow), p. 19–29 (in Russian).

Vaschilov, Yu. A., Gainanov, A. G., 1970, On the nature of gravity anomalies of the Kuril Islands arc zone, in: *Marine Gravimetric Investigations, 5*, Fedynsky V. V., ed., Moscow: Moscow State University, p. 28–31 (in Russian).

Vassilkovsky, N. P., 1967, On the geological nature of the Pacific mobile belt, *Tectonophysics*, v. 4, p. 583–593.

Veldyaksov, F. F., Ivanov, V. A., Peskov, E. G. and Ryabov, A. V. 1970, Main features of tectonics and development of the Okhotsk massif, in: *Tectonics of Siberia*, v. 4. Bogolepov K. V., ed., Moscow: Nauka, p. 82–91 (in Russian).

Vereschagin, V. N., ed., 1970, *Sakhalin Island, Geological Description*, v. 33, Moscow: Nedra, 431 p. (in Russian).

Verzhbitzky, E. B., Langseth, M. G. and Suzyumov, A. E. 1976, Geophysical investigations during RV *Vema* cruise N 32–13, *Oceanol.* v. 16 p. 181–184 (in Russian).

Vlasov, G. M., ed., 1964, *Kamchatka, the Kuril and Komandor Islands, Part 1, Geological Description*, v. 31, Moscow: Nedra 734 p. (in Russian)

Watanabe T., Langseth, M. G. and Anderson, R. N. 1977, Heat flow in back-arc basins of the western Pacific, in: *Island Arcs, Deep-Sea Trenches and Back-Arc Basins*, Talwani, M., and Pitman III, W. C., *Maurice Ewing Ser*, v. 1, Washington, D.C.: American Geophysical Union p. 137–161.

Yasui M., Hashimoto, Y. and Uyeda, S. 1967, Geomagnetic and bathymetric studies of the Okhotsk Sea (1). Report of the expedition in the Okhotsk Sea (1), 1966. *Oceanogr. Mag.*, v. 19, p. 73–85.

Yasui M., Nagasaka, K., Hashimoto, Y., and Anma, K. 1968, Geomagnetic and bathymetric study of the Okhotsk Sea (2), *Oceanogr. Mag.*, v. 20, p. 65–72.

Zverev, S. M., and Tulina, Yu. V., eds., 1971, *Deep Seismic sounding of the Earth's Crust of the Sakhalin–Hokkaido–Primorie Zone*, Moscow: Nauka, 286 p. (in Russian).

Chapter 9

SEA OF JAPAN AND OKINAWA TROUGH

Kazuo Kobayashi

Ocean Research Institute
University of Tokyo
1-15-1 Minamidai, Nakano-ku
Tokyo, 164 Japan

I. INTRODUCTION

The northeastern margin of the Asiatic continent is fringed by trench–arc–back-arc systems. Best known is a system consisting of the Japan Trench–Nankai Trough, Japanese island arcs, and the Sea of Japan. The Sea of Japan is, as a whole, a marginal sea with an area of about 1 million km^2, bounded on the north by Sikhote Alin eastern Siberia, on the west by the Korean Peninsula, and on the east and south by Honshu and Hokkaido, Japan (Fig. 1.).

In the central portion of the Sea of Japan there exists a topographic high called the Yamato Ridge. It extends southwestward and is connected with the shelf of Honshu by the Oki Bank. The Korea plateau is located in the western part of the Sea of Japan. The deep basin of the Sea of Japan is divided into the Japan Basin, the Yamato Basin, and the Tsushima Basin as shown in Fig. 2.

The Sea of Japan is underlain by an inclined slab of oceanic lithosphere subducted from the Pacific Ocean along the Japan Trench. The slab is extremely long, amounting to about 1,000 km and dips with an angle of approximately 30° to the horizontal plane. It is recognized by a zone of deep seismicity as well as by a slab of high seismic velocity and high Q (Utsu, 1971). Deep earthquakes with depths of 400–700 km are observed beneath

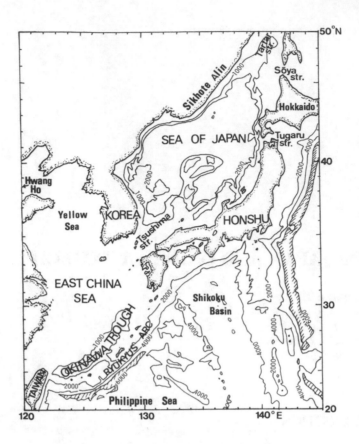

Fig. 1. Index map of the Sea of Japan and Okinawa Trough based upon the GEBCO chart of the NW Pacific. Depth contours of 1000 m and 2000 m only are shown in the back-arc regions and those of 2000 m, 4000 m, and 6000 m only are illustrated in the Pacific side, for simplicity.

the eastern margin of the Asiatic continent. Volcanic activity is now very high on northeastern Honshu, caused by upwelling magmas and heat from the asthenosphere overlying the subducted slab. The basins, however, exhibit no evidence of present crustal extension related to current ocean floor spreading and, therefore, may possibly be old back-arc basins that ceased to open some ages ago.

To the southwest of Kyusyu there exists a chain of islands called the Ryukyus or Nansei-shoto (Fig. 1). The lithosphere of the Philippine Sea is being subducted northwestward beneath this island arc along the Ryukyu Trench. The marginal sea between Ryukyu and the Asiatic continent is the East China Sea. Most of the East China Sea is covered by thick (greater than 1 km) sediment supplied by rivers and wind. Water depths over the thick-sediment area do not exceed 200 m (Emery et al., 1969). A deeper elongated basin, the Okinawa Trough, exists behind the Ryukyu Arc,

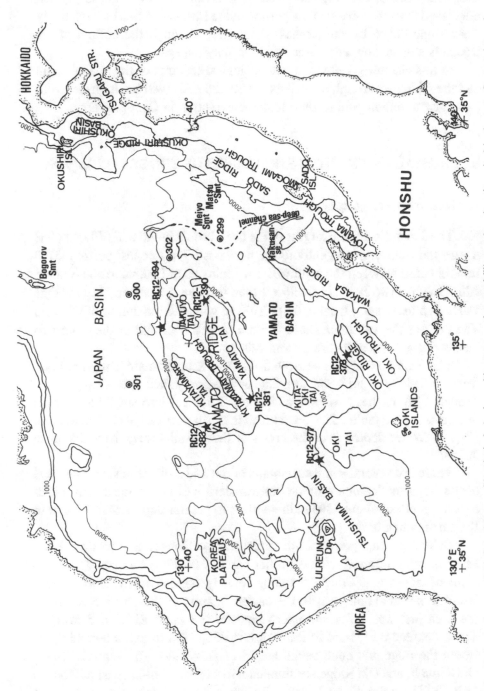

Fig. 2. Bathymetric map of the Sea of Japan with principal physiographic names and some drilling and coring sites cited in the text. Contour interval: 1000 m.

bounded by the arc to the south and east and by the eastern margin of the East China Sea to the north and west. The Okinawa Trough is sharply distinguished from the very shallow portion of the East China Sea by a relatively steep slope edged by the Senkaku-retto Arc, although the bottom of the trough is still shallow compared to other marginal seas.

In this chapter geophysical and geological features of the Sea of Japan and the Okinawa Trough as well as the Ryukyu Arc will be described with particular attention paid to their tectonic evolution in Cenozoic time.

II. CRUSTAL AND MANTLE STRUCTURES OF THE SEA OF JAPAN

A. Submarine Topography

The Japan Basin, the largest of the deep basins in the Sea of Japan, has a very flat floor dipping gently to the northeast. The deepest portion (3700 m) is situated between the southwestern peninsula of Hokkaido and Sikhote Alin. The Yamato Basin is shallower than the Japan Basin and has average water depths ranging from 2500 to 2700 m with the maximum of 2970 m. The floor of the Tsushima Basin is more rugged than that of the other two basins. Its water depths are around 2200 m.

The Yamato Ridge is separated into the Kita–Yamato Tai (bank) and Yamato Tai by a NE–SW trending depression called the Kita–Yamato Trough. The northeastern part of Yamato Tai is cut into small blocks by a few crosscutting faults. The largest high in this part is called the Takuyo Tai (Fig. 2). Water depths over the crests of these banks range from 200 m to 500 m.

There are several small seamounts in the midst of the basins. The crest of the Bogorov Seamount in the northwestern part of the Japan Basin is at a water depth of about 1300 m in contrast to greater depths than 3500 m in the surrounding basin.

Submarine topography in the area near to Honshu and western Hokkaido is quite different from the other parts of the Sea of Japan. The alignment of ridges and troughs trending roughly parallel to the northeastern Honshu is remarkable near to the eastern margin of the Yamato Basin. In the area just west of northeastern Honshu the Sado Ridge and Mogami Trough extend northward to the Okushiri Ridge, the trough intervening between the ridge and continental margin of Hokkaido. The Wakasa Ridge, Oki Trough, and Oki Ridge are situated north off the San-in coast of Honshu (Fig. 2). The trend of the Toyama Trough is different from the others. It is located along the northern seaward extension of the Itoigawa–Shizuoka tectonic line in Honshu and cuts some other tectonic structures. The Toyama

Trough may, therefore, be of a tectonic origin and younger than most of the other topographies, although it has acted as a channel of turbidity currents since the early Pliocene.

B. Crustal Structure

Depths to the Moho boundary and seismic structure of the crust in the Sea of Japan have been determined by two-ship refraction and sono-radio buoy methods (Kovylin *et al.*, 1966; Den, 1972; Murauchi, 1972; Ludwig *et al.*, 1975). These analyses have almost unanimously indicated that the Moho depth of the Japan, Yamato, and Tsushima basins is about 12 km and that their crustal structure is similar to the standard oceanic crust. The structure of the Kita–Yamato Trough has not been clearly shown but some preliminary results (Murauchi, 1972) suggest that it, too, is oceanic.

In the Yamato Tai, Kita–Yamato Tai, and Takuyo Tai, in contrast to the deep basins, the depth to the Moho boundary is much greater, possibly greater than 20 km. Beneath these submarine banks a layer with *P*-wave velocity of about 6 km/sec has been recognized by sono-radio buoy observations and is often considered to be a granitic layer characteristic of the continental crust. However, Ludwig *et al.* (1975) mentioned that the major part of the topography of these banks (the Yamato Ridge as a whole) is built of layers of *P*-wave velocity 3.5 and 5.0 km/sec and that the thickness of the 6.0 km/sec layer beneath them does not significantly differ from that in the Yamato Basin. Yoshii (1979) analyzed available seismic data and showed that an anomalous second layer with *P*-wave velocity of about 3.5 to 3.8 km/sec and average thickness greater than 1 km exists at depths of about 2 km beneath the Yamato Basin, whereas no such a layer is found in the crust of the Japan Basin and other marginal basins (Fig. 3). This layer with relatively low seismic velocity may be composed of a type of altered porous volcanogenic tuffs and lavas that are similar to the so called "green tuff" occurring on land of the inner side of Honshu.

C. Thickness and Nature of the Lithosphere

Dispersion of surface waves (both Rayleigh and Love waves) has indicated that the thickness of the lithosphere beneath the Sea of Japan is, an average, only 30–40 km. (Abe and Kanamori, 1970; Evans *et al.*, 1978). Such a thin lithosphere is consistent with the high heat flow in the Sea of Japan basins (Yasui *et al.*, 1968; Uyeda and Vacquier, 1968). The simple mean of the heat flow values measured at 224 stations is 2.21 HFU (1 HFU = 42 mW/m^2). The averaged value of the Japan Basin is 2.15 HFU (46 stations), and that of Yamato Basin (66 stations) and Tsushima Basin (18

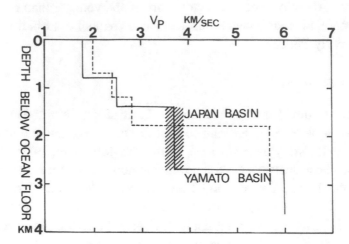

Fig. 3.   Anomalous second layer in the crust beneath the Yamato Basin (Yoshii, 1979). Regional averages of the *P*-wave velocities of the Yamato and Japan basins are shown by solid and dotted lines, respectively.

stations) is 2.15 and 2.35, respectively. Yoshii (1973) obtained the average thickness of lithosphere of about 30 km from these heat flow values.

Shots of 5 tons each of dynamite were fired on July 28 and 30, 1976 at a depth of 175 m in the Japan and Yamato basins about 250 km off northern Honshu. Seismic signals (*Pn* arrivals) were observed at 103 temporary and permanent seismological stations located in Hokkaido, Honshu, Sado and Oki islands (Fig. 4). The horizontal directional dependence of *P*-wave velocity was thus derived as shown in Fig. 5. Considering the distances between the shot points and observation stations, these values provide the *P*-wave velocities of the lithosphere just beneath the Moho layer under the Japan and Yamato basins. The velocities show a distinct directional anisotropy approximated by a sinusoidal curve with an average of 7.94 km/sec and an amplitude of about 0.4 km/sec (5% of the average). The direction of the maximum velocity (8.20 km/sec), N141°E, is roughly perpendicular to the general trend of northeastern Honshu and the magnetic lineations in both basins (Okada *et al.*, 1978).

Similar horizontal anisotropy in the upper mantle was reported from the East Pacific (Raitt *et al.*, 1969) and the northwestern Pacific (Shimamura *et al.*, 1977). In the eastern Pacific the axis of the maximum velocity is roughly parallel to the direction of seafloor spreading. The anisotropy seems to exist in the asthenosphere just below the bottom of the oceanic lithosphere in the northeastern Pacific. Such an anisotropy may possibly originate from a statistical alignment of the crystal axes of olivine, which composes most of the oceanic lithosphere that is formed by the seafloor spreading and that

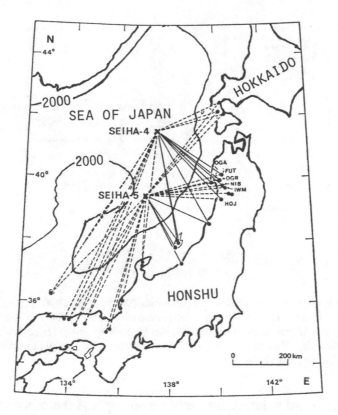

Fig. 4.   Shot points and seismographic stations in the Sea of Japan experiment (Okada *et al.*, 1978). Wave paths with the average interval velocities higher and lower than 7.95 km/sec are shown by solid and dashed lines, respectively.

gradually thickens by cooling as it moves away from the spreading ridge. The mechanism of crystal alignment has still to be elucidated. Existence of the horizontal anisotropy implies, whatever its physico-chemical origin, that the Japan and Yamato basins were formed by a spreading process similar to the normal ocean basins and that direction of the Sea of Japan opening was roughly N141°E.

D. Gravity Anomalies

Surface gravity in the Sea of Japan has been extensively measured mostly by the Japanese research vessels of the Hydrographic Department, Ocean Research Institute and Geological Survey of Japan and compiled by Tomoda and Fujimoto (1982). Free-air anomalies in the basins are roughly zero, indicating that the basins are in an isostatic compensation. The banks have free-air anomalies amounting to 40 milligals and Yoshii (1973) estimated

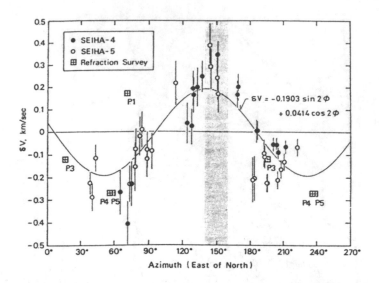

Fig. 5. Velocity anisotropy in the Sea of Japan basins (Okada *et al.*, 1978). Deviations of *P*-wave velocities from the regional average 7.94 km/sec are plotted versus azimuths of wave paths. Rectangles with crossmarks indicate velocities obtained by seismic refraction in the sea (Murauchi, 1972). The shaded area is a direction perpendicular to magnetic lineations. Error bars in velocities are shown in the figure.

from the areal distribution of the gravity anomaly that the Moho depth of the Yamato Ridge (a group of banks) is about 23 km. These banks are out of isostatic balance owing to their small sizes.

Depth-age relationships in the marginal basins have been determined using the DSDP results and seismic reflection data. It is remarkable that the basement depth (water depth plus sediment thickness) in the marginal basins is about 1000 m greater than that of the normal oceans with the same ages (Yoshii, 1972; 1973), except for very young basins that are not isostatically compensated.

E. Magnetic Anomalies

Anomalies in the total force of the geomagnetic field have been measured by a number of Japanese investigators. Results prior to 1973 were summarized by Isezaki and Uyeda (1973) and Kobayashi and Isezaki (1976). The Geological Survey of Japan has added much data measured by R V *Hakurei Maru* (Honza, 1978; 1979). Features of local magnetic anomalies in the Sea of Japan so far identified are as follows;

> (*i*) Peak-to-trough amplitudes of the anomalies in the basins of 100 to 300 gammas (nannotesla, nT). Their wavelengths range from 20 to 40 km.

(*ii*) The linearity of the anomalies is clearly recognized in the central part of the Japan Basin. The trend of linear patterns is nearly N60°E and appears to be parallel to the general trend of southwestern Honshu.

(*iii*) In the western portion of the Japan Basin and the Yamato Basin the linearity is less clear than in the central Japan Basin but still recognizable. The trend is also roughly N60°E. The northeastern portion of the Japan Basin off Hokkaido has linear anomalies trending NE–SW to NNE–SSW.

(*iv*) Magnetic anomalies in the Tsushima Basin are irregular and their linearity is not identified.

Isezaki (1975) analyzed the observed magnetic anomalies and examined their symmetry relative to a postulated axis. One axis of symmetry was found in both the Japan and Yamato basins. Furthermore, offsets of the symmetry axes were shown. If they are really relicts of the past spreading centers, there should be some indications in the bottom and basement topographies, but available information is not sufficient to test their possible occurrence. Isezaki (1979) further attempted to correlate the Japan Basin anomalies with the Heirtzler model of magnetic reversals and indicated that the highest correlation is found with the axial age of 30.6 Ma and rate of spreading of 3 cm/yr. However, correlation with other ages can not be excluded.

In conclusion, no convincing identification of the local magnetic anomalies of any basins with magnetic reversal time seals has been successful in the Sea of Japan. Linearity may be disrupted by many small transform faults. Anomalies may be weakened by overlying lava flows with alternate polarities (Johnson and Hall, 1976) and possible off-ridge magmatism as demonstrated in the Shikoku Basin (Klein *et al.*, 1978; Klein and Kobayashi, 1981) and in the Nauru Basin (Larson *et al.*, 1981). If basaltic layers beneath the Yamato Basin are badly altered by hydrothermal activity as suggested by existence of 3.7 km/sec layer, their magnetic intensity may be much reduced or their original magnetic polarity may be partly lost causing difficulty in the identification of the linearity.

It must be noted here that, in spite of the difficulty in age identification of anomalies, their most plausible origin is no other than the spreading during reversals. As the average frequency of polarity reversals in Tertiary is 1 to 3 per million years (Cox, 1976), correlation of peaks to the normal polarity periods provides the average with wavelengths of 20 to 40 km rate of spreading as 1 to 6 cm/yr (most likely about 3 cm/yr). Considering that the width of the Japan Basin is roughly 300 km, the time interval needed to form the whole basin was probably about 5 Ma, unless a number of parallel spreading axes are assumed.

Another distinct magnetic feature of the Sea of Japan is its long-wavelength (~2000 km) anomaly. It has been reported that a wide area, including the whole region of the Sea of Japan, has the total field intensity approximately 200 gammas (nT), smaller than expected from the global field model (e.g. IGRF) (Isezaki, 1973; Nomura, 1979). Nomura (1979) attributed the source of this type of anomalies with long-to-intermediate wavelengths to a high induced magnetization situated at depths of 50 to several hundred km below sea level and postulated a cold downgoing slab in the subduction zone slightly above its Curie temperature. Hansen *et al.* (1983), based upon data of the Geological Survey of Japan, have proposed the importance of a relatively shallow Curie temperature isotherm beneath the Sea of Japan to explain the regional negative anomaly. Final conclusion on the validity of these models will need further testing in various regions with similar anomaly features.

F. Sedimentary Structure and Deep-Sea Drilling

A great number of single-channel seismic reflection profiles across the Sea of Japan have revealed layered structures of the sediment. In both the Japan and Yamato basins the sedimentary layer can be divided into an upper section of highly stratified sediments and a lower weakly stratified section (Hilde and Wageman, 1973; Ludwig *et al.*, 1975; Honza, 1979). The upper section is often called the opaque layer, because it reflects seismic waves from many stratified surfaces so as to make the reflection record very dark. This section has a thickness of about 500 m over the entire basin and is undeformed.

Deep-sea drilling by the *Glomar Challenger* during Leg 31 (Sites 299–301) indicated that the upper opaque layer consists of diatomaceous silty clay deposited since early Pliocene time (about 5 Ma BP). It contains many thick turbidites, particularly in Sites 299 and 300 situated close to the Toyama deep-sea channel (Fig. 6). The Tartar Strait between Hokkaido and Sikhote Alin is probably another source of turbidites, since many turbidites are found in the Pleistocene sections of Site 301 located in the Japan Basin.

Hole 302 (length 531.5 m) situated at the northeastern flank of the Yamato Ridge (water depth 2400 m) penetrated about 250 m of thick zeolitic clay–claystone underlying 71 m of silts (opaque layer) and about a 200 m thick diatom ooze (acoustically transparent layer). From the bottom of the hole latest Miocene (5.1 to 6.2 Ma) zeolitic clay with pale green, volcanogenic silty sand and fragments of green colored tuff (Karig, Ingle *et al.*, 1975) were

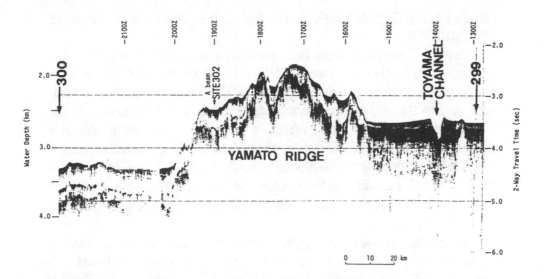

Fig. 6. *Glomar Challenger* seismic reflection profile from Site 299 to 300 through Site 302. Note continuity of the pelagic section of the Japan Basin with that on the Yamato Rise flanks (Karig *et al.*, 1975).

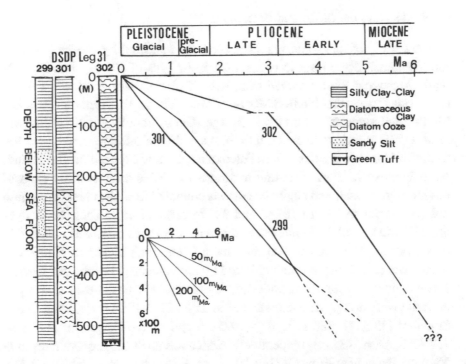

Fig. 7. Lithology and depth versus age diagram of Deep Sea Drilling Project Leg 31 Sites 299, 301, and 302 sediments. Slopes of lines give approximate rates of sedimentation (Koizumi, 1975b; Karig *et al.*, 1975).

recovered (Fig. 7). This level seems to correspond to the acoustic basement of Yamato Tai.

Several short cores have been recovered by piston coring from the Yamato Ridge. One of them (RC12-383, 38°55'N; 133°48'E, with a length of about 10 m at water depth of 1,437 m) contains early late Miocene (6.2 to 9.5 Ma) marine diatom ooze. In this sediment core a hiatus between about 3 Ma BP and recent is recognized. The result indicates that age of the first submergence of the Yamato Ridge is older than 9 Ma. Although there is continuity of acoustic reflectors from the Yamato Ridge to basin, the initial sedimentation age (i.e., possibly age of formation of the ocean floor rock) in the Yamato Basin seems to be much older than 9 Ma.

Occurrence of fragmented green colored tuff at the flank of the Yamato Ridge implies that at least a part of the ridge is composed of altered volcanogenic tuffs that may extend toward floor of the Yamato Basin. The age and petrological nature of the rock, however, must await more samples of complete blocks that may hopefully be dredged or drilled in future from this area.

G. Rocks from the Ridge and Seamounts

A variety of rocks and gravels have been collected by dredge hauls on the Yamato Ridge. Granitic rocks with K-Ar ages of 197 Ma and 220 Ma (early Mesozoic) were dredged from Kita–Yamato Tai and Takuyo Tai, respectively (Iwabuchi, 1968). Sr-Rb age of the Takuyo Tai granodiorite is 227 Ma in good agreement with its K-Ar age (Ueno, unpublished). As Ludwig *et al.* (1975) mentioned, the granitic rocks of Yamato Ridge may be related to the Hida metamorphics of late Paleozoic age, now exposed on Oki Island, Noto Peninsula, Hida Mountain and a part of Siberia. A boulder each of olivine augite basalt and augite basalt was collected from the crest of Yamato Tai (water depths 347 to 385 m and 320 m respectively). Their K-Ar ages are 21.6 Ma and 19.3 Ma but the real age may perhaps be slightly older due to alteration. The Yamato Tai may have been much closer to Honshu at these ages than at present and affected by arc volcanism before the Yamato Basin opened. Hypersthene andesite, augite hypersthene andesite and augite andesite were dredged at Hakusan-se (38°31'N, 137°04'E, D = 325 m), Matsu Seamont (39°32'N, 138°11'E, D = 975 m) and Meiyo Seamount (39°36'N, 137°43'E, D = 1435 m), respectively, which are all at the eastern margin of Yamato Basin near Honshu (Fig. 2). Their K-Ar ages are 7.7 Ma, 4.2 Ma, 13.4 Ma, respectively (Ueno *et al.*, 1971). Their source was most possibly arc magma of Honshu.

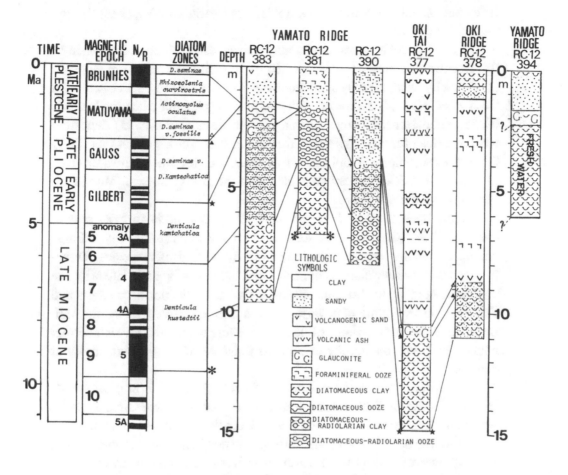

Fig. 8. Lithology and ages of six piston cores from the Sea of Japan. RC12-394 is from Burkle and Akiba (1977). The others are from Koizumi (1978).

III. PALEOENVIRONMENT OF THE SEA OF JAPAN IN THE LATE NEOGENE AND QUATERNARY

A. The Late Neogene Paleoenvironments

A 6 m long piston core, RC 12-394, taken from the northeastern flank of Yamato Ridge (40°19'N, 136°14'E, D = 2340 m, Figs. 2 and 8). contained more than 3 m of dark grey diatomaceous clay of freshwater origin at its core bottom (Buckle and Akiba, 1977). Such freshwater deposits have been found neither from any DSDP cores nor from other piston cores in the Sea of Japan. A late Miocene age was suggested based upon micropaleontology of the overlying marine sediment and the pollen data, but Koizumi (1977b;

1978) postulated much older ages (older than early late Miocene) for the freshwater diatoms.

Several possible explanations may be suggested for the occurrence of this freshwater deposit; (*1*) transported by trubidity currents from Honshu, (*2*) blown by wind, (*3*) deposited when the Yamato Ridge was an island with freshwater lakes, (*4*) deposited with continental sediment when the Yamato Ridge joined Honshu before opening of the Yamato Basin, (*5*) deposited when the Sea of Japan itself was a freshwater lake. Burkle and Akiba (1977) favored explanation (*5*) and tended to speculate that the freshwater environment in the Sea of Japan coincided with the latest Miocene (Messinian) Mediterranean dessication.

It has been concluded by Koizumi (1978) that the lowermost part of core RC 12-383 (late Miocene, about 6.2 to 9.5 Ma BP) from Yamato Ridge contains a diatom assemblage dominated by warm-water species. Dominance of warm-water diatoms indicates that the Sea of Japan in late Miocene was connected by the Tsushima Strait as well as by the northern straits and that an extensive influx of a warm-water current effectively raised the temperature of the surface water while the upwelling of the nutrient-rich bottom water due to the topography of Yamato Ridge caused high productivity of diatoms.

In contrast the latest Miocene to Pliocene (6.2 to 1.8 Ma) sediments recovered by DSDP holes and piston cores contain predominantly cold-water species. This indicates that the only continuous inflow of open ocean water into the Sea of Japan came through the northern straits at that period, although intermittent influx of warm water from the southern straits caused sporadic occurrences of warm-water diatom floras (Koizumi, 1975a, b).

B. Enclosure of the Sea of Japan in the Glacial Periods

During the Pleistocene glacial periods the sea level of the oceans was lowered by about 120 m. The Sea of Japan was cut off from the Pacific Ocean by sills at several straits where the present water depths are less than 140 m. This enclosure of the sea resulted in the cessation of the inflow of oxygen-rich ocean water and the draining of anaerobic bottom water causing a stagnant basin. Kobayashi and Nomura (1972) showed that the lower section of 12 m cores taken from the Japan Basin is so reduced as to alter magnetite and hematite (iron oxides) to pyrite and pyrrhotite (iron sulfides) by the stagnant and reducing conditions during the glacial periods. Miyake *et al.* (1968) postulated that an increase in the uranium content of the Sea of Japan deposits during glacial periods also resulted from the anaerobic condition.

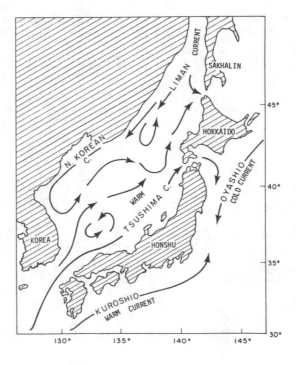

Fig. 9.   Schematic view of the present surface current systems in the Sea of Japan.

The present Sea of Japan is, in a sharp contrast to its environment during glacial periods, under an aerobic condition over its entire area and depth. The water-mass characteristic of the Sea of Japan existing below a depths of 200 to 300 m has a very uniform temperature (0.1 to 0.3°C, predominantly 0.2°C), low salinity (34.0 to 34.1 0/00) and high content of dissolved oxygen (5 to 6 ml/l). There exists a distinct thermocline at 100 to 300 m between the surface Tsushima current water (Fig. 9) and the Sea of Japan water. The cold oxygen-rich bottom water is formed at the surface and cooled in winter by the Siberian cold air sinking to the bottom and accumulating in the deep basins. The calcium carbonate compensation depth (CCD) is thus only 2000 m in the present Sea of Japan, much shallower than that of the normal oceans (approximately 4500 m). Tests of foraminifera are quickly dissolved before sufficient accumulation at depths greater than 2000 m.

DSDP hole 299 (D = 2600 m) in the Yamato Basin has an exceptionally well-preserved Pleistocene planktonic foraminiferal sequence probably owing to burial and protection by rapidly accumulated submarine fan deposits in the Toyoma channel. Ingle (1975) analyzed the biospecies and obtained a middle through late Pleistocene climatic record with a pattern similar to the adjacent North Pacific and world ocean. Koizumi (1975b) examined the diatom species in the same cores and showed consistent climatic changes.

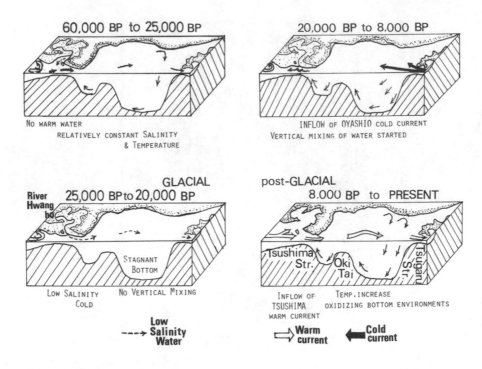

Fig. 10. Simplifed paleo-oceanographic history of the Sea of Japan since the last glacial period.

Oba *et al.* (1980) and Arai *et al.* (1981) published detailed analyses of
the paleo-oceanographic record in late Quaternary using tephrochronolog-
ically dated piston cores collected from Oki Tai (water depth of approximately 1000 m). Salinity and water temperature in the past were reconstructed from the ^{18}O ratios in benthic and planktonic foraminiferal tests
contained in the cores. Notable variations in the surface water of the Sea
of Japan have been recognized as follows (Fig. 10):

> (*i*) Salinity (33 to 34 per mil) and temperature (8 to 10°C) were relatively
> constant in the period between 60,000 yr BP and 25,000 yr BP.
> Some Pacific water was flowing into the Sea of Japan.

> (*ii*) A conspicuous and continuous decrease in salinity took place in
> the period from 25,000 yr BP to 20,000 yr BP, as inferred from
> decrease in the ^{18}O ratio of planktonic foraminiferal tests. Lowering
> of the sea level during the glacial period prevented inflow of the
> open sea water, and inflow of freshwater from the Hwang Ho River
> caused a decrease in salinity.

> (*iii*) At about 20,000 yr BP the ^{18}O value suddenly increased. Subse-
> quently a benthic foraminiferal fauna similar to that living in the
> present shallow water of the northwestern Pacific coast appeared

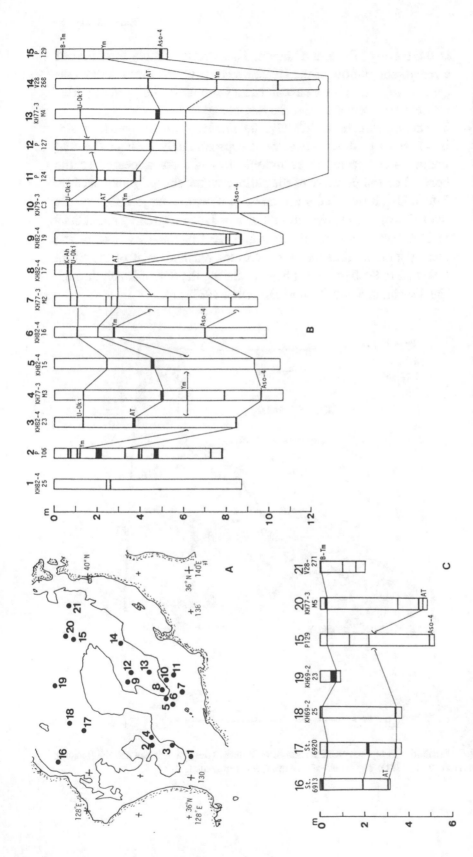

Fig. 11. Occurrence and correlation of tephra layers in piston cores collected from the Sea of Japan (Arai *et al.*, 1981). Ah, Akahoya; K, Kikai; AT, Aira–Tn; Ym, Yamato; U-Oki, Ulreung do-Oki; B-Tm, Baegdusan–Tomakomai ash. (a) Map showing core locations. Identification of cores—1:KH82-4-25; 2:P106; 3:KH82-4-23; 4:KH77-3-M3; 5:KH82-4-15; 6:KH82-4-16; 7:KH77-3-M2; 8:KH82-4-17; 9:KH82-4-19; 10:KH79-3-C3; 11:P124; 12:P127; 13:KH77-3-M4; 14:V28-268; 15:P129; 16:6913; 17:6920; 18:KH69-2-25; 19:KH69-2-23; 20:KH77-3-M5; 21:V28-271. (b) Intercore correlation of tephra layers based on petrographic analysis (southern part of the Sea of Japan). (c) Intercore correlation of tephra layers based on petrographic analysis (northern part of the Sea of Japan).

in the cores of the Sea of Japan. These results imply the onset of a remarkable inflow of the Oyashiro cold current through the Tsugaru Strait, transporting a water-mass similar to the present northwestern Pacific water into the Sea of Japan.

(iv) The coiling direction of *Globigerina pachyderma* changed from sinistral (cold) to dextral (warm) at approximately 8000 yr BP. and warm-water species of planktonic foraminifera appeared in the cores. The temperature of the surface water abruptly increased by 7 to 8°C. These results are consistent with a large influx of the warm Tsushima current into the Sea of Japan beginning about 8000 yr BP. Two cores, on the east and west of the Tsushima Strait, clearly indicate that the warm current reached west of the Strait long before 8000 yr BP but began entering the Sea of Japan through the Tsushima Strait 8000 yr BP (Oba, 1983a, b).

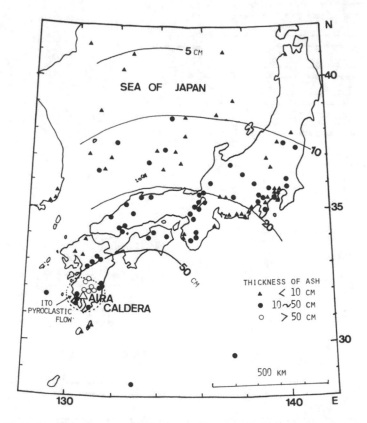

Fig. 12.   Distribution of tephras erupted from Aira Caldera in the Sea of Japan and the Japanese islands (Arai *et al.*, 1981). Thickness of tephra layers is indicated.

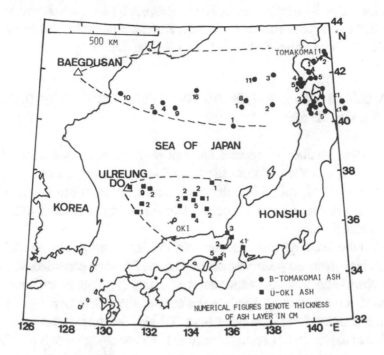

Fig. 13.   Distribution of the Baegdusan–Tm ash and Ulreung–Oki ash (Arai *et al.*, 1981).

C. Tephrochronology of the late Quaternary Sediments in the Sea of Japan

Several distinct layers of volcanic glass shards and volcanogenic minerals are recognized in the late Quaternary sediment cores recovered from the Sea of Japan (Machida, 1981; Arai *et al.*, 1981; Machida *et al.*, 1981; Machida and Arai, 1983) (Fig. 11 a, b, c). Six marker tephras have been identified; three are correlated with precisely dated widespread tephra originating from gigantic caldera-forming eruptions in Kyushu (Fig. 12). The tephras are named as the Akahoya Ash (from Kikai Caldera, eruption at 6300 yr BP), the Aira–Tn ash (from Aira Caldera, eruption at 21,000–22,000 yr BP) and the Aso-4 ash (Aso Caldera, eruption approximately 80,000 yr BP).

Sources of two other marker tephras, the Oki Ash and Yamato Ash are located on Ulreung-do Island in the western part of the Sea of Japan. The Oki Ash lies stratigraphically between the Akahoya and the Aira-Tn tephras. The ^{14}C ages of the eruption are around 9300 yr BP. The Yamato Ash is between the Aira-Tn and Aso-4 tephras and its age must be in a range 25,000 to 35,000 yr BP, although no reliable radiocarbon age has been obtained.

One tephra distributed in the northern part of the Japan Basin to the northern Honshu and Hokkaido is called the Tomakomai Ash (B–Tm) and is distinguished by occurrence of alkali feldspar. Its source volcano has been identified as Baegdusan in the northern Korean peninsula. The age of eruption seems to be about 800 to 1000 yr BP (Fig. 13).

IV. MODELS AND AGE CONSTRAINTS FOR FORMATION OF THE SEA OF JAPAN

A great number of speculative models have been postulated for the origin and age of formation of the Sea of Japan (Kaseno, 1975, 1980). The models may be classified into three; (*1*) oceanization or crustal thinning (*2*) entrapment of ocean basin, (*3*) ocean floor spreading associated with relative continental drift.

A hypothesis of oceanization was originally proposed by Beloussov (1968). However, no positive evidence has so far been provided either from field observations or by experiment and theory. The entrapment model seems to have difficulty, because the ages of the oldest portion of Honshu and the Yamato Ridge are older than 200 Ma, whereas the deep basins of the Sea of Japan do not seem to be so old. The opening models are the most popular but the ages and modes of opening are still in debate.

Uyeda and Miyashiro (1974) proposed that the opening of the Sea of Japan was caused by subduction of the Kula–Pacific Ridge, presumably in the Cretaceous. The extreme contrast is Tamaki *et al.* (1981) who postulated that the opening periods of the Yamato Basin to be 4.5 to 9.0 Ma on the basis of the DSDP data and paleontological ages of sediment samples dredged from the outcrops that appear to be continuous with the lowest acoustic horizon in the deep basin. Since DSDP hole 299 penetrated the early Pliocene (about 4.0 Ma) sediment in the Yamato Basin and the acoustic basement is situated at several hundred meters below its base, the postulated age of 4.5 Ma seems to be the minimum estimate of the opening age.

Hilde and Wageman (1973) suggested based upon the thickness and structure of the sedimentary strata in the seismic reflection records that the Japan Basin opened at a period prior to the opening of the Yamato Basin. Their conclusion is consistent with the water depths of the Japan and Yamato basins. It is well known that depth of the ocean floor (igneous layer) increases with its age t in proportion to $\sqrt{t}$. A fairly precise $\sqrt{t}$ curve has been drawn for normal oceans such as the Pacific, Atlantic and Indian oceans. However, this relationship needs a modification when applied to the marginal seas. Depths of the marginal basins are about 1000 m greater than those of the normal oceans with the same age (Yoshii, 1972, 1973). As the basement depths of the Japan and Yamato basins are about 5500 m and 4000 m, the $\sqrt{t}$ law provides ages of approximately 35 to 45 Ma for the Japan Basin and

10 to 15 Ma for the Yamato Basin, if the isostatic adjustment of the sediment load is neglected. In the Yamato Basin if the effect of the low velocity altered second layer is taken into account, slightly older ages (20–25 Ma) may be suggested.

Other constraints for the ages of the Sea of Japan are provided from the adjacent lands. Ichikawa (1972) suggested from correlation of the geological strata in southwestern Japan with those in Korea that the opening of the Sea of Japan was probably older than 25 Ma but younger than 60 Ma BP. Sillitoe (1977) investigated the zonal distribution of the metal belts, i.e. the fluorite–tungsten–molybdenum zone, lead–zinc zone and copper–tungsten zone in southwestern Japan and South Korea and concluded that the offset of these zones was related to the southward drift of Southwestern Japan relative to the Korean Peninsula and postdates the formation of these zones. The opening of the Sea of Japan is, therefore, younger than 46 Ma, the age of formation of the youngest Mo belt.

Paleomagnetic results indicating possible rotations of southwestern and northeastern Japan have been published by Kawai (1961) and Kawai *et al.*(1969, 1971). It has been suggested that southwestern Japan rotated clockwise by nearly 40° relative to northeastern Japan at some period younger than the Cretaceous. Yaskawa (1975) summarized 120 paleomagnetic measurements from southwestern Japan and compared them with the paleomagnetic directions of Korea. It was concluded that southwestern Japan drifted southwards by about 4° in latitude and rotated clockwise by about 24° relative to Korea (Fig. 14). The motion occurred at some period probably post-Cretaceous and pre middle Tertiary but the age resolution in the paleomagnetic analysis was not sufficiently high to provide a more detailed chronology.

Otofuji and Matsuda (1983) reported results of their paleomagnetic and fission track analyses of rhyolitic and dacitic pyroclastics occurring in San-in district, in southwestern Japan and concluded that the clockwise rotation of southwestern Japan occurred at some period younger than 28 Ma and prior to 12 Ma BP. Torii (1983) and Otofuji *et al.* (1985) further investigated paleomagnetic directions of igneous and sedimentary rocks in southwestern Japan and concluded that the rotation of southwestern Japan occurred between 15 and 12 Ma or probably between 15 and 14 Ma. If these ages are correct, the opening of the Yamato Basin occurred immediately after the Shikoku Basin ceased to open and ended at the same time as possible rifting in northeastern Japan associated with Kuroko mineralization was aborted.

Paleomagnetic information showing the motion of northeastern Japan was mostly from granitic rocks in which the bedding correction is difficult. Sasajima (1981) postulated that northeastern Japan rotated anticlockwise by

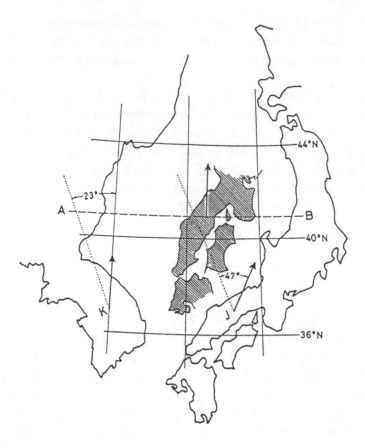

Fig. 14.   Reconstruction of paleogeography of southwestern Japan in the Cretaceous period
(Yaskawa, 1975). The coordinate of the map refers to the paleomagnetic latitudes obtained at
the reference point of Korea K (37°N,128°E). J denotes the reference point of southwestern
Japan (35°N,134°E). Arrows indicate paleomagnetic directions. The angles shown represent
deviations of paleomagnetic directions from the present meridian (dotted lines).

about 50° relative to Korea, on an assumption that northeastern Japan tilted
toward the Pacific with a tilting axis parallel to the trend (N63°E) of the
present Kuril Trench. However, this assumption seems to be unrealistic.
Ito and Tokieda (1974) pointed out that the rocks on the Sea of Japan side
are tilted westwards, whereas the rocks on the Pacific side are tilted eastward
by several tens of degrees. The observed paleomagnetic directions can be
explained by these tiltings alone. No horizontal rotation of northeastern
Japan was thus detected by paleomagnetic methods. If this is correct, the
opening of the northern Japan and Yamato basins occurred with parallel drift
of northeastern Japan or at least a part of it. If opening of the southwestern
portion of the Japan Basin was also associated with parallel drift of southwestern Japan, it may not be recorded in the paleomagnetic directions of

the southwestern Japan and can be older than the ages of rotation inferred from paleomagnetism of dated strata.

V. STRUCTURES AND TECTONICS OF THE RYUKYU ARC AND THE OKINAWA TROUGH

A. Ryukyu Arc-Trench System

1. *Geographical Setting*

The northwestern margin of the Philippine Basin is fringed by the Ryukyu Islands and trench aligned from Kyusyu in the northeast to Taiwan in the southwest, a length of about 1,200 km (Figs. 1 and 15). They are convex toward the Philippine Basin and bend westward at their southern end. The Ryukyu Arc is composed of more than 100 small islands, the largest of which is Okinawa Island with an area of 1211 km^2 situated at approximately 26°30'N 128°E.

The Ryukyu Arc is divided into the northern, central and southern segments by the Tokara Channel and Kerama Gap (Fig. 15). The latter was previously named Miyako depression (Konishi, 1963) but the present name is preferred because of its geographical affinity to Kerama Islands. Both the Tokara Channel and Kerama Gap seem to be left-lateral strike-slip faults. The depth contours of 2000 m and 1000 m are also bordered by the northwestern extension of the faults.

2. *Radiometric Ages of the Islands*

The oldest radiometric age so far known in the Ryukyu Islands is the 202 Ma of muscovite separated from crystalline schist exposed in the northern part of Ishigaki-jima (Rb-Sr age given by Shibata *et al.*, 1968). The K-Ar age of the same sample is 178 Ma. Slightly younger K-Ar ages of 175, 167, 161 and 159 Ma have been obtained with muscovite separated from schists exposed in the same island (Nishimura *et al.*, 1983). Presumably late Paleozoic formations containing limestone, chert and igneous rocks are exposed in Motobu Peninsula of Okinawa Island and other islands but their radiometric ages have not been determined (Konishi, 1963; Ujiie, 1983).

Granitic instrusions found at Tokunoshima and Amami–Oshima islands have K-Ar ages of 61, 56, 55 and 46 Ma (Shibata and Nozawa, 1966). Altered volcanic rocks of possibly Eocene age have been described at Iriomote-jima (Saito *et al.*, 1973; Tiba and Saito, 1974) but their radiometric dating has not

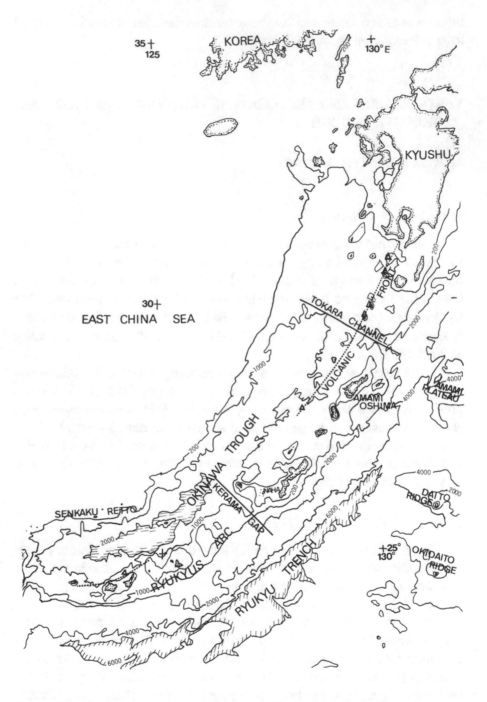

Fig. 15.   Bathymetric map of the Okinawa Trough and Ryukyu Trench region. Depth contours of 200 m, 1000 m, 2000 m, 4000 m, and 6000 m are shown based upon Hydrographic chart No.1009, Hydrographic Department, Maritime Safety Agency of Japan (1978). Location of the volcanic front is indicated by dotted line. Solid lines represent Tokara Channel and Kerama Gap.

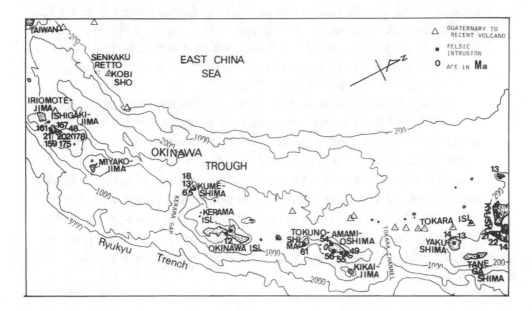

Fig. 16.   Radiometric ages of igneous and metamorphic rocks in the Ryukyu Islands. Sources of data are cited in the text.

been successful. The Fission-track age of a rhyolite dyke intruding schist located on northern Ishigaki-jima island is about 48 Ma (Matsumoto, 1983).

For the Neogene period a K-Ar age of 21 Ma has been reported with granitic rock collected at Ishigaki-jima Island (Kawano and Ueda, 1966). Bowin and Reynolds (1975) published a ^{40}Ar/^{39}Ar age of 12 Ma with an intrusive biotite dacite obtained from Okinawa Island. K-Ar age of the same rock is also 12 Ma (Shibata *et al.*, 1979). Both K-Ar and ^{87}Sr/^{86}Sr ages of biotite granite from Yaku-shima are 14 Ma (Shibata and Nozawa, 1967; Shibata and Ishihara, 1979). Ages of altered andesites occurring at Kume-shima are 18, 13 and 6.5 Ma (Nakagawa and Murakami, 1975).

Pliocene to early Pleistocene volcanism is found at Kume-shima (basalt and andesite) and Okinawa (rhyolitic tuff) but their radiometric ages have not been obtained. Volcanic activity in this period was generally weak and most of the Ryukyu Islands subsided under the ocean in which the Shimaziri Formation was thickly deposited. Volcanoes of the late Pleistocene to Recent are aligned along the Tokara Islands situated about 100 km west of the main island chain in the northern and central segments (Fig. 16). Volcanic rocks of these ages are mostly two-pyroxene andesite. No active volcanoes are found in the southern segment of the Ryukyus except for submarine volcanic activity 30 km north of Iriomote-jima. At Kobi-sho (a circular conical island with diameter of about 1 km) of Senkaku-retto northwest of the Okinawa Trough recent volcanic activity is recognized but its geochemical nature is that of alkali olivine basalt, which is quite different from that of recent andesitic volcanoes in the volcanic front of the northern and central Ryukyus (Matsumoto and Nohara, 1974).

These ages of igneous rocks indicate that the Ryukyu Arc has experienced at least five magmatic episodes, each of which may possibly be correlatable to a particular stage of individual cycles of subduction of the oceanic lithosphere from the Ryukyu Trench, although the earliest (Paleozoic to Mesozoic) episode is likely to be a mid-oceanic magmatism forming an ophiolitic complex later accreted to the continental margin of the Eurasian plate. Pre-Miocene sequences of sedimentary strata exposed in the Ryukyu Islands are intensely folded and often intruded by igneous bodies, whereas the unconformably overlying Miocene to Recent strata are relatively undisturbed.

3. *Ryukyu Trench*

The topographic axis of the Ryukyu Trench (also called the Nansei-shoto Trench in some Japanese literature but this nomenclature is not used in this article) is situated about 120 km southeast of the Ryukyu Ridge axis. The maximum water depth is about 7500 m located at 24°35'N 127°25'E, approximately 180 km south of Okinawa Island. The axial depth in other portions is around 6500 m. The trench topography is interrupted by two topographic highs; Amami Plateau and Kyushu–Palau Ridge. Between Amami Plateau and Kyushu–Palau Ridge the trench is a narrow oval-shaped trough with the maximum water depth of only 5500 m, which is still about 1000 m deeper than its seaward basin.

The free-air gravity anomaly is negative at the trench axis in the southern and central segments. The minimum amounts to -100 to -140 mgal. However, another trough of free-air gravity is clearly recognized along the continental slope about 80 km landward from the trench axis. (Tomoda and Fujimoto, 1981; Watts, 1976). Distance of the axis of this gravity minimum from the island shelf is less than 40 km. In the northern segment north of the Tokara Channel the landward minimum extends toward a basin east off Kyushu (Hyuga-nada) and the Bungo Strait between Kyushu and Shikoku (Fig. 17).

Seismic refraction study of the northern segments of the Ryukyus by Ludwig *et al*. (1973) has shown that the landward gravity minimum is underlain by an enormously thick (probably exceeding 4 km) sediment that fills a crustal trench having no surface expression. (Fig. 18). The accumulated sediment may have been transported from Kyushu by river and airfall of volcanic ashes. Beneath the thick sediment cover there exists an oceanic

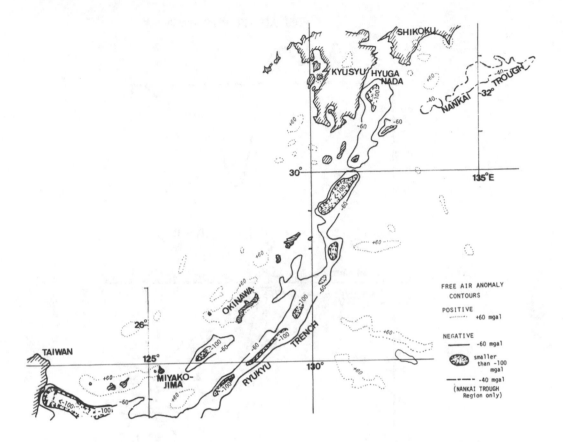

Fig. 17.   Free-air gravity anomaly in the Ryukyu region (Tomoda and Fujimoto, 1981).

crust that continues from the ocean floor of the north Amami Basin with a gentle northwestward slope.

In the southern and central segments the crustal structure of the landward (northwestward) slope of the Ryukyu Trench appears to be in a marked contrast to that of the northern. Seismic refraction profile (line 17 of Murauchi *et al.*, 1968) recorded along a deep-sea terrace on the trench slope indicates the existence of a thick (possibly thicker than 6 km) layer with V_p = 5.0 km/sec underlying a 3 km thick layer with V_p = 3.0 km/sec, that is covered by a thin (less than 1 km) soft sediment with V_p = 2.0 km/sec (Fig. 19). Seismic reflection profiler records also show that this terracelike platform is a sediment-filled trough situated between the main arc and the mid-slope high.

The landward minimum of the free air gravity anomaly is situated on the deep-sea terrace. The mid-slope high has a free-air anomaly about 100

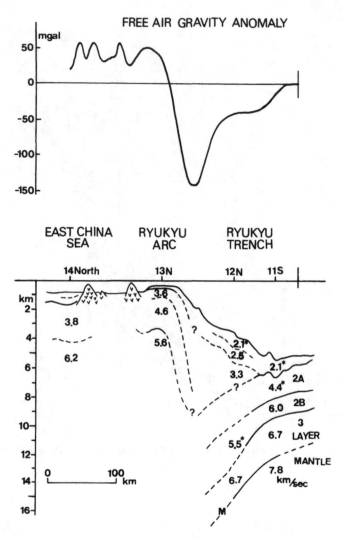

Fig. 18.   Crustal structure section and free-air gravity anomaly profile across the northern
segment of the Ryukyu trench–arc system (compiled from Ludwig *et al.*, 1973, and Tomoda
and Fujimoto, 1981). Numerical figures indicate *P*-wave velocity. Asterisks denote assumed
velocity.

mgal higher than that of the trench and the terrace, indicating its crust is
composed of rocks with comparatively high density. They may be either a
part of continental crust similar to the Ryukyu Arc crust or accreted oceanic
basalts.

B. Subduction Zone at the Ryukyu Trench

Seismicity along most of the Ryukyu arc-trench system is not so high
as along the Japan Trench and no earthquakes with epicenters deeper than

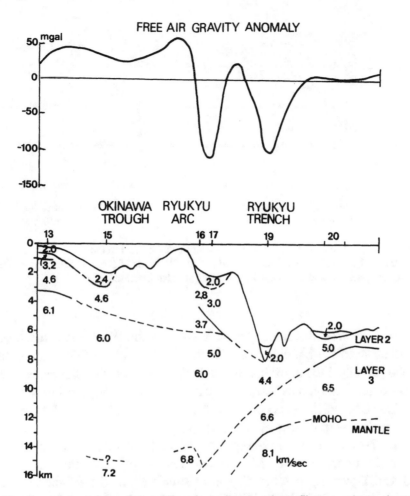

Fig. 19.   Crustal structure section and free-air gravity anomaly profile across the south central segment of the Ryukyu trench–arc–back-arc system close to the Kerama Gap (compiled from Murauchi *et al.*, 1968, and Tomoda and Fujimoto, 1981).

300 km have been observed in this area. However, the Wadati–Benioff zone of intermediate depth earthquakes can be recognized along the entire region of the Ryukyus (Katsumata and Sykes, 1969). The direction of motion of the Philippine Sea plate relative to the Eurasian plate is roughly NW, as indicated by the focal mechanism solution of shallow earthquakes occurring in the Ryukyu Arc. The oceanic lithosphere of the Philippine Sea is subducted under the Ryukyu Arc toward this direction.

A sharp contrast seems to exist in shape and internal stress of the descending lithosphere between the north and south central segments of the Ryukyus separated by the Tokara Channel (Shiono *et al.*, 1980). In the northern segment the Wadati–Benioff zone bends sharply at a depth of about 70 km and dips steeply at a high angle amounting to 70° at greater depths. The

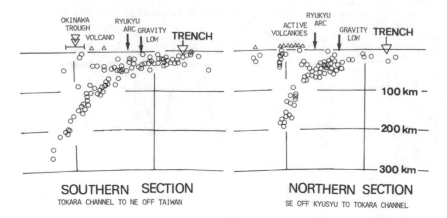

Fig. 20.   Hipocenters of selected ISC records projected normal to the north and south central segments of the Ryukyu Arc (Shiono *et al.*, 1980). Positions of Ryukyu Trench axis, free-air gravity low, main islands, and active volcanoes are also indicated.

minimum of free air anomaly is situated over the W–B zone with a gentle slope close to the bending line. The maximum depths of deep-focus earthquakes is only 200 km in this segment (Fig. 20). Focal mechanisms of these deep-focus earthquakes show a prevalence of down-dip tension in the subducting slab. Volcanism is active in the overlying slab above the deep-focus earthquakes with depths of 100 to 150 km.

In the south and central segments of the Ryukyus, on the contrary, the angle of subduction is 40°–50°. The maximum depth of earthquakes is about 270 km. Down-dip compression is predominant in the subducting lithosphere. Volcanic activity in this portion of the arc is low and confined to a submarine volcano situated above deep-focus earthquakes with depths of about 100 km. Back-arc opening seems to be underway in the south and central segments as described in the following sections. The origin and/or consequences of this geodynamic contrast have not been clarified but may be correlated with the difference in the rate of convergence of two plates greater in the south than in the north.

Several great earthquakes have occurred in the Ryukyu Trench. The epicenter of the June 15, 1911 earthquake (magnitude 8.2) is near Kikai-jima close to the Tokara Channel Fault (Mogi, 1969). The focal depth is about 160 km. No focal mechanism solution is available with this earthquake due to lack of an instrumental observation network at that time. The epicenter of the June 10, 1938 earthquake (magnitude 7.7) is located east of Miyako-jima Island. Its focal depth is 38 km, which is in the upper surface of the subducted lithosphere. The focal mechanism deduced using old data appears to be a normal fault dipping at 65° toward N35°W. The southeastern end of

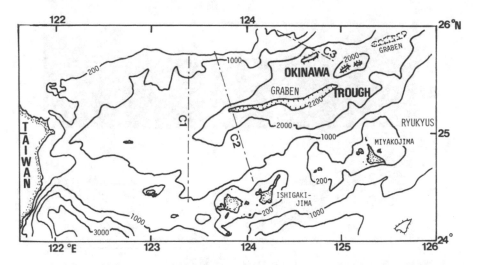

Fig. 21. Detailed bathymetric map of the southwestern part of the Okinawa Trough indicating position of the central grabens (Lee *et al.*, 1980). Seismic reflection profiles C1, C2, and C3 are shown by broken lines.

the Ryukyu arc-trench system east of Taiwan is another site of major earthquakes. Occurrence of these earthquakes seems to be related to dislocations of the subducted lithosphere.

Many medium to small earthquakes occurred in the Hyuganada Basin east of Kyushu. They are attributed to the stress concentration in the overlapping subducted lithosphere caused by the sharp change in geometry of the trench axis from the Ryukyu Trench to the Nankai Trough.

C. Okinawa Trough

1. *Submarine Topography*

The Okinawa Trough is distinguished from the surrounding areas by the ‘1000 m and 2000 m. isobaths. General trend of the trough thus defined is arcuate, convex toward the Pacific and roughly parallel to the trend of the Ryukyus Islands (Figs. 1 and 15). The trough, a graben, is considerably shallower than the other back-arc basins. Its deepest part, only 2270 m deep, is situated in the southwestern part of the trough about 100 km north of Ishigakijima, in the southwestern part of the Ryukyu chain (Fig. 21). A seamount with the crestal depth of about 1,200 m exists northeast of the graben. Many E–W trending grabens, similar to but not so prominent as the southwestern one, have been found in the other parts of the trough (Kimura, 1983).

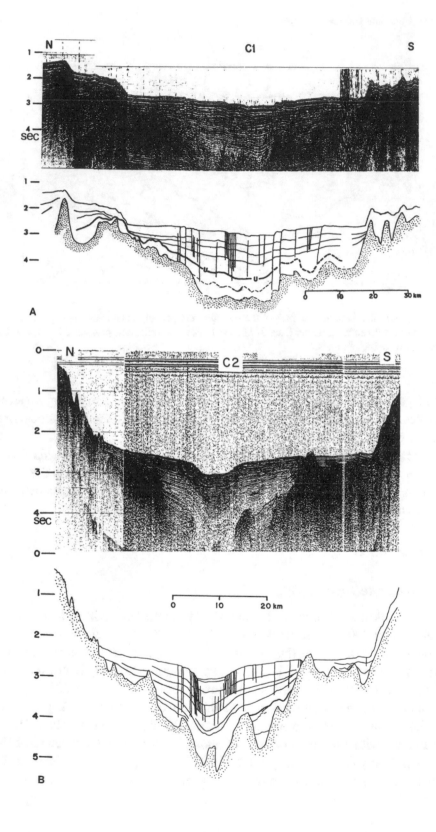

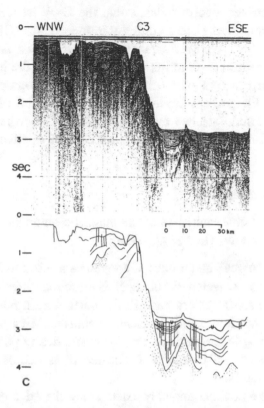

Fig. 22.   Seismic profiler records from the Okinawa Trough with interpretations by line drawings. U in the drawings denotes proposed unconformity (Herman *et al.*, 1978). (a) C1, (b) C2, (c) C3. Positions of profiles are given in Fig. 21.

The trough shoals gradually northeastward toward Kyushu. The northeastern limit of the 2000 m isobath is located along the northwestward extension of the Miyako Depression or Kerama Gap. The 1000 m isobath is limited on the northeast by the Tokara Channel (Fig. 15). In the northeast of the Tokara Channel the trough is barely distinguishable by the 500 m contour, although some E–W trending grabens seem to trend toward central Kyushu.

2. *Sedimentary Structures*

A number of seismic reflection profiles have been obtained by several investigators (Wageman *et al.*, 1970; Kagami, 1975; Honza *et al.*, 1976; Herman *et al.*, 1978; Lee *et al.*, 1980). Sediment cover over the Okinawa Trough is generally very thick, amounting to 1 to 3 km except for an E–W trending ridge and some diapirs in the trough. Most of sediments are acoustically well-stratified. Two layers are distinguished; the upper unfolded but faulted layer unconformably overlies a highly deformed layer (Fig. 22).

Seismic profiler records indicate that the E–W trending grabens were formed by normal faults clear cutting the upper layer (C1, C2 in Fig. 22) and suggesting their younger spreading origin. Normal faults are also remarkable in the lower layer at the extreme margins of the trough (C3 in Fig. 22). In the northeastern part of the trough the topographic and seismic expressions of the grabens are less distinct. The central rifts are in several places intruded and extruded by volcanic plugs. Nevertheless, the general patterns of deformation of sediments show currently active extensional tectonics.

3. *Magnetic Anomalies*

The general characteristics of the magnetic anomalies in the Okinawa Trough are as follows (Lee *et al.*, 1980):

(*i*) Peak to peak amplitudes of anomalies are 200 to 300 gammas (nT) and their wavelengths along N–S direction range from 20 to 50 km in the southwestern part. In the northeastern portion their amplitudes are lower and wavelengths shorter (Miyazaki *et al.*, 1976).

(*ii*) The magnetic anomalies are linear and nearly parallel to the trend of grabens and trough itself. The lineations are offset by many transform faults.

(*iii*) A large positive anomaly exists along the edge of the East China Sea Shelf. Large anomalies associated with volcanic plugs are observed. In contrast the anomalies are very low along the Ryukyus Islands and trench.

4. *Heat Flow*

Heat flow values in the Okinawa Trough are very variable, ranging from 0.36 to 10.4 HFU (15.1 to 437 mW/m²) (Yasui *et al.*, 1970; Herman *et al.*, 1978) (Fig. 23). The scatter in the heat flow values implies existence of hydrothermal circulation in the Okinawa Trough. High average value (4.06 HFU or 170 mW/m²) indicates that the trough is very young and possibly still active. All the data cited above indicate that at least a part of the Okinawa Trough has been formed by a process of back-arc opening and it is still being rifted although its water depth is too shallow to be an oceanic basin.

5. *Seismicity*

Shallow earthquakes beneath the Okinawa Trough are not frequent and their magnitudes are all smaller than 6. Eguchi (1982) reported focal mechanism of one moderately large earthquake (magnitude 5.5) that occurred at the northern margin of 1000 m isobath of the Okinawa Trough (29.6°N

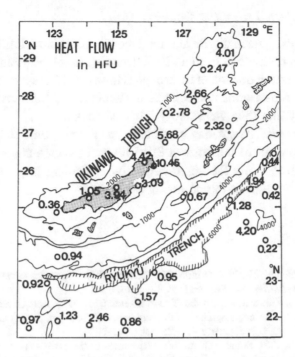

Fig. 23.   Heat flow values in the Okinawa Trough (complied from Yasui *et al.*, 1970, and Herman *et al.*, 1978).

127.9°E, D = 39 km). Stress responsible for the earthquake is extensional in a direction of NNW–SSE, implying that this part of the trough is now rifted in this direction. The direction is oblique to the trend of Tokara Channel along which pre-Miocene strata in the Ryukyu Ridge appear to be offset. Rifting of the Okinawa Trough may have occurred at two major episodes since Miocene, one of which proceeds at present.

6. *Crustal Structure beneath the Trough*

Murauchi *et al.* (1968) conducted a seismic refraction survey of the Okinawa Trough and found a thick crustal layer similar to that of the continental crust, although the Moho depth beneath the trough was not determined. Later work by Lee *et al.* (1980) indicated that the crustal thickness beneath the central rift of the Okinawa Trough is about 15 km, intermediate between continental and oceanic.

Free-air gravity anomalies in the Okinawa Trough are positive (20 to 60 mgal, Tomoda and Fujimoto, 1982), indicating that the trough is non-isostatic and can be several hundred meters shallower than isostatically compensated oceanic basin.

7. Rocks Collected from the Trough

Biotite-rich quartz diorite and metamorphosed pillow basalts typical of island arcs were dredged from an E–W trending ridge between two grabens in the southwestern portion of the trough (Honza, 1976). Herman *et al.* (1978) regarded them as remnant blocks of an ancient arc but seismic profiling records imply that they were newly intruded rocks.

The Okinawa Trough may possibly be in an embryonic stage of opening in a similar manner to the African Rift Valley. In such a stage the region is upheaved, rifted along the central zone but intruding magma is generally more felsic than that of the normal oceanic ridge.

REFERENCES

Abe, K. and Kanamori, H., 1970, Mantle structure beneath the Japan Sea as revealed by surface waves. *Bull. Earthquake Res. Inst.*, v. 49, p, 1011–1021.

Arai, F., Oba, T., Kitazato, H., Horibe, Y. and Machida, H., 1981, Late Quaternary tephrachronology and paleo-oceanography of the sediments of the Japan Sea, *Quat. Res. (Tokyo)*, v. 20, p. 209–230. (in Japanese with English abstracts).

Beloussov, V. V., 1968, Some problems of development of the earth's crust and upper mantle of oceans, in: *The Crust and Upper Mantle of the Pacific Area.*, Knopoff, L., Drake, C., and Hart, P., eds. *Am. Geophys. Union Monogr.* 12, p. 449–459.

Bowin, C. and Reynolds P. H., 1975, Radiometric ages from Ryukyu Arc region and an ^{40}Ar/^{39}Ar age from biotite dacite on Okinawa, *Earth Planet. Sci. Lett.*, v. 27, p. 363–370.

Burckle, L. H., and Akiba, F., 1977, Implications of late Neogene freshwater sediment in the Sea of Japan, *Geology*, v. 6, p. 123–127.

Cox, A., 1976, The frequency of geomagnetic reversals and the symmetry of the nondipole field, *Rev. Geophys. Space Sci.*, v. 13, p. 35–51.

Den, N., 1972, Crustal structures in the western Pacific Ocean, in: *The Crust and Upper Mantle of the Japanese Area, Part I.*, Geophys., Jpn. Com. UMP, p. 57–68.

Eguchi, T., 1982, On the opening of the Okinawa Trough, *Abstr. Seism. Soc. Jpn.*, A75, p. 77, (in Japanese).

Emery, K. O., Hayashi, Y., Hilde, T. W. C., Kobayashi, K, Koo, J. H., Meng, C. Y., Niino, H., Osterhagen, J. H., Reynolds, L., M., Wageman, J. M., Wang, C. S., and Yang, S. J., 1969, Geological structure and some water characteristics of the East China Sea and the Yellow Sea, *United Nations ECAFE Tech. Bull.*, v. 2, p. 3–43.

Evans, J. R., Suyehiro, K. and Sacks, I. S., 1978, Mantle structure beneath the Japan Sea— A re-examination, *Geophys. Res. Lett.*, v. 5, p. 487–490.

Hansen, R. O., Okuro, Y., Graf, R. J., Tsu, H. and Ogawa, K., 1983, Nationwide Curie point depth analysis of Japan, *Abst. Soc. Explor. Geophys. Fall Meeting, Las Vegas*, p. 218–223.

Herman, B. M., Anderson, R. N., and Truchan, M., 1978, Extensional tectonics in the Okinawa Trough. in: *Geological and Geophysical Investigations of Continental Margins*, Watkins, J. S., Montadert, L. and Dickerson, P. W., eds., *Am. Assoc. Pet. Geol. Mem.* 29, p. 199–208.

Hilde, T. W. C. and Wageman, J. M., 1973, Structure and origin of the Japan Sea, in: *The Western Pacific: Island Arcs, Marginal Seas, Geochemistry*, Coleman, P. J., ed., University of Western Australia Press, Nedsland, p. 415–434.

Honza, E., ed., 1976, *Ryukyu Island (Nansei Shoto) Arc. GH 75-1 and GH 75-5 Cruises*, Geological Survey of Japan, Tsukuba, p. 20–22; p. 25–26.

Honza, E., ed., 1978, Geological investigation of the Okhotsk and Japan Seas, off Hokkaido, June–July 1977 (GH 77-3 Cruise). *Geol. Surv. Jpn. Cruise Pap.* 11, 72p.

Ichikawa, K., 1972, On reconstruction of the ancient Sea of Japan region, *Kagaku*, v. 42, p. 630–633. (in Japanese).

Ingle, J. C. Jr., 1975, Pleistocene and Pliocene foraminifera from the Sea of Japan, leg 31 DSDP, in: *Initial Reports of the Deep Sea Drilling Project*, v. 31. Karig, D. E. and Ingle, J. C. Jr. *et al.*, Washington DC: United States Government Printing Office, p. 693–702.

Isezaki, N., 1973, Geomagnetic anomalies and tectonics around the Japanese islands, *Oceanogr. Mag.*, v. 24, p. 107–158.

Isezaki, N., 1975, Possible spreading centers in the Japan Sea. *J. Mar. Geophys. Res.*, v. 4, p. 53–65.

Isezaki, N., 1979, Ages of magnetic anomalies in the Sea of Japan, *Nihon-Kai.* v. 10, p. 111–119 (in Japanese).

Ito, H., and Tokieda, K., 1974, Tilting of Hokkaido Island and region, *Geol. Soc. Jpn J.*, v. 81,; Koizumi, I., 1975a, Granitic rocks, *Rock Mag. and Paleogeophys.* (Tokyo), v. 2, p. 54–58.

Iwabuchi, Y., 1968, Submarine geology of the southeastern part of the Japan Sea, *Tohoku Univ. Inst. Geol. Paleontol. Contr.*, v. 66, p. 1–76. (in Japanese).

Johnson, H. P. and Hall, J. M., 1976, Magnetic properties of the oceanic crust, considerations from the results of DSDP Leg 34, *J. Geophys. Res.*, v. 81, p. 5281–5293.

Kagami, H., ed., 1975, *Preliminary Report of the Hakuho Maru Cruise KH 72-2*, University of Tokyo: Ocean Research Institute, 144p.

Karig, D. E., Ingle, J. C., Jr., *et al.*, 1975, *Initial Reports of the Deep Sea Drilling Project*, v. 31, Washington, D.C.: United States Government Printing Office, 974p.

Kaseno, Y., 1975, *Enigma of the Sea of Japan*, Tsukiji Shokan, Tokyo, 169p, (in Japanese).

Kaseno, Y., 1980, *Nihon-kai (Sea of Japan)* research circular, Kanazawa Japan, (in Japanese).

Katsumata, M., and Sykes, L., 1969, Seismicity and tectonics of the western Pacific. Izu–Mariana–Caroline and Ryukyu–Taiwan regions, *J. Geophys. Res.*, v. 74, p. 5923–5948.

Kawai, N., 1961, Deformation of the Japanese islands as inferred from rock magnetism, *Geophys. J.*, v. 6, p. 124–129.

Kawai, N., Hirooka, K. and Nakajima, T., 1969, Paleomagnetic and potassium-argon age informations supporting Cretaceous Tertiary hypothetic bend of the main island Japan, *Paleogeogr. Paleolim. Paleoecol.*, v. 6, p. 277–282.

Kawai, N., Nakajima, T. and Hirooka, H., 1971, The evolution of the island arc of Japan and the formation of granites in the circum-Pacific belt, *J. Geomagn. Geoelectr.*, v. 23, p. 267–293.

Kawano, Y. and Ueda, Y., 1966, K-A dating of Japanese igneous rocks (V)-granitic rocks in the southwest Japan, (in Japanese), *J. Jpn. Assoc. Petrol. Miner. Econ. Geol.*, v. 56, p. 191–211.

Kimura, M., 1983, Formation of Okinawa Trough grabens, in: *Geohistory of the Ryukyu Islands*, Kizaki, K., Nakagawa, H., Konishi, K. and Ujiie, H., eds., *Mem. Geol. Soc. Jpn.*, 22, p. 141–157. (in Japanese with English abstracts.)

Klein, G. D., Kobayashi, K., Chamley, H., Curtis, D. M., Dick, H. J. B., Echols, D. J., Fountain, D. M., Kinoshita, H., Marsh, N. G., Mizuno, A., Nisterenko, G. V., Okada, H., Sloan, J. R., Waples, D. M. and White, S. M., 1978, Off-ridge volcanism and seafloor spreading in the Shikoku Basin, *Nature*, v. 273, p. 746–748.

Klein, G. D. and Kobayashi, K., 1981, Geological summary of the Shikoku Basin and northwestern Philippine Sea, Leg 58, DSDP/IPOD drilling results, *Oceanol. Acta, Spec. Issue*, p. 181–192.

Kobayashi, K., and Nomura, M., 1972, Iron sulfides in the sediment cores from the Sea of Japan and their geophysical implications, *Earth Planet. Sci. Lett.* v, 16, p. 200–208.

Koizumi, I., 1975a, Late Cenozoic diatom biostratigraphy in the circum-Pacific region, *Geol. Soc. Jpn. Jour.*, v. 81, p. 611–627.

Koizumi, I., 1975b, Neogene diatoms from the western margin of the Pacific Ocean, leg 31, DSDP, in: Initial Reports of the Deep Sea Drilling Project, v. 31, Karig, D. E., Ingle, J. C. Jr. *et al.* eds., Washington, D.C.: United States Government Printing Office, p. 779–819.

Koizumi, I., 1977a, Deep-sea sediments and history of the Sea of Japan, *Kagaku*, v. 47, p. 45–51, (in Japanese).

Koizumi, I., 1977b, Diatom biostratigraphy in the North Pacific region, in: *Proceedings of the First International Congress on Pacific Neogene Stratigraphy*, Saito, T. and Ujiie, H. eds. Tokyo: Science Council of Japan, p. 235–254.

Koizumi, I., 1978, Neogene diatoms from the Sea of Japan, *Mar. Geol.* v. 26, p. 231–248.

Konishi, K., 1963; Pre-Miocene basement complex of Okinawa and the tectonic belts of the Ryukyu Islands, *Sci. Rep. Kanazawa Univ.*, v. 8, p. 569–602.

Kovylin, V. M., Karts, B. Ya., 1966, Structure of the crust and sedimentary layer of the Sea of Japan based on seismic data, *Dokl. Akad. Nauk SSSR*, v. 168, p. 1048–1051. (in Russian).

Larson, R., Schlanger, S. O., *et al.*, 1981, *Initial Reports of DSDP, v. 61*, Washington, D.C.: United States Government Printing Office, 885p.

Lee, C. S., Shor, G. G. Jr., Bibee, L. D., Lu. R. S., and Hilde, T. W. C., 1980, Okinawa Trough: origin of a back-arc basin, *Mar. Geol.*, v. 35, p. 219–241.

Ludwig, W., Murauchi, S. and Houtz, R. E., 1975, Sediments and structure of the Japan Sea, *Geol. Soc. Am. Bull.*, v. 86, p. 651–664.

Ludwig, W. J., Murauchi, S., Den, N., Bull, P., Hotta, H., Ewing, M., Asanuma, T, Yoshii, T., and Sakajiri, N., 1973; Structure of East China Sea–West Philippine Sea margin off southern Kyushu, Japan, 1973, *J. Geophys. Res.*, v. 78, p. 2526–2536.

Machida, H., 1981, Tephrochronology and Quaternary studies in Japan, in: *Tephra Studies*, Gelf, S. and Sparks, S. J., eds., D. Reidel Publishing Company, Dordrecht, p. 161–191.

Machida, H. and Arai, F., 1983, Extensive ash falls in and around the Sea of Japan from large late Quaternary eruptions, *J. Volcanol. Geotherm. Res.*, v. 18, p. 151–164.

Machida, H., Arai, F. and Moriwaki, H., 1981, Two Korean tephras, Holocene markers in the Sea of Japan and the Japan Islands, *Kagaku*, v. 51, p. 562–569. (in Japanese).

Matsumoto, Y., 1983, The Cenozoic volcanism in the Ryukyu Islands, Japan, *Mem. Geol. Soc. Jpn.*, 22, p. 81–91 (in Japanese).

Matsumoto, Y. and Nahara, T., 1974; Volcanic rocks in Kobi-sho, Senkaku-retto, *Mem. Dept. Liberal Arts, Nagasaki Univ., Nat. Sci.*, v. 15, p. 21–35. (in Japanese).

Miyake, Y., Sugimura, Y., and Matsumoto, E., 1968, Ionium-thorium chronology of the Japan Sea cores, *Oceanogr. Works Jpn. Rec.*, v. 9, p. 189–195.

Miyazaki, T., Tamaki, K. and Murakami, F., 1976, Geomagnetic survey, in: *Ryukyu Island (Nansei-Shoto) Arc, GH 75-1 and GH 75-5 Cruises*, Honza, E., ed., Kawasaki-shi: Geological Survey of Japan, p. 52–54.

Mogi, K., 1969; Relationship between the occurrence of great earthquakes and tectonic structures, *Bull. Earthquake. Res. Inst., Univ. Tokyo*, v. 47, p. 429–451.

Murauchi, S., 1972, Crustal structure of the Japan Sea *Kagaku* v. 42, p. 367–375, (in Japanese).

Murauchi, S., Den, N., Asano, S., Hotta, H. Yoshii, T., Asanuma, T., Hagiwara, K., Ichikawa, K., Sato, T., Ludwig, W. J., Ewing, J. I., Edgar, N. T. and Hontz, R. E., 1968, Crustal structure of the Philippine Sea, *J. Geophys. Res.*, v. 73, p. 3143–3171.

Nishimura, Y., Matsubarra, Y. and Nakamura, E., 1983, Zonation and K-Ar ages of the Yaeyama metamorphic rocks, Ryukyu Islands *Mem. Geol. Soc. Jpn*, 22, p. 27–37. (in Japanese with English abstracts).

Nomura, M., 1979, Marine Geomagnetic Anomalies with Intermediate Wavelengths in the Western Pacific Region, *Bull. Ocean Res. Inst. Univ. of Tokyo*, No. 10, p. 1–42.

Oba, T., 1983a, Paleoenvironment of the Sea of Japan since the last glacial period, *Chikyu*, v. 5, p. 37–46. (in Japanese).

Oba, T., 1983b, Oxygen isotopic analysis, in: *Preliminary Report of the Hakuho Maru Cruise KH 82-4*, Kobayashi K., ed. University of Tokyo: Ocean Research Institute, p. 140–141.

Oba, T., Horibe, Y. and Kitazato, H., 1981, Paleoenvironmental analysis of a period since the last glacial using two sediment cores form the Sea of Japan, *Kokogaku-to-Shizenkagaku*, v. 13, p 31–49, (in Japanese).

Okada, H., 1978, Sedimentary patterns in apparent back-arc basins: A case study of the Neogene sequence in Northwestern Hokkaido, Japan, *J. Phys. Earth*, v. 26S, p. 477–490.

Otofuji, Y. and Matsuda, T., 1983, Paleomagnetic evidence for the clockwise rotation of Southwest Japan, *Earth Planet. Sci. Lett.*, v. 62, p. 349.

Otofuji, Y., Hayashida, A., and Torii, M., 1985, When did the Japan Sea open?—paleomagnetic evidence from southwest Japan, in: *Formation of Active Ocean Margins*, Nasu, N., Kobayashi, K., Uyeda, S., Kushiro, I., and Kagami, H. eds., Terra Science, Tokyo and D. Reidel, Dordrecht, p. 230–255.

Raitt, R. W., Shor, G. G., Jr., Francis, J. G. and Morris, G. B., 1969, Anisotropy of the Pacific upper mantle, *J. Geophys. Res.*, v. 74, p. 3095–3109.

Saito, Y., Tiba, T. and Miyagi, H., 1973, Geology of Iriomote-jima, Ryukyu Island. *Mem. Nat. Sci. Mus. (Japan)*, v. 6, p. 9–22.

Sasajima, S., 1981, Pre--Neogene paleomagnetism of Japanese Islands (and vicinities). in: *Paleoreconstruction of the Continents, Geodynamics Series 2*, McElhinney, M. W., and Valencio, D. A., eds., American Geophysical Union-Geological Society of America, p. 115–128.

Shibata, K., and Nozawa, T., 1966, K-Ar ages of granites from Amami–Oshima, Ryukyu Island, Japan, *Bull. Geol. Surv. Jpn.*, v. 17, p. 430–435.

Shibata, K., Konishi, K. and Nozawa, T., 1968, K-Ar age of muscovite from crystalline schist of the northern Ishigaki-shima, Ryukyu, Islands, *Bull. Geol. Surv. Jpn.* v. 19, p. 529–533.

Shibata, K. and Ishihara, S., 1979, Initial 87Sr/86Sr rations of plutonic rock from Japan, *Contrib. Mineral. Petrol.*, v. 70, p. 381–390.

Shimamura, H., Asada, T. and Kumazawa, M., 1977, High shear velocity layer in the upper mantle of the western Pacific. *Nature*, v. 269, p. 680–682.

Shiono, K., Mikumo, T., and Ishikawa, Y., 1980, Tectonics of the Kyushu–Ryukyu Arc as Evidenced from seismicity and focal mechanism of shallow to inter-mediate-depth earthquakes, *J. Phys. Earth*, v. 28, p. 17–43.

Sillitoe, R. H., 1977, Metallogeny of an Andeantype continental margin in South Korea: Implications for opening of the Japan Sea, in: *Island Arcs, Deep Sea Trenches and Back-Arc Basins*, Talwani, M. and Pitman. W. C. III, eds. *American Geophysical Union Maurice Ewing Series* v. 1, p. 303–310.

Tamaki, K., 1982, Consideration on age and of formation of the Sea of Japan, *Abstr. Seism. Soc. Jpn.*, v. A73, p. 75.

Tiba, T. and Saito, Y., 1974, A note on the volcanic rocks of Iriomote-jima, Ryukyu Island, *Mem. Nat. Sci. Mus., (Japan)*, v. 7, p. 25–30.

Tomoda, Y., and Fujimoto, H., 1982, Maps of gravity anomalies and bottom topography in the western Pacific, *Bull. Ocean Res. Inst. No. 14*, p. 1–158.

Torii, M., 1983, Final age of the clockwise rotation of the southwest Japan paleomagnetism of the Tosho and Shozushima formations in Kagawa Prefectures, *Abstr. Soc. Terr. Mag. Electr. Japan* Tokyo, p. 143, (in Japanese).

Ueno, N., Kaneoka, J., Ozima, M., Zashu, S., Sato, T., and Iwabuchi, Y., 1971, K-Ar age, Sr isotopic ratio and K/Rb ratio of the volcanic rocks dredged from the Japan sea, (in Japanese). in: *Island Arc and Marginal Seas*, Asano, S. and Udintsev, G. B., eds., Tokyo: Tokai University Press, p. 305–309.

Ujiie, H., 1983, Submarine geology west off the Okinawa Island in relation to the Ryukyu Arc development *Mem. Geol. Soc. Jpn.*, 22, p. 131–140, (In Japanese with English abstracts).

Ujiie, H. and Ichikawa, M., 1973, Holocene to uppermost Pleistocene planktonic foraminifera in a piston core from off Sanin district, Sea of Japan, *Trans. Proc. Paleontol. Soc. Jpn.* v. 91, p. 137–150.

Utsu, T., 1971, Seismological evidence for anomalous structure of island arcs with special reference to the Japanese region, *Rev. Geophys. Space Phys.*, v. 9, p. 839–890.

Uyeda, S. and Miyashiro, A., 1974, Plate tectonics and the Japanese Islands; A synthesis, *Geol. Soc. Am. Bull.*, v. 85, p. 1159–1170.

Uyeda, S. and Vacquier, V., 1968, Geothermal and geomagnetic data in and around the island of Japan. in: *The Crust and Upper Mantle of the Pacific Area*, Knopoff, L., Drake, C. and Hart, P., eds., *Am. Geophys. Union Mongr.*, 12, p. 349–366.

Wageman, J. M., Hilde, T. W. C., and Emery, K. O., 1970, Structural framework of East China Sea and Yellow Sea. *Am. Assoc. Pet. Geol. Bull.*, v. 54, p. 1611–1643.

Watts, A. B., 1976, Gravity field of the northwest Pacific Ocean basin and its margin; Philippine Sea, *Geol. Soc. Am. Spec. Map & Chart Ser. MC-12*.

Yaskawa, K., 1975, Paleolatitude and relative position of southwest Japan and Korea in the Cretaceous. *Geophys. J. R. Astron. Soc.*, v. 43, p. 835–846.

Yasui, M., Kishii, T., Watanabe, T. and Uyeda, S., 1968, Heat flow in the Sea of Japan. in: Knopoff, L., C. Drake and P. Hart (eds.) *The Crust and Upper Mantle of the Pacific Area*, Knopoff, L., Drake, C., and Hart, P., eds., *Am. Geophys. Union Mongr.*, v. 12, p. 3–16.

Yasui, M., Epp. D., Nagasaka, K. and Kishii, T., 1970, Terrestrial heat flow in the seas around the Nansei Shoto (Ryukyu Islands), *Tectonophysics*, v. 10, p. 225–234.

Yoshii, T., 1972, Features of the upper mantle around Japan as inferred from gravity anomalies, *J. Phys. Earth*, v. 20, p. 23–34.

Yoshii, T., 1973, Upper mantle structure beneath the North Pacific and the marginal seas, *J. Phys. Earth*, v. 21, p. 313–328.

Yoshii, T., 1977, Crust and mantle structures of the Northeastern Japan, *Kagaku*, v. 47, p. 170–176. (in Japanese).

Chapter 10

THE BONIN ARC

E. Honza and K. Tamaki

Geological Survey of Japan
1-1-3 Higashi, Yatabe
Ibaraki, 305 Japan

I. INTRODUCTION

The Bonin Arc is a well-developed island arc, associated with a deep trench, an active volcanic chain, and a back-arc basin. The Bonin Arc is approximately 1100 km long, extending north–south from latitude 35°N to 25°N within longitude 139°E and 145°E, facing the Pacific plate in front and the Shikoku Basin behind (Fig. 1). At its northern limit, the Bonin Arc forms part of a triple junction of trench–trench–transform fault or trench–trench–trench with the Tohoku and the Seinan (southwest) Japan arcs. The Mariana Arc continues the line of the Bonin Arc to the south. There is no marked morphological boundary between the Bonin and the Tohoku arcs, but one is inferred by the depression along the Sagami Trough (see Fig. 13). The southern boundary of the Bonin Arc is even less well defined than the northern margin. The Bonin Arc extends to the Mariana Arc as part of the same morphological sequence from north to south. These three arcs face the Pacific plate.

Two island chains are distinguished in the Bonin Arc. One is along the volcanic chain and the other, restricted to the southern part, lies along the fore-arc basement high.

In this chapter the geology and geophysics of the Bonin Arc are discussed, with particular reference to the data obtained by the *Hakurei-Maru*

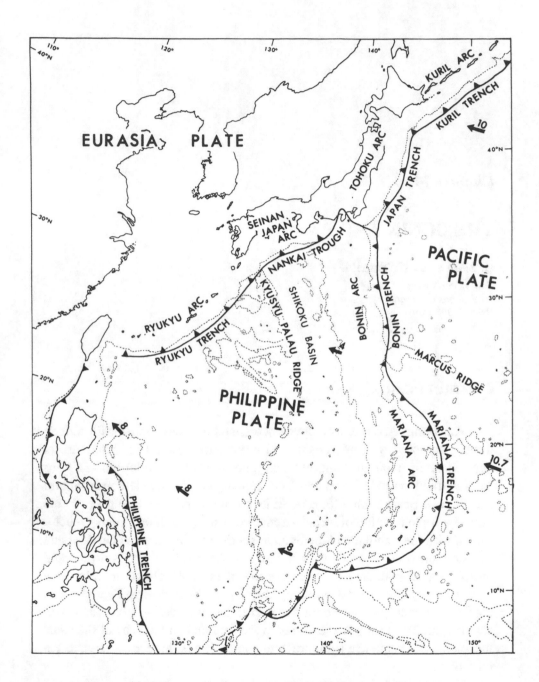

Fig. 1.   Plate environment around the Bonin Arc based on Nishiwaki (1981). The thick solid line shows present plate boundary; the arrows show absolute motion of each plate with values in cm/yr.

cruises in 1979, which covered the whole offshore area of the arc (Honza
et al., 1981; Honza *et al.*, 1982; Yuasa *et al.*, 1982).

II. BOTTOM TOPOGRAPHY

The physiography of the northern part of the Bonin Arc is different from
the southern part, but both are bounded by an E–W trending fore-arc depression just east of Sofugan Island (Fig. 2). The southern Bonin Arc is characterized by the presence of the Bonin Ridge in the fore-arc area and the
less prominent development of the Shichito Ridge (active volcanic chain).

Several ridges and troughs extend north–south parallel to the trench,
but the volcanic Shichito (Iwo Jima) Ridge is the dominant ridge extending
throughout the arc. The Bonin (Ogasawara) Trench, which also extends
throughout the arc, is interrupted by the Ogasawara Plateau at the southern
end. The trench to the south of the plateau is the Mariana Trench. To the
north of the Bonin Trench is the Japan Trench bordered by a seamount on
the axis of the trench (Fig. 2).

Several terms have been used for the topographic features in the Bonin
Arc (Hess, 1948; Mogi, 1972; Karig and Moore, 1975; Honza *et al.*, 1981).
The term Izu–Ogasawara or Ogasawara is used for the arc, trench, and fore-
arc ridge (Mogi, 1972). Ogasawara is the Japanese name for Bonin. The term
Bonin is also used for the same arc, trench and fore-arc ridge (Karig and
Moore, 1975). The term Shichito–Iwoto Ridge (Mogi, 1972) or Iwoto Ridge
(Karig and Moore, 1975) is used for a volcanic ridge in the central part of
the arc. The name Nishi–Shichito Ridge is used for the ridge west of the
volcanic ridge (Mogi, 1972).

Here, we use the term Bonin (Ogasawara) for arc, trench, fore-arc ridge
and fore-arc trough, instead of the term Izu–Ogasawara; Shichito (Iwo-Jima)
Ridge for the volcanic ridge; and Izu Ridge instead of the Nishi–Shichito
Ridge for the ridge to the west of the Shichito Ridge. The Nishinoshima
Trough is used for a trough between the Shichito and the Izu ridges in the
southern arc (Fig. 2).

Small highs arranged discontinuously immediately to the east of the
Shichito Ridge are called the Shinkurose Ridge in this chapter. In the northern arc, the fore-arc continental slope is remarkably smooth and gentle, with
some minor irregularities in the trench-slope break area where small highs
are observed.

The existence of the Bonin Ridge and the Bonin Trough defines the
nature of the southern Bonin Arc. The Bonin Ridge has a steep slope on the
west side and a gentle slope in the east side. The northern extension of the
ridge consists of small highs on the trench slope break. The Bonin Trough,

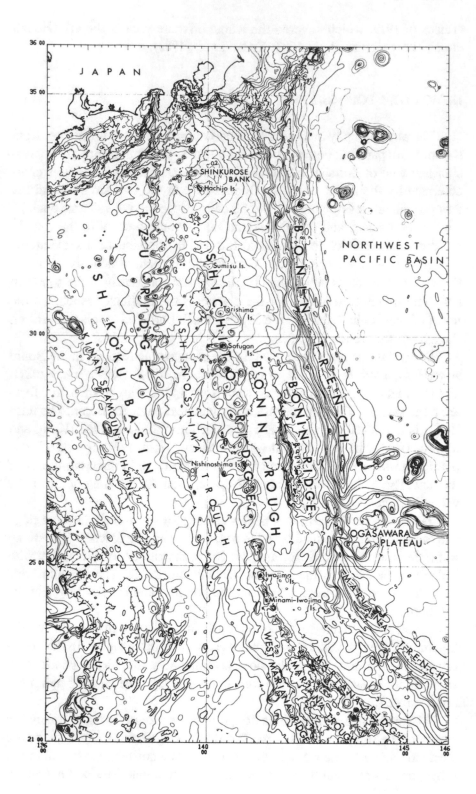

which has a deep and smooth bottom, extends to a relatively shallow and gentle continental slope in the northern Bonin Arc.

There are several canyons on the northern fore-arc slope perpendicular to the slope contour. Some of them have ponded terminals, but have an erosional character throughout the channel itself.

Most of the Bonin Trench has a depth of more than 9000 m with steep walls in the inner trench slope and a relatively gentle slope on the outer trench slope. Generally, the trench has a V-shaped profile with a narrow flat bottom along the axis. The bottom is wider (approximately 18 km) in the northern part and narrower in the south. It is shallower with deformed features along the contact zone with the Ogasawara Plateau at the southern end of the arc. The rugged topography of the outer trench slope suggests horsts and grabens as is commonly observed in other trenches.

Topographic depressions are observed just behind the Shichito Ridge and Tamaki *et al.* (1981a) tentatively called them "back-arc depressions." The width of the depressions range from less than 20 km to 60 km, and their depth varies from 500 m to 1300 m. Small depressions are not shown on the bathymetric map of Figure 2.

The Izu Ridge along the western margin of the arc is one of the prominent topographic features in the back-arc region. It consists of some highs on the mid-slope of the western flank of the Shichito Ridge.The Izu Ridge is more prominent in the northern than in the southern Bonin Arc, and the ridge is broken up into several en echelon segments arranged in a northeast to southwest direction as shown on the bathymetric map (Fig. 2). The Izu Ridge consists of small irregular highs in the southern arc and the number of highs decreases to the south. The Izu Ridge can be discontinuously traced to southwest of Minami–Iwo Jima Island, and does not extend to the West Mariana Ridge. The area between the Shichito Ridge and Izu Ridge is rugged with many small highs and troughs in the northern arc. Some of these highs and troughs also trend in a northeast to southwest direction (Mogi, 1972; Karig and Moore, 1975; Honza *et al.*, 1982). Karig and Moore (1975) postulated compression of the arc along its trend for origin of the ridge and trough topography and concluded that this compression could be attributed to the southward movement of Japan.

Many knolls are distributed in the back-arc depressions. We call the area "back-arc knolls zone." The back-arc knolls zone represents a rise

Fig. 2. Bathymetric map of the Bonin Arc and the northern end of the Mariana Arc. Depth values are shown in kilometers. Depth contours are based on digitized contour-line data of the Japan Oceanographic Data Center of Maritime Safety Agency of Japan, excluding the area south of 24°N.

feature that extends in NE–SW direction with a left stepping en echelon arrangement.

A trough is developed along the eastern margin of the Izu Ridge. Honza *et al.*, (1981) called it the Nishinoshima Trough after Nishinoshima Island. it is more prominent in the southern Bonin Arc than in the northern arc. In the northern arc, the trough has the same en echelon pattern as the Izu Ridge.

III. GEOPHYSICAL DATA

A. Gravity Anomalies

Free-air gravity anomalies are well correlated with the topographic features. The topographic highs such as ridges are associated with high anomalies, and topographic lows such as troughs and trenches with low anomalies (Ishihara *et al.*, 1981, Ishihara *et al.*, in press). The high anomalies occur along the Shichito Ridge and the Bonin Ridge and the low anomalies along the Bonin Trench and on the Bonin Trough (Fig. 3).

High free-air anomalies of 200 mgal are observed to the east of Hachijo Island where there are some shallow banks. The high free-air anomaly of 380 mgal that occurs over the northern part of the Bonin Ridge is one of the largest anomalies in the world (Watts *et al.*, 1976; Ishihara *et al.*, 1981).

There are two interpretations of the high anomalies over the Bonin Ridge, one is in terms of crustal thinning and the other is of a high density crust. It is established that the Bonin Ridge has boninite, which is believed to have been formed by cooling of magma from the mantle. Such a magma most probable could account for high gravity anomalies.

If the upper mantle density and the mean crustal density are assumed to be 3.3 and 2.67 g/cm^3, the the Moho Discontinuity is calculated to lie approximately 8 km below sea level (Ishihara, 1981). If the crustal thickness is 12 km, as in the case of the trench-slope break area (Hotta, 1970), the mean crustal density will be 2.89 g/cm^3 (Ishihara, 1981).

Gravity minima along the trench are located a few kilometers to the inner side of the trench axis. The distance between the trench axis and gravity minimum is larger in the Japan Trench by more than 10 km (Nishimura and Murakami, 1977). The shorter horizontal distance between the gravity minimum and the trench axis in the Bonin Trench may reflect a steeper angle of subducted Pacific plate under the arc. This is deduced from

Fig. 3.   Free-air gravity anomaly map. The contour lines are based on a digitzed chart of Ishihara (in press). The solid lines show positive anomalies, and the dashed lines show negative anomalies. Values are represented by mgal. The stippled area shows trench depth.

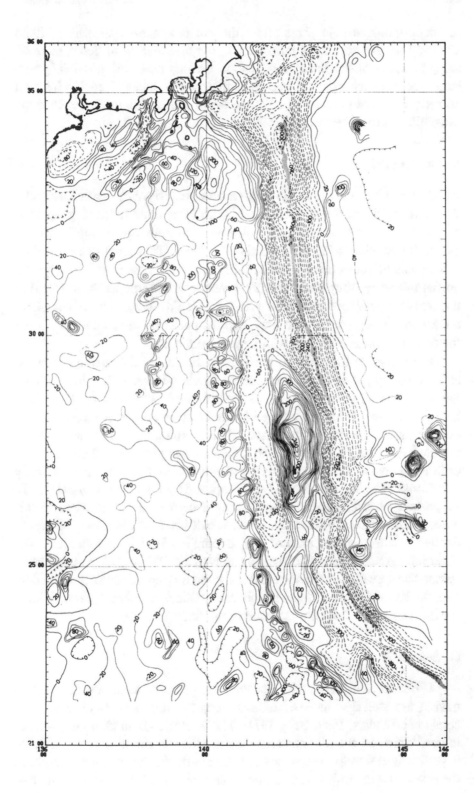

the deep-focus seismic plane under the arc (Katsumata and Sikes, 1969; Utsu, 1974). An additional possibility is the distribution of dense material along the foot of the inner trench slope. In this case, the possibility of a large accretionary prism is less than that in the Japan Trench where an accretionary prism is developed in the lower half of the inner trench slope (Scientific Party, 1980; Honza, 1981a).

B. Geomagnetics

The Shichito and the Bonin ridges are characterized by marked magnetic anomalies with short wave length and with large amplitudes (Fig. 4). Relatively high positive anomalies are observed over the topographic highs just east of the Shichito Ridge (Miyazaki et al., 1981). These highs lie over older volcanoclastic rocks (Honza et al., 1982). Anomalies over the Izu Ridge are not as intensive, they show the same anomaly patterns as those found in the area of rugged topography between the Shichito and the Izu Ridges where the anomalies appear to have a overlapped pattern. Anomalies found over the frontal slope are smooth with long wave length.

Anomalies in the Shikoku Basin show lineations that strike northwest approximately parallel to the trend of the Shichito Ridge and of the Kyushu–Palau Ridge (Watts and Weissel, 1975; Kobayashi and Nakada, 1979). The lineation pattern is best developed in the western part of the basin. In the eastern part of the basin, the basement morphology is rough and complex and magnetic anomalies cannot be identified unequivocally. There are two different interpretations regarding the identification of the magnetic anomaly lineations of the Shikoku Basin. One, proposed by Watts and Weissel (1975), suggests a spreading pattern with the ages increasing to the west. In this case, the anomalies 7-5E in the western half are identified, but the identification of those of the eastern half, anomalies 5E-5 is hypothetical. The other interpretation is by Kobayashi and Nakada (1979); their pattern shows symmetric spreading. With the center of symmetry approximately coincident with the Kinan Seamont chain, which extends along the center of the Shikoku Basin. Anomalies 7-5D are identical in this interpretation.

C. Seismicity

Differences in the spatial distribution of intermediate and deep earthquakes are well distinguished in arcs along the northern Pacific rim (Katsumata and Sykes, 1969; Utsu, 1974). The Wadachi–Benioff zone is steeper in the Bonin and the Mariana arcs than in other neighboring arcs. At depth it gradually becomes steeper toward the south. In the northern margin of the Mariana Arc, however, a smaller angle is observed. Shallow earthquakes

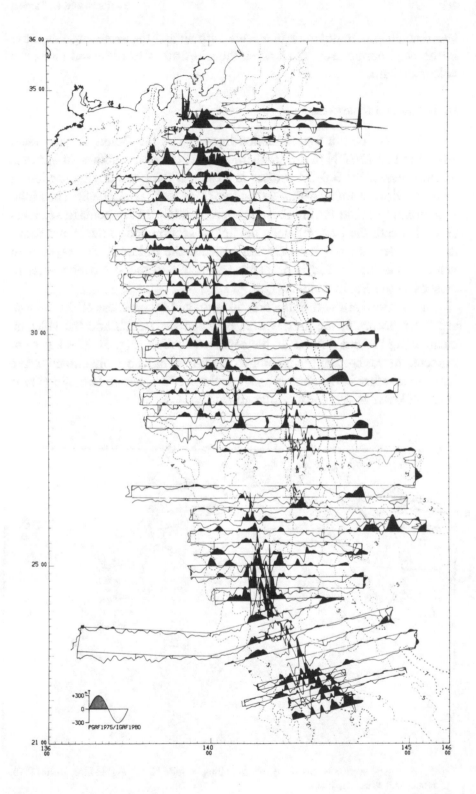

Fig. 4. Geomagnetic anomaly data profile along the ship's tracks. PGRF1975 and IGRF1980 are used as a reference field.

less than 100 km in depth lie in a plane with almost the same angle as those in the neighboring arcs. Shallow earthquakes are also observed along the volcanic chain.

D. Refraction Measurement

There are only a few refraction lines across the Bonin Trench. One, along the line 23°30'N in the southermost part of the arc shows an increase in thickness of the 3.0–3.5 km/sec layer and a thick underlying section of material with a velocity of 5.5–6.0 km/sec (Murauchi *et al.*, 1968). The Moho Discontinuity in the Pacific Basin is approximately 13 km beneath sea level (7 km beneath the bottom) along the line of section and in the Bonin Ridge is 17 km beneath sea level (15 km beneath the bottom). The layer with velocity 3.0–3.5 km/sec is probably correlated to the Paleogene volcani-clastics in the Bonin Ridge.

In the northern part of the arc, an appreciable thickness of the 5.4 km/sec layer occurs in the narrow area between the trench and the volcanic chain along the line 32°N as in the southern section (Fig. 5). The layer not reflected in the bottom topography in the north may possibly extend to the fore-arc member in the southern part of the arc. The Moho Discontinuity is also shallower in the north (Hotta, 1970).

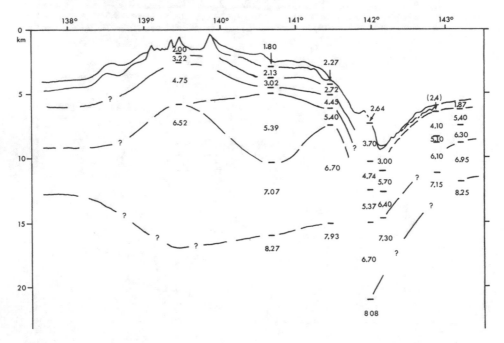

Fig. 5.   A crustal section across the Bonin Arc along latitude 32°N (revised after Hotta, 1970, and Houtz and Windish, 1980).

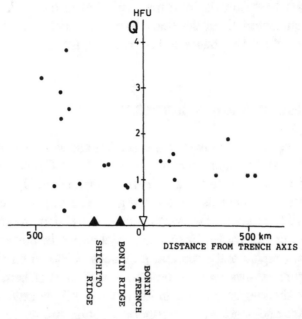

Fig. 6. Heat flow values between latitudes 25°N and 31°N plotted against the distance from the trench axis (Matsubara, 1981). Heat flow values in 10^{-6} cal/cm^2 sec in the Bonin Arc.

The layer on the foot of the inner trench slope has a velocity with 3.7 km/sec, which is slightly higher than that in the Japan Trench. (2.5–2.8 km/sec; cf. Nagumo *et al.*, 1980; Murauchi and Ludwig, 1980). A relatively lower velocity at the foot of the Japan Trench is interpreted to reflect lower formation of an accretionary prism that consists of a melange overlain by younger sediments (Scientific Party Leg 56 and 57, 1980; Honza, 1981a). The slightly higher velocity at the foot of the Bonin Trench is considered to reflect that there is small accretionary prism in it (Honza *et al.*, 1982; Yuasa *et al.*, 1982).

E. Heat Flow

In the southern Bonin Arc, a minimum heat flow is observed just landward of the trench axis. Uniform, and relatively low values are observed on the Pacific side. Rather lower values [less than 1.5 HFU (10^{-6} cal/cm^{2}sec)] occur in the fore-arc area. Heat flow values gradually increase toward the volcanic chain and are independent of the morphology of the Bonin Ridge and the Bonin Trough. In the back-arc area, heat flow values vary greatly from place to place (Fig. 6). Low values may result from hydrothermal circulation caused by the rough topography and the figures may have a low reliability (Matsubara, 1981; McKenzie and Sclater, 1968).

The lower heat flow as compared with that in the back-arc area is a remarkable characteristic of the Bonin Trough. This low heat flow refutes the possibility of active modern spreading in the trough.

IV. STRATIGRAPHY AND STRUCTURE

The observations presented in this section are based on single channel seismic profiles and bottom sampling data of GH79-2, GH79-3, GH79-4, and GH80-3 cruises of the Geological Survey of Japan (Fig. 7). Sound source for the seismic profiling was a 300 in³ airgun. The ship's speed during the survey was 10 to 11 knots. Two-way acoustic travel time in seconds is used in discussions of the profiles for sediment thickness. The nomenclature used for the physiographic and structural provinces is shown on the profile of Figure 8. Figure 9 shows a seismic profile record of the northern end Mariana Arc which is the longest traverse in the cruises. All the seismic profiles of the Bonin Arc and some of the northern Mariana Arc are summarized in Figure 10.

The stratigraphy of the Bonin Arc is summarized in marine geological maps by Honza *et al.* (1982) and Yuasa *et al.* (1982) that were made by the synthetic study of seismic profiles, bottom sampling data, and extrapolation from regional land geology (Fig. 11).

A. Trench and Trench Slopes

The Bonin Trench is generally deeper than 9000 m and occasionally exceeds 9500 m (Fig. 2) in the northern Bonin Arc. The depth of North-western Pacific Ocean floor along the Bonin Trench, over the outer gravity high, is about 5500 m. Thus the trench bottom is about 3000 m deeper than the Pacific Ocean floor. The trench axes show offsets of 10 to 20 km in several places (Fig. 12).

Flat-lying turbidites deposits are commonly observed along the trench bottom in the northern Bonin Arc, while to the south and in the northern Mariana Trench the axis is V-shaped and free of sediments. The width of the trench floor filled by turbidite deposits is generally 5 to 10 km with a maximum of 18 km on Profile 49 (Fig. 10). The maximum thickness of the trench fill, which reaches 1.5 sec, is also observed on Profile 49. Profile 49 is located at the triple junction where the Japan and Bonin trenches, and the Sagami Trough join. The trench fill diminishes to the north and to the south of Profile 49. This shows that the thick sediments fill is derived through several submarine canyons developed at the junction area (Fig. 2).

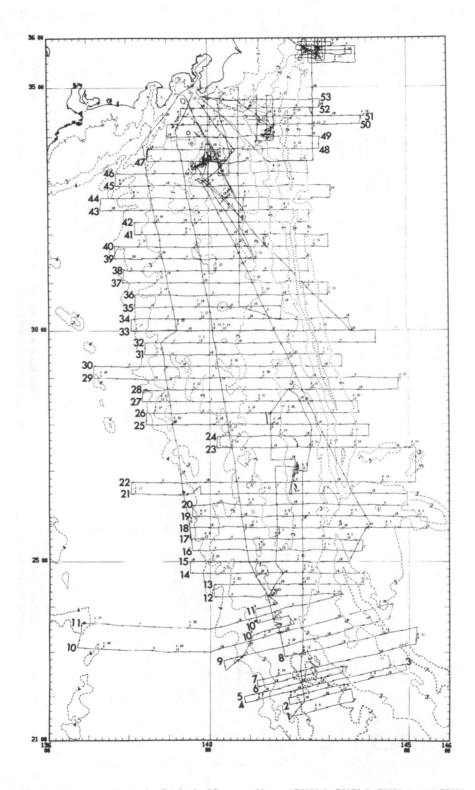

Fig. 7. Survey tracks by the Geological Survey of Japan (GH79-2, GH79-3, GH79-4, and GH80-3 cruises of RV *Hakurei Maru*). Numerals show numbers of traverse lines.

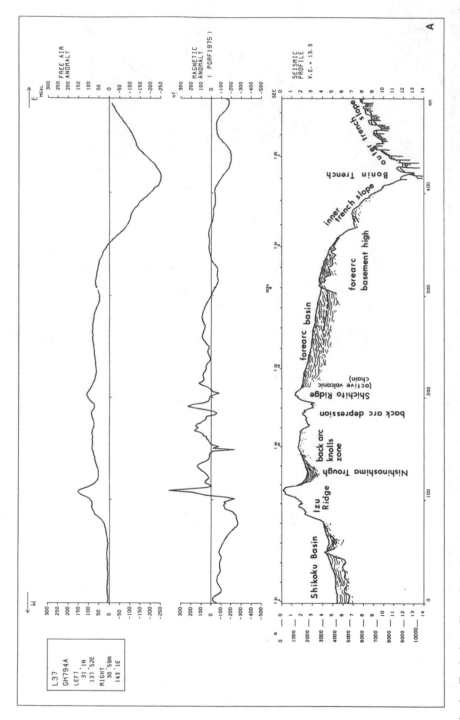

Fig. 8. Typical profiles of the Bonin Arc and the northern end of the Mariana Arc, with free-air gravity anomaly and geomagnetic anomaly profiles, and interpretations of the seismic profiles. Traverse lines for the profiles are shown in Fig. 7: (a) Profile 37; (b) Profile 25; (c) Profile 19.

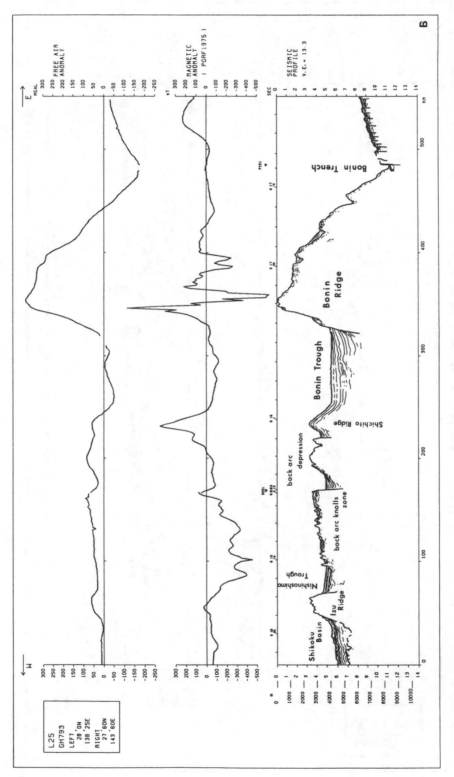

Fig. 8. (*continued*)

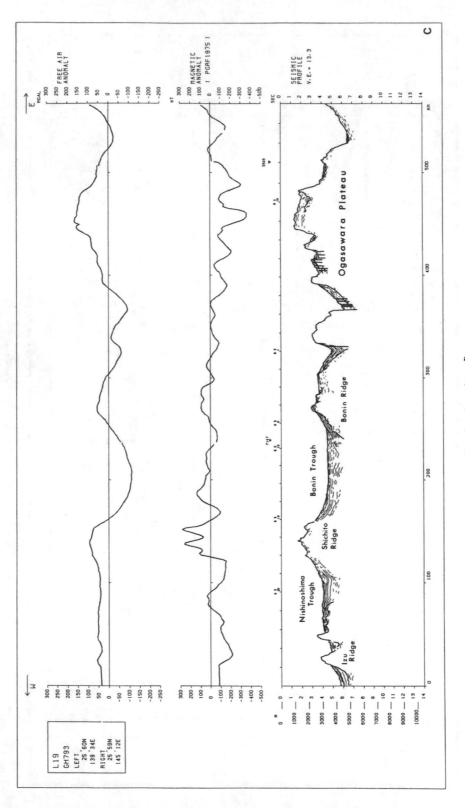

Fig. 8. (*continued*)

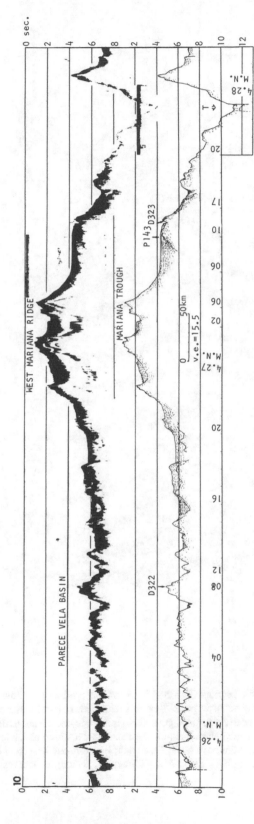

Fig. 9. Typical seismic profiles with interpretation of the Bonin Arc and the northern end of the Mariana Arc along Profile 10. The vertical scales are two-way acoustic travel time in seconds.

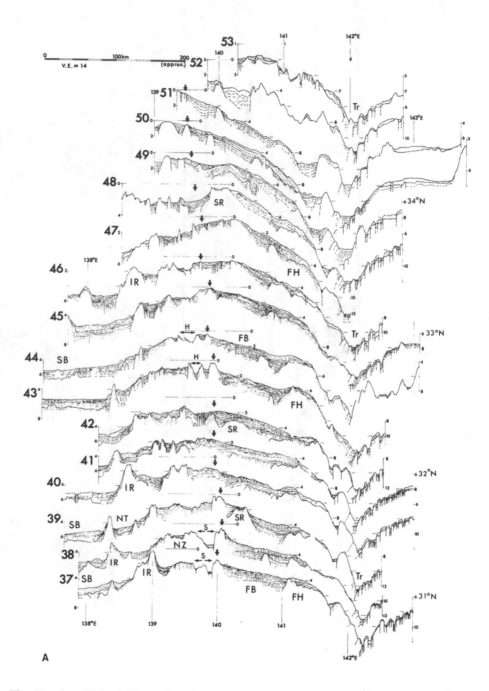

Fig. 10. Compiled seismic profiles. Traverse lines for the profiles are shown in Fig. 7. The vertical scales are two-way acoustic travel time in second. Horizontal distances are based on Mercator projection, and each profile is aligned along longitude lines. The vertical arrows show the volcanic front. Paired arrows show back-arc depressions. Physiographic and structural provinces are shown in abbreviations as follows; Tr, trench; FH, fore-arc basement high; BR, Bonin Ridge; FB, fore-arc basin; SR, Shinkurose Ridge; H, Hachijo Depression; S, Sumisu

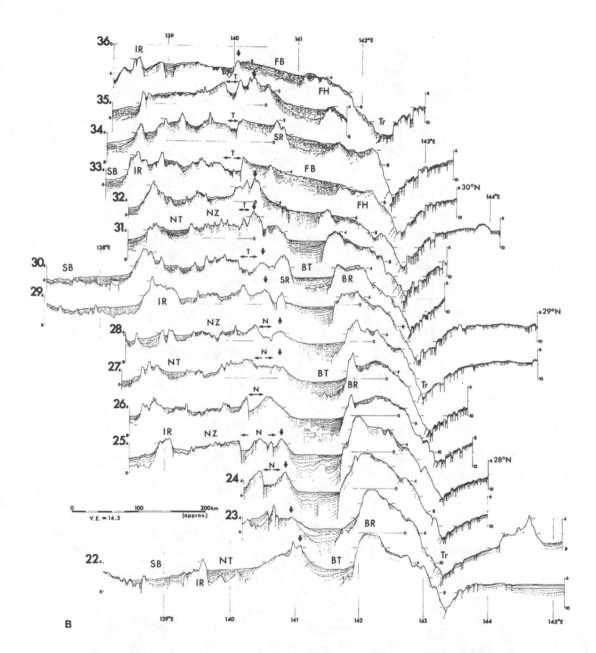

Depression; T, Torishima Depression; N, Nishinoshima Depression; NZ, back-arc knolls zone; NT, Nishinoshima Trough; IR, Izu Ridge; SB, Shikoku Basin; MR, Mariana Ridge; MT, Mariana Trough; and WR, West Mariana Ridge. (a) Profiles 53–37. (b) Profiles 36–22. (c) Profiles 21–4.

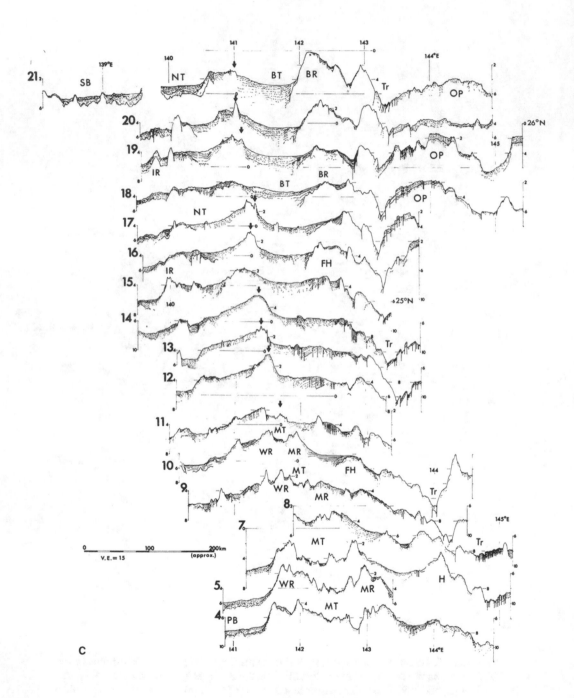

Fig. 10.   (*continued*)

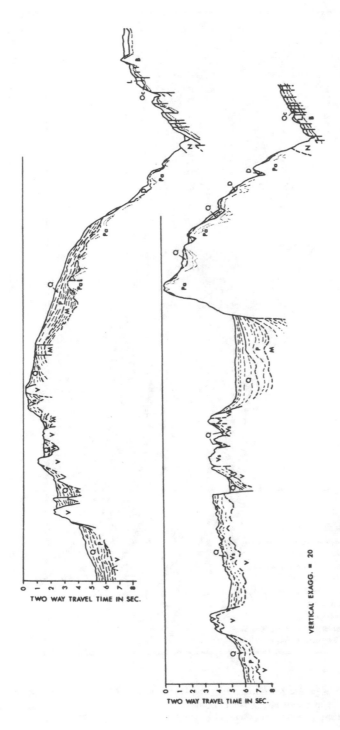

Fig. 11. Stratigraphy of the Bonin Arc illustrated in the typical profiles (Honza *et al.*, 1982; Yuasa *et al.*, 1982). Q, Quaternary; P, Pliocene; M, Miocene; Pa, Paleogene and older layers; D, Neogene deformed layer; N, deformed layer in the inner trench slope; V, volcanic rocks; Vs, volcanoclastics; Oc, Pelagic sediments; L, Layered basement in the Pacific floor; B, oceanic basement. Upper, Profile 40 (northern Bonin Arc), Lower, Profile 25 (southern Bonin Arc).

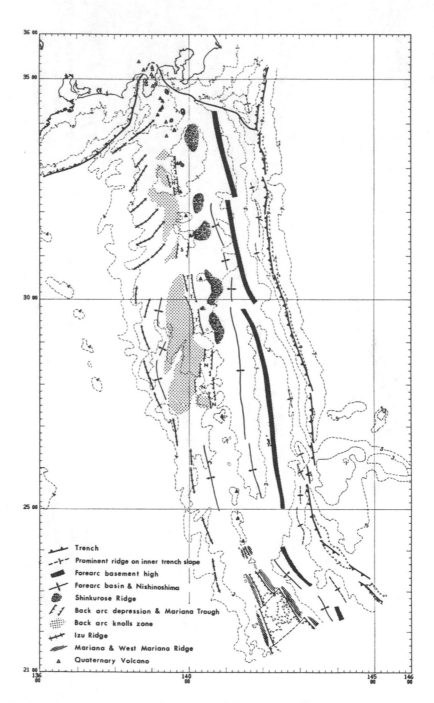

Fig. 12. Summary of geological structure of the Bonin Arc and the northern Mariana Arc. Names of back-arc depressions are shown in abbreviations as follows; H, Hachijo Depression; S, Sumisu Depression; T, Torishima Depression; and N, Nishinoshima Depression.

A distinctive feature of the outer trench slope is the well-developed horst and graben structure and normal step faults, commonly observed on the outer slopes of many other trenches. These structural features apparently indicate the tensional stress field of the outer trench slope caused by downward bending of the subducting lithosphere.

The faults are seldom observed on the Pacific Ocean floor outside the trench slope area. Fault development is concentrated on the slope area and large displacements of the faults are observed in the mid-slope to down slope area. The maximum displacement of the faults attains 1200 m on Profile 36. Here, it is anticipated that oceanic basement rocks crop out with a thickness of 1000 m, and hence it should be possible to dredge samples of the lower part of Oceanic Layer 2. The faults on the inner trench slope of the Bonin Trench show a larger displacement than those of the Japan Trench where faults on the outer trench slope are also well developed but with displacements never exceeding 300 m (Tamaki et al., 1977).

The difference of the fault displacement between the Japan Trench and the Bonin Trench may be related to the difference of the dip angle of the shallow part of the downgoing slab; 15 degrees along the Japan Trench (Nakatsuka et al., 1981) and 30 to 35 degrees along the northern Bonin Trench (Horiuchi, 1977). The more strongly bent slab along the Bonin Trench results in the development of more conspicuous faults than along the Japan Trench.

The fault distribution pattern suggests that faulting is initiated at the top of outer trench slope and reaches a maximum in the mid-slope area. The trend of the faults cannot be precisely defined here because our survey tracks have a separation of 15 nautical miles, but the faults appear to be parallel or slightly subparallel (diverging southward) to the trench axis. The distance between the faults is generally a few kilometers to 10 kilometers on the profiles. Some of the grabens are filled with turbidites that are believed to be triggered by fault scarps collapse.

The inner trench slope materials are acoustically massive and opaque while at the foot of the inner trench slope they are commonly acoustically transparent. The development of benches and ridges on the inner trench slope is common. A prominent mid-slope terrace or ridge is observed in the northern Bonin Arc. The ponded sediments behind the mid-slope ridge generally dip westwards (Profiles 39, 40, 43, and 44 in Fig. 10) suggesting the relative uplift of the mid-slope ridge.

It is difficult from the seismic data to interpret uniquely the relative uplift of the ridge. One explanation is that the ridge is thrust up by compressional stress. The alternative interpretation is that the ridge has undergone relative uplift by the subsidence of the fore-arc basement high in the tensional stress field.

Minor benches and ridges are developed at the foot of the slope near the trench axes. These minor features which are acoustically transparent, appear to be accretionary sedimentary prisms caused by subduction of oceanic lithosphere. The feature of the minor benches and ridges near the trench axes on the inner trench slope on the seismic profiles suggests active tectonic movement in the trench axis (Profiles 36, 43, 46, 48, and 50, for example). Active accretionary sedimentary prisms seem to be predominant in the northern Bonin Arc.

In the collision area of the Ogasawara Plateau and the Bonin Arc, the trench bottom is V-shaped and shallow (4500 m Profiles 17–21). The outer trench slope (trenchward slope of the Ogasawara Plateau) is segmented by many faults that are inferred to be mostly normal faults of the same type as the faults on the outer trench slope of the other parts of the Bonin Trench. If this is the case, the Ogasawara Plateau is bent concordantly with the underlying slab and tensional stress is predominant on the trenchward slope of the plateau. On the inner trench slope (eastward slope of the Bonin Ridge) in the collision area, a large ridge occurs and depositional structures behind it suggest that it is formed by uplift. The ridge may have been uplifted by mass concentration beneath the inner trench wall by undercoating or subaccretion of the crust of the subducted Ogasawara Plateau.

B. Fore-Arc Basin and Fore-Arc Basement High

The physiographic expression of the trench-slope break is remarkably different in the northern Bonin Arc and in the southern Bonin Arc. The fore-arc basement high along the trench-slope break is well developed in the northern Bonin Arc as shown in Fig. 10. In the southern Bonin Arc, the ridge is an outstanding topographic feature with extraordinarily high gravity anomaly of 380 mgal.

The Bonin Ridge is a shallow basement high on the extension of the northern trench slope break clearly shown on the seismic profiles (Fig. 10). Honza (1981a) called the basement high the "fore-arc basement high" and emphasized that it is a common basic component of island arc systems. The Bonin Ridge does not continue smoothly to the basement high along the trench slope break of the northern Bonin Arc. The slight discontinuity of the fore-arc basement highs between the northern and the southern Bonin arcs observed from Profile 32 to 33 in Figure 11 (b), shows a right lateral offset. The offset coincides with the E–W trending topographic depression that is obvious on the free-air gravity anomaly map (Fig. 3). The geomagnetic signal from the fore-arc basement high of the northern Bonin Arc is similar to that of the Bonin Ridge. Both basement highs have appreciable magnetic anomalies (Fig 4). This suggests that volcanic rocks are the main constituent of the basement although no rock samples have been recovered from the fore-arc basement high in the northern Bonin Arc. The fore-arc basement high and the Bonin Ridge are considered to be Paleogene in age (Fig. 11).

A fore-arc sedimentary basin is well developed between the fore-arc basement high and the active volcanic chain throughout the whole Bonin Arc, while, in the northern Mariana Arc, the development of fore-arc sedimentary basin is rather poor. The fore-arc basin of the northern Bonin Arc continues southward into the Bonin Trough in the southern Bonin Arc.

The sediments of the fore-arc basin of the northern Bonin Arc exceed 2.0 sec. in thickness and the basement underlying the sediments cannot be detected by single channel seismic profiles in areas with the thickest sedimentary pile. Thicker sediments are observed in the Bonin Trough. Sediment thickness of the Bonin Trough generally exceeds 3.0 sec and on some profiles exceeds 4.0 sec (Profiles 26–23).

The sediments clearly onlap eastward on to the fore-arc basement high and the Bonin Ridge, but show a different depositional pattern westward to the Shichito Ridge, the active volcanic chain. To the west, the sediments of the fore-arc basin increase in acoustic opaqueness and change continuously to the acoustic basement of the Shichito Ridge without any reflecting boundary. This means that the sediments of the basin were derived from the Shichito Ridge and associated with its volcanic activity. The above observations indicate that the fore-arc basin sediments are mostly volcanogenic deposits that have been shed from the active volcanic chain and have been trapped behind the fore-arc basement high.

The sedimentary layers in the fore-arc basin are correlated with whole Neogene–Quaternary strata (Fig. 11). In the southern Bonin Arc, where the sediments are much thicker than in the northern arc the age of the lower sedimentary layer is unknown (Yuasa et al., 1982).

Several basement highs are observed just east (fore-arc side) of the Shichito Ridge (Profiles 42–38, 34, 30, and 29 in Fig. 10). These basement highs, here called the Shinkurose Ridge, continue to the Shinkurose Bank to the north. This Ridge is onlapped by the sediments of the fore-arc basin. This depositional feature means that the Shinkurose Ridge is not volcanically active and has been inactive during the deposition of the sediments of the fore-arc basin. The Shinkurose Ridge is assigned a Miocene to Paleogene in age by Honza et al., (1982). (Fig 11).

The northern Bonin Arc represents physiographically a single arc not associated with frontal land mass (frontal arc) and it is different from the double arcs common around Japan. The Shinkurose Ridge of the Bonin Arc may represent a poorly developed frontal arc.

The uplift of the Bonin Ridge can be deduced from the sedimentary structure of the Bonin Trough. The sediments of the Bonin Trough, which exceed 3.0 sec in thickness, tilt toward the west with the lower horizons dipping more steeply than the upper horizons. This tilt feature suggests the uplift of the Bonin Ridge, while, in the northern Bonin Arc, sedimentary structure around the trench-slope break does not show any appreciable movement of the fore-arc basement high.

The Bonin Ridge is considered to be an anomalously uplifted fore-arc basement high at the trench-slope break in the southern Bonin Arc. This is in sharp contrast to the fore-arc basement high in the northern Bonin Arc, which was stable and was not uplifted. The collision of the Ogasawara Plateau against the Bonin Arc is the most plausible cause of the uplift of the Bonin Ridge. Subaccreted or undercoated oceanic plateau beneath the inner trench wall results in the uplifted ridge as shown on Profiles 21–18.

The Ogasawara Plateau is a large tablemount that belongs to the W-NW trending Marcus–Wake Seamount Chain in the Pacific Basin and there was, most probably a northwestward extension of the seamount chain west of the present Ogasawara Plateau. If these subducted seamounts west of the Plateau undercoated the inner trench wall of the Bonin Arc, the succession of ridge uplift on the inner trench wall will occur progressively from north to south. The Bonin Ridge could have been constructed by such a mechanism. According to this hypothesis, the middle Bonin fore-arc discontinuity is interpreted as having been formed by the initiation of the collision of the Marcus–Wake Seamounts Chain against the Bonin Arc.

C. Active Volcanic Chain and Back-Arc Regions

The Shichito Ridge, an active volcanic chain, is topographically obvious on the bathymetric chart (Fig. 2). The Mariana Ridge and the West Mariana Ridge meet at Minami–Iwo Jima Island and extend northward as the Shichito Ridge. The topographically complex back-arc region (west of the Shichito Ridge) is subdivided into a back-arc depression, a back-arc knolls zone and the Nishinoshima Trough, and Izu Ridge. The back-arc region, where so many seamounts, knolls, and ridges occur, is associated with marked geomagnetic anomalies. This indicates that volcanic rocks prevail here and that the topographic highs are due to submarine volcanic activity.

The Nishinoshima Trough, which is developed along the eastern margin of the Izu Ridge, is filled by rather thick sediments. The average thickness of the sediments in the trough is 1.0 sec, and, in a few profiles, it reaches 2.0 sec (Profiles 40, 43 and 45). In the southern Bonin Arc, the sediments of the Nishinoshima Trough continues to the basement of the Shichito Ridge while they onlap the Izu Ridge as shown in Profiles 22, 20, 19, 17, and 16 (Fig. 10).

The sediments of the Nishinoshma Trough were derived mainly from the Shichito Ridge with its volcanic activity and it appears that the Izu Ridge is considerably older than the Shichito Ridge. In the northern Bonin Arc, however, such a depositional contrast is not observed in the Nishinoshima Trough. The lower sediments of the Nishinoshima Trough overlap both the Shichito Ridge to the west and the eastern basement of back-arc knolls zone (Profile 41 of Fig. 10). The sediments are thought to be Quaternary to Pliocene in age, but the age of the lower layer is unknown (Honza *et al.*, 1982; Yuasa *et al.*, 1982).

The back-arc depressions, which are nearly free of sediments and often flanked by scarps on both sides, occur discontinuously just behind the volcanic chain, from west of Hachijo Island to northwest of Nishinoshima Island. They are not found between Nishinoshima Island and Minami–Iwo Jima Island. Profile 37 (Figs, 8, 9 and 10) implies that the back-arc depressions are extensional basins younger than the Shikoku Basin which was generated by back-arc spreading during late Oligocene to middle Miocene, because the depression is nearly free of the sediments and has a central high that resembles a spreading center. The possibility of the active back-arc spreading in the back-arc depressions is discussed in Section V of this chapter.

The back-arc knolls zone is covered by thin sediments less than 1.0 sec in thickness. The presence of magnetic anomalies here implies that the knolls are all small submarine volcanoes. The thin sedimentary cover on the knolls indicates that the volcanic activity here is older in age than the volcanic activity along the volcanic front. The back-arc knoll zone is not observed in the arc south of 22°N (to the south from profile 22). South of 22°N, the Nishinoshima Trough between the Shichito Ridge and the Izu Ridge contains a thick synclinal sedimentary trough of 2.0 sec in thickness.

At the eastern margin of the Shikoku and Parece Vela basins, the sediments are thicker than in the central part reaching 1.0 to 2.0 sec. The sediments increase in reflectivity and thickness to the east as shown in Profile 30. This feature shows that the sediments were dominantly derived from the eastern topographic highs of the Bonin Arc. White *et al.* (1980) on the basis of the sedimentary analysis of IPOD Leg 58 documented that the sediment was fed from the Bonin Arc into the eastern Shikoku Basin. The lower sediments in the eastern Shikoku Basin increase in reflectivity to the east and continue to the acoustic basement of Izu Ridge. This continuity indicates that the lower sediments are synchronous with the volcanic activity of the Izu Ridge. These lower sediments are assigned a Miocene age by Honza *et al.* (1982) based on IPOD Leg 58 results.

D. Northern Mariana Arc and Its Northern Continuation to the Bonin Arc

The fore-arc basement high at the trench-slope break is also observed in the northern Mariana Arc (Profiles 4–10 in Fig. 10), although its structure is complicated and its continuity from profile to profile is very poor compared to the Bonin Arc. The depositional structure of the sediments of the fore-arc basin indicates that they were derived from the Mariana Ridge and trapped behind the basement high of the trench-slope break and that the basement high was uplifted after deposition of the lower sedimentary layer, causing a slight disconformity between the upper and lower sedimentary layers in the fore-arc basin.

The northern Mariana Trough shows a stepwise decrease in width to the north, and suddenly disappears just south of Minami–Iwo Jima Island. The northernmost part of the Mariana Trough is observed on Profile 11' where it has a width of about 9 km. There is, however, no sign of trough features on Profiles 10" and 11, located south of Profile 11'. In the eastern half of the Mariana Trough on Profile 10 (Figs. 8, 9 and 10), where the width of the Trough is 20 km, a depressional feature with width of 7 km is observed along the foot of the Mariana Ridge. The sediment-free depression marked is in contrast with the western sedimentary pile in the trough (Profile 10). We infer that the depression represents a very young extensional feature in the Mariana Trough. Miyazaki et al. (1981) identified the magnetic anomaly lineations of the northern Mariana Trough from anomaly 1 to 2A. This identification indicates that the trough increases in age to the west and that the eastern end is probably the active extensional zone. The appearance of an active extensional depression on the eastern half of the northernmost Mariana Trough on Profile 10, agrees well with the identification of the magnetic anomaly lineations.

The bottom of the northern Mariana Trough is rugged and deepens southward from 1700 m (Profile 10) to 4200 m (Profile 1). The trough bottom is generally free of sediments except in a few depressions. The maximum thickness observed in the trough was 0.3 sec in small depressions among the knolls. The eastern end of the trough appears to be generally characterized by thin sediment deposition. This observation is also consistent with the identification of magnetic anomaly lineations by Miyazaki et al, (1981).

V. POSSIBILITY OF ACTIVE BACK-ARC SPREADING

The Mariana Arc, the southern extension of the Bonin Arc, has a well documented active back-arc basin, the Mariana Trough which is one of the most typical active back-arc basins in the world. The Mariana Trough diminishes in width to the north and closes just south of Minami–Iwo Jima Island. The bathymetry shows that the Mariana Trough does not extend into the back-arc area of the Bonin Arc. The Shikoku Basin, an inactive back-arc basin of the Bonin Arc, was active from 30 Ma to 15 Ma (deVries Klein and Kobayashi, 1980). We now discuss the existence of the active back-arc spreading of the Bonin Arc.

The features of the back-arc depression of the Bonin Arc on Profile 37 (Fig. 8a) strongly suggest that the back-arc depression may be a back-arc basin younger than the Shikoku Basin. The back-arc depressions first were observed by Mogi (1968) and then Hotta (1970) noted that the sediment thickness in the depression was remarkably low. Karig and Moore (1975) showed the distribution of similar depressions throughout the Bonin Arc and argued that the depressions were "young extensional basins" or active back-arc basins on the basis of the topographic and the seismic reflection data. Tamaki *et al.* (1981b) presented a different distribution of the back-arc depressions and discussed the possibility of their activity based on seismic profiles and bottom sampling data.

The distribution of the back-arc depressions based on Tamaki (1981b) is summarized in Figure 12. From this distribution, the following three prominent characteristics can be deduced:

(*i*) The back-arc depressions do not show any continuous features behind the volcanic chain, but rather appear as a discontinuous distribution of four segmented depressions. We call the four segmented depressions as follows; from north to south, the Hachijo Depression, the Sumisu Depression, the Torishima Depression, and the Nishinoshima Depression after neighboring islands.

(*ii*) No back-arc depression is observed in the area between Nishinoshima Island and Minami–Iwo Jima Island. The four segments mentioned above lie north of Nishinoshima Island. South of Minami–Iwo Jima Island, the Mariana Trough is developed. It is very interesting that the area where no back-arc depressions appear is located just west of the collision zone of the Ogasawara Plateau and the Bonin Arc.

(*iii*) The back-arc depressions are located just behind (west of) the active volcanic chain. The distance between the eastern margin of the back-arc depression and the young volcanoes on the volcanic front is only about 10 km.

If the back-arc depressions constitute an active spreading zone, the characteristics mentioned provide a very important constraint on the origin of the back-arc spreading.

The depth of the depressions increases southwards. The maximum depth of the Nishinoshima and the Torishima depressions exceeds 3500 m, while the Sumisu Depression and the Hachijo Depression are about 2200 m and 1300 m deep respectively. This variation reflects the general northward shallowing of the Bonin Arc. The depth of the depression surface below the surrounding seafloor has a similar tendency, decreasing northward from 1000–2000 m in the Nishinoshima and the Torishima depressions to just less than 1000 m in the Sumisu Depression, and about 700 m in the Hachijo Depression. The width of the depressions is generally 20 to 30 km and the maximum width is 60 km at the Nishinoshima Depression.

Topographic highs are observed in the center of the Sumisu Depression and Torishima Depression (Profiles 37 and 25). Basaltic pillow lava fragments were dredged from two sampling sites on the central high of the Sumisu Depression. K/Ar data show that the quantity of the Ar^{40} is below the apparatus resolution level (below 0.002 $sccAr^{40}Rad/g \times 10^{-5}$) (Tamaki et al., in preparation). This result indicates the young age of the samples and suggests volcanic activity in the back-arc depression. Two possible submarine eruptions in the Sumisu Depression are shown on the distribution map of active volcanoes around Japan compiled by Ono et al. (1981). The submarine eruptions also strongly support the theory of active back-arc depressions. Karig and Moore (1975) have also sampled "fresh basalt" in the Torishima Depression.

The sediment in the depression are thin compared to the surrounding basin area, but reaches a thickness of 500 m in the widest part of the Nishinoshima Depression (Profile 25) and 200 m in the Torishima Depression (Profile 30). Tamaki et al. (in preparation) obtained a 3 m core sample in the Sumisu Depression consisting wholly of massive acidic tuff. Thus, the thick sediments in the depressions are considered to be composed mainly of volcanic products. The origin of the volcanic sediments is considered to result from submarine eruptions in the depressions or volcanoes at the volcanic front or both. Rapid deposition of the volcanic products is plausible, and the local distribution of thick deposits in the depressions does not contradict the possibility of active extensional tectonics of the back-arc depressions.

The distribution of back-arc depressions on Figure 12 implies that they are a northern extension of the Mariana Trough. Also, it shows that the back-arc depressions represent the initial stage of the formation of the back-arc basin. It is very interesting that there is no back-arc basin in the area where the Ogasawara Plateau collides with the Bonin Arc. The Ogasawara Plateau seems to cause compressional stress in the Bonin Arc as shown by the westward deviation of the Bonin Trench. It is reasonable to consider that the compressional stress continues into the back-arc area and that no extensional depressions are observed in the back-arc area west of the Oga-sawara Plateau.

The back-arc depressions, as discussed above, are considered to be very young tectonic features. The present field observations, however, do not uniquely indicate active extensional tectonics in the back-arc depression. Additional data in support of the contention can be deduced from Uyeda and Kanamori's (1979) observations.

Uyeda and Kanamori divided the arc–back-arc systems of the world into two types from a comparative study of all the Recent island arcs. The two types are the Mariana type and the Chilean type; the Mariana type being characterized by tensional tectonics of the arc and the presence of active back-arc spreading, whereas the Chilean type is characterized by the com-pressional tectonics of the arc and the lack of active back-arc basins. The difference between the two types depends upon the degree of mechanical coupling between the subducting plate and the plate at the back side of the arc. If the two plates are decoupled tensional tectonics prevail in the arc area and cause back-arc spreading. Uyeda and Kanamori proposed that the Bonin Arc belongs to the Mariana type because of the lack of very large earthquakes, the poor development of outer gravity high, and the steep dip of the Wadachi–Benioff zone of the Bonin Arc. The data suggest a tensional stress regime in the Bonin Arc. The west-northwestward retreat of the Phil-lipine Sea plate (Seno, 1977) supports the tensional stress field in the Bonin Arc. These data agree with the active back-arc spreading of the Bonin Arc. Direct evidence of the active spreading in the back-arc depressions is still not available. A detailed survey, including heat flow measurements and deep-sea camera observation is needed to document active back-arc spread-ing in the Bonin Arc.

VI. TECTONICS OF THE JUNCTION AREA IN THE NORTHERN MARGIN

The Bonin Arc is bordered by the Sagami and the Suruga troughs at the northern margin where it contacts the Tohoku (northeast Japan) Arc to the north and with the Seinan Japan (southwest Japan) Arc to the west. Boundaries of these arcs form a triple junction of trench–trench–trench or trench–trench–transform fault. This is also the boundary of the Eurasia and the Philippine Sea plates. The Philippine Sea plate is subducting under the Eurasia plate along the Sagami, the Suruga, and the Nankai troughs (Sug-imura, 1972; Shimazaki et al., 1981).

Bottom topography of the south (Bonin Arc) side of these troughs is low, gentle, and smooth as compared with that on the north (Tohoku and Seinan Japan arcs) side where it is high, steep, and rough, with ridges, troughs, and canyons (Fig. 13). Many banks with canyons among them are arranged in a northwest to southeast direction on the northeast side of Sagami Bay where the Sagami Trough starts to extend southeastward in the central part of the bay. Seismic reflection profiles (Fig. 14) in the trough show faults along the foot of the southwest side of the banks, and sediments thickening toward the bank where the topographic low runs slightly southwest of the faults. Basement, that probably constitutes the Izu Peninsula, tends to deepen toward the banks and is cut by the faults at the foot of the bank (Fig. 14). These faults develop discontinuously in a zone. The banks that consist dominantly of Paleogene and Miocene sediments are not arranged on a line, but arranged in a zone associated with faults along the southwest foot of the bank (Kimura, 1975).

There are steep scarps to the immediate north of the Sagami Trough which lies southeast of the Boso Peninsula. Many canyons run east to southeastward among highs on the north side of the scarps. These highs consist dominantly of Miocene and older sedimentary layers (Tanahashi and Honza, 1983). The bottom topography on the south side of the trough is low, gentle and smooth, generally deepening toward northeast or east where it is overlain by thick, presumable Quaternary, sediments. Here, the deepest depression is located a little south of the scarps (Fig. 15), as in Sagami Bay.

Topographical features quite similar to those observed in Sagami Bay are seen in the Suruga Trough. A few banks associated with steep scarps on the east side and many canyons arranged on the west side of the trough, whereas the bottom topography is low, gentle, and smooth on the east side of the trough. The basement that constitutes the Izu Peninsula deepens toward the west, and is overlain by the deformed layers associated with westward dipping thrusts (Fig. 16). The deformation could be a manifestation of subduction of the east side under the westside. This is commonly observed in the southwestern extension of the trough where the Shikoku Basin plate is being subducted under the Seinan Japan Arc along the Nankai Trough.

Structural features of Sagami Bay may extend onshore, at least to the eastern foot of Mt. Fuji. The structural line that borders the Philippine and the Eurasia plates may pass along the southern foot of the Mt. Fuji, but no marked structural change is observed. There may be a collision of both plates along the foot of the Mt. Fuji as deduced from the fact that there is no surficial impression in the area (Shimazaki et al., 1981).

The Philippine Sea plate is approaching the Eurasia plate with a velocity of 3–5 cm/year on a bearing of $310° \pm 5°$ in this area. This is calculated on

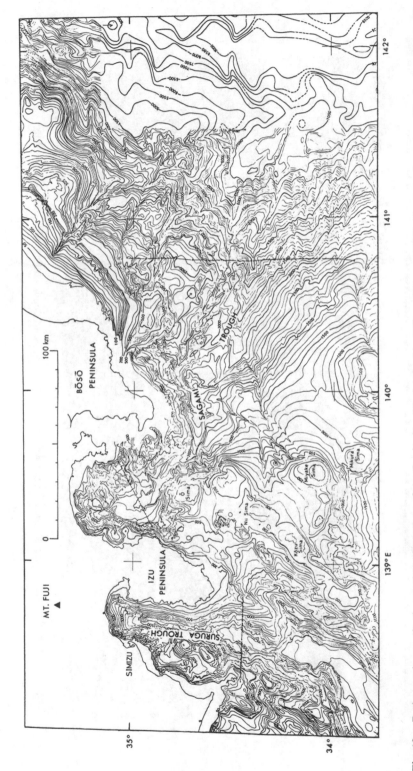

Fig. 13. Bathymetric map in the northern boundary of the Bonin Arc (Hydrographic Office of Japan, 1979). Contours in meters. The broken lines with dots are seismic reflection lines.

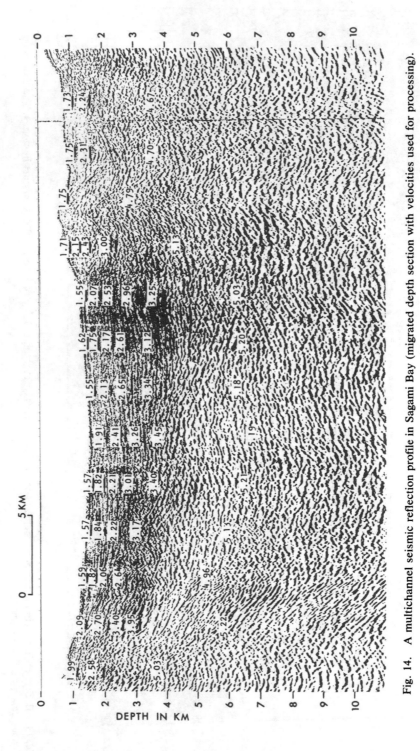

Fig. 14. A multichannel seismic reflection profile in Sagami Bay (migrated depth section with velocities used for processing).

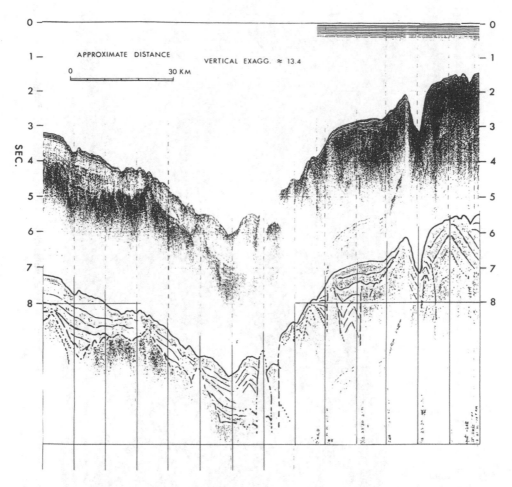

Fig. 15   A single-channel seismic reflection profile in the Sagami Trough southwest offshore of the Bōsō Peninsula.

the basis of displacement and direction of earthquake first-motion studies along the boundaries between the Philippines and the Eurasia and between the Philippine Sea and Pacific plates during the last several decades (Seno, 1977). The direction of maximum crustal shortening during the last ten years based on geodetic measurement (Nakane and Fujii, 1979; Simazaki *et al.*, 1981) is analyzed in detail and the result is consistent with the subduction of the Philippine Sea plate northwestwards under the Eurasia plate (Fig. 17).

The uplift along the Eurasia plate side of the troughs is also well demonstrated onshore. The Boso and Miura peninsulas are uplifted along the

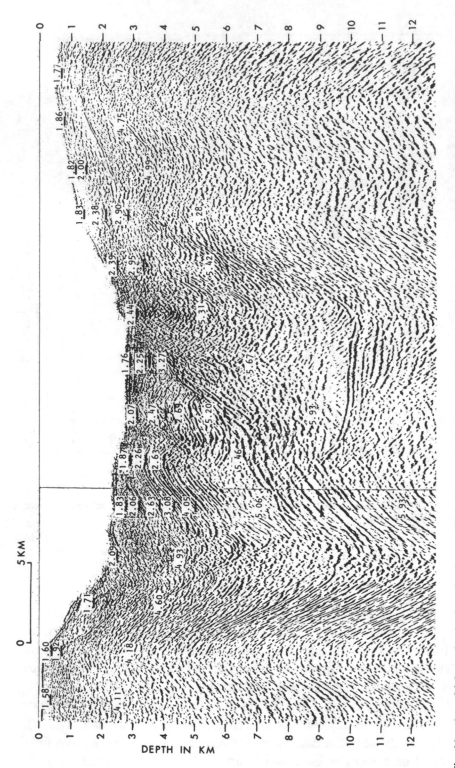

Fig. 16. A multichannel seismic reflection profile in Suruga Bay (migrated depth section with velocities used for processing). Subduction is toward west and the formation of the subduction complex in the west side slope are observed.

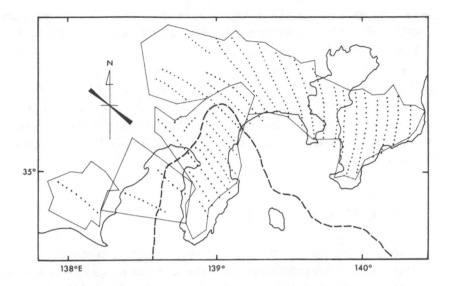

Fig. 17. Contraction lines after the 1923 Kanto and the 1930 Kita–Izu earthquakes (after Nakane and Fujii, 1979, and Shimazaki *et al.*, 1981).

southern margins and have relative undergone subsidence at the northern foot of the both peninsulas, indicating a tilt toward the north. The uplift of the southern margin of the Boso Peninsula reached 3 m per 1000 years during Holocene on the basis of terrace analysis. This is considered to be accumulated by thrust movements of earthquakes, an example of which is the 1703 Kanto earthquake (Matsuda *et al.*, 1978).

The same feature is suggested on the west side onshore of the Suruga Trough. Tilt movements to the northwest in the coastal area and in offshore banks have been taking place over the last 0.3–0.7 Ma (Nasu *et al.*, 1968).

The rate of uplift may be consistent with the regional compressional field generated by the collision of the Izu Peninsula with the Eurasia plate during the Quaternary (Matsuda, 1979).

At the junction with the Bonin Trench, the features of the trench bottom are somewhat different from the neighboring trenches. The trench bottom is wider with a thicker sediment fill as is shown in the earlier section. This indicates a very large sediment supply along the canyon and suggests a right lateral displacement of 50 km in horizontal component (Shimazaki *et al.*, 1981).

The fault scarps on the Eurasia plate side of the Sagami Trough called the Sagami Bay faults, strike approximately parallel to the subduction direction of the Philippine Sea plate in this area. This fact suggests a temporal transformation to an eduction boundary from oblique subduction that may have occurred in the earlier stage (Sugimura, 1972). In reflection profiles, they appear to be normal faults, however, from the analysis of displacement of earthquakes, they are interpreted as thrust faults with steep dip near the surface (Ando, 1971; Kanamori, 1971).

The development of thrusts in areas slightly landward from the maximum topographic low is introduced as a concept of a dynamic zone the boundary of plates along the fault zone displaced from the topographic low (Shimazaki et al., 1981). This is commonly observed in the trench area where the dynamic boundary is thought to be located slightly landward from the trench axis.

VII. DISCUSSION AND SUMMARY

Two possibilities can be considered for the initial position of the Bonin Arc. One is along the position of the Kyushu–Palau Ridge in which the arc shifted eastwards forming the Shikoku Basin behind (Karig 1975; Karig and Moore, 1975; Seno and Maruyama, 1984) and the other is with the arc fixed in the same position throughout its development (Honza, 1978; Matsuda, 1979). In the latter case, the arc may have been initiated as an aseismic ridge or a semicontinent along the boundary of the Kula and the Pacific plates (Uyeda and Ben-Avraham, 1972; Hilde et al., 1977).

The islands in the Bonin Arc consist of volcanic and, some sedimentary rocks. Volcanism in the Bonin Arc has been intermittent since middle Eocene, although the data obtained are only from the islands that occupy small areas on the ridge (Kaneoka et al., 1970; Hanzawa, 1974; Iwasaki and Aoshima, 1970; Matsumaru, 1974; Ujiie and Matsumaru, 1977).

The first sign of arc formation is the volcanism on the Bonin Ridge from middle to late Eocene. The Bonin Ridge in the southern arc extends to a high in the trench-slope break. However, there is a slight possibility that the northern extension of the ridge consisted of a different material for the magnetic anomalies observed in the high of the trench-slope break show a somewhat different pattern, although it is covered by sediments.

The next pulse consists of volcanism in the Bonin Ridge during late Oligocene, which occurred in the same place as in the initial stage. The volcanic rock is calc-alkalic with a higher content of silica and incompatible elements whereas in the initial stage the volcanics are arc tholeiites characterized by low contents of incompatible elements at all levels of silica enrichment (Shiraki et al., 1978; Kuno, 1968; Miyashiro, 1974).

The last stage of volcanism, which formed the volcanic chain of the modern Bonin Arc, occurred along a different line from that of the earlier stages in the Bonin Ridge. The Bonin Ridge forms a fore-arc basement high in the modern arc. The volcanic chain (Schichito Ridge) is approximately 100 km west of the earlier volcanic chain (Bonin Ridge) and separated by the intervening Bonin Trough.

The Bonin Trough is an inactive trough located in the fore-arc area. The trough has a thick sedimentary sequence which shows a complicated development with unconformities.

Studies of the magnetic anomalies in the Shikoku Basin show northwest striking lineations parallel to the volcanic chain of the Bonin Arc. The lineation pattern is identified as a sequence of anomalies 7 (27 Ma) through 5 (10 Ma) between the base of the Kyushu–Palau Ridge and the east margin of the basin. The identification of the eastern side is tentative because of the rough and complex basement morphology (Watts and Weissel, 1975). The other identification is based on a symmetrical anomaly profile with a spreading center along the center of the basin (Kobayashi and Nakada, 1979).

Spreading of the Shikoku Basin thus appears to have begun in late Oligocene and continued until middle to late Miocene. On the other hand, no volcanism is known in the Bonin Arc from Miocene to early Pliocene. An alternating occurrence of arc volcanism with marginal sea spreading and also activity related to the subduction complex in the trench have been explained by a convection current under the arc formed by frictional heating along the upper surface of the subducted oceanic plate (Honza, 1981b; Honza 1983). Low level volcanism occurs in the volcanic chain while the marginal sea is opening and slight spreading is occurring in the marginal sea when volcanism is active in the volcanic chain.

There are some common features associated with arcs. They are an active volcanic chain, a fore-arc basement high and a trench, even if there are discontinuities in some of the features. A marginal sea is also one of the common features in the NW Pacific rim. These features form the fundamental geomorphologic elements which constitute an arc (Honza, 1981a).

On the basis of the above discussions, tectonic evolution of the Bonin Arc is summarized as follows.

During the first stage of development the arc in middle to late Eocene, there was an arc with a volcanic chain in the Bonin Ridge. During this period, a fore-arc basement high which cannot now be seen in the modern Bonin Arc lay east of the volcanic chain. (Fig. 18).

The Bonin Trough opened in a quiet period of intermittent volcanism during early to middle Oligocene and ceased opening during late Oligocene when the second stage volcanism occurred in the same ridge. The complicated sediment deposition associated with unconformities in the trough suggest a sedimentary history modified by the formation of a new ridge in the later stages.

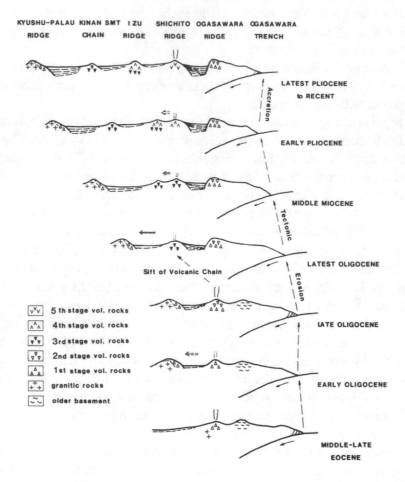

Fig. 18. Summary of evolution of the Bonin Arc, based on the model proposed by Honza (1979, and 1983).

The final stage is accompanied by the appearance of a new volcanic chain west of the older stage volcanism. However there may have been a few more volcanic stages before the final one.

A few ridges developed parallel to the modern volcanic chain in the back-arc area. One is the Kinan Seamount Chain in the central part of the Shikoku Basin. The second is the Izu Ridge at the eastern margin of the basin and the third is the Nishinoshima Knolls between the Izu and the Shichito ridges. Although samples have not been obtained, these ridges and highs consist of material that seem to be volcanic rocks on the basis of their magnetic anomalies. The Kinan Seamount chain is located between anomalies 5B and 5C (approximately 16 Ma), the Izu Ridge is younger than anomaly 5 (10 Ma) in the tentative analysis of magnetic lineations by Watts and Weissel (1975). The Upper Miocene and the Pliocene layers rest on some of the highs in the Izu Ridge, which suggest a greater age of the ridge.

Therefore, it is inferred that volcanism occurred during Miocene between the spreading of the Shikoku Basin and the back-arc area of the arc.

ACKNOWLEDGMENT

We wish to thank Dr. Y. Shimazaki, Geological Survey of Japan, for his comments and critical reading of the manuscript.

REFERENCES

Ando, M., 1971, A fault-origin model of the Great Kanto Earthquake of 1923 as deduced from geodetic data, *Bull. Earthquake Res. Inst.*, v. 49, p. 19–32.

deVries Klein, G. and Kobayashi, K. *et al.*, 1980, *Initial Reports of the Deep Sea Drilling Project*, v. 58 Washington, D.C.: United States Government Printing Office.

deVries Klein, G. and Kobayashi, K., 1980, Geological summary of the northern Philippine Sea—Drilling Project Leg 58 results, in: *Initial Reports of the Deep Sea Drilling Project*, v. 58, Washingtonm, D.C.: United States Government Printing Office, p. 951–961.

Hanzawa, S., 1974, Eocene foraminifera from Haha-jima (Hillsborough Island), *J. Paleontol.*, v. 21, p. 254–259.

Hess, H. H., 1948, Major structural features of the western north Pacific, an interpretation of H. O. 5485 bathymetric chart, Korea to New Guinea, *Geol. Soc. Am. Bull.*, v. 59, p, 417–446.

Hilde, T. W. C., Uyeda, S. and Kroenke, L., 1977, Evolution of the western Pacific and its margin, *Tectonophysics*, v. 38, p. 145–165.

Honza, E., 1978, Geological development in the junction of the Seinan (SW) Japan and Ryukyu Arcs—preliminary concluding remarks, in: *Geological Investigations in the Northern Margin of the Okinawa Trough and the Western Margin of the Japan Sea, Cruise Report 10, Kawasaki-shi:* Honza, E., ed, Geological Survey of Japan, 79p.

Honza, E., 1981a, Subduction and accretion in the Japan Trench, *Oceanol. Acta.*, Sp., p. 251–258.

Honza, E., 1981b, The geological settings of the Ogasawara and the northern Mariana Arcs, in: *Geological Investigation of the Ogasawara (Bonin) and Northern Mariana Arcs, Cruise Report 14, Kawasaki-shi:* Honza, E., et al., eds. Geological Survey of Japan, 170p.

Honza, E., 1983, Evolution of arc volcanism related to marginal sea spreading and subduction at trench, in: Tectonics of Arc Volcanism, Shimozuru D., ed., p. 177–189.

Honza, E., Inoue, E. and Ishihara, T. eds., 1981, *Geological Investigation of the Ogasawara (Bonin) and Northern Mariana Arcs, Cruise Report 14, Kawasaki-shi:* Geological Survey of Japan,

Honza, E., Tamaki, K., Yuasa, M., Tanahashi, M. and Nishimura, A., 1982, *Geological Map of the Northern Ogasawara Arc*, Marine Geol. Map Ser. 17, Geological Survey of Japan.

Horiuchi, S., 1977, Deformation of sinking slab in consequence of absolute plate motion of island-side Izu–Bonin–Mariana Arc, *Zisin (J. Seismol. Soc. Jpn.)* v. 30, p. 435–447, (in Japanese with English abstract).

Hotta, H., 1970, A crustal section across the Izu–Ogasawara Arc and Trench, *J. Phys. Earth*, v. 18, p. 125–141.

Houtz, R. and Windisch, C., 1980, Changes in the crust and upper mantle near the Japan–Bonin Trench, *J. Geophys. Res.*, v. 85, p. 267–274.

Hydrographic Office of Japan, 1979, *Bathymetric Chart of the Area Adjacent to Sagami and Nankai Trough*, 1:500,000, Hydrographic Office of Japan.

Ishihara, T., Murakami, F., Miyazaki, T. and Nishimura, K., 1981, Gravity survey, in: *Geological Investigation of the Ogasawara (Bonin) and Northern Mariana Arcs, Cruise Report 14*, Honza, E. *et al.*, eds., Kawasaki-shi: Geological Survey of Japan, p. 45–78.

Ishihara, T., Gravity field around Japan—sea gravimetry by the Geological Survey of Japan, *Mar. Geod.*, (in press)

Iwasaki, Y. and Aoshima, M., 1970, Report on geology of the Bonin Islands, in: *Nature of the Bonin Islands*, p. 205–219, Educational Agency of Japan, Tokyo (in Japanese).

Kanamori, H., 1971, Faulting of the Great Kanto Earthquake of 1923 as revealed by seismological data, *Bull. Earthquake Res. Inst.*, v. 49, p. 13–18.

Kaneoka, I., Isshiki, N. and Zashu, S., 1970, K-Ar ages of the Izu–Bonin Islands, *Geochem. J.*, v. 4, p. 53–60.

Karig, D. E., 1975, Basin Genesis in the Philippine Sea, in: *Initial Reports of the Deep Sea Drilling* Project, v. 31 Washington D.C.: United States Government Printing Office.

Karig, D. E. and Moore, G. F., 1975, Tectonic complexities in the Bonin Arc System, *Tectonophysics*, v. 27, p. 97–118.

Katsumata, M. and Sykes, L. R., 1969, Seismicity and tectonics of the western Pacific: Izu–Mariana–Caroline and Ryukyu–Taiwan regions, *J. Geophys. Res.*, v. 74, p. 5923–5948.

Kimura, M., 1976, *Submarine Geological Map of Sagami-nada and Its Vicinity*, 1:200,000, Marine Geol. Map Ser. 3, Geological Survey of Japan.

Kobayashi, K. and Nakada, M., 1979, Magnetic anomalies and tectonic evolution of the Shikoku inter-arc basin, in: *Geodynamics of the Western Pacific*, Uyeda, S., *et al.*, eds., Tokyo: Japan Scientific Societies Press. p. 391–402.

Kuno, H., 1968, Differentiation of basalt magmas, in: *Basalts*, Hess, H. H. and Poldervaart, A., eds., Wiley, New York, v. 2. p. 623–688.

Matsubara, Y., 1981, Heat flow measurements in the Bonin Arc area, in: *Geological Investigation of the Ogasawara (Bonin) and Northern Mariana Arcs, Cruise Report 14*, Honza, E., *et al.*, eds., Kawasaki-shi: Geological Survey of Japan, p. 130–136.

Matsuda, T., 1979, Collision of the Izu–Bonin Arc with central Honshu: Cenozoic tectonics of the Fossa Magna, Japan, in: *Geodynamics of the Western Pacific*, Uyeda, S., *et al.*, eds., Tokyo: Japan Scientific Societies Press, p. 409–421,

Matsuda, T., Ota, Y., Ando, M., and Yonekura, N., 1978, Fault mechanism and recurrence time of major earthquakes in southern Kanto District, Japan, as deduced from coastal terrace data, *Geol. Soc. Am. Bull.*, v. 89, p. 1610–1618.

Matsumaru, K., 1974, The transition of the larger foraminiferal assemblages in the western Pacific Ocean—especially from the Tertiary period, *J. Geogr. Soc. Jpn.*, v. 83, p. 281–301.

McKenzie, D. P. and Sclater, J. G., 1968, Heat flow inside the island arcs of the northwest Pacific, *J. Geophys. Res.*, v. 73, p. 3173–3179.

Miyashiro, A., 1974, Volcanic rock series in island arcs and active continental margins, *Am. J. Sci.*, v. 274, p. 321–355.

Miyazaki, T., Murakami, F., Nishimura, K., and Ishihara, T., 1981, Geomagnetic survey, in: *Geological Investigation of the Ogasawara (Bonin) and northern Mariana Arcs, Cruise Report 14*, Honza, E., *et al.*, eds., Kawasaki-shi: Geological Survey of Japan, p. 79–82.

Mogi, A., 1968, The Izu Ridge, in: *Fossa Magna*, Hoshino, M., ed., Tokyo: Geological Society of Japan, p. 217–221, (in Japanese).

Mogi, A., 1972, Bathymetry of the Kuroshio region, in: *Kuroshio—Its Physical Aspects*, Yoshida, K., ed., University of Tokyo Press, p. 53–80,

Murauci, S. and Ludwig, W. J., 1980, Crustal structure of the Japan Trench: the effect of subduction of ocean crust, in: *Initial Reports of the Deep Sea Drilling Project*, v. 56, Washington, D.C: United States Government Printing Office p. 463–469.

Murauchi, S., Den, N., Asano, S., Hotta, H., Yoshii, T., Asanuma, T., Hagiwara, K., Ichikawa, K., Sato, T., Ludwig, W. J., Ewing, J. I., Edgar, N. T., and Houtz, R. E., 1968, Crustal structure of the Philippine Sea, *J. Geophys. Res.*, v. 73, p. 3143–3171.

Nagumo, S., Kasahara, J., and Koresawa, S., 1980, OBS airgun seismic refraction survey near sites 441 and 434(J-1A), 438 and 439(J-12) and proposed site J-2B: Legs 56 and 57, in: *Initial Reports of the Deep Sea Drilling Project*, v. 56, 57 Washington, D.C.: United States Government Printing Office, p. 459–462.

Nakane, K. and Fujii, Y., 1979, On accumulation and release of the earth's strain in the south Kanto district, *J. Geod. Soc. Jpn.*, v. 25, p. 26–37.

Nakatsuka, T., Horikawa, Y., Nakai, J., and Ono, Y., 1981, Dip of subducting oceanic crust off the Sanriku Coast, northeast Japan, inferred from geomagnetic anomalies, *U.N. ESCAP, CCOP Tech., Bull.*, v. 14, p. 53–60.

Nasu, N., Tuchi, R. and Honza, E., 1968, Geological structure in the west side of Suruga Bay, in: *Fossa Magna*, Hoshino, M., ed., *Proc. Symp. 75th An. Meeting Geol. Soc. Japan*, p. 191–195, (in Japanese).

Nishimura, K., and Murakami, F., 1977, Gravity measurements, in: *Geological Investigation of Japan and Southern Kurile Trench and Slope Areas, Cruise Report 7*, Honza, E., ed., Kawasaki-shi: Geological Survey of Japan, p. 21–42.

Nishiwaki C., ed., 1981, *Plate Tectonic Map of the Circum-Pacific Region—Northwest Quadrant*, Circum-Pacific Council for Energy and Mineral Resources, American Association of Petroleum Geologists.

Ono, K., Soya, T., and Mimura, K., 1981, *Volcanos of Japan*, (second ed.), 1:2,000,000 Map. Ser. 11. Geological Survey of Japan.

Scientific Party Legs 56 and 57, 1980, Initial Reports of the Deep Sea Drilling Project v. 56 and 57, Washington D.C.: United States Government Printing Office.

Seno, T., 1977, The instantaneous rotation vector of the Philippine Sea plate relative to the Eurasian plate, *Tectonophysics*, v. 42, p. 209–226.

Seno, T., and Maruyama, S., 1984, Paleogeographic reconstruction and origin of the Philippine Sea, *Tectonophysics*, v. 102, p. 53–84.

Shimazaki, K., Nakamura, K., and Yonekura, N., 1981, The Suruga Trough and the Sagami Trough: movements deduced from geodesy and morphology and their relation to plate movement, *The Earth Monthly* (Tokyo), v. 31, p. 455–463, (in Japanese).

Shiraki, K., Kuroda, N., and Maruyama, S., 1978, Evolution of the Tertiary volcanic rocks in the Izu–Mariana Arc, *Bull. Volcanol.*, v. 41, p. 548–562.

Sugimura, A., 1972, Boundaries of plates around Japan, *Kagaku*, v. 42, p, 192–202, (in Japanese).

Tamaki, K., Inouchi, Y., Murakami, F., and Honza, E., 1977, Continuous seismic reflection survey, in: *Geological Investigation of Japan and Southern Kurile Trench and Slope Areas, Cruise Report 7*, Honza, E., ed. Kawasaki-shi: Geological Survey of Japan, p 50–71.

Tamaki, K., Tanahashi, M., Okuda, Y., and Honza, E., 1981a, Seismic reflection profiling in the Ogasawara (Bonin) Arc and the northern Mariana Arc, in: *Geological Investigation of the Ogasawara (Bonin) and Northern Mariana Arcs, Cruise Report 14*, Honza, E., *et al.*, eds., Kawasaki-shi: Geological Survey of Japan, p. 83–91.

Tamaki, K., Inoue, E., Yuasa, M., Tanahashi, M, and Honza, E., 1981b, Possible active back-arc spreading of the Bonin Arc, *Abstract of Symposium on Geotectonics of Sagami Trough–Suruga Trough Junction Area (First French-Japanese Symposium on Japanese Subduction Program)*, June 04–05, Tokyo.

Tamaki, K., Inoue, E., Yuasa, M., Tanahashi, M., and Honza, E., Possible active back-arc spreading of the Bonin Arc, (in preparation).

Tanahashi, M., and Honza, E., 1983, Geological map off the Boso Peninsula, *Mar. Geol. Map Ser.*, Geological Survey of Japan.

Utsu, T., 1974, Seismicity of Japan, *Kagaku*, v. 44, p. 739–746, (in Japanese).

Uyeda, S., and Ben-Avraham, Z., 1972, Origin and development of the Philippine Sea, *Nature*, v. 40, p. 176–178.

Uyeda, S., and Kanamori, H., 1979, Back-arc opening and the mode of subduction, *J. Geophys. Res.*, v. 84, p. 1049–1061.

Ujiie, H., and Matsumaru, K., 1977, Stratigraphic outline of Haha-jima (Hillsborough Island), Bonin Islands, *Mem. Nat. Sci. Mus. (Japan)* v. 10, p. 5–18.

Watts A. B., Talwani, M., and Cochran, J. R., 1976, Gravity field of the northwest Pacific Ocean Basin and its margin, in: *Geodynamics of the Pacific Ocean Basin and Its Margin, Geophysical Monograph Series* 19, Washington D.C.: American Geophysical Union., p. 17–34,

Watts, A. B., and Weissel, J. K., 1975, Tectonic history of the Shikoku marginal basin, *Earth Planet, Sci. Lett.*, v. 25, p. 239–250.

White, S. M., Chamery, H., Curtis, D., deVries Klein, G., and Mizuno A., 1980, Sediment synthesis: Deep Sea Drilling Project Leg 58, Philippine Sea, in: *Initial Reports of the Deep Sea Drilling Project*, v. 58, United States Government Printing Office, p. 963–1013.

Yuasa, M., Honza, E., Tamaki, K., Tanahashi, M., and Nishimura, A., 1982, *Geological Map of the Southern Ogasawara and Northern Mariana Arcs*, Mar. Geol. Map Ser. 18, Geological Survey of Japan.

Chapter 11

TAIWAN: GEOLOGY, GEOPHYSICS, AND MARINE SEDIMENTS

Biq Chingchang

Geology Department
Chinese Culture University
Yangmingshan, Taiwan

C. T. Shyu and J. C. Chen

Institute of Oceanography
National Taiwan University
Taipei, Taiwan

and

Sam Boggs, Jr.

Department of Geology
University of Oregon
Eugene, Oregon 97403

I. INTRODUCTION

The ocean margin island of Taiwan is a geodynamic body of young and complex build. Occupying an area of 36,000 km² between the Chinese part of the Eurasian continent capping one lithospheric plate and the northwestern part of the Philippine Sea floored by another (Figs. 1, 7), this island, as a compression-plus-shear product, is elevated to a maximal height of almost 4000 m, higher than any other fold mountains on the northwest coast of the Pacific Ocean. It is an arcuate island extending its shorter arm eastward

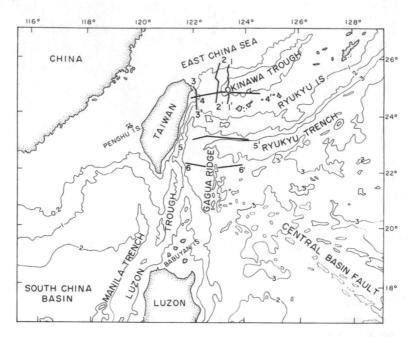

Fig. 1.   Map of Taiwan and offshore areas. Isobaths (depth in 1000 m) are based on Chase *et al.* (1971). Heavy lines are locations of seismic reflection profiles.

to the Ryukyus and its longer arm southward to the Philippines. The crestal zone of this mountainous island is the Central Range, which is fringed on its west side by the Foothill Zone and separated on its east from the Coastal Range by the corridor known as the Longitudinal Valley (Fig. 2). West of the Foothill Zone is a vast coastal plain with the very shallow Taiwan Strait farther west; east of the Coastal Ranage is the deep Philippine Sea whose floor exhibits arcs and trenches like the North Luzon Ridge and Trough. The dependent islands of Taiwan include the Penghu Islands in the Taiwan Strait and the islands of Lutao and Lanhsü off the southeast coast. It must be added that, in the less tightly compressed northeastern and southwestern parts of the mountain complex of the Central Range and Foothill Zone, there are, respectively, the Ilan Plain and the Pingtung Valley, each in the form of an intramontane trough wedging from the sea into the island.

The ocean immediately east of Taiwan is separated into three compartments by two island arcs. East of the northeast coast of Taiwan and north of the Ryukyu Arc is a part of the East China Sea, and east of the southeast coast and west of the North Luzon Ridge is a part of the South China Sea, both on the Eurasian plate; the remaining compartment is the northwestern part of the Philippine Sea floored by the plate named after it (Fig. 7). The Ryukyu Trench is on the south side of the Ryukyu Arc and

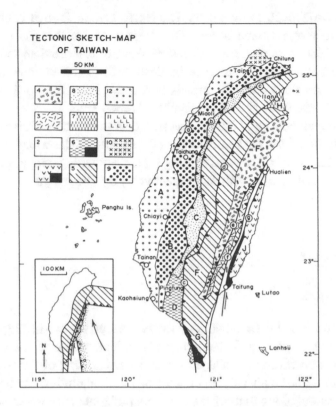

Fig. 2.   Tectonic sketch map of Taiwan. Geographic units: A, Coastal Plain; B, Outer Foothill Zone; C, Inner Foothill Zone; D, Pingtung Valley; E, Central Range (Hsüehshan Range); F, Central Range (main range); G, Central Range (Hengchun Peninsula); H, Ilan Plain; I, Longitudinal Valley; J, Coastal Range. Tectonic units: 1, Impinging allochthon folded of late Tertiary arc massif and subduction complex, mélange in solid black; 2, Fault valley underlain by basement complex and Tertiary cover rocks; 3, Mesozoic subduction complex reconstituted into basement complex; 4, Mesozoic arc massif reconstituted into basement complex; 5, Slate belt folded of Eocene–Miocene cover rocks; 6, Intraplate subduction zone buried beneath Pleistocene and Recent sediments, Pliocene mélange plastered on slate belt in solid black; 7, Interarc basin filled with Pleistocene and Recent sediments; 8, Inner Foothill Zone folded of Miocene cover rocks; 9, Outer Foothill Zone folded of Miocene cover rocks, Pliocene–Pleistocene flysch, and Pleistocene molasse; 10, Volcanic terrane of Pleistocene arc magmatism; 11, Undisturbed area covered with Pleistocene flood basalt; 12, Coastal plain and coastal terraces underlain by less disturbed Cenozoic rocks. Fault lines (indented on overriding side), a, Boundary thrust of Outer Foothill Zone; b, Boundary thrust of Inner Foothill Zone; c, Chaochow–Chüchih Fault; d, Lishan Fault; e, Shoufeng Fault; f, Central Range Fault; g, Coastal Range Fault. The inset map shows the present-day plate-tectonic setting of Taiwan. The obliquely ruled area includes the Central Range of Taiwan and the Ryukyu Arc, both on the edge of the Eurasian plate; the stippled area includes the Coastal Range of Taiwan and the Luzon Arc, both on the edge of the Philippine Sea plate. Their convergent plate boundary is depicted with saw-teeth lines indented on the overriding side. The long arrow indicates the general direction of the movement of the Philippine Sea plate.

the Okinawa Trough to the north. The North Luzon Trough is on the west side of the North Luzon Ridge. The deepest of the three is the Ryukyu Trench, which at its western end, east of Taiwan, is more than 4500 m deep. The easternmost offshore area to be dealt with in this chapter is the submarine Gagua Ridge, which extends northward from the northeast corner of Luzon to the west end of the Ryukyu Trench (Fig. 1).

To give an outline of the various surface and crustal features of Taiwan and the floor of the seas surrounding it, as briefly described above, this chapter consists of three parts dealing separately with geology, geophysics, and marine sediments, written, respectively, by Biq, Shyu, and Chen and Boggs.

II. GEOLOGY

A. Stratigraphy

The stratigraphic framework of Taiwan comprises two geologic columns that combine to exemplify a eu/miogeosynclinal couple in classic geology or an arc-upon-continent collage in new tectonics. One of the two rock sequences underlies Taiwan proper west of the Longitudinal Valley and the other crops out along and off the coast east of that morphotectonic break.

The rock sequence underlying Taiwan proper is readily divisible into a pre-Tertiary basement of a great but unknown stratigraphic thickness and a Cenozoic cover more than 10 km thick. Cropping out in the form of a 240-km-long inlier on the east flank of the Central Range, the basement is mainly a schist–marble complex with minor amounts of ortho- and paragneisses, amphibolites, etc. Also found in this assortment of metamorphics are exotic blocks of ophiolitic and other ocean floor rocks, including those converted into glaucophane schist. The age of the basement rocks, as indexed by fusulinids from the marble, is Permian (Yen *et al.*, 1951), although it is possible that there are older Paleozoic and/or Mesozoic beds in some other parts of the sequence. These rocks experienced during the Mesozoic Yenshanian movement the earliest phase in their polymetamorphism, of an amphibolite facies (Ernst *et al.*, 1981), before their late Cenozoic retrogradation to a greenschist facies (Liou, 1981a). Worth noting but not fully understood yet is the metamorphic history of a small number of exotic rocks. It seems that the blueschist metamorphism overprinted on them represents an earlier deep-reaching event that took place before transport as tectonic blocks to a low-pressure environment where they were jumbled with other exotics in a mélange (Biq, 1978; Liou, 1981a). The polymetamorphism has left on the basement rocks a wide spread of isotopic dates between 86 and 5 Ma (Jahn and Liou, 1977; Jahn *et al.*, 1981; Juan *et al.*, 1972; Yen and Rosenblum, 1964).

Above a seldom observed unconformity often obscured by longitudinal thrusts (C. H. Chen, 1979; Tsan, 1977) lies the mighty Cenozoic sequence covering the basement. It is convenient to divide the cover rocks, widely distributed in different morphotectonic belts, into a Central Range Group and a Foothill Zone Group. Of a multikilometer thickness, the former group is a succession principally of slate and phyllite with subordinate quartzite exposed mainly on the crest and west flank of the Central Range, forming the slate belt of the island. The protoliths of these low-grade metamorphics were mainly shelf sediments, additional to which were sediments forming deepsea fans. This lithologically monotonous succession is divisible into a number of formations ranging in age from Eocene through Miocene. The post-Miocene metamorphism, that made this Tertiary cover in the Central Range a sequence of slate and quartzite and retouched the basement with a retrograde overprint, was of greenschist and lower pressure and temperature facies. The grade of metamorphism gradually decreases westward, regardless of the stratigraphic position and the formation boundaries of the rocks affected.

The Foothill Zone Group of the cover rocks is a late Tertiary and Quaternary record of recurrent short-lived marine transgressions and regressions. The early cycles are recognized from interruptions of the marine sandstone–shale sequence by three Miocene coal-bearing formations, that pass laterally into marine strata in the area south of central-west Taiwan. The deposition of the last coal-bearing formation was followed by deepening of the sedimentary basin, which allowed a new transgression leading to the invasion of mainly Pliocene and partly Pleistocene flyschoid sediments of great thickness. The flysch sedimentation was succeeded in middle Pleistocene time by the accumulation of a large volume of predominantly nonmarine conglomeratic deposits, typically molassic, that represent the last regress of the Cenozoic sea from western Taiwan. Probably coeval with the Pliocene–Pleistocene flysch is a 2000-m-thick mélange fringing the west coast of the peninsular southern part of the island (Biq, 1977; Huang *et al.*, 1983). Entombed in this unconsolidated and structureless argillaceous mass are exotics of basic and ultrabasic rocks in addition to blocks of local sedimentary and exotic acid igneous rocks (Tsan, 1974). In the form of a prism tapering westward, the cover succession of western Taiwan is 8000 m thick in the Foothill Zone but thins rapidly to only 500 m on the Penghu Islands. Practically all the cover rocks are free from regional metamorphism, except probably those in the zeolite facies zone marginal to the slate belt (Liou, 1981b). The cover succession is topped with Pleistocene volcanics in two widely spaced areas of unrelated magmatic activity: the island arc andesites on and off the north coast and the flood basalt on the Penghu Islands. The youngest sediments in western Taiwan are gravels and sands that underlie the outbuilding coastal plain and form high coastal terraces in uplifted areas of neotectonic interest (Ku, 1963).

Built up during the last 20 m.y. and cropping out in a narrow coastal strip only 150 km long, the rock sequence of the domain east of the Longitudinal Valley is by no means less important in the dual-column stratigraphy of Taiwan despite the shortness of the history it reflects and the smallness of the area it underlies. This sequence is exposed in the Coastal Range and in part on the offshore islands of Lutao and Lanhsü. It is a eugeosynclinal pile in classic geology parlance as expressed by its two-stage accumulation of andesitic volcanic and volcanogenic rocks occupying the lower part and typical flysch sediments occupying the upper. The volcanics consist of a great volume of agglomerate and lesser amounts of lava and ash and contain, as a product of geosynclinal cannibalism, a conglomerate composed wholly of water-sorted andesite pebbles. This volcanic and volcanogenic group, totalling some 1500 m in thickness, was invaded by a stock of comagmatic diorite and is covered with patches of shallow-water limestone generally less than 10 m in thickness. The isotopic dates from the diorite (Ho, 1969) and fossils from the limestone (Chang, 1967) both indicate that the invaded and underlying andesitic group is of Miocene age.

The upper stage of the stratigraphic framework of eastern Taiwan is characterized by an amazing flysch buildup. The thickness of this flysch sequence, 3000 to 4000 m, contributes one half of the aggregate thickness of all the stratigraphic units of the Coastal Range. This sequence of grade-bedded rocks consists mainly of tuffaceous sandstone and low-rank graywacke, often turbiditic, and polymict conglomerate (Teng, 1979, 1980). The nature of the flysch-volcanic contact in the Coastal Range is not everywhere clear, but it is certain that the two rock groups are in thrust contact in many places. The age of the flysch is mainly Pliocene but may range from the Upper Miocene (Chang, 1967, 1968, 1969) to the middle Pleistocene (Chi *et al.*, 1981).

The last sediments in the flysch buildup in eastern Taiwan are preserved in the form of mélange, which was presumably an argillaceous mass of black flysch with the incorporation of a multitude of exotic blocks, forming a wildflysch-like chaotic body. It is the first mélange recognized in Taiwan (Biq, 1956) and has continued to be the favorite and, inevitably, debatable subject of Coastal Range geologists (e.g., Biq, 1969, 1971b; T. L. Hsu, 1976; Page and Suppe, 1981; Teng, 1981; Wang, 1976). The countless random exotics in this voluminous and chaotic body of pervasively sheared scaly clays are of various lithotypes and of different sizes. Particularly significant among the exotics are ophiolites and those of hill mass dimensions. The mélange crops out around the south end of the Coastal Range and continues northward for a distance of 70 km along the tectonic break between this range and the Longitudinal Valley. An indication that it is not simply a product of some nontectonic surficial movement, i.e., an olistostromal sheet, is its unexpectedly great thickness, which exceeds 1000 m, in a stratigraphic test well near the south end of the Coastal Range that did not reach its base (Meng and Chiang, 1965). It has been suggested that the mélange inter-tongues with some parts of the above mentioned flysch sequence (Page and Suppe, 1981). This is not as clear as the fact recognized at many localities that the mélange is often overriden by thrust plates of other rock groups. Both the exotic blocks and clay matrix yield fossils of various Tertiary stages, but mostly reworked except probably those of Pliocene age. This has led to the common belief that the rock mass was emplaced as a mélange during Plio–Pleistocene tectonic movement.

In the southernmost part of the Longitudinal Valley, immediately west of the mélange terrane, there is a Pleistocene conglomerate of pebbles almost wholly derived from the metamorphics of the Central Range. Albeit only locally developed, it attains a thickness in excess of 1000 m. Since this conglomerate is in immediate contact with the mélange, it is convenient to look on it as an additional segment to the stratigraphic column of the Coastal Range.

It is clear from all the above that the duality of the stratigraphic frame-work of Taiwan is in reality the duality of the tectonic framework. Repre-senting one petrotectonic milieu, the cover succession of the greater part of the island west of the Longitudinal Valley is an immense volume of terri-genous sediments amassed on an extensive shelf in an inactive Mesozoic convergent zone; representing the other, the well-developed eugeosynclinal sequence forming the coastal strip east of the Longitudinal Valley was an assemblage of volcanic products and foretrough fill of a short segment of a west-facing arc–trench system (Fig. 3). Largely coeval but quite different in lithofacies, these Cenozoic sequences must have accumulated separately in two distant crustal domains. It was the northwestward movement of the arc-carrying oceanic plate that, at the end of the Tertiary, brought the shelf and the arc together.

B. Structural Framework

Characterized by a westward orogenic polarity, the tectonic design of Taiwan is almost wholly a strong response to the stress imposed upon the Cenozoic rocks by the end of the Tertiary continent-arc collision that still continues today. The collision remobilized the Mesozoic tectonized base-

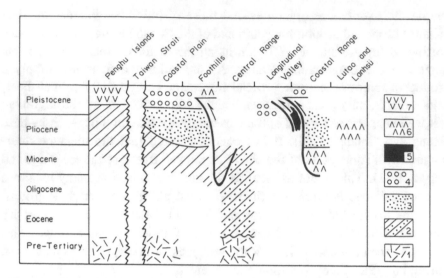

Fig. 3.   Diagram showing the distribution in time and space of rock groups of Taiwan. Rocks
built on the pre-Tertiary basement and forming a complete Cenozoic sequence widely distributed
in the areas west of the Longitudinal Valley are miogeosynclinal sediments on the passive
continental margin of Eurasia; the late Cenozoic sequence forming a separate stratigraphic
column in the coastal areas east of the valley is a eugeosynclinal pile on the leading edge of
the Philippine Sea plate. Rock masses subsequently displaced westward as nappes are marked
with heavy lines beneath. KEY: 1, basement complex; 2, shelf sediments, partly converted into
low-grade metamorphics (stippled); 3, flysch and flyschoid sediments; 4, molasse and other
post-orogenic sediments; 5, ophiolitic mélange; 6, arc andesites; 7, flood basalt.

ment rocks and refolded them on various scales. Though fusiform in shape,
the island of Taiwan is structurally a continent-facing arc convex toward the
northwest, with its east arm recurved offshore to be the westernmost segment
of the ocean-facing Ryukyu Arc. Moreover, if the structural features on the
nearby seafloor are examined, it will be realized that the overall tectonic
pattern of the region is in the form of a double free virgation tightened in
the central part of Taiwan, from which divergent structural lines run out
southward and northward. Manifestly the virgation is caused by a 60-km-
wide basement high extending from the China coast eastward to underlie
the alluvial central part of western Taiwan, where west-vergent folds are
tightly bundled into a remarkable recess (Biq, 1978).

Vigorous uplift brought what is the structurally lowest part of Taiwan
to a lofty position in the Central Range, which is orographically distinguishable into a main range and the crescent Hsüehshan Range immediately northwest of it. Only on the east flank of the main range is the innermost anatomy
of the island exposed through the Eocene to Miocene slate–quartzite mantle.
The marble-bearing and often migmatized portion of the basement schists
has been deformed and reshaped into a series of mostly overturned and

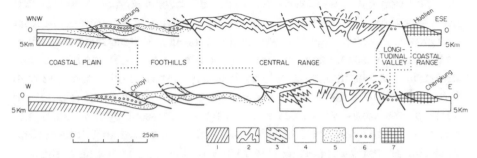

Fig. 4. Generalized structure sections of Taiwan. The lower section is a composite of two sections, the easternmost part of the section passing through Chiayi being some 25 km north of the westernmost part of the section through Chengkung. KEY: 1, Mesozoic foreland-basement; 2, pre-Tertiary metamorphic complex; 3, Slate Belt Tertiary; 4, Foothill Zone Miocene; 5, Pliocene flyschoid sediments; 6, Pleistocene molasse and other post-orogenic sediments; 7, Tertiary and Pleistocene arc-massif and subduction complex. Heavy lines are thrust faults.

occasionally fan-shaped anticlines and synclines in consequence of three or four generations of flow and shear folding (Fig. 4). These rocks are replaced, at the east foot of the main range, by the other group of the basement rocks that contain ophiolitic exotics. Such an abrupt facies junction between the two groups indicates a crust-shortening tectonic break. This is the Shoufeng Fault (Fig. 2) which is considered a Mesozoic plate boundary, the rocks of one facies underlying the magmatic arc on the west side of the boundary and those of the other facies filling the trench on the east side (Biq, 1971a; Liou, 1981a). The steeply sloping easternmost part of the slate massif of the Central Range has been eroded to a tectonically interesting depth of at least 10 km to uncover a fossil subduction zone. The remaining parts of the Central Range, underlain by the slaty cover, have their own structural style characterized by a widespread shear folding imprinted on generally not tight but occasionally recumbent folds (H. T. Chu, 1982). Such variation in tightness is perhaps due largely to the disharmonic folding of the whole sequence resulting from the difference in competence of the slaty and quartzitic beds.

The uplift of the Central Range was due to block thrusting that has produced two boundary faults, each being an outward thrust probably with some sinistral motion. One is the Central Range Fault along the east foot of the range and the other is the Chaochow–Chüchih Fault (Fig. 2) between the range and the Foothill Zone west of it. In between, there is another thrust, the Lishan Fault, which branches from the Chaochow–Chüchih Fault to separate the Hsüehshan Range from the main range. The deep valley along this fault widens in the north to form an actively subsiding and thickly alluviated intramontane trough, the deltaic Ilan Plain, on the northeast coast of the island.

The Foothill Zone is typically a zone of cover tectonics. The fact that no rocks older than the late Tertiary are involved in this fold-and-thrust complex is a telltale sign that the foothill tectonics is *décollement* in style. In the Foothill Zone, each *décollement* surface is a west-directed listric sole thrust, above which, and beneath as well, asymmetrical folds combine with thrusts in close association to show a westward displacement (Fig. 4). In some places where such folds and thrusts are crosscut by tear faults, the structural details on the two sides of each fault reveal that transcurrent buckling played a supporting role in the folding and thrusting (Tsan and Keng, 1962). It is noteworthy in the Foothill Zone that Pliocene–Pleistocene sediments, never found in the Central Range and the foothills immediately west, suddenly appear as a flysch sequence thousands of meters thick in the hills farther west. This has led to distinguishing of an inner or east zone and an outer or west zone in the foothills according to the absence or presence of the Pliocene and younger rocks. Since the outher hills are usually separated from the inner by westward thrusts, it is believed that the inner or eastern part of the deep flysch trough was overridden by masses of Miocene strata from the area farther east, that have now been carved into the inner hills. These parautochthonous slabs are very probably gravity nappes from the Central Range tectonically denuded shortly before it was upthrust by the Chaochow–Chüchih Fault (Fig. 3). When entering the trough, these plunging rock masses, acting like bulldozers, deformed the Pliocene and younger sediments into the superficial folds and thrusts of the outer hills (Biq, 1972b). The forefront of this group of "epidermal" structural features lies where the westwardly polarized cross-island tectonic drive finally died out along the west coast in the northern part of the island and along the east limit of the coastal plain in the southern part (Fig. 4).

Folds and thrusts are less closely packed in the southern half of Taiwan than in the northern half. The extra area in this less tightly bundled part of the Taiwan virgation is occupied by the extensive Pingtung Valley that separates the south segment of the Foothill Zone from the Central Range. This intramontane trough onshore is filled with thick recent sediments and represented offshore by a drowned valley. Geologic and geophysical observations (Hsieh, 1970) both permit the conclusion that the Pingtung Valley has been sinking since Pleistocene time as a concomitant trough of the upthrust Central Range. Between the two is the Chaochow–Chüchih Fault which has a large throw. In the west coast areas of the Hengchun Peninsula, which is the southernmost segment of the Central Range, the Pliocene mélange is exposed for a distance of some 60 km. Its presence between the Pingtung Valley and the Central Range suggests that the actively subsiding valley is virtually the northernmost segment of a present-day east-dipping subduction zone (Biq, 1972a). Almost collinear with the Pingtung Valley is the Manila Trench where this subduction zone has attained its fullest development.

The structure of the allochthonous Coastal Range impinging upon the autochthonous part of Taiwan seems disproportionate to the plate–junction setting. The surface structure has the form of a west-vergent series of simple folds and thrusts all trending north-northeast. Nevertheless, the left-handed echelon pattern into which these folds and thrusts are arranged reveals that the Coastal Range owed its tectonization to coast-parallel shearing in addition to the west-directed compression during the collision. The major anticlines are cored with andesitic rocks, the dioritic subvolcanic part of which crops out only in deeply eroded areas of axial culmination. Parts of these arc-magmatic products at both high and low levels are mineralized·with porphyry copper characteristic of such a metallogenic setting. It is noteworthy that the Plio–Pleistocene ophiolitic mélange is not involved in the folding and thrusting within the Coastal Range but is deformed in a quite different style. It forms a belt of badlands along the southern half of the west foot and around the southern end of the range. The simplest interpretation of the structure of the Coastal Range, the map pattern of which resembles to some extent a klippe resting upon the mélange, is that the fold-and-thrust part higher in altitude is an arc–massif nappe that slid downward and westward into the adjacent closing trench undergoing mélange tectonics (Biq, 1969, 1971b). Almost wholly submarine, the southern extension of the Coastal Range is represented by the two islands of Lutao and Lanhsü perched on the volcanic arc *in statu quo*. This offshore equivalent of the Coastal Range is now traveling the last 50 km of its overseas journey to the east coast of Taiwan before finally running aground to become an allochthon south of Taitung.

Very important in east coast tectonics because of its great magnitude and long-continued activity is the Coastal Range Fault, a crustal break between the arc range and the Longitudinal Valley. It is an oblique-slip fault, along which the entire body of the Coastal Range has been moving as an upthrust massif and as a north-translated block. Besides the left-handed pattern of the en echelon folds and thrusts in the Coastal Range, the evidence for the sinistral motion of the fault includes the offset of streams flowing eastward across the Longitudinal Valley (Biq, 1965; York, 1976) and the strike slip of 2 m, in addition to a dip slip of 1.3 m, produced along the fault by the 1951 earthquake (T. L. Hsu, 1962). The conclusion is strengthened by a recent retriangulation in an area covering the northern parts of the Coastal Range and the Longitudinal Valley, where this presently active fault has accumulated during a period of some 60 years a net left-lateral strike slip of some 3.7 m (C. Y. Chen, 1974). This horizontal movement has caused the north end of the Coastal Range to collide with the westernmost segment of the Ryukyu Arc (Biq, 1981). Located exactly at the right-angle junction of the impinging range and the passive arc is the collision product in its embryonic form. It is an isolated hill mass of disturbed Holocene marine sediments, at the foot of which on the seashore side is constructed the Hualien harbor immediately north of the seaside city. This freshly emerging neotectonic body has been uplifted at the extraordinary rate of 9 mm/yr during the last several millennia (Konishi *et al.*, 1968). Such unusually great crustal mobility underlying the locale of the incipient collision is also reflected by the extremely dense concentration of shallow earthquakes in the area with Hualien at its center (Tsai *et al.*, 1977).

C. Tectonic Evolution

Little is known about pre-Permian Taiwan, simply because this part of Asia was at that time far from the forefront of the early growth of the continent. Furthermore, even the Variscan Orogeny is undiscernible in the island's earliest Permian rock record. It was not until the late Mesozoic when this region became the leading edge of Cathaysia during a long period of continental accretion that Taiwan began its tectonic evolution patterned by episodic subduction and collision. The earliest recognized orogeny in Taiwan was of Cretaceous age, at 86 Ma as radiometrically registered in the basement (Yen and Rosenblum, 1964). It represents in Taiwan the sweeping Yenshanian movement of China; moreover, its tract in the Central Range is actually a segment of the west-dipping subduction zone underlying Yenshanian China. This Taiwan segment of the subduction zone is marked by trench fill; and paired with the trench is a magmatic arc immediately to the west that constitutes an Andean-type cordillera, hundreds of kilometers wide, including maritime and inland China (Biq, 1979). It is noteworthy that this fossil subduction zone, exposed over a narrow zone through a thick cover in Taiwan, is the outcrop of the Yenshanian arc–trench junction not found throughout eastern China. Indeed, only the Yenshanides on alpinotype Taiwan are orthotectonic in nature while those on the platform-built mainland without exception are paratectonic and free from regional metamorphism.

While nearly the whole length of the Yenshanized continental margin of eastern Asia continued to undergo subduction tectonics in Cenozoic time, only the Taiwan sector enjoyed an early to middle Tertiary noncompressional interval. This was due to the inception of a new east-dipping subduction zone no less than 1000 km southeast of Taiwan replacing the old zone beneath the Central Range. The north end of this intra-Pacific subduction zone must have been linked to the junction of inactive Taiwan and the active Ryukyu Arc by a sinistral trench–trench transform fault of which the former length of some 1000 km has now been reduced to nil in consequence of the continuous southeastward subduction of the off-Taiwan oceanic crust. Before Taiwan was reactivated in a new convergent-zone setting, its western part and the area further west, together with the area behind the Ryukyu Arc, had undergone, in early Tertiary time, interarc rifting resulting in a number of small half grabens (Sun, 1982). Following the diminution of crustal extension, the area was transformed into an extensive continental shelf during a quiescent Miocene interval. Beneath the 10-km-thick shelf were buried the abandoned Mesozoic arc and trench of Taiwan. Contemporaneous with the shelf buildup was the development of the distant west-facing Luzon Arc and its subduction zone. Inching ever closer toward Taiwan, at a velocity of about 5 cm/yr (Seno, 1977), this arc, in late Pliocene time, finally touched at its northern extremity the shelf at a point near the Taiwan–Ryukyu junction. This began the continent-arc collision that remobilized the already stabilized structure of Taiwan into a new framework. As a result of the lateral variation in timing along the shelf edge, this collision shows at the present time an ensemble of structural features in various stages of development that reveal an overall southward progressive decrease in age and intensity of tectonization.

The diachronism of the Taiwan collision was due to the precollision geography in which the trend of the arc was meridional while that of the shelf edge was north-northeast. It was this obliquity which led the north tip of the northwest-vectored arc to impinge first upon the shelf and made the vicinity of Hualien the point of initial contact of the diachronous collision. The gradually south-extending continent-arc suture has now reached Taitung after its 150 km growth (Fig. 2, inset map) during the Quaternary. The Coastal Range with an equal length is the collision product on the arc side, and is, as already described, a west-vergent fold-and-thrust package of the arc itself with the closely related sediment masses including the ophiolitic mélange. The transformation of the last trench fill into the mélange was an immediate consequence of the collision. In this grand finale of the arc-trench tectonics, pieces of ophiolitic rocks are believed to have been peeled off from the top of the latest engulfed oceanic crust and mingled with strongly disturbed sediments in the closing trench to become the most significant among the countless exotics in the mélange.

South of Taitung, the untouched part of the arc emerges from the sea as Lutao and Lanhsü, which, plus the Batan Islands collinear with them to the south, form a flottila of volcanic arc islets known as the North Luzon Ridge. It is quite significant that the andesitic volcanism on Lanhsü took place between 13 Ma and 6 Ma (Ho, 1975), younger than the same arc magmatism in the Coastal Range but older than that on the Batan Islands. This clearly demonstrates that the cessation of the arc magmatism kept pace with the collision in which the arc has been involved. Immediately west of Lutao and Lanhsü is the deep-sea furrow forming the Taiwan segment of the North Luzon Trough, which extends south-southwestward to become the West Luzon Trough, if not running southward into the Lingayen Gulf on the northwest coast of Luzon. In the Cenozoic plate tectonic picture of Taiwan, the North Luzon Trough is the precollision part of the subduction zone, above which are not only the volcanic arc partly tectonized into the Coastal Range, but also the submarine Gagua Ridge genetically related to the subduction (R. S. Lu and Liaw, 1979). Eastward subduction still active in the North Luzon Trough will eventually cause the overlying volcanic islands to edge westward across the remnant ocean basin and then to be thrust ashore in another episode of the incomplete collision. However, since the West Luzon Trough is considered a fore-arc basin (Ludwig, 1970; Ludwig *et al.*, 1967), it is still little understood how the North Luzon Trough changes its status in extending to the west of Luzon.

At the same time when the normal stress component of the impact of the northwest-advancing Luzon Arc compressed the passive continental shelf and the arc itself into two fold-and-thrust belts, the shearing stress component put Taiwan under a tectonic regime of longitudinal sinistral faulting. Although the strike-slip movement was distributed among many faults on land and offshore, the greatest displacement was along the plate boundary Coastal Range Fault. This megashear is still operative today, at a velocity of at least 6 cm/yr according to the previously mentioned retriangulation results. It is virtually the velocity of the north-moving Philippine Sea plate, which descends beneath the Ryukyu Arc but leaves the nonsubductible Coastal Range to collide in a scene of flake tectonics with the arc at Hualien (Biq, 1981).

The collision product on the continental side is the foreland fold-and-thrust belt that comprises the deep-rooted Central Range and the superficially deformed Foothill Zone. Also included in this Eurasian plate portion of Taiwan as negative elements are intramontane and extramontane basins. The collision not only folded the continental shelf and refolded the basement, but also resulted in the greenschist metamorphism that transformed great parts of the shelf sediments into slaty rocks and caused the retrograde metamorphism of the Yenshanian aged basement rocks. Depending upon how deep from the surface and how far from the plate junction they were at the time of collision, the folded and refolded rocks responded to the impact with different tectonic styles distinguishable at three structural levels. The remobilized basement becomes the integral infrastructural part of the new orogen; parts of the Miocene strata free from the metamorphism and flexural-slip in fold style represent the suprastructure; and, between these two levels, all the slaty Tertiary rocks constitute a thick transition zone characterized by shear folding (Biq, 1966).

The foreland folding was at its peak in Pliocene time when the Central Range began to rise and a collateral foredeep was created in the immediate west of it. The latter reached its maximum development when the hinge between the upbuckled and downbuckled parts of the crust was stretched into the deep-reaching Chaochow–Chüchih Fault Zone which runs the full length of the island (Biq, 1972a). Such shortening of foreland crust, usually a sequel to continental collision, is in fact the renewal of the subduction halted in the collision. Accordingly, the Pliocene foredeep of western Taiwan is virtually an Ampferer subduction zone (Bally and Snelson, 1980), born an intraplate substitute for, or subsidiary to, the already partly inactive interplate subduction zone on the other side of the Central Range (Biq, 1982). However, after running southward out of Taiwan via the sediment-filled Pingtung Valley and the mélange-plastered west coast of the Hengchun Peninsula, this subduction zone loses its Ampferer self to become the Manila Trench, an ordinary subduction zone of the Benioff type. With the Central Range as an ample source of huge volumes of orogenic sediments, the foredeep became a flysch trough in Pliocene time and a molasse trough during the Pleistocene. The rising cordillera later lavishly shed gravity nappes into the foredeep when uplifted high enough to facilitate the westward downsliding of the suprastructural slabs over the discordantly folded slaty rocks immediately lower in structural level (Fig. 3). The folding in western Taiwan shows that the tectonization of the Foothill Zone on the foreland was not an immediate consequence of the continent-arc collision but in reality a corollary of the secondary subduction of the Ampferer type taking place in the foreland itself.

The main part of Taiwan, although less tectonically active than the adjacent arc, is still a zone of high mobility, characterized by vertical and horizontal movements both of notable vigor. The Central Range, of which 227 peaks exceed 3000 m and beneath which the crust has been thickened from 24 km to 35 by the collision (C. P. Lu and Wu, 1974), is the most strongly uplifted tract in this part of the island. It is being uplifted at an average rate of 5 mm/yr (Peng et al., 1977) following a period of uplift of 10 mm/yr during late Pleistocene time (T. K. Liu, 1982). Concurrently active is the subsidence of two intramontane troughs. The Pingtung Valley is sinking as a part of a subduction zone collinear with the Manila Trench. The Ilan Plain, as a part of an interarc basin, is the southwest end of the Okinawa

Trough behind the Ryukyu Arc, which has extended onto northeastern Taiwan as a result of the northward subduction of the northwestern part of the Philippine Sea plate after the collision of its leading edge with Taiwan. Horizontal movement is very clearly demonstrated in the present century by three earthquake faults in the Tainan–Chiayi and Taichung–Miaoli areas of western Taiwan. They are east-northeast-striking dextral surface faults with comparable lengths of some 10 km and with strike slips of 1.3 m to 2.4. Interestingly, each of them is a miniature version of transform faults of the trench–trench type (Biq, 1976a). Evidently these faults represent, in the present-day stress field, one of the two sets of shear fractures forming a conjugate system, the other set being represented by the north-northeast-striking sinistral faults of much greater magnitude that have placed postcollision Taiwan, especially eastern Taiwan, under the dominance of wrench tectonics.

The tectonic activity of Taiwan between its revival from the early Tertiary anorogenic quiescence and the establishment of the present arc-alongside-continent framework of the island can be appreciated as the interaction between two lithospheric plates. The northwest corner part of the Philippine Sea plate is subducted northward beneath one part of the Eurasian plate and at the same time obducted eastward, in the manner of a continent-arc collision, onto another part. Hence, Taiwan as a geodynamic body is remarkable for its duality and uniqueness: dual for its being underridden in one direction but also overridden in another by the Philippine Sea plate, and unique for its being the only tectonic product of such duality on the boundary between this plate and the Eurasian plate. However, the uniqueness of Taiwan does not prevent it from being in good comparison with such Tertiary collision products as New Guinea (Biq, 1978) and the Himalayas (Biq, 1976b) and even with more ancient ones such as the Caledonides of Scotland and Ireland (Mitchell, 1981). Besides what is shown by the various styles of folding and faulting, there is a greater structural complexity of Taiwan induced by the heavy impact and diachronism of the collision. This impingement has produced an intraplate subduction zone on the cratonic foreland to renew the crustal consumption after having converted the most vulnerable part of the interplate zone into a continent-arc suture. Such subduction-on-foreland tectonics means the diffusing of the collision impact over a larger area in order to make the subduction not only continue but also more efficient. Each of the two subduction zones, Ampferer in type whether from the beginning or after the collision, has an active southern Benioff segment. Their coexistence brings out the imcomplete state of the collision, in which tectonic strips offshore are in an earlier stage of orogenic growth and more widely spaced from one another than their counterparts on land.

III. GEOPHYSICS

Taiwan, together with parts of the Ryukyu Islands and the Philippines, is a region (Fig. 1) of remarkable tectonic complexity. The geophysical work done in Taiwan in the past two decades has revealed to a considerable extent the major crustal features of the region. Although much remains to be understood, existing geophysical data, combined with results from geologic and geochemical researches, have already set up certain constraints for the interpretations of Taiwan tectonics. The following sections are a summary of the seismicity, gravity, geomagnetism, heat flow, and seismic reflection and refraction of Taiwan and nearby areas.

A. Seismicity

The distribution of earthquake epicenters between 1901 and 1974 in the Taiwan region is shown in Fig. 5, and the depth distribution of the hypocenters in different sectors of the region in the same period are depicted with profiles in Fig. 6. It is clear from the latter figure that, as the earthquake zone extends northward from northern Luzon to Taiwan, the foci become shallower. Focal depths are less than 70 km throughout the length of the Coastal Range of eastern Taiwan between latitude 23° and 24.2°N, but become greater in areas north and south of this tract. Both shallow and deep earthquakes are comparatively rare in the area off southern Taiwan, and how the tectonic boundary on the island is related to the Manila Trench is therefore not very clear. Based on focal mechanism solutions, a short dextral transform fault of the trench–trench type is tentatively delineated along latitude 21°N as a small segment of the tectonic boundary extending from eastern Taiwan to the Manila Trench (Lin and Tsai, 1981). The location of the plate boundary in eastern Taiwan is unquestioned and the submarine segments south of it according to the intermediate hypocenters between Taiwan and Luzon, form a 50-km-thick tabular zone dipping 55° eastward to a depth of 170 km. Considered in conjunction with the geology of eastern Taiwan, this is conceivably a Benioff zone underlying the volcanic arc islands of Lutao and Lanhsu, the inactive northern extension being now represented by the crustal suture between the Coastal Range and the Longitudinal Valley following the end of the Tertiary collision of the Luzon Arc and the Eurasian portion of Taiwan.

Three major seismic zones can be distinguished in the Taiwan region (Tsai *et al.*, 1977) (Fig. 5). In the northeast the seismic zone occupying the area east of longitude 121.5°E and north of latitude 24°N, is primarily associated with the northward subduction of the Philippine Sea plate underneath the Eurasian plate. The increasingly deeper hypocenters define a

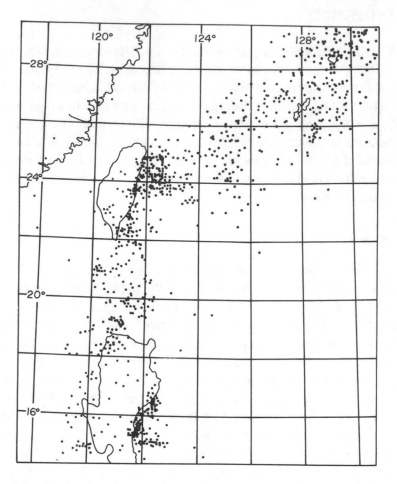

Fig. 5. Map showing the epicenters of the earthquakes occurring between 1904 and 1974 in the Taiwan region including parts of the Ryukyus and the Philippines. (Redrawn from Wu, 1978.)

planar Benioff zone about 50 km thick and dipping from its surface trace along latitude 24°N at an angle of 45°–50° to a depth of some 130 km. The west side face of the subducting Philippine Sea plate lies beneath north Taiwan, its surface projection striking along longitude 121.5°E. Microseismicity and geothermal activity giving rise to hot springs confirm this boundary (Tsai *et al.*, 1977). This hypocenter-defined subduction zone appears to trend parallel to the Ryukyu Trench but is offset by a north–south dextral transform fault at about longitude 123°E (Wageman *et al.*, 1970; Biq, 1972c; Wu, 1978). Some of the earthquakes in this northeast seismic zone have epicenters near the Quaternary volcanic centers at the north end of Taiwan, an obvious onshore continuation of the Ryukyu magmatic arc. Microseismicity

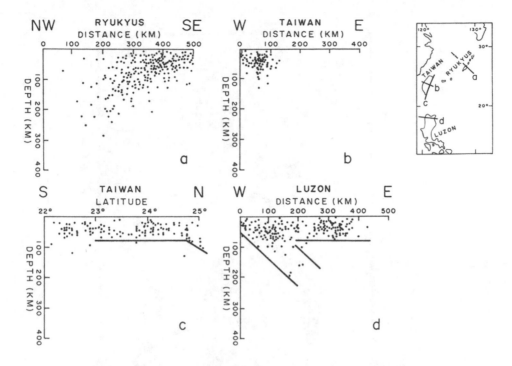

Fig. 6.   Profiles showing the distribution of the hypocenters of the earthquakes occurring between 1904 and 1974 in the Taiwan region including parts of the Ryukyus and the Philippines.
The map shows the locations of the profiles. (Redrawn from Wu, 1978.)

in the Ilan area (Tsai *et al.*, 1975) has delineated a zone of vertical faulting
that extends into the Okinawa Trough; a focal mechanism not far east of
Ilan showed normal faulting (Wu and Lu, 1976). The volcanic ridges are
traceable from northeastern Taiwan to the Okinawa Trough (C. S. Lee and
Lu, 1976). Based on these observations and some other geophysical data,
R. S. Lu *et al.* (1977) and Bowin *et al.* (1978) suggested, following the common belief, that extensional opening of the Okinawa Trough had continued
into Taiwan to form the alluvial Ilan depression. The east seismic zone
includes the Longitudinal Valley, the Coastal Range, and the offshore area
further east. The west limit of this zone lies west of, and in parallel with,
the Longitudinal Valley, and has a boundary surface dipping 50°–55° eastward down to a depth of 50 km or so. This boundary is generally accepted
as a collision zone as well as a major sinistral fault between the Eurasian
plate and the Luzon Arc of the Philippine Sea plate (Allen, 1962; Wu, 1970,
1972, 1978; Biq, 1972a,c; Fitch, 1972; Tsai, *et al.*, 1977; Bowin *et al.*, 1978).
This continent-arc suture, is no longer an active Benioff zone at the present

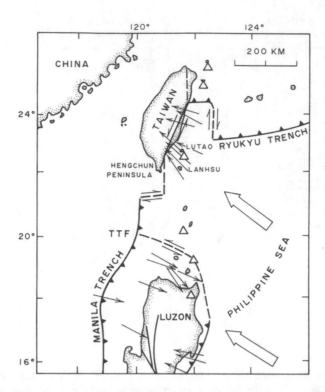

Fig. 7.   Map showing the plate boundaries in the vicinity of Taiwan, dash-lined where inferred. Triangles are active volcanoes; thin arrows, slip vectors of shallow earthquakes; large arrows, direction of the movement of the Philippine Sea plate relative to the Eurasian. (After Lin and Tsai, 1981.)

day. The west seismic zone covers most of the Hsüehshan Range, the Foothill Zone, and the coastal plain of western Taiwan. Hypocenters in this zone are all confined in the crust, which is subjected to a compression from the southeast due to the collision in eastern Taiwan. Seismic activities in southern Taiwan are generally related to a system of longitudinal faults, such as the Chukou, Liukuei, and Chaochow faults. Between the Pingtung Plain and the Manila Trench, a few focal mechanism solutions indicate a right-lateral strike-slip fault (Bashi Fault) near the Hengchun Peninsula and a left-lateral one near the Babuyan Island (Figs. 1 and 7). The overall picture of Taiwan seismicity has been depicted by Lin and Tsai (1981) with a model based on focal mechanism solutions (Fig. 7). Showing that Taiwan owes its tectonic framework mainly to the northwestward movement of the Philippine Sea plate, this model supports with more seismic data the general opinion that the Luzon Arc collided obliquely with the Eurasian continental margin at Taiwan and, in consequence of this collision, the northward subduction of the Philippine Sea plate at the Taiwan–Ryukyu junction and the strong uplift of the island of Taiwan took place separately on the two sides of the plate boundary.

B. Gravity

Figure 8 is the free-air gravity anomaly map of the Taiwan region published by Bowin *et al.* (1978). The map is characterized by generally north-trending isogals and a pronounced gradient. Gravity highs are found over the east flank of the Central Range and prolonged to the south tip of Taiwan, and also over the Lutao–Babuyan Ridge which follows the Luzon Arc. The east-trending, elongated gravity lows extend from the northeast coast of Taiwan coincide with the Ryukyu Trench and the associated upper slope basin. The coincidence delineates approximately the boundary between the Philippine plate and the Eurasian plate. On the other hand, elongated gravity highs lie mainly over the Ryukyu Arc where solid rocks are exposed. Again, as discussed in the previous section on seismicity, the free-air gravity anomalies show that the Ilan Plain in northeastern Taiwan is the southwestward extension of the Okinawa Trough. The gravity lows overlying the western margins of the Lutao–Babuyan Ridge correspond to the low-density sediment masses in the north segment of the Manila Trench and in the Luzon Trough.

For a closer study of the basement configuration and geologic structures on land, a Bouguer gravity anomaly map was compiled by Hsieh and Hu (1972) (Fig. 9). Some significant parts of the map are briefly described as follows. The lowest gravity anomalies in the Taichung–Tachien area indicate that this part of the island may be underlain by a thick mass of trough sediment. The southwestern low overlying the Pingtung Plain is in good accordance with the downbuckling of the crust in the immediate west of the south segment of the uplifted Central Range. The gravity high over the Peng-hu Islands extending eastward to Peikang corresponds in a striking manner, both in trend and areal extent, to the flood basalt on these islands and the 500-m-deep Mesozoic basement beneath the Peikang area. The north-trending gravity high in eastern Taiwan is separated by a relatively lower belt over the Longitudinal Valley into two highs, one corresponding to the Central Range and other to the Coastal Range. The Coastal Range gravity high results from its high-density magmatic arc rocks and the still denser oceanic crust beneath. As calculated from the Bouguer anomalies (C. C. Liu and Yen, 1975), the greatest crustal thickness in Taiwan is 40 ± 3 km in the northern central part of the island, the least being 27 ± 2 km in the southwestern and southernmost coastal areas. Another calculation (C. P. Lu and Wu, 1974) shows that the crust is thickened to 35 km beneath the Central Range from 24 km beneath the Coastal Range.

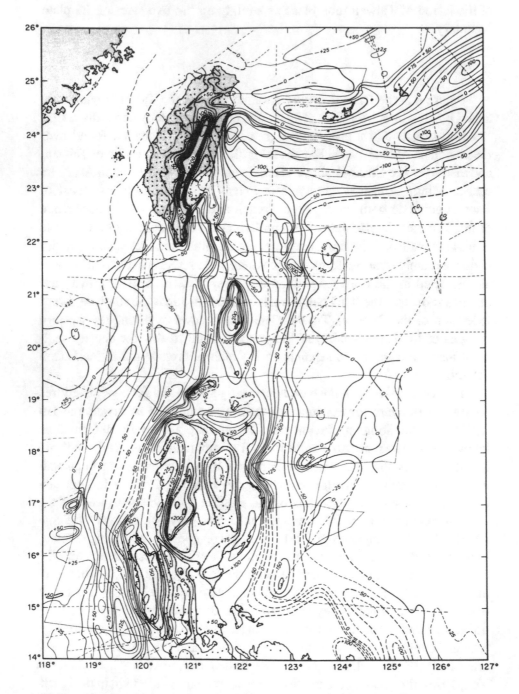

Fig. 8.   Map showing the free-air anomalies of the Taiwan–Luzon region. Isogal interval is 25
mgals. Dots show locations of on-land measurements. (After Bowin *et al.*, 1978, with permission
of the American Association of Petroleum Geologists.)

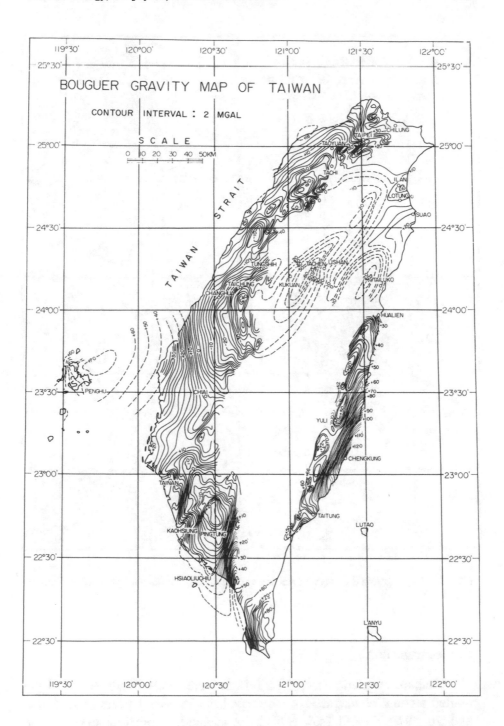

Fig. 9.   Map showing the Bouguer anomalies over Taiwan. (After Hsieh and Hu, 1972.)

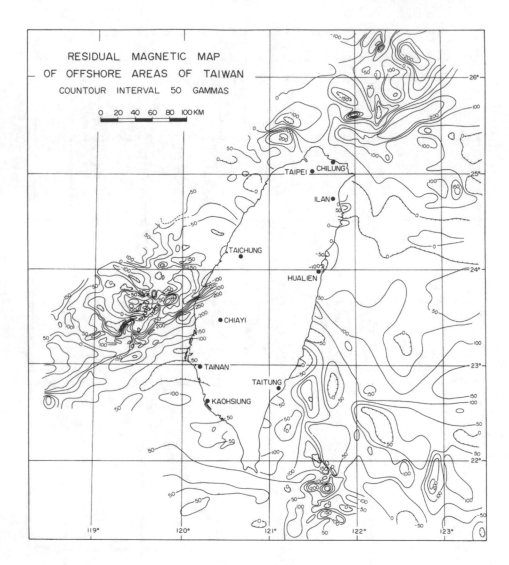

Fig. 10.   Map showing the magnetic anomalies of areas off the Taiwan coasts. (After Hu and Lu, 1979.)

C. Geomagnetism

Onshore magnetic surveys in Taiwan were rarely conducted until the limited amount of magnetic work carried out in recent years (e.g., Hsiao and Hu, 1978; Yu and Tsai, 1979a,b). A magnetic anomaly map of Taiwan is therefore not available at present. The following discussions are based mainly on a residual magnetic map of the offshore areas of Taiwan (Fig. 10) (Hu and Lu, 1979). In that map, three zones of conspicuous magnetic anomalies, trending east-northeastward from the offshore areas northwest of Taipei, southwestward from the west coast between Changhua and Chiayi, and southward from the east coast between Taitung and Hualien, are believed to reflect increasing magnetic susceptibilities of intrusive and extrusive igneous rocks. The anomalies in the first zone are caused by the Pleistocene volcanic activity of the Tatun volcanoes, the Chilung volcanoes, etc. These volcanoes are inferred to have resulted from island arc magmatism related to the northward subduction of the Philippine Sea plate beneath the Eurasian plate. The anomalies of the second zone are related to the Pleistocene basalt capping the Penghu Islands. Those of the third zone are considered to be related to the subduction of the Eurasian plate beneath the Philippine Sea plate and the subsequent collision of the former plate and the Luzon Arc of the latter plate.

Other interesting magnetic features that are correlated well with the geology and/or gravity are: (1) the east-trending magnetic high from Penghu to Peikang that is associated with a tectonic dam trapping a thick sequence of Neogene and Quaternary sediments in northwestern Taiwan (C. S. Lee et al., 1973); (2) the positive anomalies on the edge of the East China Sea shelf that correspond to the Taiwan–Sinzi folded zone (Wageman et al., 1970; C. S. Lee et al., 1980); (3) the east-trending and northeast-trending anomalies in the southwestern and northeastern parts, respectively, of the Okinawa Trough that are associated with the volcanic ridges in the trough; and (4) the magnetic anomalies of the Ilan Plain correlated with those of the offshore area that suggest that the magmatic Ryukyu inner ridge may extend southwestward onto the plain (Yu and Tsai, 1979a).

D. Heat Flow

More than 100 geothermal surface indications, including hot springs and steam vents, have been found on Taiwan. They are distributed mainly in metamorphic and Miocene volcanic terranes but also in Pliocene or Miocene sedimentary basins, and are localized generally along faults, fractured zones, and other structural lines (Fig. 11). However, owing to the high cost of drilling, shallow holes for temperature measurements and geoelectric resistivity surveys were bored only in a limited number of areas with surface thermal indications such as hot springs and fumaroles. Detailed information of this type in northern and eastern parts of the island may be found in a series of reports issued between 1969 and 1980 by the Mining Research and Service Organization. Heat flow values measured on Taiwan and off its coasts are not available now, except those from the southwestern part of the Okinawa Trough. Here the values are widely scattered between 0.36 and 10.49 HFU, among which only low values have been found at the southwest

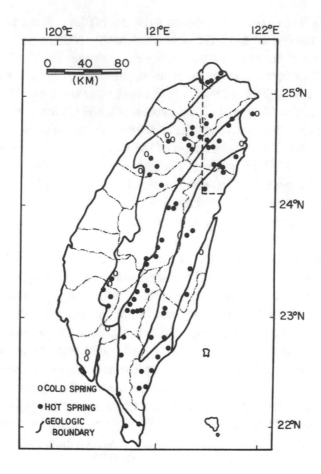

Fig. 11.   Hot springs on Taiwan. (After Yen, 1955.)

end of the trough adjacent to northeastern Taiwan (R. S. Lu *et al.*, 1981). The variable low heat flow values are probably associated with areas of hydrothermal recharge. On the other hand, the high heat flow values suggest that the age of the Okinawa Trough may be as young as 2 m. y. (Watanabe *et al.*, 1977), i.e., the trough began its rifting in early Pleistocene time. It must be added that it seems unrealistic to relate the southwestward extension of the Okinawa Trough to the Iland Plain and the nearby geothermal fields merely on the basis of the limited heat flow data alone.

E. Seismic Reflection and Refraction

Various results from the studies on the seismicity (M. T. Hsu and Yang, 1969; C. P. Lu, 1975; Y. C. Lee and Tsai, 1978; Yeh and Tsai, 1981), explosion seismology (Sato *et al.*, 1970; Pan and Hsiao, 1971; Tsai *et al.*, 1974;

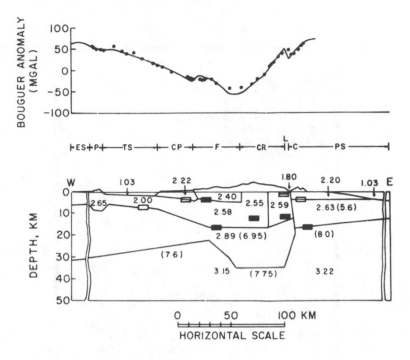

Fig. 12. An east–west profile across central Taiwan, showing Bouguer anomalies (above) and the possible crustal structure based on seismic data (below). Mainly after C. P. Lu and Wu, 1974. Refer to Fig. 13 for the location of the profile. Figures in the lower profile are velocities (in parentheses) and densities. ES, East China Sea; P, Penghu Islands; TS, Taiwan Strait; CP, coastal plain; F, Foothill Zone; CR, Central Range; L, Longitudinal Valley; C, Coastal Range; PS, Philippine Sea.

J. S. Chen *et al.*, 1974), gravity (Hsieh and Hu, 1972; C. P. Lu and Wu, 1974; C. C. Liu and Yen, 1975), and deep wells on Taiwan have combined to reveal a general picture of the crust and upper mantle beneath the island. These results are generally in agreement with what has been shown by Wu (1978) (Fig. 12). Most notable in the crustal profile across Taiwan is the marked difference between the structure beneath the Coastal Range and that beneath the area west of the Longitudinal Valley.

Some interesting submarine seismic profiles (Fig. 13) obtained in the vicinity of Taiwan may be described briefly as follows. The seismic profiles across the Okinawa Trough (Fig. 14a,b,c) are clearly comparable with one another. This indicates that the southwest end of the trough has extended as far as the northeast coast of Taiwan. In this part of the trough, flat-lying sediments thicken toward the center of the trough (Fig. 14c), and rough topographic features indicating local volcanic intrusions and extrusions generally occur on the southeast margin (Fig. 14a,b,d) (C. S. Lee *et al.*, 1980). The low ridge found in an area off the eastern Taiwan coast (Fig. 14e) is

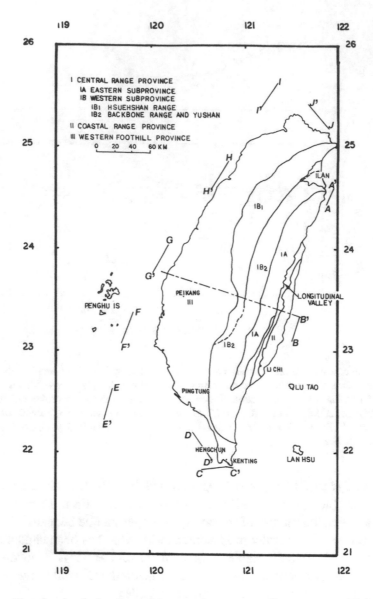

Fig. 13.   Map showing the locations of the crustal structure profile across central Taiwan (dash line) (Fig. 12) and of some seismic reflection profiles (solid line) (Fig. 15).

probably an accretionary wedge on the west end of the Ryukyu Arc. The topographic peak in Fig. 14f is the north end of Gagua Ridge, and the flat-lying sedimentary layers indicate that there is no subduction east of the Luzon Arc. Some other examples of structural complexities in the areas off Taiwan are: (1) A submarine canyon associated with an onshore river is found in a shallow basement (Fig. 15a). (2) Strongly folded zones (Fig. 15b)

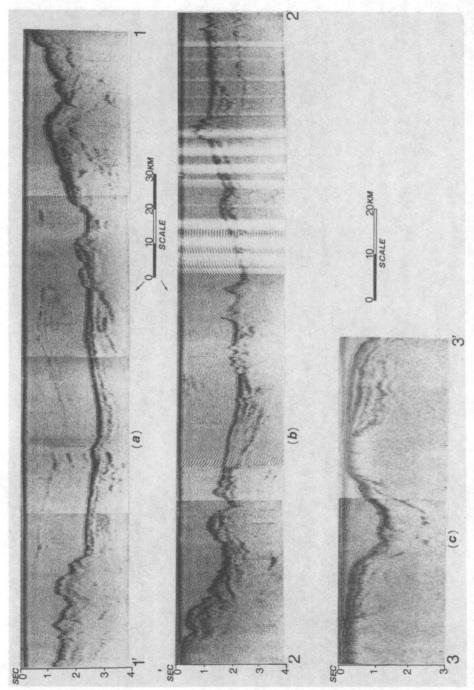

Fig. 14. Four seismic reflection profiles (profiles 1 through 4 in Fig. 1) of the southwestern part of the Okinawa Trough, showing the sediment thicknesses and igneous activities. Profiles 5 and 6 off the eastern Taiwan coast showing the conspicuous ridges of the Ryukyu Arc and the Gagua Ridge, respectively. Refer to Fig. 1 for the locations of the profiles.

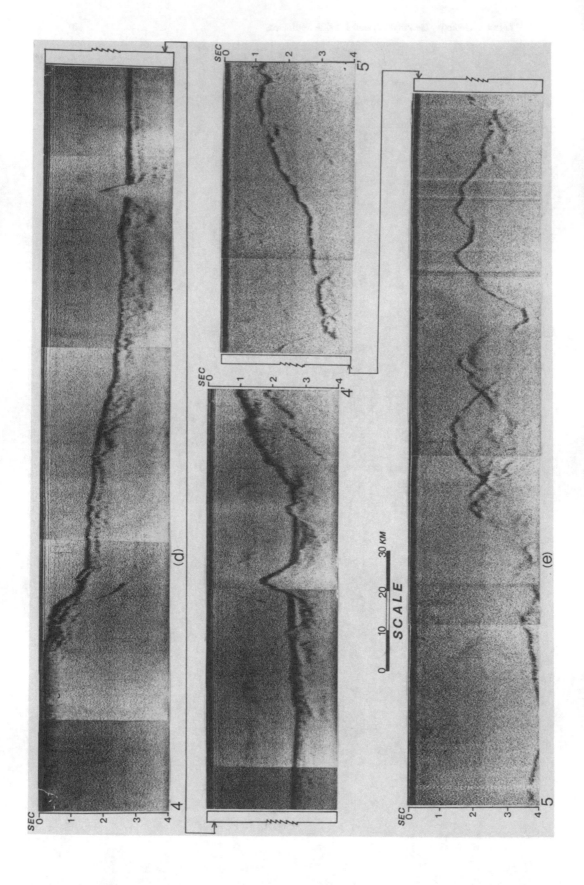

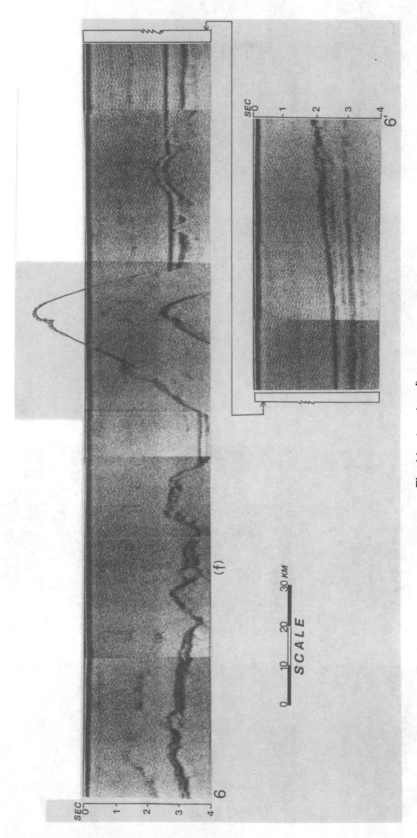

Fig. 14. (continued)

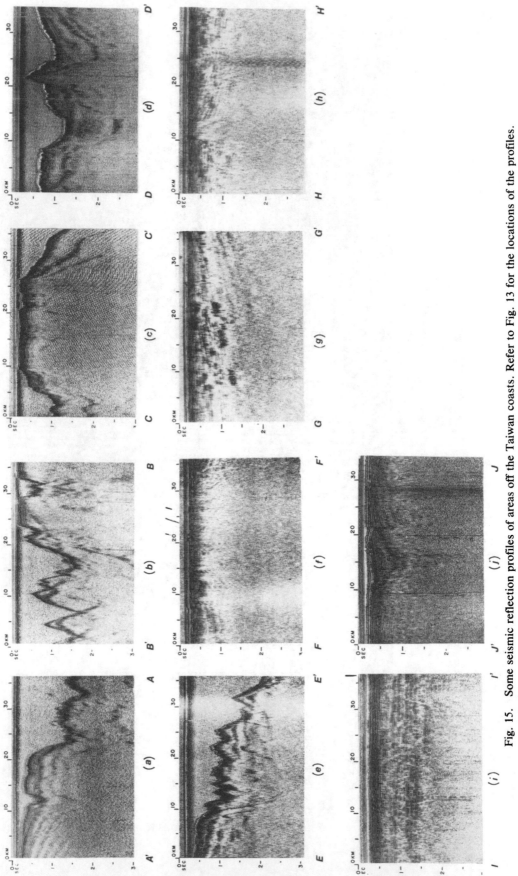

Fig. 15. Some seismic reflection profiles of areas off the Taiwan coasts. Refer to Fig. 13 for the locations of the profiles.

are caused by the north-northwestward movement of the Philippine Sea plate. (3) Well-stratified sediments overlie the folded basement (Fig. 15c). (4) Growth faults and igneous extrusions (Fig. 13d) can be found in many places off the coast of southern Taiwan and may be taken as an evidence for the subduction of the South China Sea part of the Eurasian plate. (5) Folded sediments, faults, and igneous intrusions occur on the edge of the continental shelf (Fig. 15e). (6) Folded sediments, faults, and igneous intrusions are found on each side of the Peikang–Penghu basement high (Fig. 15f). (7) A sedimentary sequence plunges northward from the Paikang–Penghu basement high to the area off the Taichung coast (Fig. 15g). (8) Folded and faulted sedimentary layers overlie the Miaoli swell (Fig. 13h). (9) Sediments in the Hsinchu Basin (Fig. 15i) correspond in attitude to the onshore Neogene and Quaternary sediments (C. S. Lee *et al.*, 1973). (10) A high-angle thrust (Fig. 15j) is found off the north coast of Taiwan and is identified as the continuation of the Hsinchuang thrust onshore.

R. S. Lu *et al.* (1977) compiled an insopach map from the reflection data of the offshore areas of Taiwan and Luzon (Fig. 16). Based on this map and the bathymetry of the map area, they support the general belief that the mélange in the Kenting area at the south end of Taiwan is of the same origin as the probably age-equivalent ophiolitic mélange in the Coastal Range, each being a product of tectonization in a trench setting.

F. Correlation with Geology

It is clear from the foregoing descriptions of the various geophysical aspects of Taiwan that, when there are sufficient data, they are generally in good accordance with a subduction-plus-collision model, based mainly upon surface geology, for the Cenozoic tectonics of this region. The better understood seismicity of the island is characterized by a distribution of epicenters in three zones of fundamental tectonic significance. Following the Ryukyu Arc, the northeastern zone enters northern Taiwan and connects at Hualien with the south-trending epicenters of the eastern zone, which extends as far as northern Luzon. These two zones combine to delineate the right-angle boundary of the Philippine Sea plate at its northwest corner. In crustal profiles, the boundary of the northeastern zone is a Benioff zone dipping beneath the Eurasian plate; the other boundary of the eastern zone dips beneath the Philippine Sea plate. Adding more complexity to the subduction system of the region is the postcollision conversion of the northernmost 150 km of the latter Benioff zone into a continent-arc suture, beneath which the only hypocenters are those of shallow shocks. The western zone, west of the suture, is another zone of shallow quakes that although fewer in number, are occasionally very strong. Tectonically, this zone occupies the foreland fold-

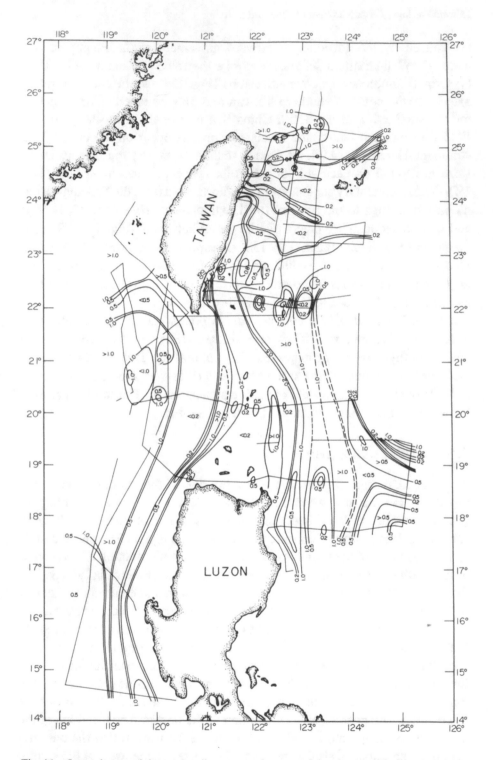

Fig. 16.   Isopach map of the total sedimentary column overlying the acoustic basement in the Taiwan–Luzon area. Isopachs are based on two-way reflection time in seconds. (After R. S. Lu *et al.*, 1977.)

and-thrust belt that is highly strained as a result of the continent-arc collision. Despite their incompleteness, the gravity data of Taiwan also reflect the crustal structure believed to underlie the island. Prominent gravity highs coincided with thinly sediment-covered areas of the cratonic foreland or with the island arc areas along the oceanic east coast; on the other hand, prominent lows are found in areas above thick sediment masses in areas of active subsidence or over deep mountain roots in isostatic uplift. The distribution of the numerous geothermal surface indications on Taiwan also show tectonic control. These occur in the areas either of arc magmatism or of foreland tectonization and regional metamorphism. In summary, although remaining to be refined with more data, the generally accepted collision model for the tectonic framework of Taiwan seems to fit the various geophysical phenomena understood to date.

IV. MARINE SEDIMENT DISTRIBUTION

A. Introduction

Taiwan is located at the junction of the Eurasian and Philippine Sea plates. The continental shelf in the Taiwan region is asymmetrical and the seafloor drops away abruptly to depths of more than 1000 m at a distance less than 10 km from the eastern shore of Taiwan. However in the western shelf, water depth is generally less than 60 m and the continental shelf extends continuously for 140 to 200 km to the mainland of China.

Investigations of the bottom sediments around Taiwan have been carried out by Niino and Emery (1961), J. C. Chen and Chen (1971), Chou and Wu (1971) and Boggs et al. (1979). Niino and Emery (1961) suggested that the coarse sediments on the Asian continental shelf represent relict littoral deposits left from Pleistocene time of glacially lowered sea level.

The ocean currents surrounding Taiwan consist of the northeast and southwest monsoon drift currents, the Kuroshio current, and the China coastal current (T. Y. Chu, 1971). These currents may have dominant influences on the sediment dispersal pattern.

B. Shelf Sediments

Shelf and slope sediments around Taiwan were collected by gravity corer, box corer and dredge at more than 170 stations (Fig. 17) during a cruise of RV *Chiu-Lien* in July 1973 (Boggs et al., 1979). The bottom sediment facies have been investigated by these authors (Fig. 18). Sand is the dominant surface sediment in the Taiwan Strait but muddy sediment covers

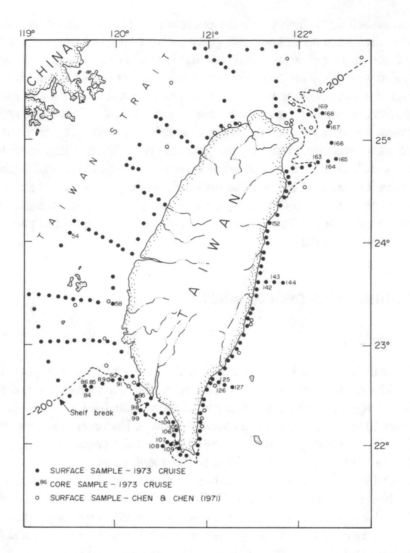

Fig. 17. Sample site map showing location of surface samples and cores collected in 1973 (Boggs *et al.*, 1979) and location of additional surface samples furnished by J. C. Chen and Chen (1971).

a segment of the strait up to 85 km side, that extends northward from the central part of the western Taiwan coast.

Muddy sediment predominates beyond the shelf break (200 m deep) and in the shallow trough that lies along the southwest coast of Taiwan. The large patch of muddy sediment that lies across the northern part of the Taiwan Strait is located on the north side of a prominent ridge 20–40 m high.

Calcareous shells occupy about 2 to 82 percent of the sandy fractions of the sediments in the Taiwan Strait. In general, the content of shell material

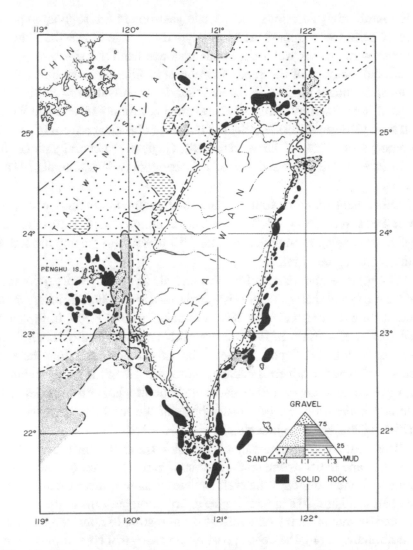

Fig. 18.   Distribution of sand, mud, gravel, and rock outcrops on the Taiwan continental shelf and upper slope. (After Boggs *et al.*, 1979.)

shows an inverse relationship to the mud content of the samples, that is, the greater the mud content the lower the content of calcareous shells. Shell material is abundant on the northern Taiwan shelf, in the southeast part of the Taiwan Strait, south of the Penghu Islands, and in the area off the southern tip of Taiwan.

Lewis (1975) identified shells and shell fragments belonging to foraminifera, ostracods, molluscs, sponges, diatoms, radiolarians, bryozoans, corals, and echinoderms as well as fish remains in the shelf and slope sediments sampled around Taiwan. Molluscs, principally pelecypods, scaphopods, benthonic gastropods and pelagic gastropods (pteropods) appear to increase seaward to a maximum just beyond the edge of the shelf and then decrease with greater water depth and distance from the shore.

Modal analyses of the sandy fraction of sediments show that quartz is the most abundant mineral in the shelf sands. It constitutes 10% to 70% of the total light fraction of the sands, averaging around 50% in the western shelf area (Taiwan strait). K feldspar ranges in abundance from 0% to 20%, averaging around 7% on the western shelf. Plagioclase generally ranges from 0% to 28% in the shelf sediments, it is commonly less abundant than K feldspar.

Other light minerals found in the shelf sediments include chert, muscovite and glauconite. Rock fragments are common constituents of all of the shelf sand ranging in abundance from 11% to 85%, averaging around 40% in the western shelf area.

The relative abundance of heavy minerals (specific gravity greater than 2.95) in the sandy shelf sediments is summarized in Fig. 19. The heavy mineral content ranges from 0.05% to 20%, averaging around 8.5%, in the eastern shelf sands. However, in the western shelf sand the heavy minerals only average about 0.6%. Figure 19 shows that on the northern shelf there is an area in which heavy minerals make up more than 4% of the sandy sediments.

Opaque minerals including magnetite, ilmenite, hematite, and leucoxene generally make up one-third to two-thirds of the total heavy mineral composition of the sandy sediments.

The most common nonopaque minerals in the shelf sand are hornblende, clinopyroxene hypersthene epidote, zircon, garnet, topaz, olivine and tourmaline. Other minor heavy minerals include monazite, sphene, rutile, biotite, and apatite. Figure 20 summarizes the distribution patterns of the most common and abundant heavy minerals. It is interesting to note that pyroxene is most abundant along the central part of the eastern shelf and hornblende is a dominant constituent of the northern shelf sands. Zircon is abundant around Penghu Islands and the southern shelf but is much less abundant on the northern shelf and virtually absent from sands of the central part of the eastern shelf. Topaz occurs mainly on the western and southern parts of the shelf and tourmaline is relatively abundant only on the western part.

The compositions of the marine sediments may be influenced by the following factors: (1) type of source rock; (2) weathering processes in the source areas; (3) transport agents; (4) extent of mixing in the marine environment of materials from different sources.

Rock fragments in the Taiwan shelf sediments indicate multiple source-rock types. On the eastern shelf, the rock fragments are mainly volcanic and metamorphic (andesite, schist, phyllite, and slate) with lesser sedimen-

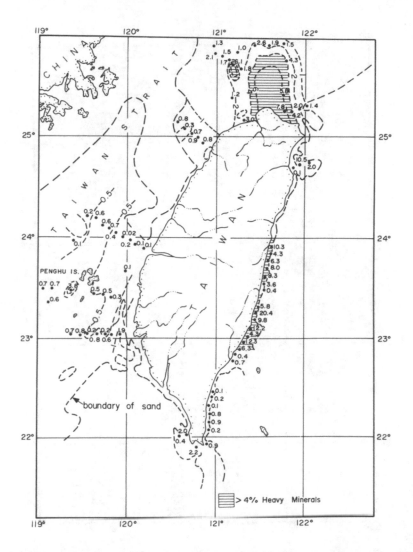

Fig. 19.   Heavy mineral abundance in Taiwan shelf sand. Contours in weight percent of total heavy minerals in sand fraction of samples; contour interval variable. (After Boggs *et al.*, 1979.)

tary rock fragments (shell and sandstone), indicating the dominance of Miocene volcanic rocks in the eastern Coastal Range and metamorphic rocks in the Central Range.

Rock fragments in sediments of the northern and western parts of the shelf represent complex mixtures of sedimentary, metamorphic, and volcanic rocks, indicating a diverse suite of source rocks for the sediments of these areas.

The presence of abundant pyroxenes in the sandy sediments of the eastern shelf show a predominant volcanic derivation (probably from the Coastal

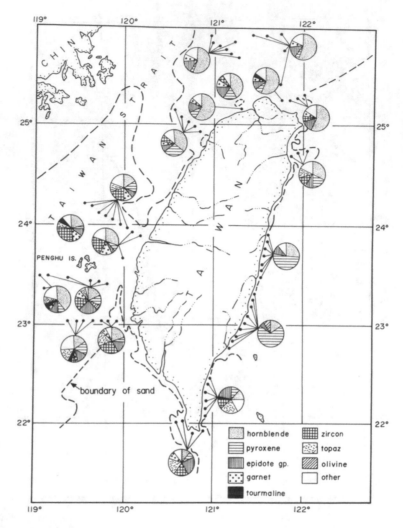

Fig. 20.   Distribution of principal nonopaque heavy minerals in sandy shelf sediment. (After Boggs *et al.*, 1979.)

Range). On the other hand, the Tatun–Chilung volcano group of northern Taiwan are the main sources for the hornblendes commonly found in the northern shelf sandy sediments.

The variety of heavy mineral species is greater on the western and southern shelves. According to Chou (1973), the source rocks for the Miocene clastic sequence in western Taiwan are acid igneous rocks including pegmatites, basic igneous rocks, various low-grade metamorphic rocks, and sedimentary rocks in the now-buried Taiwan strait and the China mainland; and the source rocks for the Pliocene and Pleistocene clastic sequences in western Taiwan are the Paleogene low-grade metamorphic rocks and, the Miocene sedimentary rocks in Taiwan. Correlation of these rock types with the shelf sediment indicates that most of the shelf sediments other than those surrounding the Penghu Islands were derived from Taiwan.

Shuford (1977) calculated the sedimentation rates on the eastern slope ranging from 2.8 to 4.5 cm/1000 yr for terrigenous mud. These low sedimentation rates are in contradiction with the high stream discharge and the extremely high rates of denudation in eastern Taiwan (Li, 1976) and indicate that much of the fine sediment discharged onto the shelf by eastern Taiwan rivers must have been carried away by the Kuroshio current beyond the inner slope.

Mineralogical data indicate that the near-shore sediments from the Taiwan Strait have a higher percentage of terrigenous sediments than those of the South China Sea (from Hainan to the Tung-sha Islands) (J. C. Chen and Chen, 1971). This is probably due to the high erosion rate on Taiwan (Li, 1976). In addition, the near-shore sediments are characterized by a lower number of total organisms, higher quartz/K feldspar ratio, and higher chlorite and illite contents as compared with shelf sediments of the South China Sea. The lower number of organisms resulted in the lower CaO content in the near-shore sediments. The relative enrichment of chlorite in the near-shore sediments from the Taiwan Strait is due probably to the weathering in temperate environments that resulted in the preservation or formation of chlorite.

The shelf sediments of the South China Sea (from Hainan Island to Tung-sha Island) are characterized by a high number of total organisms, low quartz/K feldspar ratio, high CaO and Sr contents, and low MgO, ΣFe_2O_3, K_2O, Rb and Li contents. The low quartz/K feldspar ratio suggests that the South China Sea shelf sediments were derived mostly from southern China (Kwangtung), where K feldspar-rich granite and granodiorite of Mesozoic age are abundant.

The decrease of quartz/calcite ratio with distance off the coast of the China mainland indicates that the depositional environment of the shelf sediments becomes more biogenic toward the shelf edge. It should be mentioned that Wang (1960) and Emery et al., (1969) concluded that sediments of the continental shelf of the East China Sea exhibit a simple pattern of silt and clay on the inner half and sand on the outer half of the shelf. This distribution pattern may be interpreted as due to the sea level change during and after the Pleistocene glacial ages.

Niino and Emery (1961) considered the shelf sediments of the East China Sea to be relict sediments that have remained unburied on the continental shelf since Pleistocene times of glacially lowered sea levels.

C. Pelagic Sediments

Revelle *et al.* (1955) divide the pelagic sediments of the Pacific into four principal types: clay, calcareous ooze, radiolarian ooze, and diatom ooze. Their distribution map shows that the western Philippine Basin is largely covered with clay. Drilling results of Leg 31 of Deep Sea Drilling Project indicate that at site 293 (20°21.25′N, 124°05.65′E) the stratigraphic column consists of 244 m of late Pliocene–Pleistocene sand-silt turbidites, 156 m of Pliocene distal mudstone turbidites, and 29 m of brown mudstone with re-worked late middle-Eocene nannofossils overlying 46.5 m (or more) of Miocene basaltic breccia (Ingle, 1973).

The pelagic sediments cored at seven stations in the western part of the Philippine Sea in the vicinity of the Central Basin Fault consist dominantly of reddish brown to yellowish brown clays. Mean grain size ranges from 10.38 ϕ (0.79 μm) to 7.05 ϕ (7.6 μm). Average mean grain size ranges from 9.52 ϕ (1.4 μm) to 7.76 ϕ (4.6 μm). These size values are slightly higher than those reported by Horn *et al.* (1974) from the north Pacific. The latter authors report a range in mean grain size of red clays in 49 cores from 0.50 μm to 2.45 μm with an average mean size of 0.97 μm. The larger mean grain size of the Philippine Sea pelagic sediments is clearly the result of local contributions of volcanic ash.

Quartz, feldspar, illite, chlorite + kaolin, and montmorillonite are the dominant minerals in the Philippine Sea pelagic sediments as identified by X-ray diffraction. The coarse fraction (> 62 μm) consists mainly of glass shards, plagioclase, amphibole, and volcanic rock fragments with rare terrigenous quartz grains. In some samples there are layers composed almost entirely of glass shards.

J. C. Chen and Boggs (1975) reported that in the pelagic sediments near the Central Basin Fault the clay minerals have the following composition: illite 60%, chlorite and kaolinite 20%, and montmorillonite 20%.

The source of the silicate minerals in pelagic sediments of the Pacific has been discussed by several authors. The quartz in the North Pacific pelagic sediments was considered to be largely of aeolian origin by Arrhenius (1963) and Bonatti (1963). Peterson and Goldberg (1962) concluded that the feldspar in South Pacific pelagic sediments is mainly of volcanic derivation.

Griffin and Goldberg (1963) suggested that illite and "stripped" chlorite in the North Pacific are primarily brought to the ocean environment by aeolian paths. However, they indicated that the predominance of montmorillonite in surface sediments of the South Pacific arises from the alteration of volcanic materials. Based on the relative abundance of quartz, feldspar, clays, zeolites, and montmorillonite, Primm *et al.* (1971) concluded that some brown clays in the Northwest Pacific are almost entirely of aeolian origin, other brown clays have a mixed aeolian and volcanic contributions, and still others are almost entirely of volcanic origin.

The range and average abundance of the elements in 31 pelagic sediment samples cored from the western Philippine Sea (22°00'N, 124°30'E) are : SiO_2, 48.09–63.85%, avg 56.44%; Al_2O_3, 15.80–18.98%, avg 17.93%; $\sum Fe_2O_3$, 5.6–8.28%, avg 7.66%; MgO, 2.00–2.94%, avg 2.62%; CaO, 1.10–5.06%, avg 3.28%; Na_2O, 1.36–2.11%, avg 1.56%; K_2O, 2.66–3.65%, avg 3.19%; TiO_2, 0.21–0.92%, avg 0.67%; MnO, 0.10–2.68%, avg 0.34%; Cd, 0.039–0.360 ppm, avg 0.107 ppm; Co, 9–61 ppm, avg 32 ppm; Cr, 35–100 ppm, avg 70 ppm; Cu, 55–125 ppm, avg 81 ppm; Li, 53–72 ppm, avg 64 ppm; Ni, 14–118 ppm, avg 40 ppm; Pb, 32–76 ppm, avg 49 ppm; Rb, 96–145 ppm, avg 129 ppm; Sr, 18–195 ppm, avg 98 ppm; Zn, 91–113 ppm, avg 101 ppm.

When compared with shelf sediments of areas adjacent to Taiwan (J. C. Chen and Boggs 1977), the above pelagic sediments from the western Philippine Basin are relatively higher in Fe, Mn, Ni, Cu, and Co, which may be due to the adsorption of heavy metals by clays and the occurrence of ferromagnesian minerals, such as amphibole, in these pelagic sediments.

It is known that calcium carbonate decreases sharply in abundance in pelagic sediments below the carbonate compensation depth (4000–5000 m), owing to the increased solubility of $CaCO_3$ with increasing hydrostatic pressure. The average $CaCO_3$ content (3.28%) in the western Philippine Basin pelagic sediments is much lower than that of shelf sediments (21.45%) from the Taiwan Strait and South China Sea reported by J. C. Chen and Chen (1971). The calcium in the pelagic sediments is probably contained mainly in feldspar because biogenic calcium carbonate is lacking.

The average Sr content of western Philippine Sea pelagic sediments (98 ppm) is lower than that of shelf sediments (447 ppm) reported by J. C. Chen and Chen (1971). The positive correlation between strontium and calcium and the lack of biogenic calcium carbonate in the pelagic sediments indicate that strontium probably replaces calcium in plagioclase and other calcium-bearing minerals such as montmorillonite.

The pelagic sediments from western Philippine Basin have an average silica content of 56.44%, which is similar to the average of 12 argillaceous pelagic sediments from all major oceans (55.34%) reported by El Wakeel and Riley (1961). The average aluminum content (17.93%) in the western Philippine Basin pelagic sediments is also similar to that reported by El Wakeel and Riley.

Microferromanganese nodules have been found in some cores indicating locally high positive Eh present during deposition of the sediments. It should be mentioned that the precipitation of manganese nodules is favored by both local oxidizing Eh and catalytic adsorption of manganese. The reaction 2 $Mn^{+2} + 2H_2O + O_2 \rightarrow 2MnO_2 + 4H^+$ will continue provided that sufficient O_2, H_2O and Mn^{+2} flow forward the deposit (nodules) and H^+ flows away from it (Crerar and Barnes, 1974).

Based upon the analyses of the cores from western Philippine Basin, the pelagic sediments appear to have both a continental source and a local volcanic source. The abundance of illite and presence of chlorite and kaolin indicate an aeolian and/or current contribution of the sediments. However, the occurrence of glass shards, volcanic rock fragment and amphibole in the coarse fraction of the pelagic sediments indicate there is substantial evidence of volcanic derivation of some of the sediments. The presence of montmorillonite in all of the samples is also indicative of volcanic derivation.

REFERENCES

Allen, C. R., 1962, Circum-Pacific faulting in the Philippine–Taiwan region, *J. Geophys. Res.*, v. 67, p. 4795–4812.

Arrhenius, G., 1963, Pelagic sediments in: *The Sea* v. 3, Hill, M. M., ed., New York: Wiley-Interscience, p. 657–727.

Bally, A. W., and Snelson, S., 1980, Realms of subsidence, *Can. Soc. Pet. Geol. Mem.* v. 6, p. 9–94.

Biq, Chingchang, 1956, The tectonic framework and oil possibilities of Taiwan, *National Taiwan University 10th Anniversary Commemoration Volume*, Taipei: Taiwan University, p. 95–105.

Biq, Chingchang, 1965, The East Taiwan Rift, *Pet. Geol. Taiwan*, no. 4, p. 93–106.

Biq, Chingchang, 1966, Tectonic styles and structural levels in Taiwan, *Proc. Geol. Soc. China*, no. 9, p. 3–9.

Biq, Chingchang, 1969, Role of gravitational gliding in Taiwan tectogenesis, *Bull. Geol. Surv. Taiwan*, no. 20, p. 1–39.

Biq, Chingchang, 1971a, A fossil subduction zone in Taiwan, *Proc. Geol. Soc. China*, no. 14, p. 146–154.

Biq, Chingchang, 1971b, Comparison of mélange tectonics in Taiwan and in some other mountain belts, *Pet. Geol. Taiwan*, no. 9, p. 79–106.

Biq, Chingchang, 1972a, Dual-trench structure in the Taiwan–Luzon region, *Proc. Geol. Soc. China*, no. 15, p. 65–75.

Biq, Chingchang, 1972b, Western Taiwan thrusts, active or inactive? *Acta Geol. Taiwan.*, no. 15, p. 63–75.

Biq, Chingchang, 1972c, Transurrent buckling, transform faulting and transpression: their relevance in eastern Taiwan kinematics, *Pet. Geol. Taiwan*, no. 10, p. 1–10.

Biq, Chingchang, 1976a, Western Taiwan earthquake faults as miniature transform faults, *Bull. Geol. Surv. Taiwan*, no. 25, p. 1–7.

Biq, Chingchang, 1976b, A tale of two orogens, *Bull. Geol. Surv. Taiwan*, no. 25, p. 149–166.

Biq, Chingchang, 1977, The Kenting mélange and the Manila Trench, *Proc. Geol. Soc. China*, no. 20, p. 119–122.

Biq, Chingchang, 1978, Taiwan vis-à-vis New Guinea: a comparison of their continent-arc collisions, *Acta Oceanogr. Taiwan.*, no. 8, p. 22–42.

Biq, Chingchang, 1979, Sinokorea–Cathaysia in its Mesozoic–Cenozoic convergence-zone setting, *Mem. Geol. Soc. China*, no. 3, p. 73–79.

Biq, Chingchang, 1981, Collision, Taiwan-style, *Mem. Geol. Soc. China*, no. 4, p. 91–102.

Biq, Chingchang, 1982, Another type of subduction zones: examples from China, *Sp. Pub. Central Geol. Surv. (Taipei)*, no. 1, p. 147–155.

Boggs, S. Jr., Wang, W. C., Lewis, F. S., and Chen, J. C., 1979, Sediment properties and water characteristics of the Taiwan shelf and slope, *Acta Oceanogr. Taiwan.*, no. 10, p. 10–49.

Bonatti, E., 1963, Zeolites in Pacific pelagic sediments, *Trans. N. Y. Acad. Sci.*, v. 25, p. 938–948.

Bowin, C., Lu, R. S., Lee, C. S., and Schouten, H., 1978, Plate convergence and accretion in Taiwan–Luzon region, *Bull. Am. Assoc. Pet. Geol.*, v. 62, p. 1645–1672.

Chai, B. H. T., 1972, Structure and tectonic evolution of Taiwan, *Am. J. Sci.*, v. 272, p. 389–422.

Chang, L. S., 1967, A biostratigraphic study of the Tertiary in the Coastal Range, eastern Taiwan, based on small Foraminifera (I. southern part), *Proc. Geol. Soc. China*, no. 10, p. 64–76.

Chang, L. S., 1968, A biostratigraphic study of the Tertiary in the Coastal Range, eastern Taiwan, based on small Foraminifera (II. northern part), *Proc. Geol. Soc. China*, no. 11, p. 19–33.

Chang, L. S., 1969, A biostratigraphic study of the Tertiary in the Coastal Range, eastern Taiwan, based on small Foraminifera (III. middle part), *Proc. Geol. Soc. China*, no. 12, p. 89–101.

Chase, T. E., Menard, H. W., and Mammerickx, J., 1971, *Topography of the North Pacific*, Geological Data Center, Scripps Institution of Oceanography, La Jolla, California.

Chen, C. H., 1979, Geology of the east–west cross-island highway in central Taiwan, *Mem. Geol. Soc. China*, no. 3, p. 219–236.

Chen, C. Y., 1974, Verification of the north-northeastward movement of the Coastal Range, eastern Taiwan, by retriangulation, *Bull. Geol. Surv. Taiwan*, no. 24, p. 119–123.

Chen, J. C., and Boggs, S. Jr., 1975, Mineralogical, chemical and textural characteristics of pelagic sediments form the Philippine Sea, *Acta Oceanogr. Taiwan.*, no. 5, p. 19–36.

Chen, J. C., and Boggs, S. Jr., 1977, Comparative studies of shelf sediments from the eastern and western offshore areas, *Pet. Geol. Taiwan*, no. 14, p. 249–262.

Chen, J. C., and Chen, C., 1971, Mineraology, geochemistry and paleontology of shelf sediments of the South China Sea and Taiwan Strait, *Acta Oceanogr. Taiwan.*, no. 1, p. 33–54.

Chen, J. S., Chou, J. N., Lu, Y. C., and Chou, Y. S., 1974, Seismic survey conducted in eastern Taiwan, *Pet. Geol. Taiwan*, no. 11, p. 147–163.

Chi, W. R., Namson, J., and Suppe, J., 1981, Stratigraphic record of plate interactions in the Coastal Range of eastern Taiwan, *Mem. Geol. Soc. China*, no. 4, p. 155–194.

Chou, J. T., 1973, Sedimentology and paleogeography of the Upper Cenozoic system of western Taiwan, *Proc. Geol. Soc. China*, no. 16, p. 111–143.

Chou, J. T., and Wu, F. T., 1971, Sediments of the Taiwan Basin (part I), *Pet. Geol. Taiwan*, v. 9, p. 30–41.

Chu, H. T., 1982, Geologic structure of the Yüshan block; a preliminary report, *Abs. 1982 Ann. Meeting Geol. Soc. China*, p. 11.

Chu, T. T., 1971, Environmental study of surrounding waters of Taiwan, *Acta Oceanogra. Taiwan.*, no. 1, p. 15–31.

Crerar, S. A., and Barnes, H. L., 1974, Deposition of deep-sea manganese nodules, *Geochim. Cosmochim. Acta*, v. 38, p. 279–300.

El Wakeel, S. K., and Riley, J. P., 1961, Chemical and mineralogical studies of deep-sea sediments, *Geochim. Cosmochim. Acta*, v. 25, p. 110–146.

Emery, K. O., Hayashi, Y., Hilde, T. W. C., Kobyashi, K., Koo, J. H., Meng, C. Y., Niino, H., Osterhagen, J. H., Reynolds, L. M., Wagemen, J. M., Wang, C. S., and Yang, S. J., 1969, Geological structural and some water characteristics of the East China Sea and the Yellow Sea, *C. C. O. P. Tech Bull.*, v. 2, p. 3–43.

Ernst, W. G., Liou, J. G., and Moore, D. E., 1981, Multiple metamorphic events recorded in the Tailuko amphibolites and associated rocks of the Suao–Nanao area, Taiwan, *Mem. Geol. Soc. China*, no. 4, p. 391–441.

Fitch, T. J., 1972, Plate convergence, transcurrent faults and internal deformation adjacent to southeast Asia and western Pacific, *J. Geophys. Res.*, v. 77, p. 4433–4460.

Griffin, J. J., and Goldberg, E. D., 1963, Clay mineral distributions in the Pacific Ocean, in: *The Sea*, v. 3, Hill, M. N., ed., New York: Wiley-Interscience, p. 728–741.

Ho, C. S., 1969, Geological significance of potassium-argon ages of the Chimei igneous complex in eastern Taiwan, *Bull. Geol. Surv. Taiwan*, no. 20, p. 63–74.

Ho, C. S., 1975, *An Introduction to the Geology of Taiwan: Explanatory Text of the Geologic Map of Taiwan*, Taipei: Ministry of Economic Affairs.

Horn, D. R., Delach, M. N., and Horn, B. M., 1974, Physical properties of sedimentary provinces, North Pacific and North Atlantic Oceans, in: *Deep Sea Sediments, Physical and Mechanical Properties, Marine Science*, v. 2, Inderbitzen, A. L., ed., New York: Plenum Press, p. 417–441.

Hsiao, P. T., and Hu, C. C., 1978, Geomagnetic study of the Changhua Plain, Taiwan. *Pet. Geol. Taiwan*, no. 15, p. 241–253.

Hsieh, S. H., 1970, Geology and gravity anomalies of the Pingtung Plain, Taiwan. *Proc. Geol. Soc. China*, no. 13, p. 76–89.

Hsieh, S. H., and Hu, C. C., 1972, Gravimetric and magnetic studies of Taiwan, *Pet. Geol. Taiwan*, no. 10, p. 283–321.

Hsu, M. T., and Yang, C. Y., 1969, P-wave velocity in the surface layer of the earth's crust in Taiwan, *Bull. Taiwan Weather Bur.*, v. 15, p. 22–32.

Hsu, T. L., 1962, Recent faulting in the Longitudinal Valley of eastern Taiwan, *Mem. Geol. Soc. China*, no. 1, p. 95–102.

Hsu, T. L., 1976, The Lichi mélange in the Coastal Range framework, *Bull. Geol. Surv. Taiwan*, no. 25, p. 87–96.

Hu, C. C., and Lu, R. S., 1979, Downward continuation of magnetic field and the magnetic anomalies of offshore Taiwan, *Acta Oceanogr. Taiwan.*, no. 1, p. 1–8.

Huang, T. C., Ting, J. S., and Müller, C., 1983, A note on Pliocene microfossils from the Kenting mélange, *Proc. Geol. Soc. China*, no. 26, p. 57–66.

Ingle, J. C. Jr., 1973, Western Pacific floor, *Geotimes*, v. 18, p. 22–25.

Jahn, B. M., and Liou, J. G., 1977, Age and geochemical constraints of glaucophane schists of Taiwan, *Mem. Geol. Soc. China*, no. 2, p. 129–140.

Jahn, B. M., Liou, J. G., and Nagasawara, H., 1981, High-pressure metamorphic rocks of Taiwan: REE geochemistry, Rb-Sr ages and tectonic implications, *Mem. Geol. Soc. China*, no. 4, p. 479–520.

Juan, V. C., Chow, T. J., and Lo, H. J., 1972, K-Ar ages of the metamorphic rocks of Taiwan, *Acta Geol. Taiwan.*, no. 15, p. 113–118.

Konishi, K., Omura, A., and Kimura, T., 1968 [234]U-[230]Th dating of some late Quaternary coraline limestones from southern Taiwan, *Geology and Paleontology of Southeast Asia*, Tokyo: University of Tokyo Press, v. 5, p. 211–224.

Ku, C. C., 1963, Photogeologic study of terraces in northwestern Taiwan, *Proc. Geol. Soc. China*, no. 6, p. 51–60.

Lee, C. S., and Lu, R. S., 1976, Significance of the southewwestern section of Ryukyu Inner Ridge in the exploration of geothermal resources in Ilan area, *Min. Tech. Dig.*, v. 14, p. 114–120.

Lee, C. S., Shor, G. G., Jr., Bibee, L. D., Lu, R. S., and Hilde, T. W. C., 1980, Okinawa Trough: origin of a back-arc basin, *Mar. Geol.*, v. 35, p. 219–241.

Lee, C. S., Shyu, C. T., and Leu, F. J., 1973, Structure of eastern Taiwan Strait, *Acta Oceanogr. Taiwan.*, no. 3, p. 117–140.

Lee, Y. C., and Tsai, Y. B., 1978, Crustal structure of Taiwan from P-wave arrival times, *Proc. Geol. Soc. China*, no. 21, p. 111–127.

Lewis, F. S., 1975, *Texture, Clay Mineralogy and Biogenic Composition of Taiwan Shelf and Slope Sediments*, Masters thesis, University of Oregon, Eugene, Oregon, 166 p.

Li, Y. H., 1976, Denudation of Taiwan island since the Pliocene Epoch, *Geology*, v. 4, p. 105–107.

Lin, M. T., and Tsai, Y. B., 1981, Seismotectonics in Taiwan–Luzon area, *Bull. Inst. Earth Sci. Academia Sinica*, v. 1, p. 51–82.

Liou, J. G., 1981a, Petrology of metamorphosed oceanic rocks in the Central Range of Taiwan, *Mem. Geol. Soc. China*, no. 4, p. 291–341.

Liou, J. G., 1981b, Recent high CO_2 activity and Cenozoic progressive metamorphism in Taiwan, *Mem. Geol. Soc. China*, no. 4, p. 551–581.

Liu, C. C., and Yen, T. P., 1975, Bouguer anomaly, surface elevation and crustal thickness in Taiwan, *Pet. Geol. Taiwan*, no. 12, p. 97–108.

Liu, T. K., 1982, Tectonic implication of fission track ages from the Central Range, Taiwan, *Proc. Geol. Soc. China*, no. 25, p. 22–37.

Lu, C. P., 1975, *Tectonics, Crustal and Upper Mantle Structure of Taiwan*, Ph.D. Dissertation, State University of New York at Binghamton, Binghamton, New York.

Lu, C. P., and Wu, F. T., 1974, Two dimensional interpretation of a gravity profiles across Taiwan, *Bull. Geol. Surv. Taiwan*, no. 24, p. 125–132.

Lu, R. S., Lee, C. S., and Kuo, S. T., 1977, An isopach map for the offshore area of Taiwan and Luzon, *Acta Oceanogr. Taiwan.*, no. 7, p. 1–9.

Lu, R. S., Pan, J. J., and Lee, T. C., 1981, Heat flow on the southwestern Okinawa Trough, *Earth Planet. Sci. Lett.*, v. 55, p. 299–310.

Ludwig, W. J., 1970, The Manila Trench and West Luzon Trough, III. Seismic refraction measurements, *Deep Sea Res*, v. 17, p. 553–571.

Ludwig, W. J., Hayes, D. E., and Ewing, J. I., 1967, The Manila Trench and West Luzon Trough, I. Bathymetry and sediment distribution, *Deep Sea Res.*, v. 14, p. 533–544.

Meng, C. Y., and Chiang, S. C., 1965, Subsurface data from wildcat S5-1, Shihshan, Taitung, *Pet. Geol. Taiwan*, no. 4, p. 283–286.

Mitchell, A. H. G., 1981, The Grampian orogeny in Scotland and Ireland: almost an ancient Taiwan, *Proc. Geol. Soc. China*, no. 24, p. 113–129.

Murphy, R. W., 1973, The Manila Trench–west Taiwan foldbelt, a flipped subduction zone, *Bull. Geol. Soc. Malaysia*, v. 6, p. 27–42.

Niino, H., and Emery, K. O., 1961, Sediments of shallow portions of East China Sea and South China Sea, *Geol. Soc. Am. Bull.*, v. 72, p. 731–762.

Page, B. M., and Suppe, J., 1981, The Pliocene Lichi mélange of Taiwan: its plate-tectonic and olistostromal origin, *Am. J. Sci.*, v. 281, p. 193–227.

Pan, Y.S., and Hsiao, P. T., 1971, Seismic exploration in Taiwan, *Pet. Geol. Taiwan*, no. 9, p. 145–165.

Peng, T. H., Li, Y. H., and Wu, F. T., 1977, Tectonic uplift rate of the Taiwan island since the early Holocene, *Mem. Geol. Soc. China*, no. 2, p. 57–70.

Peterson, M. N. A., and Goldberg, E. D., 1962, Feldspar distributions in south Pacific pelagic sediments, *J. Geophys. Res.*, v. 67, p. 3477–3497.

Primm, A. C., Garrison, R. E., and Boyce, R. E., 1971, Sedimentary synthesis: Lithology, chemistry and physical properties of sediments in northwestern Pacific Ocean, *Initial Reports of the Deep Sea Drilling Project*, v. 6, p. 1131–1252.

Revelle, R., Bramlette, M., Arrhenius, G., and Goldberg, E. D., 1955, Pelagic sediments of the Pacific, *Geol. Soc. Am. Spec. Pap.*, v. 62, p. 221–236.

Sato, K., Mong, C. T., Kurihara, S., Kamota, S., Obayashi, H., Inoue, E., and Hsiao, P. T., 1970, Reports on the seismic refraction survey on land on the western part of Taiwan, Republic of China, *Pet. Geol. Taiwan*, no. 7, p. 281–293.

Seno, T., 1977, The instantaneous rotation vector of the Philippine Sea plate relative to the Eurasian plate, *Tectonophysics*, v. 42, p. 209–226.

Shuford, M. E., 1977, *Biostratigraphy and Sedimentation Rates of the Taiwan Continental Slope*, Master thesis, University of Oregon, Eugene, Oregon, 126 p.

Sun, S. C., 1982, The Tertiary basins of offshore Taiwan, *Proceedings of the 2nd ASCOFE Conference*, Manila, p. 125–135.

Teng, L. S., 1979, Petrographic study of the Neogene sandstones of the Coastal Range, eastern Taiwan (I. northern part), *Acta Geol. Taiwan.*, no. 20, p. 129–155.

Teng, L. S., 1980, Lithology and provenance of the Fanshuliao Formation, northern Coastal Range, eastern Taiwan, *Proc. Geol. Soc. China*, no. 23, p. 118–129.

Teng, L. S., 1981, On the origin and tectonic significance of the Lichi Formation, Coastal Range, eastern Taiwan, *Ti-Chih*, v. 2, p. 51–61.

Tsai, Y. B., Feng, C. C., Chiu, J. M., and Liaw, H. B., 1975, Correlation between microearthquakes and geologic faults in Hsintien–Ilan, *Pet. Geol. Taiwan*, no. 12, p. 149–167.

Tsai, Y. B., Hsiung, Y. M., Liaw, H. B., Lueng, H. P., Yao, T. H., Yeh, Y. H., and Yeh, Y. T., 1974, A seismic refraction study of eastern Taiwan, *Pet. Geol. Taiwan*, no. 11, p. 165–182.

Tsai, Y. B., and Liu, H. L., 1977, Spatial correlation between hot springs and microearthquakes in Taiwan, *Pet. Geol. Taiwan*, no. 14, p. 263–277.

Tsai, Y. B., Teng, T. L., Chiu, J. M., and Liu, H. L., 1977, Tectonic implications of the seismicity in the Taiwan region, *Mem. Geol. Soc. China*, no. 2, p. 13–41.

Tsan, S. F., 1974, Stratigraphy and structure of the Hengchun Peninsula, with special reference to a Miocene olistostrome, *Bull. Geol. Surv. Taiwan*, no. 24, p. 99–108.

Tsan, S. F., 1977, Remarks on the Suao section of the Central Range of Taiwan, *Mem. Geol. Soc. China*, no. 2, p. 141–146.

Tsan, S. F., and Keng, W. P., 1962, The strike-slip faulting and the concurrent or subsequent folding in the Alishan area, *Proc. Geol. Soc. China*, no. 5, p. 119–126.

Wageman, J. M., Hilde, T. W. C., and Emery, K. O., 1970, Structure framework of the East China Sea and Yellow Sea, *Bull. Am. Assoc. Pet. Geol.*, v. 54, p. 1641–1643.

Wang, C. S., 1960, Sand fraction study of the shelf sediments off the China Coast, *Proc. Geol. Soc. China*, no. 4, p. 33–49.

Wang, C. S., 1976, The Lichi Formation of the Coastal Range and arc-continental collision in eastern Taiwan, *Bull. Geol. Surv. Taiwan*, no. 25, p. 73–86.

Watanabe, T., Langseth, M. G., and Anderson, R. N., 1977, Heat flow in the back-arc basin of the western Pacific, in: *Island Arcs, Deep Sea Trench and Back-Arc Basin*, Talwani, M., and Pittman, W. C., III, eds., Washington, D.C.: American Geophysical Union, p. 137–161.

Wu, F. T., 1970, Focal mechanisms and tectonics of Taiwan, *Bull. Seismol. Socl. Am.*, v. 60, p. 2045–2056.

Wu, F. T., 1972, The Philippine Sea plate: a sinking towel? *Tectonophysics*, v. 14, p. 81–86.

Wu, F. T., 1978, Recent tectonics of Taiwan, in: *Geodynamics of the Western Pacific, Phys. Earth Suppl. Issue*, Uyeda, S., Murphy, R. W., and Kobayashi, K., eds., p. 265–299.

Wu, F. T., and Lu, C. P., 1976, Recent tectonics of Taiwan, *Bull. Geol. Surv. Taiwan*, no. 25, p. 97–111 (in Chinese).

Yeh, Y. H., and Tsai, Y. B., 1981, Crustal structure of Taiwan from inversion of P-wave arrival times, *Bull. Inst. Earth Sci. Acadamia Sinica*, v. 1, p. 83–102.

Yen, T. P., 1955, Thermal springs in Taiwan, *Q. J. Bank of Taiwan*, v. 7, p. 129–147 (in Chinese).

Yen, T. P., and Rosenblum, 1964, Potassium-argon ages of micas from the Tananao Schist terrane of Taiwan, a preliminary report, *Proc. Geol. Soc. China*, no. 7, p. 80–81.

Yen, T. P., Sheng, C. C., and Keng, W. P., 1951, The discovery of fusiline limestone in the metamorphic complex of Taiwan, *Bull. Geol. Surv. Taiwan*, no. 3, p. 23–26.

Yen, T. P., Sheng, C. C., Keng, W. P., and Yang, Y. T., 1956, Some problems on the Mesozoic formation of Taiwan, *Bull. Geol. Surv. Taiwan*, no. 8, p. 1–14.

York, J. E., 1976, Quaternary faulting in eastern Taiwan, *Bull. Geol. Surv. Taiwan*, no. 25, p. 63–72.

Yu, S. B., and Tsai, Y. B., 1979a, Geomagnetic anomalies of the Ilan Plain, Taiwan, *Pet. Geol. Taiwan*, no. 16, p. 19–27.

Yu, S. B., and Tsai, Y. B., 1979b, Geomagnetic investigations in the Pingtung Plain, Taiwan, *Bull. Inst. Earth Sci. Academia Sinica*, v. 1, p. 189–208.

Chapter 12

NEW GUINEA AND THE WESTERN MELANESIAN ARCS

John Milsom

Department of Geological Sciences
University College
University of London
London, WC1E 6BT England

I. INTRODUCTION

The area described in this chapter amounts to about 3,000,000 sq. km, divided almost equally between land and sea and dominated by the large island of New Guinea (Fig. 1). The Indonesian province of Irian Jaya (Irian) makes up the western half of the island, but most of the remainder of the area falls within the limits of the independent state of Papua New Guinea. Prior to independence in 1974, Papua New Guinea was administered as a single unit by Australia, although legally divided into the colony of Papua in the south and east and the Mandated Territory of New Guinea, which included a number of large offshore islands such as New Britain, New Ireland, Manus, and Bougainville, in the north. Bougainville, although politically part of Papua New Guinea, is geologically a part of the Solomon Islands chain and will not be considered directly in this chapter. Also, the south-

NOTE: In references to western New Guinea the current Indonesian geographical names are used. However, since much of the available literature uses the former Dutch names, these are given in brackets after the first use of the Indonesian name. The term Papua New Guinea is used throughout to refer to the eastern part of the island and its politically associated offshore islands, even where reference is being made to work done prior to the establishment of the independent state.

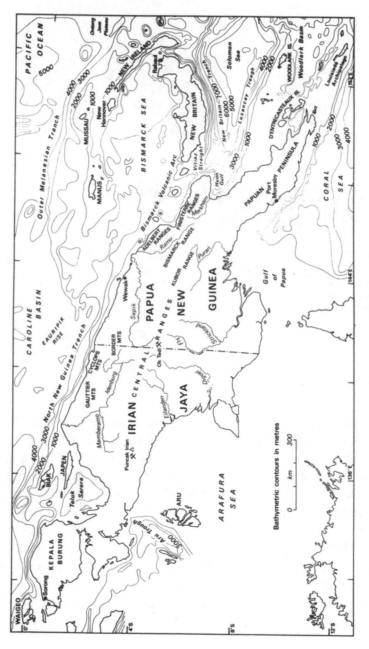

Fig. 1.   The New Guinea region.

ernmost part of New Guinea, south of the central mountain ranges, may be regarded as a northern extension of the stable Australian Platform and is therefore not discussed in detail. The Coral Sea Basin, which lies immediately to the south of the Papuan Peninsula, is considered in Volume 7B of this title.

Even when restricted in this way, the area considered includes a remarkable diversity of geomorphological units and geological provinces. Elevations range from depths of more than 8 km below sea level in the New Britain Trench to a height of more than 5 km above sea level on the glaciated peak of Puncak Irian (Carstenz Top) in the central ranges of Irian. Although none of the geological provinces into which the area can be divided has evolved in complete isolation from its neighbors, because of the complexity of the area and the uncertainty as to the nature of some of the tectonic events, it is in practice necessary to describe each individually before attempting to fit all into a common geological framework. It is also useful to consider the history of geological exploration in the area, and this is briefly reviewed in Section III.

II. MAJOR SUBDIVISIONS

The geological subdivisions upon which the discussions in this chapter are based are shown in Fig. 2. The most obvious physiographic distinction in mainland New Guinea is between the flat, swampy plains in the south and the more northerly mountain masses of the Central Orogen and of the north coast region. The two elevated blocks are separated by a series of broad, flat depressions at or near sea level that are characterized by sluggish, meandering rivers, large lakes, and extensive swamps. These depressions also roughly mark the zone of suture between the island-arc units that make up the north coast ranges and which are not older than late Cretaceous, and the deformed and partly metamorphosed continental margin sequence of the Central Orogen. The eastern (Papuan) and western peninsulas are considered separately from the central part of the orogen since although they also once formed parts of the continental margin, they have been rifted away from the main continental mass and now differ in important ways from the central nucleus.

The large island of New Britain to the northeast of New Guinea is geologically a direct continuation of the immediately adjacent parts of the north coast ranges and is here considered together with them, while the central and western parts of the north coast, which have been more intensively deformed, are discussed separately as the North New Guinea Province. Further north, Manus, New Hannover, New Ireland and some smaller is-

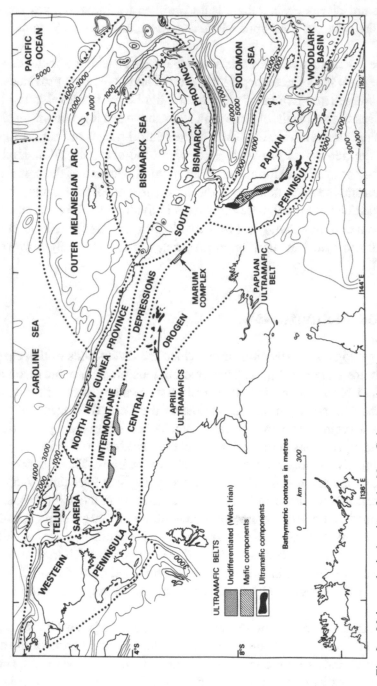

Fig. 2. Major geological units of the New Guinea region. The ultramafic belts are treated in this chapter as a single unit although they do not form one continuous domain.

lands, together with the intervening marine areas and a trench facing the Pacific, form the Outer Melanesian Arc, which includes active volcanoes and Paleogene island arc rocks, which is convex northwards and which is clearly a single unit. There are five distinct marine basins. The smallest, Teluk Sarera (Geelvink Bay), separates the north coast ranges from the Western Peninsula and is almost closed in the north by groups of islands. The Caroline Basin to the north and east of Teluk Sarera seems to be an Oligocene marginal basin defined by subsea ridges, while the Bismarck and Solomon seas and the Woodlark Basin, all of which are probably even younger, are separated from the open ocean and from each other by linear island chains.

III. HISTORY OF GEOLOGICAL INVESTIGATIONS

A. Volcanology

The existence of a considerable number of active volcanoes in east New Guinea and on the adjacent island groups (Fig. 3) has been known since the earliest days of European exploration and discovery. Their study became a matter of immediate concern when the administrative center of north-east New Guinea was established at Rabaul, on New Britain. The superb natural harbor that led to the choice of this site proved to be a volcanic caldera ringed with satellite vents, some of which are active. A volcanological survey service was instituted at Rabaul after eruptions in 1937 that cost several hundred lives, and was first seriously employed in monitoring a series of eruptions that began in 1939 and continued after the destruction of the survey center at the time of the Japanese occupation. The observatory was rebuilt after the war but although the New Guinea and Papua administrations were amalgamated in 1945, the observatory staff remained concerned only with the volcanoes of the former New Guinea mandate until Mt. Lamington in Papua erupted in 1951 with great loss of life. Prior to the eruption the mountain had not even been recognized as a volcano (Taylor, 1958). Since 1951 there have been numerous eruptions in the Papua New Guinea offshore islands, but none on the mainland.

During the past two decades there has been increasing interest in the volcanoes not merely as potential hazards but also as indicators of tectonic processes. A considerable amount of geochemical and petrological work has been done and has included not only the active centers but also older volcanoes and volcanic provinces throughout Papua New Guinea. Extensive studies have been made of the shoshonitic and calc-alkaline volcanoes of the highlands of central New Guinea, of the rather similar volcanoes of the Papuan Peninsula and the D'Entrecasteaux Islands, of the subduction-re-

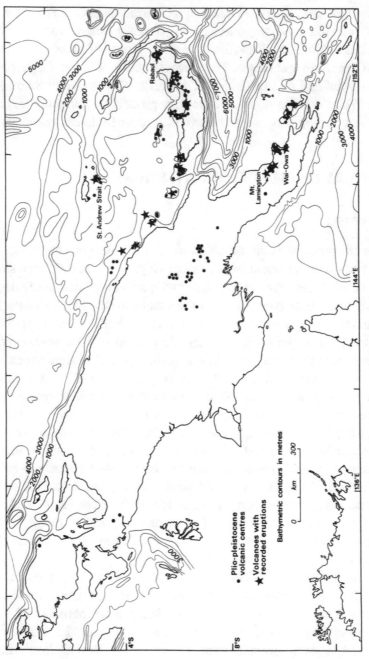

Fig. 3.   Late Cenozoic volcanism in the New Guinea region.

lated centers of the Bismarck Arc and of the possibly 'hot spot' generated volcanics of the St. Andrew Strait region southeast of Manus (Johnson, 1979). One of the most interesting facts to emerge is that many of the volcanic rocks are highly potassic, a characteristic that has apparently persisted through most of the Cenozoic. The widespread continuing eruption of "subduction-type" volcanics in areas in which no present or Recent subduction zone can be identified has led to theories being proposed that involve the reactivation of subduction-modified mantle as a volcanic source by processes that do not require simultaneous crustal consumption (Johnson *et al.*, 1978).

In contrast to the wide distribution of volcanism in Papua New Guinea, there are no active volcanoes in Irian, although Hermes (1974) emphasizes the presence of Quaternary tuffs near the international border, Robinson and Ratman (1978) note the existence of a well-preserved cone in the northern part of the Western Peninsula and Dow and Barlow (1982) report evidence of Recent activity south of Teluk Sarera.

B. Seismology

The study of the seismicity of an area often goes hand-in-hand with study of its volcanology, and in New Guinea the first recordings of earthquake waves were made at the volcanological observatory at Rabaul, where seismographs were established to monitor the ascent of magmas towards the surface (Myers, 1976). Seismological research is now also carried out at an observatory in Port Moresby, where records can be obtained that are not affected by any local volcanic activity. This observatory forms part of the World Standard Seismograph Network and is also a center where focal depth and focal mechanism determinations are made, using WSSN data (cf. Ripper, 1982). Invaluable plots of the seismicity of the whole of the area covered in this chapter have recently been prepared by observatory staff (Ripper and McCue, 1982), updating and expanding the work of Denham (1969).

The most seismically active part of the area is undoubtedly New Britain, where intense stress release is associated with the steeply dipping earthquake zone that outcrops in the New Britain Trench (Fig. 4). The western extension of this arc along the north coast ranges of the mainland is also seismically active but to a lesser degree. Earthquake foci are distributed throughout the mainland and it is difficult to allocate them to any very definite zones, although it has been suggested that the highlands volcanoes are underlain by an old subducted slab (Johnson *et al.*, 1978). Offshore, a narrow but well-defined belt of shallow seismicity across the Bismarck Sea (Fig. 8) has been thought to mark either an extension zone or a zone of transcurrent faulting (Taylor, 1979), while a more diffuse belt in the Woodlark Basin (Fig. 14) has

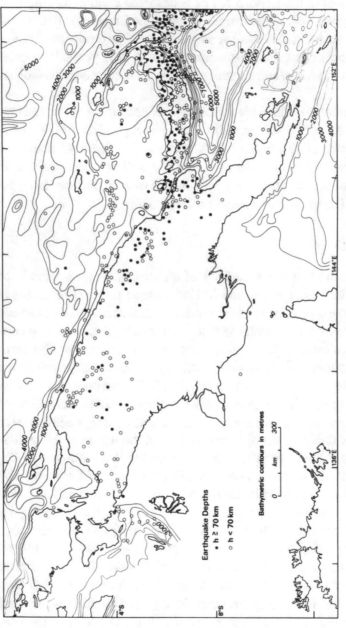

Fig. 4.   Seismicity of the New Guinea region (after Denham, 1969).

also been regarded as an indicator of active seafloor spreading (Weissel *et al.*, 1982).

Although almost no part of the area is entirely free of seismic activity, it is noticeable that there are very few shocks, and certainly no definable Benioff zone, associated with the subdued trench and dying volcanoes of the Outer Melanesian Arc (Fig. 8).

C. Geological Mapping

The original impetus for geological work in the area was economic. There were a number of gold rushes to parts of Papua New Guinea in the nineteenth century and important alluvial deposits were exploited by dredging high in the mountains of the Papuan Peninsula both before and after the Second World War. More systematic geological mapping was initiated following the discovery of a number of oil seeps in both the Dutch and British colonies. Two major publications resulted from the work of consortia headed by Shell in Irian (Visser and Hermes, 1962) and by British Petroleum in Papua New Guinea (Australasian Petroleum Co, 1961) and these provided a framework for much of the later exploration activity. Indeed, the Irian report is still the major published source of information on the geology of the western half of the island. Shell's operations were, from an economic point of view, only poorly rewarded. Three small oil fields were discovered, but only one of these, in a Miocene reef in the Salawati Basin, in the extreme west of Irian, was of much importance. Since the transfer of Shell's interests to Pertamina, the Indonesian state oil company, in 1965, a number of similar fields have been found in the same general area (cf. Froidevaux, 1978). Source rock studies have indicated that the Mesozoic and Tertiary sediments outcropping in the area are thermally immature and that the source of the hydrocarbons is probably to be found in the abundant black shales of the Palaeozoic section (Dow and Barlow, 1982).

British Petroleum's exploration efforts in Papua New Guinea have proved even less successful, but the Phillips group have discovered large, although presently uneconomic, gas reserves in the Gulf of Papua.

Metallic minerals have assumed renewed importance in Melanesia in the last two decades and in Papua New Guinea many of the former goldfields have been reinvestigated for their copper potential (Grainger and Grainger, 1974). The discovery of a very large copper–gold porphyry deposit at abandoned gold workings on Bougainville stimulated exploration and a second deposit, at Ok Tedi near the Irian border (Fig. 1) is now being prepared for production. Other cupriferous porphyries with considerable potential are known and nickel and chrome deposits associated with ultramafic rocks are also being assessed.

Alongside the work of private companies, great progress was made between about 1965 and 1975 in the regional mapping of Papua New Guinea at 1:250,000 scale by government agencies. The key to the rapid improvement in the rates of coverage lay in the use by the Australian Bureau of Mineral Resources (BMR) of small helicopters to provide logistic support. This technique was also used by the Geological Survey of Papua New Guinea (GSPNG), which was established in 1972. So successful did this prove that by 1975 the reconnaissance fieldwork had been virtually completed and the GSPNG was able to concentrate on more specific areas and projects.

In Irian also, copper and nickel have been the two principal metals sought. A copper deposit very similar to Ok Tedi was found high in the central ranges near Puncak Irian in 1936 by a scientific expedition and was brought to production by Freeport Indonesia in 1973 (Adams, 1973). Nickeliferous laterites on islands off the Western Peninsula have been prospected and a report on the mineral potential of the northeastern part of the peninsula has been published (D'Audretsch et al., 1966). Between 1978 and 1983 Irian was the subject of a major Australian aid program of geological and geophysical mapping carried out in conjunction with the Indonesian Geological Research and Development Centre (GRDC). Most of the Australian geologists engaged in this project had previously worked in Papua New Guinea and similar techniques were used. The project covered the Western Peninsula and a small part of the Central Orogen.

The considerable disparity in geological coverage of the two parts of New Guinea makes preparation of a combined study difficult. In particular, a single named unit in Irian may be subdivided into several distinct formations in Papua New Guinea. Only a few representative formations, chiefly those referred to in the text, are shown on the summary stratigraphic columns (Fig. 7 and 9). A virtually complete list, with brief descriptions, has been compiled for Papua New Guinea by Skwarko (1978).

D. Geochronology

Although southern New Guinea shares a common continental basement with Australia, only limited correlations have been possible since so many of the island's outcropping rocks belong to the Tertiary era which is only poorly represented further south. The subdivision of the Tertiary rocks has traditionally been based on a series of faunal zones in turn based on foraminifera. The precise correlation of these zones with standard sequences elsewhere in the world is only now being established. Shell Oil Company geologists in Irian used a different system based on benthonic larger foraminifera and pelagic foraminiferal zones (Visser and Hermes, 1962). One of the crucial points in the establishment of a more firmly based chronological system was the radiometric dating of one of the stages as middle Miocene (Page and McDougall, 1970). This was rather younger than had been thought and indeed a common feature of the use of radiometric methods in the Neogene has been the downward revision of ages.

Once some synchronism between radiometric dates and fossil stages had been established, it became possible to place many of the igneous and metamorphic events in their tectonic and evolutionary contexts. Particular attention was given to determination of the ages of the mineralized intrusions in the porphyry copper provinces and Ok Tedi, at 1.1 Ma, was shown to be very young indeed (Page and McDougall, 1972). Age determinations were subsequently made on rocks collected throughout the Papua New Guinea highlands as an aid to regional correlation (Page, 1976). There are far fewer published radiometric ages for rocks in Irian, but dating has been an important element in the joint BMR–GRDC project (cf. Ryburn, 1978).

Another aspect of the chronological work has been the relating of processes seen to have occurred in what is now the land area to seafloor-spreading phases in the surrounding oceans. Dateable magnetic lineations have been identified in the Woodlark, Caroline and possibly Bismarck Sea basins, all discussed in this chapter, as well as in the Coral Sea and the Pacific and Indian oceans, where major spreading episodes have affected New Guinea.

E. Geophysical Surveys

The gravity method was the first regional geophysical technique to be widely used in New Guinea, measurements being made in Papua New Guinea from 1939 onwards (Australasian Petroleum Company, 1961) and in Irian as early as 1936 (Visser and Hermes, 1962). In both cases readings were taken as part of oil exploration programs. Work of this type ceased in Irian in 1960, by which time some very detailed surveys had been made on the Western Peninsula and regional reconnaissance had been completed along the southern coast, extending in some places up to 100 km inland. Some work had also been done on the north coast and reconnaissance traverses along the Mamberamo River and its tributaries had indicated that large positive anomalies and steep gradients were associated with some of the ultramafic rocks on the northern side of the Central Orogen.

In the mid-1960s a major advance in regional coverage of Papua New Guinea was made by the University of Tasmania's Papua Project. Whereas previous surveys had been relatively detailed but had been confined to areas of possible hydrocarbon potential, the university took advantage of the helicopter-assisted triangulation surveys to establish stations throughout the Papua New Guinea mainland (St. John, 1967). As a part of this work, the isolated oil company surveys were integrated into a common system. Very large positive anomalies were shown to be associated with the Papuan Ultramafic Belt and these were further investigated by the BMR (Milsom, 1973). From this beginning, modifications of the helicopter-based, aneroid barometer levelled methods that had proved successful in Australia were applied in various areas and about half the country had been covered at a reconnaissance level by 1975. Although levelling accuracy was reduced by the nature of the terrain and although very large topographic corrections were needed, geologically useful maps were produced. The BMR also extended the Australian base station network throughout Papua New Guinea and into the Solomon Islands. A regional Bouguer anomaly map based on this work and complementary marine surveys has been published, covering Papua New Guinea and the Solomons (Bureau of Mineral Resources, 1979).

Gravity work recommenced in Irian with the joint GRDC–BMR mapping project, and links have been made to the Indonesian and Australia–Papua New Guinea base systems. The results of an initial orientation survey have been published (Untung and Barlow, 1981) and also a generalized Bouguer anomaly map of all of the Western Peninsula (Dow and Barlow, 1982).

Aeromagnetic work came relatively late in the exploration of Papua New Guinea, but a major BMR contract survey during the period 1967–69 covered most of eastern Papua and the immediately adjacent offshore areas (cf. Milsom and Smith, 1975), complementing unpublished work by oil companies further west. There have been no attempts to integrate the results of these surveys into a single map. An aeromagnetic survey flown over the southern plains of Irian in 1953 indicated the possible presence there of a deep sedimentary basin (Visser and Hermes, 1962).

Seismic surveys have been carried out in most of the main offshore and onshore sedimentary basins in the course of hydrocarbon exploration. Results have been poor in the highland areas because of structural complexity and the presence of thick limestones that form a karst topography in which much of the seismic energy is reflected and refracted at shallow depths. In addition to the detailed and systematic offshore work, the Gulf Oil research vessel *Gulfrex* collected reconnaissance data throughout New Guinea waters in 1970. No general report on the *Gulfrex* cruise has been published, but some information has appeared in various review articles (cf. Hamilton, 1979). In addition to the commercially oriented work, the BMR and several research institutions have obtained seismic profiler and other geophysical data around New Guinea. The BMR surveys, carried out under contract by Compagnie Generale de Geophysique, formed part of a general Australian continental margin project and were particularly intensive in the Bismarck Sea (Connelly, 1974). The University of Hawaii (Hawaii Institute of Geophysics) was active from the middle 1960's in seismic refraction studies in the Bismarck and Solomon seas (Furumoto *et al.*, 1968), and this work culminated in a major land based refraction study in eastern Papua, organized by the BMR but involving the Hawaiian and a number of other institutes (Finlayson *et al.*, 1976, 1977). There has been no equivalent work in any part of Irian. Some heat flow observations have been made in the marine areas (Halunen and von Herzen, 1973), but these have been too few in number for any firm conclusions to be drawn.

IV. GEOLOGICAL EVOLUTION

A. Central New Guinea Orogen

Geological mapping at 1:250,000 scale has been completed in the Papua New Guinea section of the Central Orogen and many of the maps have been published. The geology of the interior of Irian is less well known and much of the published geological map is based on airphoto interpretation (Visser and Hermes, 1962). Visser (1968) notes that the observations made during a later foot traverse by Dow (1968) were broadly in agreement with this interpretation, except that the Miocene volcanic sequences had not been recognized on the photos and were included with the Mesozoic sediments. The least known part of the orogen in Irian is the metamorphic belt along its northern margin, which was not mapped in detail since the rocks have no hydrocarbon potential. The greater apparent simplicity of the composite geological map (Fig. 5) in the western part of the Orogen is thus to some extent unrealistic, but it does seem that the wider eastern part is also somewhat more complex and has been more intensively faulted.

The orogenic belt consists of a mountain range culminating in peaks more than 4000 m in height, with numerous broad valleys at altitudes of about 1000 m. The range separates the stable Australian continental platform to the south from the island arc units to the north. Despite the more advanced state of geological mapping in Papua New Guinea, early Palaeozoic rocks have so far been reported only from Irian, where Visser and Hermes (1962) note the presence of limestone pebbles containing fossils of Silurian and Devonian age in streams draining the southern slopes of the highlands. They also assign outcrops of hard silty clays and volcanics near the international border to the late Cambrian on the basis of their supposed lithological similarity to rocks of this age in northern Australia. Large areas in the southern part of the orogen in Irian have been mapped as Palaeozoic on the basis of photogeological interpretations.

The Palaeozoic and supposed Palaeozoic rocks in Irian show little or no evidence of metamorphism, but the oldest rocks in Papua New Guinea

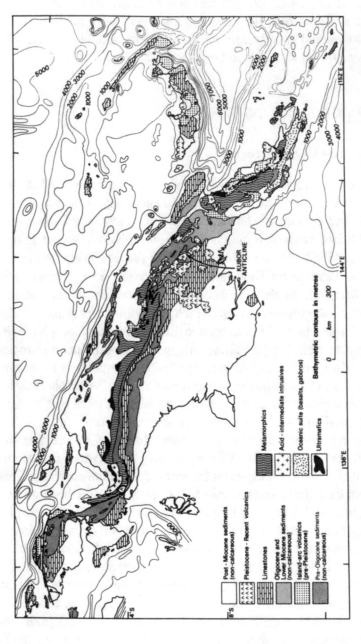

Fig. 5. Generalized geological map of the New Guinea region (after Bain *et al.*, 1972 and Visser and Hermes, 1962, with some later amendments).

are metamorphics (Omung Metamorphics) that are exposed, together with the intruding Kubor Granodiorite, in the core of the Kubor Anticline (Fig. 5). The intrusives have been isotopically dated as Permian or Triassic (215–244 Ma; Page, 1976) and resemble rocks of similar age on the Cape York Peninsula, the northernmost part of Australia (Bain *et al.*, 1975).

In Irian, Permo–Triassic rocks have been reported from numerous localities along the southern margins of the central ranges, where they are exclusively terrestrial in origin. A complex unit that includes Triassic and older rocks which outcrops just east of the base of the Western Peninsula is referred to by Visser and Hermes (1962) as the "overthrust mass." Its relationship to the surrounding younger rocks has not been determined in detail but the structural style it represents may be repeated in the less well-mapped areas to the east.

In Papua New Guinea, the crystalline rocks in the core of the Kubor Anticline are unconformably overlain by reef limestones of the late Triassic Kuta Formation. A little further to the north, slightly younger shallow marine sediments (Jimi Greywacke) and overlying acid–intermediate volcanics (Kana Volcanics) are exposed in fault wedges at the northeastern margin of the orogen (Dow *et al.*; 1972). The volcanic activity was followed by a period of emergence and erosion, but there was a general and long lasting return to marine conditions in the middle Jurassic (Burns and Bein, 1980).

The post-Triassic sedimentary history of New Guinea seems to have had two aspects, characterized by a stable shelf association in the south and by eugeosynclinal conditions in the north. The boundary between these "Fly" and "Sepik" associations (Brown *et al.*, 1980) is not simply equivalent to the present-day distinction between the north coast ranges and the units to the south, since rocks of both associations are faulted together throughout the Central Orogen, at least in Papua New Guinea. Brown *et al.* (1980) consider that the northern Australian margin developed through the Mesozoic in a series of phases of subduction and marginal basin formation, the details of which have yet to be established.

The Jurassic–Cretaceous succession has been most thoroughly mapped in western Papua New Guinea, where rocks of both the arenaceous shelf facies (Jurassic Kuabgen Group and Cretaceous–Eocene Feing Group) and the argillaceous slope facies (Jurassic Om Beds and Cretaceous Salumei Formation) outcrop extensively (Davies and Norvick, 1974). Further east only the slope facies of the Jurassic (Maril Shale and Sitipa Shale) is exposed. Mesozoic rocks in the Irian section of the Central Orogen have been generally mapped as undifferentiated Kembelangan Formation; the type sections for this formation are all on the Western Peninsula, where A, B, C and D members are distinguished. The A member seems generally equivalent to

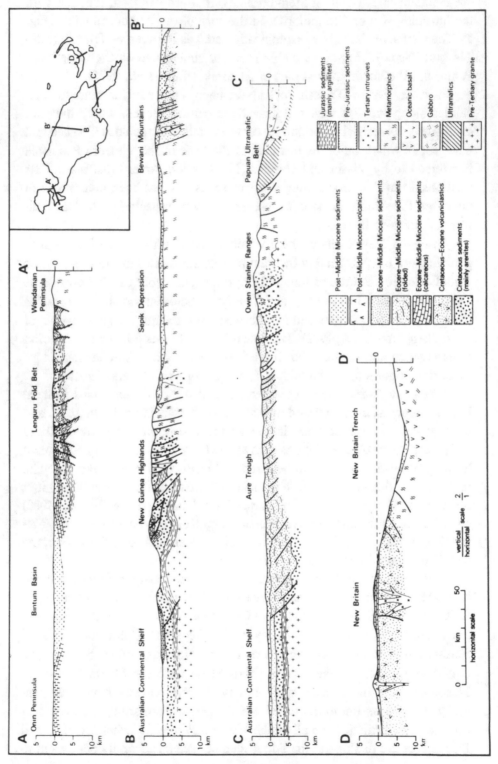

Fig. 6. Hypothetical geological cross sections (after Visser and Hermes, 1962, and Dow, 1977).

the Om Beds and the B and C members to the Kuabgen and Feing Groups, although Visser and Hermes (1962) caution against the use of these subdivisions away from the vicinity of the type localities. The Kembelangan D member has been dated as uppermost Cretaceous and Paleocene.

Subsidence seems to have been sufficiently slow during the Jurassic for shallow water conditions to be maintained almost everywhere, but was more rapid during the Cretaceous, when much thicker sediments were deposited along the northern and eastern margins of the orogen, intercalated with basaltic agglomerates, pillow lavas and associated tuffs (Kumbruf Volcanics and Kondaku Tuff). These rocks are thought to record the existence of a continental-margin volcanic arc that formed north and east of the platform in the early Cretaceous and which continued to contribute volcanic detritus to sediments well into the late Cretaceous (Brown *et al.*, 1980). Polarity criteria for this arc have been obscured by Tertiary deformation but the evidence available suggests that it faced north to the north of the platform and east to the east of it, and was separated from it by back-arc basins and marginal seas. Neither volcanic rocks nor volcanic detritus of this age have been recorded in Irian, but the area where outcrops are most likely to occur has not been mapped at more than reconnaissance level.

At the start of the Tertiary, the stable shelf to the south emerged above sea level but marine conditions prevailed along the line of the orogenic belt and there was widespread deposition of massive shoal limestones (Darai Limestone in Papua New Guinea; New Guinea Limestone Group in Irian), until a major orogeny in the Oligocene that resulted in metamorphism, folding, faulting, and finally uplift of the mountain range. Metamorphism has been generally most intense along the outer margin of the former trough and faulting most intense along the inner margin. Dow *et al.* (1972) describe two types of metamorphic rock. In addition to the high temperature Ambunti Metamorphics in the north (Section 4.4), a high-pressure/low-temperature glaucophane-bearing assemblage (Gfug Gneiss and metamorphic phase of the Salumei Formation) is locally developed near changes of strike in a major transcurrent fault zone. In one area the transition is seen over a distance of about 8 km between hard, indurated but essentially unmetamorphosed Salumei Formation of Eocene age and well cleaved slates and phyllites. Limestones in the same zone are recrystallised to marble but massive volcanics are almost unaltered. Visser (1968) notes that in Irian there is a rather sudden transition from black shales into their mildly metamorphosed equivalents but the abruptness could be due to fault juxtaposition, as is commonly the case in Papua New Guinea.

Clastic sedimentation in the Oligocene and Miocene was particularly intense in the Aure Trough, which separates the Central Orogen from the orogenic belt of the Papuan Peninsula and which probably developed at the same time as the Coral Sea Basin to the southeast. The trough sediments have been folded and uplifted to form rugged, north-trending mountain ranges. Since the Oligocene, the orogen has been characterized by continuing uplift with concomitant rapid erosion, accelerated by frequent landslips, often initiated by earthquakes, and by molasse-type sedimentation into terrestrial flanking basins. The uplift seems to have commenced with a short-lived but violent episode of island arc volcanism during the middle Miocene (Dow, 1977). The volcanics are often highly potassic and this seems to be a quite general feature of Tertiary volcanism in the New Guinea region. The structural pattern has been complicated and confused in Papua New Guinea by the Pleistocene and Recent eruption of large quantities of potassic lavas from central vents scattered through the orogen. The area is seismically active, with most shocks at shallow depths and with no clearly defined Benioff zone (Ripper and McCue, 1982). The rare deeper earthquakes are believed by Johnson et al. (1978) to be related to the presence of subduction remnants that also cause the volcanic activity.

Structurally, the orogen is dominated by strike-slip faulting in the north and by folding in the south; drilling results clearly show that, as in most foreland thrust belts, the folds are interrupted by thrust faults at depth. The main trend of transcurrent faulting is southeasterly in the east, north of the Kubor Anticline, but becomes more easterly in the west. Dow (1981) notes that at the western end of the orogen, near the base of the Western Peninsula, platform sediments are separated from metamorphosed fine-grained sedimentary rocks by a major fault zone that trends slightly south of east. The sediments in this area are folded into symmetrical anticlines separated by narrow zones of vertically dipping strata.

Although the highest mountains and the oldest rocks of New Guinea are found in Irian, St. John (1967) notes that the pattern of gravity anomalies indicates much thicker crust east of the border. The explanation for the greater elevations in the west may lie in the nature of the rocks exposed; Dow (1968) notes that with only limestones and friable sandstones outcropping in the highest parts of Irian, the rivers lack coarse detritus to abrade their beds and generally flow at grade. The patterns of exposure of the older rocks in fact roughly parallel those of northern Australia, where a late Palaeozoic orogenic belt in the east is backed by the Mesozoic of the Great Artesian Basin, with Cambrian sediments still further to the west.

There are permanent snowfields on several of the highest peaks in Irian. The highest parts of Papua New Guinea lie just below the permanent snow line, but young glacial features are common at high altitudes on both sides of the border.

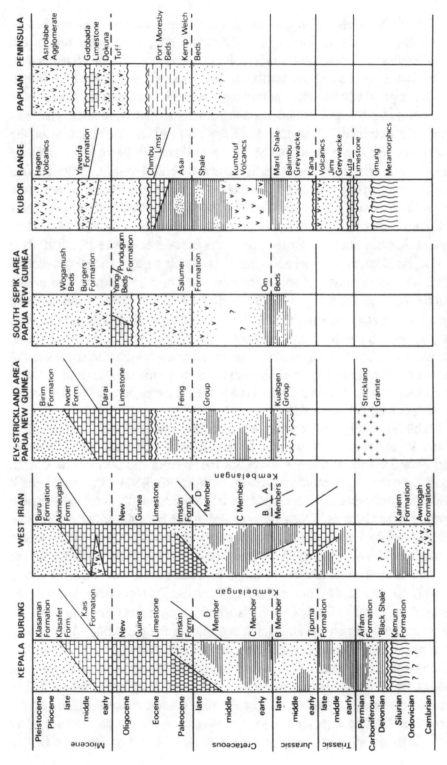

Fig. 7. Simplified stratigraphic columns; Central Orogen and Western and Papuan Peninsulas.

B. The Outer Melanesian Arc

The Outer Melanesian Arc, which marks the transition from the Pacific Ocean Basin to the archipelagoes of New Guinea, is the only one of the units discussed in this chapter that includes significant areas both above and below sea level (Fig. 8). Its northern, outer, margin is defined by a trench that has a generally subdued morphology but which reaches depths of more than 6 km in a few places and has a cross section that is, in some places at least, characteristic of subduction (Hamilton, 1979). To the south of the trench are two ridges separated by a deep sedimentary basin and capped by islands. The islands of the outer ridge are generally small and are formed of Cenozoic, mainly Neogene, limestones and volcanic rocks. There are a few solfatara fields but no other indications of current volcanic activity. Nor is there much seismic activity associated with the arc (Fig. 8) and the few shocks that have been recorded during the past decade have all taken place at shallow depths. Weissel and Anderson (1978) suggest that the trench is either extinct or active at only very slow rates of convergence. The trench and the volcanic arc merge with the North New Guinea Trench and northern New Guinea near the Irian/Papua New Guinea border; at their eastern ends they appear to continue into the Solomons without any intervening break. De Broin *et al*. (1977) suggest that the sedimentary trough that separates the volcanic arc from the inner ridge also continues and links with the "Slot" basin between the two branches of the Solomon Archipelago, but this seems to be less well substantiated.

The inner ridge, which marks the northern limit of the Bismarck Sea, protrudes above sea level to form the large islands of Manus, New Hannover and New Ireland. It follows a smooth curve from the north New Guinea coast near the Sepik mouth to a deep bathymetric hollow that separates it from the northern end of the Solomon Islands Group. The best geologically mapped of the islands is New Ireland, which has been described by Hohnen (1978). The basement rocks, which are mainly found in the broad eastern part of the island and which have their equivalents on New Hannover at the western end of the "dumbell", are Oligocene island arc volcanics (Jaulu Volcanics) very similar to the Baining Volcanics of New Britain. Scattered outcrops of Jaulu Volcanics occur also on the long narrow neck that extends between the wider blocks and which is mainly built up of Miocene limestone (Surker Limestone and Lelet Limestone) and of volcanic rocks of the late Miocene Rataman Formation. It is thought that the latter were extruded in the course of rifting and transcurrent movement along faults subparallel to the present trend of the island. The faulting is seen as having moved New Hannover and, to a lesser extent, the eastern part of New Ireland, away from formerly adjacent positions at the northeast end of the New Britain

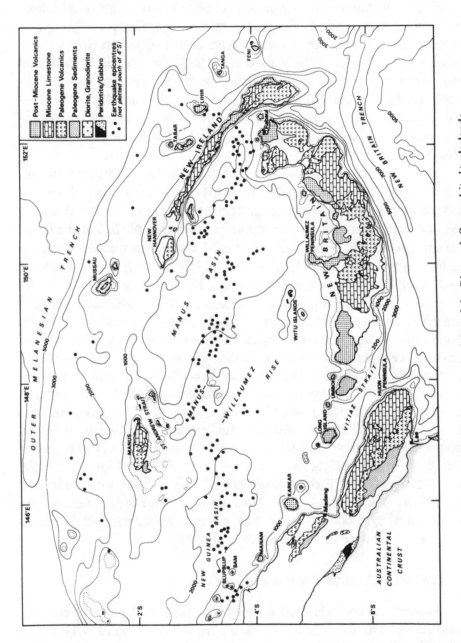

Fig. 8. Geology, bathymetry, and seismicity of the Bismarck Sea and its borderlands.

Arc. The large transcurrent movements this implies are by no means inconceivable since it is generally accepted that the entire region has been subjected to pervasive left-lateral shear consequent on the movement of the Pacific past the northern margin of the Australian plate (Johnson and Molnar, 1972). Most of the critical evidence lies offshore, but faults with the appropriate trends and sense of movement have been mapped on New Ireland. Other authors (Taylor, 1979; Weissel *et al.*, 1982) believe that New Ireland has moved much further and that during the Oligocene it occupied the position, relative to New Britain, now occupied by Bougainville. In this case it is supposed that New Ireland moved northwestwards into its present position during the Neogene, with the Solomon Islands moving into place behind it.

Other volcanic rocks, younger than those of the Rataman Formation, may be associated with continuing shear movements. Although the level of seismic activity is low, much of the present-day movement of the Pacific past northern Australia must be accommodated near the line of the Outer Melanesian Trench. Some element of subduction seems necessary in the eastern part of this trench but there can be little or no convergence in the west, where it must mark a mainly transform boundary. DeBroin *et al.* (1977) note that not all the seismic profiles across the trench indicate subduction and that in some cases tensional stress is suggested. Perhaps the most curious feature of the area is the position of the young volcanoes, since they lie between the trench and the nonvolcanic ridge in a reversal of the normal relationship. The chemistry of the volcanic rocks is also anomalous; they are mainly low-silica, highly potassic basalts that seem to have no counterparts in any other island arcs (Johnson *et al.*, 1976). Although the total amount of volcanic rock exposed is actually quite small, activity seems to have spanned a rather long period of time. Johnson *et al.* (1976) believe it may date from the Oligocene, when the North Solomons Trench was the site of subduction of Pacific lithosphere prior to collision with the thicker crust of the Ontong Java Plateau. Most of the arc's anomalous features can be explained by supposing shear to have been dominant and subduction to have been very slow through most of its history.

C. The South Bismarck Province and New Britain

Present-day north-directed subduction in the New Britain region is associated with a trench along the northern margin of the Solomon Sea reaching to depths of more than 8 km, a near-vertical Benioff zone extending to a depth of about 600 km and a line of active volcanoes along the north coast of the island (Johnson, 1976). The present volcanic line and also the pre-

Pliocene lithologies continue almost undeviated along the north coast of New Guinea as far west as the mouth of the Sepik River (Fig. 8), but the Benioff zone is profoundly modified.

The oldest rocks outcropping in the province are the Eocene Baining Volcanics of New Britain and the Gusap Argillite of the mainland (Dow, 1977). Both these formations consist of intermixed sediments and island arc volcanics. On New Britain there may have been a nonvolcanic interlude in the early Oligocene, perhaps during the time of formation of the Solomon Sea, but on the mainland the argillite grades upwards via a steadily increasing volcanic component into the highly potassic Finisterre Volcanics (Jaques and Robinson, 1977). Intrusive rocks are rare on the mainland but comagmatic intrusions are common within the volcanic sequences of New Britain.

Volcanism ceased abruptly throughout the province in the earliest Miocene and was followed by a period of calcareous sedimentation that was ended in the later Miocene by uplift that progressed gradually from west to east. This uplift seems to have been accompanied by volcanism in the extreme east of New Britain (Sigule Volcanics), but not elsewhere. Island arc volcanism returned to the province in the Pliocene with the development of the Bismarck Arc volcanoes and was swiftly followed in the west by collision with the continental block (Jaques and Robinson, 1977). One result of this collision seems to have been a considerable change in chemistry of the arc volcanoes. Johnson (1976) shows that, in contrast to the usually recognized pattern, changes in alkali content along the strike of the western part of the arc are of considerably greater magnitude than the across-strike variations. He also comments on the much reduced deep-focus earthquake activity. Almost the only deep shocks in the last decade have been in a roughly cylindrical zone that extends vertically beneath Long Island.

The frontal arc exposed on the mainland of New Guinea falls naturally into two sections, east and west of Madang. These sections seem to be offset from each other by a right-lateral fault marked by a north-trending section of coastline. A similar, although rather less obvious, offset occurs in the strait that separates New Britain from the mainland. The arc segment between New Britain and Madang (Finisterre Ranges; Fig. 1) has been interpreted as having suffered "thin-skin" thrusting of the frontal zone across the continental margin (Milsom, 1981). It is noticeable that the offset does not affect the continuity of the line of active volcanoes to the north, which is related to processes at much greater depths.

Another interesting but still unexplained offset, along the Willaumez Peninsula on New Britain, affects only the youngest volcanism, which reappears further north in the Witu Islands. The peninsula is built up of the products of volcanic eruptions along the fracture associated with the offset.

D. The North New Guinea Province

From the mouth of the Sepik River in Papua New Guinea to Teluk Sarera in Irian, the north coast of New Guinea is flanked by mountain ranges in which a Cretaceous–early Tertiary basement of igneous and metamorphic rocks is exposed in a complex jumble of fault-bounded blocks and is overlain by a thick sequence of later Tertiary nonvolcanic and mainly clastic sediments. As with the South Bismarck Province, the history of the area has been interpreted in terms of collision between an island arc and the continental margin, but it is thought that this collision occurred earlier, possibly in the early Miocene, and that the units involved have been distorted by subsequent movements (Jaques and Robinson, 1977). Brown *et al.* (1980) regard this as merely the latest in a series of subduction related tectonic events that affected the area during the Mesozoic. In contrast to the north coast region east of the Sepik, there are no active volcanoes offshore, nor is there any evidence for volcanic activity since the early Miocene. The basement igneous rocks are known as the Bliri Volcanics and the Torricelli Igneous Complex in Papua New Guinea (Hutchison and Norvick, 1978) and are equivalent to the lower part of the Auwewa Formation in Irian (Visser and Hermes, 1962). Not all of the outcrops shown on the Visser and Hermes (1962) geological map are necessarily part of this province, however. The authors note (p. 101) that "the common presence of volcanic rocks is a good criterion for recognition" of the Auwewa but in view of the complexity known east of the international border this would seem likely to be an oversimplification. South of the arc, late Cretaceous to Eocene sediments were regionally metamorphosed at the subduction zone, mainly to greenschists although blueschists were formed locally, and were intruded by diorite (Amanab Metadiorite). The greenschists in Papua New Guinea are known as the Ambunti Metamorphics (Section 4.5) but no formal name has been given to their equivalents in Irian.

A further consequence of the collision was the upfaulting of complexes of mafic and ultramafic rocks within the north coast ranges. The largest of these bodies may lie in the poorly known Gauttier Mountains (Fig. 1) but a better known, slightly smaller mass forms the Cyclops Mountains which extend for 50 km along the Irian coast just west of the border. A similar body, the Mt. Turu Igneous Complex, 50 km southeast of Wewak (Fig. 1), has been described by Hutchison and Norvick (1978).

Collision and the cessation of subduction were followed by rapid uplift of the central axis of the ranges and by sedimentation in a number of well-defined troughs formed by subsidence of the adjacent blocks (Grund, 1975). This area was presumably the fore-arc region during the middle Miocene when island arc volcanism in the central orogenic belt was associated with

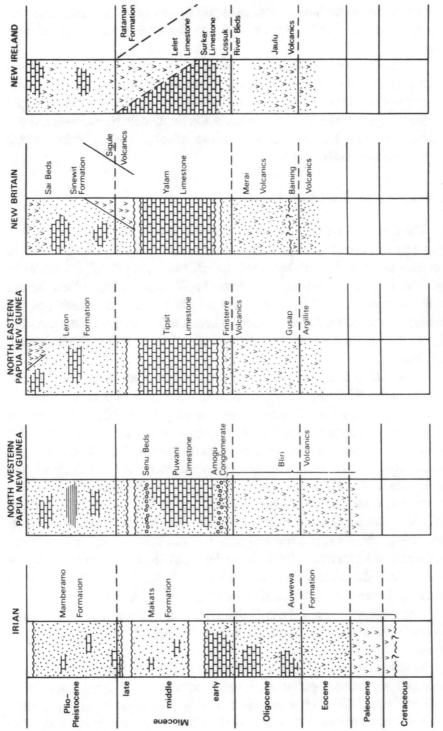

Fig. 9.   Simplified stratigraphic columns; northern New Guinea and the island arcs.

the trench which is still visible off the north coast. The short-lived volcanic episode seems not to have contributed greatly to the trough sediments, the bulk of the volcaniclastics presumably being deposited closer to the volcanic islands.

A further change in the late Miocene to a tectonic regime that did not involve subduction may have dictated the change at the same time to shallow-water deposition. Continued crustal compression was accommodated by a major system of east-trending strike-slip faults with thrust components. The continuing tectonism in the area is demonstrated by the continuing high levels of seismic activity (Fig. 4).

E. The Intermontane Plains

The Central Orogen is separated from the North New Guinea Province by a series of broad swampy depressions within which the land surface is at or near sea level. In Irian the area is termed the "Lake Plain" (Meervlakte, Fig. 10a); it is separated from the broader Sepik Plains of Papua New Guinea by the Border Mountains. East of the Sepik, southward thrusting of the Finisterre block has reduced the gap between the northern and central mountains to little more than the immediate width of the valleys of the Ramu and Markham rivers.

Basement outcrops are rare within the troughs and correlations are difficult to make. There has been some exploration for hydrocarbons (Grund, 1975) and in Papau New Guinea aeromagnetic, gravity and seismic surveys and a limited amount of drilling have provided some clues to concealed basement structures. In some places at least, basement is gabbroic (Jaques and Robinson, 1977) but elsewhere basement rocks have been found that can be correlated with those outcropping in both the North New Guinea Province and the Central Orogen. According to Dow et al. (1972), these Ambunti Metamorphics underlie most of the Sepik Depression. They range from low-grade slate, phyllite and schist in the east to amphibolite facies rocks in the west and are believed to be the regionally metamorphosed equivalents of the Mesozoic and Eocene sediments in the Central Orogen. They are certainly pre-middle Miocene since overlying sediments of this age have not been metamorphosed. Page (1976) reports a number of age determinations that place the metamorphic event in the earliest Miocene or late Oligocene.

In Irian, Visser and Hermes (1962) believe that the Auwewa Formation, the island arc volcano-sedimentary unit, underlies most of the Lake Plain. Their evidence, based on a very few isolated outcrops, is not conclusive and it seems rather unlikely that a rock unit of this type should constitute basement over such a considerable area.

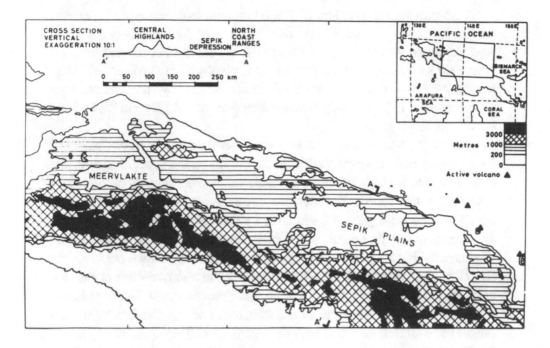

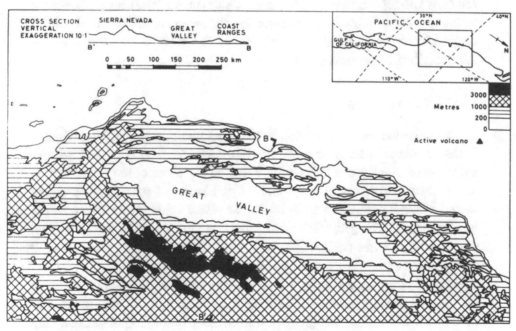

Fig. 10. (a) Physiography of the north coast region of New Guinea and the intermontane depressions. (b) Physiography of California.

One outstanding problem of the intermontane depressions is the discontinuity at the international border. Because of their inaccessibility and rugged topography, the Border Mountains are poorly known and their relationship with the disparate units around them is not fully understood. They are reported to consist of metamorphic rocks intruded by late Permian granites (Norvick, 1975) but only isolated boulders of the plutonic rocks have been found (Mackenzie, 1977).

An interesting analogy can be drawn between the northern plains of New Guinea and the Great Valley of California, on the opposite side of the Pacific Ocean. In California also, low mountains with peaks everywhere less than 2000 m high, border the coast and are separated from the main ranges, which include large granitic batholiths, by the Great Valley, which in turn has rather similar dimensions to the plains in New Guinea. The impression of similarity is strengthened by the remarkable resemblance between the shapes of the two coastlines (Fig. 10) and by the fact that each of the two areas represents a break in the almost continuous circum-Pacific volcanic ring. It also appears that the Franciscan rocks of the Californian coastal ranges have much in common with the island arc collision complex of northern New Guinea (Section 4.4), while the Californian inland ranges, like those of central New Guinea, are notable for the presence of thrust sheets of ultramafic rock along their seaward flanks. The New Guinea thrust sheets are described in the next section.

F. The Ultramafic Belts

In common with many other active continental margins, New Guinea is characterized by abundant outcrops of ultramafic rocks (Fig. 2). The largest of these, the Papuan Ultramafic Belt (PUB), is one of the world's most spectacular ophiolite complexes and forms a series of subsidiary mountain ranges along the northern flank of the mountain spine of the Papuan Peninsula (Fig. 11). The geological history and composition of the belt have been discussed by Davies (1971, 1980) and the geochemistry of the various rock suites of the association has been described by Jaques and Chappel (1980). The complex consists of a basal ultramafic tectonite layer that is principally harzburgite and is typically 4 to 8 km thick, overlain by a few hundred meters of cumulus ultramafics that grade upwards into gabbroic cumulates that may also be several hundreds of meters thick. These are in turn overlain by 3 to 4 km of noncumulus gabbros, a sheeted dyke complex that has so far been observed in a few places only and by some 4 km of basaltic lava and pillow lava intercalated with cherts and limestones. The gabbros have been dated as Jurassic by the K-Ar method and a basalt sample as early Cretaceous.

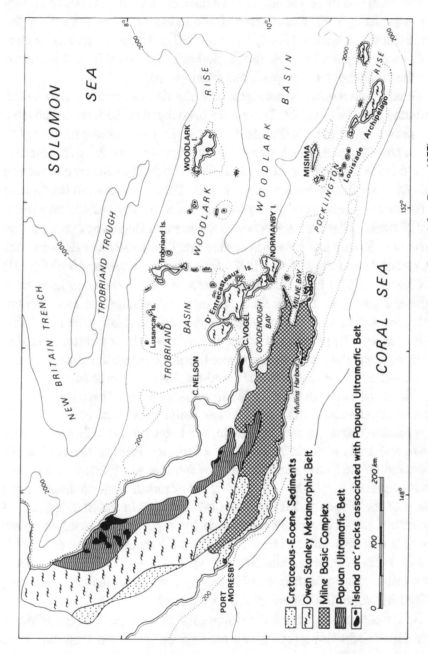

Fig. 11. Simplified pre-Miocene geology of eastern Papua (after Dow, 1977).

Associated sediments have been assigned a late Cretaceous faunal age but all dates are based on only a very small number of samples.

The PUB and its surroundings have been covered by a series of gravity surveys (Milsom, 1973) and its deep structure has been investigated by the seismic refraction method (Finlayson *et al.*, 1977). The geophysical evidence in both cases supports the geological hypothesis that the PUB has been thrust on to the sialic rocks of the peninsula from the north and east.

Tonalite intrusions and comagmatic andesitic volcanics are also widely distributed through the PUB. These have been dated to the Eocene and have been shown to be genetically distinct from the mafic/ultramafic complex (Jaques and Chappell, 1980). They thus provide some of the strongest tangible evidence for the hypothesized origin of the PUB as the frontal section of an island arc that collided with the Papuan Peninsula between the Eocene and the first appearance of ultramafic detritus in nearby sedimentary rocks in the Miocene. Island arc volcanics of Eocene and Oligocene age are found also on New Britain and it seems possible that this island represents another fragment of the same arc (Karig, 1972). However, as Davies and Smith (1971) and Connelly (1979) have both emphasized, the PUB seems to have been emplaced from the east rather than the north and relating it to New Britain, which now strikes east to northeast, implies considerable rotation of that island; Davies (1980) prefers to regard the New Britain volcanism as having developed separately in response to continued convergence after the collision further south between the PUB and the Papuan Peninsula.

The actual extent of the PUB is somewhat in doubt. There are no major outcrops of ultramafic rocks on the New Guinea mainland east of 149° E, but peridotite masses, some quite large, are known in the D'Entrecasteaux Islands and some ultramafics have been reported from as far east as the eastern end of the Louisiade Archipelago (Smith *et al.*, 1973). It seems possible that east of 149° E the root zone of the ultramafic belt continues along the line of the Woodlark Rise and that the outcrops in the islands south of this feature, which are not accompanied by major gravity anomalies, are fragments of a dismembered thrust sheet. On the mainland between 148° E and 149° E the main ultramafic outcrops seem to be of just this type. They occur within and on both sides of a gravity low, while the corresponding gravity high lies in the alluvial plains to the north.

As its northern end the PUB disappears beneath the young sediments of the Huon Gulf. Scattered gravity readings on the Huon Peninsula have indicated a possible continuation beneath the Finisterre Ranges thrust sheet (St. John, 1967; Milsom, 1981). Some 200 km further west another gravity high is associated with the Marum Complex, which consists dominantly of mafic and ultramafic plutonic rocks. Basalts make up only a very small part of this complex and seem in most cases to have been overthrust by the ophiolitic mass rather than to be part of it. Jaques *et al.* (1978) present analyses of the lavas that show them to be enriched in lithophile elements and compositionally more akin to basalts from oceanic islands than to those extruded at mid-ocean ridges or island arcs. They have a distinctly different chemistry from the plutonic rocks of the complex (Jaques, 1981), and may have formed at a site hundreds of kilometers away from them before being brought into contact by faulting and crustal absorption.

It seems reasonable to class the Marum Complex as an ophiolite because the mafic and ultramafic components are both present and in their expected structural relationship (Jaques, 1981). Still further west and as far as the Irian border, numerous ultramafic bodies ranging from a few meters to 50 km in length have been mapped under the general name of April Ultramafics (Dow *et al.*, 1972). Typically all are elongated parallel to the WNW structural grain, the smaller masses being intensively sheared and largely serpentinized while the larger ones are irregular and consist of almost unaltered dunite and peridodite. Small blocks of gabbro are encased in serpentinite in fault zones at the margins of many of the larger bodies, but these are really volumetrically too insignificant to allow the term ophiolite, which implies a more equal association, to be used. It seems that the April Ultramafics continue along the northern front of the ranges in Irian as far as Teluk Sarera, but they have not been mapped in any detail. Some of these bodies are associated with gravity highs and may therefore be deeply rooted, but in other cases only the expected decrease in Bouguer anomaly towards the thickened crust beneath the central ranges is seen (Visser and Hermes, 1962). Dow (1981) notes that just east of Teluk Sarera the metamorphic belt of the Central Orogen is flanked by a thick belt of ultramafic rocks and that (Dow and Barlow, 1982) this continues as part of the Weyland thrust sheet south of the bay.

In New Guinea the transition can thus be observed from massive, thrust-emplaced ophiolite in the east via a smaller but still essentially coherent mafic/ultramafic mass to the disrupted and partially serpentinized bodies in western Papua New Guinea and in Irian. All have a common relationship to the central New Guinea ranges and all were emplaced at some time in the early to middle Tertiary, although it seems scarcely possible that emplacement took place simultaneously along the whole of the 2000 km belt. Throughout this belt there is also evidence for early Tertiary collision between island arc systems and the Australian continental margin. Where this collision was most complete the ultramafic sheets are most shattered and distorted; where, as with the PUB, convergence may have been halted at an early stage because of collision elsewhere, the ophiolite has retained its original form and the igneous stratigraphy has been preserved. It may, however, equally well be the case that it was the northerly trend of the Papuan Peninsula that allowed the PUB, emplaced from the east, to retain much of its original coherence. Further west the ultramafic sheets may have been more severely deformed because they were being emplaced from the east on a continental margin with an easterly trend.

G. The Papuan Peninsula

The existence of the Papuan Peninsula as an identifiable geological unit dates from the late Paleocene, when a part of the continental slope turbidite sequence of northeastern Australia was rifted away from the continental shelf by the opening of the Coral Sea (62–56 Ma; Weissel and Watts, 1979). This event seems to be marked by a change in the character of the sediments along the southern coast of the peninsula, from the mainly clastic Kemp Welch Beds to the cherty Port Moresby Beds (Pieters, 1978). The partly metamorphosed slope complex now makes up the core of the peninsula and similar rocks reappear in the D'Entrecasteaux Islands and, far to the east, on the islands of the Louisiade Archipelago (Davies and Smith, 1971). The extreme east of the peninsula, however, and much of the southern foothills zone is built up of a monotonous sequence of Cretaceous and Eocene submarine basalts and associated gabbroic intrusives whose relationship to the sialic rocks is still not understood (Smith and Davies, 1976). It has been variously suggested that the mafic rocks (Milne Basic Complex, Fig. 11) were thrust across the sialic core, were extruded on top of it or were intruded and extruded adjacent to it during the extension of the Coral Sea. It is not clear whether the absence of Paleocene dates indicates a real hiatus or simply a lack of information. The main intrusive body, the Saddowa Gabbro, has been tentively dated as Oligocene (Pieters, 1978).

The third major pre-Miocene unit (Fig. 11) is the Papuan Ultramafic Belt, described in the previous section, which was probably emplaced during the Oligocene as part of a southwest-facing island arc. In addition to the Eocene tonalites and their volcanic equivalents that intrude the mafic rocks, Oligocene boninites on Cape Vogel have also been ascribed to the island arc suite (Jaques and Chappel, 1980).

Subduction and collision were probably responsible for metamorphism on the peninsula. As with the ophiolite emplacement, evidence for age of metamorphism is indirect, being based on the presence or absence of metamorphic clasts in dated sediments. Various ages have been suggested, from lowermost Eocene to Lower Miocene (cf. Brown, 1977). Overall, the evidence points to a time-transgressive metamorphism, which accords well with progressive absorption in an advancing subduction zone. Metamorphic grade increases steadily eastward and northward towards the PUB and the Mesozoic and Eocene sediments near the coast are virtually unaltered. In most areas the sialic rocks are separated from the basal ultramafics by a thin sheet of metamorphics of basic composition that are thought to be altered lavas (Pieters, 1978). The grade of metamorphism attained provides another argument in favour of the subduction hypothesis; it seems too high to have been attained simply by underthrusting a relatively thin ophiolitic nappe.

An additional argument that has been advanced in favor of the frontal arc collision hypothesis is that it accords well with the pattern of Oligocene collision elsewhere in New Guinea. There are obvious geometrical problems in supposing almost simultaneous collision along the entire north New Guinea margin, especially since it seems that the PUB was originally emplaced from the east rather than the north (cf. Connelly, 1979), but most of the evidence does point to widespread collision during this era.

The Papuan Peninsula evidently participated in the general southward-directed phase of subduction indicated by middle Miocene intrusion and volcanism throughout New Guinea (Dow, 1977). The actual areas of exposure of igneous rocks of this age are not large on the peninsula and almost negligible on the outer islands, but larger masses may well be concealed beneath the sea and beneath younger sediments and volcanics. The most easterly known outcrops are the andesites in the western Louisiade Archipelago (Smith, 1976, 1982) and the highly potassic rocks of the shoshonite association on Woodlark Island (Ashley and Flood, 1981), both dated to about 11 Ma.

In the extreme west of the peninsula the Morobe Granodiorite has been dated to 15 to 12 Ma (Page, 1976) and the Talama Volcanics have been placed in the middle Miocene on palaeontological grounds (Brown, 1977). Between these rocks and the outcrops in the islands the pattern is complicated by abundant younger volcanism, but potassic basalts and intrusives in the eastern part of the peninsula have also been radiometrically dated to the middle Miocene (Smith, 1972).

Miocene subduction was accompanied by formation of a deep, fault-bounded basin between the Trobriand Trough and D'Entrecasteaux Islands, with an eastern extension into the Woodlark Basin. The basin has been explored for petroleum; the two holes drilled both bottomed in middle Miocene tuffaceous basement that probably dates the initial extension (Tjhin, 1976). The middle Miocene was also a time of considerable uplift throughout the peninsula.

Extension, vertical tectonism and volcanism all seem to have suffered a hiatus between about 9 and 6 Ma but then recommenced with sometimes spectacular results. Extension associated with opening of the Woodlark

Basin was probably responsible for at least the peralkaline rhyolites of the D'Entrecasteaux Islands (Smith, 1976). Milne Bay, Goodenough Bay and Millins Harbour in the eastern part of the peninsula either formed or expanded in the Pliocene. Subsidence in these areas was accompanied by dramatic uplift in some others, notably the D'Entrecasteaux Islands and the adjacent parts of the mainland (Ollier and Pain, 1980). There was also a resurgence of island arc type volcanism that may point to reactivation of previously subducted lithosphere or of subduction modified mantle (Johnson *et al.*, 1978). Activity continues to the present day with recorded eruptions at Mt Lamington and Wai-Owa (Taylor, 1958). Although earthquakes are quite common in eastern Papua (Fig. 14), they occur almost exclusively at shallow depths. The only possible Benioff zone is a weak feature defined by a very few intermediate-depth shocks in the extreme west of the peninsula, near the Aure Trough (Fig. 2) and far removed from most of the sites of volcanic activity (Ripper, 1982).

H. The Western Peninsula

The "Bird's Head" and "Bird's Neck" areas of western New Guinea form a westward continuation of the Central Orogen, distinguished by the presence of oceanic crust to both north and south (Fig. 12). The peninsula can be subdivided into Kepala Burung (the Vogelkop or Bird's Head), the Lenguru Foldbelt, which occupies most of the neck west of Teluk Sarera, and the Weyland overthrust, which dominates the base of the neck to the south.

The geology of Kepala Burung is better described than that of any other part of Irian. It contains all the province's known oilfields and was therefore discussed in detail by Visser and Hermes (1962). Also, a report has been published on the economic geology of its northeastern part that includes many later observations (D'Audretsch *et al.*, 1966). More recently, a number of preliminary reports have been published by the BMR/GRDC teams (cf. Robinson and Ratman, 1978; Dow and Barlow, 1982).

The oldest rocks of the peninsula, and possibly the oldest exposed anywhere in New Guinea, belong to the Kemum Formation, which is at least partly Upper Silurian and which consists of a thick and remarkably uniform succession of thinly-bedded sands, silts and clays with widespread intercalations of basic igneous rocks and abundant evidence of submarine slumping. The Kemum Formation was intensively folded and metamorphosed to phyllitic slate in an orogeny that is usually regarded as pre-middle Carboniferous. Small granite plutons were thought by D'Audretsch *et al.* (1966) to have been intruded in the closing stages of this orogeny but Dow (1980)

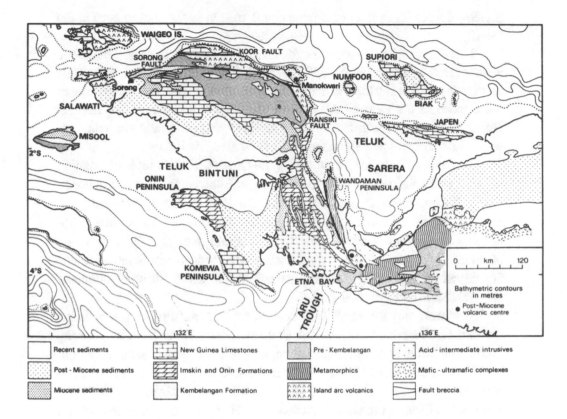

Fig. 12.   Geology of the western peninsula region of New Guinea (after Visser and Hermes, 1962; Robinson and Ratman, 1978; and Dow and Barlow, 1982).

notes that radiometric dates span the time interval from Devonian to Permian.

The metamorphic mountains were eroded during the Permian and sediments of the Aifam Formation were laid down over a wide area. This formation, which includes both marine and terrestrial deposits, seems to be absent in eastern Kepala Burung, where younger Mesozoic rocks rest directly on the Kemum.

Shelf conditions prevailed throughout the Mesozoic in Kepala Burung as elsewhere in New Guinea. The Tipuma Formation was deposited during the Triassic and Jurassic as a terrestrial to littoral clastic succession derived from the north and northeast, and is locally absent where removed by erosion following epeirogenic uplift in the early Triassic. The subsequent marine transgression did not reach the more strongly uplifted western part of the peninsula until the Eocene. Thus the overlying Kembelangan Formation, in which marls and limestones are an important although subordinate component, decreases in thickness from about 1000 m in the east to zero in the west.

There is no sharp break at the Mesozoic–Tertiary boundary. The clastic element in the sediments gradually declines, however, and aphanitic pelagic limestones of the Imskin Formation are in places the lateral equivalents of the clastics of the Upper Cretaceous–Paleocene D member of the Kembelangan Formation. Both the Imskin and Kembelangan grade gradually upwards into the massive shoal facies New Guinea Limestone.

Tectonic activity increased during the Oligocene, with rapid alternation of uplift and subsidence, prior to collision with the island arcs to the north. The collision suture is now marked by the Ransiki Fault (Robinson and Ratman, 1978) and the middle Miocene Kais Formation, the youngest subdivision of the New Guinea Limestone, is found overlying older rocks on both sides of this fault. Also in the Oligocene, the Bintuni and Salawati basins developed south of the basement massif, and began to fill with sediments derived from the north and east.

Robinson and Ratman (1978) suggest that at the time of the collision the Ransiki Fault had an orientation similar to that of the present-day Sorong Fault. The latter is certainly not a simple feature. It was originally thought by Visser and Hermes (1962) to be transcurrent because of the presence of large exotic blocks, but there is little direct evidence for transcurrent movement in the form of offsets of watercourses. Nor can it be a simple collision suture since large outcrops of Kemum Formation and Palaeozoic granites occur on the northern side as well as on the southern. Recent mapping has defined an important fault, the Koor Fault, which runs parallel to the Sorong Fault and some twenty kilometers to the north and which does appear to separate the island-arc and continental environments (Dow, 1981). Another important recent observation is that middle Miocene andesitic volcanics and volcaniclastics overlie the basement rocks between the two faults (Dow, 1980). Volcanics of this age are not known in the peninsula south of the Sorong Fault but are widely distributed in the Central Orogen and on the Papuan Peninsula.

Vertical movements intensified during the Plio–Pleistocene and resulted in gentle folding of the Miocene sediments and in places in their wholesale removal by erosion. There was also some intermediate volcanism east of the Ransiki and north of the Sorong Fault zones. An agglomerate cone about 50 km west of Manokwari might be the most recently active eruptive center in Irian (Robinson and Ratman, 1978). The total post-Miocene uplift in the northeast of the peninsula has amounted to about 2000 to 3000 m, while the Wandaman Peninsula, which protrudes into Teluk Sarera further south, has risen at least 1000 m.

The general geological trends in Kepala Burung appear to be east–west, but north–south trends predominate in the Lenguru Foldbelt. The belt consists of complexly interfolded sediments of varying ages that have suffered varying degrees of metamorphism. The oldest rocks, exposed in the cores of the anticlines, belong to the Triassic–Jurassic Tipuma Formation, but the folding also affected the Jurassic–Cretaceous Kembelangan Formation and the Paleocene–Miocene New Guinea Limestone (Dow, 1981). The Wandaman peninsula is separated from the foldbelt by the Wandaman Fault Zone, a series of straight faults marked by zones of intense shearing (Dow, 1981). Amphibolites have been mapped on the peninsula and probably include equivalents of the Kembelangan A member, older sediments and Permo–Triassic granites; Ryburn (1978) reports metamorphic dates of only 3 to 4 Ma.

Visser and Hermes (1962) note that the foldbelt is a sharply bounded, localized feature that narrows rapidly to the north, is cut off by the coastline in the south and disappears beneath alluvial plains in the southeast. Throughout the area it occupies, the folds are disturbed by both longitudinal and cross-faults. Fold patterns are irregular but the western flanks are usually steeper and are sometimes overturned. The most tectonically complicated area is at the head of Teluk Bintuni, where the deep Bouguer anomaly low

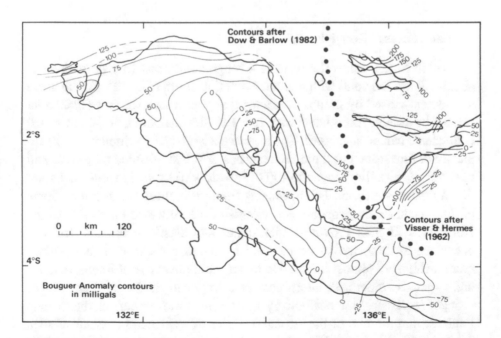

Fig. 13.   Generalized Bouguer anomaly map, western New Guinea (after Visser and Hermes, 1962, and Dow and Barlow, 1982).

that underlies most of the Foldbelt reaches its minimum value of −75 mgal (Dow and Barlow, 1982). Just north of this region, the belt ends in an unusual sigmoidal bend.

The fold structures and the gravity low both die away in the region south of Teluk Sarera (Fig. 13). Recent mapping has shown that most of this area is occupied by rocks that have been thrust at least 35 km southwards over platform sediments of the Australian continental margin. The overthrust consists of metamorphics and ultramafics similar to those found along the northern margin of the Central Orogen, intruded by a middle Miocene granitic batholith that predates the thrusting (Dow and Barlow, 1982). The toe of the thrust forms an isolated, arcuate mountain range with peaks 400 m high. Some of the associated faults displace Recent alluvial fans and are probably still active. A very young date for the thrust is probable and is consistent with the young metamorphic dates from the Wandaman Peninsula. Another interesting indication of Recent tectonism has been the discovery of two probably Holocene andesitic cones near where the western margin of the overthrust abuts against the southern margin of the Lenguru Foldbelt (Dow and Barlow, 1982).

V. MARINE BASINS

A. Teluk Sarera (Geelvink Bay)

Teluk Sarera takes the form of a rough equilateral triangle with sides about 250 km long and with one apex pointing south (Fig. 12). To the north it is almost closed by groups of islands that Visser and Hermes (1962) note cannot be readily correlated with those of the mainland, although rough equivalents can be suggested. The northern islands (Biak–Supiori group) are mainly of limestone but there are outcrops of schist intruded by basalts and overlain by basaltic lavas and tuffs. A wide variety of igneous rocks are found further south, on Japen, ranging from peridotites to andesites. Some of the volcanics are probably equivalents of the Auwewa Formation of the North New Guinea Province. Limestones and clastic sediments that also occur are all reported to be intruded by igneous rocks and sedimentary relationships are almost impossible to determine because of intensive faulting. Japen is sometimes interpreted as a fragment of the northern ranges brought to its present position by transcurrent movement on the Sorong Fault. It may be one of the strongest pieces of evidence in favour of such movements, although it lies rather to the south of the trend of this fault as extrapolated from Kepala Burung. Dow (1981) notes that the gravity pattern in Teluk Sarera (Fig. 13) is not consistent with the hypothesis that Japen marks the boundary between continental and oceanic crust.

Teluk Sarera itself is generally more than 1500 m deep and reaches to more than 2000 m in some places. Below the 500 m isobath it forms a closed, more or less flat-bottomed depression whose northern and western slopes rise more steeply than those to the south and east. Although there has been some exploration for oil and a number of holes have been drilled offshore near the eastern end of Japen (Rutherford and Qureshi, 1981), the bay is not generally well known and deductions as to its structure and origins have to be made on the basis of the rather scantily known geology of its margins. The western shore seems to be controlled by the southeast-trending Ransiki Fault, which Robinson and Ratman (1978) have described as an early Miocene collision suture. They believe that at the time of collision the fault had an east–west trend and involved sinistral transcurrent movement as well as thrusting, and also that it has assumed its present NNW orientation as a consequence of rotation of Kepala Burung. It passes offshore about 70 km south of Manokwari and is not seen to cross the "Bird's Neck" at the southern apex of Teluk Sarera, unless it can be regarded as having its continuation in the Wandaman Fault Zone, which separates the metamorphic rocks of the Wandaman Peninsula from the Lenguru Foldbelt.

The eastern shore of the bay is largely occupied by young sediments in the area of a deep (-115 mgal) linear gravity minimum (Visser and Hermes, 1962). Near the southern apex of the bay the minimum seems to be offset to the east by the high associated with the Weyland thrust, but as this area lies at the limit of the presently available regional gravity coverage (Dow and Barlow, 1982), the pattern is not yet completely defined (Fig. 13).

Most reconstructions of the history of Teluk Sarera show the bay forming as a closed feature as a result of the rotation of Kepala Burung (Visser and Hermes, 1962; Robinson and Ratman, 1978; Hamilton, 1979). Such a rotation would require a considerable amount of crustal shortening to have taken place in some areas to the north and east, and it seems possible that the sinuosity of the neck and the formation of the Lenguru Foldbelt arose not from compression from the Banda Arc region but from compression due to the Oligocene–early Miocene collision at the north New Guinea margin. Visser and Hermes (1962) date the folding to the middle Miocene. In central New Guinea the forces involved in this collision would necessarily have been accommodated by thrust shortening, but west of the limits of the main continental block a narrow tongue of continental crust could have responded by shearing southwestwards along the continental margin. This slip zone could now be marked by the eastern shore of Teluk Sarera and the parallel gravity low (which implies at least 6 km of light sedimentary infill), by the dislocations at the southern end of the foldbelt indicated by the Recent volcanism and the Weyland thrust sheet and by the Aru Trough. Although the latter has been widely regarded as a subduction trench (cf. Hamilton, 1979), Bowin *et al.* (1980) show seismic reflection profiles in which there are no obvious subduction characteristics, and suggest that at the present time the trough is an extensional feature. It seems quite possible that it was initially the site of a major transcurrent fault along the continental margin, which now merges with the Banda Arc collision system. Some southward movement of the lithosphere that formerly lay south of the "Bird's Neck" could have been accommodated in the Banda system. When the southerly movement that opened Teluk Sarera to the rear of the advancing crustal segment could no longer be absorbed in this way, the advance might have been halted with the piling up of the asymmetric folds of Lenguru, and possibly later with thrusting at the base of the Bird's Neck.

B. Caroline Basin

The Caroline Basin is discussed in detail elsewhere in Volume 7B of this title, and is considered here only insofar as processes within it may have affected the New Guinea borderland. The sea is divided into an eastern and western subbasin by the north-trending Eauripik Rise. Bracey (1975) has suggested that the crust of the basin was formed in an extensional phase beginning at about 42 Ma and lasting until 25 Ma, with later extension in the far northwest, well away from the New Guinea margin. Weissel and Anderson (1978) have discussed the evidence for the existence of the basin as an independent plate of lithosphere and note that both magnetic lineations and Deep Sea Drilling Project results strongly suggest that the entire basin crust was formed during the Oligocene. They view the Eauripik Rise, which at a depth of 2.5 km stands 2 km above the general floor of the basin and has approximately double the normal oceanic crustal thickness, as having been formed by magmatism at a "leaky" transform fault at right angles to the spreading ridge. They support the suggestion, also made by Bracey (1975), that there was subduction at the trough along the northern New Guinea coast in the Miocene. Partial subduction of the thickened crust of the rise might explain some of the anomalous features of the onshore geology in the vicinity of the international border, notably the discontinuity in the line of intermontane lows. Hamilton (1979) presents seismic reflection profiles across the trough that strongly support the idea of a subduction origin.

C. Bismarck Sea

The Bismarck Sea (Fig. 8) is possibly the best known of the small ocean basins that surround New Guinea. It consists of two distinct subbasins, the

Manus and New Guinea basins, separated by the NW-trending Willaumez–Manus Rise (Taylor, 1979). These three units are enclosed within the loop formed by the nonvolcanic section of the Outer Melanesian Arc to the north and the large islands of New Guinea and New Britain to the south. The most striking geophysical feature of the sea is the narrow zone of earthquake foci that crosses it from east to west. Fault plane solutions on this "Bismarck Earthquake Lineament" (BEL) indicate left-lateral movement and have been interpreted as showing that much of the shear of the Pacific past the northern edge of the Australian plate is occurring along this line (Johnson and Molnar, 1972). The ambiguity in direction inherent in fault plane solutions would also allow these solutions to relate to shear on transform faults crossing a spreading axis; this is an attractive alternative since it is now known that the earthquake zone is not a single linear feature but that it is divided into three or four straight-line sections of which the easternmost is entirely separate from the others and parallels the southern coast of New Ireland. Such a pattern is difficult to reconcile with well-established transcurrent faulting but could be explained by a combination of spreading and transcurrent movement. However, even the strongest advocates of spreading have agreed that all the fault plane solutions obtained to date have recorded almost pure strike-slip motion (cf. Taylor 1979).

Magnetic anomalies in the Manus Basin were claimed by Connelly (1974) to run parallel to the BEL and to have a degree of symmetry about it. Taylor (1979), using additional data, plotted magnetic lineations trending ENE within the basin, which the BEL enters from the west on an ESE trend. The BEL seems to terminate within the basin and reappears further north, near New Ireland. Taylor (1979) dates the magnetic anomalies to the period from about 3.5 Ma up to the present day, deducing a fast total-opening rate of about 13 cm/yr.

The hypothesis of local spreading in the Manus Basin is supported by the almost complete absence of sediments in its central region, but encounters considerable obstacles when viewed in a regional context. There seems, for example, to be no corresponding offset in the smooth curves of New Britain or of the Outer Melanesian Arc and Trench, nor does the evidence for major strike-slip displacements on New Ireland (Section 4.3) seem to have been integrated into discussions of basin development. The role of the Willaumez Rise is also rather uncertain (cf. Johnson et al., 1974). This feature seems to have a very sharp boundary with the Manus Basin but a much more gradual one with the New Guinea Basin, where magnetic lineations have not yet been recognized. One notable feature of the rise is the present-day activity at each end; the Willaumez Peninsula of north New Britain is entirely volcanic in origin and in the northwest there have been recent eruptions in the isolated volcanic region of St. Andrew Strait. Basalts were erupted from the older centers but the younger volcanoes, which may mark the site of a developing ring fracture, are very silicic and moderately alkaline (Johnson and Smith, 1974). The activity may be the consequence of a local mantle plume ("hot spot") that has produced melts from a number of crustal levels (Johnson *et al.*, 1978).

In the New Guinea Basin there are thick sediments along the epicenter line, particularly near the north coast of New Guinea. These may have been very rapidly deposited by turbidity currents originating in the Ramu–Sepik Delta, or there may have been no recent spreading in this area. Whether as a spreading center or as a transcurrent fault zone, the BEL certainly divides the Bismarck Sea into two small plates. The thickness of the Recent sediments deposited on the southern plate increases southwards towards the land and, as might be predicted, is particularly great near the volcanoes of the Bismarck Arc. Sedimentation increases northwards on the northern plate towards Manus, New Hannover and New Ireland, but these relatively small islands have been less significant as sediment sources than have the large land masses along the sea's southern margin. Interestingly, although seismic refraction work has failed to locate any material with a velocity in the continental 5.8 to 6.5 km/sec range, the estimated Moho depths of 15 to 20 km are much greater than would be expected for oceanic crust (Connelly, 1976).

The BEL converges with the New Guinea coast in an area of strong shallow seismic activity just west of the Sepik River mouth. It is possible that the shear, if such it is, continues along the north coast ranges and links with the Sorong Fault Zone in Kepala Burung.

D. Solomon Sea

The Solomon Sea is bounded to the northwest by the New Britain Trench and to the northeast by the North Solomons Trench, each of which is the surface trace of a very active subduction zone. There is also a linear zone of deeper water, the Trobriand or Lusancay Trough, along the sea's southern margin. It has been variously suggested that this trough is a former subduction trench, a tensional graben or a locus of strike-slip faulting; all these solutions could be made compatible with the information at present available. There has been no dating work done on samples from the sea's basaltic floor, unless the Cretaceous basalts of the Papuan Ultramafic Belt can be regarded as such. Heat flow is moderately high (Halunen and von Herzen, 1973) and sediment cover is thin, suggesting a younger, perhaps Miocene or Oligocene, age.

The sea has not been systematically surveyed by geophysical methods but a certain amount of refraction work has been done. In the center and east the crust is about 10 km thick, which is close to the thickness found in normal oceans and well within the range in recognized marginal basins (Furumoto *et al.*, 1968). In the west, however, crustal thicknesses of as much as 30 km have been interpreted from refraction data, supported by the existence of a considerable area with a free-air anomaly of −150 mgal or less (Fig. 14). Areas further east with similar water depths have regional free-air anomalies of +177 mgal and Finlayson *et al.* (1976) emphasize that the difference cannot be explained simply as the effect of a thick pile of young sediments. Although there is a thick wedge of deltaic sediments in the extreme west of the sea, the low-gravity area lies east of this and has a rugged bottom topography. Moreover, it is traversed by a canyon that feeds sediment from the delta to the New Britain Trench but is not itself a site of present-day sediment accumulation.

Several suggestions have been made to explain the presence of thick crust and of these one of the most promising and certainly the most interesting is that offered by Johnson and Molnar (1972) and supported by Johnson (1976). In the first of these papers, which discussed the seismo-tectonics of much of the southwest Pacific, the existence of a very complex triple junction at the western end of the Solomon Sea was noted and the observed seismic pattern was explained by supposing that ocean crust dips almost vertically or is even being overturned as it encounters the New Britain subduction system. Johnson (1976) points to the abrupt disappearance of deep focus earthquakes in the New Britain Benioff zone west of about 150° E and the accompanying abrupt change in chemical character of the arc volcanics. He also draws attention to the splitting of the New Britain Trench into two branches at its western end, separated by a block of relatively elevated seafloor, that, he suggests, may be the surface expression of a major imbricate wedge. Such a wedge might well have been produced under the complex compressional regime consequent on active subduction close to a continental mass. Effectively, a new fault may be developing south of the present surface trace of the New Britain Trench that will merge with the subduction zone at depth. The intervening slice of crust and uppermost mantle will thus be only partially subducted. A hypothesis of this nature could explain the frequent presence of a thin sheet of metamorphosed basic rocks beneath ophiolite overthrusts.

If the Solomon Sea was largely formed in the Oligocene or Miocene, New Britain was presumably adjacent to the Papuan Peninsula prior to that time. A young age for the sea is in fact necessary to any theory that supposes the Papuan Ultramafic Belt was the frontal part of the arc whose major volcanic development is recorded in the Paleogene igneous complexes of New Britain. The main objection to this hypothesis lies in the 600 km depth

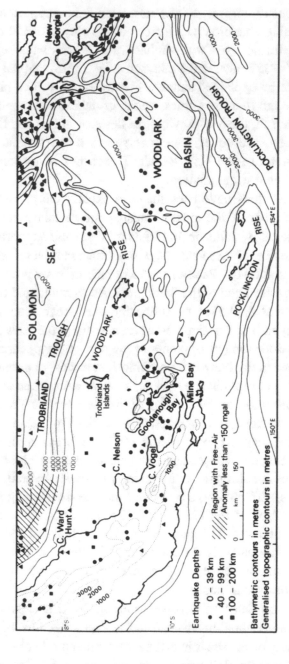

Fig. 14. Seismicity of the Papuan Peninsula and the Woodlark Basin. (Area of low free-air anomaly after Bureau of Mineral Resources, 1979).

of the Benioff zone beneath New Britain. Unless the deeper shocks are associated with relics of earlier subduction, some 800 km of seafloor must have been created (more, if allowance must also be made for southward subduction in the middle Miocene) and 700 km must then have been absorbed beneath New Britain in a relatively short time. Known rates of spreading in various parts of the world and estimates of present-day convergence in the New Guinea region (10 cm/yr; cf Ripper, 1982) are only just compatible with such a sequence of events.

E. Woodlark Basin

The Woodlark Basin lies between the Woodlark and Pocklington rises which together may be regarded as the largely submerged extensions of the Papuan Peninsula (Fig. 14). The islands of the Louisiade Archipelago along the Pocklington Rise south of the basin are made up mainly of metamorphic rocks very similar to those forming the core of the western part of the peninsula (Smith, 1972), while islands closer to the mainland are formed of basaltic pillow lavas identical to those of the eastern part. The geology of the Woodlark Rise to the north is less well known since noncoralline rocks have been described only from Woodlark Island itself. These are early Tertiary low-K basalts and middle Miocene high-K shoshonites, both of which have counterparts on the Peninsula (Ashley and Flood, 1981).

Refraction surveys by the University of Hawaii in 1966 showed that the crust of the western part of the basin is less than 10 km thick and is underlain by low-velocity mantle (Furumoto et al., 1968), while detailed bathymetric data collected by the Royal Australian Navy in 1967 showed the basin floor to be extremely rugged with probably only thin sediment cover. These observations, together with the line of weak seismicity along the basin axis, led to a revival of Carey's (1958) proposal that the basin was of rift origin (cf. Milsom, 1970).

The results of a geophysical cruise in 1971 have been reported by Luyendyk et al. (1973). They concluded from the very thin sediment cover that they observed in the central part of the basin, from the block faulting, from the pattern of magnetic anomalies and from the distribution of earthquakes that seafloor spreading has occurred in the area and is continuing. There seems to have been an early stage of rifting followed by a period of near or complete quiescence, since an older sediment pile was seen on some records to have been rifted apart by more recent movements. This original rift may have formed in the middle Miocene, perhaps as a basin marginal to a volcanic arc in eastern Papua. Certainly a graben formed at this time north of the D'Entrecasteaux Islands (Tjhin, 1976). The present spreading phase was seen as occurring during the last 3 Ma. Weissel et al. (1982), using additional data gathered since 1971, elaborated on this conclusion, postulating an initiation of spreading at some time before 3.5 Ma in the east and progressively later towards the west.

The eastern and western ends of the Woodlark Basin are both of considerable interest. In the east or northeast the basin trends at right angles to the Solomon chain, interrupting the deep trench that is elsewhere a feature of the southwestern flank of these islands. Where it is developed, the trench is associated with a well-developed Benioff zone and there seems to be no doubt that both the Solomon and Coral seas are being subducted. The tectonics of the Woodlark Basin–Solomon Islands triple junction near the very volcanically active island of New Georgia have been discussed in some detail by Weissel *et al.* (1982).

In the west the basin narrows towards its "pole of rotation". Before this is reached, however, the main basin is replaced by a number of smaller isolated features such as Milne Bay and Goodenough Bay, both of which are very deep for such landlocked basins. They are clearly rift structures, although it seems unlikely that there has been any actual generation of oceanic crust so close to the Papuan mainland. The structure and morphology of Milne Bay have been discussed by Jongsma (1972), who shows that subsidence has been particularly rapid during and since the Pleistocene, although the basic structure may be much older.

VI. PLATE TECTONIC EVOLUTION

With so many unresolved problems in even the Tertiary in New Guinea, it is obvious that reconstruction of the pre-Tertiary geological history will be extremely difficult. The sequence of tectonic events in Fig. 15 is highly speculative and, in particular, the periods of island arc activity are suggested solely because of the occurrence of intermediate volcanics of the appropriate age. The present volcanism of the New Guinea mainland demonstrates however that such rocks can be produced without simultaneous subduction.

One of the most striking geological features of the area is the abrupt change from the northerly trends of northern Australia to the strong west to northwest trends of the New Guinea mainland. Hamilton (1979, p. 235) argues strongly that parts of Southeast Asia are fragments of the Australian continent that have been displaced thousands of kilometers westwards by spreading in the Pacific Ocean during the Triassic and early Jurassic. Support has recently been provided for this hypothesis by the identification of Permian faunas in Kepala Burung that have some affinities with assemblages from Western Australia but much closer links with Thailand (Archbold *et al.*, 1982). The distribution of Permian granitic rocks and Palaeozoic met-

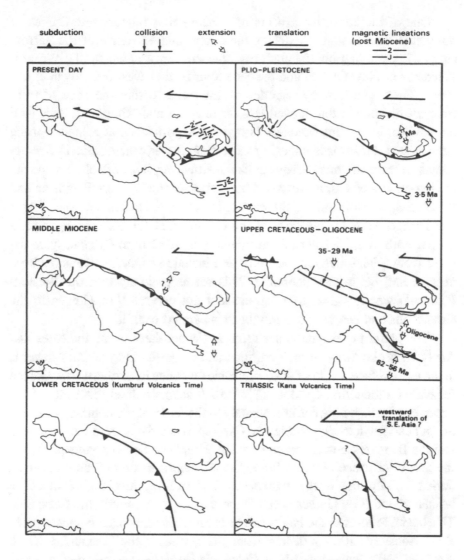

Fig. 15.   Hypothetical tectonic evolution of the New Guinea region.

amorphics can also be seen as evidence for displacement of this sort. The outcrops in the Kubor Range are almost directly on strike from their analogues in eastern Australia, while further to the northwest there are thought to be Permian granites in the Border Mountains. Still further north and west, granites in Kepala Burung have been definitely dated as Permian or earliest Triassic, and intrude metamorphics that in places contain Silurian fossils. Conversely, the Palaeozoic rocks of the Central Orogen in Irian have been only very weakly metamorphosed and in this and some other respects resemble Cambro–Ordovician rocks in the parts of Australia directly to the south.

One of the attractive aspects of a theory that relates New Guinea to Southeast Asia is that it explains the anomalous existence of the narrow strips of Australian continental crust that are represented by the Western Peninsula of New Guinea and the Sula Spur further west (cf. Bowin *et al.*, 1980). These can best be regarded as fragments within the zone of transcurrent movement that moved less far than the main Southeast Asia land masses. An interesting corollary is that the hypothesis that Kepala Burung later moved northwards under pressure from the advancing Banda Arc becomes unnecessary and perhaps unlikely. However, it is difficult to reconcile the hypothesis of simple westward translation of the Western Peninsula and the Sula Spur with some of the reported Cretaceous facies patterns. Visser and Hermes (1962) are clearly of the opinion that the Kembelangan sediments, both in the Western Peninsula and in the Central Orogen, were derived from a land mass to the south and include sediments deposited both near to and far from a shoreline. This seems to imply that the Western Peninsula was marginal to Australia and not central New Guinea in the Cretaceous and has only recently been separated from it.

The pattern of magnetic lineations in the southern ocean indicates that Australia broke away from Antarctica and began its long drift northwards in the late Cretaceous, but there is volcanic evidence in northern New Guinea for earlier subduction episodes (Fig. 15). It seems unlikely that any of the ocean crust which now lies immediately to the north of the continent's active margin existed in the Mesozoic; the Caroline Basin dates from the Oligocene and the Bismarck and Solomon seas are thought to be even younger, while the first brief phase of extension in the Woodlark Basin probably occurred during the Miocene. Rather unexpectedly, therefore, the oldest of the ocean basins around New Guinea seems to be the one to the south, the Coral Sea. This began to form in the Paleocene or earliest Eocene in a movement probably associated with, but distinct from, the opening of the Tasman Sea which removed New Zealand and New Caledonia from the Australian continental margin. Little can be deduced about former ocean basins west of New Guinea because the continental crust of Australia and New Guinea is in virtually direct contact with the Banda Arc collision zone and any formerly intervening material has presumably been destroyed. It seems that the only fragments of Mesozoic oceanic crust that can be identified near New Guinea may be the ophiolite masses that have been thrust over the island from the north and east.

Evidence of late Cretaceous to Oligocene island arc activity is preserved along the north coast ranges of the New Guinea mainland and on the islands that fringe the Bismarck Sea. It is not yet clear whether the arc faced south and absorbed the oceanic crust of the Australian plate or faced north and was separated from the continental mass by only a narrow marginal basin. The first of these alternatives requires the arc to have formed autonomously in mid-ocean but otherwise seems to present the fewest difficulties. If subduction had developed southwards beneath the Australian margin, it seems likely that the initial volcanism would have taken place within the continental mass, as in the Andes, and that the volcanic belt would only later have separated away as a marginal basin developed. The Kumbruf volcanics of Papua New Guinea might be remnants of such a phase of activity, but are generally considered to be early rather than late Cretaceous in age (Dow, 1977). There is no evidence for the incorporation of Australian continental basement in to any of the islands of the Bismarck Archipelago. Furthermore, when arc activity ceased in the Bismarcks in the earliest Miocene, it was replaced by true continental margin subduction which is attested by Miocene intermediate volcanics throughout central New Guinea and by a now inactive trench at the southern margin of the Caroline Sea. If this merely replaced a southward subducting arc further north, its origin might be difficult to explain, but if it developed after collision of the continent with a southward-facing arc that could then no longer accommodate the northward drift of the Australian plate, its role is obvious. Finally, if the Papuan Ultramafic Belt and the other ophiolite masses of northern New Guinea are to be regarded as consequences of collision, whether as parts of actual frontal arcs or as fragments of ocean crust caught up in a collision process, a destructive margin absorbing crust to the south or west seems necessary at least during the Eocene–early Miocene interval. The development of the Caroline Basin during the Oligocene presumably accelerated the convergence of the arc systems with Australia, and so hastened the collision.

It is difficult to understand why the subduction zone establshed along the New Guinea margin at the beginning of the Miocene ceased to be active at about the close of the same era, but it seems likely that the cause lay in a change of spreading regime in the Pacific. Developments in the Solomon Islands, where there may have been a reversal of arc polarity due to collision with the Ontong Java Plateau, may also have influenced the tectonic style in New Guinea. Since that time the New Guinea region as a whole seems to have been subjected to pervasive left-lateral shear which has shattered it into a number of microplates and complex imbricate zones. The present disposition of the tectonic elements around the island can best be understood if they are seen as very temporary features in a system that is undergoing major readjustment. Analyses of magnetic lineations in the Woodlark and Manus basins seem to indicate that these small oceans began to develop almost simultaneously, at about 3 to 4 Ma. It is also a striking fact that the New Britain Arc which, in terms of release of seismic energy and depth of

Benioff zone, is one of the world's most active plate margins (Hedervari and Papp, 1977), acts merely to absorb the relative motions of two units, the South Bismarck and Solomon Sea plates, which are of quite trivial size in terms of global tectonics.

One other anomaly of present-day New Guinea is the wide distribution of intermediate volcanic activity which cannot be regarded as being related to any existing subduction process. The very alkaline volcanic centres of the eastern D'Entrecasteaux Islands may be the results of extensional movements but it is still not clear whether the activity is a direct consequence of rifting or a by-product of the thermal conditions that caused it. It does seem conceivable that the recently active volcanoes of the Papuan Peninsula, together with those of the Central Highlands, are caused by reactivation of mantle modified by an earlier subduction episode, the trigger perhaps being the changes in the directions of major plate convergence that initiated the spreading in the Manus and Woodlark basins. Alternatively, the mainland volcanoes, as well as those of St. Andrew Strait, may be related to processes similar to those that produced Tibesti in the middle of the African continent and Hawaii in the middle of the Pacific. In New Guinea we may be seeing the results of the same processes occurring in a continental margin environment.

VII. FURTHER WORK

The summary of the geological history of New Guinea emphasizes the major gaps that still exist in present knowledge. One of these, the onshore geology of Irian, has received a considerable amount of attention in recent years and much new information on the Western Peninsula region should reach publication in the near future. Detailed studies of the Mesozoic facies belts should help to define the pre-Tertiary position of the peninsula. Correlations between New Guinea and Misool, west of Kepala Burung, where a nearly complete Mesozoic sedimentary sequence is preserved (Pigram *et al.*, 1982) are likely to prove especially important.

Two major new "transect" programs were proposed at the second SEA-TAR workshop in Bandung in 1978 (CCOP–IOC, 1980; pp. 157–161). The first is to run parallel to and just to the east of the international border in New Guinea, beginning in the Caroline Sea north of the junction of the North New Guinea and Outer Melanesian trenches and extending from there through the central highlands to the south coast. The second, a double transect, is designed to have one branch crossing the Solomon Sea from the Papuan Peninsula to Bougainville and thence to the Ontong Java Plateau and another beginning in the Solomon Sea, crossing New Britain and the

Manus Basin and ending in the Caroline Basin north of the Outer Melanesian Trench. Any work done on these transects will obviously be very valuable but in the absence of any plans for deep drilling in the Bismarck or Solomon seas, or in Teluk Sarera, the prospects for a significant improvement in geological control in the offshore areas seem poor. Deep drilling in these regions might well bring considerable benefits in terms of our understanding of the evolution of active continental margins and collision zones, and of the mechanisms of emplacement of ophiolite masses.

ACKNOWLEDGMENTS

I thank Ian Ripper of the Geophysical Observatory, Port Moresby, for providing invaluable plots of earthquake epicenter locations and copies of his forthcoming publications in draft form.

REFERENCES

Adams, B. W., 1973, Ertsberg project, *Min. Mag.*, v. 129, p. 310–323.

Archbold, N. W., Pigram C. J., Ratman, N. and Hakim, S., 1982, Indonesian Permian brachiopod fauna and Gondwana–Southeast Asia relationships, *Nature*, v. 296, p. 556–558.

Ashley, P. M. and Flood, R. H., 1981, Low-K tholeiites and high-K igneous rocks from Woodlark Island, Papua New Guinea, *J. Geol. Soc. Aust.*, v. 28, p. 227–240.

Australasian Petroleum Co, 1961, Geological results of petroleum exploration in western Papua, *J. Geol. Soc. Aust.*, v. 8, p. 1–133.

Bain, H. J., Mackenzie, D. E., and Ryburn, R. J., 1975, Geology of the Kubor Anticline, Central Highlands of Papua New Guinea, *Bur. Miner. Resour. Aust., Bull.* 155.

Bain, J. H., Davies, H. L., Hohnen, P. D., Ryburn, R. J ., Smith, I. E., Grainger, R., Tingey, R. J. and Moffat, M. R., 1972, Geology of Papua New Guinea (1:1,000,000 map), *Bur. Miner. Resour. Aust.*, Canberra.

Bowin, C., Purdy, G. M., Johnston, C., Shor, G., Lawver, L., Hartono, H. M. S. and Jezek, P., 1980, Arc-continent collision in Banda Sea region, *Bull. Am. Assoc. Pet. Geol.*, v. 64, p. 868–915.

Bracey, D. J., 1975, Reconnaissance geophysical survey of the Caroline Basin, *Geol. Soc. Am. Bull*, v. 86, p. 775–784.

Brown, C. M., 1977, Explanatory notes, Sheet SC/55-2, Yule (with 1:250,000 geological (map), *Bur. Miner. Resour. Aust.*, Canberra.

Brown, C. M., Pigram, C. J., and Skwarko, S. K., 1980, Mesozoic stratigraphy and geological history of Papua New Guinea, *Palaeogeogr. Paleoclimatol. Palaeoecol.*, v. 29, p. 301–322.

Bureau of Mineral Resources, 1979, Gravity Map of Melanesia (1:5,000,000), Geophysical Branch, Canberra.

Burns, B. J., and Bein, J., 1980, Regional geology and hydrocarbon potential of the Mesozoic of the western Papuan Basin, Papua New Guinea, *Aust. Pet. Explor. Assoc. J.*, v. 20, p. 1–15.

CCOP-IOC, 1980, Studies in east Asian tectonics and resources (SEATAR) United Nationsl ESCAP, *CCOP Tech. Pub.* 7A.

Carey, S. W., 1958, The tectonic approach to continental drift, in: *Continental Drift, a Symposium* Carey, S. W., ed., Tasmania: University of Hobart, p. 1–177.

Connelly, J. B., 1974, Structural interpretation of magnetometer and seismic profiler records in the Bismarck Sea, Melanesian Archipelago, *J. Geol. Soc. Aust.*, v. 21, p. 459–469.

Connelly, J. B., 1976, Tectonic development of the Bismarck Sea based on gravity and magnetic modelling, *Geophys. J. Astron. Soc.*, v. 46, p. 23–40.

Connelly, J. B., 1979, Mode of emplacement of the Papuan Ultramafic Belt, *BMR J. Aust. Geol. Geophys.*, v. 4, p. 57–65.

D'Audretsch, F. C., Kluiving, R. B. and Oudemands, W., 1966, Economic geological investigations of the NE Vogelkop (Western New Guinea), *Verh. K. Ned. Geol. Mijnbouwk. Genoot.* v. 23.

Davies, H. L., 1971, Peridotite-gabbro-basalt complex in eastern Papua: an overthrust plate of oceanic mantle and crust, *Bur. Miner. Resour. Aust.*, Bull. 128.

Davies, H. L., 1980, Crustal structure and emplacement of ophiolite in southeastern Papua New Guinea, *Collo. Int. C.N.R.S.* No. 272, p. 17–33.

Davies, H. L. and Norvick, M., 1974, Explanatory notes, Sheet SB/54-7, Blucher Range (with 1:250,000 geological map), *Bur. Miner. Resour. Aust.*, Canberra.

Davies, H. L. and Smith, I. E., 1971, Geology of eastern Papua, *Geol. Soc. Am. Bull.*, v. 82, p. 3299–3312.

DeBroin, C. E., Auberton, F. and Ravenne, C., 1977, Structure and history of the Solomon–New Ireland region, in: *Proceedings of the International Symposium on Geodynamics in the Southwest Pacific*, Paris: Editions Technip, p. 37–50.

Denham, D., 1969, Distribution of earthquakes in the New Guinea–Solomon Islands region, *J. Geophys. Res.*, v. 74, p, 4290–4298.

Dow, D. B., 1968, Geological Reconnaissance in the Nassau Range, West New Guinea, *Geol. Mijnbouw*, v. 47, p. 37–46.

Dow, D. B., 1977, Geological synthesis of Papua New Guinea, *Bur. Miner. Resour. Aust. Bull.*, no. 201.

Dow, D. B., 1980, Irian Jaya geological mapping project, in: *Geological Branch Summary of Activities*, 1979, Canberra: Bureau of Mineral Resources, Report 222, p. 231–236.

Dow, D. B., 1981, Irian Jaya geological mapping project, in: *Geological Branch Summary of Activities, 1980*, Canberra: Bureau of Mineral Resources Report 230, p. 277–292.

Dow, D. B., Smit, J. A. J., Bain, J. H. G., and Ryburn, R. J., 1972, Geology of the South Sepik region, New Guinea, *Bur. Miner. Resour. Aust. Bull.*, no. 133.

Dow, D. B., and Barlow, B. C., 1982, Irian Jaya geological mapping project, in: *Geological Branch Summary of Activities, 1981*, Canberra: Bureau of Mineral Resources. Report 239, p. 231–241.

Finlayson, D. M., Muirhead, K. J., Webb, J. P., Gibson, G., Furumoto, A. S., Cooke, R. J. S. and Russel, A. J., 1976, Seismic investigation of the Papuan Ultramafic Belt, *Geophys. J. R. Astron. Soc.*, v. 44, p. 44–59.

Finalyson, D. M., Drummond, B. J., Collins, C. D. M., and Connelly, J. B., 1977, Crustal structures in the region of the Papuan Ultramific Belt, *Phys. Earth Planet. Inter.*, v. 14, p. 13–29.

Furumoto, A. S., Hussong, D. M., Campbell, J. F., Sutton, G. H., Malahoff, A., Rose, J. C., and Woolard, G. P., 1968, Crustal and upper mantle structure of the Solomon Islands as revealed by seismic refraction survey of November–December 1966, *Pac. Sci.*, v. 24, p. 315–332.

Froidevaux, C. M., 1978, Tertiary tectonic history of the Salwati area, Irian Java, Indonesia, *Bull. Am. Assoc. Pet. Geol.*, v. 78, p. 1127–1150.

Grainger, D. J. and Grainger, R. L., 1974, Explanatory notes on the 1:2,500,000 mineral deposits map of Papua New Guinea, *Bur. Miner. Resour. Aust. Bull.*, no. 148.

Grund, R. B., 1975, North New Guinea Basin, in: *Economic Geology of Australia and Papua New Guinea, Vol 3, Petroleum*, (Knight, C. L., ed., Monograph 7, Australian Institute of Mineralogy and Metallurgy, p. 499–506.

Halunen, A. J. and von Herzen, R. P., 1973, Heat flow in the western equatorial Pacific Ocean, *J. Geophys. Res.*, v. 78, p. 5195–5208.

Hamilton, W., 1979, Tectonics of the Indonesian Region. *U. S. Geol. Surv. Prof. Pap 1078.*

Hedervari, P. and Papp, Z., 1977, Seismicity maps of the New Guinea–Solomon Islands region, *Tectonophysics*, v. 42, p. 261–281.

Hermes, J. J., 1974, West Irian, in: *Mesozoic–Cenozoic Orogenic Belts*, Spencer, A. M., ed., *Geol. Soc. London Spec. Pub. No. 4*, p. 475–490.

Hohnen, P. D., 1978, Geology of New Ireland, *Bur. Min. Resour. Aust. Bull.*, v. 194.

Hutchison, D. S., and Norvick, M., 1978, Explanatory notes, Sheet SA/54–16 Wewak (with 1:250,000 geological map), *Bur. Miner. Resour. Aust.* Canberra.

Jaques, A. L., 1981, Petrology and petrogenesis of cumulate periodotites and gabbros from the Marum ophiolite complex, northern Papua New Guinea, *J. Petrol.*, v. 22, p. 1–40.

Jaques, A. L. and Robinson, G. P., 1977, Continent/island arc collision in northern Papua New Guinea, *BMR J. Geol. Geophys. Aust.*, v. 2, p. 289–303.

Jaques, A. L., and Chappell, B. W., 1980, Petrology and trace-element geochemistry of the Papuan Ultramafic Belt, *Contrib. Mineral. Petrol.*, v. 75, p. 55–70.

Jaques, A. L., Chappell, B. W. and Taylor, S. R., 1978, Geochemistry of LIL enriched tholeiites from the Marum ophiolite complex, northern Papua New Guinea, *BMR J. Aust. Geol. Geophys.*, v. 3, p. 297–310.

Johnson, R. W., 1976, Late Cenozoic volcanism and plate tectonics at the southern margin of the Bismarck Sea, Papua New Guinea, in: *Volcanism in Australia*, Johnson, R. W., ed., Amsterdam: Elsevier, p. 101–106.

Johnson, R. W., 1979, Geotectonics and volcanism in Papua New Guinea, a review of the late Cenozoic, *BMR J. Aust. Geol. Geophys.*, v. 4, p. 181–207.

Johnson, R. W. and Smith, I. E., 1974, Volcanoes and rocks of St. Andrew Strait, Papua New Guinea, *J. Geol. Soc. Aust.*, v. 21, p. 333–351.

Johnson, R. W., Mutter, J. C. and Arculus, R. J., 1974, Origin of the Willaumez–Manus rise, Papua New Guinea, *Earth Planet. Sci. Letts.*, v. 44, p. 247–260.

Johnson, R. W., Wallace, D. A., and Ellis, D. J., 1976, Feldspathoid-bearing volcanic rocks and associated types from volcanic islands off the coast of New Ireland, Papua New Guinea, in *Volcanism in Australasia*, Johnson, R. W., ed., Amsterdam: Elsevier, p. 297–316.

Johnson, R. W., Mackenzie, D. E., and Smith, I. E., 1978, Delayed partial melting of subduction modified mantle in Papua New Guinea, *Tectonophysics*, v. 46, p. 197–216.

Johnson, R. W., and Jaques, A. L., 1980, Continent-arc collision and reversal of polarity: new interpretations from a critical area, *Tectonophysics*, v. 63, p. 111–124.

Johnson, T. and Molnar, P., 1972, Focal mechanisms and plate tectonics of the southwest Pacific, *J. Geophys. Res.*, v. 77, p. 5000–5032.

Jongsma, D., 1972, Marine geology of Milne Bay, eastern Papua, in: *Geological Papers, 1969*, *Bur. Miner. Resour. Aust. Bull.*, no. 125, p. 35–54.

Karig, D., 1972, Remnant Arcs, *Geol. Soc. Am. Bull.*, v. 83, p. 1057–1068.

Luyendyk, B. P., MacDonald, K. C. and Bryan, W. B., 1973, Rifting history of the Woodlark Basin in the southwest Pacific, *Geol. Soc. Am. Bull.*, v. 84, p. 1125–1134.

Mackenzie, D. E., 1977, North Sepik Project, in: *Geological Branch Summary of Activities, 1976*, Canberra: Bureau of Mineral Resources Australia, Report 196, p. 141–143.

Milsom, J. S., 1970, Woodlark Basin, a minor center of seafloor spreading in Melanesia. *J. Geophys. Res.*, v. 75, p. 7335–7339.

Milsom, J. S., 1973, Papuan Ultramafic Belt: gravity anomalies and the emplacement of ophiolites, *Geol. Soc. Am. Bull.*, v. 84, p. 2243–2258.

Milsom, J. S., 1981, Neogene thrust emplacement from a frontal arc in New Guinea, in: *Thrust and Nappe Tectonics* (McClay, K. and Price, N. J., ed.), London: Geological Society of London, p. 417–426.

Milsom, J. S. and Smith, I. E., 1975, Southeastern Papua: generation of thick crust in a tensional environment? *Geology*, v. 3, p. 117–120.

Myers, N. O., 1976, Seismic surveillance of volcanoes in Papua New Guinea in: *Volcanism in Australasia*, Johnson, R. W., ed., Amsterdam: Elsevier, p. 91–99.

Norvick, M., 1975, North Sepik project, in: *Geological Branch Summary of Activities, 1974*, Canberra: Bureau of Mineral Resources, Report 189, p. 85–86.

Ollier, C. D. and Pain, C. F., 1980, Actively rising surficial gneiss domes in Papua New Guinea, *J. Geol. Soc. Aust.*, v. 27, p. 33–44.

Page, R. W., 1976, Geochronology of igneous and metamorphic rocks in the New Guinea highlands, *Bur. Miner. Resour. Aust. Bull.* no 162.

Page, R. W. and McDougall, I., 1970, Potassium-argon dating of the Tertiary fl-2 stage in New Guinea and its bearing on the geological time scale, *Am. J. Sci.*, v. 269, p. 321–342.

Page, R. W. and McDougall, I., 1972, Ages of mineralization of gold and porphyry copper deposits in the New Guinea highlands, *Econ. Geol.*, v. 67, p. 1065–1074.

Pieters, P. E., 1978, Explanatory Notes, Sheets SC/55-6, -7 and -11 Port Moresby, Kalo, Aroa, (with 1:250,000 geological map), *Bur. Miner. Resour. Aust.*, Canberra.

Pigram, C. J., Challinor, A. B., Hasibuan, F., Rusmana, E. & Hartono, U., 1982, Lithostratigraphy of the Misool archipelago, Irian Jaya, Indonesia, *Geol. Mijnbouw* v. 61, p. 265–279.

Ripper, I. D., 1982, Seismicity of the Indo-Australian/Solomon Sea plate boundary in the southeast Papua region, *Tectonophysics*, v. 87, p. 355–370.

Ripper, I. and McCue, K. F., 1982, Seismicity of the New Guinea region, 1964–1980, computer plots. *Papua New Guinea Geol. Surv. Rep. 1982/10* (unpublished).

Robinson, G. P. and Ratman, N., 1978, Stratigraphic and tectonic development of the Manokwari area, Irian Jaya, *Bur. Miner. Resour. J. Aust. Geol. Geophys.*, v. 3, p. 19–24.

Rutherford, K. J., and Qureshi, M. K., 1981, Geothermal gradient map of southeast Asia, *Indonesian Petrol. Assoc.*, Jakarta

Ryburn, R. W., 1978, Irian Jaya Project, in: *Geological Branch Summary of Activities, 1977*, Canberra: Bureau of Mineral Resources, Report 208, p. 187–191.

Skwarko, S. K., 1978, Stratigraphic tables, Papua New Guinea, Canberra: Bureau of Mineral Resources, Report 193.

Smith, I. E., 1972, High-potassium intrusives from southeastern Papua, *Contr. Mineral. Petrol.*, v. 34, p. 167–176.

Smith, I. E., 1973, Geology of the Calvados chain, southeastern Papua, *Bur. Miner. Resour. Aust. Bull.*, no. 139, p. 59–66.

Smith, I. E., 1976, Peralkaline rhyolites from the D'Entrecasteaux Islands, Papua New Guinea, in: *Volcanism in Australasia*, Johnson, R. W., ed., Amsterdam: Elsevier, p. 275–285.

Smith, I. E., 1982, Volcanic evolution in eastern Papua, *Tectonophysics*, v. 87, p. 315–334.

Smith, I. E., and Davies, H. L., 1976, Geology of the southeast Papuan mainland, *Bur. Miner. Resour. Aust. Bull.* no. 165.

Smith, I. E., Pieters, P. and Simpson, C. J., 1973, Notes to accompany a geological map of Rossel Island, southeastern Papua, in: *Geological Papers, 1970–1971, Bur. Miner. Resour. Aust. Bull.*, no. 139, p. 75–78.

St. John, V. P., 1967, Gravity Field of New Guinea, Ph.D. dissertation, University of Tasmania (unpublished).

Taylor, G. A., 1958, The 1951 eruption of Mt. Lamington, Papua, *Bur. Miner. Resour. Aust. Bull.*, no. 38.

Taylor, B., 1979, Bismarck Sea: evolution of a back-arc basin, *Geology*, v. 7, p. 171–174.

Tjhin, K. J., 1976, Trobriand basin exploration, *Aust. Pet. Explor. Assoc. J.*, v. 16, p. 81–90.

Untung, M. and Barlow, B. C., 1981, Gravity field in eastern Indonesia, *Geol. Res. Dev. Cen. Spec. Pub.*, no. 2, p. 53–63.

Van der Wegen, G., 1966, Contribution of the Bureau of Mines to the geology of the central mountains of west New Guinea, *Geol. Mijnbouw*, v. 45, p. 249–261.

Visser, W. A., 1968, Discussion of a paper by D. B. Dow 'Geological reconnaissance in the Nassau Range', *Geol. Mijnbouw*, v. 47, p. 47–48.

Visser, W. A. and Hermes, J. J., 1962, Geological Results of Exploration for Oil in Netherlands New Guinea, *Verh. K. Ned. Geol. Mijnbouwk. Genoot.*, v. 20, p.

Weissel, J. K. and Anderson, R. N., 1978, Is there a Caroline plate ? *Earth Planet. Sci. Letts.*, v. 41, p. 143–158.

Weissel, J. K. and Watts, A. B., 1979, Tectonic evolution of the Coral Sea Basin, *J. Geophys. Res.* v. 84, p. 4572–4582.

Weissel, J. K., Taylor, B. and Karner, G. D., 1982, Opening of the Woodlark Basin, subduction of the Woodlark spreading system and the evolution of northern Melanesia since mid-Pliocene time, *Tectonophysics*, v. 87, p. 253–277.

Chapter 13

SOLOMON ISLANDS

Frank I. Coulson*

British Geological Survey
Keyworth, Nottingham NG12 5GG, England

I. INTRODUCTION

Solomon Islands form part of a complex of Melanesian island arcs and marginal basins that extend in a southeasterly direction across the southwest Pacific from New Ireland, New Britain and Bougainville in Papua New Guinea, through Vanuatu and Fiji to Tonga and the Kermadec Islands (Fig. 1). This region has also been called the Melanesian Reentrant (Coleman, 1970) and the Melanesian Borderlands.

The major islands within the western and central Solomon Archipelago extend in a northwest to southeast belt across some 850 km of the southwest Pacific between the latitudes 6°35'S and 11°50'S and longitudes 155°30'E and 162°20'E. Six main islands form an en echelon double island chain, closed in the west by the island of Bougainville, and in the east by San Cristobal.

All the islands show some degree of NW–SE lineation although this is most pronounced in Santa Isabel and least pronounced in Guadalcanal, which is slightly sigmoidal and more stumpy in shape. Malaita, which trends north-northwesterly is noticeably "out of line" with the rest of the island chain.

Separating the northern island chain of Choiseul, Santa Isabel and Malaita from the southern island chain of New Georgia, Guadalcanal, and San

---

* Present address: Directorate of Mineral Resources, Jl. Diponegoro 57, Bandung 40122, Indonesia

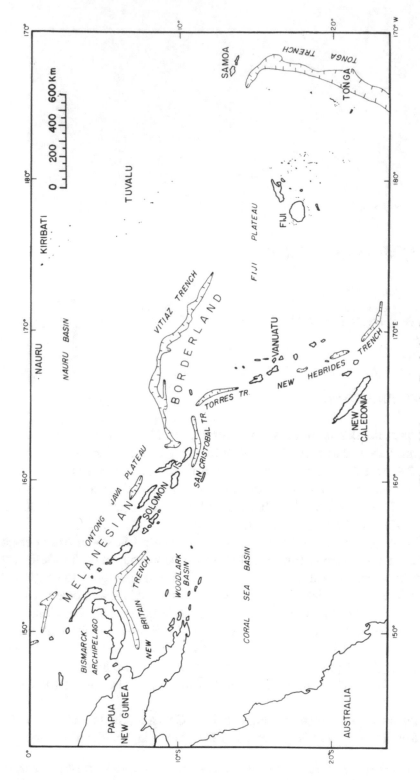

Fig. 1.   Melanesian Borderland.

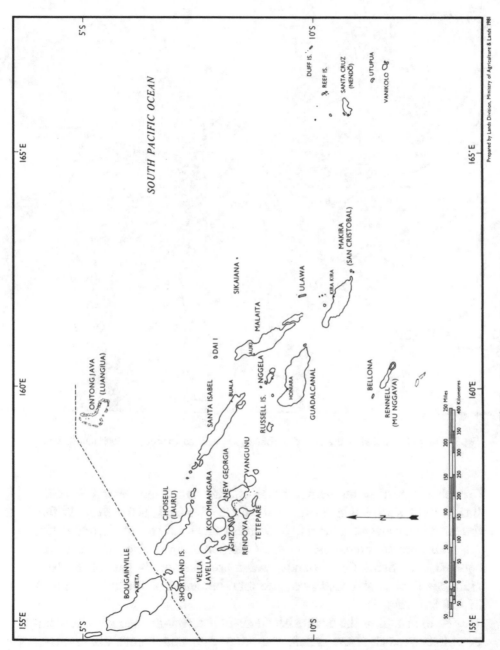

SOUTH PACIFIC OCEAN

Prepared by Lands Division, Ministry of Agriculture & Lands 1981

Fig. 2. Map of the Solomon Islands.

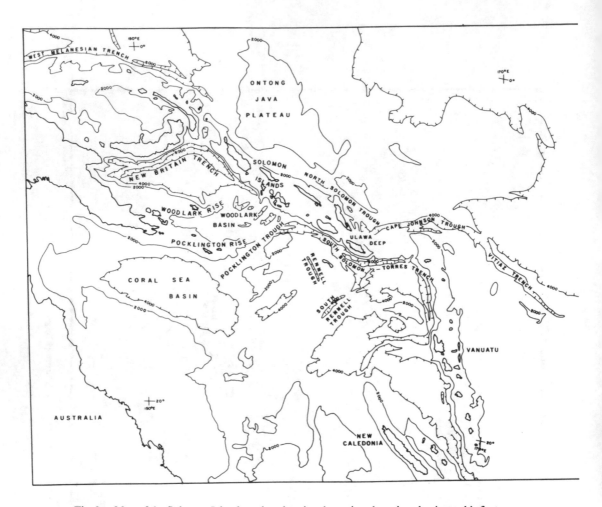

Fig. 3.   Map of the Solomon Islands region showing the main submarine physiographic features.

Cristobal is an elongate marine basin—herein called the Central Solomon Basin—that is 350 km in length, up to 90 km wide, and 1800 m deep. Within the main chain and peripheral to it are many small islands, atolls, and reefs.

The political boundaries of the Solomon Islands also include in the southeast the Santa Cruz Islands, which are structurally part of the New Hebrides (Vanuatu) Island Arc. The total landmass of the group is about 27,750 km^2 (Fig. 2).

To the south of the double island chain is a well-defined but discontinuous linear trench (New Britain and South Solomon Trench) that reaches depths of over 6000 m at a point 80 km south of San Cristobal and over 9000 m in the Planet Deep of the New Britain Trench. The trench almost disappears south of the New Georgia Group where a northeast-trending submarine ridge, the Woodlark Rise abuts onto the Solomon chain from the eastern end of the Papuan promontory.

The limits of the Solomon segment are clearly defined by sharp reentrants in the southern trench system, in the northwest in the New Britain Trench, and in the southeast where the South Solomon Trench meets the Torres (New Hebrides) Trench (Fig. 3). To the south of the trench is a complex area of troughs, basins and ridges of the Coral Sea.

To the north and east of the Solomon Islands a less well-defined trench system (West Melanesian Trench, North Solomon Trough, Cape Johnson Trough, Vitiaz Trench) reaches depths of 6000 m at the eastern end of the Solomon Islands where it forms a sharp reentrant, the Ulawa Deep. To the north of this trench system is the broad anomalously shallow area of the Pacific known as the Ontong Java Plateau.

The main islands, which are largely volcanic in origin, are rugged, jungle covered and experience an equatorial oceanic climate. The population, predominantly of indigenous Solomon Islanders, numbers about 214,000 and is forecast to rise to 435,000 by the year 2000. The capital, Honiara (pop. 18,000), which lies on the north coast of Guadalcanal, the largest island in the group, is some 4,430 km northeast of Sydney and 2,415 km northwest of Suva. Communications are not well developed. Domestic sea and air services link the main islands, but a limited road network makes travel within the islands difficult. On 7th July, 1978, the Solomon Islands became an independent member of the British Commonwealth.

II. HISTORY OF INVESTIGATIONS

A. Geological Investigations

Prior to the establishment of the Geological Survey in 1950, the Solomon Islands were, geologically speaking, "terra incognito." The Geological Survey, which began with one geologist and later (1954) increased to three, initiated a program of reconnaissance geological mapping at a scale of 1:200,000, carried out as a collaborative effort between the University of Sydney (Australia) and the Geological Survey. The first University of Sydney expedition commenced work on Guadalcanal, Malaita and Santa Isabel in 1951 (Coleman, 1960; Rickwood, 1957 and Stanton 1961), followed by a survey of Guadalcanal initiated in 1954 as a joint program between the University of Sydney and the Geological Survey (Pudsey-Dawson and Thompson, 1958; Coleman, 1960).

In 1956, reconnaissance surveys of both San Cristobal (Thompson and Pudsey-Dawson, 1958) and the Florida Islands (Thompson, 1958) were completed. Work on Choiseul began in 1957 (Coleman, 1960) and in New Georgia the following year (Stanton, 1961). In 1962 this early reconnaissance work was compiled into the first geological map of the Solomon Islands (Coleman, 1965), the culmination of some 12 years of collaborative effort.

In 1963, following the provision of accurate 1:50,000 scale topographic maps of the Protectorate, a 1:50,000 scale regional geological mapping program was begun, initially on the island of Guadalcanal (Dennis and Hackman, 1977; Hughes, 1977a, 1977b; Hackman, 1977, 1979, 1980; Hackman and Turner, 1977). Regional mapping continued during the period 1963–1975, although the effort was to be considerably reduced and even stopped altogether by the Geological Survey's commitment to the UNDP Aerogeophysical Survey Project during the period 1965–1968. By the end of 1975, south Malaita (Hughes and Turner, 1976), the Eastern Outer Islands (Hughes et al., 1981), northwest San Cristobal (Jeffery, 1977), Savo (Proctor and Turner, 1977), the Russell Islands and Mborokua (Danitofea, 1981), western Florida (Taylor, 1977) and Ulawa (Danitofea, 1978) had been mapped at 1:50,000 scale. Commencing in 1976, the regional mapping program was greatly accelerated as a British Technical Cooperation Project, with senior staff supplied by the Institute of Geological Sciences, UK. By 1979 the Shortland Islands and Choiseul had been completed (Ridgway and Coulson, in preparation) and the New Georgia Group was completed in 1983. (Dunkley and others, in preparation).

Approximately 66% of the Solomon Islands have been surveyed in detail (1:50,000 scale mapping) with the completion of the New Georgia Mapping Project in 1983 (Fig. 4). Santa Isabel, Malaita, San Cristobal, and the eastern Florida Islands remain to be surveyed and constitute a significant gap in our knowledge of the land geology. In particular, the detailed mapping of Santa Isabel where the Pacific and Central Province (Coleman, 1975) appear to be juxtaposed, would reveal valuable information on the relationships between these two contrasting geological domains.

Complementing the mapping program over the years, special interests have been pursued by various workers in the region, e.g. in palaeontology (Coleman, 1965, Coleman and McTavish, 1964; Hughes, 1977), and ultrabasic rocks (Thompson, 1960).

B. Geophysical Investigations

During the period 1965–1968, a major aerogeophysical survey project of Solomon Islands was mounted as a joint venture between the United Nations Development Program and the Government of British Solomon Islands. In addition to the aerogeophysics itself, the project also included photogeological interpretation, reconnaissance stream sediment geochem-

Additional material from *The Ocean Basins and Margins,*
ISBN 978-1-4612-9440-5 (978-1-4612-9440-5_OSFO2),
is available at http://extras.springer.com

istry and limited ground follow-up geological and geophysical investigations (ABEM 1967). The United Nations contracted ABEM Company of Stockholm, Sweden to undertake the aerogeophysical survey which was to combine airborne magnetometer, electromagnetometer and scintillometer surveys.

The survey was carried out using two Piper Aztec aircraft flying in tandem, the leader aircraft carrying all instrumentation except the electromagnetic transmitter, which was mounted in the following aircraft. The instruments used included a total-field magnetometer (Barringer Nuclear Precession Magnetometer Model AM - 101), an ABEM Rotary Field System for the electromagnetic survey and a Nuclear Enterprise Mark XII Scintillation Counter for the radiation survey.

A total of about 25,000 line miles (40,000 line km) was flown with the three geophysical methods combined, and about 1750 line miles (2800 line km) of inter-island magnetometer survey were added in order to establish the regional magnetic field. The mean line spacing was mostly 400 m, in some areas 800 m. Mean terrain clearance was 125 m.

The results of the aeromagnetic survey were presented as a series of contoured maps of magnetic total field without correction for regional gradient. The contour interval varied from 20 to 100 gammas, depending upon the magnetic gradient. The magnetic total field is characterized by generally high relief, and magnetic gradients of 0/5–2.0 gamma per meter at the average flying height of 125 m were common (ABEM, 1967). The magnetic pattern shows an E–W grain, interrupted by more complex zones of high magnetic relief that correspond to intense faulting and Pliocene to Recent vulcanicity. Magnetic highs tend to correspond with blocks of oceanic basalt; magnetic lows with intervening sedimentary basins (Fig. 5).

In a regional interpretation of the results of the project (Winkler, 1968), it was concluded that the central trough along the axis of the island chain contains very large amounts of fill, most of it of volcanic origin. Normal block faulting, most of it high angle, dominates the present structure.

The first gravity survey was made in 1960, across the plains of north-central Guadalcanal (Coleman and Day 1965). Some 200 stations were surveyed and tied into the newly established base station at the Geological Survey Department in Honiara. This survey discovered the "Tetere High," a broad, gravity maximum striking northwards across the plains with gradients of up to 9.5 mgal/km on its eastern flank—this was at that time, the steepest gradient ever recorded (Fig. 6). The Tetere High was interpreted as an upfaulted basement horst flanked by sedimentary basins that, in the east, attained a thickness of 10,000 m.

In 1961, T. S. Laudon of the University of Wisconsin linked the Solomon airfields into the world gravity survey network and subsequently undertook

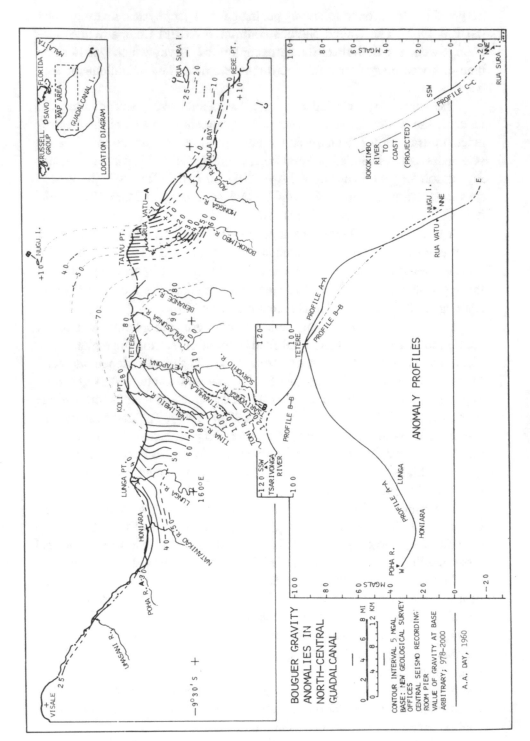

Fig. 6. Bougeur gravity anomalies in north-central Guadalcanal.

a regional land gravity survey involving approximately 2000 stations and using three La Coste and Romberg geodetic gravity meters (Laudon, 1968). The regional Bougeur anomaly map (Fig. 7) as interpreted from the land gravity surveys of the individual islands, shows the Solomon Islands as a somewhat symmetrical regional low, elongated northwest–southeast, parallel to the trend of the islands, a feature that corresponds to a major bulge in the geoid and which causes deflection of orbital satellites. This regional low flanks the northeast side of a large composite Bougeur positive that extends over the Coral Sea to the south.

Each major island or group of islands is expressed as a smaller regional high superimposed on the flanks of the major gravity low. On the larger islands, closed highs are usually associated with basement rock exposures in the centers of the islands, often at high elevation. Over the Russell and New Georgia islands, which are composed of Pliocene to Recent volcanic piles, gravity highs are not closed, but are expressed as sharp northward deflections of the northwest trending isogals.

The minimum value associated with the axial basins are probably between 0 and −50 mgal. Exceptionally steep gravity gradients on the southern flanks of the Indespensable Strait Basin lead to the highest positive reading of about +250 mgal in the southeast of San Cristobal. The islands are characterized by large free-air and Bougeur anomalies, and considerable local departure from isostasy is indicated by the regional association between elevation and the Bougeur anomaly value. These departures from isostasy can be attributed to high density crustal material, especially ultrabasic intrusives within the basement or to upwarping of the base of the crust, or both. Local large anomalies in Pliocene to Recent volcanic areas are attributable to the crust's supporting accumulated volcanic piles without isostatic compensation.

The Solomon Islands constitute an island arc that has evolved through the complex interaction of the major Australian and Pacific plates. At the present time the Australian plate, moving north-northwest with an absolute motion of 7 cm/yr is being consumed beneath the Pacific plate along a trench system flanking the southwest side of the arc. The Solomon Islands sit on the leading edge of the Pacific plate which is moving northeastward with an absolute motion of 10.7 cm/year (Minster and Jordan, 1978). The resultant high rate of oblique convergence—generally thought to exceed 10 cm/yr (e.g. Johnson and Molnar, 1972)—is probably being accommodated through a combination of subduction and sinistral strike-slip shearing along the length of the arc (e.g. Coleman, 1975).

The Solomon Islands Arc is characterized by intense seismicity along the northwestern and southeastern parts of the arc where the southern trench

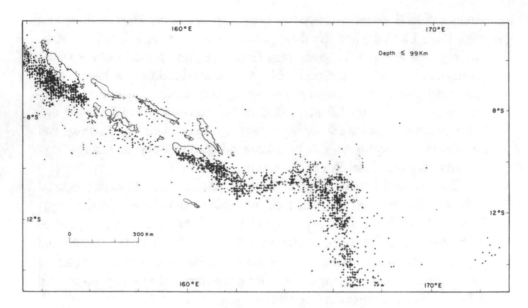

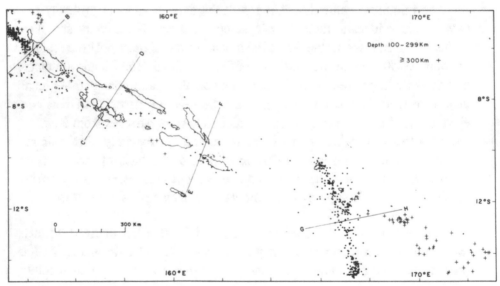

Fig. 8.   Earthquake epicenters for the Solomon Islands, 1962–1977.

system is well developed (Fig. 8). Clearly defined Benioff zones dip steeply
northeastward beneath the arc in the region of Bougainville and the Santa
Cruz Islands (Fig. 9). In the San Cristobal region, the Benioff zone is less
well defined and probably steeply inclined (Fig. 9). However, focal mechanism solutions indicate that the Australian plate underthrusts the Pacific
plate in this region (Johnson and Molnar, 1972). In contrast, along the central

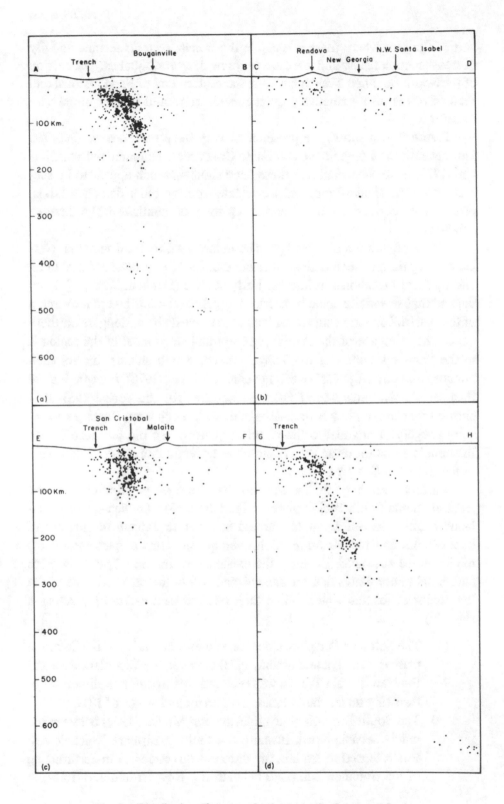

Fig. 9.   Distribution of earthquake depths in the Solomon Islands.

part of the arc, where there is a gap in the trench system, seismic activity decreases dramatically and there are very few deep earthquakes. The section of the South Solomon Trench between the eastern end of San Cristobal and Santa Cruz is also a relatively quiet seismic zone and is interpreted as a transform.

Figure 9 also shows the presence of very deep earthquakes (500–700 km) beneath the Bougainville and Santa Cruz areas. Halunen and von Herzen (1973) have suggested that these deep shocks, which appear to be separate from the Benioff zone, may originate from an older detached lithospheric slab derived from an earlier episode of southwesterly directed subduction.

The configuration of crustal plate boundaries within the Solomons–New Guinea region has been defined through studies of earthquake distribution and by focal mechanism solutions. Early studies (Denham, 1969, 1971) indicated that seismicity could be ascribed largely to the northward movement of the Australian plate coupled with an east-to-west shear along its northern edge. The influence of the Pacific plate is small since most of the region is in the "shadow zone" of the Tonga Trench, which shields the Solomon Islands from any large east–west movement. Ripper (1970), interpreting the focal mechanism analyses of large earthquakes for the period 1963–1967 showed that in the New Britain–Bougainville Trench area, the slip vectors were roughly orthogonal to the trench, a pattern that did not accord with the simple situation of an Australian plate underthrusting the Pacific plate in the Solomon Sea area.

Further seismicity studies showed that the zone of interaction of the Australian and South Pacific plates in the Papua New Guinea–Solomon Islands region was not a simple one and in order to explain the pattern of hypocenters and the direction of slip vectors in terms of plate tectonics, models were advanced involving the existence of three and possibly more additional minor plates that are sandwiched within the collision zone of the two major plates and which derive their relative motion from the collision (Fig. 10):

(*i*)   The Solomon Sea plate (Johnson and Molnar, 1972) or the Solomon plate (Luyendyk and others, 1973) between the Woodlark Rise and the New Britain Trench was required by the different directions of thrusting under the Solomon Arc north and south of 10°S.

(*ii*) The South Bismark plate (Johnson and Molnar, 1972) between the mildly seismic North Bismarc Arc and the Bismark Volcanic Arc was required to explain the different directions of underthrusting of the Solomon Sea plate beneath the New Britain and Solomon Arc.

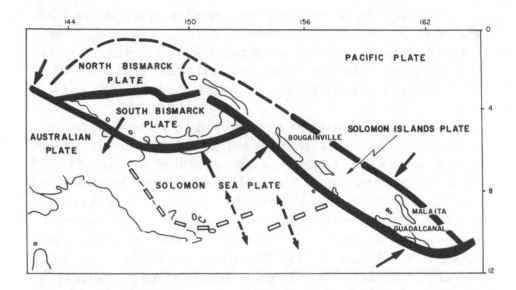

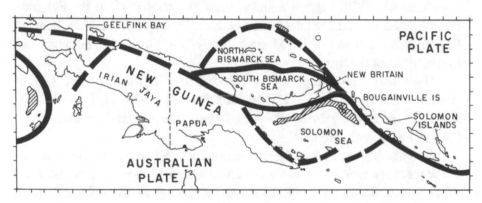

Fig. 10. Plate boundary models for the Solomon Islands–New Britain region.

(*iii*) The North Bismark plate (Johnson and Molnar, 1972) between the
       North Bismarck Arc and the West Melanesian Trench is indicated
       by an E–W belt of epicenters near 4°S, showing left lateral strike-
       slip motion between this plate and the South Bismark plate.

(*iv*) The New Britain plate (Curtis, 1973) which is almost synonymous
       with the South Bismark plate of Johnson and Molnar (1972).

(*v*)   The Manus plate (Curtis, 1973) introduced tentatively between the
       New Britain and Solomon Sea plate and the West Melanesian
       Trench. It includes much of the Solomon Arc, its northern bound-
       ary coinciding with Coleman's (1965) Central Province–Pacific
       Province boundary. This plate also appears as the Solomon plate
       in some texts.

Denham (1975), in considering the distribution of underthrust lithosphere in the Solomon Islands Arc, reports a very complex situation including current underthrusting of the Pacific plate beneath San Cristobal, deep remnant lithospheric slabs beneath the central part of the island chain and underthrusting of the Pacific plate beneath Bougainville and San Cristobal (Fig. 11).

Crustal structure in the Solomon region was first investigated by Rose *et al.* (1968). In the central Solomon Islands crustal thickness was estimated to vary from about 27 km under Indespensable Strait to 17 km southwest of Guadalcanal. From five other profiles across the entire Solomon–northern New Hebrides region, crustal thickness was estimated to vary from 9 to 29 km, the higher figures being associated with New Georgia Sound and Indespensable Strait. Rose *et al.* (1968) concluded that the Solomon Islands appear to have little or no "root" and that crustal thickness is similar to that of the oceanic region to the north. Seismic refraction studies by Furumoto *et al.* (1970) suggest a linear, blocklike character for the Solomon Islands, with crustal thickness varying from 15–20 km. The crust beneath Ontong Java Plateau is up to 35 km but only 10–12 km in the Solomon Sea to the south of the arc.

High heat flow is reported in the Woodlark Basin and Solomon Sea (Halunen and von Herzen, 1973; Macdonald *et al.*, 1973, Taylor and Exon, 1983). There appears to be low heat flow in the Central Solomon Basin (Taylor and Exon, 1983).

Geophysical surveys in the surrounding seas have been carried out over the years since 1964 by a variety of vessels including naval, institutional and oil company. Table I attempts to list these surveys in chronological order, but some of the earlier records are poor and the list may be incomplete. Track charts for some of these surveys are given in Figs. 12–16. Although there has been some integration of the marine seismic data into regional syntheses, (Katz, 1980; Maung and Coulson, 1983) no work of a comparable nature has been attempted for the magnetic and gravity data.

Marine geophysical surveys in the Solomon Islands began in the mid-1960s. Following a short test run of a shipboard gravity meter abroad the USS *Wandank* in 1964 (Rose *et al.*, 1966), a three-month marine gravity survey was completed by HMS *Dampier* in 1965 (Fig. 12). In 1966, the Hawaii Institute of Geophysics launched a three-ship refraction seismic survey in the Solomon Sea (Woolard *et al.*, 1967). The RV *Conrad* and the RV *Vema* also cruised the region in 1966/7 (Ewing *et al.*, 1969; Houtz *et al.*, 1968). The early findings of these cruises gave impetus to a greatly expanded investigation of the Ontong Java Plateau in 1967, 1968, 1970 and 1971 (Kroenke, 1972, Fig. 13).

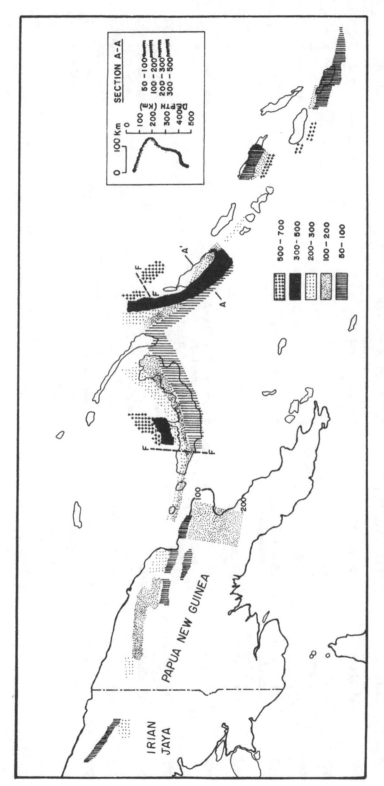

Fig. 11. Distribution of underthrust lithosphere in the Solomon Islands Arc.

TABLE I

Marine Geophysical Surveys in Solomon Islands Waters, 1958–1982

Year	Vessel	Operator	Area	Length (km)	Type
1958	HMS *Cook*	Royal Navy	Transit passage and New Georgia Sound		Magnetic
1963	HMS *Cook*	Royal Navy	Blanche Channel, New Georgia		Magnetic
1965	USS *Wandank*	US Navy	Reconnaissance		Magnetic
1966	HMS *Dampier*	Royal Navy	Reconnaissance		Magnetic
1966	MV *Taranui*	Hawaii Institute of Geophysics	Ontong Java Plateau		Seismic
1966	*California*	Hawaii Institute of Geophysics			
1966	MV *Machias*				
1966	RV *Conrad*				
1967	RV *Vema*				
1969	MV *Teledex IV*	Magellan Petroleum (Aust)	Shortlands, Bougainville Manning, and Indespensable Straits	769	Seismic
1969	DS *Glomar Challenger*	NSF	Ontong Java Plateau (DSDP Site 64)		Seismic, gravity, magnetics
1971	RV *Chain*	Woods Hole Cruise 100, Leg 8	Woodlark Basin, Rennell, Bellona		Seismic, gravity, magnetics
1971	MV *Petrel*	Shell International	Reconnaissance	2764	Seismic, gravity, magnetics
1971	MV *Teledex IV*	Southern Pacific Petroleum, NL	Manning Strait	161	Seismic, magnetics

Year	Ship	Organization	Area	Line-km	Survey
1971	MV *Teledex IV*	Teledyne Exploration	Reconnaissance	2458	Seismic, magnetics
1972	*Fred H. Moore*	Mobil Oil	Reconnaissance	4447	Seismic, magnetics, gravity
1972	MV *Gulfrex*	Australian Gulf Oil	Reconnaissance	2408	Seismic, magnetics, gravity
1972	HMS *Hydra*	Royal Navy	New Georgia Sound, Manning, and Bougainville Straits		Magnetics, gravity
1973	DS *Glomar Challenger*	NSF	Transit, Leg 30 (DSDP Sites 287–289)		Seismic, gravity
1973	RV *Coriolis*	Western Geophysical ORSTOM	Reconnaissance	4389	Seismic
1975	MV *Western Islander*	Austradec III Pacific Energy and Minerals	Reconnaissance		Seismic, magnetics
1978	MV *Machias*	CCOP/SOPAC	North Guadalcanal	100	Seismic
1979	MV *Machias*	CCOP/SOPAC	New Georgia Sound	4064	Seismic
1982	RV *S.P. Lee*	CCOP/SOPAC	Rennell	1272	Seismic, magnetics
1982	RV *Kana Keoki*	CCOP/SOPAC	Reconnaissance	3700	Seismic, magnetics, gravity
1982			Woodlark Basin, New Georgia		Seismic, magnetics, gravity

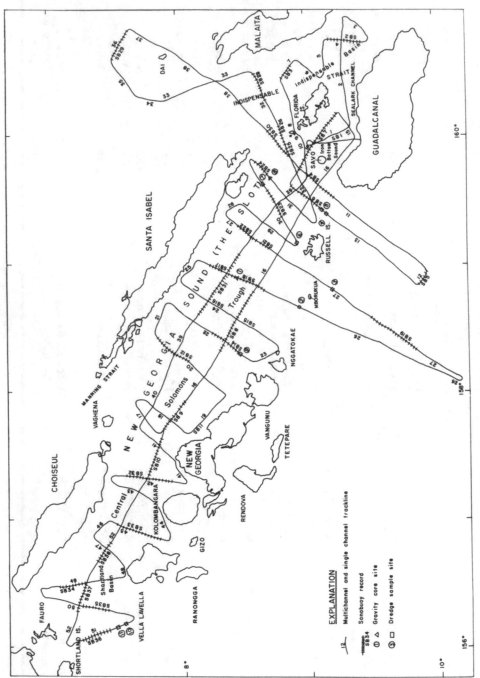

Fig. 15.   RV *S.P. Lee* geophysical survey of Solomon Islands waters, 1982.

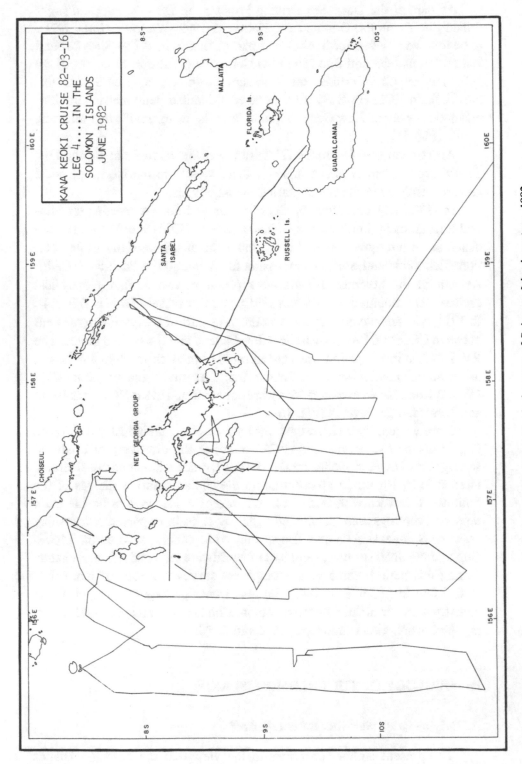

Fig. 16. RV *Kana Keoki* geophysical survey of Solomon Islands waters, 1982.

As part of the Deep Sea Drilling Project, the DS *Glomar Challenger* drilled Site 64 on the Ontong Java Plateau in 1969 (Winterer *et al.*, 1971), to be followed by two further holes as part of Leg 30 in 1972—Sites 287 and 288 on the plateau and Site 286 in the Coral Sea (Andrews *et al.*, 1975). In 1971, the RV *Chain* cruised the Woodlark Basin region, (Leg 8 of Cruise No. 100). In 1972, HMS *Hydra* conducted a detailed bathymetric–gravity–magnetic survey of New Georgia Sound, and the Bougainville and Manning straits (Fig. 12).

Apart from the *Austradec III* seismic survey carried out by the OR-STOM group from Noumea in 1975, there was little marine geophysical research cruising in the region during the mid-1970s.

In 1972, the Committee for Coordination of Joint Prospecting for Mineral Resources in South Pacific Offshore Areas (CCOP/SOPAC) was established as an intergovernmental body under the auspices of the United Nations Economic and Social Commission for Asia and the Pacific (ESCAP). As part of the Solomon Islands work program, two marine geophysical cruises were mounted in 1979 and 1981 using the vessel *Machias* (Fig. 14). In 1982, two seismic cruises were mounted as part of a tripartite agreement between CCOP/SOPAC, Australia, New Zealand and the United States. The RV *S.P. Lee* investigation was concentrated in the Central Solomon Basin, using multichannel reflection seismic, refraction, gravity and magnetics (Fig. 15). RV *Kana Keoki* investigations were concentrated in the Woodlark Basin and New Georgia regions (Fig. 16).

Marine geophysical investigations by the petroleum industry took place largely during the period 1969–1973, following the discovery of crude oil seepages on Tonga in 1968. The Tongan occurrence stimulated considerable interest in hydrocarbon prospecting on the shallow marine shelves of the Southwest Pacific, with buried carbonate reef considered to be the main prospect. Surveys were conducted either by ships of opportunity transiting between Southeast Asia and Tonga–Fiji, while others, such as the Mobil, Gulf and the Shell surveys, were part of worldwide reconnaissance progams. During this period, some seven companies surveyed a total of more than 17,000 line km (Coulson, 1981) Fig. 14. Investigations by the petroleum industry were terminated by the Solomon Islands Government in 1973, pending the introduction of modern petroleum legislation.

III. GEOLOGY OF THE SOLOMON ISLANDS

A. Origin—An Enigmatic, Ensimatic Arc?

The present author subscribes to the view that the Solomon Islands have evolved since Eocene times as an island arc in response to the complex interactions between the Australian and the Pacific plates. The account of the geology of the islands that follows attempts to integrate the various geological disciplines—volcanic and sedimentary stratigraphy, petrochemistry, geophysics—into a geological history related to two periods of subduction involving a reversal of arc polarity. Nevertheless, the author having declared his faith, is bound to point out that the Solomon Arc is enigmatic. Some features put forward in support of an arc origin are ambiguous and can equally support a transcurrent fracture origin. Given the highly oblique interaction between the Pacific and Australian plates, strike-slip motion across the length of the arc must be considered as a tangible structural element in the evolutionary process, but its relative importance is conjectural. This being so, it is appropriate to review the history of thought on this subject and to highlight some of the problem areas.

It was not until the late 1960s, nearly 20 years after the establishment of the Geological Survey that sufficient geological and geophysical data had been gathered by the Geological Survey, academic institutions and international aid agencies on which to base speculation on the origin of the Solomon Islands. Plate tectonic theory was at that time in its infancy and was, with good cause, being applied with some caution to the complex and enigmatic arcs of the southwest Pacific. To quote one of the most eminent authors in the region "I view plate tectonic theory and its handmaiden, seafloor spreading, as an elegant youthful working theory" (Coleman, 1975).

The enigmatic style of the Solomon Islands Arc has encouraged considerably speculation on its origin. The early models emphasized the taphrogenic nature of the island chain (Coleman, 1965, 1975, 1976, Coleman and Packham, 1976, Hackman, 1973) and proposed that evolution of the islands may have been controlled largely by major transcurrent fractures along which magmas may have arisen and upon which there has been substantial strike-slip movement to accommodate convergence between the Australian and Pacific plates. The models envisaged the islands arising as geanticlinal welts that, in the late Mesozoic, may have approached the linear, but which were progressively broken up by tension and sheared into the present en echelon arrangement. Terms such as "fractured arc" (Coleman, 1970; Hackman, 1973) and "nonarc" (Coleman, 1975) were used to describe the chain, Taylor (1975) emphasized the importance of Carey's Tethyan Shear System which includes the Solomon megashear (Carey, 1958).

Features that have been cited in favor of such a nonarc origin have been summarized by Coleman (1975), and include:

(i) the disposition of the islands, which are linear and en echelon rather than arcuate;

(ii) the pattern and distribution of rock types, the comparative dearth of calc-alkaline rocks and the absence of a tholeiite–calc-alkaline–high-K progression;

(iii) the absence of a trench and the seismicity gap between eastern Bougainville and Guadalcanal, i.e., bordering the volcanic province;

(iv) the anomalous near-trench disposition, abnormal lava geochemistry, and high heat flow in the New Georgia Group;

(v) the absence of Pliocene to Recent volcanism in the East Guadalcanal–San Cristobal region associated with the South Solomon Trench.

Several models describing the Solomon Islands as an island arc have been proposed since Coleman (1966) drew attention to several arclike features of the region. In seeking to explain the Solomon Islands as an arc, and in accepting that the region is presently subject to northeast directed subduction, it is implict that the arc has reverse polarity. It follows therefore that during the Palaeogene, the greater part of the Solomon Islands (Coleman's Central Province) was built by south-dipping subduction bordering the northeastern flank. While it has been suggested (Curtis, 1973; Neef, 1978) that the Solomon Islands might be the site of double subduction, most arc models involve a polarity reversal, probably within the last 10 m.y. (Kroenke, 1972; Karig and Mammerickx, 1972; Packham, 1973; Falvey, 1975; de Brion *et al.*, 1977, Coleman and Kroenke, 1982; Dunkley, 1983).

The author subscribes to the view that no two island arcs are the same and that the Solomon Arc is therefore not "abnormal". It is unique, complex and not fully understood. However, considerable progress in unravelling the enigmatic features of this arc have been made in recent years, e.g., the ridge subduction phenomenon and its by product, the New Georgia Group. Important elements of the arc that require further elucidation include:

- the "volcanic gap" adjacent to the South Solomon Trench in the region of east Guadalcanal and San Cristobal.
- the origin and evolution of the Central Solomon Basin.
- the Solomon segment of the Inner Melanesian arc.
- the nature and origin of the Pacific Province and its relationship to the Central Province.

B. An Account of the Geology and Geological History of the Solomon Islands

The absence of allochthonous continental material in the geological column indicates that the Solomon Islands has evolved in an entirely oceanic environment and for much of its history has developed through the complex interaction between the Australian and Pacific plates.

Following an initial phase characterized by the generation of mid-ocean ridge basalts and the deposition of pelagic sediments, two distinct and separate periods of arc volcanism are recognized and are thought to have resulted from a reversal in the polarity of the arc. A pre-Miocene oceanic basement, characterized by the absence of terriginous material but extensively tectonized and metamorphosed contrasts with an overlying sedimentary column marked by rapid horizontal facies changes and effected by relatively simple post-Oligocene tectonics with normal faulting and only minor gentle folding. The sedimentary column is dominated by volcaniclastic sediments although reef and platform carbonates formed extensively during periods of volcanic quiescence. A generalized geological map of the Solomon Islands is shown at Fig. 17, and a generalized stratigraphic column at Fig. 18. Turner and Hughes (1982) have described the sedimentary rocks in terms of nine facies and four facies associations.

Coleman (1965, 1970) subdivided the modern Solomon Islands Arc into a number of geological provinces (Fig. 19). The value of Coleman's concept of geological provinces as a means of identifying contrasting geological domains within the modern island arc has been accepted by many subsequent workers, and the terms are now firmly entrenched in the geological literature. However, with increasing understanding of the geology of the Solomon Islands brought about by a broader data base and the application of modern theories of island arc formation, the limitations of the concept, especially when viewed in a temporal sense, need to be recognized. For instance it is now generally accepted that in Cretaceous and Eocene time, the Solomon Islands was a region of oceanic lithosphere generated at an oceanic ridge, the location of which can only be surmised and that areas now designated as Central and Volcanic provinces could all be considered as Pacific Province. Oceanic Phase (Hackman, 1980) would seem an eminently suitable term for this period. In like manner, the Volcanic Province, at least in its development on Guadalcanal and Choiseul, rests upon a Central Province substrate.

Three of Coleman's Provinces, the Central Pacific and Volcanic provinces, constitute the greater part of the Solomon Islands.

Within the *Central Province* (Choiseul, southwestern side of Santa Isabel, Florida Islands, Guadalcanal and San Cristobal), the islands are characterized by intensely faulted cores of pre-Miocene basic lavas, and associated gabbros and dolerites in part metamorphosed to a low grade (greenschist or amphibolite facies). Bodies of serpentinized ultrabasic rocks are widespread within the Central Province "basement." These are overlain by a varied sedimentary succession ranging in thickness from 5 km in east-central Guadalcanal to less than 700 m on San Cristobal. The sediments

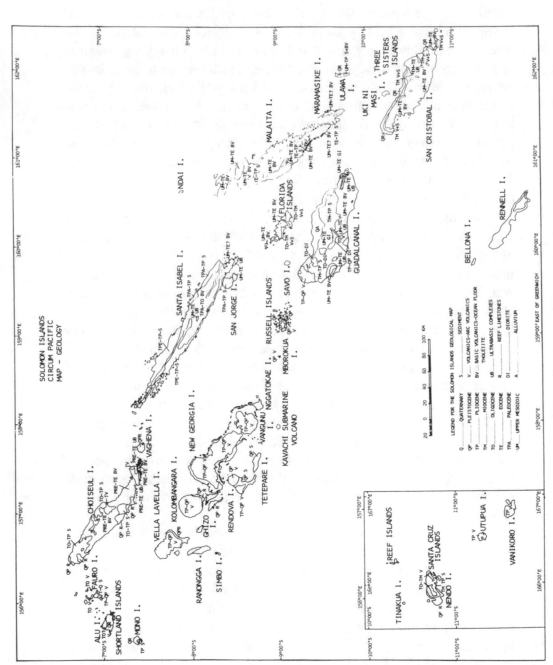

Fig. 17. Geological map of the Solomon Islands.

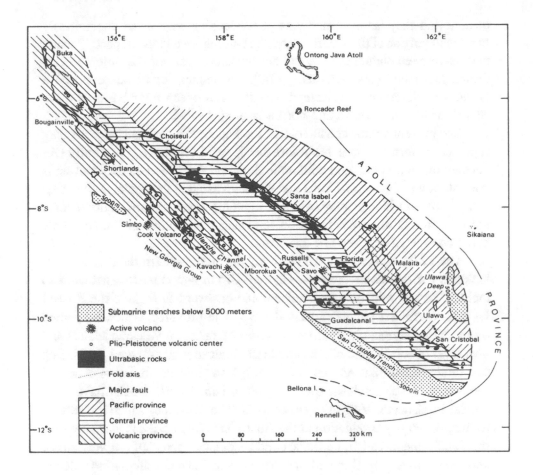

Fig. 19. Geological provinces of the Solomon Islands.

include biogenic limestones, calcarenites and arenaceous and volcaniclastic material ranging in age from Lower Miocene to Holocene. In general, the sedimentary piles display shallow dips, extensive block faulting and gentle "drape" folds, the latter usually reflecting underlying basement structures. Within the postbasement successions on several islands within the Central Province, tholeiitic basalts of late Oligocene to Lower Miocene age together with intrusive complexes of diorite are recognized as an initial phase of arc volcanism.

The *Pacific Province* includes the whole of Malaita, Ulawa and the northeastern flanks of Santa Isabel. This province also has a basement of oceanic basalt, but it is not metamorphosed. The overlying sediments include about 1200 m of chalky pelagic carbonates that, in contrast to the Central Province, range in age from Cretaceous to Recent and are folded along northwest-trending axes. The rocks of the Pacific Province are an oceanic sequence that may have formed part of the leading edge of an anomalously thickened portion of the Pacific plate (the Ontong Java Plateau) part of which may have been obducted onto the northwestern front of the Solomon Arc during Miocene times (Kroenke, 1972). Subsequent uplift has exposed a broad, anticlinorium of deformed pelagites and ocean floor basalts on the islands of Malaita, Ulawa and northeast Santa Isabel.

The present author considers that the Pacific Province successions may represent deformed and uplifted fore-arc deposits that have, since early Oligocene times, accumulated within the Solomon Arc. Such an explanation is consistent with the depositional history as revealed in the geological successions, most of which display evidence of an increasing incursion of volcanic detritus and shoaling conditions. Palaeomagnetic studies on Malaita now being completed may resolve this issue.

The *Volcanic Province* which forms the southwestern flank of the arc, includes the islands of the New Georgia Group and extends northward to include certain volcanics on Choiseul and eastward to include the Russell Islands, northwest Guadalcanal and Savo. The province, which is typified by the New Georgia Group, forms a series of emergent volcanic islands and lava piles of subalkaline basalts and lesser andesites surrounded by fringing and offshore reef that provide a framework for the accumulation of varied sediments of volcaniclastic and biogenic origin. Diorite stocks have been emplaced at depth. Volcanics rocks within the province, which are related to the present day subduction of the Australian plate northeastward beneath the Pacific plate, constitute the second recognized phase of arc volcanism, and range in age from Upper Miocene to Recent. Kavachi submarine volcano in the New Georgia Group is currently active and in total there are nearly thirty well-preserved centers and two active or quiescent volcanoes within this province.

In addition to the three main Provinces, the *Atoll Province* includes Lord Howe and Sikaiana on the Ontong Java Plateau to the north of the main island chain, while to the south the uplifted atolls of Rennell and Bellona lie on a sinuous ridge that links New Caledonia with Papua. While these islands form a distinct geographical grouping, their probable geological evolution is diverse.

1. *Cretaceous–Lower Oligocene; Oceanic Basement*

The oldest rocks exposed in the main islands surrounding the Central Solomon Basin consist of ocean floor tholeiites of Cretaceous to Paleocene age. These rocks form a basement for later arc volcanism and sedimentation in the Shortland Islands, Choiseul, Santa Isabel, Floridas, San Cristobal and Guadalcanal. It is probable that a similar basement now lies buried beneath the later volcanics in the New Georgia Group and Bougainville. The basement succession is best observed in outcrop on the island of Malaita where it is unmetamorphosed.

Typically, the basement sequence consists of deep-water nonvesicular basaltic pillow lavas and associated volcaniclastics together with intercalations of pelagic limestone lenses and local cognate intrusions of gabbro and dolerite of late Cretaceous age. On the islands of the Central Province the basalts pass at depth into dynamometamorphically altered metabasalts of greenschist to amphibolite facies, which on Guadalcanal and Choiseul, exhibit distinct sets of schistosity (Hackman, 1980; Stanton and Ramsay, 1975; Arthurs, 1980). Thrusting, faulting and deep erosion has subsequently juxtaposed schists and unmetamorphosed basalts in fault or thrust-contact shuffled blocks.

On all the islands of the Central Province, alpine-type ultrabasic complexes, consisting predominantly of harzburgites, variously serpentinized, have been tectonically emplaced as pods into the axial zones of geanticlinal structures or as thrust sheets onto basement volcanics. With the exception of the ultrabasic sheet on east Choiseul, and the Suta Ultrabasic Body on Guadalcanal, the larger ultrabasic bodies of the Solomon Islands are arranged in a linear fashion, parallel to northwesterly trend of the major islands.

The age of emplacement of the ultrabasic complexes is imprecisely known. On Guadalcanal, Eocene to Upper Miocene ages have been invoked (Thompson, 1960; Coleman, 1965; Hackman, 1980). On Choiseul, the Siruka sheet was thrust over already deformed and metamorphosed basement but is probably no later than middle Miocene (Ridgway and Coulson in press). Remobilization of some ultrabasic bodies occurred in middle Miocene times. This "pre-Miocene basement," consisting more or less of strongly folded and metamorphosed oceanic basalts, cognate gabbroic intrusions and "alpine-type" ultrabasics, while not conforming in every respect to the classical ophiolite sequences described in the geological literature, nonetheless does display enough of the necessary features so that authors have been prompted to use the term ophiolite in describing the basements of Choiseul (Coleman, 1977), Florida (Neef and Plimer, 1978), and Santa Isabel (Stanton and Ramsay, 1975).

It is believed that the tectono-metamorphic event that resulted in the widespread metamorphism, thrusting and block shuffling of the basement and the emplacement of ultrabasic thrust sheets, was a response to the initiation of subduction of the Pacific plate in a southwesterly direction beneath the Australian plate (Dunkley, 1983). Radiometric dating of the metamorphism on Choiseul gives an age range of 32–51 Ma with a mean of 44 Ma (Webb *et al.*, 1966) and on the Florida Islands, an age of 35–44 Ma (Neef and McDougall, 1976), corresponding to the Upper Eocene–Lower Oligocene. This trench system, which lay to the northeast of the present island chain and which was subsequently blocked off in mid-Miocene times, has its remnants today in the Vitiaz Trench–Cape Johnson Trough to the east of the Solomon Islands and the West Melanesian Trench to the north of Bougainville.

On Guadalcanal Hackman (1980) has subdivided the pre-Miocene basement into the Mbirao Group and the Guadalcanal ultrabasics. The Mbirao Group is essentially a thick sequence of basic volcanics, subsidiary limestones and doleritic sills intruded (at depth) by gabbro. Within the group, the Mbirao volcanics are distinguished as being predominantly of relatively unaltered pillow lavas, some massive lavas and included minor bodies of white and pink limestone. Measured thicknesses of up to 1200 m of pillow lavas occur on the Weather Coast block. Minor, but discrete bodies of limestone (Tetekanji Limestone) up to 200 m thick are also recognized.

Bodies of gabbro (Guadalcanal Gabbro) occur in sheared and faulted contact with both the Mbirao Metabasics and the Guadalcanal Ultrabasics, but are intrusive into the Mbirao Volcanics. A belt of metamorphic rocks (the Mbirao Metabasics) form a major east–west axial zone of eastern Guadalcanal, consisting of altered brecciated and schistose metabasics. Hackman (1980) considers the Mbirao Metabasics as the metamorphosed equivalent of the Mbirao Group altered to greenschist grade.

Three roughly linear belts of ultrabasic rocks outcrop on Guadalcanal— the Marau, Suta and Ghausava–Itina Ultrabasics. All are predominantly of harzburgite, variously serpentinized, and were probably emplaced during Eocene or possibly Oligocene time.

On Choiseul, Coleman (1960, 1962, 1965) described the basement consisting of an older series—the Choiseul Schists, overlain by later Voza Lavas. More recent work (Ridgway and Coulson, in press) have confirmed this main subdivision, but have shown that the Choiseul Schists are the more intensely metamorphosed and deformed equivalent of the Voza Lavas, the two rock types being folded and now juxtaposed in thrust and fault-shuffled blocks.

The Voza Lavas form a pile of pillowed, massive, brecciated and sheared ocean floor tholeiitic lavas that display all gradations between unaltered lavas to massive hornblende amphibolites. The Choiseul Schists, composed mainly of amphibole schists with a dominant northwest to southeast trending foliation, are extensively exposed in central and eastern Choiseul. Three periods of deformation are recognizable in the schists. Intrusives into the basement include dolerite dikes and an extensively altered microgabbro (the Oaka metamicrogabbro) that invades the Voza Lavas of east-central Choiseul. In east Choiseul a thrust-emplaced slab of serpentenized harzburgite up to 560 m in thickness (the Siruka Ultramafics), rests on both the Voza Lavas and Choiseul Schists.

On Santa Isabel, Stanton (1961) and Stanton and Ramsay (1975) describe the Central Province basement as a well developed ophiolite sequence. The basalt–gabbro sequence, with a combined thickness of 6.5 km, is particularly well developed.

Ultramafic rocks are present as the San Jorge Ultramafites, a subcircular outcropping body on the island of San Jorge, that appears pluglike in form. On the mainland, the Kolomola Ultramafites form a series of linear, fault-controlled bodies of harzburgite grading locally into dunite. Layered micro-gabbro (the Vitoria Microgabbro), which in places attains amphibolite grade metamorphism, occurs in faulted and intrusive contact with the harzburgites. A thickness in excess of 3 km is estimated for the microgabbro. Overlying the ultramafites–gabbros are the Sigana Volcanics consisting of a 3.5 km pile of well-formed pillow lavas, thin lava flows and rare volcaniclastics. These are assigned an Upper Cretaceous to Palaeocene age on the basis of a 66 m.y. radiometric date. (Snelling, personal communication in Hackman 1980).

Thompson (1958), Taylor (1977), and Plimer and Neef (1979) have described the basement rocks of the Florida Islands as an ophiolitic sequence. An essentially continuous pile of mafic pillow lavas of pre-Miocene age is divided into an older Kasika Metabasics and a younger Naghoto Volcanics. The lavas are partly metamorphosed and extensively intruded by dolerite dike swarms. Intruding the Kasika Metabasics are the Vatilau Gabbros which are associated with various ultramafic rocks (Nggela Ultramafics).

The basement of San Cristobal has been described by Thompson and Pudsey-Dawson (1958), Coleman (1965) and Jeffries (1977), the latter intro-ducing the term San Cristobal Basement Complex. Pre-Miocene pillow bas-alts, massive lavas and pods of gabbro (the Warahito Lavas) containing lenses of pelagic limestone predominate and show evidence of extensive fracturing, shearing and low-grade metamorphism. Discrete limestone masses up to 200 m thick (Ravo Limestones) are recognized resting on the lavas. Ultrabasic complexes occur in fault-bounded enclaves in eastern San Cristobal and appear to be tectonically emplaced. The widespread dynamic alteration of the basement lavas and the emplacement of the ultrabasics is believed to have occurred during Eocene–Oligocene times.

In the Shortland Islands, volcanic "basement" is exposed on all three main islands. However the age of the volcanics is uncertain and the chem-istry of the lavas indicates that only the basement lavas on Fauro may be oceanic tholeiites (Turner 1978; Ridgway and Coulson in press). On Fauro, over 500 m of basement lavas (the Masamasa Volcanics) consist of massive, brecciated and pillowed lavas. On Alu, a basement of over 400 m of altered, massive and brecciated lavas is overlain by up to 450 m of Plio–Pleistocene calcareous sediments. The chemistry of these lavas indicates that they are probably early island arc tholeiites. On Mono, a small area of "basement" pillow lavas are both admixed with and are overlain by pelagic limestones containing tuffaceous mudstones, a sequence very similar to that seen in southern Malaita. Chemically however, these lavas are altered hawaiites and benmoreites, and are of a type found on oceanic islands.

There is no "basement" exposed in the Eastern Outer Islands, the volcanics being considered to be arc tholeiites.

2. *Late Oligocene to Late Miocene: Early Arc Volcanism and Sedimentation*

The essentially "oceanic phase" of the Solomon Islands development, which was characterized by the extrusion of mid-ocean ridge basalts and open ocean pelagic sedimentation was succeeded, during the late Eocene and early Oligocene, by a period of tectonism. A regional metamorphic event imparted a greenschist to amphibolite facies on the basement rocks of the Central Province. On Guadalcanal and on Choiseul, detailed mapping has shown that at least two major phases of tectonism each imprinted a set of metamorphic *S*-surface on the basement successions. On Guadalcanal, the Suta and Marau ultrabasics were emplaced into axial locations, their dispositions conforming with that of the axial plane schistosity (Hackman, 1981).

Dating of this metamorphic event is only approximate. The Choiseul Schists (the metamorphosed basement Voza Lavas) radiometric dates range from 32.4 to 51.5 Ma, with a mean value of 44 $\pm$ 18 Ma (Webb *et al.*, 1966), while on Small Nggela in the Florida Islands greenschists and a metagabbro have yielded radiometric ages of 44.7 $\pm$ 2.1 and 35.2 $\pm$ 0.7, respectively (Neef and McDougall, 1976), corresponding to an Upper Eocene to basal Oligocene age. A basal Oligocene age of 35 Ma has also been reported for the metamorphism of the ophiolitic basement on Pentecost Island in Vanuata (Mallick and Neef, 1974).

It is believed that this widespread dynamothermal metamorphic event occurred in response to the initiation of subduction of the Pacific plate in a southwesterly or westerly direction beneath the Australian plate and the creation of a nascent Solomon Arc. A process of downbuckling of the oceanic lithosphere with concomitant increase in the depth of burial is considered to have facilitated the metamorphism. The onset of subduction and the creation of an island arc environment led, probably through a process of differential block faulting, to the emergence of islands of pre-Miocene rocks.

Hackman (1980) refers to the emergence of the Guadalcanal–Mbirao block possibly during the early Oligocene. Emergence of the Choiseul basement probably occurred in the late Oligocene (Ridgway and Coulson, in press). Jeffries (1977) refers to the emergence of a proto-San Cristobal as a series of horst blocks in pre-Miocene times. In the Florida Islands, basement was probably emergent in the late Oligocene and early Miocene (Neef, 1979).

Subduction also led to the production of the first island arc volcanic episode with the eruption of island arc theoleiites and the intrusion of igneous complexes. The main activity appears to have been on Guadalcanal where a minimum thickness of 2500 m of basaltic andesite lavas with intercalations of volcaniclastic and carbonate sediments (Suta and Marasa volcanics) of Late Oligocene to early Miocene age are now exposed.

Arc volcanism of this period is also evident: in the Florida Islands, the Soghonara Lavas (Taylor 1977); the Alu Basalts and Masamasa Volcanics of the Shortland Islands (Turner and Ridgway 1981); and in the Eastern Outer Islands the Mergalue and Nolua volcanics of Nondo (Hughes *et al.*, 1981). The Kieta Volcanics of Bougainville (Blake and Miezitis, 1967) are also of Oligocene–Lower Miocene age and probably form part of this episode.

The volcanics of this period consist mainly of lavas with intercalations of volcaniclastics and carbonate sediments. Pillow lavas are common and most of the volcanics are probably of submarine origin. Basalts and basaltic andesites predominate (Fig. 20) with andesites and more salic rocks accounting for only a small proportion of the total volume (80% of the total population contains less than 55% SiO_2). The suite is tholeiitic and is considered typical of the island arc tholeiitic series (Dunkley, 1983).

On Guadalcanal in the late Oligocene the Suta Volcanics were intruded by the Poha Diorite dated at 24.4 ± 0.3 Ma (Chivas and McDougall, 1978). The Lungga Diorite is probably of similar age. These intrusives are generally more siliceous than their contemporaneous volcanics, ranging in composition from gabbro to tonalite with quartz diorite predominating; in contrast to the volcanics they are calc-alkaline in character. Thus began a period of arc volcanism and sedimentation that was to lead to the accumulation of up to 6 km of Neogene and Quaternary rocks on the pre-Miocene basement.

The early arc volcanism, which appears from the geological record preserved in the island successions to be localized and relatively short-lived, had virtually ceased by the early Miocene. The widespread deposition of limestones, calcarenites and foram-rich marls in the early Miocene across the shelves of Guadalcanal, the Florida Islands and San Cristobal attest the cessation of volcanism. On Bougainville, Lower Miocene reefal limestones (the Keriata Limestone) unconformably overlying the Oligocene Kieta Volcanics (Blake and Miezitis 1976), also indicates an early Miocene termination of volcanism.

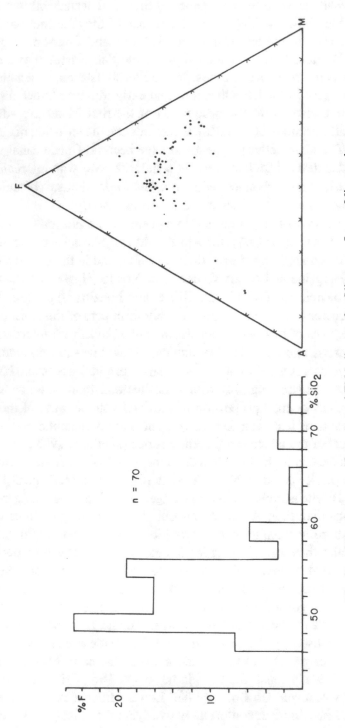

Fig. 20.   Silica frequency histogram and FMA diagram for Oligocene–Lower Miocene lavas.

The absence of volcanic rocks in the Miocene stratigraphy of the Solomon Islands is also reflected throughout the rest of Melanesia, from the Fiji Group in the east, to the islands of Bougainville and New Guinea in the west. The Miocene sedimentary column in the Solomon Islands is dominated by the deposition of volcanogenic clastic deposits in deep-water basins and the extensive development of carbonates in the shallow water shelf areas. Rapid erosion of the newly emerged but tectonically unstable island blocks led to the denudation of both the pre-Miocene basement and the early arc volcanics, producing voluminous volcanogenic clastic deposits, essentially of the greywacke type, that were deposited in fault-bounded basins and troughs.

On Choiseul, up to 2500 m of sediments of the Mole Formation lie with pronounced unconformity on the basement. At the base of the sequence a rudaceous facies (the Koloe Breccias) form a basal breccia or conglomerate that has been recognized at several localities across western Choiseul. Its erratic distribution, localized nature and marked thickness variations (up to a maximum of 250 m) suggest that it infills hollows or small fault-controlled basins within the basement. The breccia consists of angular basaltic clasts set in a basaltic grit matrix, the lithology being controlled by the character of the underlying basement, from which they are derived. By a decrease in the proportion of included lithic fragments and a concomitant increase in matrix, the Koloe Breccias grade upwards into the well-bedded and color banded sandstone–siltstone–mudstone sequences that are characteristic of the bulk of the Mole Formation.

Fringing coralgal reefs became established upon the flanks of the possibly partly subaerial Choiseul ridge and are now evident as the Mount Vuasa Limestone. These Lower to middle Miocene limestones, which form a series of discontinuous lenses up to 50 m thick in western Choiseul, include calcisiltites, calcarenites, reef limestones and biocalcirudite and represent both reef and reef-slope deposits.

The bulk of the Mole Formation consists of repeated sequences of interbedded microbreccias, sandstones, siltstones and mudstones, the succession becoming generally more fine-grained and calcareous upwards. Clasts within the breccias are mainly derived from the basaltic and schistose basement, although limestone, mudstone and siltstone lithics occur higher in the succession. Color banding of the finer grained parts of the succession is very well developed. A variety of sedimentary structures are evident including graded and cross-bedding, load-casts, ripple marks and scour. A turbidite origin for much of the Mole Formation seems likely although some parts may be estuarine.

The Mole Formation is generally of Miocene age, although paleontological evidence indicates that deposition in central-north Choiseul commenced in late Oligocene times, but did not begin until the middle to late Miocene in northwest Choiseul where the beds are only a few hundred meters thick.

On Guadalcanal, a similar early history of sedimentation is recognized, with a variety of sedimentary rocks laid down upon a heterogenous irregular basement. In southwest Guadalcanal, late Oligocene to Miocene greywackes and poorly calcareous clastic sediments accumulated in the deep basinal areas, having been deposited as marine turbidite and debris flows derived from adjacent high volcanic terrains. Such deposits include the Kavo Greywacke Beds and Lungga Beds.

The Kavo Greywacke Beds, now exposed in the Itina River Basin and Kavo Ranges form a series of thick (minimum thickness 2500 m), poorly sorted sandstones and conglomerates. Within this sequence Hill (1960) has described dark massive shales, often black in color that indicate a euxinic environment. The greywackes both overlie and interdigitate with the Oligocene–Miocene Suta volcanics from which they are principally derived. They are succeeded by the Lungga Beds which outcrop extensively over western Guadalcanal. They comprise a succession of arenites and wackes derived primarily from volcanic sources with subsidiary conglomerates, mudstones and andesitic lava flows. The variety of lithic fragments is greater than in the Kavo Greywackes and local rapid facies changes are characteristic and indicative of the constantly changing sedimentary environments created by active tectonism during accumulation. The succession, which in part rests upon oceanic basalt (Turner and Hackman, 1977) and in other areas on biostromal limestone, reaches a thickness of 1200–1500 m (Wright, 1968, Hughes, 1977) and ranges in age from middle Miocene to Upper Pliocene(?).

Deep marine clastic turbidite deposits, possibly with deep marine carbonates may also have been accumulating at this time along the northern slope of the rising ridge of Guadalcanal. Such deposits would now lie buried beneath later sediments exposed in northern Guadalcanal.

In the shallower eastern parts of the basin and along the basinal margins of the west, biogenic limestones and interreefal calcarenites (the Mbetilonga Group) form the predominant lithology and are mainly of Lower Miocene age although ranging into the Piocene in eastern Guadalcanal. From west to east across the island these include the Mbonehe Limestone, Mbetilonga Limestone, Tina Calcarenite, Lake Lee Calcarenite and the somewhat younger Valasi Limestone. These deposits, which rest unconformably upon a variety of older rocks, range in minimum thickness from 100 to 400 m, with a maximum estimated thickness of 1000 m for the Valasi Limestone. The group is composed essentially of carbonate rocks with a variety of texture, grain size and visible organic content, and characteristically containing a high admixture of terriginous material. Biostromal limestones, consisting of off-white pure carbonate rocks are massive and usually recrystallized to varying degrees; such facies constitute up to 70% of the Mbonehe Limestone and are well represented in both the upper half of the Mbetilonga Limestone and the basal beds of the Lake Lee Calcarenite. A coralgal reef facies comprises the western part of the Valasi Limestone.

Foraminiferal biocalcarenites, usually thickly bedded, may be intercalated with or grade laterally into the more massive recrystallized limestones. Such facies are common thoughout the Lake Lee Calcarenite and the eastern part of the Valasi Limestone. The Tina Calcarenite is mainly composed of well-bedded flaggy calcarenites with a high terriginous admixture including occasional carbonized wood fragments. In both the Mbetilonga and the Valasi limestones, the terriginous content generally increases downsection, with basal conglomerates containing clasts of demonstrably local origin.

For the most part, the Mbetilonga Group sediments are impure shelf carbonate deposits. Foraminiferal assemblages suggest water depths in the range of 40–80 m. However, the cleaner biostromal limestones must have accumulated in quieter back-reef conditions or as fringing reefs sheltered from contamination.

The island of Santa Isabel has not been subject to detailed mapping and in consequence, little detail is known of the Tertiary sedimentary cover overlying the ophiolitic basement. The distribution of sedimentary facies on the island is complicated by the juxtaposition of the Central and Pacific provinces.

Stanton (1961) informally described all the sediments as belonging to the Tanakau Group, a term that implied no stratigraphic or facies connotation, although the essential contrasts of the sedimentary successions on either side of the axial volcanic spine were established in the early reconnaissance work. Along the northeast coast a "Malaitan-type" succession containing an abundance of pelagic limestones was distinguished from a "Choiseul-type" sequence in the south containing terrigenous material but no limestone.

Subdivision of the Tanakau Group in southern Santa Isabel has since been attempted (Coleman, 1965). The oldest sediments are the Loguhutu Beds (San Jorge) and the Bero Beds of southwest Santa Isabel, consisting of Lower Miocene coarse lithic volcanic sandstones. The succession in southwest Santa Isabel shows a minimum thickness of 2.2 km of volcanogenic greywacke type sediments consisting predominantly of volcanic sandstones with interbedded shales and mudstones and occasional sandstones and conglomerates. While it has been suggested that these rocks are the approximate equivalents of the Kavo Greywackes of Guadalcanal or the Mole Formation of Choiseul, they may not be of local provenance, but possibly derived from the Choiseul or New Georgia region (Stanton, 1961). The Pliocene is represented, at least in part, by tuffaceous sandstones and shales that may be highly calcareous with abundant foraminifera.

In the Florida Islands, detailed mapping is incomplete. Coleman (1965) describes the sedimentary succession commencing with coarse-grained calcareous lithic sandstones of Lower Miocene age; these are clearly derived from erosion of the basement although they contain no ultrabasic detritus. Calcarenites, fine-grained volcanic sandstones and siltstones succeed the basal sediments and in turn, give way to the Anuha Calcarenite, a prominent horizon of coarsely bedded foraminiferal calcarenite of Lower Miocene age; correlation of the Anuha Calcarenite with the Mbetilonga Group of Guadalcanal is probable.

The Lower Miocene sediments are overlain by middle to Upper Miocene calcareous sandy siltstones and fine-grained tuffaceous sandstones, indicative of rapid accumulation on an unstable shelf. These become increasingly calcareous upwards and, in the west, grade into the Florida Limestone, a Pliocene fringing reef deposit.

Taylor (1977), in detailed mapping of western Florida, has given the name Mboli Beds to the Oligocene–Pliocene sedimentary succession that rests unconformably on the basement lavas. Within this succession, which has a minimum thickness of 600 m, the Kombuana Sandstone forms the dominant member, being an essentially turbidite sequence of volcaniclastic arenites, lutites and epiclastic rudites. The Mboli Beds are probably the equivalent of the Florida Limestone and Anuha Calcarenite to the east.

On Small Nggela, Neef (1979) has divided the Lower Miocene sediments into the Siota Beds, consisting of up to 850 m of uniform massive sandstone with occasional mudstones and separated by an ophiolitic wedge from the Ghumba Beds to the southwest which comprise a more diverse sequence of arenites, rudites and lenses of pillow lavas, the whole sequence attaining a thickness of 1500 m and becoming progressively finer upwards.

The geology of San Cristobal is probably the least known of the major islands. The sedimentary succession that is believed, for the most part, to lie upon the basaltic basement has been named the San Cristobal Group (Coleman 1965). It represents one of the thinnest post-Miocene covers in the islands, having a total thickness of only 700 m.

Within the San Cristobal Group in the east of the island, the Hariga Conglomerates both interdigitate with and overlie the basement pillow lavas. They include coarse- and fine-grained conglomerates and sandstones with a calcareous, tuffaceous matrix. The latter contain Lower Miocene foraminifera although the conglomerates probably range in age from Lower to Upper Miocene. They have a partial stratigraphical equivalent in foraminiferal calcarerites that occur in southeast San Cristobal, and may represent the equivalent of the Mbetilonga Group on Guadalcanal (Hackman, 1980).

Jeffery (1976) in detailed mapping of northwestern Santa Isabel describes the Miocene succession as calcareous mudstones, wackes and volcaniclastic slump breccias (the Ruawai Beds), with the development of a pelagic limestone at the base of the succession (the Hautarau Limestone). A minimum thickness for this sequence is 410 m. A volcanic sequence (the Waihada Volcanics), which is in part equivalent to and in part younger than the Ruawai Beds, consists of tuffs, agglomerates and pillow breccias that show evidence of submarine reworking and slumping.

The waning of first phase arc volcanism by early Miocene times, enabling the extensive development of reef and platform carbonates during the Miocene period, is believed to have been caused by arrival at the Vitiaz–West Melanesian trench system, of the Ontong Java Plateau, a collision process that gradually choked off subduction and its volcanic by-product. However Coleman and Packham (1976) attribute this quiescent period to orientation of the Melanesian arcs in line with a direction of transform motion, and suggest that the encroachment of the Ontong Java Plateau upon the Central Province may have been through transform motion.

Kroenke (1972) and Coleman and Kroenke (1982) describe the Ontong Java Plateau as an anomalously thick oceanic crust (40 km) and lithosphere that probably originated through a process of abnormally slow spreading at a mid-ocean ridge prior to 67 m.y. ago. The plateau is believed to have begun to impinge upon the North Solomon Trench about 10 m.y. ago, causing tensional ruptures along the crest of the outer trench flexural high; progressive blocking and obliteration of parts of the trench gradually led to the choking off of the subduction process and its related volcanism. By about 8 m.y. ago, deformation from the collision folded and faulted the sediments along the margin of the plateau, this deformation being most intense in the Malaita region (Fig. 13).

3. Pliocene to Recent: Second-Stage Arc Volcanism and Continued Sedimentation

With subduction at the North Solomon Trench cut off by the Ontong Java Plateau, it is thought that further convergance of the Pacific and Australian plates was accommodated by a subduction flip and the formation of a new trench system on the southwest side of the arc along which the Australian plate is now being subducted in a northeasterly direction beneath the

Pacific plate. The establishment of the subduction process led to a resurgence of volcanism along the southwest side of the arc. Pliocene to Recent calc-alkaline activity built a series of volcanoes along a belt stretching from the Shortland Islands to northwest Guadalcanal, and was accompanied by the emplacement of several intrusive complexes.

The second phase of volcanism, which gave rise to Coleman's Volcanic Province, was preceded by tectonism in the middle and late Miocene, believed to be associated with the development of incipient northeastward subduction under the southwestern flank of the arc.

On Guadalcanal, renewed anticlinal horst formation heralded a period of instability and intense alluviation centered on the Gold Ridge area, while the Mbirao Metabasics and the Guadalcanal Gabbro were tectonized along fault lines. Thompson (1960, 1968) considers that the Ghausava–Itina Ultrabasic Bodies, initially emplaced in the Oligocene, were thrust higher into the crust at this time. In the early Pliocene, the Miocene sediments responded to reactivation of basement faults, forming a series of gentle open folds along northwesterly axes. Within the basement, earlier schistosities were folded and S_3, S_4 surfaces were imprinted on the basement.

The renewal of subduction created a new pattern of sedimentation. In the region of the volcanic axis rapid uplift and instability brought reef growth to an end, and high erosional energy levels led to the shedding of huge quantities of coarse greywacke-type sediments into adjacent marine basins. Further behind the arc however, sheltered open-marine conditions allowed the deposition of extensive platform carbonates.

(a) Pliocene to Recent Arc Volcanism—the "Volcanic Province". Volcanics related to the second period of subduction occur mainly in a belt along the southwest side of the arc, stretching from Bougainville to Savo and northwest to Guadalcanal. However, two Plio–Pleistocene volcanoes on Choiseul are also included within the Volcanic Province. A number of high level intrusive complexes also occur in these calc-alkaline suites. Coleman and Kroenke (1982) have described an approximately 75 km spacing of volcanoes along the southwest side of the arc along an axis stretching from Bougainville to Savo. Off-trend "displaced" volcanism is evident in the location of some of the New Georgia volcanoes, which are anomalously near-trench.

In general, within the volcanic piles of the Volcanic Province lavas and epiclastic breccias derived mainly from lavas predominate over pyroclastic deposits. Ash flow tuffs and pumice flows of intermediate composition are of limited areal extent, which on some quiescent volcanoes, are associated with andesite and dacite domes.

Compositionally, the volcanics of this second period are variable in character. In contrast to the earlier Oligocene–Lower Miocene volcanics,

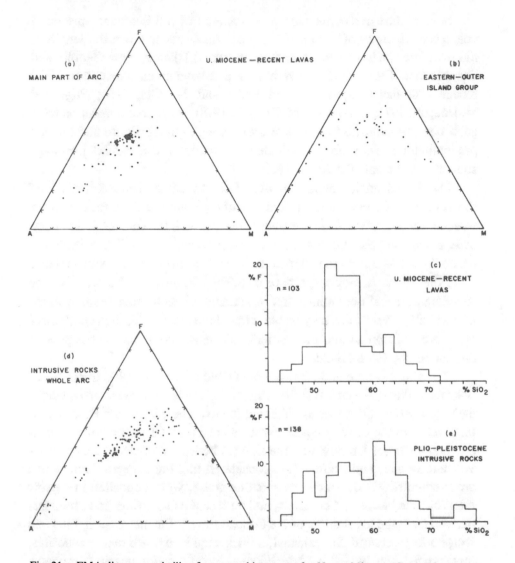

Fig. 21. FMA diagram and silica frequency histograms for Upper Miocene–Recent lavas and intrusions (excluding the New Georgia Group).

those in the main part of the arc are typically calc-alkaline (Fig. 21), whereas those in the Eastern Outer Islands are tholeiitic with low TiO_2 and belong to the island arc tholeiitic series. A further variation occurs in the New Georgia Group where large volumes of olivine rich basalts and picrites have been erupted; however these are not thought to be related to normal sub-duction processes. Excluding the anomalous New Georgia volcanism, the Upper Miocene to Recent volcanics are generally more salic compared to the early arc eruptives and range in composition from basalt to rhyodacite with basaltic andesites and andesites predominating (Dunkley, 1983).

Several intrusive complexes are associated with this second episode of volcanism. A series of irregularly shaped stocks occur down the length of the main arc, including Bougainville, Shortland Islands, New Georgia and Guadalcanal. These stocks, several of which have given Pliocene to Pleistocene radiometric ages (Well in Blake and Miezitis, 1967; Page and McDougall, 1972; Chivas and McDougall, 1978), range in composition from gabbro to granodiorite to quartz monzonite with quartz diorite and tonalite predominating. They are calc-alkaline in character and exhibit late-stage sodium enrichment (Dunkley, 1983).

On Guadalcanal, igneous activity reached a climax towards the end of the Pliocene, and was concentrated in, although not confined to, the northwest corner of the island. In northwest Guadalcanal, the Gallego Lavas comprise plugs and volcanic cones composed predominantly of hornblende andesite lavas that are up to 900 m in thickness; aprons of pyroclastic breccias flank the lavas. A single age date of 6.39 ± 1.95 m.y. or Lower Pliocene (Snelling personal communication in Hackman, 1980) indicates that much of the Gallego Volcanics may be older than Pleistocene although geothermal areas are widespread and intermittent vulcanicity continues to the present day on nearby Savo Island.

The Gallego-type andesites are blanketed by considerable thickness of volcanic rudite (the Tiaro Tuff Breccias) which are considered as an epiclastic facies probably laid down as subaerial lahars. The Tiaro Tuff Breccias interfinger with the cone-complex pyroclastic facies near to the volcanic centers and also pass laterally into the Lungga Beds, composed of sedimentary volcanic wackes and rudites. It is considered that the Lungga Beds, which are predominantly of sandstone-siltstone grade, were deposited in narrow intervolcanic basins and channels, and represent ash showers and sediments derived from the breakdown of the Gallego Lavas. Further east, in the Gold Ridge area of central Guadalcanal, a thick sequence of extrusive andesites, now eroded and buried by late Pliocene paralic sedimentation, probably represents an isolated calc-alkaline volcanic center.

The Koloula Diorite, which outcrops in south-central Guadalcanal, is associated with this volcanic phase. Radiometric age dating (Chivas and McDougall, 1978) indicates a polyphase emplacement that extended through Pliocene times.

To the north, the island of Savo represents a quiescent volcano of the Pelean type. Proctor and Turner (1977) describe the main rock type as a hornblende andesite, the volcanic pile being composed predominantly of agglomeratic and tuffaceous deposits and their slope-wash derivatives.

The Russell Islands, which lie midway between Guadalcanal and eastern New Georgia, form an emergent island volcano composed of basaltic andesite breccia (the Pavuvu Breccias) overlain by basalt lavas (the Banika Lavas); the volcanic core is encircled by siltstones and reef limestones of Pleistocene age (Danitofea and Turner, 1981). Mborokua Island, situated between the Russell Islands and New Georgia, forms an extinct volcano composed of massive basaltic lavas and volcanic breccias (Turner, 1975).

On the island of Choiseul, two volcanic piles comprising the mountains of Maetambe and Komboro are composed predominantly of andesitic pyroclastic deposits. Petrochemically these are calc-alkaline rocks, and are considered to form part of the Volcanic Province. The Maetambe Volcanics, which have a minimum estimated thickness of 500 m, consist predominantly of andesitic tuffs, ashes and breccias. Although the volcanics for the most part overlie the basement and the Mole Formation, some interbedding of the latter with the Maetambe volcanics indicates contemporary volcanism and sedimentation at least in the early stages. Maetambe volcanism possibly began in middle Miocene times and may have continued into the Pleistocene. Geothermal springs on Mt. Maetabe suggest the waning of relatively recent volcanism. The Komboro Volcanics consist of a sequence of andesitic breccias and tuffs. No direct evidence for the age of these volcanics is available, but ash-fall material from Komboro occurs in nearby sediments of early Pliocene age.

The islands of the New Georgia Group represent by far the most extensive and voluminous development of the second arc volcanic period. While the New Georgia Group, which consists of a complex of emerging and coalescing island volcanoes encircled by a complex of fringing reefs and lagoons, is, in its geographical expression, typical of the Volcanic Province, in the location of the volcanoes, which are anomalously near-trench, and in the chemical composition of the lavas, it is, in fact, an atypical development. From Pliocene to Recent times, the New Georgia Group has been the site of voluminous eruptions of volcanic rocks of unusual composition, some of which are among the most basic lavas known to occur in island arcs.

The volcanics consist of large volumes of highly porphyritic olivine basalt and picrite basalt lavas and breccias with minor basaltic hornblende and basaltic andesites and andesites. The basalts and picrites are hypersthene normative and on an FMA diagram show a trend to magnesium enrichment (Fig. 22), whereas the minor andesites with which they are both temporally and spatially associated, show a broad calc-alkaline trend more akin to the "normal" volcanics occurring elsewhere in the arc. (Dunkley, 1983)

It is difficult to define the petrogenetic character of the New Georgia suite. It is subalkaline and has a low TiO_2 content associated with arc volcanics, but the suite is high in potash with basalts containing up to 2.6%

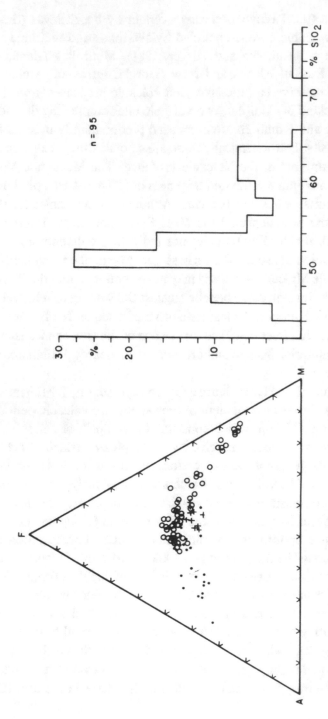

Fig. 22.   FMA diagram and silica frequency histogram for Upper Miocene to Recent lavas from the New Georgia Group.

K$_2$O. Typically, the volcanic piles display a lateral facies change from lavas to volcaniclastic rocks, the latter showing an increasing degree of reworking away from the eruptive centers. Slope outwash material is a significant facies within the volcanic edifices (Dunkley, personal communication).

Extensive development of sedimentary rocks is only evident in the fore-arc region, where on the islands of Tetepare and Rendova, deep water silt-stones and sandstones containing chaotic melanges of rudaceous clastic deposits together with volcanic, plutonic and coral reef debris that have probably slumped off the main arc to the north. The unusual composition of the New Georgia Lavas and the anomalously near-trench position of the volcanoes is attributed to subduction beneath New Georgia of the Woodlark Rift which is considered to be an active spreading center within the Woodlark Basin to the west and southwest of the New Georgia Islands. The Woodlark Basin is described in Section IV.

The subduction of the Woodlark spreading center and the adjacent young crust is also believed to be the cause of the seismicity gap between Guadalcanal and the Shortland Islands. The virtual absence of intermediate and deep earthquakes associated with this section of the Benioff zone is belived to be the result of the subduction of very young, warm and ductile oceanic lithosphere that is likely to deform plastically.

An additional effect of subducting young, bouyant oceanic crust is the considerable uplift of the arc and fore-arc region, a feature that is clearly manifest in New Georgia where deep water fore-arc sediments have been elevated to varying heights above sea level on the islands of Tetepare and Rendova.

Intrusives related to this second period of arc volcanism include gabbro to tonalite stocks in the Mase and Hube areas of New Georgia together with several high-level andesite plugs.

In the Shortland Islands, younger calc-alkaline suites comprising both intrusive and extrusive rocks have been described. In the Fauro Group, two intrusive bodies (the Tauna Microdiorite and the Fauro Dacite) have been emplaced into the basement lavas, while further north a high-level horn-blende andesite intrusive forms Oema Island.

The horseshoe shaped bay of north Fauro is believed to have been formed by the caldera collapse of the main Fauro calc-alkaline volcano that produced at least 1500 m of (reworked) crystal tuff and minor lavas (the Koria Sandstones) overlain by 600 m of volcanic breccias, tuffs and minor lava flows (the Togha Pyroclastics).

In the Alu Group, the dioritic Hisiai Complex intrudes the basement lavas. The Kamaleai Pyroxene Andesite may be the extrusive equivalent of the Hisiai Complex.

In the Eastern Outer Islands, volcanics of this period are considered to make up most of the group, forming submarine and subaerial piles of lavas and volcaniclastics of Pliocene to Recent age. Only Nendo, the largest island in the group, and composed predominantly of Lower Miocene volcanics, appears to belong to the earlier arc period. The volcanics are predominantly tholeiitic (Fig. 21), and belong to the island arc tholeiitic series (Dunkley, 1983).

One of the often-commented upon enigmatic features of the Solomon Arc is the absence of post-Miocene volcanics associated with the supposed subduction along the line of the South Solomon Trench between East Guadalcanal and the South Solomon–Torres Trench reentrant. While part of this volcanic gap may be attributed to transform rather than subductive plate motion, the total absence of volcanism where seismicity describes a Benioff zone requires explanation.

Coleman and Kroenke (1983) have suggested that subducted Australian plate is in direct contact with cold, depleted, abnormally thick Ontong Java lithosphere which in the absence of an intervening asthenospheric wedge, inhibits magma production. However it has been argued by Dunkley (1983) that such contact is unlikely as the Benioff zone beneath eastern Guadalcanal is nearly vertical and that the Ontong Java lithosphere is presumably confined to the region north and northeast of the Vitiaz Trench. The presence of Ontong Java lithosphere south of this line would imply successful subduction of the plateau.

Dunkley (1983) considers the volcanic gap not to be an unusual feature when the spacing of active or recently active volcanoes is viewed across the entire arc, especially if the abnormal ridge–subduction related volcanism in New Georgia is excluded. He concludes that volcanism within the arc is very much on the wane due to a combination of steeply subducted lithosphere, which would reduce the overlying asthenospheric wedge, and a refractory mantle that has now been "double-cooked" through two periods of subduction.

However, there are a number of objections to this proposal. In considering the lack of second stage arc volcanism in east Guadalcanal and San Cristobal, it is not valid to use the distribution of only active and recently active volcanoes as a yardstick when the later arc volcanism spans the period from Pliocene to Recent. There appears to be a total lack of volcanics of this period in east Guadalcanal and San Cristobal yet Pliocene to Recent volcanoes are numerous to the west. Furthermore, while there might be grounds for excluding the anomalous New Georgia volcanics in this comparison, it is not unreasonable to assume that had subduction conditions been "normal" in this section of the arc, "normal" volcanoes would have been created.

The present author considers the "volcanic gap" in east Guadalcanal and San Cristobal not yet satisfactorily explained. A solution may have to await the detailed geological mapping of San Cristobal and focal mechanism studies in the South Solomon Trench. The case for waning volcanism is also doubtful. If it is accepted that the enormous volume of New Georgia Volcanics is due to abnormal subduction conditions, then they should be excluded. This would drastically reduce the volume of Plio–Pleistocene volcanics to a level more comparable to present day volcanism. In proceeding back through time, it should be borne in mind that there was a period of several million years in the Miocene when there appears to have been no volcanic activity, and first stage, Oligocene–Lower Miocene arc volcanism has only been positively identified on Guadalcanal and the Florida Islands. The present author is inclined to the view that, apart from the abnormal New Georgia volcanism, the Solomon Islands has never been a particularly "active" arc, but then what constitutes an "active" arc?

(b) Pliocene to Recent Sedimentation. While the second phase of volcanism was producing the Volcanic Province along a volcanic arc belt that extended from Bougainville to Guadalcanal, and also in the Eastern Outer Islands sedimentation continued throughout the region in a variety of conditions. In the southwest, uplift of the frontal arc, coupled with active volcanism created high rates of sedimentation with contemporary volcanism contributing detritus to adjacent sedimentary basins. Elsewhere low-energy environments were created in which fine-grained calcareous deposits accumulated with only the occasional incursion of contemporary volcanic detritus. By early Pliocene times, Guadalcanal, now forming part of the frontal arc, was being subject to increasing uplift, and in the west, to renewed volcanism. The uplift was accompanied by increased faulting, upheaval and an increased sedimentation.

Pliocene sedimentary successions are characterized by rapid lateral facies changes developed around a complex of emergent and volcanic islands, basins and interisland channels; Reef limestones were no longer able to flourish except on narrow shelves protected from the intense alluviation. The effect of this upheaval, coupled with the vulcanicity in western Guadalcanal, was to unite the Oligo–Miocene sediments and volcanics with the Mbirao basement block to form one coherent landmass, leaving the Lungga Basin as an almost landlocked focus of sedimentation. To the north, the deep and subsiding Mbokokimbo Basin received a steady supply of terrestrial detritus throughout the Pliocene.

Detailed mapping of Guadalcanal (Hackman, 1980) has outlined a number of basinal sedimentary formations spanning the late Miocene to Pleistocene period. Several of these form lithosomes with complex intertonguing

relationships. Commonly they are of rudaceous to arenaceous grade material, derived principally from adjacent volcanic sources and display structures indicative of unstable, moderate to high sedimentation.

The Mbokokimbo Formation consists of a sequence of poorly sorted
arenites containing abundant foraminifera of Pliocene age. The formation,
which takes the form of a lithosome with complex intertonguing relationships
with adjacent rocks, has a maximum thickness of 4000 m. The deposit ranges
in grade from coarse sand to clay, although fine-grained sandstone and siltstone predominate. Sporadic conglomerate layers and lenses occur throughout the sequence, often as cut and fill structures. Carbonized wood and other
plant remains occur in the coarse silt and sandstones and dark siltstones
contain pyritic concretions indicative of anaerobic conditions.

Sedimentary structures are indicative of accumulation in a fluviatile or
neritic environment, with the continued turbid deposition of debris, mainly
of volcanic provenance. Phases of marked instability, possibly seismically
generated, produced turbiditic, chaotic slump sedimentation. Hackman
(1980) has suggested a parallelism between the Mbokokimbo Formation and
the Pliocene Pemba Siltstones of northwest Choiseul. Further westward, the
Toni Formation consists of a series of volcaniclastic rudites and arenites
with subsidiary pyroclastics, extrusives and biogenic limestone that range
in age from middle Miocene to Upper Pliocene. Considerable lateral facies
change characterizes the Toni Lithosome, although it is composed predominantly of rudaceous material, mostly of volcanic origin. It includes both
paraconglomerates and orthoconglomerates, and volcanic and lithic arenites
that may be locally calcareous and grade into micrites and marls. Lenses of
coralgal limestone (the Kombusoe Limestone) occur, but nowhere exceed
100 m in thickness.

The Toni Formation was deposited on a tectonically unstable marine
shelf with slumping and turbid flow transporting shallow-water sediment into
deeper water areas. The majority of the phenoclasts in the conglomerates
are from the Gold Ridge Volcanics and this period of intense alluviation may
have derived its energy from the active volcanism of the Gold Ridge area.
The equivalent of the Toni Formation in Western Guadalcanal are the
Lungga Beds which comprise a succession of arenites and wackes derived
principally from volcanic sources, with subsidiary conglomerates and mudstones. A similar age range to the Toni Formation—middle Miocene to
Upper Pliocene or Pleistocene—is indicated.

In contrast to conditions in central and west Guadalcanal, in northeast
Guadalcanal quieter conditions permitted the growth of fringing reefs (the
Valasi Limestone). These are succeeded by the Vatumbulu Beds, a sequence
of up to 700 m of shallow water conglomeratic sediments of Upper Pliocene–
Pleistocene age, derived mainly from pre-Miocene basement volcanics.

By the end of the Pleistocene, Guadalcanal has almost attained its modern shape with the calcareous Honiara Beds forming as a northern fringe complex of reef and shallow marine and estuarine sediments.

On Choiseul, by late Miocene times, a low energy environment was becoming established in the north, which enabled widespread organic colonization to take place. The upper Mole Formation is increasingly calcareous and is succeeded by the Pemba Formation, a 400 m thick Pliocene sequence of calcarenites overlain by marls. The calcarenites, which form the lower part of the Formation (200–250 m thick) are well bedded, often displaying slump and current bedding and containing noncalcareous sandstone and paraconglomerate horizons, which possibly represent turbidites seismically generated from the slopes of the active Maetambe Volcano. The calcarenites grade upwards into uniform calcisiltites that have a maximum thickness of 120 m. There are no sedimentary structures and an increasing fineness upwards may indicate a gradual transition to deeper water conditions in the middle to Upper Pliocene.

Similar environmental conditions prevailed in southern Choiseul where the Lower Pliocene Vaghena Formation was being laid down. The Vaghena Formation, which lies unconformably upon the basement, consists of up to 530 m of calcareous arenites, siltstones and mudstones containing varying amounts of crystal debris, derived from the nearby active Komboro Volcano. There is no sedimentary record spanning the middle to late Pliocene in southeast Choiseul.

Following the deposition of the Pemba Formation, Choiseul underwent uplift, tilting and erosion. Pleistocene coral–algal–foraminiferal reefs (the Nukiki Limestone) were deposited on Mole and Pemba formations as well as over the basement down the length of Choiseul, and now preserved as thick (up to 150 m) isolated blocks along the southwestern side of the island.

In the Shortland Islands, Plio–Pleistocene sediments are represented on the islands of Alu and Mono and record a general transition from the pelagic deposition of volcanic detritus to shallow-water reef limestone deposits.

On Alu, the Alu siltstones consist of about 300 m of silts, clays and fine sandstones of Pliocene age. They are considered to represent deposition on an open marine shelf receiving detritus from Pliocene volcanoes on Bougainville and possibly Fauro. In the late Pliocene, uplift and shallowing of the seafloor allowed the deposition of up to 150 m of calcareous sandstones and chalky limestones (the Lofang Limestone). Late Pleistocene gravel deposits (the Toapina Conglomerates) represent alluvial or littoral gravels eroded from the emerging island.

On Mono, some 250 m of siltstones (the Mono Siltstones) represents a pelagic deposit, the material being largely of terriginous origin and probably representing ash-falls from Bougainville volcanism. Shoaling occurred in the late Pliocene and the Kohele Limestones were deposited, representing a reef limestone facies within the Mono Siltstones. Gradual emergence in the early Pleistocene gave rise to reef limestone deposits (Bare Reef Limestone and Toloko Reef Limestone) and lagoonal siltstones (Soanatalu Siltstone).

Plio–Pleistocene sediments of the Florida Islands are confined to residual patches of the Florida Limestone, consisting of uniform calcarenites with a minimum thickness of 250 m (Thompson, 1958; Neef, 1979). The Florida Limestone in Small Nggela rests upon folded Ghumba Beds, and represents a patch reef deposit.

On San Cristobal, Plio–Pleistocene deposits are confined to the northern and western coastal fringes where the Arosi Beds, of Pleistocene age form a reef facies including fore-reef coralgal limestone and back-reef calcarenites. Jeffery (1977) has suggested that San Cristobal was essentially emergent during the Pliocene and that Pleistocene sea level variations led to the establishment of fringing reefs along wave cut benches on the northern coast.

In the Pacific Province, the Plio–Pleistocene was an important tectonic period. In response to northeast directed compressive stress, faulting and folding of the Cretaceous to Pliocene successions took place concomitant with the emergence of the islands. The pre-Pliocene pelagic successions are succeeded by deposits showing increasing evidence of shoaling and the influx of terrestrial detritus, culminating in the deposition of shallow-water conglomerates, sandstones and siltstones and Pleistone reef limestones.

In north Malaita, the pelagic Suaba Chalk, the upper part of which may be of Pliocene age, is overlain by some 250 m of Toomba Silts composed of foraminiferal and tuffaceous debris. The Silts are overlain by Pleistocene reef limestones (Rickwood, 1957). In south Malaita the Are'are limestones are unconformably overlain by the Hauhui Conglomerates, comprising a shallow-water sequence of interbedded conglomerates and sandstones of Pleistocene age (Hughes and Turner, 1976). In small Malaita, the Haruta Calcisiltites pass upwards into the Hada Calcisiltites a sequence of calcilutites, calcisiltites and minor calcarenites and containing brown noncalcareous mudstone horizons, the latter representing incursions of volcanic detritus. The sequence, which is up to 200 m thick, is of late Miocene and Pliocene age. It is unconformably overlain by shallow-water reef limestones of Pleistocene age (Rokera Limestone).

In Ulawa, rapid shoaling in Pliocene times is evident in the deposition of up to 330 m of calcareous mudstones and slumped conglomerates (the Holohau Mudstones). They are probably the equivalent of the Hada Calcisiltites and Toomba Silts on Malaita (Danitofea, 1978), and are overlain by Pleistocene reef limestone (Ngorangora Limestone).

4. *The Pacific Province—Malaita, Ulawa and Northwest Santa Isabel*

Rocks of the Pacific Province form the northeastern flank of the modern Solomon Island Arc. They form a broad anticlinorium of ocean floor basalts and deep sea pelagic sediments. Kroenke (1972) and Coleman and Kroenke (1983) suggest that these rocks form a part of the southwest leading edge of the Ontong Java Plateau overthrust onto the Solomon Arc. Collision with the arc probably occurred in early Miocene times and subsequent tilting and uplift in response to vertical tectonics in the Plio–Pleistocene has now exposed rocks of the Pacific Province on the islands of Malaita, Ulawa and the northeastern rim of Santa Isabel.

The geology of the Pacific Province has been described by a number of workers including Coleman (1965, 1970), Rickwood (1957), Hughes and Turner (1976, 1977), Stanton (1961) and Danitofea (1978). However little of the province has been mapped in detail and in particular, the extent of the province on Santa Isabel and its relationship to the Central Province require further elucidation. A generalized stratigraphic column for the Pacific Province successions on Malaita and Ulawa is shown in Fig. 23.

Rickwood (1957), in describing the geology of the northern half of Malaita, distinguished an older basic lava and volcaniclastic sequence (the Alite volcanics) and a younger limestone and mudstone succession (the Malaita Group). The Alite Volcanics consist predominantly of unmetamorphosed pillow lavas and dolerites (the Fiu Lavas) and the Fo'ondo Clastics which comprise up to 600 m of black, finely bedded tuffs that appear to be the fragmented tuffaceous equivalent of the Fiu Lavas. The volcanics are of probable Cretaceous age. The Kwara'ae Mudstones are found overlying the volcanic basement in many areas, particularly in the north. They consist of up to 270 m of siliceous mudstones, representing a lithified deep sea siliceous ooze deposited at depths in excess of 4300 m below sea level.

The bulk of the sedimentary succession, common to all areas, consists of up to 2000 m of pelagic limestone, predominantly biocalcilutites, calcarenites and peperite breccias with thin beds of nonorganic pelagic mudstone and radiolarian chert. The sequence ranges in age from Upper Cretaceous to Pliocene. In the north, the lower half has been termed the Alite Limestone, consisting of up to 900 m of bedded limestone with chert horizons, and the upper, more friable part, the Suaba Chalk, comprising up to 760 m of pale grey, thickly bedded chalk. In the far north, the Suaba Chalk is conformably overlain by at least 250 m of Toomba Silts containing foraminiferal and tuffaceous debris.

In the south of Malaita and on Small Malaita, more detailed work (Hughes and Turner 1976, 1977) has revealed a similar succession. In south Malaita, early Cretaceous oceanic tholeiitic basalts (Malaita Volcanics) are overlain by the Kwara'ae Mudstone which become increasingly calcareous upwards to be succeeded by the Are'are Limestones, a pelagic limestone with chert that contains peperite horizons and is of Upper Cretaceous to Eocene age. Further south on Small Malaita, the Kwara'ae Mudstones are wedged out by the Apuloto Limestones, consisting of up to 550 m of pelagic limestone with chert that rest conformably upon the basement tholeiites (the "older basalts"). This limestone, which is of Upper Cretaceous to Eocene age, contains thickly interbedded horizons of oceanic basalt (the "younger basalts").

The Apuloto Limestone, which represents a lithified marine calcareous ooze deposited in a open ocean floor environment is overlain by the Haruta and Hada calcisiltites containing an increasing number of brown mudstone horizons, possibly representing distal turbidite deposits or the primary product of relatively proximal volcanism. An increasing development of sedimentary structures in the later sediments attests to a shallowing of the seas and an increasing influx of terrestrial debris in the late Miocene and Pliocene. Coleman and Nixon (1978) have described a brecciated alnoite volcanic center in north-central Malaita that is of deep seated origin and contains xenolithic suites that normally characterize kimberlites in continental cratonic settings. The unique occurrence of garnet lherzolite xenoliths (unmodified mantle material) is especially unusual in a supposed island arc situation. The breccia appears to form a pipelike intrusion into Kwara'ae Mudstones (?) and is overlain by the Haruta Calcisiltites. An age of about 34 m.y. is suggested for the intrusion (Nixon and Boyd, in press). Calculations based upon the pyroxene geotherm is suggestive of a lithospheric thickness of 110 km.

It is probable that Malaita sediments were extensively folded and that the land rose above sea level during the late Pliocene. The Tomba Silts in north Malaita and the Hauhui Conglomerates on the southwest coast, the latter being composed of 180 m of interbedded conglomerates, sandstones and siltstones, are terriginous shallow water deposits of probable Pliocene age. Uplifted coral terraces (the Rokera Limestones) are prominent in Small Malaita and are thought to be of Pleistocene age.

On Ulawa, Danitofea (1978) describes a similar succession to that exposed on Malaita. A basement of massive and pillowed unmetamorphosed oceanic tholeiitic basalts (the Oroa Basalts) are of pre-Upper Cretaceous age and probably correlate with the Fiu Lavas of north Malaita (Rickwood, 1957) and the "older basalts" of south Malaita (Hughes and Turner, 1977). There is no equivalent of the Kwara'ae Mudstones on Ulawa, the basement lavas being conformably overlain by the Arau Limestone, a pelagic limestone with chert sequence that is up to 400 m thick and of Upper Cretaceous to Upper Eocene age.

A gradual change from quiet, deep-water oceanic sedimentation to shallower water and an increasing influx of volcanic detritus is apparent in the overlying Waipaina Calcisiltites, a 380 m thick succession of calcisiltites and calsilutites containing thin bands of mudstone and intercalations of pillowed and massive alkalic basalts (the Haumela Basalts). The Waipaina Calcisiltites, which are of lower Oligocene to Upper Miocene age, are probably the lateral equivalent of the Haruta Calcisiltites of South Malaita (Hughes and Turner, 1976).

Rapid shallowing is evident in Pliocene times with the formation of up to 330 m of calcareous mudstone and slumped conglomerates (the Holohau Mudstones) that may correlate with the Hada Calcisiltites and the Toomba Silts on Malaita. The late Pliocene deformation that affects the Pacific Province rocks has resulted in mild folding along NNW–SSE axes and predominantly normal faulting of the Cretaceous to Pliocene succession. Ulawa probably emerged in the Pleistocene and is now encircled by up to 80 m of Pleistocene reef limestones (the Ngorangora Limestones), that rest unconformably upon the older rocks.

Little detailed information is available on the Pacific Province successions of Santa Isabel. Reconnaissance mapping by Stanton (1961) predated Coleman's concept of geological provinces and the full realization that the Pacific and Central Provinces were juxtaposed within the island along a tectonic front. Nevertheless Stanton (1961) recognized that his Basement Complex on the one hand the Sigana Volcanics and Tanakau Group sediments on the other, constituted the two basic structural units on Santa Isabel, and that the sedimentary sequences to the north and south of the axial spine of Sigana Volcanics were fundamentally different in character.

He describes a basal sequence of pillow lavas (the Sigana Volcanics) overlain by a sedimentary succession dominated, especially in the north, by pelagic limestones, thrust southwestward along the line of the Kaipito–Korigole Fault Zone, over a pre-Miocene basement complex of matamorphosed lavas, dolerites and gabbros and arenaceous sediments metamorphosed to schists and amphibolite—the Central Province.

Coleman *et al.* (1978) has described the Sigana Volcanics as a sequence of deep-water pillow lavas, massive lava and tuffs, the lavas being chemically oceanic tholeiitic basalts. An age determination of 66 ± 3 Ma is given for an intrusive basalt in the Sigana Volcanics (Snelling, personal communication in Hackman, 1980). The sedimentary succession north of the Sigana Volcanics, which probably has a thickness in excess of 900 m, is folded and crumpled, especially lower down in the succession (Coleman, 1965). The sedimentary pile is variable in composition. the sediments that are clearly intercalated with the Sigana Volcanics occur as lenses or more massive units of pelagic carbonates containing chert. Planktonic foraminifera indicate the pelagites range in age from late Palaeocene to early Oligocene. This part of the succession resembles the lower part of the Malaita Group; it is succeeded by tuffaceous calcarenites containing Lower Miocene foraminifera.

The sediments that appear to overlie the bulk of the Sigana Volcanics are more varied in composition and range from massive pink chert horizons, infrequent breccias, bedded calcisiltites and calcilutites and rare tuffaceous sandstones. They are probably conformable with the Sigana Volcanics although in some areas postdepositional sliding has given rise to apparent unconformity. Volcanic wackes, consisting of an admixture of volcanic sandstones and siltstones make up the rest of the succession. They are noncalcareous and contain plant remains representing the incursion of terriginous material, a feature that occurs as early as late Oligocene in the Santa Isabel successions (Coleman et al., 1978).

While the possibility of detailed stratigraphic correlations will have to await detailed mapping of Santa Isabel, the succession can be matched in their faunas and lithologies with counterparts in the Malaitan pelagic successions to the east. According to Kroenke (1972), the Pacific Province represents the leading edge of the Ontong Java Plateau, which was emplaced on the fore-arc as on obducted slab during Miocene times. The collison of the Ontong Java Plateau with the Solomon Arc is believed to have commenced about 10 m.y. ago, leading to deformation and obduction, a process that began about 8 m.y. ago and continued to 6 m.y. ago.

Subsequent uplift has exposed the Ontong Java flake as a sequence of deformed pelagites and ocean floor basalts on the islands of Malaita, Ulawa and northeast Santa Isabel, where, in the latter case, the Ontong Java Plateau is thought to be thrust against the Central Province successions along the line of the Kaipito–Korigole Thrust (Stanton, 1961).

While the theory of an obduction origin for the Pacific Province is well reasoned and based upon good acoustic stratigraphy and drill core evidence, it has posed a number of questions which have yet to be convincingly addressed:

(1) It is not unreasonable to surmise that the obductive process would have infilled or otherwise obliterated the subduction zone or trench, particularly as the plugging of the subduction zone by the Ontong Java Plateau ans the supposed obduction of a tectonic flake must be related events and approximately coeval. Yet there remains to the north of the Solomon Arc a linear deep—the North Solomon Trough which is up to 4300 m deep off northeast Malaita, and its salient, the Ulawa Deep which is 6000 m deep.

(2) The absence of the "other half" of the Ontong Java Plateau cannot be ignored. If it is accepted that the Ontong Java Plateau was the result of

abnormally slow spreading at a mid ocean ridge, then the spreading axis has to be south of the southern margin of the plateau because it "youngs" to the south. (In the obduction model, this margin is the Kaipito–Korigole thrust on Santa Isabel and its extension along a boundary somewhere to the southwest of Malaita.) The "other half" of the plateau therefore also has to be south of this line.

(3) If the Ontong Java Plateau is a spreading phenomenon created prior to 65 m.y. ago, it is implicit that prior to its collison with the Solomon Arc 8–10 m.y. ago, it lay far to the east-northeast of its present position, presumably in a truly oceanic environment as reflected in the cores of the three DSDP drill holes (Sites 64, 288 and 289) which shown purely pelagic sedimentation.

However, while the Pacific Province successions do show general similarities to the Ontong Java drill cores, Hughes and Turner (1976) highlighted three significant points of difference between the two sequences:

(1) No basalts younger than Cretaceous were encountered north of Malaita, i.e. no equivalents of the "younger basalts" and peperites of southern Malaita were encountered in the drill holes.

(2) No brown mudstones have been found in the drillhole pelagites. The brown mudstones of the Haruta and Hada calcisiltites on Malaita are interpreted as incursions of terrigenous material, either as the products of distal turbidites or relatively close submarine volcanism (Turner and Hughes, 1982).

(3) A major hiatus at DSDP Sites 288 and 289 show the Eocene and part of the Lower Oligocene missing. This has not been observed on Malaita where McTavish (1966) recognized foraminifera representing a complete succession from Upper Eocene to Pliocene.

However, a recent reexamination of McTavish's material in the light of the Ontong Java drill cores has revealed that stratigraphic breaks are in fact present in the Malaita sequences with parts of the Eocene and Oligocene not recorded (Resig, personal communication).

The present author, while accepting that the Ontong Java Plateau did collide with the arc and shut off subduction, considers that it may not be necessary to invoke the complications of an obduction origin for the Pacific Province, and puts forward the suggestion that the Pacific Province may represent deformed and uplifted fore-arc, an interpretation that is both simpler and equally, if not more consistent with several of the known features of the region.

The upward increase in the amount of terrestrial material is a significant feature of the Pacific Province successions. Whether interpreted as distal turbidites or as the product of relatively close volcanism, it does imply that proximity of the Pacific Province to an eroding and/or volcanically active landmass since the late Oligocene times. Guadalcanal, Florida and San Cristobal were emergent from late Oligocene times onwards and are a logical source for this detritus. However, the addition of such terrigenous material from the Solomon Arc to the Pacific Province would only have been possible if the Pacific Province seafloor were on the south side of the North Solomon Trench, that is, in a fore-arc position.

5. *The Eastern Outer Islands*

The Eastern Outer Islands comprise a group of small scattered islands some 450 km east of the main Solomon Island chain. They lie some 130 km north of the New Hebrides and are structurally the northern extension of that island arc system. Tinakula, Nendo, Utupua and Vanikoro form a well-defined chain in the west of the group, while the Duff Islands, Anuta and Fatutaka form a less well-defined chain to the east; Tikopia lies midway between the two groups. Eight of the nine islands are volcanogenic and volcanic cones are well preserved on Vanikoro, Utupua and Tikopia, as well as the active volcano of Tinakula. The Reef Islands, to the northeast of Nendo, which are coral and sand cays, are probably founded upon a volcanic pedestal.

The Eastern Outer Islands rise steeply from a northwest extension of the Fiji Plateau (also called the North Fiji Basin), which is an extensive back-arc basin of average water depth 3,600 m. Some 95 km off the west coast of Nendo, the reentrant of the South Solomon–New Hebrides (Torres) Trench system reaches depths of about 7000 m and forms a western boundary to this region. Some 270 km north of Anuta, the southeast trending Vitiaz Trench, reaching depths of 6150 m forms the northeast boundary of the plateau, this boundary feature extending westwards as the Cape Johnson Trough (Fig. 24).

Seismic data appears to define a Benioff zone reflecting the eastward subduction of the Australian plate along the line of the Torres Trench (Fig. 9). There are limited seismic data for the central area except for a few deep epicenters south of Tikopia, and considered by Kroenke (personal communication in Hughes 1978) to represent a detached fragment of lithospheric material. The Vitiaz Trench is seismically inactive, and is thought to be a fossil trench representing a former south dipping subduction zone. The Eastern Outer Islands were mapped in detail by the Solomon Islands Geological Survey during the period 1968–1973 (Hughes, 1981).

The islands form distinct volcanic piles of submarine and subaerial lava flows and volcaniclastic rocks, ranging in age from Oligocene to Recent.

Basalts predominate and include tholeiitic, high alumina and alkaline types, together with high and low silica andesites and dacitic rocks. The distribution of analyzed rock types in each island suggests that predominantly tholeiitic basalts are found in the western chain of islands and also in samples from the Vitiaz Trench. In the eastern chain and Tikopia however, high alumina and alkali basalts are common, together with andesites and dacites (Hughes, 1978). On the basis of this distribution (Fig. 25) and on possible differentiation trends, Hughes (1978) has proposed an evolutionary arc model involving two subduction polarity reversals between the Vitiaz and Torres Trench subduction sites. Five stages of volcanism are recognized; the western tholeiitic basaltic island chain are ascribed to two periods of arc volcanism resulting from eastward subduction of the Australian plate at the Torres Trench, while the eastern island lavas are thought to be produced by westward subduction of the Pacific plate at the Vitiaz Trench. Jesek *et al.* (1977) reports a 2.2 m.y. (Upper Pliocene) age date for high alumina tholeiitic lavas on Fatutaka and considers they originate from Vitiaz Trench subduction. However, Falvey (1978) reports a 30° clockwise rotation of the New Hebrides Arc, commencing in the late Miocene (6 m.y. BP) associated with the formation of the North Fiji Plateau about a northerly spreading axis. Carney and Macfarlane (1978) have suggested that extension of this basin was accompanied by a subduction polarity reversal in the New Hebrides Arc, from the eastern side of the New Hebrides Ridge (Vitiaz Trench) to the western side (Torres or New Hebrides Trench).

The evolutionary model of Hughes (1978) seems to the present author to be unnecessarily complex and the geology of the Eastern Outer Islands is more simply explained in terms of the single subduction reversal model that is proposed in the adjacent areas of the main Solomon chain and the New Hebrides. Dunkley (1983) has suggested that the high alumina tholeiitic basalts on Fatutaka are unlikely to be related to subduction processes because they contain much higher TiO_2 (1.4–1.9%) and Al_2O_3 (up to 20.5%) and lower K_2O than arc volcanics found on the same island; their origin is ascribed to plateau extension.

6. *Late Pliocene–Pleistocene Uplift*

Evidence for considerable uplift in late Pliocene and Pleistocene times has been found on several islands within the main chain, notably on those where detailed mapping is complete.

On Choiseul, late Pliocene uplift caused erosion of the Pemba Formation, prior to the deposition of the Pleistocene Nukiki Reef Limestones. These limestones, which now form a series of isolated blocks flanking the

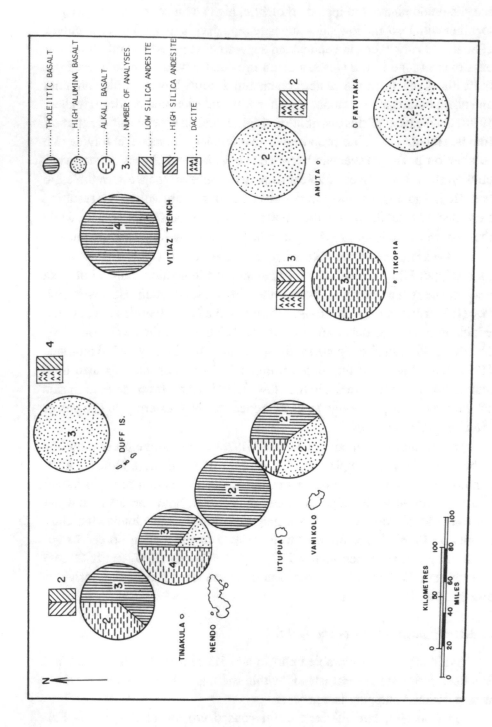

Fig. 25.   Distribution of analyzed rock types of the Eastern Outer Islands.

southwest coast of the island, are now raised to heights of 800 m as little as 7 km inland from the sea.

In the New Georgia Group, uplift of the fore-arc region is most pronounced. Pleistocene to Recent coral reefs are raised to 800 m above sea level. This uplift increases southwestward toward the site of subduction, and islands nearest the projected line of the trench contain basal Pleistocene sediments now raised to 800 m above sea level. Studies of foraminiferal depth assemblages in these sediments indicate deposition in water depths of 1.5–2.0 km (Hughes, 1981), implying an uplift of 2–3 km during and since the Pleistocene. (Dunkley, 1983). Late Pliocene and Quaternary sediments dredged from the New Georgia Platform northwest of the Vella Lavella during the 1982 cruise of the RV *S.P. Lee* (Fig. 15) contained more than one biofacies of different age: paleobathymetry indicates uplift of at least 600 m in 0.5 m.y. (Resig, 1983).

On Guadalcanal, the Pleistocene Honiara Reef limestones form a spectacular series of raised coral terraces, climbing to a height of 200 m above sea level. Hackman (1980) reports Pleistocene marine erosion surfaces at heights of up to 800 m in north Guadalcanal.

On the south coast, the Koloula Diorite Complex, the youngest phase of which has been dated at 1.4 m.y. (Chivas and McDougall, 1978), and which forms a high level intrusion into the Suta Volcanics, is now elevated to form a 1000 m plus mountain range adjacent to the south coast.

IV. MARINE GEOLOGY

Because of the complexities of the Solomon Islands Arc, it is not possible to describe the marine basins that flank the islands in island arc terminology. Three broad structural elements are apparent. The median sedimentary basin herein called the Central Solomon Basin, the complex area of the Coral Sea to the south of the main arc, and the Ontong Java Plateau to the north.

A. Central Solomon Basin

In the western and central Solomon the main islands form a double chain; the northern chain of Choiseul, Santa Isabel and Malaita and the southern chain of New Georgia, Gaudalcanal and San Cristobal semienclose an elongate, northwest-trending marine basin, closed in the west by Bougainville and the Shortland Islands and in the east by Ulawa, the Ulawa Trough and San Cristobal. The basin has a maximum width of 110 km between New Georgia and Santa Isabel and an overall length of some 850 km.

Various geographical names have been given to parts of this semienclosed elongate sea; the waters between the Shortland Islands and western Choiseul are known as the *Bougainville Straits*, the main body of water between New Georgia, Santa Isabel, the Russell Islands and the Florida Islands is known as *New Georgia Sound (the Slot)*, the waters between Guadalcanal, Florida Islands and Savo are known as *Iron Bottom Sound* and the waters between Florida and Malaita are known as the *Indispensable Strait*.

In the geological literature, de Brion *et al.* (1977) describe the overall structural pattern of the Solomon–New Ireland region in terms of a number of subparallel northwest-trending structural elements:

(*i*) a southern ridge, which in the Solomon Arc corresponds to the southern island chain of San Cristobal to New Georgia—passing to the southwest of Bougainville and to New Ireland;

(*ii*) a northeastern ridge, which in the Solomon Islands corresponds to the northern island chain of San Cristobal to Choiseul—thence to Bougainville and extending to form an upper axis bounding the New Ireland Basin;

(*iii*) a median sedimentary basin separating these two ridges stretching from the New Ireland Basin some 1600 km southeastward to where it is closed off at its eastern end by the Florida Islands.

This median sedimentary basin, which de Brion *et al.* (1977) describe as the main structural feature of the region is called the Solomon Basin where it corresponds to the New Georgia Sound (the 'Slot') bathymetric depression, where it is up to 70 km wide and contains 4–5 km of sediments, mainly of volcanic origin but with some carbonates. It should be noted that the sedimentary basin of de Brion *et al.* (1977) does not fully correspond to a bathymetric basin.

Katz (1980) in his study of this median basin, informally divided the region into an oceanic–pelagic Malaita Province (corresponding to the Pacific Province of Coleman) and a main Solomon Province (Coleman's Central and Volcanic provinces combined). New Georgia Sound, which Katz (1980) named the Central Solomon Trough, was described as a continuous deep sedimentary basin some 500 km in length and some 30–70 km in width. Some internal structural complexity is evident, perhaps even fragmentation where a broad fault zone in Manning Strait appears to have produced an en echelon arrangement of two partly separated basins. Katz (1980) describes the basin as shallowest (7–800 m) north of Kolombangara, deepening to 1500 m south of Bougainville Strait, and to 1800 m northwest of Guadalcanal. Sediment thickness increases correspondingly with depth and the trough is underlain by 2.5–4.5 km of sediments. Deformation within the sediments is slight and mainly confined to the upturning of sediments on the faulted basin margins, sometimes accompanied by the formation of local anticlines. Although fold structures are unconformably overlain by later sediments and are therefore probably of Pliocene age, most normal faulting towards the uplifted basin margins have disturbed the seafloor topography indicating they have remained active until Recent times (Fig. 26).

In a more recent study of the median basin. Maung and Coulson (1983) (Fig. 27) have proposed the term Central Solomon Basin for the entire semienclosed sea between the main island chain, stretching from Bougainville in the west some 850 km southeastward to be pinched out by Ulawa, Ulawa Trough and the eastern end of San Cristobal. Preliminary results from the 1982 RV *S.P. Lee* geophysical cruise indicate the presence of four subbasins within the Central Solomon Basin (Vedder, Tiffin *et al.*, 1982). From west to east these are the Shortlands subbasin, the Russell subbasin, Iron Bottom subbasin and the Indespensable subbasin (Fig. 28).

The Shortlands subbasin, which is approximately 140 km long and 60 km wide, is bounded by the Shortland Islands in the northwest, Choiseul in the northeast and Vella Lavella in the south-southeast. The slope of the basin is generally very steep (200–1000 m), becoming broad and gently inclined at about 1000 m; maximum depth is 1400 m.

This subbasin is a graben, bounded by approximately east–west trending faults on the north and south. It contains up to 5.2 km of well-laminated, mostly conformable sediments which occur in three sequences, separated by unconformities (Vedder Tiffin *et al.*, 1983). The lowest sequence is well-bedded, south-dipping and is folded into broad anticlinal folds that trend WNW–ESE. This is succeeded unconformably by a sequence that both dips and thins northward off the platform extending west from Vella Lavella. The topmost sequence, which is up to 0.8 seconds thick, is horizontal or gently north-dipping and overlaps the middle sequence.

This succession reflects three well-defined events. Following initial sedimentation, uplift on the north side of the Basin tilted the earlier Oligo–Miocene (?) sediments southward. After subduction polarity reversal, a buildup of volcanics in the south produced a north-dipping volcanoclastic apron of Pliocene (?) age. The final infilling of the basin is probably composed of Pleistocene and Recent sediments.

Separating the Shortland and Russell subbasins is an isolated, mid-basin complex of topographic highs recently named "Fatu O Moana" and located in New Georgia Sound northeast of Kolombangara. Originally inferred to be a submarine volcanic complex (Katz, 1980; Coleman and Kroenke, 1982), recent work has shown that it is an upfaulted sequence of folded and faulted sediments. Pliocene sandstone has been dredged from one of the steep northern faces. Low heat flow was recorded (Taylor and Exon, 1982).

8

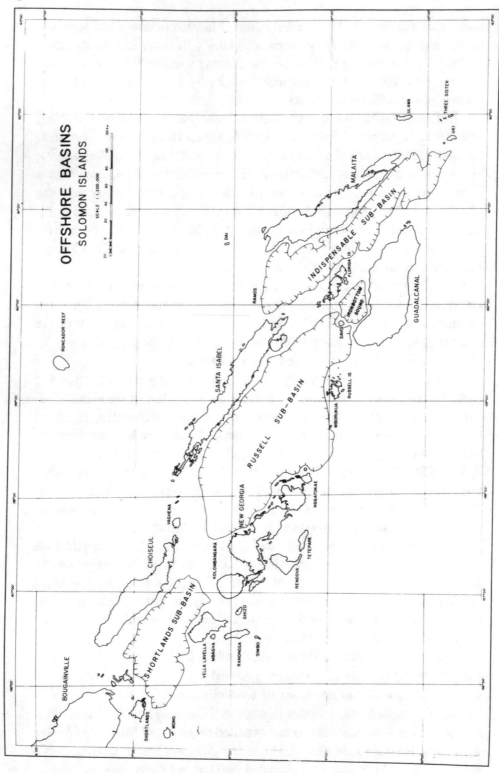

Fig. 28.   Subbasins within the Central Solomon Basin.

The Russell subbasin, which forms the central subbasin and coincides approximately with the area known as New Georgia Sound, is the largest subbasin, being 350 km long and 100 km wide. Like the Shortlands subbasin, it is steeply inclined from 200 m to 1000 m, becoming broad at the latter depth; the maximum depth of the basin however is 1800 m.

The Russell subbasin is a graben with numerous step faults on its northern and southern edges. Several north-northeast-trending faults occur in the northwestern part of the subbasin and reflect similar fault trends apparent in the New Georgia Group; the majority of faults affect seafloor topography and have been active in recent times. Refraction analysis indicate that up to 5.7 km of sedimentary section lie directly on the oceanic crustal layer. Several anticlinal structures occur, striking mostly in a northwesterly direction parallel to the faults. The seismic character of the Russell subbasin varies from east to west, reflecting differences in the type of sediments.

In the west axial sediments up to 0.5 seconds thick onlap against volcaniclastics (?) in the east Choiseul to Manning Strait area to the north. On the south side of the basin, volcanics (?) persist in the deep section as far eastward as the Russell Islands, giving way in the north to well-laminated sediment that thicken and spread southward toward the eastern end of the basin. Structure is confined to the area around Santa Isabel where some folding, possibly related to faulting can be seen.

In the central part of the basin, axial deposits become thicker and better laminated with onlap relations to the south and interfinger with coarse steeply dipping material in the north. In the eastern part of the basin, well-laminated sediments are widespread. Sediment layers from north and south interfinger or onlap alternately in mid-basin (Vedder, Tiffin *et al.*, 1983).

Velocities of the crustal rocks beneath the sedimentary section range from 4.8 to 7.5 km/s. The acoustic basement, thought to be volcanic and volcaniclastic rocks, in part metamorphosed, has velocities of 4.8 to 5.8 km/s; a deeper layer, which has a velocity of 6.4 to 7.5 km/s, is 3–5 km below the acoustic basement. These higher velocity rocks may include deeper crustal igneous rocks similar to those observed in island outcrops in the Central Province.

In the Russell subbasin there is low heat flow in the north (ca. 30 mK/m) which gradually increases towards the New Georgia Group where a value of 96 mK/m was recorded (Taylor and Exon, 1982).

Iron Bottom Basin, the area between Guadalcanal, Savo Island and the Florida Islands, constitutes the third and smallest of the subbasins. It is approximately 40 km long and 30 km wide and has a maximum depth of only 600 m. It is connected with the Russell subbasin by a narrow depression north of Savo Island in which as much as 3 km of sediment may have accumulated. Several reefal buildups within this basin have been identified on seismic sections (Maung and Coulson, 1983) and may range in age from Miocene to Recent. The Iron Bottom Basin is structurally complex. Large folds are present. Several are buried in the deeper sediment, although some affect the ocean floor. Well-defined and continuous sedimentary reflectors are present throughout the section in most places. (Vedder, Tiffin *et al.*, 1983)

The Indispensable subbasin in the southeast is bounded by the Florida Islands in the west, Malaita in the northeast, Guadalcanal in the south-south-west and San Cristobal in the southeast. This is the narrowest subbasin with a width of 40 km and a length of about 280 km. Within this subbasin are three small basins having depths of between 1600 and 2000 m. This basin, which almost coincides with the Indispensable Straits, is bounded by the deeply buried, tightly folded southwestern flank of the Malaita anticlinorium and the faulted northeast margin of the Florida platform; the northwestern edge of the basin abuts Santa Isabel. Up to 2.2 km of sediment may be present in this basin forming a southwestward dipping triangular shaped wedge with its deepest side against the fault. There is some evidence of anticlinal structures formed by draping of sediments over preexisting basement highs. The major part of this basin lies within the Pacific Province.

The stratigraphy, age and origin of the Central Solomon Basin must, at this stage, remain speculative. The Basin is essentially extensional in character with uplifted basin margins characterized by normal faults, many of which affect the seafloor topography. There is generally little deformation of the sedimentary pile although gentle folding affects as much as 2 km of sediments low in the pile, such folds being unconformably overlain by 1500–200 m of later sediment. (Maung and Coulson, 1983), Katz (1980) has suggested that the folding is late Pliocene in age and that the unconformable cover is Pleistocene. Attempts to gravity core and dredge the basin have yielded only Pleistocene and late Pliocene deposits (Tiffin *et al.*; 1983, Resig, 1983).

The islands that semienclose the basin testify to a history of rapid uplift accompanied by deep erosion of the volcanic basement, punctuated by periods of volcanic outpourings, episodes that probably enhanced the erosional energy levels. The author is therefore inclined to Katz's (1980) view that the volcanogenic ridges that flank the basin have, since late Oligocene time, repeatedly supplied large volumes of detritus to the basin and that the sediments therein correlate, at least in principle, with the late Oligocene–Pleistocene successions exposed within the islands. Facies variation, both laterally and vertically would be expected to reflect the local provenance and the temporal character of the detritus being shed into the basin. By analogy with the island successions, it is likely that reef limestones and open marine carbonates have developed in shelf areas around the emergent horsts while the deeper marine areas received continuous clastic turbidite sedimentation.

The basin probably deepened rapidly in the late Miocene and Pliocene as detritus from adjacent uprising land area rapidly filled the basin with several thousand meters of sediment; bathyal turbidites and local chaotic slump deposits would be characteristic of the deeper parts of the basin (Katz, 1980). Such deposits might be expected to resemble those seen in the various Pliocene lithosomes on Guadalcanal, e.g. the Mbokokimbo Formation, Toni Formation and Lungga Beds.

The extensional tectonics that characterize the basin, intensified in the late Pliocene and Pleistocene culminating in the taphrogenic collapse of the basin through a series of steeply inclined step faults along the basin margins. Downthrowing of the basin by 2000 m possibly within the last 1.0 m.y. is suggested (Katz, 1980). A similar Plio–Pleistocene taphrogeny is considered to have shaped the Central Basin in the New Hebrides Arc (Carney and McFarlane, 1980). Indeed, de Broin et al. (1977) have pointed out that these two basins closely correspond in their "intra-arc" positions.

The Central Solomon Basin is an extensional basin within the arc—the median sedimentary basin of de Broin et al. (1977). Any attempt to further define or classify the basin in island arc terminology is frustrated by the complexity of the arc, i.e., it does not fit the textbook model of an island arc basin. Since late Miocene times it could be considered to form a back-arc basin that has formed behind the Bougainville–west Guadalcanal volcanic arc. However the amount of sedimentary fill and lack of a central magnetic anomaly would set it apart from the normally accepted back-arc basin type.

The earlier Oligo–Miocene evolutionary history of the basin is even more speculative. If the earlier north Solomon Arc was represented by a volcanic axis through San Cristobal–Guadalcanal–Bougainville (Kroenke, 1983), then was the basin at that time possibly a part of the fore-arc?

B. Coral Sea

To the south of New Britain–South Solomon Trench system—the Outer Melanesian Arc of Carey (1958)—a largely inactive marginal zone is characterized by a complex seafloor topography of ridges, basins, troughs and plateaus and by a diversity of crustal structure. It includes what may be segments of the Inner Melanesian Arc (Carey, 1938; Glaessner, 1950). The unravelling of the tectonic history of this complex region is far from complete.

1. *The Rennell Arc*

To the south of the island of Guadalcanal, Rennell and Bellona, Indispensable Reefs and the adjacent Rennell trough form a northwest-trending structural unit about 200 km in length (Recy *et al.*, 1975). The Rennell Trough, which is generally less than 4,500 m deep is flatfloored and partially filled with horizontal sediments. To the northeast, it abuts against a 1300 m deep submarine plateau crowned by the atolls of Rennell and Bellona. Landmesser (1974) was the first to suggest that the Rennell Trough and Ridge were vestiges of a former northeast subduction zone along an intermediate plate boundary, at one time contiguous with New Caledonia. The asymmetrical trench morphology suggestive of a former northeast subduction zone is evident in the reflection profiles across the Trough (Recy *et al.*, 1977) and is shown at Fig. 29.

Correlation of the ponded sediments in the trough, (which appear to have accumulated after subduction ceased) with DSDP data is uncertain and indicates that this subduction zone may have been active in Eocene or Miocene times. Kroenke (1983) argues for Eocene subduction coeval with adjacent subduction in New Guinea and New Caledonia. If this interpretation is correct, then the Rennell Arc represents the earliest subduction period in the Solomon Islands, i.e., that associated with northward subduction of the Australian plate in Eocene time.

2. *The Pocklington Trough*

Lying to the north of the Louisiade Plateau and stretching from the Louisiade Islands almost to Guadalacanal, having a thick sedimentary fill and a large negative gravity anomaly (Coleman and Packham, 1972) is the Pocklington Trough. Seismically inactive, it has been interpreted as a relic subduction zone by Karig (1972). Recy *et al.* (1977) consider that the gross morphology of the trench supports the contention that this is a fossil north-directed subduction zone. If correct, this feature may constitute another fragment of the earlier Inner Melanesian Arc.

3. *The South Rennell Trough*

Lying to the south of the Rennell Arc, is a northeast-trending deep some 700 km in length and up to 30 km wide. It reaches depths of 5000 m. Larue *et al.* (1976) suggest that on the basis of magnetic spreading anomalies, gravity data and bottom sampling that this trough is the remains of a spreading rift of possible Oligocene age (Fig. 30).

4. *Santa Cruz Basin*

Little is known about the Santa Cruz Basin. Luyendyk *et al.* (1974) and Ravenne *et al.* (1977) describe subhorizontal, well-bedded sediments with a

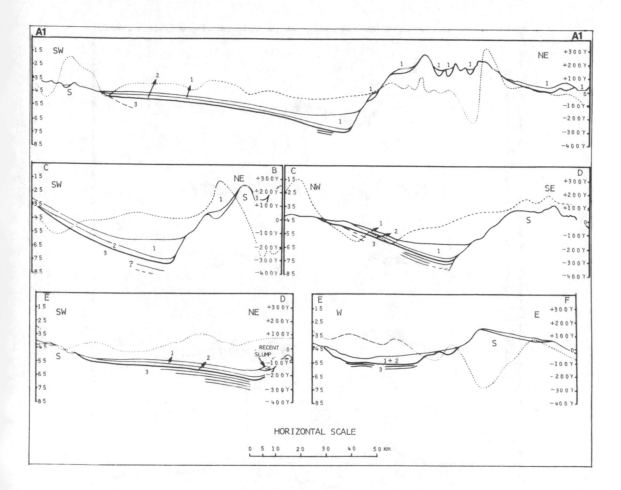

Fig. 29.   Reflection profile across the Rennell Trough.

total thickness of less than 1000 m. Ravenne *et al.* describe steep eastern slopes in the eastern part of the basin and consider it representative of an outer swell approaching the subduction zone. Klein *et al.* (1975), reporting on the *Glomar Challenger* Leg 30 DSDP results, describes a 650 m middle Eocene to Pleistocene sedimentary succession in the southern Santa Cruz Basin; turbidites formed the lower two-thirds of the succession, overlain by pelagic sediments.

5. *The Woodlark Basin*

The Woodlark Basin, which lies to the southwest of the Solomon Arc and adjacent to the line of the Bougainville–South Solomon Trench, forms a 3–4 km deep basin bounded in the north by the Woodlark Rise and in the south by the Pocklington Rise. Sediment thickness varies from less than 50 m in

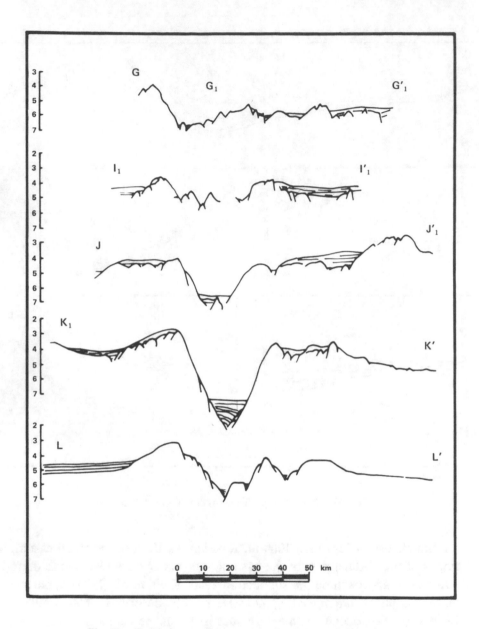

Fig. 30.   Reflection profile across the South Rennell Trough.

the center of the basin to local ponds containing in excess of 500 m of sediment near the margins (Taylor and Exon, 1983). The area is characterized by high heat flow, at least 3 to 4 H.F.V., reflecting the young age of the oceanic crust (Halunen and von Herzen, 1973; Taylor and Exon, 1983) and is floored by basalts flows and pillow lavas that appear to be typical mid-ocean ridge type (Taylor and Exon, 1983).

Following Carey's (1958) suggestion of an extensional origin for the Woodlark Basin, plate boundary models were advanced by Luyendyk *et al.* (1973) on the basis of a few magnetic seafloor spreading anomalies and by Curtis (1973) on the basis of seismicity. Weissel *et al.* (in press), on the basis of additional magnetic and seismic data, further defined the spreading axis, which extends from immediately south of the New Georgia Group west-southwest to the D'Entrecasteaux Islands (Fig. 31). Recent investigations of the Woodlark Basin during the *Kana Keoki* cruise 82-03-16 Leg 4 (Fig. 16) have shown that the spreading center itself is characterized by a axial rift valley about 10 km wide and with 500 m to 1 km of relief. The pattern of seafloor spreading is characterized by east–west-trending magnetic lineations sinistrally offset by north–south fracture zones. Simbo Ridge, which marks a major fracture zone, offsets the anomalies by some 65 km (Taylor and Exon, 1983).

The initial separation of the Woodlark and Pocklington rises, which form the margins of the basin, occurred approximately 4 m.y. BP. Since then seafloor spreading has occurred at an average total opening rate of 6–7 cm/yr. Spreading however, has been asymmetric, being generally twice as fast on the northern limb.

During the 4.0 m.y. opening history of the Woodlark Basin, the spreading ridge and crust so formed has been subducted at the South Solomon Trench during the oblique convergence of the major Pacific and Australian plates. There is no apparent flexure of Woodlark Basin crust into the New Georgia subduction zone, the basin crust simply abuts the fore-arc lower slope and there is no bathymetric trench. Seismicity is low.

This situation, whereby an active spreading system is entering an oceanic subduction zone with a high angle between the ridge segments and the trench, is probably unique at the present day. The considerable influence this has within the Solomon Arc, on its volcanism, petrochemistry and tectonics is considerable and goes a long way to explaining some of the enigmatic features of the arc.

C. The Ontong Java Plateau

To the northwest of the main Solomon Islands chain lies the Ontong Java Plateau. The Plateau, which strikes northwest–southwest, parallel to the Solomon chain, is over 1600 km long by 80 km wide. It is anomalously shallow with average depths of less than 2000 m over the central portion. Large submarine plateaus such as Ontong Java, Shatsky and Manihiki, are enigmatic features of the Pacific Basin. The Plateau has been surveyed in detail (Fig. 33) and the geology and acoustic stratigraphy of the Plateau has been described by Kroenke (1972) who demonstrated that it is covered by

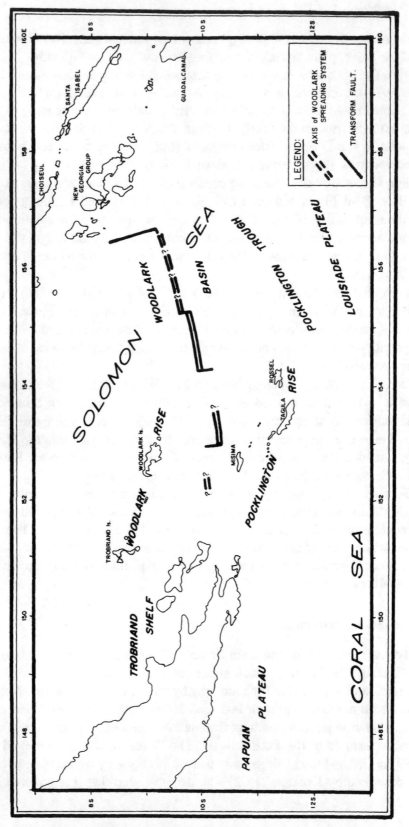

Fig. 31. The Woodlark spreading system.

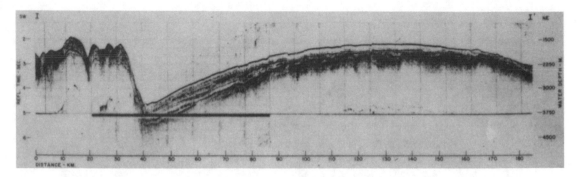

Fig. 32.   Reflection profile showing normal faulting on the Stewart Arch north of Malaita.

a uniformly thick (+ 1000 m) highly stratified conformable sedimentary blanket that is, together with the underlying basement, highly deformed on its southwestern edge adjacent to the Solomon Island. The sediments range from Lower Cretaceous to Recent in age. The crustal structure has been described by Furumoto *et al.* (1976), Murauchi *et al.* (1973) and Hussong *et al.* (1979). The crust is anomalously thick (43 km); layer velocities are similar to normal oceanic crust, but each layer is grossly and abnormally thickened. Kroenke (1974) has suggested that the Plateau arose about a spreading axis, but that spreading rates were abnormally low. Such massive outpourings of flood basalt (6 million km^3) may mark the beginning of continent formation.

The southwestern flank of the plateau, adjacent to the Solomon Islands, is dominated by the Roncador Homocline–Stewart Arch (Fig. 32), a lithospheric flexure believed to be the result of severe bending developed during attempted subduction of the plateau (as part of the Pacific plate) at the North Solomon Trench prior to late Miocene arc reversal. This flexure, which is characterized by submarine slumps and slides, terminates at the base of the Malaita Foldbelt in a series of en echelon troughs. The troughs and ridges that comprise the foldbelt form a large anticlinorium, most intense over Malaita where it is characterized by tight and even overturned folds. Steep submarine scarps indicate intense faulting. The folding diminishes in intensity along the flank to the northwest, becoming mild opposite Santa Isabel and represented only as a single northwest-trending trough opposite Choiseul.

The stratigraphy of the Plateau has been established by the Deep Sea Drilling Project which drilled Sites 64, 288 and 289 on the Plateau (Andrews *et al.*, 1975; Winterer *et al.*, 1971). The correlation of sedimentary horizons at Site 64 with laterally continuous seismic reflectors that can be traced almost to the shoreline of Malaita led Kroenke (1972) to correlate the DSDP 64 site with the succession established for north Malaita by Rickwood (1957).

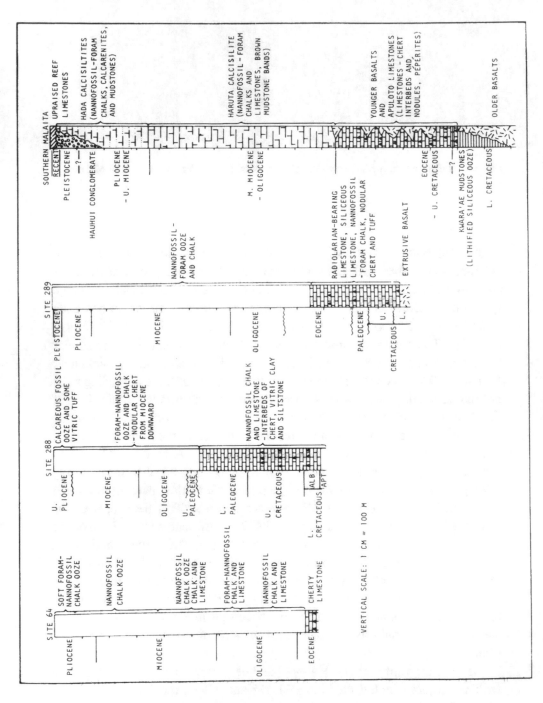

Fig. 33. Stratigraphic successions in the Deep Sea Drilling Project, Ontong Java Plateau.

In all three boreholes (Fig. 33), cherty limestones have been found toward the base of the succession and at Site 289, these overlie Lower Cretaceous basalts. The younger sedimentary deposits tend to be nannofossil–foram limestones, chalk and ooze. All three boreholes successions have obvious similarities with the succession found in north Malaita although the Miocene–Pleistocene successions in south Malaita, Santa Isabel and Ulawa differ from the ocean floor successions by containing more terriginous debris derived from an area of active erosion. Also within the older limestone in the three ocean floor boreholes, no basaltic flows were encountered. However, Kroenke (1972) recognized seismic reflectors within the sedimentary succession to the south of the plateau which he interpreted as lava flows or sills and which could be the lateral equivalent of the younger basalts and peperites of southern Malaita. Nixon (1980) argues that some of them may represent alnoite pipes and may be similar to those found in north Malaita.

The DSDP drilling results show that a major dislocation—a fracture zone—is present between DSDP Sites 288 and 289, offsetting the plateau to the south. With supporting evidence from foraminiferal ages of sediments on northeast Santa Isabel and Malaita, the line of this dislocation is postulated to lie between these two islands (Coleman and Kroenke, 1983).

The deformation of the southwestern margin of the plateau together with the apparent correlation of the Malaitan and plateau geological successions has led Kroenke (1972) to postulate that Malaita and the northwestern margin of San Cristobal (Coleman's Pacific Province) constitute an obducted slice of the Plateau rim, tectonically welded to the Solomon Arc during the unsuccessful attempt to subduct the plateau when it collided with the arc in late Miocene times.

REFERENCES

A.B.E.M., 1967, *Report on an Airborne Geophysical Survey in the British Solomon Islands*, Stockholm: Aktiebdag Elektrisk Malmletning 1965–1966, v. 1, 160 p., and vol. 2, 156 p.

Andrews, J. E., and Packham, G. H., 1975, *Initial Reports of the Deep Sea Drilling Project*, v. 30, Washington, D.C.: United States Government Printing Office.

Arthurs, J. W., 1981, The geology of the Mbambatana area, Choiseul. An explanation of 1:50,000 scale geological map sheet CH5, *Brit. Tech. Coop. West. Solomons Map. Proj. Rep.* No. 5.

Blake, D. H., and Miezitis, Y., 1976, Geology of Bougainville and Buka Islands, New Guinea, *Aust. Bur. Miner. Resour. Geol. Bull., Geophys.* v. 93, p.

Broin, C. E., de Aubertin, F., and Ravenne, C., 1977, *Structure and History of the Solomons–New Ireland Region, International Symposium on Geodynamics in Southwest Pacific*, Paris: Editions Techniques p. 37–50.

Cargy, S. W., 1958, *Continental Drift. A Symposium* Hobart: University of Tasmania, p. 177–355.

Carney, J. N., Mc Farlane, A., 1980, A sedimentary basin in the Central New Hebrides Arc, *United Nations, ESCAP, CCOP/SOPAC Tech. Bull.* v. 3, p.

Chivas, A. R., and McDougail I., 1978, Geochronology of the Koloula porphyry copper prospect, Gaudalcanal, Solomon Islands, *Econ. Geol.,* v. 73, p. 678–679.

Coleman, P. J., 1960a, North-central Guadalcanal—an interim geological report *Br. Solomon
Is. Geol. Rec.* v. 1, p. 4–13.

Coleman, P. J., 1960b, An introduction to the geology of the island of Choiseul in the Western
Solomons, 1957. *Br. Solomon Is. Geol. Rec,* v. 1, p. 16–26.

Coleman, P. J., 1962, An outline of the geology of Choiseul, British Solomon Islands. *J. Geol.
Soc. Aust.,* v. 8, p. 135–58.

Coleman, P. J., 1965a, Stratigraphical and structural notes on the British Solomon Islands with
reference to the first geological map, *Br. Solomon Isl. Geol. Rec.,* v. 2, p. 16–17.

Coleman, P. J., 1965b, Tertiary assemblages of larger foraminifera in the Solomon Islands and
New Hebridges archipelago. *Contr. New Hebrides Annu. Rep. Geol. Surv.,* p. 48–51.

Coleman, P. J., 1966, The Solomons as an island arc *Nature (London),* v. 211, p. 1249–1251.

Coleman, P. J., 1970, Geology of the Solomon and New Hebrides Islands, as part of the Melanesian reentrant. *Pac. Sci.,* v. 24, p. 289–314.

Coleman, P. J., 1975a, On Island Arcs, *Earth Sci. Rev.,* v. 11, p. 47–80.

Coleman, P. J., 1975b, The Solomons as a non-arc, *Bull. Soc. Geol. Explor. Geophys.,* v. 6,

Coleman, P. J., 1976, A re-evaluation of the Solomon Islands as an arc system, in: *Marine
Geological Investigations in the Southwest Pacific and Adjacent Areas* Glasby, G. P. and
Katz, H. R., eds. *UNESCAP Tech. Bull.,* v. 2, p. 134–139.

Coleman, P. J., and Day, A. A., 1965, Petroleum possibilities and marked gravity anomalies
in north-central Guadalcanal. *Br. Solomon Is. Geol. Rec.* v. 2, p. 112–119.

Coleman, P. J., and Kroenke, L. W., 1983, Subduction without Volcanism, *Geomarine Lett.,*

Coleman, P. J., and McTavish, R. A., 1964, Association of larger and planktonic foraminifera
in single samples from middle Miocene sediments Guadalcanal, Solomon Islands, Southwest Pacific, *J.R. Soc. West Aust.,* v. 47, p. 13–24.

Coleman, P. J., and Packham, G. H., 1976, The Malanesian Borderlands and India–Pacific
plates' boundary *Earth Sci. Rev.,* v. 12, p. 197–233.

Coleman, P. J., McGowran, B., and Ramsay, R. W., 1978, New early Teritiary ages for basal
pelagites, northeast Santa Isabel, Solomon Islands, *Bull. Aust. Soc. Explor. Geophys.,* v.
9, 110–114.

Coulson, F. I., 1981, The history of hydrocarbon prospecting in Solomon Islands and an index
to oil company and other marine data., *Geol. Surv. Solomon Is. Rep.* (unpublished).

Curtis, J. W., 1973, The spatial seismicity of Papua New Guinea the Solomon Islands, *J. Geol.
Soc. Austr.,* v. 20, p. 1–20.

Daniel, J., Jouannic, C., Larve, B. M., and Recy, J., 1978, Marine geology of (eastern margin
of Indo–Australian plate north of) New Caledonia *South Pac. Mar. Geol. Notes,* v. 1,

Danitofea, S., 1978, The geology of Ulawa. *Geol. Surv. Solomon Is. Bull.,* no. 15 (unpublished).

Danitofea, S., 1981, The geology of the Russell Islands and Mborokua. *Geol. Surv. Solomon
Is. Bull.,* no. 12 (unpublished).

Denham, D., 1969, Distribution of earthquakes in the New Guinea–Solomon Islands region, *J.
Geophys. Res.,* v. 74,

Denham, D., 1971, Seismicity and tectonics of New Guinea and the Solomon Islands *R. Soc.
N.Z. Bull.,* v. 9, p. 31–8.

Denham, D., 1975, Distribution of underthrust lithospheric slabs and focal mechanisms–Papua
New Guinea and Solomons Island region *Bull. Aust. Soc. Explor. Geophys.,* v. 6,

Dennis, R. A., and Hackman, B. D., 1977, The geology of the Cape Esperance area, Guadalcanal
Geol. Surv. Solomon Is. Bull. no. 5 (unpublished).

Dunkley, P. N., 1983, Volcanism and the Evolution of the Ensimatic Solomon Islands Arc, in:
Arc Volcanism, Physics and Tectonics, Yokoyama, I., Shimozuru, D., eds. Terro Scientific
Publications and Reidel Press.

Ewing, J., and Houtz, R., 1969, Mantle reflections in airgun sonobuoy profiles. *J. Geophys.
Res.,* v. 74, p. 6706–6709.

Falvey, D. A., 1975, Arc reversal and a tectonic model for the North Fihi Basin. *Bull. Aust. Soc. Explor. Geophys.*, v. 6, p. 47–49.

Falvey, D. A., 1978,

Furumoto, A. S., Hussong, D. M., Cambell, J. F., Sutton, G. H., Malahoff, A., Rose, J. C., and Woolard, G. P., 1970, Crustal and upper mantle structure of the Solomon Islands as revealed by seismic regraction of November–December, 1966. *Pac. Sci.*, v. 24, p. 315–332.

Glaessner, M. F., 1950, Geotectonic position of New Guinea *Bull. Am. Assoc. Pet. Geol.*, v. 34, p. 856–881.

Hackman, B. D., 1973, The Solomon Islands fractured arc, in: *The Western Pacific, Island Arcs, Marginal Seas, Geochemistry*, edited by Coleman, P. J. ed. University of Western Australia. p. 179–191.

Hackman, B. D., 1977, The geology of the Tiaro Bay area, Guadalcanal, *Geol. Surv. Solomon Is. Bull.* no. 8, (unpublished).

Hackman, B. D., 1979, The geology of the Honiara area, Guadalcanal. *Geol. Surv. Solomon Is. Bull.* no. 3, (unpublished).

Hackman, B. D., 1980, The geology of Guadalcanal, Solomon Islands, *Overseas Mem. Inst. Geol. Sci.*, no. 6, H.M.S.O. London.

Hackman, B. D., and Turner, C. C., 1977, The geology of the Beaufort Bay area, Guadalcanal. *Geol. Surv. Solomon Is.* Bull No. 9 (unpublished).

Halunen, A. J. Jr., and Von Herzen, R. P., 1978, Heatflow in the western equatorial Pacific *Aust. Bur. Miner. Geol. Geophys. Bull.* v. 194, p.

Hill, J. H., 1960, Further exploration in the Betilonga area of Guadalcanal *Br. Solomon Is. Geol. Rec.*, v. 1, p. 81–94.

Houtz, R., Ewing, J., and Le Pichon, X., 1968, Velocity of deep sea sediments from sonobuoy data, *J. Geophys. Res.*, v. 73, p. 2615–2641.

Hughes, G. W., 1977a, The geology of the Itina Basin area, Guadalcanal *Geol. Surv. Solomon Is. Bull.* no. 7 (unpublished).

Hughes, G. W., 1977b, The geology of the Lungga Basin, Guadalcanal. *Geol. Surv. Solomon Is. Bull.* no. 6 (unpublished).

Hughes, G. W., 1978, The relationship between volcanic island genesis and the Indo–Australian Pacific plate margins in the Eastern Outer Islands, Solomon Islands, southwest Pacific, *J. Phys. Earth.* v. 26, Suppl., S123–S128.

Hughes, G. W., Sedimentary basins of the Solomon Islands region—their extent, biostratigraphic correlation and a revised Pliocene zonation, in: *Stratigraphic Correlation between Sedimentary Basins of the ESCAP Region. Min. Res. Dev. Ser.*, United Nations, New York. (in press)

Hughes, G. W., and Turner, C. C., 1976, Geology of Southern Malaita, *Geol. Surv. Solomon Is. Bull.*, v. 2, p. 80.

Hughes, G. W., and Turner, C. C., 1977, Upraised Pacific Ocean floor, southern Malaita, Solomon Islands, *Geol. Soc. Am. Bull.*, v. 88, p. 412–424.

Hughes, G. W., Craig, P. M., and Dennis, R. A., 1981, Geology of the Easter Outer Islands, *Geol. Surv. Solomon Is. Bull.*, no. 4, p. 108.

Hussong, D. J., Wipperman, L. K., and Kroenke, L. W., 1979, The Crustal structure of the Ontong Java and Manihiki oceanic plateaus *J. Geophys. Res.*, v. 84, p. 6003–6010.

Jeffery, D. H., 1977, Geology of northwestern San Cristobal, Uki ni Masi and Pio and the Three Sisters, *Geol. Surv. Solomon Is. Bull.*, no. 10 (unpublished).

Jesek, P. A., Bryan, W. B., Haggerty, S. E., and Johnson, H. P., 1977, Petrography, petrology and tectonic implications of Mitre Island, northern Fiji Plateau, *Mar. Geol.*, v. 24, 123–148.

Johnson, T., and Molnar, P., 1972, Focal mechanisms and plate tectonics of the southwest Pacific, *J. Geophys. Res.*, v. 77, p. 5000–5032.

Karig, D. E., 1972, Remnant arcs, *Geol. Soc Am. Bull.*, v. 83, p. 1057–1068.

Karig, D. E., and Mammerickx, J., 1972, Tectonic framework of the New Hebrides Island Arc, *Mar. Geol.*, v. 12, p. 187–205.

Katz, H. R., 1980, Basin development in the Solomon Islands and their petroleum potential *United Nations ESCAP, CCOP/SOPAC Tech. Bull.* v. 3, p.

Klein, G. de V., 1975, Sedimentary tectonics in southwest Pacific marginal basins based on Leg 30, Deep Sea Drilling Project. Cores from the South Fiji, Hebrides and Coral Sea Basins. *Geol. Soc. Am. Bull.*, v. 86, p. 1012–1018.

Kroenke, L. W., 1972, Geology of the Ontong Java Plateau, *Hawaii Inst. Geophys. Rep.*, HIG-72-5.

Landmesser, C. W., 1974, Submarine Geology of the Eastern Coral Sea Basin, Southwest Pacific, M.S. Thesis, University of Hawaii, Honolulu, p. 64.

Larue, B. M., Daniel, J., Jouannic, C., and Recy, J., 1977, *The South Rennell Trough; Evidence for a Fossil Spreading Zone, International Symposium on Geodynamics in Southwest Pacific*, Paris: Editions Technip p. 51–62.

Laudon, T. S., 1968, Land gravity survey of the Solomon and Bismark islands, in: *The Crust and Upper Mantle of the Pacific Area. Mongr. Am. Geophys. Union* no. 12, p. 279–295.

Luyendyk, B. P., MacDonald, K. C., and Bryan, W. B., 1973, Rifting history of the Woodlark Basin in the Southwest Pacific, *Geol. Soc. Am. Bull.*, v. 84, p. 1125–1134.

Luyendyk, B. P., Bryan, W. B., and Jezek, P. A., 1974, Shallow structure of the New Hebrides arc, *Geol. Soc. Am. Bull.*, v. 85, p. 1287–1300.

Mallick, D. I. J., and Neef, G., 1974, Geology of Pentecost, *New Hebrides Geol. Surv. Reg. Rep.*,

Maung, T and Coulson, F. I., 1983, Assessment of petroleum potential in Central Solomons Basin, *UNDP-CCOP/SOPAC Tech Rept. No. 26.*

McTavish, R. A., 1966, Planktonic foraminifera from the Malaita Group, British Solomon Islands, *Micropalaeontology*, v. 12, p. 1–36.

Minster, J. B., and Jordan, T. H., 1978, Present day plate motions, *J. Geophys. Res.* v. 83, p. 5331–5345.

Murauchi, S., Ludwig, W. J., Den, N., Hotta, H., Asanuma, T., Yoshii, T., Kubotera, A., and Hagiwara, K., 1973, Seismic refraction measurements Ontong Java Plateau northeast of New Ireland, *J. Geophys. Res.*, v. 78, p. 8653–8663.

Neef, G., 1978a, A convergent subduction model for the Solomon Islands, *Bull. Aust. Soc. Explor. Geophys.*, v. 9,

Neef, G., 1978b, Cenozoic stratigraphy of Small Nggela Island, Solomon Islands—early Miocene deposition in a fore-arc basin followed by Pliocene patch reef deposition, *N.Z. J. Geol. Geophys.* v. 22, p. 53–70.

Neef, G., and McDougall, I., 1976, Potassium-argon ages on rocks from Small Nggela Island, British Solomon Islands, *Pac. Geol.*, v. 11, p. 81–85.

Neef, G., and Plimer, I. R., 1979, Ophiolite complexes on Small Nggela Island, Solomon Islands: Summary, *Geol. Soc. Am. Bull.*, v. 90, p. 136–138.

Nixon, P. H., 1980, Kimberlites in the southwest Pacific, *Nature*, v. 287, p. 718–720.

Nixon, P. H., and Boyd, F. R., 1979, Garnet-bearing herzolites and discrete nodules from the Malaita alnoite, Solomon Islands, S.W. Pacific, and their bearing on oceanic mantle composition and geotherm, in: *The Mantle Sample: Inclusions in Kimberlites and other Volcanics, v. 2, Proceedings of the Second International Kimberlite Conference, Santa Fe, New Mexico*, Boyd, F. R. and Meyer, H. O. A., eds. Washington, D.C: American Geophysical Union, p. 400–423.

Packham, G. H., 1973, A speculative Phanerozoic history of the Southwest Pacific, in: *The Western Pacific Island Arcs, Marginal Seas, Geochemistry*, Coleman, P. J., ed., Perth: University of Western Australia Press, p. 369–88.

Page, R. W., and McDougall, I., 1972, Geochronology of the Panguna Porphyry copper deposit, Bougainville Island, New Guinea, *Econ. Geol.* v. 67, p. 1065–1072.

Plimer, I. R., and Neef, G., 1980, Early Miocene extrusives and shallow intrusives from Small Nggela, Solomon Islands, *Geol. Mag.*, v. 117, p. 565–578.

Proctor, W. D., and Turner, C. C., 1977, Geology of Savo Island, *Geol. Surv. Solomon Is. Bull.*, no. 11 (unpublished).

Pudsey-Dawson, P. A., and Thompson, R. B., 1958, The detailed geological survey of Western Guadalcanal 1954, in: *The Solomon Islands—Geological Exploration and Research, 1953– 1956, Mem. Geol. Surv. Br. Solomon Is.*, no. 2, p. 43–56.

Ravenne, C., Pascal, G., Dubois, J., Dugas, F., and Montadert, L., 1977, *Model of a Young Intra-oceanic arc; the New Hebrides Island Arc International Symposium on Geodynamics in southwest Pacific*, Paris: Editions Technip, p. 63–77.

Recy, J., Dubois, J., Daniel, J., DuPont, J., and Launay, J., 1977, *Fossil Subduction Zones: Examples in the Southwest Pacific, International Symposium on Geodynamics in Southwest Pacific*, Paris: Editions Technip, p. 345–356.

Resig, J., 1983, Foraminiferal stratigraphy and palaeobothymetry of dredged rock, Lee 1982 Cruise, Solomon Islands, *Contrib. to R. V. S. P. Lee Cruise Report Leg 3, Solomon Is.*

Rickwood, F. K., 1957, Geology of the island of Malaita, in: *Geological Reconnaissance of Part of the Central Islands of BSIP: Colonial Geology and Mineral Resources.*, v. 10, p. 113–145.

Ridgway, J., and Coulson, F. I. E., The geology of Choiseul and the Shortland Islands, (in preparation).

Ripper, I. D., 1970, Global tectonics and the New Guinea–Solomon Islands region. *Search* v. 1, p. 226–232.

Rose, J. C., Woolard, G. P., and Malahoff, A., 1968, Marine gravity and magnetic studies in the Solomon Islands, in: *The Crust and Upper Mantle of the Pacific Area, Mongr. Am. Geophys. Union.*, no. 12, p. 379–410.

Stanton, R. L., 1961, Explanatory notes to accompany a first geological map of Santa Ysabel, British Solomon Islands Protectorate, *Overseas Geol. Miner. Resourc.*, v. 8, p. 127–149.

Stanton, R. L., and Ramsay, W. R. H., 1975, Ophiolite basement complex in a fractured island chain, Santa Isabel, British Solomon Islands, *Bull. Aust. Soc. Explor. Geophys.*, v. 6, p. 61–64.

Taylor, G. R., 1976, Styles of mineralization in the Solomon Islands—a review, in: *Marine Geological Investigations in the Southwest Pacific and Adjacent* Areas Glasby, G. P. and Katz, H. R. eds., *UNESCAP Tech. Bull.*, v. 2, p. 83–91.

Taylor, G. R., 1977, The ophiolite terrain and volcanogenic mineralization of the Florida Islands, Solomon Islands, Ph.D. dissertation University of New England, (unpublished).

Taylor, B., and Exon, N. F., 1983, Cruise report, ridge subduction in the Woodlark–Solomons region, *Kana Keoki* cruise 82-03-16, Leg 4. *CCOP/SOPAC Rep.*

Thompson, R. B. M., 1958, The geology of the Florida Group, 1956, in: *The Solomon Islands— Geological Exploration and Research, 1953–1956, Mem. Geol. Surv. Br. Solomon Is..* no. 2, p. 97–101.

Thompson, R. B. M., 1960, The geology of the Ultrabasic Rocks of the British Solomon Islands, Ph.D. dissertation, The University of Sydney, (unpublished).

Thompson, R. B. M., 1968, Southwest Guadalcanal-the Itina River Basin, 1964–1965, *Br. Solomon Is. Geol. Rec.*, v. 3, p. 9–14.

Thompson, R. B. M., and Pudsey-Dawson, P. A., 1958, The geology of eastern San Cristobal, 1955–1956, *Mem. Geol. Surv. Brit. Solomon Is.*, p. 90–95.

Turner, C. C., 1975, The geology of Mborokua, *Geol. Surv. Solomon Is. Bull.*, v. 7, p. 15.

Turner, C. C., and Hackman, B. D., 1977, The geology of the Beaufort Bay Area, Guadalcanal. *Geol. Surv. Solomon Is. Bull.*, no. 9, (unpublished).

Turner, C. C., and Ridgeway, J., 1982, Tholeiitic, calc-alkaline and (?) alkaline igneous rocks of the Shortland Islands, Solomon Islands, *Tectonophysics*, v. 87, p. 335–354.

Turner, C. C., and Hughes, G. W., 1982, Distribution and tectonic implications of Cretaceous– Quaternary sedimentary facies in Solomon Islands, *Tectonophysics*, v. 87, p. 127–146.

Webb, R. J. R., Cooper, A. W., and Coleman, P. J., 1966, Potassium-argon measurements of the age of basal schists in the British Solomon Islands, *Nature*, v. 211, p. 1251–1252.

Weissel, J. K., Taylor, B., and Karner, G. D., The opening of the Woodlark spreading system and the evolution of northern Melanesia since mid Pliocene times, *Tectonophysics*, (in press).

Winkler, H. A., 1968, *Regional Geophysical Structure of the British Solomon Islands. UN Special Development Program Aerial Geophysical Survey Project Report* Honiara: Government Printing Office.

Winterer, E. L., *et al.*, 1971, *Initial Reports of the Deep Sea Drilling Project*, v. 7. Washington D.C: United States Government Printing Office, p. 473–606.

Woolard, G. P., *et al.*, 1967, Cruise report on 1966 seismic refraction expedition to the Solomon Sea, *Hawaii Inst. Geophys. Rep* HIG-67-3, p. 31.

Wright, P. C., 1968, Western Guadalcanal—the geology of the Lungga and Tenaru River systems. *Br. Solomon Is. Geol. Rec*. no. 3, p. 25–40.

Vedder, J., Tiffin, D., and others, 1983, Draft cruise report on Leg 3 of the S.P. Lee Solomon Islands, May 1982, *Cruise Report No. 71 of PE/SI. 2, CCOP/SOPAC*.

Chapter 14

THE VANUATU ISLAND ARC: AN OUTLINE OF THE STRATIGRAPHY, STRUCTURE, AND PETROLOGY

J. N. Carney, A. Macfarlane and D. I. J. Mallick

British Geological Survey
Keyworth, Nottingham NG12 5GG, England

I. INTRODUCTION

Vanuatu (formerly the New Hebrides Anglo–French Condominium) forms one sector of a Cretaceous to Recent island arc system extending from New Britain through the Solomon Islands to Vanuatu, Fiji, Tonga and the Kermadec Islands (Fig. 1). The component arcs originally formed a volcanic chain to the south and west of a subduction zone dipping towards Australia and now seen as a discontinuous line of active and inactive trenches. Fragmentation of this earlier arc–trench system during the late Miocene resulted in the formation of new subduction zones dipping northwards and eastwards beneath the Vanuatu, Solomon Islands and New Britain arc sections.

The changeover, or "reversal," of arc polarity of the Fiji and Vanuatu sections occurs in a complex zone of marginal basins and transform faults that surround, and largely isolate, the Fiji Platform. Chase (1971) first noted the young age of this fragmentation, suggesting that by about 10 m.y. Vanuatu had begun to migrate, rotating southwestwards and opening behind it the North Fiji Basin. More recent palaeomagnetic surveys (Malahoff *et al.*, 1982a) have confirmed the onset of related rotational deformation on the Fiji Platform at about 10 m.y., while magnetic anomalies in the North Fiji Basin date from about 8 m.y. (Malahoff *et al.*, 1982b). Palaeomagnetic studies in

683

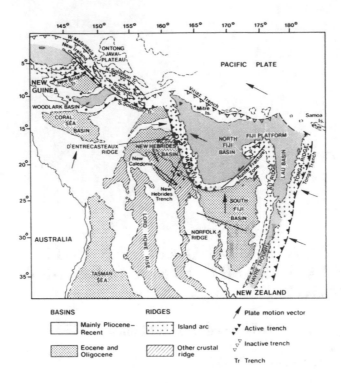

Fig. 1.   Regional setting of Vanuatu in the Southwest Pacific.

Vanuatu by Falvey (1978) have detected rotation in rocks younger than 6 m.y. but this does not rule out earlier movements associated with arc rifting.

The relatively young age of arc fragmentation in this region implies that prior to the important 10–8 m.y. period Vanuatu must have been located a few hundred kilometers further to the northeast against a former zone of convergence between the Austral–Indian and Pacific plates that is now represented by the "fossil" Vitaz Trench (Fig. 1). Postdating this period, however, the arc was "transferred" to the Pacific plate Margin (i.e., North Fiji Basin), its volcanism being generated by eastward subduction of the Austral–Indian plate.

This then, is the unusual geodynamic framework within which the geology of Vanuatu must be considered. In the account that follows, interpretative aspects are not emphasized unduly but as they are important to a proper understanding of arc evolution they are discussed where relevant to the correlation of intra-arc events with regional plate interactions.

II. STRATIGRAPHICAL AND STRUCTURAL SUBDIVISIONS

Vanuatu was subdivided (Mitchell and Warden, 1971; Mallick, 1973a) into three major volcanic provinces; these are the *Western Belt* of late Oli-

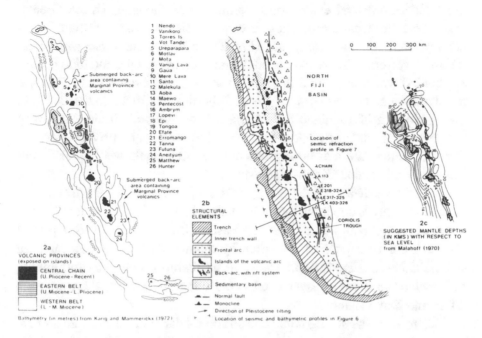

Fig. 2.   (a) Volcanic provinces of Vanuatu. (b) Structural elements of Vanuatu. (c) Crustal thickness variation in Vanuatu calculated from gravity data (after Malahoff, 1970).

gocene to middle Miocene age, the *Eastern Belt* dated at Mio–Pliocene, and the *Central Chain* active from the late Pliocene to present day. A largely submerged *Marginal Province*, of similar age to the Central Chain but now largely inactive, was proposed by Carney and Macfarlane (1982). These volcanic provinces, distributed as in Fig. 2a, are spatially separated in the central latitudes of the arc but further to the south, the Central Chain must overlie the submerged continuation of the Western Belt along the frontal part of the arc.

Subaerial volcanics are limited to the Central Chain islands. Fragmental rock sequences are mostly derived from volcanic episodes either directly, as primary volcanic breccias, or secondarily as volcaniclastics deposited by erosion or tectonic instability along the slopes of former volcanic islands. Additionally, there are appreciable accumulations of basinal clastics grading to pelagites, a present-day example being a thickness of at least 2000 m of sediment beneath the 3000-m deep basin between the Western and Eastern belts (Ravenne *et al.*, 1977; Carney and Macfarlane, 1980).

The structure of Vanuatu was first described by Karig and Mammerickx (1972) in the terms of the pattern observed in other western Pacific arc systems. Their subdivisions served as a basis for later offshore studies by Luyendyk *et al.* (1974), Dubois *et al.* (1975) and Ravenne *et al.* (1977). More recent syntheses, incorporating seismic profile results from a CCOP/SOPAC hydrocarbon potential evaluation program, are forthcoming (Katz; Greene *et al.*, both in preparation). These diverse studies all recognize that Vanuatu's present structure comprises (Fig. 2b): a *trench* and *inner wall* along the arc's western offshore margin succeeded eastwards by a horstlike *frontal arc* behind which is located the present active *volcanic arc* and, finally, a rifted and block faulted *back-arc* zone. Subordinate structures that indicate young block uplifts include the "*Central*," or "*Aoba*," basin downwarp between the frontal and back-arc systems (Ravenne *et al.*, 1977). Extending from the central latitudes northwards are the rather shallower "Banks" and "Santa Cruz" basins (Katz, in preparation).

Gravity modelling of the arc by Malahoff (1970) shows the frontal arc structural entity is underlain by a longitudinal belt of 24 to 26-km thick crust, whereas high gravity values over the Eastern Belt suggest that this part of the back-arc zone may be underlain by thinner crust or contain uplifted mantle material (Fig. 2c). Other significant features shown by the gravity map include apparent transverse left-lateral offsets of the frontal arc between Efate and Santo islands.

Comparing the stratigraphic and structural diagrams of Figs. 2a and 2b, it is apparent that certain of the major volcanic provinces also define structural entities. For example, the Western Belt of Santo and Malekula islands is coincident with the frontal arc crustal ridge (Fig. 2c), from which it is inferred that the crustal thickening process occurred during the late Oligocene to middle Miocene volcanic episode represented in the Western Belt. By contrast, the Eastern Belt and Marginal Province volcanics are both mainly located in the back-arc zone. The Central Chain, as the present volcanic arc, is in the south of Vanuatu located along the rear margin of the frontal arc and, as discussed below, has a tectonic history related to frontal arc movements.

III. OUTLINE OF THE STRATIGRAPHY

Stratigraphic columns for the four volcanic provinces are given in Fig. 3. The account that follows describes the evolution of these rock sequences on a chronological basis in order to emphasize correlations from east to west across the arc.

A. Nature of the Basement Rocks

So far there are no confirmed exposures of "sub-arc" basement stratigraphically beneath the arc-derived volcanic or clastic rock sequences shown in Fig. 3. In the Eastern Belt, however, there exists an ophiolitic

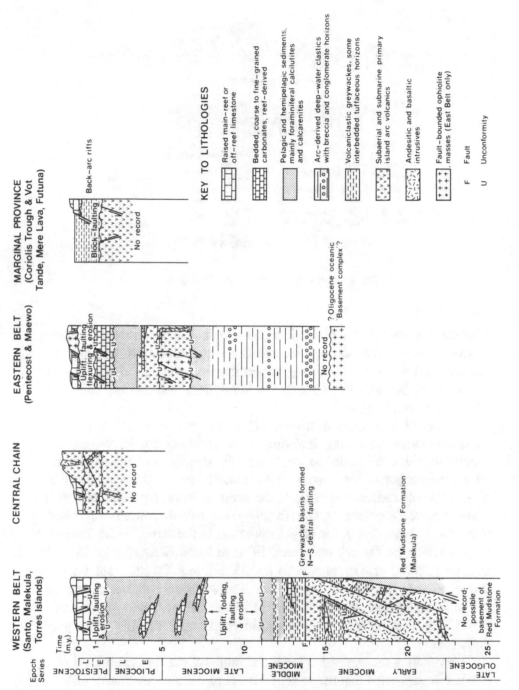

Fig. 3. Generalized stratigraphic columns of the main volcanic provinces of Vanuatu.

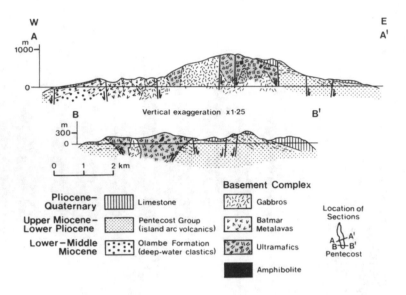

Fig. 4.   Sections across southern Pentecost.

"Basement Complex" of peridotites and serpentinized peridotites associated with amphibolite rafts, gabbro intrusives and metalavas, an ophiolite association. Its field relations suggest tectonic emplacement along the axis of a block faulted anticline structure in Mio–Pliocene island arc volcanics (Mallick and Neef, 1974).

The age of the Pentecost Basement Complex remains an unsolved problem at the time of this writing. It obviously predates basal arc-derived clastics of early to middle Miocene age since these dip steeply away from the basement complex horsts and have been downfaulted on the western side (Fig. 4). Attempts to radiometrically date the complex have, however, met with mixed success. For example, Gorton (1974) determined a range of K/Ar ages of between 13 and 6 m.y. for amphibolite rafts in the serpentinites whereas N. J. Snelling (in Mallick and Neef, 1974) deduced K/Ar ages of 28 ± 6 m.y. for gabbro intruding metalavas and 35 ± 2 m.y. for the amphibolites. If one assumes that certain systems were reset during subsequent heating by the Mio–Pliocene arc volcanism on Pentecost (Figs. 3 and 4), then only the 35 m.y. determination can be considered to approach the true age of these rocks.

Considering the age and ophiolitic affinities of the Basement Complex, it might be suggested that the various island arc volcanic sequences of Vanuatu rest directly on Oligocene oceanic crust. The formation of this crust could correlate with an early Oligocene hiatus of arc volcanism on Tonga and the Fiji Platform (e.g., Coleman and Packham, 1976), both events perhaps indicating a phase of rifting and marginal sea development within the earlier arc system located along the Tonga–Fiji–Vitiaz zone of Pacific plate subduction.

B. Latest Oligocene to the End of the Early Miocene (22–14 m.y.)

During this period an active island arc gave rise to the Western Belt volcanic sequences on Santo, Malekula and the Torres Islands which now form the frontal arc to the Vanuatu system. Radiometric and micropalaeontological studies indicate that the volcanics are mainly of early Miocene age, though they may extend back to the latest Oligocene. There are isolated K/Ar determinations of 39 ± 5 m.y. and 36.7 ± 1.0 m.y. on andesite from the Torres Islands (Greenbaum *et al.*, 1975) but these contradict early Miocene micropaleontological ages from nearby sedimentary sequences and it is therefore questionable whether they can reliably be used to indicate that the first phase of volcanism started in late Eocene or early Oligocene (as quoted by Malahoff, 1982). However, as will be seen below, the volcanic ridge now known as Vanuatu may formerly have been adjacent to an older frontal arc system along the Vitiaz Trench, and containing volcanics dating back to Eocene which are similar to those exposed on Fiji and Tonga (e.g. Carney and Macfarlane, 1978).

Estimates of 4000 to 6000 m thickness for the Western Belt volcanics (Robinson, 1969; Mitchell, 1971; Mallick and Greenbaum, 1977) are minimum values since no subvolcanic basement is exposed. Red mudstones containing fine volcaniclastic laminae and forming isolated outcrops on the northwestern margin of Malekula have depositional depths estimated at 4000 to 6000 m (Mitchell, 1971) and could, therefore, represent accumulations on the proposed Oligocene oceanic floor to the arc. Unfortunately, their age is unknown and it cannot be ruled out that the mudstones are the uplifted deep-water lateral equivalents of the Western Belt volcanics.

In lithology, the Western Belt volcanics comprise mainly basaltic and basaltic–andesite breccias, paraconglomerates and volcanogenic sandstones together with intercalated reef-derived clastic carbonates, the whole complex presenting all the features of derivation from an active, partly emergent island arc. A number of volcanic centers can be distinguished by extensive sequences of palagonitic tuffs with pillow lava and pillow breccia lenses. Dyke swarming in the vicinity of certain centers suggests large volume eruptions in extension environments (Mallick and Greenbaum, 1977; Carney and Macfarlane, in preparation).

The stratigraphic evolution of the Western Belt was marked by interruptions in the eruptive cycle, firstly in the uppermost early Miocene with the intrusion in southern Santo of gabbros, andesites and microdiorites. Vol-

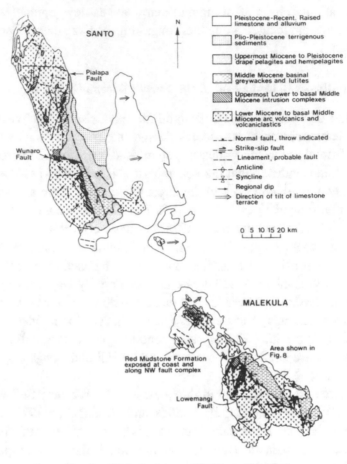

Fig. 5.   Geological map of Santo and Malekula.

canism was then resumed and continued into early middle Miocene times before being brought to a close by a further episode of faulting and intrusion.

C. Middle Miocene (14–11 m.y.)

At this time major faulting and intrusion occurred in the Western Belt, and coarse arc-derived clastic sediments were deposited in the eastern part of the arc.

Arc volcanism in the Western Belt culminated in a final episode of andesite–microdiorite intrusion along a sinuous system of right-lateral wrench faults described further in the structural section of this paper. A radical change in the stress regime then followed, resulting in the formation of narrow, north-trending graben along the former intrusive axis of Santo and Malekula (Fig. 5). The graben were rapidly filled by at least 4000 m of

volcanogenic greywackes containing horizons of slumped, reef-derived carbonates and occasional developments of fine tuff (Robinson, 1969; Mallick
and Greenbaum, 1977; Mitchell, 1971). Lingering volcanism may have produced the occasional breccia horizons in these basinal successions, while
distant pyroclastic activity certainly accounts for the tuffaceous horizons.
However, no intrusives are known and since there are few primary volcanics
associated with the basins it is concluded that the main phase of island arc
activity on the Western Belt had ceased by this time. The basinal sediments
"fine" upwards, in South Santo, to pelagic mudstones and calcilutites (Mallick and Greenbaum, 1977).

In the Eastern Belt, bedded coarse to fine-grained volcanogenic sediments form the earliest exposed rocks on Maewo and Pentecost islands.
Sedimentary structures that include graded bedding, convoluted lamination,
and mud rafting are indicative of deposition as flysch near to an unstable
slope (Carney, in preparation; Neef, 1982), while the presence of cobble
conglomerates with reef limestone clasts suggests derivation from a landmass (Liggett, 1967). Volcanic cobbles in the conglomerates comprise highly
differentiated types such as quartz dacite and rhyolite of island arc tholeiite
composition (Carney and Macfarlane, 1978), more akin lithologically and
chemically to the Eocene–Oligocene Wainimala Group of Fiji (Rodda, 1967)
than to the Western Belt volcanics of Vanuatu. Carbonate clasts in the conglomerates on Maewo contain shallow-water larger foraminifera with assemblages appropriate to late Eocene (Coleman, 1969), and early Miocene
(Carney, in preparation) ages, the youngest with derived faunas dated at
about the early–middle Miocene boundary (Carney, in preparation; Mallick
and Neef, 1974). Unfortunately, planktonic globigerinid foraminifera in the
sediments hosting these clastic carbonates are poorly preserved, preventing
an accurate determination of the depositional age of the sediments, but it is
probable that the fine-grained, often tuffaceous upper part of the sequence
may range some way into the late Miocene (Fig. 3).

Neef (1982) has shown that the Eastern Belt sediments are of a lithology
appropriate to deposition below the carbonate compensation depth, estimated at 4.25 km. It is therefore possible that they may rest directly upon
oceanic crust represented on Pentecost by the tectonically uplifted ophiolitic
"Basement Complex" although this cannot be proven. Further work to date
the planktonic component of these sediments is required to resolve whether
the fine-grained host rocks have a similar age range, of early to early–middle
Miocene, as do the younger of the included clasts.

D. End of the Middle to the Early Late Miocene (11–8 m.y.)

This period is marked by uplift and erosion of the Western Belt and
coincides with the onset of seafloor spreading which was to form the North

Fiji Basin. As noted above, the Western Belt was volcanically quiescent by about 13 m.y. save for the presence of fine tuffaceous horizons in the grey-wacke basins of Malekula Island and an uncorrelated breccia sequence on eastern Malekula dated by the K/Ar method at 10.7 m.y. (Gorton, 1974). This latter age is somewhat anomalous and worthy of further examination, as it is broadly comparable with the 12.7 m.y. recorded by Jezek *et al.* (1977) from volcanics on Mitre Island, which is located on the eastern margin of the North Fiji Basin and therefore perhaps was originally in close proximity to Vanuatu prior to inception of that marginal sea (Fig. 1).

The uplift of Santo followed a change, in the greywacke basins, to the deposition of predominantly fine-grained foraminiferal sediments dated at 11 m.y. (Mallick and Greenbaum, 1977). The magnitude of the ensuing erosion is best appreciated in the northern part of the island, where late Miocene hemipelagic sediments are observed to veneer steep underlying slopes on earlier Miocene formations, and to infill former V-shaped river valleys developed on them (Carney and Macfarlane, in preparation). Important infilling relationships of the hemipelagites occur along the Wunaro Fault, a lineament that may have a left-lateral as well as a vertical throw, and around the southern slopes of Santo. Both these westwards salients of hemipelagite (Fig. 5) suggest that valleys had been eroded along traverse faults whose movements may have been associated with the apparent left-lateral offset of Santo relative to Malekula (Fig. 2b).

In the Eastern Belt, continued deep-water sedimentation resulted in a predominantly fine-grained capping to the volcanogenic flyschoid sequence described above. Shallowing of water depths towards the end of this cycle is however suggested by palaeodepths of 2000 to 3000 m for overlying highly foraminiferal sediments (Carney, in preparation).

E. Late Miocene (8–6 m.y.)

Important unconformities and/or lithological changes at about this time occurred in the successions of the Eastern and Western belts (Fig. 3).

In the Western Belt, subsidence beneath sea level of the previously eroded Miocene volcanic topography resulted in the deposition of "drape" hemipelagic foraminiferal mudstones and calcilutites. Palaeodepths of 700 m determined in mudstones now exposed at 1100 m elevation on Santo (Taylor, 1977) suggest that submergence of the Western Belt was complete by earliest Pliocene times (about 5 m.y.).

The Eastern Belt succession evolved rapidly during the latest Miocene period in response to seafloor spreading that was to result in a change of polarity of the island arc from northeast-facing to southwest-facing. The initial event, on Maewo Island, comprised an abrupt lithological change from fine-grained arc-derived volcaniclastic sediments to highly foraminiferal mudstones containing fossil *Globigerina* oozes (Carney, in preparation). As noted above, palaeodepths of 2000 to 3000 m deduced for the microfauna of these rocks indicate a shallowing of depositional depths, but no intervening faulting or tilting is discernible between the two formations. Slightly higher in the succession however, major faulting accompanied by a 2° to 3° westwards tilting of the foraminiferal mudstones occurred as a prelude to the eruption of a voluminous series of basic island arc type pillow lavas. On Maewo, the lowest lava horizons are picritic and there is a progressive change upwards into feldsparphyric basalts (Liggett, 1967). On Pentecost, Mallick and Neef (1974) recorded at least two cycles of basic to intermediate lava eruption. These volcanics are the first products of subduction of the Austral–Indian plate beneath the developing North Fiji Basin.

F. Early Pliocene (5–4 m.y.)

Volcanic events at this time were confined to the Eastern Belt, the Western Belt remaining submerged and continuing to accumulate drape-type hemipelagi..es.

Shallowing of waters over the Eastern Belt by basal Pliocene times is deduced from the increased pyroclastic character of the volcanism and the occasional presence in volcanic breccias of reef limestone clasts. On Maewo, the changeover was accompanied by block faulting and uplift of the central part of the island (Carney, in preparation) while on Pentecost the presence of serpentinite clasts in volcanic breccias suggests that the "Basement Complex" may have been undergoing uplift to high structural levels within the volcanic carapaces (Mallick and Neef, 1974).

Termination of Eastern Belt volcanism occurred at about 3 to 4 m.y., and on Maewo there was a return to pelagite accumulation.

G. Late Pliocene to the Present Day (3–0 m.y.)

Major vertical movements occurred along the frontal arc (Western Belt) and in the back-arc zone. This period also saw the main development of the present Central Chain active volcanic arc, and of the Marginal Province volcanic belt.

In the central latitudes of Vanuatu, frontal arc uplift is reflected on Santo by a change, dating from about 2 m.y., to fluviatile and shallow water marine sedimentation in the central area (Fig. 5). Uplift with eastwards tilting continued throughout the Pleistocene when successive regressive fringing reef complexes were constructed around the flanks of the emerging landmass.

Important block faulting characterized the back-arc zone at this time. In the Eastern Belt, shallow water limestones with reef-derived clasts were deposited on Maewo at about 1.8 m.y. and further uplift throughout the Pleistocene was accompanied by westwards tilting of about 20° (Carney, in preparation). On Pentecost, uplift and unroofing of the "Basement Complex" occurred pre-late Pliocene, but further uprise of the ophiolite evidently accompanied faulting and general elevation through the Pleistocene into Recent times. Both Eastern Belt islands are mantled by highly faulted raised reef limestones, also containing atoll and lagoonal facies.

In the back-arc zone to the north and south of the Eastern Belt ridge and its submerged geophysical continuations (Fig. 2c) Pleistocene block faulting resulted in the development of a system of narrow, sinuous rifts. Island arc volcanics dated at late Pliocene to early Pleistocene (1.8 m.y.) have been recovered in dredge hauls from the Coriolis Trough (Dugas *et al.*, 1977) and are exposed on Futuna Island (Carney & Macfarlane, 1979). Volcanics dated at 3.5 m.y. are also exposed on Vot Tande Island in the north of Vanuatu. The age of these volcanics indicates that prior to the early Pleistocene, arc volcanism was continuous between the Marginal Province and Central Chain (Fig. 2b).

The Central Chain volcanic sequences expose basic lavas dated at 2.4 m.y. on Tanna and Erromango islands, and these represent the earliest known *in situ* eruptions. Three cycles of volcanism were suggested for the three southern islands of Efate, Erromango and Tanna (Carney and Macfarlane, 1982):

(*i*) 2.4 m.y. on Erromango and Tanna only;
(*ii*) 1.7–1.1 m.y.;
(*iii*) 0.2 m.y. to present-day.

North of Efate, volcanic formations are mostly rather younger in age, dating back 0.7–0.4 m.y. on Western Epi and Vanua Lava and ranging through to recent times on those and other islands. Remnants of older phases of volcanism occur in the southern Banks Islands, the older volcanics of Gaua having yielded an age of 1.8 m.y. and Merig 1.1 m.y. (Mallick and Ash, 1975).

In compositional range, the Central Chain volcanics comprise a spectrum from basalts through to andesites and dacites. Basalts and andesites predominate and are probably roughly equal in volume. The more differentiated types are subordinate, but occur as locally voluminous sequences of potash-rich rhyodacite tuffs and breccias on Efate (Ash *et al.*, 1978) and Western Epi (Warden, 1967).

Basic accumulative lavas of picritic and ankaramitic composition are common on Ambrym and Aoba volcanoes (Gorton, 1974), but are rare elsewhere. By analogy with the basal Eastern Belt picrite lavas on Maewo, the islands of Ambrym and Aoba might represent early eruptions of the Central Chain in this part of Vanuatu. These two volcanic islands are furthermore unusual in possessing low-angled shield morphologies and in being surmounted by circular calderas of ~10 km diameter containing younger resurgent volcanoes. Prominent zones of basaltic fissure volcanism are responsible for the east-northeast alignment of Aoba and the WNW alignment of Ambrym. A small caldera, with resurgent volcanic cone, is also present at the summit of Gaua island (Mallick and Ash, 1975) and a similar structure may at one time have existed also on Vanua Lava (Ash et al., 1980).

Volcanoes on the other islands generally have steeper profiles and lack significant caldera formation at their summits; uneroded examples of perfect basalt–andesite cones are the dormant island (?) of Mere Lava and the now continuously active volcano of Lopevi.

Reef limestone and associated clastic carbonates are variably developed on the Central Chain islands and reflect, in a general way, the age of the volcanism present. Fringing reefs are present around parts of most islands at the present day. Raised reefs are absent from some of the active islands (e.g. Gaua) or are present only as small developments at low altitude (e.g. on Vanua Lava and Ambrym). On older, eroded islands the raised reefs are present to higher altitudes, for example to ca. 40 m on the young island of Motlav in the Banks Islands and to between 300 and 630 m on the older parts of the southern islands of Efate, Erromango and Tanna. Radiometric determinations on Efate raised reefs indicate an average rate of uplift of almost 1 mm/year for the last 200,000 years (Bloom et al., 1978) while limestones from eroded plateaus overlying the Efate rhyodacitic tuffs have microfauna appropriate to the early Pleistocene zone N22, or 1.5 m.y. (D. Taylor in Carney, 1982). Similarly on Tanna, a raised reef limestone sequence occurs between volcanic formations dated at between 2.4 m.y. and 1.1 m.y.

IV. ARC STRUCTURE

In this section an account is given of the major structural subdivisions detailed in Fig. 2b. Surface structures such as faults and folds are also described and where relevant deformational histories of well-exposed areas are outlined.

A. Trench and Inner Wall

The New Hebrides Trench in past accounts has been described as a discontinuous feature interrupted in the latitudes of Santo and North Ma-

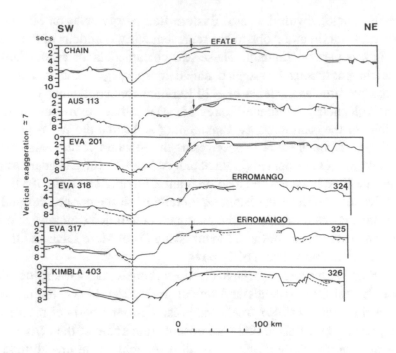

Fig. 6.   Seismic reflection profiles to illustrate the morphology of the southern part of the New Hebrides Trench between Efate and Erromango (from Daniel, 1978).

lekula. More recent studies (Greene *et al.*, in preparation) show that the trench is in fact present along most of this arc sector even though its expression is much subdued. This effect coincides with the anomalous uplift of the frontal arc at Santo and Malekula islands and has been attributed by Chung and Kanamori (1978) to buoying-up of the overlying lithosphere by the presence on the downgoing slab of a ridge of thickened crust known as the D'Entrecasteaux Fracture Zone (e.g. Daniel *et al.*, 1977).

The Morphology of the inner trench wall was initially described by Karig and Mammerickx (1972) as typically steep with a benched profile but no well-defined mid-slope basement "high." Similarities in profile between the inner slopes of the Vanuatu and Mariannas trenches were noted by Karig and Sharman (1975) who observed that the apparent absence of the mid-slope basement "high" could mean that accretion had played an insignificant role in the development of the arc. Further seismic profiling (Daniel, 1978) has, however, demonstrated the presence of a variety of inner wall morphologies of which some suggest that accretion has taken place (Fig. 6, profiles EVA 201 and AUS 113). Furthermore, on the seismic refraction profile (Fig. 7) a possible accretionary prism is outlined by the trenchward-thickening 4.1 km/sec velocity layer. Before firmer conclusions can be

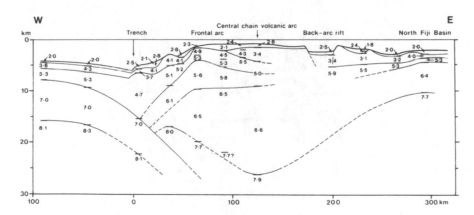

Fig. 7. Seismic refraction profile across the central part of Vanuatu (vertical exaggeration approximately × 7) (from Pontoise *et al.*, 1980).

drawn, however, more seismic work is needed to define the attitudes of reflectors in these zones of inferred accretion.

B. Frontal Arc

The frontal arc is defined by the islands of the Torres, Santo and Malekula (Fig. 2b) and on offshore seismic profiles by a horstlike ridge of acoustic "basement." Geophysically the frontal arc is expressed as a longitudinal ridge of 24 to 26-km thick crust that may be offset by transverse structures between Efate and the Torres Islands (Fig. 2c). Seismic refraction profiles across the frontal arc show (Fig. 7) a "root" of 15 km thickness comprising 6.6 km/s velocity material interpreted by Pontoise *et al.* (1980) as arc crust. Above this is a 7-km thick layer with 5.8 to 4.9 km/s velocity that is probably made up of early and middle Miocene basic arc volcanics. The uppermost layers are uplifted along the leading edge of the frontal arc into a horstlike structure in keeping with the observed elevation of Santo, Malekula and the Torres Islands. As previously noted, this positive expression of the frontal arc is also suggested by seismic reflection profiles that show a progressive thinning out westwards of sedimentary reflectors on to acoustic basement. On Santo Island, the same relationship may be demonstrated by the major erosional unconformity some 3 m.y. in duration, that intervened between early middle Miocene arc volcanics and late Miocene–Pliocene pelagic deposits. The structural evolution of the frontal arc, deduced from field relationships on Santo and Malekula islands, can be summarized as follows:

(*i*) Latest Oligocene–early Miocene. Arc volcanism and intrusion resulted in the progressive development of the 24–26-km thick crustal ridge

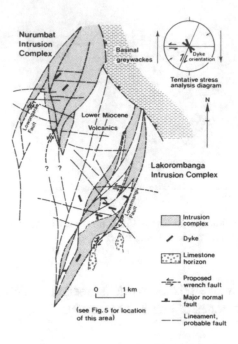

Fig. 8.   Fault control of the Nurumbat and Lakorombanga intrusion complexes, Malekula.

that now defines the frontal arc. As discussed below, the volcanism can be attributed to a pre-middle Miocene phase of westwards subduction of the Pacific plate at a former zone of convergence along what is now the north-eastern margin of the North Fiji Basin (Fig. 1).

The most uplifted part of the frontal arc, in southern Santo Island (Mallick and Greenbaum, 1977), records an early Miocene phase of intrusion of andesite–gabbro bodies accompanied by block uplifts and the formation of subsiding greywacke/breccia basins. On Malekula there is little evidence of similar happenings, possibly because the island is relatively less elevated.

(*ii*) Early/middle Miocene boundary. Island arc volcanism continued on Santo, and possibly also on Malekula and the Torres Islands.

(*iii*) Early/middle Miocene (14–12 m.y.). Important tectonic episodes at this time were associated with cessation of volcanism in the Western Belt. Recent mapping on Malekula by one of the writers (J.N.C.) has shown that firstly a number of basalt–andesite–microdiorite intrusion complexes were emplaced along regionally extensive fault systems with NW to NNE trends (Fig. 5). In the Lakorombanga intrusion complex, for example, basalts and andesites were emplaced into a braided fault zone (Fig. 8) as a series of NNE aligned slivers and later as E–W trending dykes, both sets of intrusives indicating conjugate fracturing within a prevailing N–S regime of dextral shear, in keeping with the observed displacement of a limestone horizon along the southern bounding fault of the complex. On Santo Island, Robinson (1969) suggested the operation of transcurrent faulting at this time, and recent mapping (by J.N.C. and A.M.) has shown that the fold systems on northern Santo could have an en echelon relationship to the major NNE-trending Pialapa Fault (Fig. 5).

Following this initial episode of intrusion and transcurrent faulting a change in the stress regime caused the formation of NW-trending greywacke basins located along, or slightly to the east of, the former axis of volcanism and intrusion on Santo and Malekula (Fig. 5). On Malekula the basinal faults truncated the earlier wrench fault systems in the south but further north they may have developed along former NW alignments. Similarly in north Santo the Pialapa Fault now became the margin to one of these basins. Sediments accumulating within the basins were mainly fine-grained volcanogenic sandstones and siltstones of distal turbidite facies (Robinson, 1969; Mallick and Greenbaum, 1977). Occasional horizons of slumped sediments associated with lithic and reef-derived bioclastic debris are indicative of periodic bouts of tectonism along the basin margins.

(*iv*) Late Miocene (11–8 m.y.). At some time during this period major uplift and erosion is recorded on Santo and, by inference, is suggested to have occurred throughout the frontal arc. The precise nature of the uplift mechanism cannot be determined, save that it probably imparted the steep eastward to southward dips observed in the greywacke basins (Fig. 5) and hence may have involved flexuring associated with uplift of the arc's western edge over a newly formed subduction zone along that western margin of Vanuatu (see Resumé).

A further consequence of compression along the new trench/frontal arc system may have been the formation of the E–W trending displacements that, on Santo, appear to have caused the sinistral translation of the northern part of an intrusive body along the Wunaro Fault (Fig. 5). Further related displacements as offsets to the frontal arc are believed to occur between Santo and Malekula, off south Malekula and beneath north Efate (Fig. 2c). Evidence that fault valleys corresponding to the displacements on Santo were in existence at this time is discussed above in the stratigraphic outline.

(*v*) Late Miocene (?8–7 m.y.). Foraminiferal age determinations on pelagites resting unconformably on the eroded Miocene volcanic and basinal greywacke terrains of Santo suggest that the frontal arc began to subside at this time and was probably almost completely submerged by basal Pliocene times. Again, the nature of the tectonic processes cannot be precisely determined, but events further east associated with marginal sea inception indicate that arc rifting, and hence regional tension, may have been involved (Carney and Macfarlane, 1982).

(*vi*) Late Pliocene–Present (2–0 m.y.). Final frontal arc uplift can be dated on Santo as commencing at 2 m.y., when terrestrial sediments succeeded the Pliocene pelagite accumulations (Mallick and Greenbaum, 1977). Pleistocene emergence of Santo and Malekula led to the formation of successive raised reef limestone terraces with altitudes that suggest progressive eastwards tilting by an aggregate of several degrees (Mitchell, 1971; Mallick, 1973b). Corroborating evidence that the tilting is still continuing is provided first by studies of Holocene terrace levels on Santo indicating an eastwards decrease in uplift rates from 5 mm to about 1 mm/yr (Jouannic *et al.*, 1980), and second, by the eastward tilting that accompanied uplift of Malekula during several large earthquakes in August 1965 (Benoit, 1967).

Exposed structures associated with this latest period of uplift include relatively minor lineaments cutting the Pleistocene limestones of Santo and Malekula. However, the eastwards tilting associated with uplifts of both these islands suggests that important westwards throwing faults may have been synchronously operative within the inner wall of the adjoining trench and along the present western coastline of Santo (Mallick and Greenbaum, 1977). Along the eastern downtilted margins of Santo and Malekula, offshore seismic profiles indicate the presence of east-facing monoclinal flexures and major normal faults of probable Pleistocene–Recent age, their movements being associated with corresponding deepening of the Central or "Aoba" Basin (Carney and Macfarlane, 1980). In the southern part of Vanuatu, east-throwing fault complexes on Epi, Efate and Tanna islands may also be related to frontal arc uplift and tilting.

C. Volcanic Arc

The present volcanic arc comprising the Central Chain islands is mainly confined to a belt 20–40 km width lying 130–150 km from the trench axis. Recent volcanoes outside of this zone and in a more "frontal" position at only 100 km from the trench include islands between Efate and Epi (Fig. 2a), while in the extreme south of the archipelago the recent volcanoes of Matthew and Hunter islands lie at only 75 km from the trench axis. (The youthful cone of Mere Lava in the north of the archipelago, and the nearby island of Merig, both represent recent volcanism within the back-arc zone.)

Plots of earthquake hypocenters show that the volcanic arc is underlain by a seismic plane dipping eastwards at about 70° and lying 150–200 km beneath Aneityum but increasing gradually in depth northwards to 250–300 km beneath Vanua Lava (Pascal *et al.*, 1978).

The seismic plane or "Benioff zone" beneath Vanuatu has evidently exerted a strong control over the location of the volcanic arc, most centers lying within a narrow belt that maintains a similar distance relative to the trench axis from north to south (Fig. 2b). Relative to the other structural elements of the arc, however, the volcanic alignment has a varied setting. Fig. 2b shows that south of Malekula, for example, the Central Chain islands are located astride the rearwards margin of the frontal arc crustal ridge whereas north of this they lie centrally within the axis of the longitudinal "Central" ("Aoba") and "Torres" basin downwarps (Fig. 2b). The structures and tectonic histories of the volcanic arc islets summarized below can be seen to reflect movements of these major components of the arc crust.

South of Malekula a history of progressive uplift, eastwards tilting, flexuring and faulting since the oldest exposed volcanism at 2.4 m.y. was inferred for Tanna Island (Carney and Macfarlane, 1979) with separate phases of tectonism dated at between 2.4 and 1.7 m.y. and 1.1 and 0.5 m.y. Similarly on Efate, uplift of the western part relative to the eastern side of the island, together with major northwards downfaulting along E–W displacements in the north can also be dated between about 1.1 and 0.5 m.y. Proximity of volcanic "basement" beneath Efate is suggested by the presence in Pleistocene limestones of redeposited middle Miocene carbonate clasts (D. Taylor in Carney, 1982). Similarly on Erromango a conglomerate horizon at the base of the 2.4 m.y. volcanics has yielded clasts with Mio–Pliocene radiometric ages (Colley and Ash, 1971). Recent uplift and easterly faulting/tilting in the islands of Epi (Warden, 1967), Efate (Carney, 1982) and Tanna (Carney and Macfarlane, 1979) testify to the present-day continuation of uplift mechanisms similar to those described above for the emergent part of the frontal arc at Santo and Malekula islands.

North of Malekula islands of the volcanic arc expose only lavas younger than 1 m.y. (except for remnants of 1.8 m.y. volcanics on Gaua) and hence do not have such complete tectonic histories. A possible reason for their apparent youth is that these islands mainly lie within the "Central" and "Torres" basins, which were evidently subsiding throughout the Pleistocene marginal uplifts in the frontal and back-arc zones (Carney and Macfarlane, 1980). Progressive subsidence of these islands during volcanic buildup could, for example, account for the subdued development, and in some cases the complete absence, of raised reef limestones around their coastlines. North of Gaua, however, raised limestones do occur at progressively higher elevations eastwards from Vanua Lava to Motlav and Mota islands (Fig. 2b), corresponding to a shallowing of the underlying basin evident on offshore seismic profiles. Furthermore, the limestone plateau surfaces on these islands dip westwards i.e. into the center of the basinal structure (Ash *et al.*, 1980).

D. Back-Arc Zone

The back-arc region of Vanuatu can be subdivided into two distinct structural entities, namely the longitudinal block faulted ridge of the *Eastern Belt* and, flanking this to north and south, a rifted and largely submerged *Marginal Province* (Fig. 2b).

The submerged extensions to the Eastern Belt structural system are indicated by a continuation to north and south of high Bouguer gravity anomalies that are the basis of Malahoff's calculation of a longitudinal zone of 18-km thick crust in this part of the arc (Fig. 2c). The gravity anomalies undoubtedly owe their origin to the presence at high crustal levels of tectonically uplifted ophiolite bodies visible, on Pentecost, as "Basement Complex" horsts (Mallick and Neef, 1974). It is noteworthy that the northern and southern terminations of the Eastern Belt gravity anomaly zone both coincide with major transverse crustal structures (Fig. 2c), respectively in the latitudes of Gaua and Efate islands. Northwards and southwards of these intersections, the Eastern Belt horst structures are replaced by narrow back-arc rift systems of the Marginal Province.

The back-arc rifts attain a maximum depth of 3300 m in the "Coriolis Trough", between Tanna and Futuna islands. They are aligned roughly parallel to the trench though appearing to curve inwards towards the trench towards their southerly termination (Fig. 2b). Dredging across the submerged scarps of the Coriolis Trough has recovered debris of vesicular volcanic breccia identical chemically to island arc type lava (Dugas *et al.*, 1977). Similar conclusions can be drawn from volcanics exposed on the three small islands of Futuna, Mere Lava and Vot Tande. The significance of these back-arc rifts with respect to island arc tectonic processes has been a subject of some speculation. Karig and Mammerickx (1972) initially described them as of probable extensional origin, but it has since been observed that there is no evidence of oceanic crust within them, and seismic reflection profiles indicate a rather simple rift-like structure involving only the arc crust (Dubois *et al.*, 1975). The seismic refraction profile of Fig. 7 seems to support a continuity of velocity layering across and beneath the rift zone, but there is admittedly a loss of detail in this area, making further deep crustal studies desirable. Against this "nonevidence" of crustal extension, there are certain features of the rifts that indicate that they may be incipiently developing or even "failed" marginal basins. First, there is geological evidence that, prior to formation of the rifts, island arc volcanism was coextensive between the Central Chain and Marginal Province (Carney and Macfarlane, 1979) whereas in post-early Pleistocene times, when the rifts initially formed, the eastern part of the Marginal Province became volcanically inactive and in effect passed into the "remnant arc" stage of island arc evolution. Second, geophysical studies (Dubois *et al.*, 1975) have delineated within the rift axes a central positive magnetic anomaly, larger in the northern rift system than in the southern and possibly indicative of recent axial intrusion. Finally, the same workers have also reported the presence in the Coriolis Trough of a 30 mGal axial positive Bouguer gravity anomaly. Daniel (1978) noted that in profile the rifts were comparable with back-arc basins in the Ryuku and Mariana island arcs if it were supposed that these latter examples are in a more advanced stage of development.

Structural evolution within the whole of the Vanuatu back-arc zone can be summarized into the number of stages described below. It must, however, be borne in mind that the picture is necessarily the most complete for the Eastern Belt where exposure is greatest.

(*i*) Early/middle Miocene (?15–14 m.y.). The deposition of arc-derived clastic sediments on Maewo and Pentecost suggests that the proto-Eastern Belt lay in deep water adjacent to a tectonically active landmass (Neef, 1982) (see Section III).

(*ii*) Late Miocene (?8 m.y.). Cessation of clastic sedimentation, coupled with a shallowing of water depths to 2000–3000 m (Carney, in press) occurred as a prelude to the following rapid sequence of events.

(*iii*) Late Miocene (?8–7 m.y.). Faulting and westwards tilting of pelagites deposited on the basal sediments; this preceded the effusion on to the seafloor of voluminous basic island arc type pillow lavas.

(*iv*) Basal Pliocene (5 m.y.). Block faulting accompanied a further shallowing of water depths, marking a change to part-subaerial pyroclastic volcanism in the arc. On Pentecost, serpentinite clasts in the volcanic breccias suggest that the "Basement Complex" ophiolitic rocks were undergoing tectonic uplift to high levels within the active arc.

(*v*) Mid-Pliocene–early Pleistocene (1.8 m.y.). Uplift commenced along the Eastern Belt, which may already have been emergent and partly limestone covered. Subsequent elevation throughout the Pleistocene was accompanied by an aggregate westerly tilt of 20° on Maewo. On Pentecost an anticlinal structure cored by "Basement Complex" ophiolite horsts was formed. In the Marginal Province, island arc volcanism continued at least as late as 1.8 m.y. on Futuna Island (Carney and Macfarlane, 1979) and probably into Recent times on Mere Lava (Mallick and Ash, 1975), but it is inferred that the area generally ceased to be volcanically active following the development of the back-arc rift system. These structures were thus synchronous with movements causing the Pleistocene faulting and tilting along the Eastern Belt.

V. RESUMÉ—BRIEF GEODYNAMIC HISTORY OF VANUATU

This account places the stratigraphic and structural evolution of the arc in the context of tectonic developments within the region as a whole (Fig. 9). As noted above, Vanuatu essentially evolved under two separate regimes of subduction. Prior to spreading within the North Fiji Basin, the arc was evidently located above the convergence zone of the Pacific plate beneath the Austral–Indian plate. Inception of the marginal sea marked a "reversal" of polarity of the arc to its present west-facing configuration (Fig. 1) and caused migration (with rotation) of Vanuatu by some hundreds of kilometers to the southwest.

Little is known of the *early Miocene* arc configuration. It has been inferred (Carney and Macfarlane, 1978) that the then active island arc that is now represented by the Western Belt volcanic succession originally lay behind (i.e. to the west of) a frontal arc composed of island arc tholeiites dating back to the late Eocene. This frontal arc system, named the "Vitiaz palaeoarc" by Jezek *et al.* (1977) may have comprised rock sequences similar to the Wainimala Group of Fiji, if the evidence of clasts in the basal sediments on Maewo has been interpreted correctly. It is probable that the Western Belt, as the active arc to this earlier system, occupied a deep basin formed by an Oligocene phase of ocean floor spreading immediately behind the frontal arc (Fig. 10a).

During build up of the Western Belt Arc, periodic block faulting may have caused the formation of a number of basins, now seen as greywacke sequences within the volcanics. The major tectonic events, however, occurred at the start of the middle Miocene and are suggested to have comprised two main phases:

(a) Phase I (Fig. 10a) saw an evolution of the Western Belt from an active volcanic arc to a plutonic/orogenic belt. Volcanism largely ceased following the onset of tectonism comprising north-trending dextral strike-slip faults along which sizeable basalt–andesite–microdiorite–gabbro bodies were emplaced. Location of some of these plutons at tensional nodes along a braided wrench fault system probably accounts for the sheeted nature of the larger intrusive bodies found in south Santo (Mallick and Greenbaum, 1977).

(b) Phase II followed immediately after Phase I but was a tensional event resulting in the formation of narrow but regionally extensive greywacke basins along, or to the east of, the faulted intrusion complexes (Fig. 10b).

At some time during these Western Belt tectonic episodes, volcanogenic deep water sediments bearing conglomeratic horizons were deposited along

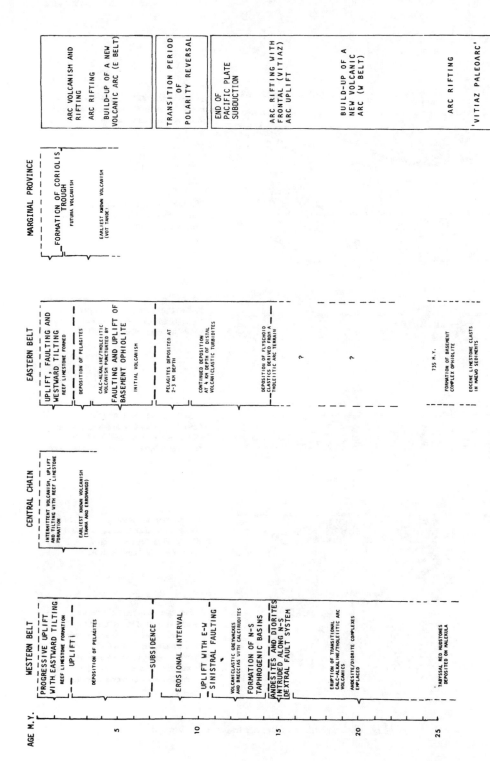

Fig. 9. A correlation of major events in the Vanuatu Arc. Interpretations are given in the right-hand column.

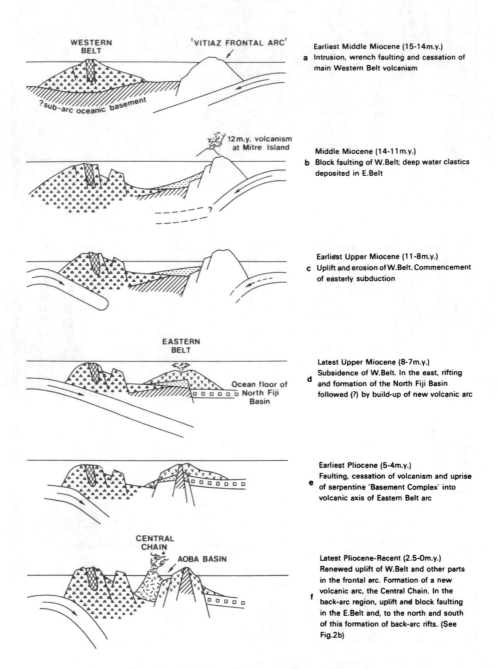

Fig. 10.   Evolution of the central part of Vanuatu.

the Eastern Belt. These sediments suggest a period of instability and erosion due to uplift along the "Vitiaz" frontal arc system (Fig. 10b). To account for this event two explanations are put forward:

(a) The uplift was caused by a collision event at the Vitiaz subduction zone; or

(b) It was essentially an intra-arc event related to arc-rifting and remnant arc formation at the Western Belt.

Of these, alternative (b) is preferred because it is difficult to reconcile collision at the trench with the essentially tensional phase of rifting observed from the formation of the greywacke basins along the Western Belt. Furthermore, the 12.7 m.y. volcanics on Mitre Island (Jezek *et al.*, 1977) suggest that subduction beneath the frontal arc system was still an active process at this time. The strike-slip faulting that immediately preceded the rifting event in the Western Belt is difficult to explain; it possibly reflects movements of the northwards-drifting Austral–Indian plate on which the Western Belt was at that time situated.

By *late Miocene* times (11–8 m.y.) significant uplift and erosion of the Western Belt was accompanied by the formation of E–W dextral faults, the latter causing the offsets observed to cross the Western Belt (Fig. 2c). The uplift and faulting are correlated with a compressional phase marking the initiation of subduction from the west (Fig. 10c). It is also inferred that Pacific plate subduction at the Vitiaz Trench had ceased by this time, perhaps because of a collision event involving converging elements of the Samoa Seamount chain (Fig. 1).

During the later stages of "polarity reversal" across Vanuatu, events occurred rapidly and are difficult to place in sequence. By 8 to 7 m.y., island arc volcanism had commenced in the Eastern Belt, whereas the Western Belt began to subside (Fig. 10d). Both these events are correlated with the onset of westwards migration and of a consequent prevailing regime of tension across the now weakly coupled arc/trench system. The prerequisite of such a configuration must have been arc rifting and inception of the North Fiji Basin (Carney and Macfarlane, 1982), for which a date of 8 m.y. is suggested from aeromagnetic anomalies immediately east of Vanuatu (Malahoff, 1982b). It is noticeable that the Eastern Belt volcanism was situated some distance further to the east than is the presently active volcanic arc; this may be of some significance to the interpretation of later tectonic developments in Vanuatu as it suggests a change in angle of the seismic zone.

By early Pliocene times the Eastern Belt volcanic arc had emerged above sea level, but then became extinct at 3 to 4 m.y., soon after block faulting had elevated the "Basement Complex" ophiolitic rocks beneath Pentecost (Fig. 10e), a phenomenon that has not yet been adequately explained. Carney and Macfarlane (1982) had attributed the Pliocene cessation of volcanism to the operation of a blocking mechanism caused by the arrival at the trench of the D'Entrecasteaux Fracture Zone (Fig. 1). Another possibility, intrinsic to the arc itself is that the ophiolites, remobilized due to heating by arc magmas, rose along faults and created unstable structural conditions wherein magma columns could not be supported. Yet a third explanation is that the Eastern Belt had evolved into a remnant arc because of incipient marginal basin formation between it and the Western Belt.

In the early Pleistocene widespread block uplifts of the frontal arc, corresponding horst formation in the Eastern Belt and back-arc rifting in the Marginal Province were synchronous, interrelated events associated with major geodynamic changes along the arc's length. The evolution in the northern and southern parts of the arc from a formerly wide volcanic front taking in the whole of the Central Chain and Marginal Province, to the present rather narrow active Central Chain arc and rifted back-arc zone is typical of migrating arc systems that periodically undergo rifting and the "shedding" of remnant arcs (e.g., Karig, 1975) even though, as previously observed, the Vanuatu back-arc rifts may not be floored by oceanic crust and in this sense are therefore only "incipient" extensional basins. Narrowing of the volcanic front during this period may have been accompanied by a steepening of the seismic zone to its present 70° eastward dip, but there is no evidence from earthquake foci for an actual rupture of the downgoing slab (Pascal *et al.*, 1978). This final Pleistocene tectonic episode correlates with uplift and basin formation in the Solomon Islands and hence is attributed to young plate interactions in the region (Carney and Macfarlane, 1980).

VI. PETROLOGICAL SURVEY OF THE VOLCANIC ROCKS

Research into major element compositions and trace element studies on rocks from the Western Belt, Eastern Belt and Central Chain have shown that each of these volcanic sequences has certain geochemical characteristics of both calc-alkaline and island-arc tholeiite magma types. The Central Chain in particular shows interisland, and in some cases intra-island, variations in potash and alkaline contents similar to those found in other Melanesian island arcs (e.g., Jakeš and White, 1969).

In the first detailed study of petrological variations in Vanuatu, Mallick (1973a) observed that between the Central Chain islands of Gaua, Merig and Mere Lava (Fig. 1) alkalies and particularly potash contents decreased eastwards, producing a negative correlation between potash content and depth to the seismic zone. In a subsequent study Colley and Warden (1974) noted that high-alumina and tholeiitic volcanics characterized the Western and

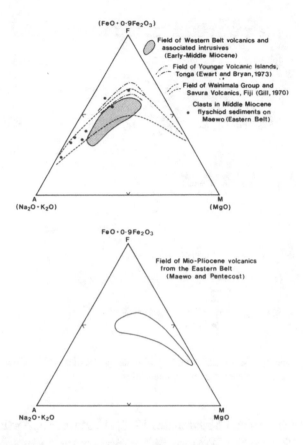

Fig. 11. (a) AFM diagram of the Western Belt volcanics (early–middle Miocene). (b) AFM diagram of the Eastern Belt volcanics (late Miocene–early Pliocene).

Eastern Belt sequences whereas the Central Chain contained a mixed tholeiitic/calcalkaline suite. The following account, necessarily abbreviated, presents an updated survey of each of the three volcanic provinces using as a basis for comparison AFM and potash–silica diagrams and drawing where possible on trace element data.

A. Early to Middle Miocene: Western Belt Volcanics, and Clasts in Eastern Belt Arc-Derived Sediments

Western Belt volcanic and hypabyssal rocks from Santo, Malekula and the Torres Islands occupy a broad field on the AFM diagram (Fig. 11a), indicating a typical calc-alkaline trend of limited iron enrichment. The K_2O–SiO_2 diagram (Fig. 12a) shows a scattering of points suggestive of "high" and "low" potash subtrends in the intermediate to acid members and both of these trends fall within the "normal-K" field of Papua New Guinea in-

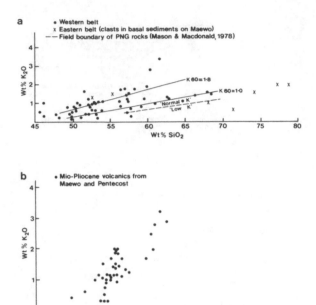

Fig. 12.   (a) K_2O–SiO_2 plot for early–middle Miocene volcanics. (b) K_2O–SiO_2 plot for late Miocene–early Pliocene volcanics of the Eastern Belt.

trusive rocks (Mason and Macdonald, 1978). Features supporting island arc tholeiite affinities for the Western Belt rocks are their generally low K_2O/Na_2O ratios, low U and Th contents and the flat patterns and low absolute abundances of REE (Gorton, 1974). In their somewhat transitional calc-alkaline/tholeiitic attributes, the Western Belt volcanics strongly resemble those of the Lau Ridge (Gill, 1976).

In the Eastern Belt, clasts from the basal arc-derived volcanogenic sediments show marked geochemical contrasts to the Western Belt suite in their notable iron-enrichment (Fig. 11a) and their highly differentiated low-potash compositions (Fig. 12a). These differences prompted Carney and Macfarlane (1978) to suggest that the clasts were not derived from the Western Belt but from an eastern "frontal" part of a former arc system that contained island arc tholeiite volcanics comparable to those from Tonga and the Wainimala/Savura Groups of Fiji.

B. Late Miocene–Early Pliocene: Eastern Belt Volcanics

On the AFM plot (Fig. 11b) the suite shows limited iron enrichment within the basaltic and andesitic members. The K_2O/SiO_2 trend (Fig. 12b) is rather steep, but conceals a stratigraphic evolution towards lower potash

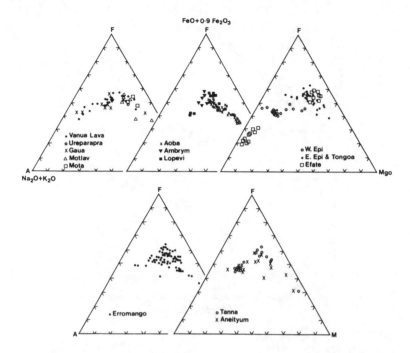

Fig. 13.   AFM diagrams for Central Chain volcanics (late Pliocene–Recent).

contents in the younger rocks noted by Gorton (1974) and Carney (in preparation). The flat REE patterns and low Th/U ratios of Pentecost basalts are suggestive of island arc tholeiite affinities (Gorton, 1974) but potash values in excess of 2% for high-silica andesites (Fig. 12b) are attributes of calc-alkaline volcanic suites.

C. Late Pliocene–Recent: Central Chain and Marginal Province

The Central Chain affords opportunities for in-depth geochemical studies of arc volcanics because variations both within and between individual centers can be discerned. A good example of this is shown in Fig. 13 where AFM diagrams for the individual islands are plotted. If these are all combined, an averaged "tholeiitic" distribution would result, concealing obvious interisland differences, for example between West Epi and East Epi/Tongoa.

The potash–silica diagram (Fig. 14a) shows diverse trends summarized as follows:

- (*i*)   High-potash "A," constituting rocks from Gaua Island (Mallick and Ash, 1978).
- (*ii*)   High-potash "B," represented by dacitic/rhyolitic sequences on Efate and Western Epi (Ash *et al.*, 1978; Warden, 1967) and by low-silica andesites on Tanna (Carney and Macfarlane, 1979).

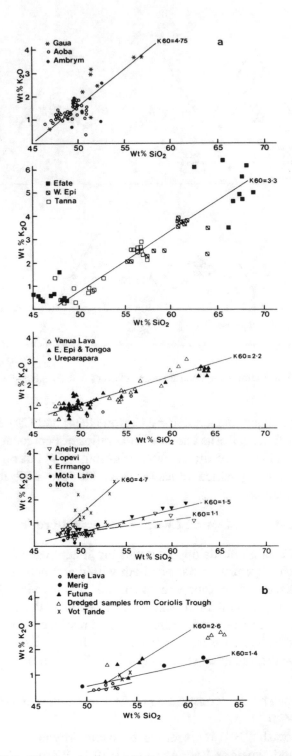

Fig. 14.   (a) K_2O–SiO_2 plots for Central Chain volcanics (late Pliocene–Recent). (b) K_2O–SiO_2 plot for Marginal Province volcanics (late Pliocene–Recent).

(*iii*) Normal potash, comprising rocks from the Central Islands (Warden, 1967) and Vanua Lava (Ash *et al.* 1980).

(*iv*) Low potash, consisting of suites from Lopevi (Warden, 1967) and Aneityum (Carney and Macfarlane, 1979).

Erromango and Ambrym islands both have interbedded lavas of contrasting high- and low-potash contents (Gorton, 1977).

Marginal Province rocks from Futuna, Vot Tande and Mere Lava islands, and from samples obtained by dredging in the Coriolis Trough backarc rift (Dugas *et al.*, 1977) show trends on AFM diagrams similar to the Central Chain suites. Potash variations in the Marginal Province (Fig. 14b) are widely variable, ranging from steeply-inclined trends for Futuna and Vot Tande to rather low, almost "tholeiitic" values for Mere Lava (Mallick and Ash, 1975).

Geochemical studies of Central Chain and Marginal Province basalts by Dupuy *et al.* (1982) have defined three main groupings of parental types:

(*i*) Island arc tholeiite, characterized by an enrichment of LILE and depletion of incompatible elements relative to MORB. Basalts of this type were found on Erromango, north Efate and the Central Chain islands (Fig. 2a).

(*ii*) A transitional tholeiitic/calc-alkaline grouping showing strongly fractionated HREE and higher contents of LREE, Rb and Sr.

(*iii*) A calc-alkaline group represented by Marginal Province basalts from Futuna. These have the highest concentrations of incompatible elements, fractionated REE patterns with high La/Yb ratios, and markedly different Ba/La, Sr/Ce and La/Ta ratios when compared with the other two groups.

Dupuy *et al.*'s study was carried out on basaltic rocks and thus throws little light on the reasons for the considerable differences in potash (and related trace element) contents between, for example, Aneityum, Tanna and Efate suites (Fig. 13a) all of which were shown to contain basalts of island arc tholeiite type. Gorton (1977) observed that fractional crystallization was inadequate to generate high-K from low-K suites on Epi though it was possible to fractionate andesite from basalt in a single suite. He suggested that basalts and andesites of the high-K suite could, however, be derived by varying the degree of partial melting of hydrous mantle at moderate depth, and proposed that arc-rifting during active Central Chain volcanism could have resulted not only in lava suites with diverse potash variations but also in the eruption of magmas in different stages of differentiation. On the other hand, Coulon (1980), studying the rhyodacitic tuffs on Efate, concluded that partial melting of acid rocks in the downgoing oceanic lithosphere crust was the most plausible explanation for the high-potash contents of these rocks.

Concerning the transverse geochemical variations across the arc between the Central Chain and Marginal Province, the presence in the latter of back-arc rifts seems to be of particular relevance. For example, potash contents of lavas of the South Banks Islands decrease eastwards, i.e. away from the trench, between Gaua and Mere Lava. This anomalous geochemical variation was first observed by Mallick (1973a) and later attributed by Ash *et al.* (1980) to the modifying influence of back-arc rifting on magmas generated at the tip of the eastward-descending lithospheric slab. Similarly, in the southern latitudes, Carney and Macfarlane (1979) noted that on Futuna Island the lavas form part of a remnant arc sequence situated east of the back-arc rifts and are of a geochemical type appropriate to their eruption in an environment transitional between, on the one hand, the deep subduction-controlled magmatism typical of volcanic arcs and, on the other, a magmatism originating at shallower levels and perhaps involving larger degrees of partial melting as the back-arc rifts developed.

Because geochemical studies to date have been concerned with relatively small island groupings, or with restricted ranges of rock types, it is clearly necessary for the whole of the archipelago, and in particular the Central Chain, to be studied using the chronological controls that have been established during Geological Survey mapping over the past twenty years. It is likely that isotope geochemistry will provide the most useful tool to unravel the relative roles that partial melting and crustal contamination and other processes may have played in the evolution of such diverse trends as are seen in the Central Chain.

VII. DISCUSSION

Stratigraphic relations within the Vanuatu arc are now reasonably well constrained as a result of systematic mapping and detailed research carried out over the past 25 years. There are, however, certain problems that remain unsolved at the time of this writing. For example, the age of the abyssal red mudstones exposed on Malekula Island, and their relationship to the Western Belt volcanics, require further investigation as this may reveal the nature of the arc basement in the west. Similarly in the Eastern Belt further geochemical and geochronological studies on clasts from the middle Miocene flyschoid sequences should be undertaken to substantiate assertions that these deposits may represent the only remaining elements of a former frontal arc system. Perhaps the most enigmatic rocks are those belonging to the Pentecost Island "Basement Complex" ophiolites. Geochronological investigations here have so far yielded ambiguous results hence the "Basement Complex" cannot yet be equated for certain with any of the various episodes of pre-Pliocene marginal basin spreading in Melanesia (see Fig. 1). Moreover, in-depth geochemical studies are needed to check the presumed marginal basin affinities of this ophiolite.

An important aspect of stratigraphic correlation to emerge from recent studies concerns the spatial separation of the major volcanic episodes within the arc. The situation in Vanuatu contrasts greatly with neighbouring arcs such as Fiji and the Solomon Islands where many of the larger islands have experienced multiphase volcanism since at least early Miocene times. In the case of Vanuatu there is little doubt that post-middle Miocene changes in the position and angle of the subduction zone have caused the observed shifts in the locus of volcanic activity, each of which was accompanied by the development of major unconformities and episodes of faulting. This does not however apply to the important middle Miocene episodes of strike-slip faulting and then taphrogenic basin formation in the Western Belt. An explanation for these earlier events may be sought for in the significant "orogenies" that took place during middle Miocene times in the formerly colinear arc systems of Fiji–Tonga, the Solomon Islands, and New Britain. These events formed the prelude to extensive late Miocene–Recent marginal sea growth and attendant fragmentation of the Outer Melanesian Arc system.

The well-documented post-middle Miocene unconformities in Vanuatu when correlated with plate tectonic events of regional important i.e. growth of the North Fiji Basin and consequent arc migration in front of this, give some insight into the response of the arc's crust to these processes. The following is a summary of the main events within the arc:

(1) Earliest late Miocene (nonvolcanic transitional period of polarity reversal). Uplift and erosion of the new frontal system (Western Belt). Static conditions further east where the former trench/frontal arc was situated.

(2) Late Miocene (Formation of new Eastern Belt volcanic province, and inception of the North Fiji Basin). Rapid subsidence at the new frontal arc leading to "drape" hemipelagite covering to the Western Belt succession. In the east, a shallowing of depth followed by block faulting and effusion of submarine arc volcanics.

(3) Latest Miocene–early Pliocene (Main migratory phase). Static, submergent conditions at the frontal arc. Block faulting along the Eastern Belt volcanic arc and eventual emergence of reef-fringed islands there.

(4) Plio–Pleistocene (Possible steepening of subduction zone to present 70° angle). Rapid uplift of frontal arc. Widespread back-arc rifting and intra-arc downwarping. Narrowing of volcanic arc width and westwards (frontal) shift in the concentration of volcanism resulting in the present well-defined Central Chain alignment.

The extent to which petrological variations within the arc can be tied in to deduced crustal movements and/or changes in attitude or direction of the subduction zone will largely depend upon the type of model called upon to account for the apparent diversity of magma type (i.e. island arc tholeiite, calc-alkaline, and transitional varieties). Such a model should also explain marked divergences in the degree of potash enrichment of suites both between and within the main volcanic provinces. Further and more refined geochemical studies will be necessary to evaluate those schemes proposed so far that include (a) varying the degree of partial melting of hydrous mantle during arc rifting and (b) partially melting acid rocks in the downgoing slab.

ACKNOWLEDGMENTS

This chapter is published by permission of the Directors of the British Geological Survey (N.E.R.C.) and of the Vanuatu Geology, Mines and Rural Water Supplies Department.

REFERENCES

Ash, R. P., Carney, J. N., and Macfarlane, A., 1978, Geology of Efate and Offshore Islands, *New Hebrides Geol. Surv. Reg. Rep.*, 49p.

Ash, R. P., Carney, J. N. and Macfarlane, A., 1980, Geology of the Northern Banks Islands, *New Hebrides Geol. Surv. Reg. Rep.*, 52p.

Benoit, M., 1967, Seismic activity, *New Hebrides Ann. Rep. Geol. Surv. 1965*, p. 19–26.

Bloom, A. L., Jouannic, C. and Taylor, F. W., 1978, Preliminary radiometric ages from the uplifted Quaternary coral reefs of Efate, *New Hebrides Geol. Surv. Regional Rep.*, p. 47–49. (Appendix to Ash, Carney and Macfarlane, 1978).

Carney, J. N., 1982, Efate Geothermal Project, Phase I: Geology and reconnaissance hydrology of the project area, *Overseas Division, Br. Geol. Surv. Rep. 82/11*, (mimeographed).

Carney, J. N., Geology and mineralisation of North and Central Malekula, *Geol. Surv. Vanuatu Rep.*, (in preparation).

Carney, J. N., Geology of Maewo, *Geol. Surv. Vanuatu Reg. Rep.* (in preparation).

Carney, J. N. and Macfarlane, A., 1978, Lower to middle Miocene sediments on Maewo, New Hebrides, and their relevance to the development of the Outer Melanesian Arc System, *Bull. Aust. Soc. Explor. Geophys.*, v. 19, p. 123–130.

Carney, J. N. and Macfarlane, A., 1979, Geology of Tanna, Aneityum, Futuna and Aniwa, *New Hebrides Geol. Surv. Reg. Rep.*, 71p.

Carney, J. N. and Macfarlane, A., 1980, A sedimentary basin in the central New Hebrides arc, in: *Petroleum Potential in Island Arcs, Small Ocean Basins, Submerged Margins and Related Areas. United Nations ESCAP CCOP/SOPAC Tech. Bull.* v. 3, p. 109–120.

Carney, J. N. and Macfarlane, A., 1982, Geological evidence bearing on the Miocene to Recent structural evolution of the New Hebrides Arc, *Tectonophysics*, v. 87, p. 147–175.

Carney, J. N. and Macfarlane, A., Geology and mineralization of North Santo, *Geol. Surv. Vanuatu Rep.*, (in preparation).

Chase, C. G., 1971, Tectonic history of the Fiji Plateau, *Geol. Soc. Am. Bull.*, v. 82, p. 3087–3110.

Chung, W. Y. and Kanamori, H., 1978, A mechanical model for plate deformation associated with aseismic ridge subduction in the New Hebrides, *Tectonophysics*, v. 50, p. 29–40.

Coleman, P. J., 1969, Derived Eocene larger foraminifera on Maewo, eastern New Hebrides, and their southwest Pacific implications, *New Hebrides Ann. Rep. Geol. Surv. 1967*, p. 36–37.

Coleman, P. J. and Packham, G. H., 1976, The Melanesian borderlands and India–Pacific plates' boundary, *Earth Sci. Rev.*, 12 p. 197–233.

Colley, H. and Ash, R. P., 1971, The Geology of Erromango, *New Hebrides Geol. Surv. Reg. Rep.*, 112p.

Colley, H. and Warden, A. J., 1974, Petrology of the New Hebrides, *Geol. Soc. Am. Bull.*, v. 85, p. 1635–1646.

Coulon, C., Maillet, P., and Maury, R., 1980, Contribution à l'étude du volcanisme de l'arc des Nouvelles Hébrides: données pétrologiques sur les laves d'Efate, *Bull. Soc. Geol. Fr.*, v. 7, p. 621–631.

Daniel, J., 1978, Morphology and structure of the southern part of the New Hebrides island arc system, *J. Phys. Earth*, v. 26, Suppl., S181–S190.

Daniel, J., Jouannic, C., Larue, B. M. and Recy, J., 1977, Interpretation of the D'Entrecasteaux Zone (north of New Caledonia). in: *Geodynamics in Southwest Pacific*, Paris: Editions Technip, p. 63–78.

Dubois, J., Dugas, F., Lapouville, A. and Louat, R., 1975, Fosses d'éffondrement en arrière de l'arc des Nouvelles-Hébrides. Méchanismes proposés, *Rev. Géogr. Phys. Géol. Dynamique*, v. 17, p. 73–94.

Dugas, F., Carney, J. N., Cassignol, C., Jezek, P. A. and Monzier, M., 1977, Dredged rocks along a cross-section in the southern New Hebrides island arc and their bearing on the age of the arc, in: *Geodynamics in Southwest Pacific*, Paris: Editions Technip, p. 105–116.

Dupuy, C., Dostal, J., Marcelot, G., Bougault, H., Joron, J. L. and Treuil, M., 1982, Geochemistry of basalts from central and southern New Hebrides arc: implication for their source rock composition, *Earth Planet. Sci. Lett.*, v. 60, p. 207–225.

Ewart, A. and Bryan, W. B., 1973, The petrology and geochemistry of the Tonga islands, in: *The Western Pacific: Island Arcs, Marginal Seas, Geochemistry*, Coleman, P. J. ed., Nedlands, W. A: University of Western Australia Press, p. 503–522.

Falvey, P. A., 1978, Analysis of palaeomagnetic data from the New Hebrides, *Bull. Aust. Soc. Explor. Geophys.*, v. 9, p. 117–123.

Gill, J. B., 1970, Geochemistry of Viti Levu, Fiji, and its evolution as an island arc, *Contr. Mineral. Petrol.*, v. 27, p. 179–203.

Gill, J. B., 1976, Composition and age of Lau Basin and Ridge volcanic rocks: Implications for evolution of an interarc basin and remnant arc, *Geol. Soc. Am. Bull.*, v. 87, p. 1384–1395.

Gorton, M. P., 1974, Geochemistry and Geochronology of the New Hebrides. Ph.D. dissertation, Australian National University, Canberra A.C.T., 300p (unpublished).

Gorton, M. P., 1977, The geochemistry and origin of Quaternary volcanism in the New Hebrides, *Geochem. Cosmochem. Acta*, v. 41, p. 1257–1270.

Greenbaum, D., Mallick, D. I. J. and Radford, N. W., 1975, Geology of the Torres Islands, *New Hebrides Geol. Surv. Reg. Rep.* 46p.

Greene, H. G., Falvey, D. A., Macfarlane, A., Preliminary Report on the geology, structure and resources potential of the central basin of Vanuatu, *Leg 2 CCOP/SOPAC S.P. Lee Cruise* (in preparation).

Jakeš, V. P. and White, A. J. R., 1969, Structure of the Melanesian arcs and correlation with distribution of magma types, *Tectonophysics*, v. 8, p. 223–236.

Jezek, P. A., Bryan, W. B., Haggerty, S. E. and Johnson, H. D., 1977, Petrography, petrology and tectonic implications of Mitre Island, Northern Fiji Plateau, *Mar. Geol.*, v. 24, p. 123–148.

Jouannic, C., Taylor, F. W., Bloom, A. L. and Bernat, M., 1980, Late Quaternary uplift history from emerged reef terraces on Santo and Malekula islands, central New Hebrides arc, *United Nations ESCAP, CCOP/SOPAC Tech. Bull.*, v. 3, p. 91–108.

Karig, D. E., 1975, Basin genesis in the Philippine Sea, in: *Initial reports of the Deep Sea Drilling Project*, v. 31, Karig, D. E., Ingle, J. C., Jr., *et al.*, eds, Washington, D.C: United States Government Printing Office, p. 1–927.

Karig, D. E. and Mammerickx, J., 1972, Tectonic framework of the New Hebrides island arc, *Mar. Geol.*, v. 12, p. 187–205.

Karig, D. E. and Sharman, G. F., 1975, Subduction and accretion in trenches, *Geol. Soc. Am. Bull.*, v. 86, p. 377–389.

Liggett, K. A., 1967, Maewo, *New Hebrides Ann. Rep. Geol. Surv. 1965*, p. 8–12.

Luyendyk, B. P., Bryan, W. B. and Jezek, P. A., 1974, Shallow structure of the New Hebrides island arc, *Geol. Soc. Am. Bull.*, v. 85, p. 1287–1300.

Malahoff, A., 1970, Gravity and magnetic studies of the New Hebrides island arc, *New Hebrides Geol. Surv. Rep.*, 67 p.

Malahoff, A., Hammond, S. R., Naughton, J. J., Keeling, D. L. and Richmond, R. N., 1982a, Geophysical evidence for post-Miocene rotation of the island of Viti Levu, Fiji, and its relationship to the tectonic development of the North Fiji Basin, *Earth Planet. Sci. Lett.*, v. 57, p. 398–414.

Malahoff, A., Feden, R. H. and Fleming, H. S., 1982b. Magnetic anomalies and tectonic fabric of marginal basins north of New Zealand, *J. Geophys. Res.*, v. 87, p. 4109–4125.

Mallick, D. I. J., 1973a, Some petrological and structural variations in the New Hebrides. In: *The Western Pacific: Island Arcs, Marginal Seas, Geochemistry*, Coleman, P. J. Ed., Nedlands, W. A: University of Western Australia Press, p. 193–211.

Mallick, D. I. J., 1973b, Santo, *New Hebrides Ann. Rept. Geol. Surv. 1971*, p. 11–12.

Mallick, D. I. J. and Neef, G., 1974, Geology of Pentecost, *New Hebrides Geol. Surv. Reg. Rep.*, 103p.

Mallick, D. I. J. and Ash, R. P., 1975, Geology of the southern Banks Islands, *New Hebrides Geol. Surv. Reg. Rep.*, 33 p.

Mallick, D. I. J. and Greenbaum, D. 1977, Geology of southern Santo, *New Hebrides Geol. Surv. Reg. Rep.*, 84p.

Mason, D. R. and Macdonald, J. A., 1978, Intrusive rocks and porphyry copper occurrences of the Papua New Guinea–Solomon Islands region: a reconnaissance study, *Econ. Geol.*, v. 73, p. 857–877.

Mitchell, A. H. G., 1971, Geology of Northern Malekula, *New Hebrides Geol. Surv. Reg. Rep.*, 56p.

Mitchell, A. H. G. and Warden, A. J., 1971, Geological evolution of the New Hebrides island arc, *J. Geol. Soc. London*, v. 127, p. 501–529.

Neef, G., 1982, Plate tectonic significance of late Oligocene/early Miocene deep sea sedimentation at Maewo, Vanuatu (New Hebrides), *Tectonophysics*, v. 87, p. 177–183.

Pascal, G., Isacks, B. L., Barazangi, M. and Dubois, J., 1978, Precise relocations of earthquakes and seismotectonics of the New Hebrides island arc, *J. Geophys. Res.*, v. 83, p. 4957–4973.

Pontoise, B., Latham, G. V., Daniel, J., Dupont, J. and Ibrahim, A. B., 1980, Seismic refraction studies in the New Hebrides and Tonga area, *United Nations ESCAP, CCOP/SOPAC Tech. Bull.*, v. 3, p. 47–58.

Ravenne, C., Pascal, G., Dubois, J., Dugas, F. and Montadert, L., 1977, Model of a young intra-oceanic arc: the New Hebrides island arc, in: *Geodynamics in Southwest Pacific*, Paris: Editions Technip, p. 63–78.

Robinson, G. P., 1969, The geology of north Santo, *New Hebrides Geol. Surv. Reg. Rep.*, 77p.

Rodda, P., 1967, Outline of the geology of Viti Levu, *N. Z. J. Geol. Geophys.*, v. 10, p. 1260–1273.

Taylor, D., 1977, The ages and environments of micropalaeontological samples from North Santo, New Hebrides. *New Hebrides Geol. Surv. Rep.*, Occ. 2/77 (mimeographed).

Warden, A. J., 1967, The geology of the Central islands, *New Hebrides Geol. Surv. Reg. Rep.*, 107p.

INDEX